CERAMIC GAS TURBINE DESIGN AND TEST EXPERIENCE

Progress in Ceramic Gas Turbine Development, Volume 1

MARK VAN ROODE
MATTISON K. FERBER
DAVID W. RICHERSON

NEW YORK ASME PRESS 2002

Library of Congress Cataloging-in-Publication Data

Ceramic gas turbine design and test experience / edited by Mark van Roode, Mattison K. Ferber, and David W. Richerson.
p. cm.
ISBN 0-7918-0181-0
1. Gas-turbines—Ceramic materials. I. van Roode, Mark. II. Ferber, Mattison K. III. Richerson, David W., 1944–
TJ778.C39 2002
621.43′3—dc21
2002005827

The front cover depicts three stages in the design and fabrication of a silicon nitride turbine rotor blade developed under the U.S. Department of Energy Ceramic Stationary Gas Turbine (CSGT) program by Solar® Turbines Incorporated. The designs were developed by Oscar Jimenez and Paul Norton. Figure arrangement by Mattison K. Ferber of Oak Ridge National Laboratory.

Cover design: Paul Moran

Table of Contents

Foreword vii

Preface ix

Acknowledgements xi

About the Editors xiii

Chapter 1
The Benefits and Challenges of the Use of Ceramics in Gas Turbines 1

Chapter 2
Evolution of Ceramic Gas Turbine Development Programs at Engine Manufacturers in the United States 11

Chapter 3
Ford's Development of the 820 High Temperature Ceramic Gas Turbine Engine 17

Chapter 4
Evolution of Ceramic Turbine Engine Technology at Honeywell Engines, Systems & Services 77

Chapter 5
Ceramic Gas Turbine Development at Rolls-Royce in Indianapolis 111

Chapter 6
Ceramic Gas Turbine Development at Hamilton Sundstrand Power Systems 133

Chapter 7
Ceramic Gas Turbine Development at Solar Turbines Incorporated 155

Chapter 8
Progress in Ceramic Gas Turbine Development Westinghouse/Siemens Westinghouse 193

Chapter 9
General Electric Company 225

Chapter 10
United Technologies Pratt & Whitney/United Technologies Research Center Experiences with Ceramics and CMCs in Turbine Engines 237

Chapter 11
Ceramic Composite Turbine Engine Test Experience at Williams International 261

CHAPTER 12
Evolution of Ceramic Gas Turbine Development Programs at Engine Manufacturers in Japan 277

CHAPTER 13
Japanese Automotive Ceramic Gas Turbine Development 283

CHAPTER 14
Development of 300kW-Class Regenerated Single-Shaft Ceramic Gas Turbine CGT301 305

CHAPTER 15
Development of the 300kW Ceramic Gas Turbine CGT302 331

CHAPTER 16
Development of 300kW-Class Ceramic Gas Turbine CGT303 361

CHAPTER 17
Application of Ceramics to a Power Generating Gas Turbine 381

CHAPTER 18
Development of Ceramic Gas Turbine for Power Plants 403

CHAPTER 19
A Design and Experimental Evaluation of Ceramic Stator and Rotor Blades for 1300°C Class Medium Gas Turbine 425

CHAPTER 20
Evolution of Ceramic Gas Turbine Development Programs at Engine Manufacturers in Western Europe 445

CHAPTER 21
German Automotive Ceramic Gas Turbine Development 453

CHAPTER 22
Ceramic Components for Automotive Gas Turbines – Experience at Volvo 499

CHAPTER 23
Ceramic Development at ONERA 513

CHAPTER 24
Ceramic Gas Turbine Component Development at Motoren Turbinen Union 539

CHAPTER 25
New Design Principles for Ceramic Hot Gas Components of Gas Turbines 555

CHAPTER 26
Rolls-Royce Experience in Ceramic Matrix Composite Demonstration 583

CHAPTER 27
'CMC-Design' A Brite Euram Program to Develop Design Methodologies for CMC Components for Gas Turbine Engines 595

CHAPTER 28
Siemens KWU – Experience with Refractory Ceramics in Stationary Gas Turbines 615

CHAPTER 29
Development of Ceramic Components for Utility Gas Turbines 635

Chapter 30
Evolution of Programs for Ceramic Gas Turbine Engine Design in
the States of the Former Soviet Union 661

CHAPTER 31
Ceramic Gas Turbine Development Experience at the Central Institute of Aviation Motors 669

CHAPTER 32
Research-Engineering Center "Ceramic Heat Engines (NIZ KTD) at the
Research-Technological Institute for Power Engineering (NITI EM)" 683

CHAPTER 33
Ceramic Component Development at Rybinsk Motors Design Bureau 709

Index 715

FOREWORD

The nation has been experiencing one of the longest periods of sustained economic growth in U.S. history. During this period, significant progress has been made in addressing some of the country's most pressing energy and environmental problems. For example, the U.S. Environmental Protection Agency reports that emissions of certain air pollutants have measurably declined since the 1980s. As a result, the once strongly held belief that it is not possible to improve the environment without slowing the economy has been proven false. One of the factors that has made these economic and environmental gains possible is greater energy efficiency, which has been brought about in part by the development and deployment of cleaner and more fuel efficient appliances, industrial equipment, transportation vehicles, and power generation devices.

As we enter the new millennium, energy problems have once again become major public concerns. Energy price spikes, electric power supply constraints, power reliability problems, and air pollution are once again national news. In fact, the need for cleaner, reliable, affordable, and more fuel efficient energy devices has never been greater.

The U.S. Department of Energy (DOE) has a long-standing commitment to supporting technological innovation aimed at increasing energy efficiency, reducing dependence on energy imports, and lowering emission of air and water pollutants from energy generation, delivery, storage, and end-use consumption. A critical energy research and development (R&D) area has been that of advanced materials for improving the efficiency of combustion processes. In fact, for more than thirty years the DOE and other government agencies have been funding the design and development of new materials for energy generation systems to improve performance and lower emissions.

The oil crises of the 1970s stimulated considerable interest in the development of more energy efficient engines for passenger vehicles. More recently, increasingly stringent environmental emissions requirements have spurred interest in low-emission electric power plants. Recent growth in worldwide power demand has led to a surge in orders for electricity generation systems, particularly combustion turbines. In fact, the worldwide demand for combustion turbines is now so great that the major manufacturers are experiencing difficulties expanding production to keep pace with customer demand.

One of the most productive R&D pathways has been the development of advanced ceramic materials in collaboration with gas turbine manufacturers and producers of ceramic materials and components as part of a concerted strategy to combat the wasting of energy resources and to promote environmentally friendly engineering practices. The Federal government initiated major programs in the 1980s, including the Advanced Gas Turbine (AGT) program and its successor, the Advanced Turbine Technology Applications Project (ATTAP). These successful programs resulted in major technology breakthroughs in ceramic component design and materials processing and properties. The industrial participants in these programs – in collaboration with government laboratories and universities – developed the basic design, fabrication, and testing tools which led to the introduction of small ceramic gas turbines for automotive and related vehicular applications.

The great strides made in this endeavor are reflected in the advancements in silicon nitride ceramics, the strength of which have almost doubled since this material was first introduced in the 1970s. In fact, as a direct result of the AGT and ATTAP programs, silicon nitride nozzles have recently been certified for auxiliary power units for commercial aircraft. Lufthansa and Alaska Airlines planes are now flying with durable and reliable ceramic parts.

In the early 1990s, the DOE launched several new initiatives to advance the development of more energy efficient gas turbines through the application of advanced ceramic materials. These included the Ceramic Stationary Gas Turbine (CSGT) program and the Continuous Fiber-reinforced Ceramic Composites (CFCC) initiative. Both programs were aimed at applying the AGT and ATTAP improvements in the design and fabrication of advanced ceramic components to improve the design and performance of medium size industrial gas turbines.

The CSGT and CFCC programs have been extremely successful. Recent tests of silicon carbide-CFCC combustor liners have now been demonstrated for over 29,000 operating hours in industrial gas turbines at commercial sites. Tests have shown NO_x emissions at these sites dropped by more than 50 percent and CO emissions have been reduced 5 to 10 fold.

In addition, microturbines – small-scale gas turbines often less than 100 kilowatts in size – are being sold commercially and have the capability to revolutionize the global power generation industry over the next ten years. To be competitive with alternative energy generation devices, the fuel efficiency and air emissions profiles of microturbines must be improved. Ceramic materials for various microturbine subsystems and components will play a key role in reaching the performance levels required for commercial success. The DOE has recently launched a five year microturbine R&D initiative in collaboration with industry to meet those technological challenges and reach the performance targets.

This book provides important perspectives on the history of ceramic gas turbine development worldwide. There are lessons that have been learned that apply to a wide variety of industry-government R&D partnerships. The book is intended to highlight the excitement of successful R&D ventures in general and detail how researchers have responded to the challenges of designing ceramic turbines, how materials suppliers have improved their processes, how scientists have developed techniques to measure properties reliably and to establish life prediction and diagnostic technologies, and the key lessons learned in each of these endeavors.

Finally, innovations require the dedication of talented, visionary individuals. Each of the three editors of this book is a major contributor to the science, engineering and impact of advanced materials on the world. Each one is clearly among the best and brightest who have worked many additional hours, uncompensated, with the desire to leave more than they take in this world. Mark van Roode ran one of the best Public-Private partnerships that the Department of Energy was engaged in during the 1990s; work that contributed to his employer winning the Malcolm Baldrige Quality Award for excellence. This book is one of many major significant contributions to Ceramics and Education that Dave Richardson has given the world.

It is my hope that readers will find many useful insights that can be applied for further improvements in ceramic materials for gas turbines and in the development of other technologies for achieving cleaner, reliable, affordable, and more efficient energy generation technologies for the 21st century.

William P. Parks, Jr.
Associate Deputy Assistant Secretary
Office of Power Technologies
United States Department of Energy

PREFACE

As ceramic components become viable design alternatives for commercial gas turbines the need will arise amongst the manufacturers of turbomachinery equipment, their component suppliers, and end users to become knowledgeable about the applicability of ceramics and their properties. For more than 50 years reports have appeared in the technical literature about design and materials development for specific gas turbine related projects. Taken together the published literature represents an impressive body of data that can be utilized to guide the materials process engineer, and the engine component and system designer.

Published conference proceedings discuss specific aspects of ceramics for gas turbine applications and represent the state-of-the-art at a certain point in time. However, at this time there exists no comprehensive text that addresses all the relevant issues involved in the application of ceramic gas turbine components and summarizes the evolution of the ceramic gas turbine development work. Ceramic gas turbine development programs have been conducted for a wide range of commercial and military applications for which the output power ranges from several tens of kilowatts to tens of megawatts in size. Large programs have been undertaken for automotive, industrial and utility size gas turbine development, each of which has its own peculiarities that need to be appreciated to obtain a complete picture of the development efforts. For example, data on component performance under typical operating conditions for automotive gas turbines is typically limited to durations from several hours to a few thousand hours at best, because the expected service life of an automotive gas turbine can be simulated by such short term testing. On the other hand, test data representing tens of thousands of hours are needed to effectively establish the life of components in an industrial gas turbine for which the end users expect service lives on the order of 30,000 hours between overhauls. And such long term data is still very scarce and has only become available over the past decade.

These and other critical issues are very evident to those who have worked on ceramic gas turbine design, materials development and engine testing. Many valuable lessons have been and are still being learned, and the emerging knowledge base starts to provide a clearer picture of the potential and limitations of designing gas turbine hot section components with ceramics.

It is against this background that a publication that consolidates the data generated and summarizes the lessons learned over the more than fifty years of ceramic gas turbine development is timely and will be helpful for those involved with designing, testing, repairing, and marketing of turbomachinery equipment with ceramic parts, the suppliers of these parts, organizations involved in materials and component testing, and the end users of equipment.

The publication in front of you starts from the premise that the contributors to the various aspects of knowledge on ceramic gas turbines themselves are the best source of information. Ceramic gas turbine design is still a developing field, and there are different design approaches. On the other hand, similar probabilistic life prediction methodologies are widely employed for a variety of applications, although there are differences in the software routines used by the various workers in the field.

The editors of the publication discussed here, when contemplating the state of the technology, feel that the best approach to become familiar with the issues of ceramic gas turbine development is to consult directly the experts that have been intimately involved with ceramic gas turbine related work at their own organizations, whether these are the gas turbine manufacturers, suppliers of ceramic materials and components, or test laboratories.

This approach also has the advantage that it brings together the view points of the contributors that have learned valuable lessons from first hand experience. As a result the text will illustrate the various approaches in which the gas turbine community has embarked upon using ceramic materials. To guide the reader there are introductory chapters that will filter the cumulative experience and distill the key lessons learned, while highlighting still unresolved issues.

The large amount of design and materials data will be valuable for those with various degrees of exposure to the field. There will be information useful to the novice as well as to the experienced gas turbine designer and materials expert. The editors hope that the data presented will shorten the iterative development cycle and provide an up-to-date perspective on recent as well as historically more distant development work in ceramic materials and components. To provide a balanced view point the applications of ceramic matrix composites are being included alongside with those of the monolithic ceramics.

This publication on ceramic gas turbines draws on the accumulated experience of those closely involved with their development in the four regions of the world where active research has been conducted, the United States, Japan, Western Europe, and Russia and other countries that previously made up the Soviet Union. Some of the data has been accessible to those that cared to search for it. However, this information has been consolidated here in one English language text for the first time. Moreover, the perspective of a worldwide body of experts with its focus on relevance adds a novel dimension to the data.

The goal of this publication is to summarize the collective experience of the international gas turbine community with ceramic gas turbine development over the past couple of decades. Representatives from various organizations (equipment manufacturers, suppliers of ceramic parts, and scientists and engineers involved with materials evaluation) from four areas in the world (U.S., Japan, Western Europe, and Russia) have contributed chapters that summarize the projects in their organizations. The emphasis is on evolution of development projects, lessons learned, successes, problems, and potential for commercialization.

The text is published in two volumes. The first volume highlights the development work carried out by the gas turbine equipment manufacturers and emphasizes design and testing of ceramic components and engines. This volume is organized by each of the four regions and there is an introductory chapter for the work in the U.S.A., Japan, Western Europe, and Russia.

The second volume summarizes the work of the component suppliers and materials laboratories. The laboratory work is predominantly organized by interest area, i.e. monolithic materials and components development, ceramic matrix composite development, materials characterization, and special challenges. This volume is concluded with a summary chapter that highlights the overall perspective.

The editors hope that the two volumes of this text will spark interest of the gas turbine design and ceramic component fabricating communities, and within other organizations that relate to these industries. Universities that educate future engineers that will be working in the gas turbine related industries will find much in this text that is valuable and intellectually challenging.

The authors retain full responsibility for the material presented in their chapters and they deserve the credit for it. The editors are greatly indebted to their dedication in presenting their material; they have merely had the pleasure of structuring the overall text and providing editorial assistance where appropriate.

Mark van Roode, Mattison K. Ferber, David W. Richerson

Acknowledgements

The initial objective of this 2-volume treatise was to document the worldwide efforts and progress involved in applying ceramic materials to gas turbine engines. Considering the diverse nature of the technologies involved and a timeline that goes back at least to the late 1960s, meeting the objective turned out to be quite a challenge. The editors are indebted to all of the authors who agreed to be part of this challenge, to the many individuals who helped us identify and recruit the authors, to the individuals and organizations who funded and managed the programs (many on a national scale), and to the hundreds of individuals who participated in the programs. In addition the editors thank ASME Press for their willingness to publish the treatise, the ASME editorial team, and all of the individuals who were willing to proofread draft chapters and provide feedback.

The editors especially acknowledge with great appreciation the support of the following individuals:

Mr. Masafumi Chikaraishi, Materials Manager of Kyocera Corporation, who was fulfilling an assignment with the Japanese Fine Ceramics Association, early on in the project invited representatives of the Japanese organizations that could contribute to this book for a meeting in Tokyo where one of the editors (Mark van Roode) was given the opportunity to explain the book project. Dr. Takane Itoh of Tokay University attended that meeting and was extremely helfpul in enlisting contributors for the various chapters, as well as being the prime author on the chapter in Volume I that introduces the Japanese ceramic programs. He further contributed as the prime author on the chapter reviewing the Japanese automotive ceramic programs. Without the help of Mr. Chikaraishi and Dr. Itoh, the Japanese contributions would not have been as numerous and complete.

Prof. Dr. Anatoly Soudarev, Assistant Managing Director at the Research Technological Institute for Power Engineering in St. Petersburg, Russia, very effectively coordinated the chapters from Russia and the Ukraine. He also organized the translation of the Russian chapter text into English, a job the editors were ill equipped to handle. The editors also recognize the helpful suggestions of Dr. Vladimir Shirmanov of Energomash, St.Petersburg, Russia, on the technical content of some of the chapters.

For the early work in Germany, the editors appreciate the comments of Dr. Wolfgang Bunk and Dr. Manfred Böhmer (retired from DFVLR), Dr. Jörgen Heinrich of the Technical University of Clausthal, Dr. Fritz Thümmler, Professor Emeritus of the University of Karlsruhe, Prof. Dr. Fritz Aldinger of the University of Stuttgart, Dr. Ernst Gugel, former Head of R&D of CFI, Dr. Gerhard Wötting of CFI, and Dr. Bjoern Schenk of Honeywell Engines, Systems & Services. The editors also thank Mr. Manfred Langer of Volskwagen AG, Dr. Klaus Mörgenthaler of DaimlerChrysler AG, Mr. Gerhard Andrees of Motoren Turbinen Union, and Dr. Ernst Gugel for their comments on the chapter on German automotive development in Volume I. Dr. Eberhard Seitz of the Research Center Jülich, Ltd., reviewed the introductory chapter on ceramic gas turbine development in Western Europe. Dr. Peter Walzer supplied supplemental literature for the automotive gas turbine development work in Germany. Prof. Dr.-Ing. Helmut Pucher of the Technical University of Berlin provided several names of researchers that were contacted to review the German ceramic gas turbine development activities.

Special thanks also to to Sandra L. Maldonado, Oak Ridge National Laboratory, and Austin K. Ferber for organizing and scanning the hundreds of drawings and photographs associated with the various figures.

The support of Dr. Michael Karnitz of ORNL throughout the book preparation project is also gratefully acknowledged.

Finally, the editors are much indebted to the support of Mary Grace Stephanchik, Acquistions Editor and Tara Smith, Production Coordinator at ASME Press who guided the project patiently through the publication preparation process.

Mark van Roode, Mattison K. Ferber, David W. Richerson

ABOUT THE EDITORS

Mark van Roode has been active in scientific and technical research and development since he received his Ph.D. in Chemistry from the University of Calgary in 1975. His varied assignments in chemistry include an assistant professorship at Memorial University of Newfoundland, Sir Wilfred Grenfell College, Corner Brook, Newfoundland, Canada, and a post-doctoral fellowship in pre-biotic chemistry at the Salk Institute in San Diego, California. He became involved in industrial research in ceramics in 1981 when he joined the Ontario Research Foundation, Mississauga, Ontario, Canada. In 1984 Dr. van Roode joined Solar Turbines Incorporated as a development engineer. He worked for a number of years on the development of ceramic and metallic coatings for oxidation and corrosion protection of gas turbine hot section components. When Solar was awarded the contract for the Ceramic Stationary Gas Turbine (CSGT) program by the U.S. Department of Energy in 1992, Dr. van Roode became the program manager with responsibility for the programmatic and technical leadership of the program. Under this program Solar and its subcontractors designed ceramic blades, nozzles, and combustor liners for the Solar Centaur 50 engine. These components were extensively tested in in-house engine testing at Solar and in field testing at several industrial sites. He is currently the Manager of Development Engineering at Solar, with responsibility for contracts in advanced materials development and technology strategy for gas turbines. Dr. van Roode has been active in the scientific community and has published extensively in the area of ceramic gas turbine development. He served on the executive of the ceramics division of the International Gas Turbine Institute of ASME from 1994 to 1999.

Mattison K. Ferber is currently employed as a Senior Research Staff Member in the Metals and Ceramic Division at the Oak Ridge National Laboratory in Oak Ridge, Tennessee. Dr. Ferber received his B.S., M.S., and Ph.D. degrees in Ceramic Engineering at the University of Illinois. From 1981 until 1984 Dr. Ferber was employed at the Oak Ridge National Laboratory as a Research Staff Member in the Structural Ceramics Group. In 1987 he joined the Department of Ceramic Engineering at the University of Illinois in Urbana-Champaign as an Associate Professor. From 1987 until 1996 he was the Group Leader of the Mechanical Characterization and Analysis Group in the High Temperature Materials Laboratory. Dr. Ferber, who is currently a member of the Structural Ceramics Group, is author or coauthor on more than 160 publications including 130 technical papers. Dr. Ferber's honorary include Tau Beta Pi, Phi Kappa Phi, Keramos, and Bronze Tablet. He is currently affiliated with the Engineering and Basic Science Divisions of the American Ceramic Society. Dr. Ferber is a Fellow of the American Ceramic Society.

David W. Richerson has been one of the long-term pioneers in efforts to develop ceramics and supporting technologies for turbines. Following his undergraduate studies in Ceramic Engineering at the University of Utah and his graduate studies in Ceramic Science at Pennsylvania State University, Mr. Richerson joined Norton Company in Worcester, Massachusetts, in 1969. There he developed the first commercial silicon nitride in the U.S. (NC-132 hot pressed silicon nitride) and fabricated many prototype components such as stator vanes and bearings. He then moved to AiResearch Manufacturing Company of Arizona (presently Honeywell Engines, Systems & Services) in Phoenix, AZ and spent the next 11.5 years guiding efforts to introduce ceramics into turbine engines. This included materials and fabrication development, characterization, design and life prediction, NDE, rig and engine testing, and failure analysis. In addition he taught courses at Arizona State University, wrote the textbook *Modern Ceramic Engineering: Properties, Fabrication, and Use in Design*, participated in National Materials Advisory Board studies, and wrote a white paper for the ASME Gas Turbine ceramics committee that became the model for the DOE/ORNL Ceramic Technology for Advanced Heat Engines Project. Mr. Richerson temporarily left gas turbine development efforts in 1985 to become Director of R&D (and later Vice President) at Ceramatec, Inc. in Salt Lake City, Utah. However, in 1991 he established his own consulting business and again focused much of his efforts on ceramics for gas turbines. Some of these efforts included being a major participant in the Solar Turbines CSGT program, supporting the DOE ATS project, helping with ceramic matrix composites development under the CFCC program, and currently supporting the DOE Microturbines programs. In addition Mr. Richerson has been active in education. Efforts have included teaching at the University of Utah, teaching numerous short courses internationally, writing a home study course on ceramics for ASM, Int., writing a second edition of *Modern Ceramic Engineering* and chapters on ceramics and composites for numerous books, establishing an education committee for the American Ceramic Society and being its chair since 1991, and writing the education outreach book *The Magic of Ceramics*.

CHAPTER

1

The Benefits and Challenges of the Use of Ceramics in Gas Turbines

Don Anson
Battelle Memorial Institute, Columbus, Ohio, USA

David W. Richerson
University of Utah, Salt Lake City, Utah, USA

INTRODUCTION

A gas turbine engine imposes severe challenges on its materials of construction, especially the materials of the rotors, nozzle guide vanes, and combustor liner. These challenges include high temperature, thermal stress, centrifugal stress, contact stress, high frequency and low frequency cyclic fatigue, static fatigue (stress rupture), creep, oxidation, and corrosion. The earliest turbines were constructed of metal alloys. However, the alloys available in the 1940s and 1950s were very limited in their temperature capability and required extreme cooling of the combustion gases, which resulted in low efficiency. Over the years there have been continuous improvements in metal alloys and in cooling schemes to allow turbine engines to operate at incremental increases in turbine inlet temperature (TIT) and thus at increased efficiency.

Ceramics have long been recognized as a potential alternate route to increased TIT and increased engine efficiency. A variety of oxides, borides, carbides, and cermets were evaluated in the 1940s and 1950s. Some ceramics had favorable strength and oxidation resistance, but none survived the severe thermal shock conditions imposed by the engines. Some ceramics could survive thermal shock and impact conditions but did not have adequate oxidation resistance and stress rupture life. During the 1960s, new ceramic materials based on silicon nitride and silicon carbide were invented that had improved properties and were more resistant to thermal shock than prior ceramics. Programs were started in the late 1960s and early 1970s to evaluate these new ceramics for turbine components.

After 30 years of continuous development, ceramic materials are just beginning to be qualified for gas turbine use (with the exception of ceramic thermal barrier coatings that have been in production use for many years). Why has it taken so long? The answer is complex and involves a number of factors:

(1) Due to their brittle fracture behavior a whole new design and life prediction methodology based on a probabilistic rather than deterministic approach needed to be developed for ceramics.
(2) 1970s vintage ceramics did not have adequate life or reliability, so many iterations of materials development and refinement have been required.
(3) Adequate property test methods, standards, and property databases did not exist in the early 1970s for candidate silicon nitride and silicon carbide ceramics.
(4) Each new improved ceramic required detailed property measurement before it could be selected as a turbine engine component candidate.
(5) No one at the ceramic suppliers or engine companies in 1970 had the technology to fabricate the required turbine component shapes. This shape-forming capability required many years to develop, followed by many more years to achieve the uniformity and properties required for turbines, followed by more years to reduce cost to viable levels (still in progress).
(6) The high price of prototype ceramic components and the long delivery times inhibited testing of ceramic hardware.
(7) Because of the low toughness and non-ductile behavior of ceramics, the critical flaw size (microstructural flaw that could cause failure) is much smaller than for metals and more difficult to detect nondestructively. New nondestructive evaluation (NDE) techniques and proof tests had to be developed and refined.
(8) Various problems such as low tolerance to impact and contact stress, high rate of slow crack growth in early materials, and inadequate oxidation resistance in the turbine environment have been encountered and required R&D to overcome.

These all turned out to be difficult challenges that were solved in an incremental fashion to progress from one plateau of technology and success to the next plateau. However, once a new plateau was reached, another problem or deficiency would be identified that had to be resolved before advancing to the next plateau. For example, early hot pressed silicon nitride was demonstrated successfully as a structural material for rotor blades attached to a metal hub for short life application, but did not have adequate high temperature properties for longer term use. It also was judged to be too expensive for commercial turbine applications, so an alternate lower cost fabrication approach with comparable properties had to be developed.

This introductory chapter provides a brief review of some basic design considerations for gas turbine engines and discusses the reasons why ceramics are important candidate materials. It then discusses some of the special design considerations for ceramics.

BASIC GAS TURBINE DESIGN CONSIDERATIONS

Thermodynamic Incentives to Increase Temperatures

In its most simple form the gas turbine (or combustion turbine) consists of a compressor, a combustor, and a turbine. The compressor delivers compressed air to the combustor. Fuel is mixed with the compressed air, and the mixture is burned in the combustor. The hot combustion products pass at high pressure and velocity into the turbine, are expanded down to near ambient pressure, and finally exhausted. The power generated by the turbine is used to drive the compressor and to provide a useful output. The corresponding ideal thermodynamic cycle (the Brayton Cycle) is characterized by isentropic compression, combustion at constant pressure, isentropic expansion, and heat rejection at constant pressure.

With a perfect working fluid, having constant thermodynamic properties, the efficiency of the Brayton cycle depends only on the pressure ratio. The pressure alone determines the relationship between

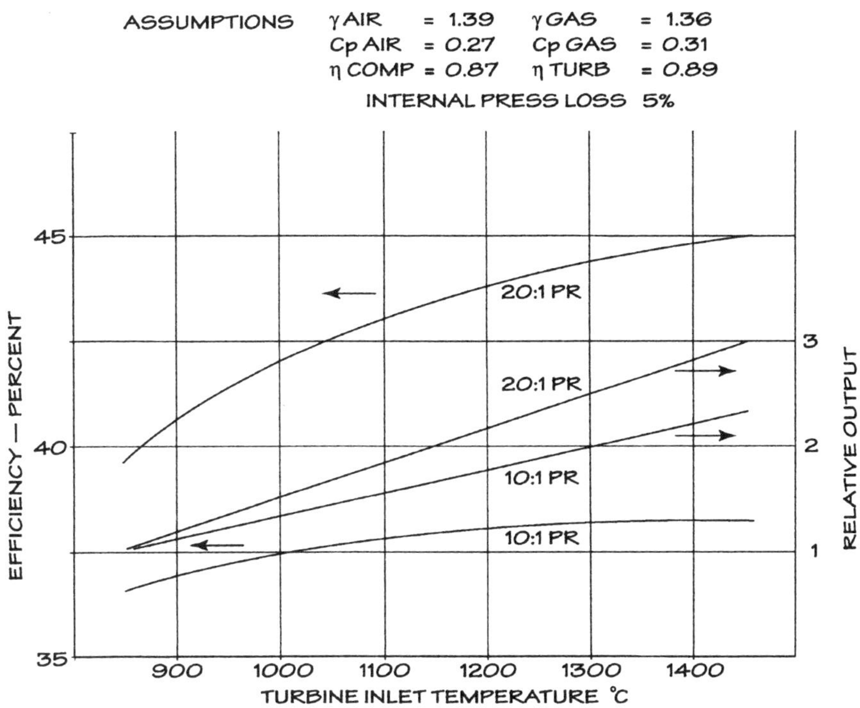

Figure 1.1 Effect of TIT on Efficiency and Specific Output

the temperatures at the beginning and end of the isentropic processes. The compressor inlet condition corresponds to the ambient, and the turbine inlet temperature (TIT) is determined by the heat released by combustion. The turbine output for a given pressure ratio depends on the amount of energy supplied by combustion. It is therefore directly related to the maximum temperature achieved in the combustor.

In practice the situation is more complex; neither the compression nor the expansion process is isentropic, there are pressure losses due to friction and turbulence in the combustion section, and the properties of the working fluid change with temperature. Figure 1.1 shows the theoretical performances of two gas turbines with different pressure ratios as functions of the turbine inlet temperature, assuming the values listed for gas properties and component efficiencies. The theoretical efficiencies are on the order of 60 percent of the ideal efficiencies when these realistic, but still approximate, values are assumed. Note that the effect on efficiency of increasing pressure ratio is more marked than the effect of increasing temperature, while the reverse is true for the effect on output. Also, for a pressure ratio of 10 the efficiency changes very little above 1200°C, and in fact it falls off at temperatures above about 1500°C. The region of peak efficiency occurs at higher temperature as pressure ratio increases, and is also a function of the initial assumptions. In the case of the ideal conditions there is no maximum. These trends explain the rapid development towards higher temperatures and pressure ratios that has occurred, especially in aircraft engines.

From the beginning the aircraft industry has been the largest user of gas turbine engines, and the course of development has been dominated by the needs of that industry. Two primary needs are a high power to weight ratio and low frontal area. Peak power is needed for short periods at takeoff and can be achieved by operating the engine at a TIT above that corresponding to peak efficiency. This allows TIT to be reduced to the most economical condition for cruise and to extend engine life. While this reasoning has propelled the development of aircraft engines, it is not necessarily the best course for the development of

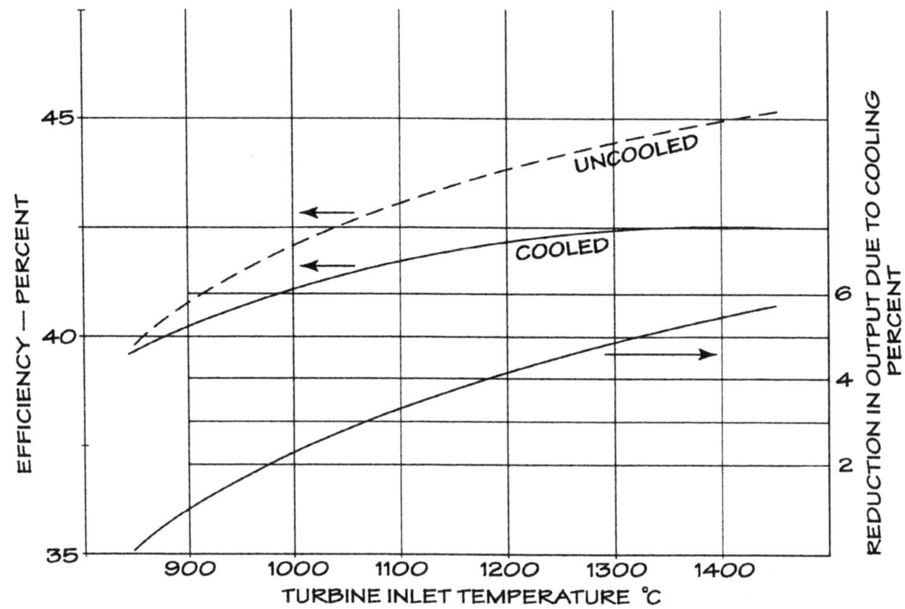

Figure 1.2 Effect of Cooling on Turbine Performance

land based engines in which there is a primary need for continuous operation near full power with high efficiency and reliability.

The high temperatures used in modern turbines require the cooling of components in the hot gas path. Cooling is effected by taking air from the compressor and ducting it to the hot components. After serving as coolant this air is returned to the turbine, but it may bypass one or more of the turbine stages and thereby incurs and efficiency penalty. Figure 1.2 illustrates the effect on turbine performance, based on the assumption that the use of air for cooling incurs a loss of one percent of turbine efficiency for each increment of 200°C in TIT above 850°C. At 1400°C this results in a drop of over two percentage points in overall efficiency and over five percent loss in output. Peak efficiency is now achieved at 1350°C. While it is still worthwhile to operate the cooled engine at 1500°C or more in an aircraft in the interests of weight reduction, it would be sensible to stop at 1350°C in a land based engine. Referring to Figure 1.1, it will be seen that this would still result in an increase in power by a factor of more than two and a gain in efficiency of some three points compared with the uncooled engine operating at TIT of 850°C. Figure 1.2 also illustrates the value of materials that can operate at high temperatures without the need for cooling.

Figure 1.3 shows how two different heavy frame power generation turbines have developed over the last two decades. The apparent superiority of turbine 'A' over turbine 'B' is not significant because of slightly different ways of defining the operating temperature. In both cases the development process has included changes in pressure ratio and mass flow rate as well as TIT. The steeper rise in combined cycle efficiency for turbine 'A' is significant and is largely attributable to the adoption of more advanced steam conditions in the bottoming cycle, made possible by increases in gas turbine exhaust temperature as TIT increased. Manufacturers of large turbines for power generation at base load tend to restrain pressure ratios in order to raise exhaust temperatures and benefit the steam cycle that uses the hot gases in combined cycle arrangements. The overall development trends of these two turbines are consistent with one another and with the foregoing discussion. The lower efficiencies quoted for the real engines compared to those of the hypothetical engines considered earlier are the result of using much more detailed analyses and including various parasitic losses including electrical generation losses.

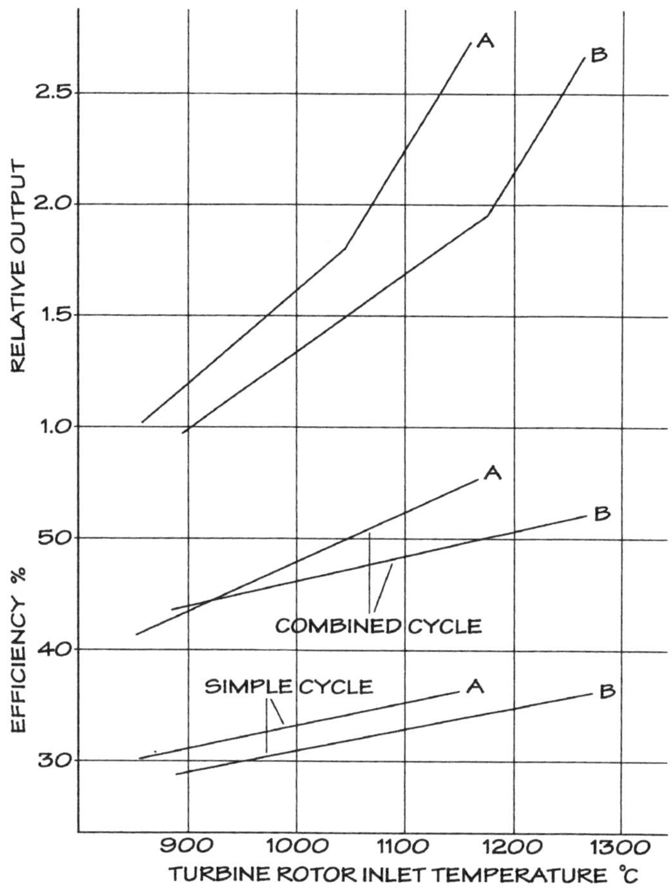

Figure 1.3 Development of Large Industrial Gas Turbines for Power Generation

Role of Material Improvements

Advances in TIT have been made possible by improvements in high temperature materials and innovative cooling techniques, especially those applicable to turbine airfoils. As might be expected these have been largely the result of pressures from the aircraft industry, often with industrial turbine manufacturers following suit at a later date after costs have been reduced and long term reliability proven. Figure 1.4 illustrates the relationships between the historical development of advanced materials and the adoption of higher values of TIT, using both better materials and cooling techniques.

In the 1950s wrought alloys were used for turbine blades to achieve the necessary strength and uniformity of quality. With the advent of new foundry techniques a decade later it became possible to cast blades with equivalent or better properties and also to make blades with cored cooling channels. There followed a series of refinements in the design of cooling techniques to exploit these possibilities, the ability to make castings with controlled orientation of the grains and finally single crystal castings. Each major metallurgical advance resulted in an increase of about 50°C in the temperature capability of the material, and the TIT values that could be used with uncooled components increased from about 800°C to 950°C or more. Advanced cooling techniques allowed even greater changes, and TIT levels well above the service temperatures of the metals became possible.

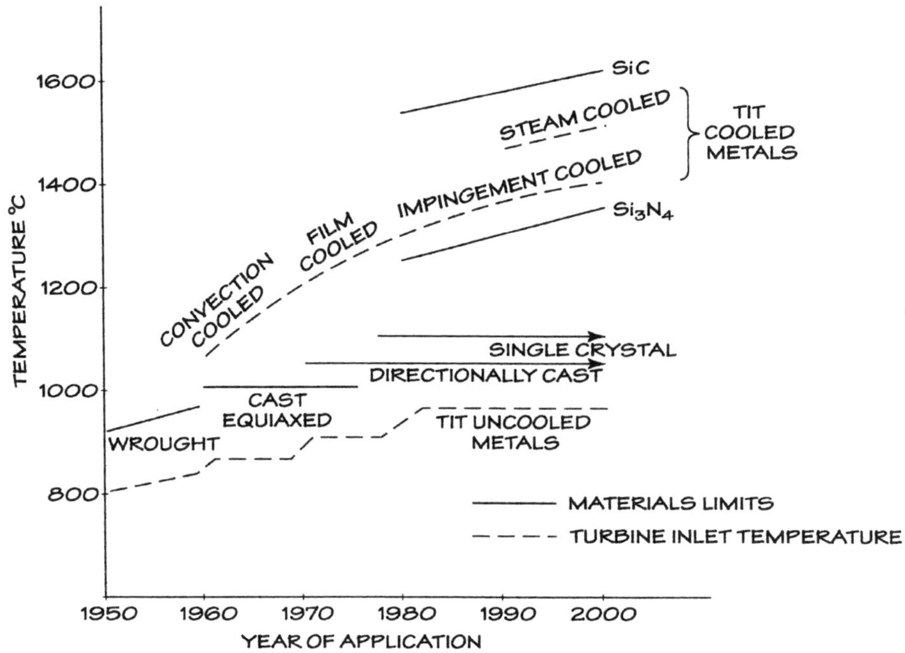

Figure 1.4 Turbine Blade Technology vs. Years of Application

As TIT increased the second and sometimes later turbine stages needed cooling and increasing amounts of air were extracted from the compressor for this purpose, with an accompanying performance penalty and a reduced amount of air available for the combustor. Steam has been used as coolant to alleviate this problem, permitting a further increase in TIT. The use of thermal barrier coatings on the metal surfaces offers another means of protection against high gas temperatures. As can be seen from Figure 1.4, ceramic materials such as silicon nitride and silicon carbide would permit the attainment of similar temperatures without cooling. Although the limits for silicon nitride are somewhat below those obtainable with cooled metals, this would be offset by the reduced cooling penalties.

DESIGNING FOR CERAMICS

For centuries ceramic materials have been used where refractory properties were required. The more common structural metals all suffer rapid deterioration as temperatures increase above about 800°C. Oxide based ceramics such as those made from natural clays retain their strength much better, although historically their properties have fallen far short of those required for most engineering applications. Recent advances in ceramic technology, including the use of synthetic materials such as silicon carbide and silicon nitride, offer the prospect of far superior physical and mechanical properties that can be retained at temperatures in the range of 1000°C to 1500°C. Such materials hold out the promise of achieving higher TIT when used for uncooled blades and nozzles and for other components in the hot gas path. Even without an increase in TIT the elimination of cooling could produce significant improvements in both efficiency and power. The use of ceramics for combustor liners offers a further advantage because the elimination of cooled surfaces facilitates the progress of chemical reactions and thus decreases the emission of pollutants.

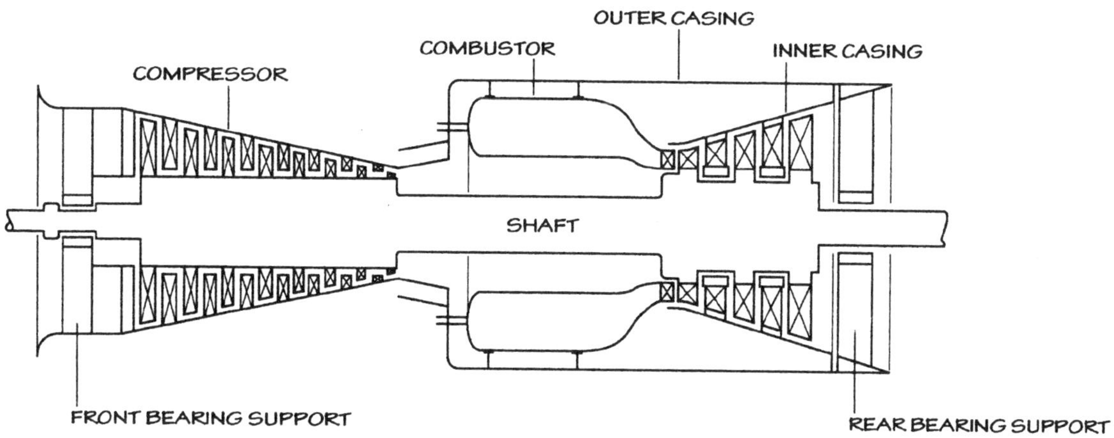

Figure 1.5 Parts of a Gas Turbine

Physical Layout and Structure and Key Issues

Figure 1.5 represents a generic single shaft gas turbine, which is about the simplest conceivable layout for a practical machine and is widely used in industrial applications. The essential functional parts are the compressor (axial flow in the illustration), the combustor, and the turbine. The compressor and turbine rotors are carried on a common shaft that is supported by two journal bearings and a thrust bearing at the compressor inlet end. The combustor is shown as an annular enclosure but alternatively it may consist of several smaller cylindrical enclosures leading into the turbine via transition sections. The outer turbine casing, which is built up of several essentially cylindrical and conical sections, constitutes the supporting structure for the whole assembly. It is protected from the hot gases leaving the combustor by the inner casing which is cooled on the outside by air at compressor delivery temperature while containing the hot gas stream from the combustor. The inner casing is supported from the outer casing, and in turn it carries the non-rotating elements of the turbine, including the vanes, interstage diaphragms, and rotor tip seal rings.

Figure 1.6 illustrates in more detail how the turbine blades, vanes, and tip seal rings relate to one another in the hot gas flow path and identifies the blade tip clearance and diaphragm/shaft seal areas that are critical features of turbine design. Ceramics are usually considered for use in the combustor, the turbine nozzle vanes and the turbine blades, which are the components contacting the hottest gases.

A fundamental principle in gas turbine layout is that the inner and outer casings and the shaft should each be able to expand and contract freely without contact between the stationary and rotating elements except at the bearings. This means maintaining strict concentricity and understanding in detail how the various components respond to temperature changes that accompany load changes. Adjoining stationary components may have telescoping joints for example to allow inner members to expand independently of the outer casing. Axial clearances between rotating parts and stationary parts must be maintained at all times. The coefficients of expansion of the various materials and their thermal conductivities become very important. Ceramics have very different thermal expansion characteristics than metals, and this can create design problems.

Where ceramic components are supported by metal components a number of problems must be addressed. These arise from differences in the rate at which the materials expand during temperature transients, differences in the extents of expansion at equilibrium, differences in temperature if the metal is cooled, and from relative movement at points of contact. The brittleness of ceramics makes them very unforgiving towards high contact stresses that may be exacerbated by thermally induced stresses if the metal is

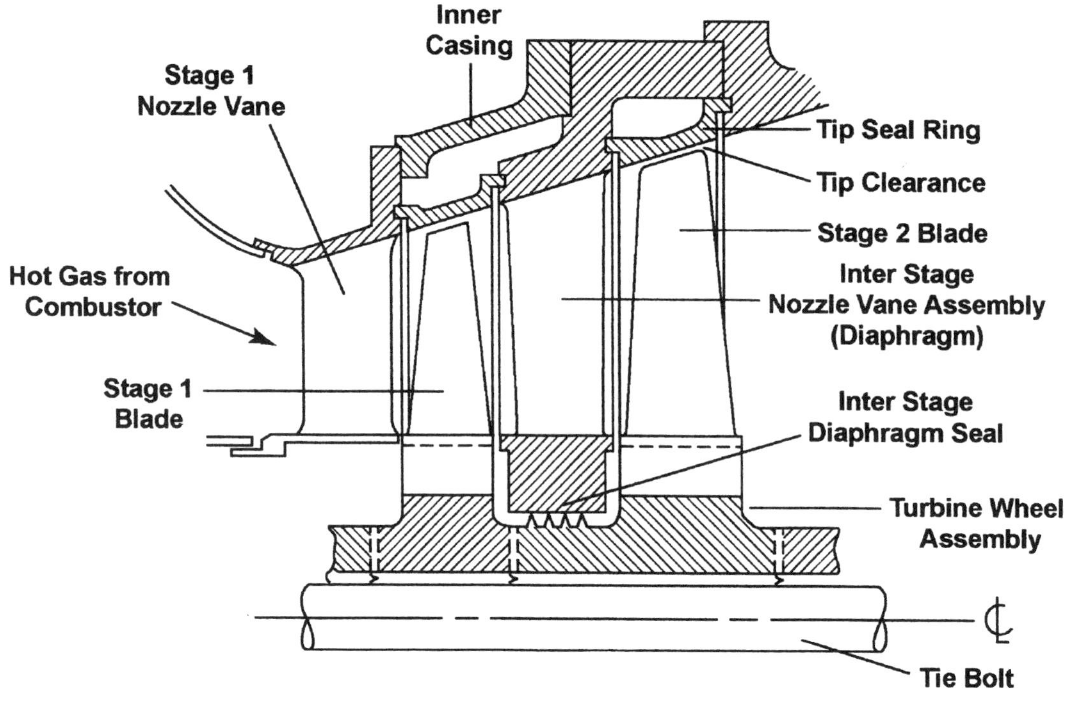

Figure 1.6 Components in Turbine Hot Gas Path

cooler than the ceramic, which is usually the case. Any distortion of the metal supporting structure may lead to a concentration of load on the ceramic component and ultimately to failure. Ideally, monolithic ceramic components should be simply supported, and rigid attachments should not be used. At high temperatures sliding joints involving ceramics generally are not acceptable because of possible chemical interaction and galling of contacting surfaces.

Clearly in the design of a hybrid structure such as a turbine interstage assembly, it is very difficult to satisfy all the above conditions. In the past, designers of all-metal gas turbines have utilized a number of devices to permit relatively free expansion of adjacent components, relying on the elastic properties and strain tolerances of the materials to accommodate minor misalignment and other imperfections. These devices often are not applicable to ceramics and the design of metal/ceramic hybrid structures has in the past called for very innovative approaches and will continue to do so.

The rotor represents a significant design and fabrication challenge. Key issues that must be resolved for a rotor to be successful include the following:

(1) The material in the hub region must withstand high centrifugal stress over a substantial volume of material, but does not require particularly high temperature capability. If this region is fabricated from ceramic, the critical flaw size for most turbine applications is less than 50μm, so highly refined fabrication and nondestructive inspection is required.
(2) An attachment is required to metal either by joining an integral ceramic rotor to a metal shaft or by inserting ceramic blades into a metal hub. Both involve high contact stress loads and challenges of stress concentration due to relative movement (resulting from the centrifugal stress and also from thermal expansion mismatch of the metal and ceramic).
(3) Precise clearances are typically required to control flow of gases and also to assure that stresses are carried in the desired path. This requires a high degree of fabrication control including expensive finish

grinding with diamond tooling and sophisticated measurement techniques. Careful design of interfaces and seals and of the tip clearance between the blades and a tip shroud are also required. The use of an abradable tip shroud coating that can tolerate a blade rub may be needed as well.

(4) The rotor blades are exposed to much higher temperatures and thermal shock than the hub and also are more difficult to fabricate. Because machining of ceramics is so expensive (and can also cause surface defects that decrease strength), the airfoil contours for small rotors must be fabricated to net shape, especially for radial rotor configurations. The ceramic materials must have superior creep resistance and stress rupture life.

(5) Rotor blades are vulnerable to occasional foreign object damage, (FOD) from debris introduced from outside the engine and also from carbon chunks that can form in the engine and break loose. Thus, the rotor blades must be designed to be robust enough to survive high- probability impact (such as from carbon) and also must be of a ceramic material of high strength and toughness.

(6) Rotor blades are vibrationally excited by the gases flowing through the nozzle guide vanes. Ceramic blades can easily fracture if they are driven at their natural flexural frequency during engine startup or operation. Great care is required in design to avoid a critical condition. Boundary conditions and damping must be better understood and controlled than for metals.

Turbine engine companies have explored many engine and rotor designs in an effort to accommodate ceramics. The various approaches are described in detail by each engine company in subsequent chapters. Contact stress in particular has been a recurring problem.

Nozzle guide vanes represent different challenges than the rotor. Attachment and control of clearances are especially important. Because the nozzle must be attached to the main structure of the engine, there is high likelihood of cantilevered motion/loading and thermal expansion mismatch, both of which cause the challenge of being sure that the loads are carried in the path desired by the designer. Other key challenges are high temperature and thermal shock. The first stage nozzle airfoils are exposed to the peak TIT (including the worst temperature spikes in the pattern factor) and to very high transient and steady state thermal stresses. The nozzle vanes also must withstand the most severe oxidation and corrosion conditions.

Numerous approaches have been taken by different engine companies to address the issues of nozzle design. These have included everything from individual airfoils piloted by slotted shrouds to one-piece nozzle rings with the airfoils and shrouds integral. These approaches are described in subsequent chapters.

Combustors vary substantially for different types of turbine engines and for different engine companies, especially in recent years as efforts have been directed towards minimizing pollution emissions. The combustors can be staged, in-line annular, in-line reverse annular, side mounted with a scroll transition, or other configuration. Ceramic linings have been tried as single cylinders, rings, slats, and refractory tiles. Both monolithic and composite ceramic liners have been designed and tested. Subsequent chapters describe a variety of ceramic combustor liner designs and test results.

Stress Analysis, Life Prediction, and Database

Because of their low strain to failure and other characteristics, ceramics require more careful stress analysis than metals. Stress analysis of complex shapes during thermal transients has required the refinement of finite element computational techniques and would not have been practical without the advances in computer technology that have occured in recent years. Codes developed at engine companies and the CARES codes developed by the National Aeronautical and Space Adminstration (NASA) have been essential to the design of ceramic components. In most cases thermally induced stresses are predominant, and a necessary step in the stress analysis process is the calculation of heat transfer rates and heat conduction within the structure. As a general rule the use of relatively thin and uniform sections is desirable in ceramic structures. This preference sometimes conflicts with the needs for mechanical strength, and several design iterations and stress analyses may be needed before an acceptable design emerges. At each step it is important to take into account all factors that contribute to stress patterns, including highly concentrated contact loads that may occur.

In predicting the endurance of ceramic components, it is also necessary to allow for time-dependent factor such as creep, slow crack growth, and oxidation/corrosion. Over extended periods these phenomena can result in significant changes in strength, so a database generated by long term testing (including testing under the best possible simulation of the engine operating environment) is essential. The interactions that must be addressed by design and analysis are complex and are still not as well understood (and accounted for) as with metals. For example while creep is usually thought of as a negative, in some cases it can lead to substantial redistribution of stress, which can be beneficial.

Impact Resistance

The brittleness and relatively low toughness of ceramics makes them liable to catastrophic failure in the event of impact by a foreign body. FOD may occur as the result of ingestion of objects with the air, by loose turbine parts, or by the detachment of carbon deposits from the combustor. If the impact causes breakage of a ceramic turbine blade, the entire blade row will likely be destroyed, as well as many downstream ceramic components. The FOD problem requires special design attention such as adoption of more robust airfoils. However, this may impair aerodynamic efficiency, so considerable design effort and iterations are typically required to address all of the tradeoffs in order to develop an acceptable design.

Scale Effects

The strength of monolithic (non-composite) ceramics is dependent on the volume and area under stress. This is because the strength is controlled by the size of defects in the volume and surface of the ceramic. Larger ceramic parts have a higher probability of containing a critical defect and thus a higher probability of fracture. This phenomenon requires a special probabilistic design approach and extra care in analysis and in fabrication/inspection. Continuous fiber-reinforced ceramic composites do not have the same flaw sensitivity as monolithic ceramics, so they are amenable to be fabricated into larger components such as combustor liners and transition structures. However, fiber-reinforced composites have highly directional architecture and properties and require a design methodology different from that for monolithic ceramics or metals.

SUMMARY

This chapter has described the characteristics of gas turbine engines and the reasons that ceramics could provide substantial benefits. It has also introduced and briefly discussed some of the challenges of trying to design and implement ceramic gas turbine components. Enormous progress has been made during the past 30 years, but there are still challenges that require further effort: cost reduction of ceramic components, improved resistance to FOD, long term resistance to contact stress at key interfaces, and resistance to accelerated oxidation/corrosion of silicon-based ceramics in moist high temperature engine environments. The remainder of this volume and the subsequent volume review the various programs that have brought the technology to its present level. The authors of each chapter describe the lessons learned and the challenges yet to overcome.

CHAPTER

2

Evolution of Ceramic Gas Turbine Development Programs at Engine Manufacturers in the United States

David W. Richerson
University of Utah, Salt Lake City, Utah, USA

Don Anson
Battelle Memorial Institute, Columbus, Ohio, USA

INTRODUCTION

Chapter 1 identified some of the challenges of implementing ceramic materials in gas turbine engines. To address and solve these challenges, engine and technology development programs have been conducted worldwide, especially since about 1970. The objective of this chapter is to provide a brief timeline and overview of the major engine programs that have been conducted in the United States. Subsequent chapters provide detailed descriptions of the individual programs and the evolution of ceramic engine technology at the various gas turbine engine companies. Many support technology (materials, property characterization, nondestructive evaluation, design codes) development programs have been conducted in parallel to the engine programs and are reviewed in Volume 2.

Timeline of U.S. Ceramic Engine Development Programs

Table 2.1 provides a timeline of key engine programs in the U.S. that have contributed to the development of ceramic components for gas turbine engines. The table identifies the name of each program, the

TABLE 2.1 TIMELINE OF KEY U.S. PROGRAMS IN THE EVOLUTION OF CERAMICS FOR TURBINES

Program	Time	Sponsor and Engine Company	Scope, Key Accomplishments, Plateaus Reached, Lessons Learned
Various programs at NACA/NASA	1944–1960	NACA/NASA	Conducted property screening for oxides, borides, carbides, cermets; conducted turbojet engine design, especially for ceramic/cermet blades inserted into a metal hub; fabrication of ceramic and cermet engine components; engine testing. Demonstrated that 1944–1960 ceramics had inadequate properties (especialy thermal shock resistance); cermets had inadequate stress rupture life and oxidation resistance; metals designs did not work for ceramics; contact stress was a problem in blade attachment and required innovative designs.
Brittle Material Design – High Temperature Gas Turbine	1971–1979	ARPA/AMMRC, Ford Motor Co., Westinghouse	Goal: to demonstrate feasibility of design of brittle materials for both small and large gas turbine engines. Established firm foundation of methods and codes for finite element analysis of ceramics; demonstrated that ceramic gas turbine components fabricated from 1975-vintage ceramics could survive under transient and steady-state engine operation conditions; substantially improved RBSN, fabricated and tested a duo-density rotor, and developed a sintered RBSN; conducted extensive material characterization, component fabrication, and rig/engine testing. Clearly demonstrated the feasibility of use of ceramics for turbine components. Identified needs for improved regenerator material and sealing technology, for materials with better oxidation resistance, for solutions to fractures initiated by contact stress, and for cost reduction of ceramic components.
10-kW Radial Turbine Development	1970–1980	Programs supported by NAVAIR and MERADCOM, conducted at Solar Turbines	Initial goal: to solve a severe dust erosion problem of nozzle guide vanes of small field generator sets, especially used adjacent to helicopter traffic. Hot pressed silicon nitride demonstrated nearly 100-fold increase in erosion life compared to metal vane materials and performed successfully in engines. Efforts were extended to other turbine components including combustor liners, transfer ducting, and bolts in efforts to increase engine performance in addition to increasing dust erosion resistance.
Hybrid Rotor Evaluation	1974–1979	DARPA, Pratt and Whitney (West Palm Beach)	Goal: Use isothermal forging to bond two AF2–1DA circumferential axial turbine disk halves around silicon nitride blades to produce a hybrid rotor. Rotor was successfully fabricated with NC-132 hot pressed silicon nitride blades and a platinum compliant layer and spin tested at 45,000 rpm at 1230°C. This was a key step in demonstrating the viability of a hybrid rotor and the need for a compliant layer.
Ceramic Gas Turbine Engine Demonstration Program	1976–1981	DARPA/Navy, AiResearch Manufacturing Co. of Arizona (became Garrett Turbine Engine Company)	Goals: retrofit ceramics into existing production engine to demonstrate increased power and reduced fuel consumption. Program expanded silicon nitride materials database (including tensile tests), extended design and life prediction codes, improved NDE methods, and demonstrated 30% increase in power and 7% decrease in fuel consumption for short duration engine operation (2–10 hours). Achieved improved understanding of the problem of contact stress. Identified need for improvement in design and ceramic materials to achieve longer life engine capability.
Ceramic Components for Turbine Engines	1977–1980	AFWAL, AiResearch Mfg. of Arizona	Goals: improve silicon nitride materials properties, reduce fabrication cost, and demonstrate feasibility of a ceramic hot section for a limited-life application such as a cruise missile. Resulted in significant improvements in the strength of slip cast RBSN from the AiResearch Casting Company, demonstration of high yield of stator vane fabrication, cost reduction of hot pressed silicon nitride rotor blades, and successful redesign and rig testing of a second generation of the DARPA/Navy engine.

Heavy-Duty Gas Turbine Engine (HDGTE)	1976–1978	NASA Lewis, Detroit Diesel Allison Div. of General Motors	Goal: Study the potential benefits of ceramic components for the Allison GT404–4 truck engine. Predicted substantial benefits and led to the CATE program.
Ceramic Applications in Turbine Engines (CATE)	1978–1983	DOE/NASA Lewis, Detroit Diesel Allison Div. of General Motors	Goals: replace metal components with ceramic components to increase the axial flow gasifier turbine rotor inlet of the 404–4 engine from 1002°C to 1241°C and demonstrate the engine in a vehicle. 7475 hours of engine testing at 1038°C were accumulated for engines containing reaction-bonded SiC vanes and tip shroud and aluminum silicate regenerator disks. This included 231 hours (6656 miles – ~10,650 km) truck demonstration without incident over streets, highways, railroad crossings, and the GM proving grounds Belgian block and truck durability road courses. Interest shifted to automotive size engines, so no testing was conducted at the original goal of 1241°C.
Advanced Gas Turbine Powertrain System Development Project (AGT)	1979–1987	DOE/NASA Lewis, Detroit Diesel Allison Div. of GM	Goal: Design, build and test a 74.6 kW (100 hp) automotive turbine engine that would implement ceramics to achieve 42.8 miles per gallon average mileage (~ 18 km/l). The Allison AGT100 design consisted of a two-shaft engine using radial ceramic rotors for both the gas generator and power turbine. Nozzle guide vanes and shrouds, the turbine inlet scroll, the combustor liner, the regenerator disks, and various transition and seal parts all needed to be ceramics to meet the TIT of 1288°C and rotor speed of 86,240 rpm to meet the efficiency goal. Much of the program focus was on working with ceramic suppliers to develop new pressureless sintered ceramics, especially the Hexology SA SiC from Carborundum. The radial rotor and scroll represented severe challenges and required many development iterations. A number of ceramic components fractured during rig and engine tests, especially rotors and scrolls. One engine accumulated nearly 100 hours, 56 start/stop cycles, and a maximum TIT of 1080°C and failed due to impact. Further ceramic material development and an impact-resistant rotor design were identified as needs.
Advanced Gas Turbine Powertrain System Development Project (AGT)	1979–1987	DOE/NASA Lewis, Garrett Turbine Engine Company (later Garrett Auxiliary Power Div. of Allied Signal Aerospace Corp.), Ford Motor Company	Goal: Same as for the Allison AGT program. The Garrett AGT101 design was also radial, but single shaft. To achieve the efficiency goal, this engine had to operate at TIT of 1371°C and rotor speed of 100,000 rpm. Due to the higher temperature and speed plus use of a foil bearing and other unproved technologies, this engine was less conservative than the Allison engine. However, it required fewer ceramic parts and was perceived to have lower cost potential. Reaction bonded and sintered silicon nitride received extensive development for the static structure components and sintered silicon nitride and hot isostatically pressed (HIPed) silicon nitride for the radial rotor. Hundreds of hours of rig testing and 230 hours of engine testing were accumulated under the program, including an engine run to 100,000 rpm and 85 hours testing of an all-ceramic engine at 1204°C. In spite of this success, the engine did not meet performance goals because of regenerator leakage, and the design and ceramics were not demonstrated to be reliable.
10 kW–30 kW Radial Turbine	1980s	Sundstrand Power Systems (former Turbomach group of Solar Turbines)	Goal: Use ceramic components to increase power of small generator set for Army and other applications. Substantial fabrication and testing of silicon nitride vanes, shrouds, and combustor liners and some experience with silicon nitride radial rotor.
Carbon/Carbon Composites for Cruise Missile	1985–1989	Air Force, Williams International	Goal: Develop a C/C composite combustor liner for the F107 cruise missile engine. Was one of the earliest substantial efforts to explore ceramic-based composites for turbine engine improvement.

(continues)

TABLE 2.1 *(Continued)*

Program	Time	Sponsor and Engine Company	Scope, Key Accomplishments, Plateaus Reached, Lessons Learned
Ceramic Matrix Composites in Short Life Turbojets	1982-present	DARPA, Air Force, Navy, Williams International	Goal: Use ceramic matrix composites to increase thrust to volume for the WR24-7 and -8 turbojet engines for Navy target drone and other applications. Has resulted in fabrication and rig testing of combustor liners, nozzle guide vanes, and axial rotors. Also has included design, fabrication, and testing of silicon nitride rotor blades.
Engine Testing of COMPGLAS® Components	1984–1996	UTRC/P&W	Goal: Explore potential turbine applications of the COMPGLAS materials developed between 1974 and 1985 at United Technologies Research Center (UTRC) under funding from ONR, AFOSR, and United Technologies Corporation. As an example, high pressure compressor inner shrouds were successfully designed and fabricated and operated without failure for > 1600 hours in F119 engines. The technology reached bill-of-materials status, but the shroud was replaced by metals for cost reasons.
Advanced Turbine Technology Applications Project (ATTAP)	1987–1992	DOE/NASA Lewis, AlliedSignal Engines	Goal: To advance the technology readiness of the ceramic automotive gas turbine engine by using the AGT101 rigs and engine as test beds to evaluate improvements in ceramic components. Subcontract with Carborundum to make transition ducts and turbine back shrouds from sintered SiC. Subcontracts with Norton/TRW and Garrett Ceramics to develop HIP silicon nitride materials for rotors and nozzle vanes and procurements from NGK Locke and Kyocera of rotors. Designed a mixed flow rotor with improved impact resistance. Achieved all-ceramic engine operation for 13 hours and 78 starts reaching 90,000 rpm and 1288°C. Substantial improvements in material properties and component reliability achieved, but the overall engine still not reliable enough and the ceramic components too expensive for automotive application.
Advanced Turbine Technology Applications Project (ATTAP)	1987–1993	DOE/NASA Lewis, Allison Gas Turbine Div. of General Motors	Goal: Same goal as the Garrett program,except Allison switched from the AGT100 engine to their axial flow AGT5 engine. Designed and validated impact-resistant silicon nitride axial gasifier rotor. Demonstrated 100-hours cyclic durability of all-ceramic gasifier power section including peak cycle condition of 1371°C TIT. Program continued under HVTE-TS project.
Hybrid Vehicle Turbine Engine Technology Support (HVTE-TS)	1994–1998	DOE, Allison Turbines (later Rolls-Royce Allison)	Goal: Demonstrate full operating AGT-5 engine and measure emissions on a chassis dynamometer. Successfully completed a 300-hours cyclic durability test of an all-ceramic gasifier turbine stage. Demonstrated fully operational all-ceramic AGT-5 automotive gas turbine engine with 1371°C TIT. Demonstrated Tier 0 federal emissions levels over the EPA urban driving cycle on a chassis dynamometer. Completed a 550-hours cyclic durability test of one-piece extruded ceramic regenerator disks under design conditions.
Ceramic Turbine Engine Demonstration Project (CTEDP)	1993–2000	DOE, AlliedSignal Engines (later Honeywell Engines and Systems)	Goal: Address several remaining barriers to commercialization by increasing maturity of support technologies, by improving ceramic component production capability, and by achieving long-term engine testing (including field testing). Switched from the AGT101 engine to the Model 331–200 (CT) auxiliary power unit. Replaced the first stage blades and nozzles with ceramics. Demonstrated refinement in design technologies, pilot production level of ceramic component fabrication, over 6800 hours of endurance testing on site, and over 2300 hours of airborne field testing in APUs in commercial service. Early in the program encountered contact stress problems with ceramic blades inserted into a metal hub, but after extensive effort achieved a successful design and compliant layer and also demonstrated an all-ceramic rotor.

Radial Inflow Turbine Advanced Material Partnership Program	1990s	DARPA, Sundstrand Power System	Goal: Demonstrate low-cost fabrication of a mixed flow rotor at Kyocera with their hybrid molded SN252 silicon nitride, bring together potential users, and verify the quality of the rotor in an engine. Other objectives were to increase the thrust of the TJ-50 turbojet engine by 50% by increasing the turbine inlet temperature from 1120 to 1423°C. Rotor successfully fabricated and then proof tested at 142,000 rpm. Engine run at about 1315°C for one hour, but vibration higher than desired. Design modifications and further engine testing in progress.
GTCP Series APU Ceramic Component Insertion Program	1992–1995	USAF, AlliedSignal Engines	Goal: Conduct endurance testing of ceramic turbine nozzles in Air Force ground power carts. Accumulated over 17,000 hours of engine operation including 7,542 hours for individual ceramic nozzles. Encountered some non-catastrophic impact damage from hard combustor-generated carbon. Solved the problem by increasing the trailing edge thickness from 0.8 mm to 1.27 mm (without a performance penalty).
GTCP85 Series APU Airborne Field Evaluation	1995–2000	AlliedSignal Engines (later Honeywell Engines and Systems)	Goal: Achieve long-term field testing of silicon nitride nozzles in APUs in MD80 aircraft. Have accumulated over 40,000 hours.
Ceramic Stationary Gas Turbine Program (CSGT)	1992-present	DOE, Solar Turbines	Goal: To retrofit ceramics inot a Solar Centaur 50S industrial turbine to improve performance (efficiency, output power) and lower emissions of NOx and CO and to operate the engine in a 4000 hour field test. Program has demonstrated at end–user test sites 948 hours of continuous engine operation with silicon nitride blades and over 33,000 hours of engine operation with SiC/SiC ceramic matrix composite combustor liners. Longest test on a single set of liners was 13,937 hours. Engine tests have shown that the ceramic materials can withstand long term exposure to engine-induced stresses, but that interaction with the engine atmosphere causes accelerated surface recession compared to predictions based on prior laboratory tests. Collaboration with Oak Ridge National Laboratory has led to improved laboratory tests. Oxide environmental barrier coatings (EBCs) have been evaluated that significantly decrease the rate of surface recession, but further work is necessary.
Cooled Silicon Nitride Vane	Early 1990s to present	DARPA, DOE, ONR, UTRC/P&W	Goal: Develop a cooled silicon nitride vane for the 25.5 MW FT8 aeroderivative engine to allow operation at increased turbine inlet temperature. Individual vanes have been designed, fabricated, and tested enough to demonstrate viability of the concept. A key 50 hour rig demonstration is scheduled for 2001. Environmental Barrier Coating (EBC) technology developed under the High Speed Civil Transport–Enabling Propulsion Materials (HSCT-EPM) Program is being integrated into this effort.
Advanced Turbine Systems (ATS)	1995–2001	Rolls-Royce Allison	Goal: To test silicon nitride vanes in a Model 501–KB5 engine at an industrial end user test site. Two separate field tests were conducted. In Test 1, 820 hours of successful engine testing were completed. Post-test examination of the vanes determined that accelerated surface recession had occurred due to attack by the combustion gases. It appears that EBCs will be required to achieve acceptable life.
Ceramic Matrix Composites for Advanced Engine Components	1999–2002	NIST, Siemens Westinghouse Power Corp.	Goal: Demonstrate the viability of oxide-oxide ceramic matrix composites for selected turbine components.

engine companies involved, government sponsorship, and a brief summary of the scope of each program and highlights of accomplishments and lessons learned.

Many challenges have been encountered and solved through interations of improved design and improved materials. Engine operation life, temperature of engine operation, and reliability all have increased. However, challenges still remain. Specifically, ceramic materials properties still need further improvement and the cost of the ceramics needs to be reduced, environmental barrier coatings (EBCs) will probably be needed for long life at higher temperatures, impact-resistant and low contact stress designs will be needed, and further field testing will be necessary to assure end users that the ceramic components are reliable. The progress, the lessons learned, and the future needs are discussed in detail in subsequent chapters.

CHAPTER

3

Ford's Development of the 820 High Temperature Ceramic Gas Turbine Engine

Dale L. Hartsock

Ford Motor Company, Dearborn, Michigan, USA

INTRODUCTION

In 1967 Ford Motor Company, already having underway a heavy-duty gas turbine program, initiated a project to build the 820 high temperature ceramic gas turbine engine for a passenger car. This was one of the projects in the Ford Research Laboratories and was under the direction of Mr. Arthur F. McLean.

As is well known by engine designers, the efficiency of an ideal thermodynamics cycle for converting heat into work is directly related to the cycle maximum temperature. Achievement of a high maximum temperature in a practical heat engine is, of course, related to materials and design (e.g. cooling methods). In the case of the gas turbine, from its inception there has been continual pressure to improve engine efficiency and power-to-weight ratio through high temperature materials development. Today's nickel-chrome, cobalt-based, directionally solidified and single crystal superalloys are all a result of this persistent push.

Reaching higher temperatures and thus higher efficiencies was also the driving force for this new engine. The initial plan was to investigate the use of ceramics for the stationary components of the hot flow path and air-cooled superalloy rotors to achieve the high turbine inlet temperature desired for full power which, in the case of the passenger car, was only required for limited time periods during acceleration. However, design and experimental studies indicated that blade cooling of such small turbine wheels was impractical. Yet, it was also recognized that in order to make the gas turbine engine efficient enough for passenger car use, the turbine inlet temperature would have to be raised significantly above the metal engine limit of approximately 980 to 1040°C (1800 to 1900°F). Thus the program was modified to include uncooled ceramic turbine rotors with a turbine inlet temperature of 1370°C (2500°F).

This program was a total system iterative approach to high temperature structural design with ceramics in highly stressed applications focusing on the development of ceramic components for the gas turbine application. The total systems approach included major efforts in ceramic design, materials development, fabrication, process development, and test and evaluation methodology. Because of the high potential, albeit high risk, of this program government support was sought. Starting in July 1971, the Defense Advanced Research Projects Agency (DARPA) sponsored a joint program with the Ford Motor Company to demonstrate and encourage the use of brittle materials in engineering designs. The 820 engine components would be used to demonstrate advancements in the technology.

By the end of the main program in 1979, design techniques, material characterization & fabrication, and testing & evaluation had all been significantly advanced. The challenges and subsequent advances are described in this chapter. The details of this work are covered in over 3000 pages of reports and technical papers [1–51]. This chapter covers some of the more interesting design and development aspects of the work. The chapter by Mr. John Mangels [52] in volume II of this work covers the materials development.

At the time of this program, the potential advantages of this engine were many:

(1) The fuel economy would be significantly above competitive powertrains.
(2) The continuous combustion of the gas turbine should have inherently lower exhaust emissions than comparable reciprocating engines.
(3) The engine would have multifuel capability.
(4) It could be made very reliable and durable.
(5) There would be low maintenance.
(6) It would be a smooth vibration-free system.
(7) It should have lower oil consumption.
(8) It would have good cold starting capabilities.
(9) It would have rapid warm up time.

TIMELINE

This ceramic gas turbine program covered many years, and some of the major events are marked on the following timeline:

1967 R&D program started at Ford to design and develop a ceramic gas turbine engine.
1970 By the end of 1970, the results of design and experimental studies indicated that the focus should be on an all-ceramic flowpath rather than using air-cooled rotors blades.
1971 Advanced Research Projects Agency (ARPA) formed joint program with Ford to encourage the use of brittle materials in engineering design [20].
1973 Over 100 hours of static testing completed on first stage ceramic stator.
1973 Ceramic turbine rotor spun to over 50,000 rpm before failure.
1974 Fifty hours of dynamic testing (with metal rotors) was completed on a full set of stationary parts.
1977 Department of Energy (DOE) provided additional funding to develop the duo-density ceramic gas turbine rotor.
1977 Ceramic turbine rotor cold spin tested to over 70,000 rpm.
1977 Ceramic rotor operated successfully for over 30 hours at 50,000 rpm and turbine inlet temperatures up to 1370°C (2500°F).
1979 DOE initiated collaborative programs between gas turbine manufacturers and the automotive industry
1979 The original Department of Defense (DOD) funded project to support brittle material design ended.
1979 Primary work on the 820 gas turbine engine stopped [53]. Separately, Ford assisted the Garrett AiResearch Company on the initial phase of a DOE funded ceramic automotive gas turbine program. Work at Ford continued on the development of ceramic technology and structural components until 1987.

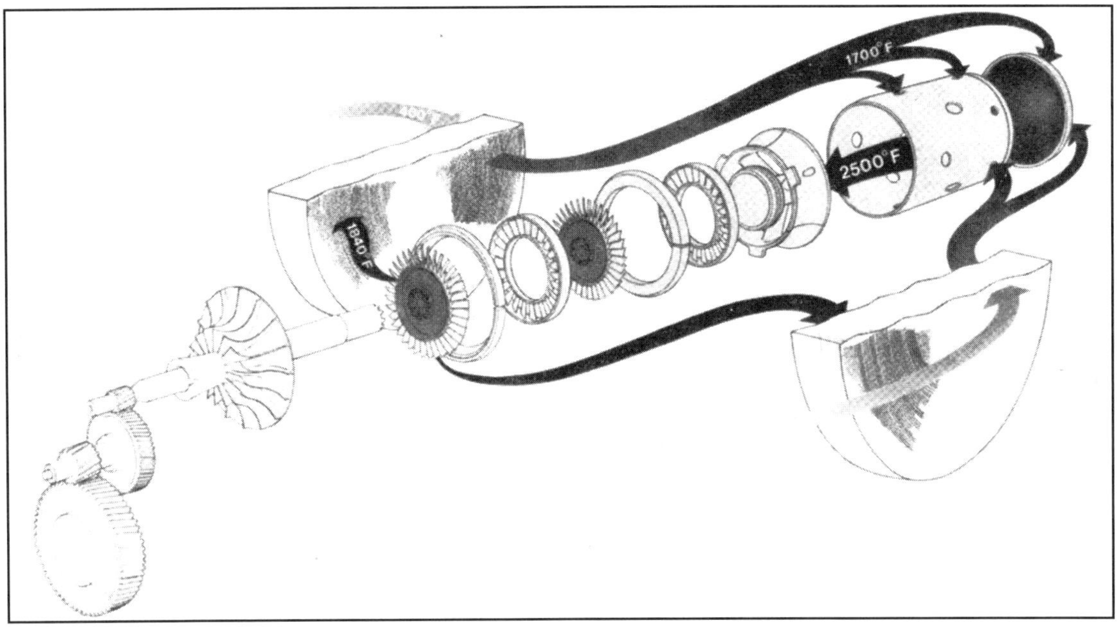

Figure 3.1 Flow Path For the 820 High Temperature Ceramic Gas Turbine Engine (Reprinted from Ref. [37])

PROGRAM APPROACH

At the program beginning, it was recognized that an interdisciplinary systems approach was required. Developments in ceramic designs, probabilistic analytical methods, materials science, ceramic processing, nondestructive evaluation, materials characterization, rig and engine testing, and statistical evaluation techniques were integrated into the following three major categories.

- Design
- Materials & Processes
- Testing and Evaluation

After assembling the team, ceramic scientists and engineers, design engineers, and development engineers were teamed together to maximize daily interfaces of the above areas and create a mutually interdependent research and development effort.

The ceramic hot flowpath is comprised of the following (see Figures 3.1 and 3.2):

- Ceramic turbine rotors
- Ceramic stationary components
  - (1) Stators
  - (2) Nose Cone
  - (3) Combustor
  - (4) Rotor Tip Shrouds
  - (5) Inner housing (referred to as the Inner Liner)
- Regenerators

The approach to develop the ceramic gas turbine components was to provide for an iterative development program as shown in the block diagram of Figure 3.3. For any major iteration, the starting point is the initiation or change in design concept. From this, design layouts are prepared and analyzed leading to

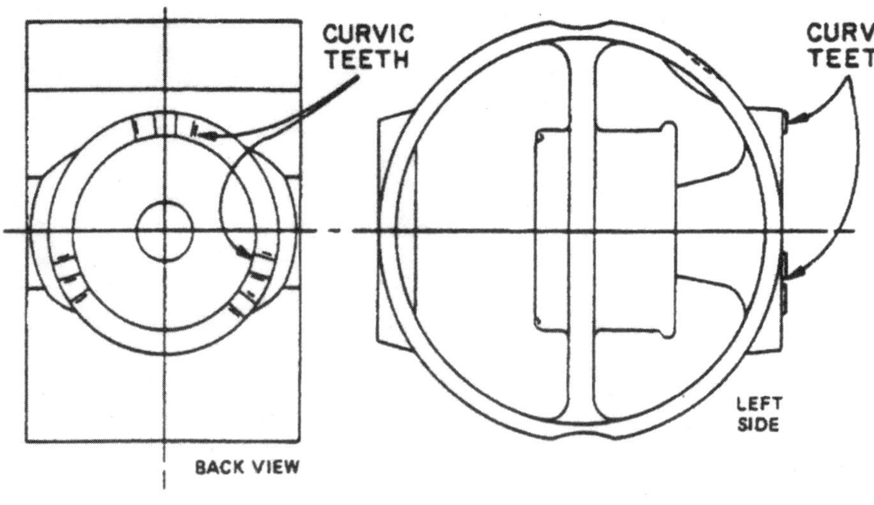

Figure 3.2 Ceramic Inner Housing Called the Inner Liner (Reprinted from Ref. [40])

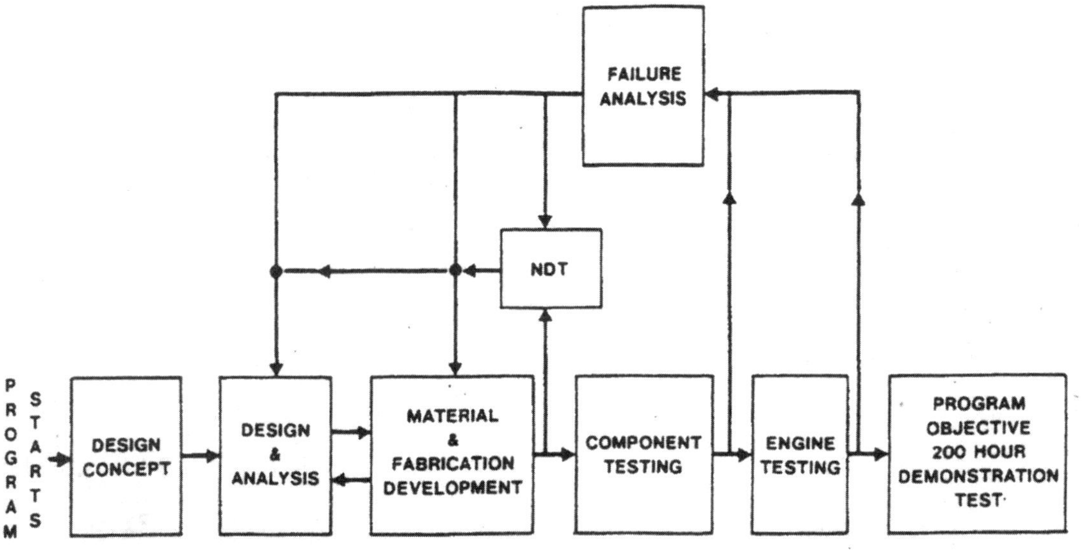

Figure 3.3 Interative Development Process (Reprinted from Ref. [35])

detailed design and analysis of the ceramic components. Ceramic materials and process development proceeds in parallel with the design phase. Material science and property characterizations are used to improve materials and processes. Feedback between design and materials prompts optimization trade-offs. Tooling is then designed and made for fabrication development of ceramic components. Non-destructive and screening tests are developed for early evaluation of ceramic components and appropriate feedback to design, and materials and process development. In parallel, all metal parts for engine rigs are made per detailed design requirements. Ceramic and metal parts are dimensionally inspected and instrumented in preparation for testing. Ceramic assembly techniques are developed and components are evaluated in engine test rigs. Failure analysis provides one of the most important feedback loops for corrective action; and or all of the factors involved in the development could be affected. This, then, was the systems approach adopted for this research program—an approach that provided broad flexibility for minor and/or major iterations.

The need to develop the specific ceramic components maintained a continual thread between the various interdisciplinary technologies of Ceramic Design, Materials and Processes And Testing and Evaluation. All aspects were generally in a state of relative infancy at the beginning of this program. The following subsections present a brief discussion of the state-of-the-art at that time along with a discussion of the approaches followed in each to meet the goals of the program.

DESIGN APPROACH

At the beginning of the program, the status of ceramic design technology could be described as follows:

(1) No set design guides for ceramics existed
(2) Limited analytical design tools were available
(3) Experience with the behavior of ceramics in an engine-type environment was limited
(4) The computer tools available were too cumbersome and costly for the detailed analyses required of brittle structures
(5) A rationale to assess the strength of ceramic component design configurations under engine loading conditions was lacking
(6) The conventional, deterministic approach to strength evaluation of ceramic components was considered inadequate due to the inductile behavior of ceramics.

As a result of these shortcomings, a basic design approach was adopted for this program which involved addressing and developing two aspects of brittle materials design technology as follows:

(1) The practical aspect, which would be composed of general guidelines and design concepts for ceramic-ceramic and ceramic-metal interfaces, e.g., attachments, assemblies, etc. (see section on Engine Design).
(2) The theoretical aspect, which would comprise the modification of existing analytical codes and generation of new analytical procedures to theoretically assess the integrity of ceramic components in a service environment, see section on Design Codes.

As the program proceeded, available analytical codes were modified and new analytical codes were generated to assist in the design of all the critical flowpath components. Component responses, such as material temperatures, stresses and deformations, were analyzed using finite element and/or finite difference computer codes.

A statistical approach was selected to determine the reliability of a ceramic component under engine loading conditions. In this connection, a probabilistic analysis based on Weibull's model [54] of brittle strength was developed along the general guidelines suggested by Dukes [55]. In addition, experimentation on design ideas or concepts, using test rigs or engines, provided important feedback for ceramic design.

The approach was to develop these new design tools and use them as a basis for the design iterations of ceramic components that took place during the program.

MATERIAL AND PROCESSING APPROACH

Near the beginning of this program, materials and process investigations had resulted in the following:

(1) Silicon nitride (Si_3N_4) and silicon carbide (SiC) were identified as viable candidate materials for ceramic turbine components. Evaluation of test samples made from several grades of these materials in test rigs confirmed these selections, although failures were encountered. There was no manufacturing source identified for Si_3N_4 for either the material or the components. Norton Company and The Carborundum Company were possible sources for the silicon carbide components.

(2) The very complex shapes of the turbine engine components led to a search for reasonable forming and processing techniques. These manufacturing process studies were designed to develop near net shape fabrication capabilities. Injection molding, isostatic pressing, and slip casting were selected as the most promising forming methods.
(3) Reaction-bonded silicon nitride (RBSN) was soon selected as the most attractive material for the stationary hot flowpath components. The low thermal expansion and relatively high strength of this material gave an advantage over other materials in the high thermal stress, low mechanical stress environment imposed upon these components. Early development work on injection molding of RBSN components having airfoils, such as stator vane segments and rotor blades with bulb-shaped root sections, established that injection molding was a very attractive forming technique but further development was needed. The initial RBSN material used for these components was of low density (2.2 g/cm^3) and low strength (83–97 MPa (12–14 ksi) MOR (modulus of rupture)). Much more development was needed on stronger materials and on improved nitriding procedures.
(4) Lithium aluminum silicate (LAS) would be used for the inner liner and the regenerator cores primarily due to it extremely low thermal expansion. The low thermal distortion would reduce sealing problems of the regenerators and reduce rotor tip clearance issues. Since the inner liner holds the stationary flow path in position relative to the turbine rotors, a low thermal distortion is critical for maintaining minimum tip clearance of the rotors to their respective shrouds. Corning Glass Works was identified as a potential supplier of this material.
(5) Very little detailed mechanical and thermo-physical characterization of turbine ceramic materials was available.
(6) Process controls to insure component uniformity needed to be developed.
(7) Critical inspection and evaluation of component quality, other than dimensional inspection, was in a state of relative infancy. Nondestructive evaluation (NDE) needed to be investigated.

Many of the design and analytical techniques available today for use on ceramic components had not yet been developed at the start of this program. Consequently, well-formulated design goals were unavailable during much of the material and process development which took place. Therefore, most of the ceramic material and process development for ceramic turbine components proceeded along the lines of 'learn by doing.'

For the case of stationary ceramic components, since operating stresses were thought to be relatively low, RBSN was selected as the principal candidate material, with RBSC (reaction-bonded silicon carbide) as the backup. Development work in both injection molding and slip casting was initiated. Isostatic pressing was tried early in the program for forming of combustor tubes but there was little opportunity to utilize this process for the other, more complex components and therefore this work was terminated. Material improvements were initiated in RBSN, resulting in several materials of increased density and strength, along with continuing effort to improve nitriding. By the end of 1979 the average strength of RBSN had increased to 380 MPa, and this was later increased to 900 MPa with the development of Sintered Reaction Bonded Silicon Nitride. The details of the RBSN improvements are discussed in more detail in Volume II [52] of this series. Work on fabrication improvements on structural components is also noted in this second volume.

Much of the program effort was directed toward the ceramic turbine rotor, due to the very high operating stresses that were unprecedented in terms of ceramic material requirements. The goal was to attempt to fabricate an all-ceramic rotor using near net shape forming techniques as much as possible, since the ultimate economics of vehicular ceramic turbine engines requires reasonable and low-cost fabrication processes. Many approaches to rotor fabrication were conceived, evaluated, and most were discarded as being unlikely to achieve success, at least within the constraints of this program. These investigations are discussed in the references and in Volume II of this series. The development of the duo-density Si_3N_4 turbine rotor utilizing RBSN for the blades and fully dense hot pressed silicon nitride (HPSN) for the hub became the leading edge of ceramic material, process, and NDE development for the program. Figure 3.4 depicts the hot pressing of the center hub with the final cross-section of the rotor shown in Figure 3.5 along with an actual finished rotor in Figure 3.6.

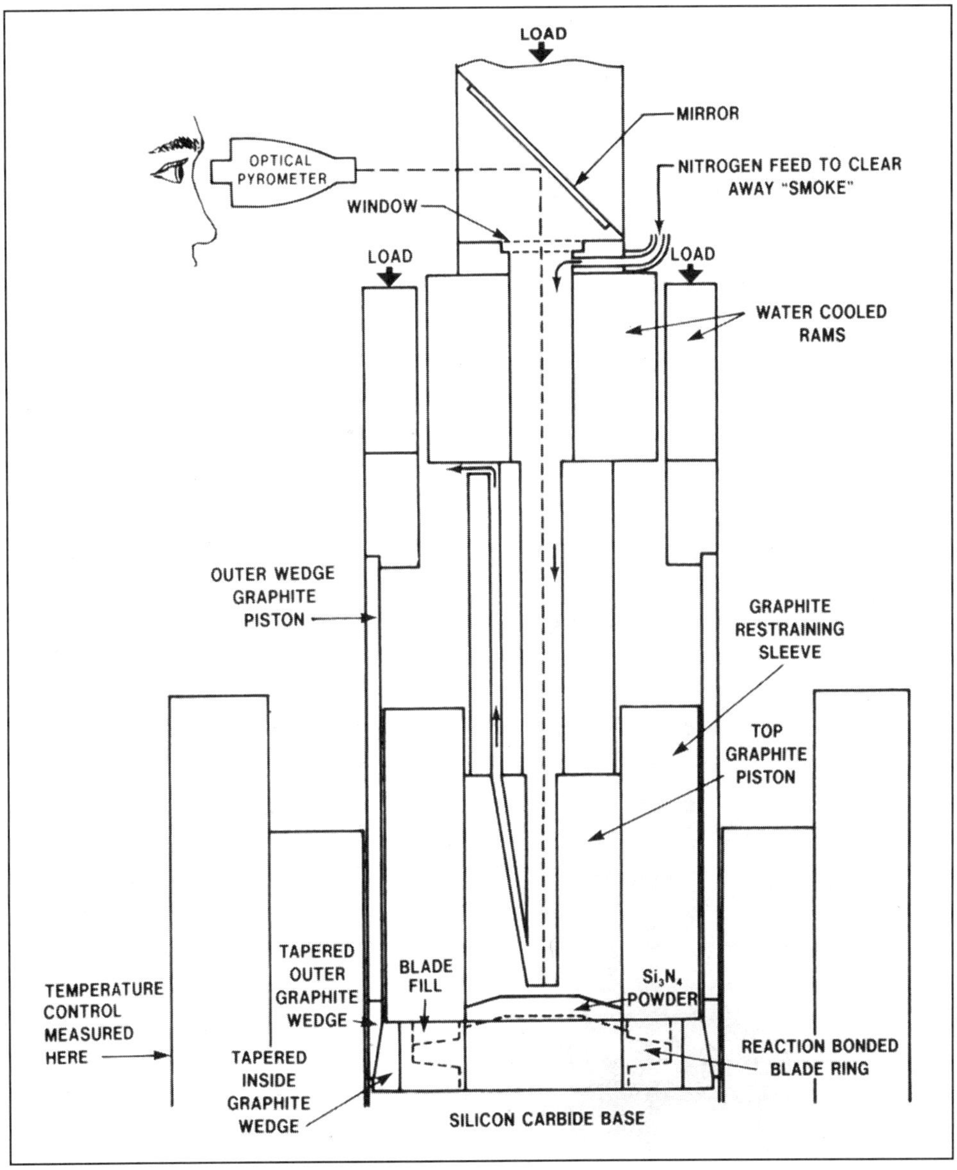

Figure 3.4 Hot Pressing of Center Hub in the Duo-Density Rotor (Reprinted from Ref. [33])

TESTING AND EVALUATION APPROACH

At the beginning of this program, many ceramic materials needed to be tested at room and elevated temperatures to evaluate relative strength, thermal shock resistance, erosion characteristics, stability, etc. As mentioned in the previous section, the majority of testing was focused on silicon nitride and silicon carbide. The approach was to determine physical and mechanical properties of both hot pressed and reaction bonded silicon nitride along with reaction-bonded silicon carbide, as required by the design activity. With respect to nondestructive evaluation (NDE), the approach was to develop new and improved NDE techniques initially for

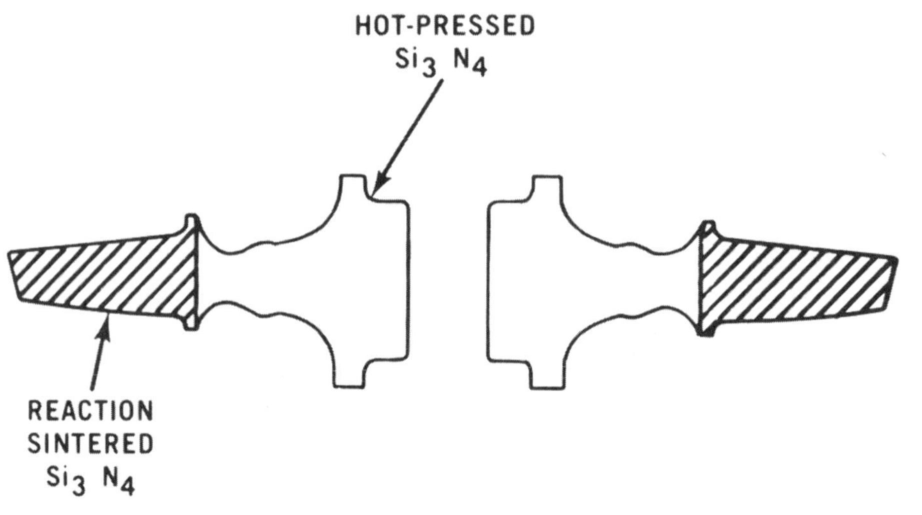

Figure 3.5 Cross-section of the Duo-density Rotor After Machining (Reprinted from Ref.[22])

Figure 3.6 Finished Duo-density Rotor After Machining (Reprinted from Ref. [35])

application to material samples and modulus of rupture bars to correlate with destructive test results generated when determining mechanical properties. Later in the program, when components were being processed, the NDE approach focused on the evaluation of macroscopic flaws in complex-shaped ceramic components such as blade rings and stator vanes as opposed to minute flaws in simpler-shaped ceramic test bars.

Testing of ceramic components was planned to play a major role in the program as a means of providing feedback to design, analysis, and material and process development. The test and evaluation of the complete ceramic engine was impossible at the onset of the program as all of the ceramic hot flowpath components were not available. In addition, it was felt that the testing of a multiplicity of undeveloped ceramic components at once was not sound engineering judgment and would make assessments of any failures very difficult, if not impossible.

The approach, therefore, was to devise several different test rigs, which utilized portions of the 820 experimental gas turbine engine to facilitate the rapid evaluation of individual ceramic components, design concepts and assembly procedures. It was also expected that testing of a ceramic component over the complete engine duty cycle would make the task of determining the critical part of the duty cycle more difficult; therefore, the planned approach was to test and evaluate components over separate portions of the duty cycle.

The testing approach devised was twofold: 1) utilize metal components along with the ceramic components under test when a full set of flowpath hardware was required. Metal turbine rotors were particularly useful as they allowed evaluation of stationary ceramic components over a major portion of the engine duty cycle prior to the availability of ceramic turbine rotors: and 2) test and evaluate individual ceramic parts in special test rigs through portions of the duty cycle. The screening of stationary ceramic flowpath components would utilize a test rig programmed to repeatably perform a sequence of cold lightoffs and hot shutdowns. Operating without turbine rotors, the test rig would reduce complexity, minimize secondary damage effects, and eliminate dynamic influences without invalidating the transient thermal evaluation of the stationary components. This approach gave clear and early definition to the modes of failure since the compounding effect of rotors was absent and the single mode of operation isolated the failure environment.

Stationary components that successfully completed the screening test were subjected to steady-state thermal environments. Initial steady-state testing identified "infant mortality" failures. The resulting iterative development feedback was of importance to the material and process development activities. Extended steady-state testing identified time-dependent problems. This iterative development feedback was of importance to the material development and reliability analysis activities. Subsequent testing of improved stationary components that successfully completed the steady-state duty cycle objectives involved testing under a combination of steady-state environments. This coupling of environments identified new situations and moved the development of stationary ceramic flow path components one step closer to successful testing over the complete duty cycle.

Ceramic turbine rotors were also screened by component testing for minimum strength and quality levels. The first step was a complete inspection. If the rotor was dimensionally acceptable, its quality was assessed using visual NDE techniques such as dye penetrant inspection and microscopic viewing. Ceramic rotors judged unacceptable for dynamic testing were used to provide feedback to the material and development activity; ceramic rotors judged acceptable were first evaluated by spin testing in a vacuum spin pit at room temperature.

The details of the many test rigs are given in the section on Ceramic Component Test Rigs.

ENGINE DESIGN

To orient the reader, at the beginning of this program, passenger cars were typically large, heavy vehicles that got less than 8.5 kilometers per liter (20 miles per gallon) of fuel. There were no emission requirements. The 820 high temperature ceramic gas turbine engine was designed for the full size passenger car, such as the 1972 Torino. This engine would utilize ceramic components for the entire hot flow path including the turbine rotors. It would be a two-stage single shaft regenerative engine coupled by a 16

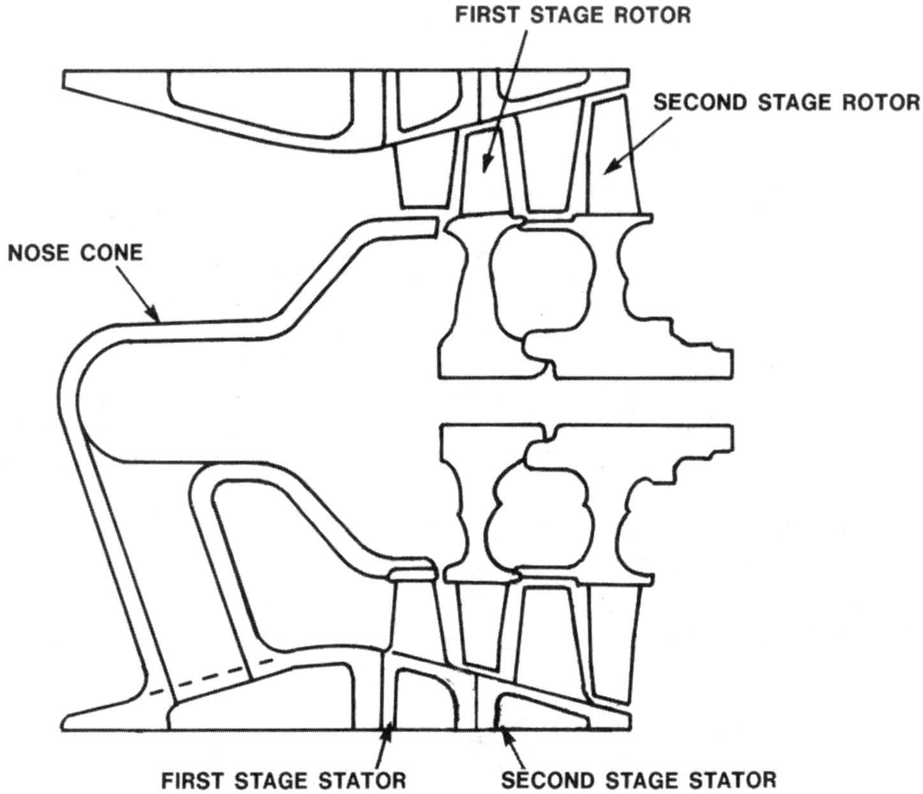

Figure 3.7 Design 'A' Flowpath Cross-section—Initial Design (Reprinted from Ref. [35])

to 1 gear train to an automatic transmission with a variable stator vane torque converter. The engine would have a maximum speed of 64,240 rpm and a 1370°C (2500°F) turbine inlet temperature producing 164 kW (220 hp), and be 20% more fuel efficient that an engine running at 1040°C (1900°F)[55]. Compared to the then-carbureted piston engine, fuel economy was projected to be about 10% better in city driving, 20% better in suburban driving and 30% better in highway driving.

An expanded view of the flow path layout is given in Figure 3.1. The cross-sections of the first (design A) and final (design D) design flow paths are given in Figures 3.7 and 3.8. The changes in the flow path were made to improve reliability of the system by reducing stresses or to permit fabrication of better quality components. The mass of the outer shrouds on the stators was minimized to reduce transient thermal stresses caused by the large cross-sectional areas of the shrouds relative to the vanes. The hub shape of the rotors was modified to reflect Weibull reliability design techniques developed and/or verified during this program. The outer shroud of the nose cone was slotted between each strut to reduce stresses caused by differential thermal growth of the struts and the outer shroud. The edges of all mating ceramic components were radiused to minimize cracking and chipping from contact stresses (see the sections on Engine and Rig Testing Results of Stationary Components and Ceramic Rotor Testing and Development for more details).

The utilization of ceramics in the engine resulted in the unique problem of addressing ceramic to metal interfaces. Since the ceramics have a much lower thermal expansion than the metals, the design must allow for relative movement at the interface. The primary areas where these interfaces occurred were the contact of the rotor/shaft and inner liner/main housing. Early attempts to use conical seats for attaching and aligning these mating surfaces were unsuccessful due to unreliable sliding resulting in misalignment and frac-

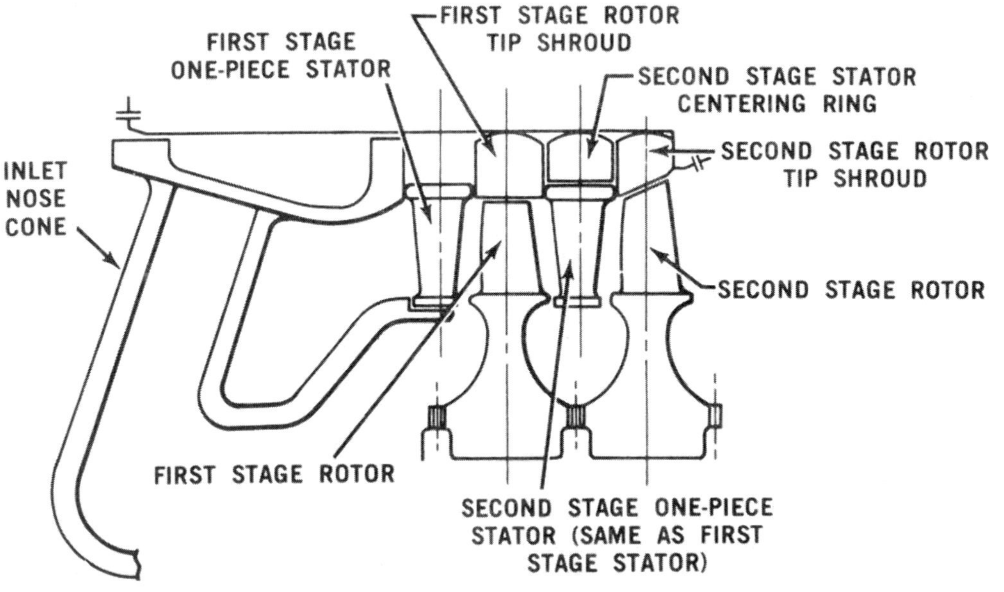

Figure 3.8 Design 'D' Flowpath Cross-section: Final Design (Reprinted from Ref. [35])

tures. The solution finally adopted was to use curvic couplings (a form of face spline) for the rotor/shaft and inner liner/main housing interfaces (see Figure 3.9). Even with this approach, long-term durability was an issue because of the high loads at the interface and limitations of the lubricants (see section on Turbine Rotor Engine Testing). The ceramic regenerator also had a thermal expansion mismatch to the iron ring gear mounted to it. The utilization of a compressed stainless steel wire mesh was used to solve this problem.

Another unique design feature that resulted from the differential thermal expansion of metals and ceramics was the rotor attachment. A special folded bolt mechanism shown in Figure 3.10 was developed. The folded bolt is made up of a tensile member of Inconel alloy 718 and a compressive member of A-286. The difference in the thermal expansion coefficients of these two materials reduced the relaxation of the clamping load due to the thermal expansion of the rotor bolt. To further control the bolt relaxation, cooling air was passed over the compressive member and through the center of the tensile member. A metallic locating adapter was placed in front of the first stage rotor to position the bolt for minimum imbalance. An installation tool was used to preload the bolt to the desired level without placing the bolt in torsion.

DESIGN CODES

A full three-dimensional analysis was required for detailed studies of thermal and structural responses of axisymmetric three-dimensional structures or certain features of these structures. In the latter category were the rotor blades, stator vanes and, most importantly, the junction of these elements with the axisymmetric body. This junction required very careful evaluation, as it was the most likely site of stress concentration.

To facilitate these analyses, computer programs were procured, modified as needed, and made compatible with each other for ease of processing. Included were two finite element codes for computing temperatures (Thermal Analysis Program, TAP) [56], and stresses (Static Analysis Program, SAP) [57], and three special purpose programs for calculating convective heat transfer coefficients and aerodynamic loading on airfoil structures (TSONIC, BLAYER, and ADFORCE) [58]. These programs functioned and communicated through the use of various buffer, geometry generation, storage, data handling and plotting routines.

Figure 3.9 Test Sample of Coupling Used on the Rotor (Reprinted from Ref. [21])

The heat transfer program, TAP, would handle both steady state and transient problems, and admit heat input by convection, conduction and heat generation internal to elements. This program was modified to allow solution of thermal problems involving temperature-dependent properties. Output from this program could be stored on tape and used as input to the stress program—SAP.

A more advanced version of the original SAP program, SAP III, was developed by Professor E. L. Wilson of the University of California at Berkeley, Division of Structural Engineering. It is a general-purpose three-dimensional static finite element stress program. The basic elements are linear eight-noded and parabolic twenty-noded isoparametric solids with elastic properties and thermal expansions fully orthotropic and varying with temperature. SAP III was modified to handle centrifugal loading, a required feature for turbine blade analysis.

The program, in its original form, would accept boundary conditions consisting of point loads, pressures and displacements. A different type of boundary problem arises, however, in the analysis of structural features that are periodic in nature; examples are the structural interactions of blades with a disk or of stator vanes and their attached shroud. These problems were more economically analyzed by the use of the "principle of repeatability," a method where a typical segment containing the repeating feature is analyzed. This method required that the displacements and rotations of the two surfaces bounding the segment be matched. Modifications to SAP III to handle this type of repetitive boundary problem failed to provide a stable solution when executed on a Honeywell H6080 computer. However, a close simulation was achieved, with the unmodified SAP III program, through coupling corresponding boundary nodal points by way of rigid trusses and executing the solution in double precision.

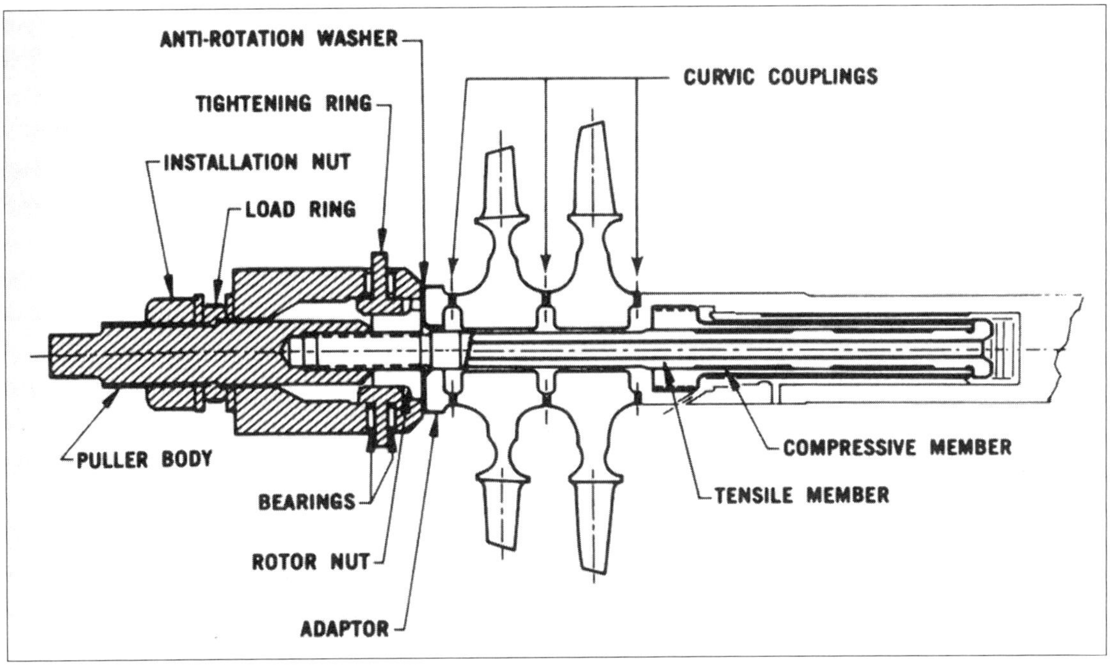

Figure 3.10 Folded Bolt Used to Control the Load on the Rotor (Reprinted from Ref. [24])

All five programs (TAP, SAP III, TSONIC, BLAYER and ADFORCE) compose the basic units of an analytical process devised for the three-dimensional analysis of turbine flowpath components. A schematic diagram of the process developed for the stator study is displayed in Figure 3.11.

RELIABILITY ANALYSIS

Two analytical methods evolved during the program for the design of structural ceramic components: (1) a method to calculate the fast fracture reliability, and (2) a method to calculate the time dependent reliability.

Fast Fracture Reliability—Analysis

Like all structural materials, ceramics have flaws; but, being brittle, ceramics lack plastic flow to relieve localized stress concentrations caused by these flaws. Since flaws are distributed randomly, the strength of ceramic components varies not only from component to component but also within the same component. If the flaws within a component could be characterized by size, shape, and location with nondestructive testing, then a deterministic type analysis could accurately predict the failure load of a structure using fracture mechanics relationships. Conversely, for a given application and desired reliability, the required accuracy of flaw detection and characterization varies with the fracture toughness of the material.

The nondestructive flaw detection techniques available at this time for ceramics lacked the capability to detect and characterize flaws of the size that often cause failure in ceramic turbine components; therefore, a probabilistic approach to the calculation of fast fracture reliability was developed and, in particular,

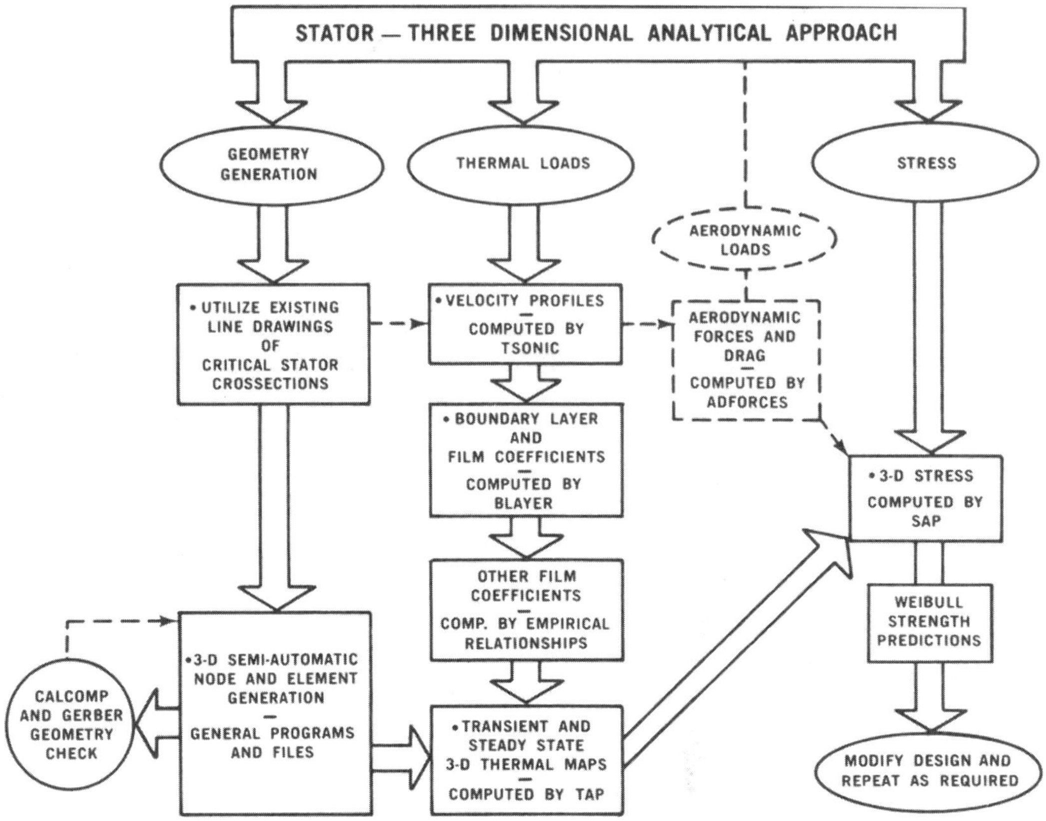

Figure 3.11 Analysis Process for the Stator Study (Reprinted from Ref. [25])

utilized the probability of failure theory devised by Weibull. [54]. The Weibull approach to predicting the probability of failure of a structure is to recognize that each element in the structure has a definite probability of failure and that the structure in its entirety could be considered to be an assembly of individual elements with an associated probability of failure, Vardar and Finnie [59] given an integral formulation of the Weibull approach:

$$P_f = 1-\exp(-B) = 1-\exp\left\{-\int_V \left[K\int_A (\sigma_n)^m dA \right] dV \right\} \tag{1}$$

The term in the parenthesis is evaluated on the surface of the unit sphere, illustrated in Figure 3.12, over the regions where the normal stress is tensile and neglecting regions where the normal stress is compressive. The reason for neglecting compressive stress is that, for ceramics, compression is not nearly as detrimental in causing fracture as is tension, and also it minimizes mathematical complexities. [54,60] The normal stress on the surface of the unit sphere in terms of the maximum principal stresses and the polar and azimuthal angles is given by:

$$\sigma_n = \cos^2\phi \left[\sigma_1 \cos^2\psi + \sigma_2 \cos^2\psi\right] + \sigma_3 \sin^2\phi \tag{2}$$

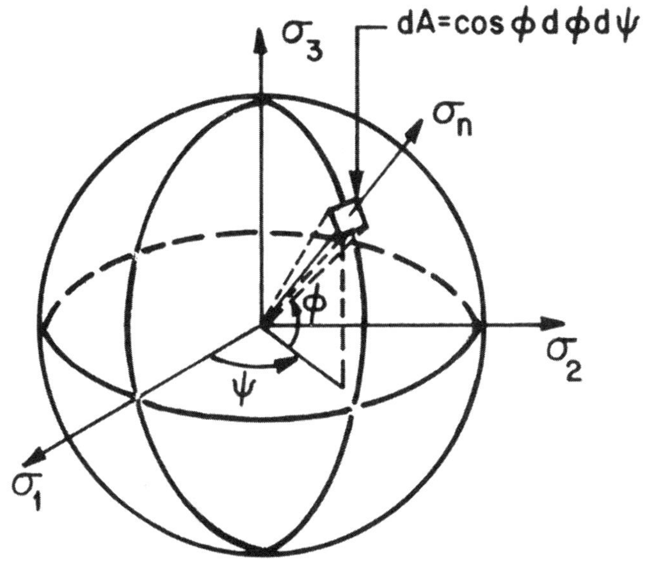

Figure 3.12 Unit Sphere Used for Weibull Analyses (Reprinted from Ref. [35])

The constant, K, is given by:

$$K = \frac{2m+1}{2\pi\sigma_0^m} \tag{3}$$

The (2m + 1)/2τ term in Equation (3) is a compatibility factor required to make the result of integrating Equation (1), using the normal stress distribution of Equation (2) for uniaxial stress cases, agree with the results obtained from the one dimensional Weibull equation:

$$P_f = 1 - \exp\left[-\int_V \left(\frac{\sigma}{\sigma_0}\right)^m dV\right] \tag{4}$$

The integration of Equation (1) over an entire structure is accomplished by dividing the structure into finite elements in which the principal stresses can be assumed to be constant. The result of integrating over the unit sphere is constant in each finite element; therefore, the risk of rupture for each element is simply the volume of the element times the integrated result over the unit sphere. This procedure is very convenient when used with finite element stress programs.

In the Weibull formulation δ_0 and m are parameters that show the probability of having a certain strength in any finite element. The values of δ_0 and m are obtained from experimental test results. Weibull showed that the characteristic strength, δ_0, is identical to the ultimate strength of the material in the classical theory as the Weibull modulus, m, increases indefinitely. This provides a physical interpretation of the characteristic strength. The Weibull modulus may be interpreted as a measure of variability or scatter in material; as m increases the variability of the strength diminishes.

The Weibull parameters δ_0 and m are obtained from strength testing. For bend tests, the stress distribution is uniaxial and the risk of rupture is given by:

$$B = \int \left(\frac{\sigma}{\sigma_0} \right)^m dV \tag{5}$$

In a bend test, the stress on the outer fiber at failure is called the modulus of rupture (MOR). In this study four-point bending, with the middle span located symmetrically to the total span, was used and Equation (5) was integrated over the volume of the bar in tension to obtain Equation (6), the expression for the risk of rupture of such a bar:

$$B = \left(\frac{MOR}{\sigma_0} \right)^m \left(\frac{bh}{2} \right) \left[\frac{L_1 + mL_2}{(m+1)^2} \right] \tag{6}$$

The data obtained from a set of test bars was fitted to Equation (7) with maximum likelihood techniques [22]:

$$P_f = 1 - \exp\left[-\left(\frac{MOR}{MOR_0} \right)^m \right] \tag{7}$$

The probability of failure of the test bars may also be represented by the Weibull integral formulation, which must be equal to Equation (7):

$$P_f = 1 - \exp\left[-\left(\frac{MOR}{MOR_0} \right)^m \right] = 1 - \exp\left\{ -\left(\frac{MOR}{\sigma_0} \right)^m \left(\frac{bh}{2} \right) \left(\frac{L_1 + mL_2}{(m+1)^2} \right) \right\} \tag{8}$$

Substituting the value of the characteristic modulus of rupture MOR_0 for the modulus of rupture MOR into Equation (8) and solving, one obtains Equation (9), the equation for the characteristic strength:

$$\sigma_0 = MOR_0 \left(\frac{bh}{2} \right)^{\frac{1}{m}} \left[\frac{L_1 + mL_2}{(m+1)^2} \right]^{\frac{1}{m}} \tag{9}$$

The parameters σ_0 and m define the strength of the material and are independent of the volume and geometry of the structure. They are the values used to determine the failure distribution of a complex structure and are generally determined from MOR bar tests. It is important to note that the MOR is not an intrinsic material parameter but depends on the geometry of the MOR bar and test fixture.

The integral formulation of the Weibull equation was programmed for a digital computer. The code used as input the principal stresses calculated with axisymmetric and three-dimensional finite element stress codes. The code evaluated the integrals in Equation (1) by numerical integration in each volume element of the finite element mesh assuming that the principal stresses were constant throughout the element. The code also used as input material strength data in the form of characteristic modulus of rupture and Weibull modulus, both as a function of temperature. The code selected the material strength data according to the temperature of the element.

As developed, the fast fracture reliability code can be used in two ways:

(1) The computer code can calculate the survival probability of a certain component under a specified load when input consists of material strength data in the form of Weibull strength parameters (Weibull parameters are the characteristic MOR strength obtained from simple four-point bend tests and the Weibull shape parameter or slope).

(2) The computer code can generate sets of Weibull strength parameters required of a material to meet a specified reliability goal, when given the desired reliability level.

Nomenclature

dA = differential area
B = risk of rupture
b = width of specimen
h = height of specimen
K = constant defined by Equation (3)
L_1 = lenght between supports
L_2 = length between load points
m = Weibull modulus
MOR = modulus of rupture
MOR_0 = characteristic modulus of rupture
P_f = probability of failure
V = volume of the structure
$\sigma_1,\sigma_2,\sigma_3$ = principal stresses
σ_n = normal stress
σ_0 = characteristic strength
ψ = azimuth angle
φ = polar angle

Utilization of Fast-Fracture Reliability Analyses

During most of the program, fast fracture reliability analyses were used to design the components or to establish what the material properties needed to be in order for the components to have an acceptable failure probability. An example of the utilization of this theory coupled with conventional finite element analyses is the design (Design C) of the first stage rotor. This rotor was analyzed for four conditions:

(1) Idle steady state,
(2) Maximum speed steady state,
(3) Start up transient to idle speed (55% of maximum speed),
(4) Maximum thermal gradient from start up transient with step change in speed to 110% of maximum speed.

The temperature contours for conditions 3 and 4 are shown in Figure 3.13. This shows that there is over a 650°C (1200°F) gradient from the bore to the blade platform which produces thermal stresses that add to the centrifugal stresses. The maximum principle stresses for condition four are shown in Figure 3.14. The highest stresses occur at the bore and the neck region of the rotor. These stresses were coupled with Weibull probability of failure analyses to generate a contour plot of the risk of rupture of the various regions of the rotor (see Figure 3.15). This showed that even though the neck and bore region had approximately the same stress level the neck region was more likely to fail. This was due to the volume effect in the Weibull theory that says that if two regions are at the same stress level the one with the higher volume will be more likely to fail.

The above analyses were conducted on a number of rotor configurations (see Figure 3.16). This figure shows the probability of failure for a number of rotor cross-sections. In all cased the probability of failure for the transient condition was much higher than for the steady state condition as was expected. The probability of failure for (1a) and (2c) are very close. Configuration (1a) was chosen for the design C rotor since it provided a ring of material just below the neck that could be used to balance the rotor.

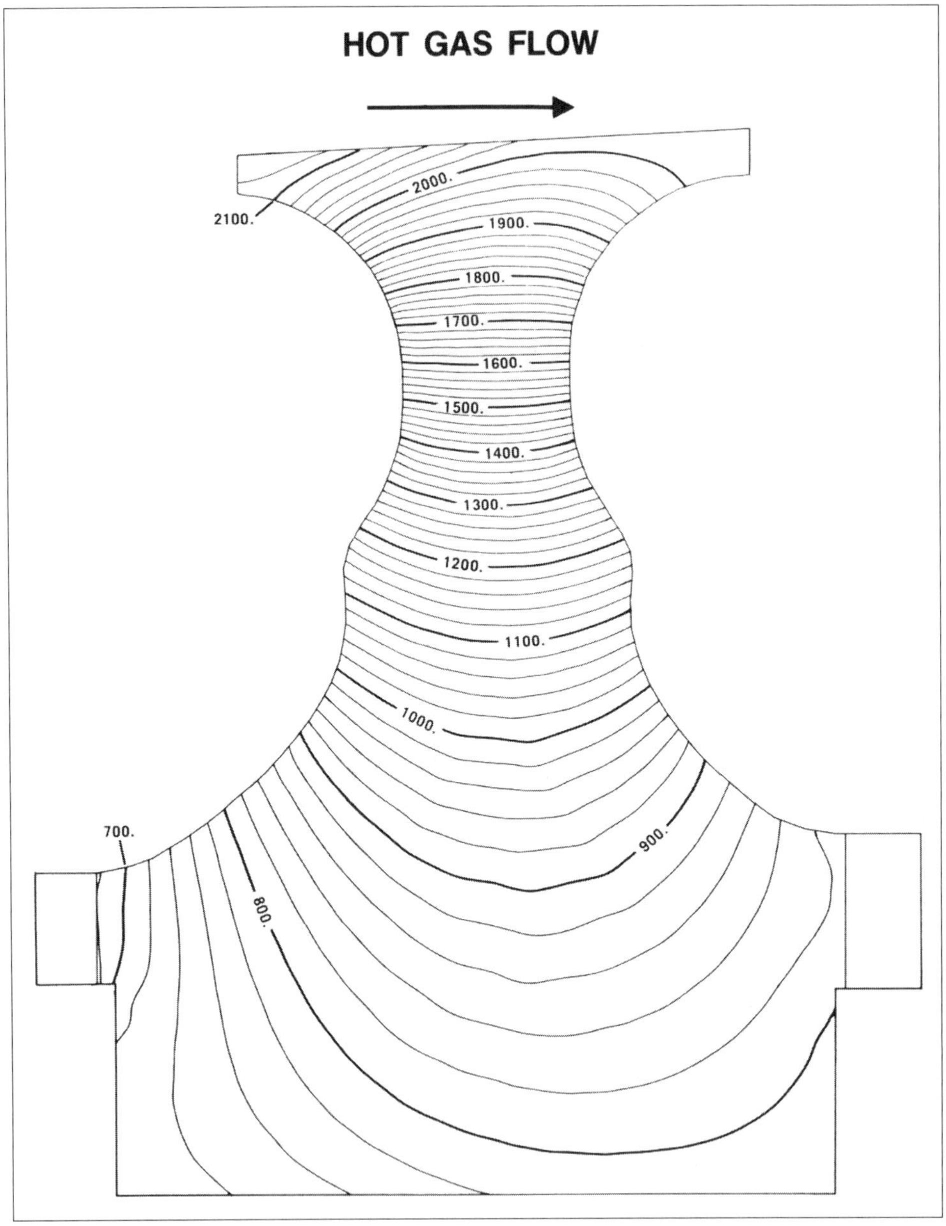

Figure 3.13 Rotor Temperatures (°F) During Cold Start Transient (Adapted from Ref. [23])

Time-Dependent Reliability Analysis

The previous section covered the development of an analytical method to calculate fast fracture reliability. This calculation of fast fracture reliability is an important first step in the calculation of time-dependent reliability, the subject of this section. Certain ceramics, including hot pressed silicon nitride [62], exhibit delayed fracture or static fatigue at elevated temperature under loads. This static fatigue is caused by existing flaws in the material growing over a period of time until they are large enough to cause failure.

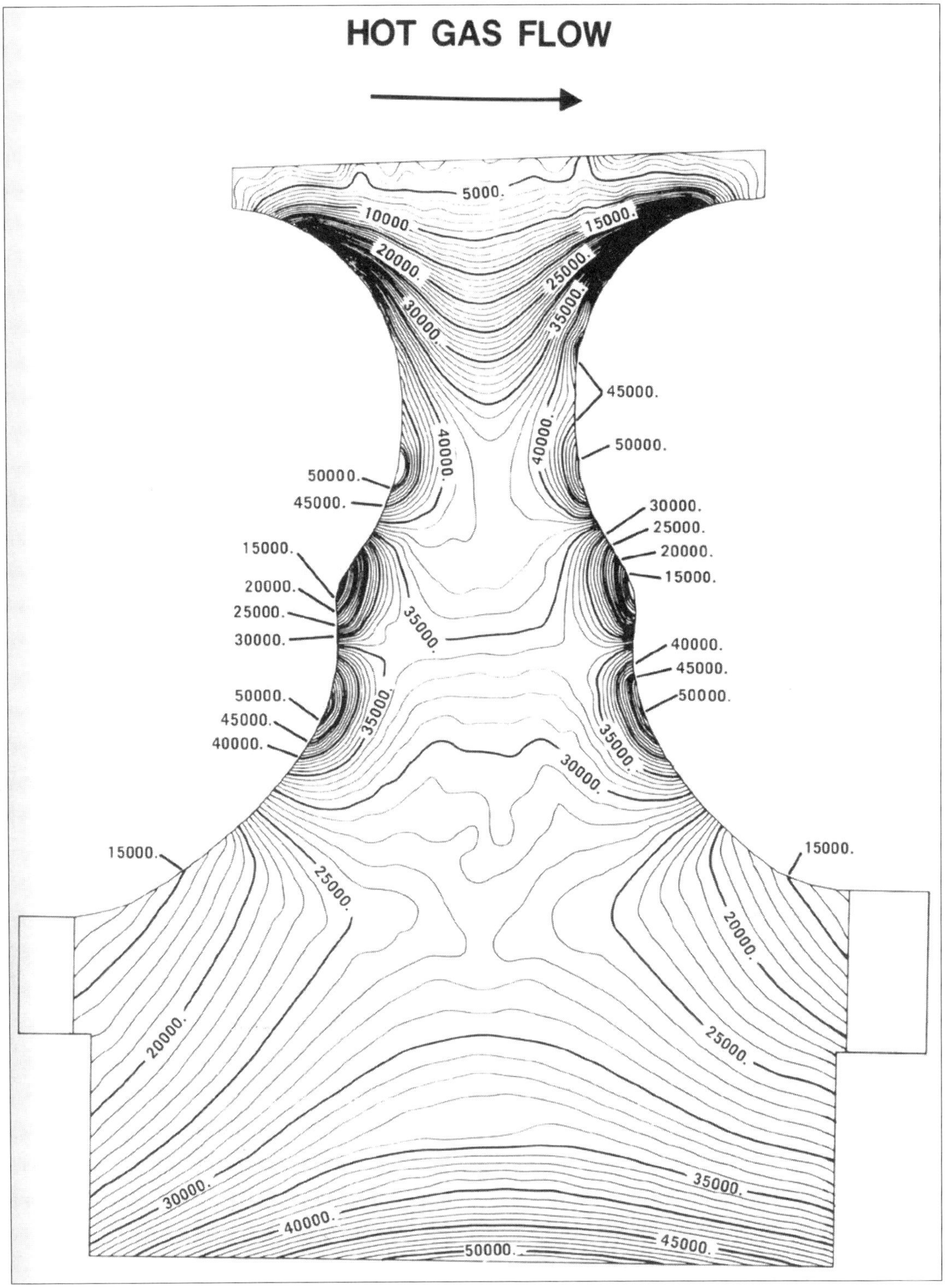

Figure 3.14 Rotor Stress (psi) for Cold Start Transient Temperature Gradient and 110% of Maximum Design Speed (Reprinted from Ref. [23])

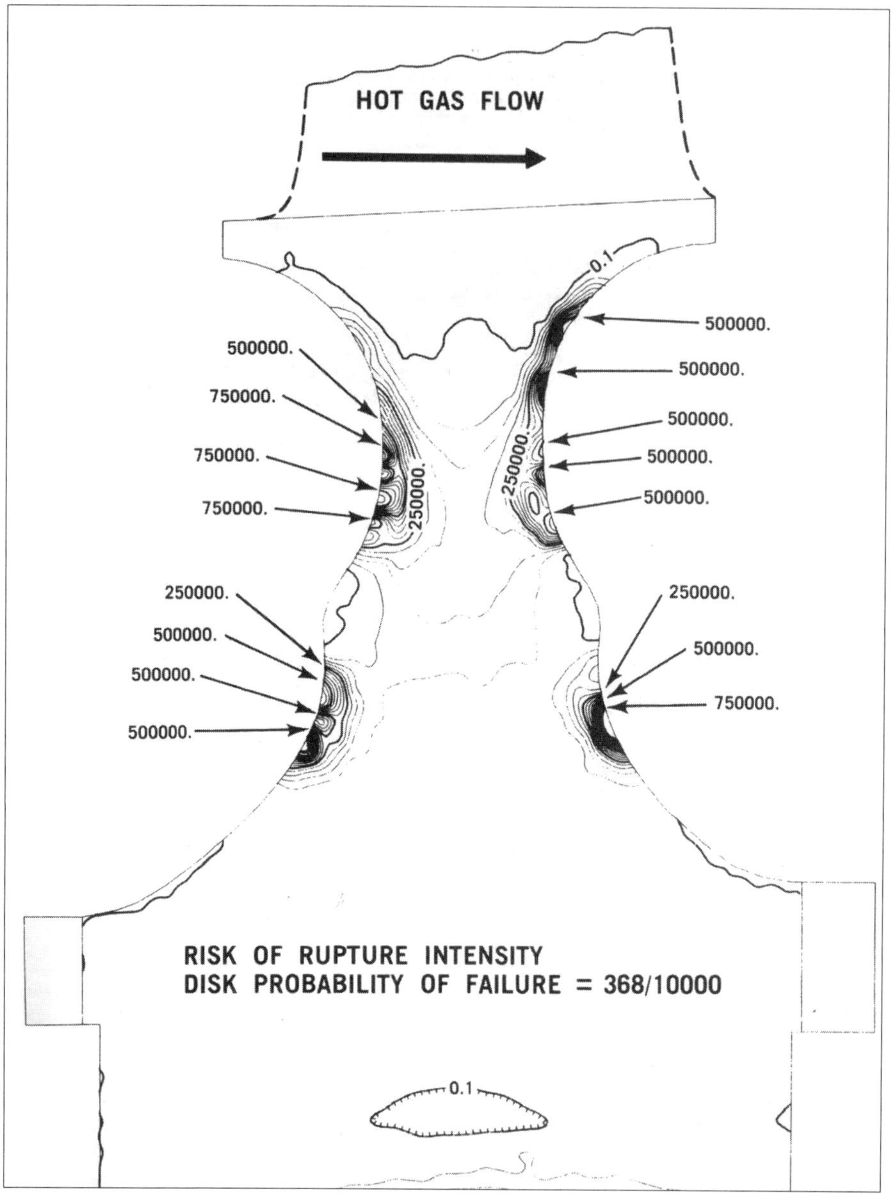

Figure 3.15 Rotor Risk of Rupture for Cold Start Transient Temperature Gradient and 110% of Maximum Design Speed (Reprinted from Ref. [24])

This phenomenon reduces the reliability of the structure with time, so the designer needs to have a method of calculating time-dependent reliability.

The basis for time-dependent reliability analysis is the experimental observation [61] that the velocity of a pre-existing crack can be described with a power law relationship:

$$V = AK_I^n \tag{10}$$

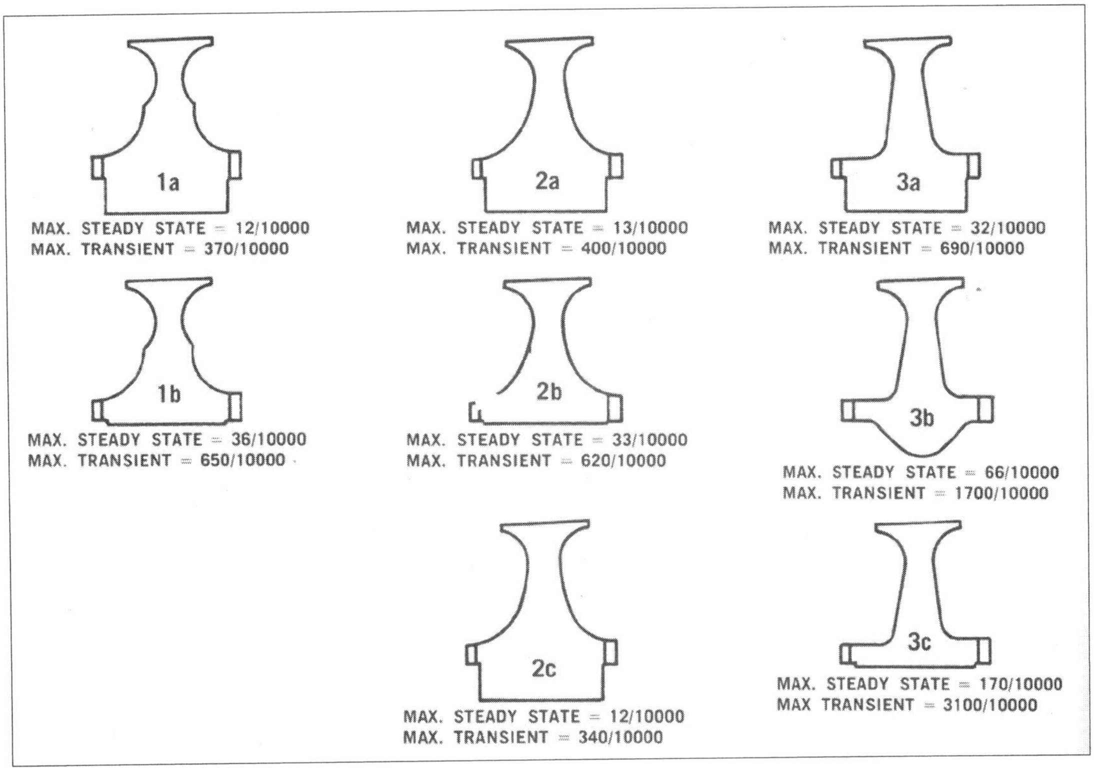

Figure 3.16 Rotor Design Options Analyzed and the Probability of Failure For Steady State and Transient Conditions (Reprinted from Ref. [24])

This equation expresses the rate of slow-crack growth caused by grain boundary sliding, typical in such materials as hot pressed silicon nitride. Paluszny and Nichols [45] combined the power law relationship for crack velocity with the Weibull relationship for fast fracture reliability to arrive at the following equation for reliability versus time:

$$R = \exp\left\{-\left[\left(\ln\left(\frac{1}{R_{ff}}\right)\right)^{\frac{n-2}{m}} - \frac{\sigma_n t}{B\sigma_\theta^{n-2}}\right]^{\frac{m}{n-2}}\right\} \tag{11}$$

In this equation, the constant B is defined by the following equation:

$$B = \frac{2}{(n-2)Y^2 A K_{IC}^2} \tag{12}$$

These equations apply to a structure under constant load, but Paluszny and Wu extended the analysis to time-dependent load by approximating the load history with a simple step function. For a finite element model, Equation (11) is applied to the volume of each element and the reliability of the entire structure is calculated with the following equation [44]:

$$R = \frac{N}{\prod_{i=1} R_i} \tag{13}$$

In extending this analysis to multi-axially stressed ceramic turbine structures, some simplifying assumptions were necessary [61]. For the convenience of the reader, these assumptions are summarized here:

(1) Omit time-dependent effects such as material instability, high-temperature oxidation, etc.
(2) Consider only the normal stress or the opening mode (K_I) of crack propagation; neglect modes K_{II} and K_{III}
(3) When the load varies with time (δ) assume that the orientation of principal stresses throughout the structure remains constant with time.
(4) Treat time to failure using Weibull's probabilistic statistical model of fracture with the premise that flaws which control strength are identical for both fast fractures and time-dependent fractures.

The time-dependent reliability equations were programmed for the digital computer. The code utilized the output of the fast fracture reliability code, comprising data on the elemental volume, fast fracture reliability, maximum principal stress, characteristic strength of the elemental structure [44], Weibull modulus, and temperature of the element. As programmed, this data was written on tape by the fast fracture reliability code in a format compatible with the time-dependent reliability code.

Two different databases were available to define the time-dependent characteristics of the material, double torsion and stress rate testing. Double torsion testing [62] was a technique of measuring the velocity of a crack under load in a specially designed double torsion specimen. Analysis of the results gave K_{IC}, n, and "A," the material parameters required for time-dependent reliability calculations. Stress rate testing measured the fracture stress of MOR specimens under varying load rates. The results were sets of fracture stresses at corresponding stress rates.

The data required for time-dependent reliability calculations were calculated with the following formulas [61]:

$$n = 1 - \frac{\ln\left(\frac{\sigma_2}{\sigma_1}\right)}{\ln\left(\frac{\sigma_{f2}}{\sigma_{f1}}\right)} \tag{14}$$

$$B = \frac{\sigma_f^3}{\sigma(n+1)} \tag{15}$$

The code incorporated the option of inputting double torsion or stress rate material data. The data was inputted in tabular form with the parameters given with their corresponding temperature. The code interpolated the table of material parameters to obtain the value of the material parameter corresponding to the temperature of the element.

As developed, the time-dependent reliability code can be used in the following modes of calculation:

(1) Survival probability after the elapse of a specified service life or time at load.
(2) Service life available for a specified reliability. This becomes an iterative process, as strength degradation in a typical turbine structure is not uniform because stresses and temperatures are seldom uniform.

Nomenclature

A = Premultiplier in crack equation
B = Constant in reliability versus time equation
K_I = Stress intensity
K_{IC} = Critical stress intensity, a material parameter
n = Crack velocity exponent
N = Number of elements in the structure
m = Weibull modulus
R = Reliability as a function of time
Ri = Reliability of ith element
R_{ff} = Fast fracture reliability
σ = Stress, MPa
σ_θ = Characteristic strength of the structure
σ_f = Fracture stress
σ = Stress rate
t = Time, hours
V = Crack velocity
Y = Stress intensity factor coefficient, non-dimensional

Utilization of Time-Dependent Reliability

The development of the time dependent reliability methods and the material properties necessary to use these theories was completed late in the program and was thus not used for design changes during most of the program. However it was used in the development of the design D rotor as discussed in the section Ceramic Rotor Testing and Development.

CERAMIC COMPONENT TEST RIGS

Testing of ceramic components played a major role in the program by providing feedback to the design, analysis, material and process development activities. As described in the Testing and Evaluation Approach, many test rigs were developed which would evaluate ceramic components, first individually and then in sets, by subjecting them to the various types of loading in separate tests. Table 3.1 shows the rigs utilized in evaluating the various ceramic components subjected to various types of loading. Generally, the more complex types of loading required more complex rigs and considerable effort was expended to devise simpler tests and rigs which would preserve the fracture surfaces intact to aid in failure analysis and corrective action in the fabrication and/or design processes.

Rotor Blade and Stator Vane Bend Testing

The primary purpose for developing the rotor blade and stator vane bend tests was to monitor the effects of processing improvements and provide timely feedback to the fabrication efforts.

However, a variety of other goals were accomplished which:

- provided the capability to correlate the analytical stress analysis model with strain gauged test results
- determined the existence of and quantified the effects of surface and subsurface flaws and determined the potential gain for unflawed components
- evaluated component strength at various stages in the fabrication process thereby identifying processing steps that affect component quality

TABLE 3.1 CERAMIC COMPONENT TEST RIGS

Test Rigs	Ceramic Components Evaluated				Types of Loading				
	Combustor	Nose Cone & Shrouds	Rotor(s)	Stator(s)	Mechanical	Thermal Shock	Thermal Steady state	Dynamic	Vibratory
Stator Vane Bend				X	X				
Rotor Blade Bend			X		X				
Stator Outer Shroud Pressure				X	X				
Cold Spin			X		X				
Thermal Shock			X	X		X	X		
Rotor Disk Thermal Stress			X				X		
Combustor	X				X	X	X		
Lightoff Qualification	X	X		X	X	X	X	X	
Ceramic Structures 1371°C (2500°F) Flowpath	X	X		X	X		X	X	
	X	X		X	X		X	X	
Hot Spin			X		X		X		
Turbine Rotor	X	X	X	X	X		X	X	X
Modified Engine	X	X	X	X	X	X	X	X	X

- evaluated processing changes on component strength
- determined effects of thermal shocking and/or oxidation on component strength
- provided a means for proof testing of as-fabricated components prior to expensive and time-consuming machining and testing
- generally preserved fracture surfaces intact to allow complete fractographic analysis to pinpoint fracture origin and cause.

The rotor blade bend test fixture [28–30], shown in Figure 3.17, subjected the airfoils to nearly pure bending with an inherent trace of torsion present due to the twisted geometry of the blade. The direction of loading shown in Figure 3.17 produced compressive stresses in the leading and trailing edges and tensile stresses at the back or convex side of the airfoil. Reversing the direction of the load produced tension at the leading and trailing edges and compression at the back. The maximum stresses for either load direction occurred at the base of the airfoils adjacent to the platform. The test fixture was designed such that both blade rings and duo-density turbine rotors in various stages of processing could be evaluated on the same equipment. Utilizing this test it was determined that the blade strength was significantly degraded during the hot press bonding of the rotor hub section to the blade ring (see Table 3.2).

The stator vane load test [27, 30, 31] is shown in Figure 3.18. The leading and trailing edges and the back of the vanes were subjected to either tensile or compressive stresses depending on whether the front or back side of the inner shroud was employed as the load surface. In either case, the maximum stresses were produced at the vane root adjacent to the outer shroud. Initially, only destructive tests were conducted to evaluate fabrication processes. Later, stator vanes were proof tested prior to machining but this was time consuming as each of the 25 vanes had to be loaded individually, first from the front and then from the back. Accordingly, the test fixture shown in Figure 3.19 was designed and built. Each of the 25 vanes was simultaneously loaded by 25 independent hydraulically loaded pins thereby reducing the proof test time by a factor of 25. By loading vanes to failure from stators that had passed light off qualification tests (see light off qualification test rig), an initial vane proof test load of 44.6 N (10 lbs) was selected. This eliminated most but not all vane failures in subsequent light off qualification tests. "Unflawed" vanes were determined to have failure loads of greater than 80 N (18 lbs) with some as high as 222 N (50 lbs). "Unflawed" here is defined as flaws less than approximately 0.07 mm (0.003 inches) in diameter. Thus the proof load was raised to 84 N (19 lb), but even this did not guarantee that the stator would pass the light off qualification test. The higher load did however weed out poorer quality stators. (NOTE: Some analysts now believe that proof testing may actually damage the part by inducing microscopic cracks.)

Stator Outer Shroud Pressure Test

The stator outer shroud pressure test was developed primarily to screen stators for defects in the outer shroud prior to engine or rig testing [30–31]. The test fixture, shown in Figure 3.20, subjected the outer shroud to a tensile tangential stress by applying an internal hydraulic pressure. The stators were tested in the finish ground condition to include any detrimental effects due to surface finish or chipping. As shown, the stator was trapped between two flat plates and sealed at the shroud inside diameter with thin, flexible rubber lip seals. The stress level in the outer shroud was controlled by a pressure control valve on the hydraulic supply line and monitored by a gage reading internal pressure in the sealed cavity. The outer shroud was subjected to 27.6 MPa (4.00 ksi) as a proof test to screen out flawed shrouds. This level of stress was sufficient to screen out low quality parts but was significantly below the strength of good quality parts which would fail at about 75.8 MPa (11.0 ksi).

Cold Spin Test

Cold spin testing was conducted in a vacuum spin pit shown schematically in Figure 3.21. The test specimen was suspended in a vacuum chamber from an air turbine capable of achieving 100,000 rpm. Photographs

Figure 3.17 Rotor Blade Bend Test Fixture (Reprinted from Ref. [35])

TABLE 3.2 EFFECT OF PRESS BONDING UPON ROTOR BLADE STRENGTH

	As Nitrided		After Hot Pressing		% Change	
Nitriding Number	Weibull Slope	Characteristic Load	Weibull Slope	Characteristic Load	Weibull Slope	Characteristic Load
		N, lbs		N, lbs		
48	9.1	401 (89.9)	6.9	343 (77.0)	–24	–14
67	11.2	382 (85.7)	6.8	236 (53.0)	–39	–38
78	9.2	355 (79.1)	15.1	275 (61.6)	+64	–22

Figure 3.18 Stator Vane Load Test (Reprinted from Ref. [35])

Figure 3.19 Stator 25 Vane Bend Load Fixture (Reprinted from Ref. [35])

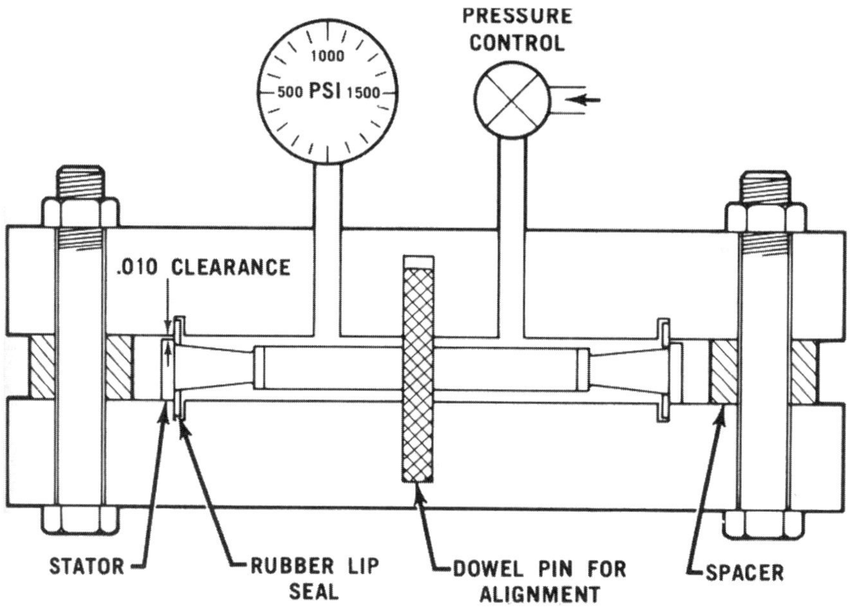

Figure 3.20 Outer Shroud Pressure Test (Reprinted from Ref. [30])

could be taken at the instant of failure through a glass plate in the bottom of the vacuum chamber. The single frame camera could be replaced by a special 16mm high-speed (3,000 frames per second) motion picture unit to study time to failure. The failure detector shown in Figure 3.21, a ceramic burst ring painted with conducting epoxy, was later replaced by a cylindrical tube of 3.8 cm (1.5 inch) wide masking tape with several windings of thin copper wire. This triggering device was less expensive, easy to fabricate and allowed failed particles to escape thereby minimizing secondary damage to the remaining rotating blades.

The spin test facility was capable of testing simple ceramic hubs, segments of blade rings or complete blade rings bonded to a metal or ceramic hub with heat-cured epoxy, and full duo-density ceramic turbine rotors.

An example of using this rig was the determination of the speed necessary to fail blades on a rotor. Several rotors were spun and the blade failures were examined for defects. It was concluded from these tests that: (1) better NDE techniques were required to detect blade flaws which result in blade failure, and (2) improvements were required in the fabrication process to eliminate these gross defects.

The duo-density rotors were cold spin tested to 10% above their planned operating speed. In many cases this would result in the failure of one or more of the rotor blades. In many cases the rest of the rotor and blades remained intact. In order to provide hardware for continued engine and rig development, these components were balanced and utilized for further tests.

Thermal Shock Test Rig

The thermal shock test rig was developed to screen candidate ceramic materials in simple bar form and evaluate changes in design and processing variations on stators and rotor blade rings [16, 22–28].

A schematic of the rig is shown in Figure 3.22. Up to eight samples could be clamped to the air actuated mechanically indexed rotary table. The cycle time could be varied from 0 to 4 minutes and table revolutions were recorded on a digital counter. Cooling air could be supplied to each of the eight samples to establish required thermal gradients by cooling the stator shrouds or rotor blade platforms.

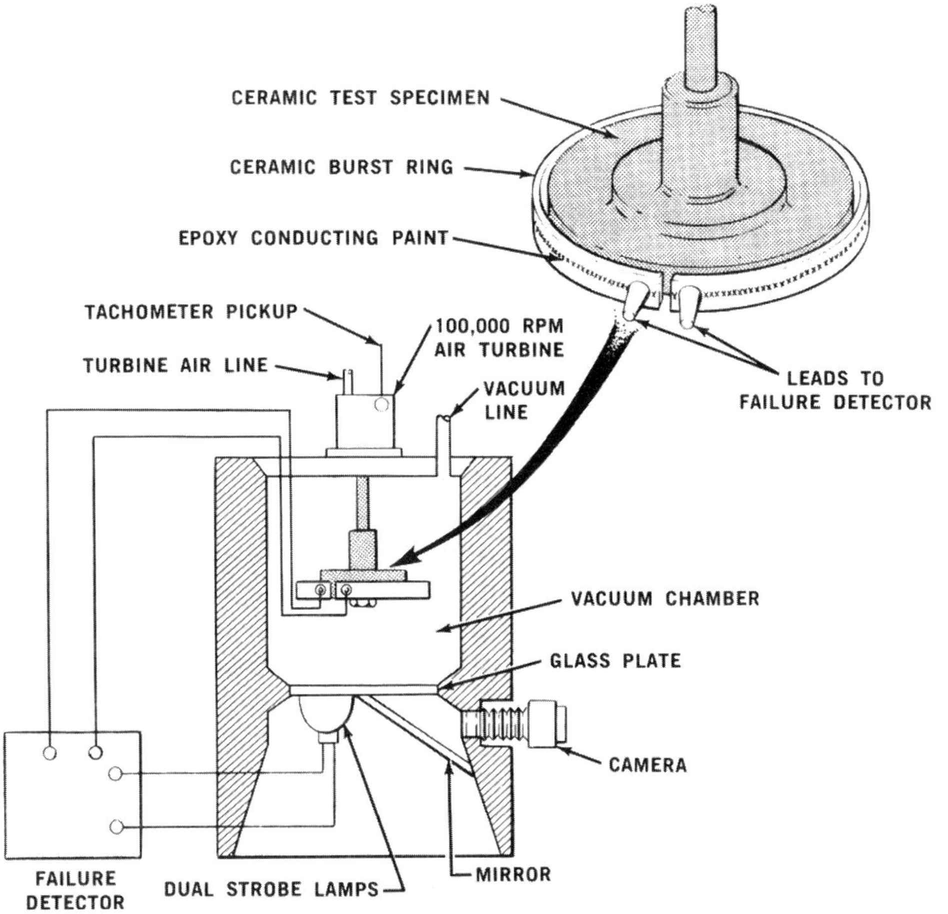

Figure 3.21 Cold Spin Test (Reprinted from Ref. [35])

Six stations supplied cool quench air that could be independently adjusted to control the downshock rate. Two mixing burners, using natural gas and oxygen, could be adjusted to a concentrated pencil flame or a more diffuse larger flame. An infrared sensing pyrometer and temperature recorder were used to set the burner and cooling airflows. Failures were automatically recorded as a step change in the recorded temperatures. Safety equipment monitored ultraviolet light from the flames and would shut off the gas and oxygen, if required.

This type of test rig offered several advantages over more extensive engine testing as follows:

- a complete stator or blade ring was not necessary as segments from damaged components could be evaluated
- thermal gradients—both upshocks and downshocks—were adjustable and repeatable
- automated testing and failure recording
- design and process changes could be evaluated quickly
- fracture surfaces were preserved intact to expedite failure analysis

This rig was extremely useful in investigating stator vane design changes to eliminate cracking which was discovered in the Light Off Qualification Test Rig (LOQTR). By adjusting the flame and quenching air, vane temperatures matching those that were measured in the LOQTR could be obtained. These setting were

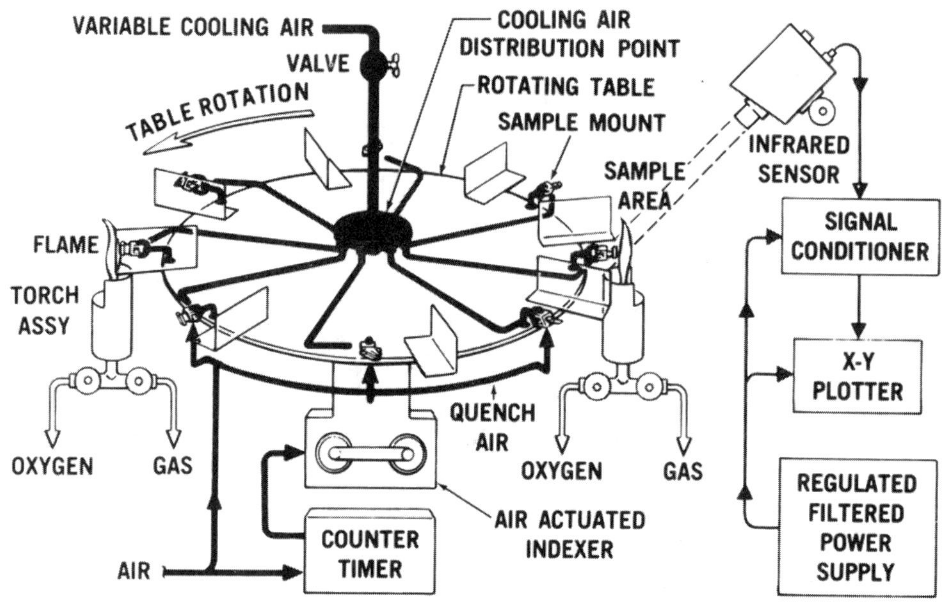

Figure 3.22 Thermal Shock Test Rig (Reprinted from Ref. [35])

then used to evaluate new vane and shroud configurations to eliminate cracking. This led to the cut back trailing edge of the vane as shown in Figure 3.36 of in the section on Ceramic Stator Testing.

Rotor Disk Thermal Stress Rig

A test rig was designed and built which subjected the turbine rotor hub to a radial thermal gradient thereby producing a tensile stress in the bore of the disk [28]. The rig consisted of a cylindrical container in which a rotor was placed between two 1000-watt electrical heaters. The heaters raised the temperature of the blades and platform while the rotor bore was cooled by a tube delivering shop air. Thermocouples and strain gauges were to be used to monitor the disk temperatures and stresses. Although the rig was shown to be capable of fracturing the rotor and preserving the fracture surfaces intact, many problems were encountered attempting to fasten thermocouples and strain gauges to the ceramic turbine disk. Since these gauges were required to control the test, the effort was de-emphasized.

Combustor Test Rig

The combustor test rig was developed to evaluate ceramic combustors independent of the other ceramic components in order to eliminate secondary damage as a result of combustor failures. The rig, shown in Figure 3.23, could operate at inlet temperatures from ambient to 980°C (1800°F), exit temperatures in excess of 1370°C (2500°F) and airflow rates up to about 0.9 kg/s (~ 2 lbs/s). Facility air could be supplied at pressures of up to 450 kPa (65 psia) and temperatures up to 590°C (1100°F). A secondary heater in the system would be utilized when inlet temperatures of 590 to 980°C (1100 to 1800°F) were desired. The quartz window observation port, shown on the right in Figure 3.23, allowed visual inspection of the exterior of the ceramic combustor. The other observation port, shown on the left in Figure 3.23, provided visual observation of both the combustion process and the interior wall of the ceramic combustor.

Figure 3.23 Combustor Test Rig (Reprinted from Ref. [35])

This rig was not only useful in determining that thin wall ceramic combustors were the best design (see section on Engine and Rig Test Results of Stationary Components), but it was also useful in doing emissions studies which started part way through the program.

Lightoff Qualification Test Rig

The lightoff qualification test rig consisted of a complete engine assembly, less the turbine rotors, as shown in Figure 3.24. The stationary ceramic components to be qualified were installed in their normal position immediately downstream of the combustor. Engine speed, and therefore airflow, was controlled by a motoring dynamometer directly coupled to the engine output shaft. Thus, simulation of the engine thermal

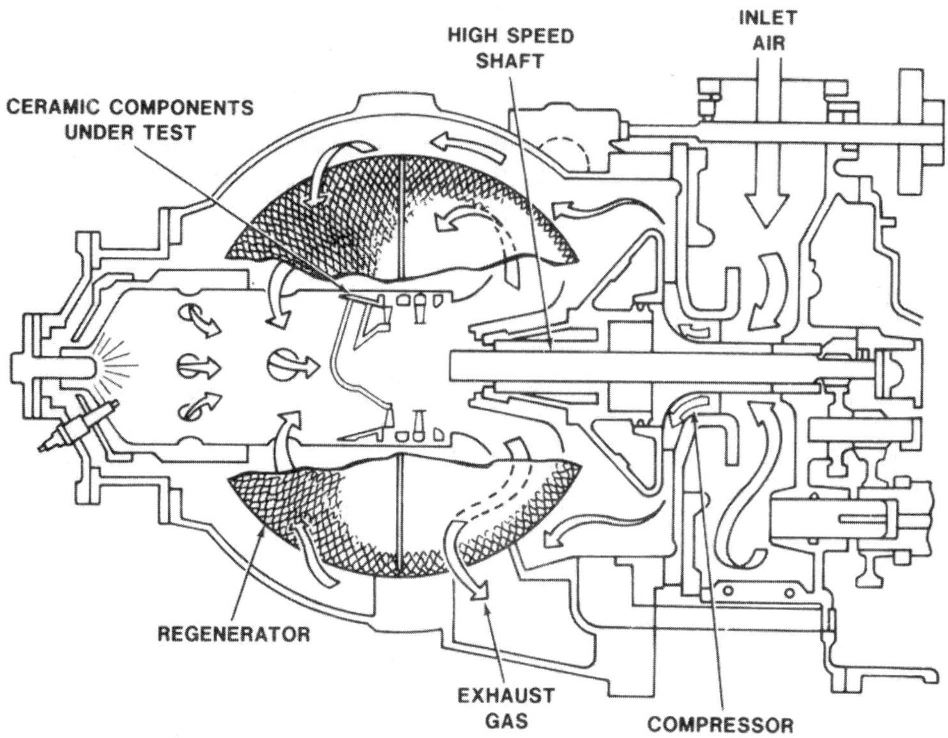

Figure 3.24 Light Off Qualification Rig (Reprinted from Ref. [35])

shock and aerodynamic loads on the stationary ceramic hot flowpath components could be achieved during engine start-up.

A control system was developed which would detect lightoff and accelerate the rig to idle speed while controlling the temperature to a predetermined schedule. Originally the lightoff signal used in the control was detected by a fast response thermocouple; however, durability problems necessitated another approach. The test rig was fitted with a quartz window in the combustor assembly, as shown in Figure 3.25. A photoelectric cell was then sighted through the window to detect combustion. Figure 3.25 also shows an infrared sensor that could be used to determine the thermal response of different areas of the nose cone and first-stage stator [35].

One of the harshest conditions for the ceramics is a cold start of the engine with the turbine inlet temperatures rapidly rising to almost 1090°C (2000°F) and then a combustor flame out a short time later. This results in rapid heating of the components and then a severe down shock. This rig would simulate this condition and accentuate the down shock by continued motoring of the rig to keep cool air flowing after the

TABLE 3.3 FLOWPATH QUALIFICATION TEST SCHEDULE

Number of the Lightoffs	Lightoff Temperature in °C (°F)	Hold Time at 1054°C (1930°F)
1	21 (70)	30 s
2–5	66 (150)	30 s
6–9	66 (150)	60 s
10	66 (150)	300 s
Total Lightoffs		Total time = 420 s

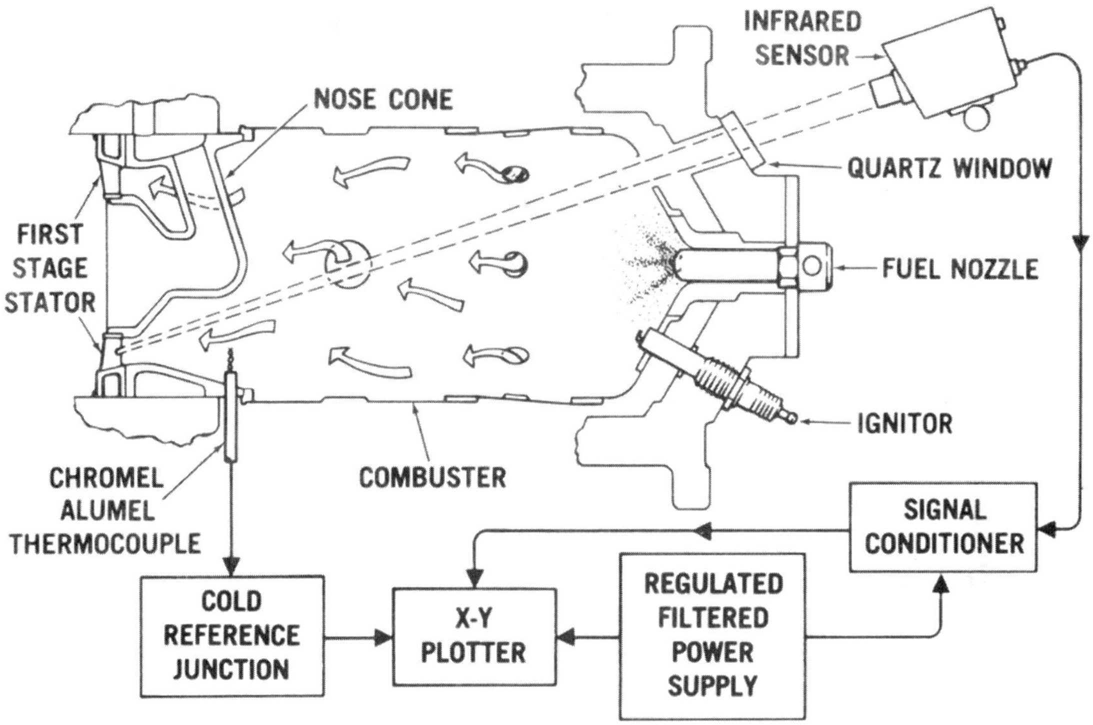

Figure 3.25 Light Off Qualification Rig with Quartz Window (Reprinted from Ref. [35])

flame out. This rig was instrumental in stationary component developments since most of the cracking problems in the ceramics occurred as a result of thermal shock. Stator vane and shroud cracks were identified in this rig as well as the nose cone shroud and bell cracks. The duty cycle is shown in Table 3.3 and Figure 3.26. This cycle resulted in measured temperature rises on the vane of over 320°C/s (580°F/s) during light off and drops of over 130°C/s (240°F/s) during the flame out.

Ceramic Structures Test Rig

The ceramic structures test rig (Figure 3.27), was configured to evaluate the stationary ceramic flowpath components for steady-state durability over the complete range of engine airflows. The maximum temperature was restricted to 1050°C (1930°F) to prevent over-temperature of the two rotary regenerators. As shown in Figure 3.27, the high-speed shafting and compressor were replaced by an external air supply that could be manually adjusted along with the fuel supply to control airflows, temperatures and pressures. A variable speed electric motor was utilized to drive the regenerator system.

This rig was instrumental in uncovering long term cracking of components. Utilizing this rig it was discovered that the silicon nitride components were gaining weight during runs at 1054°C (1930°F). In some cases, this weight gain was correlated to failures of the components, see section on Ceramic Stator Testing.

1370°C (2500°F) Flowpath Test Rig

The 1370°C (2500°F) flowpath test rig was designed to evaluate the stationary ceramic flowpath components up to temperatures of 1370°C (2500°F) over the complete range of engine airflows. Figure 3.28

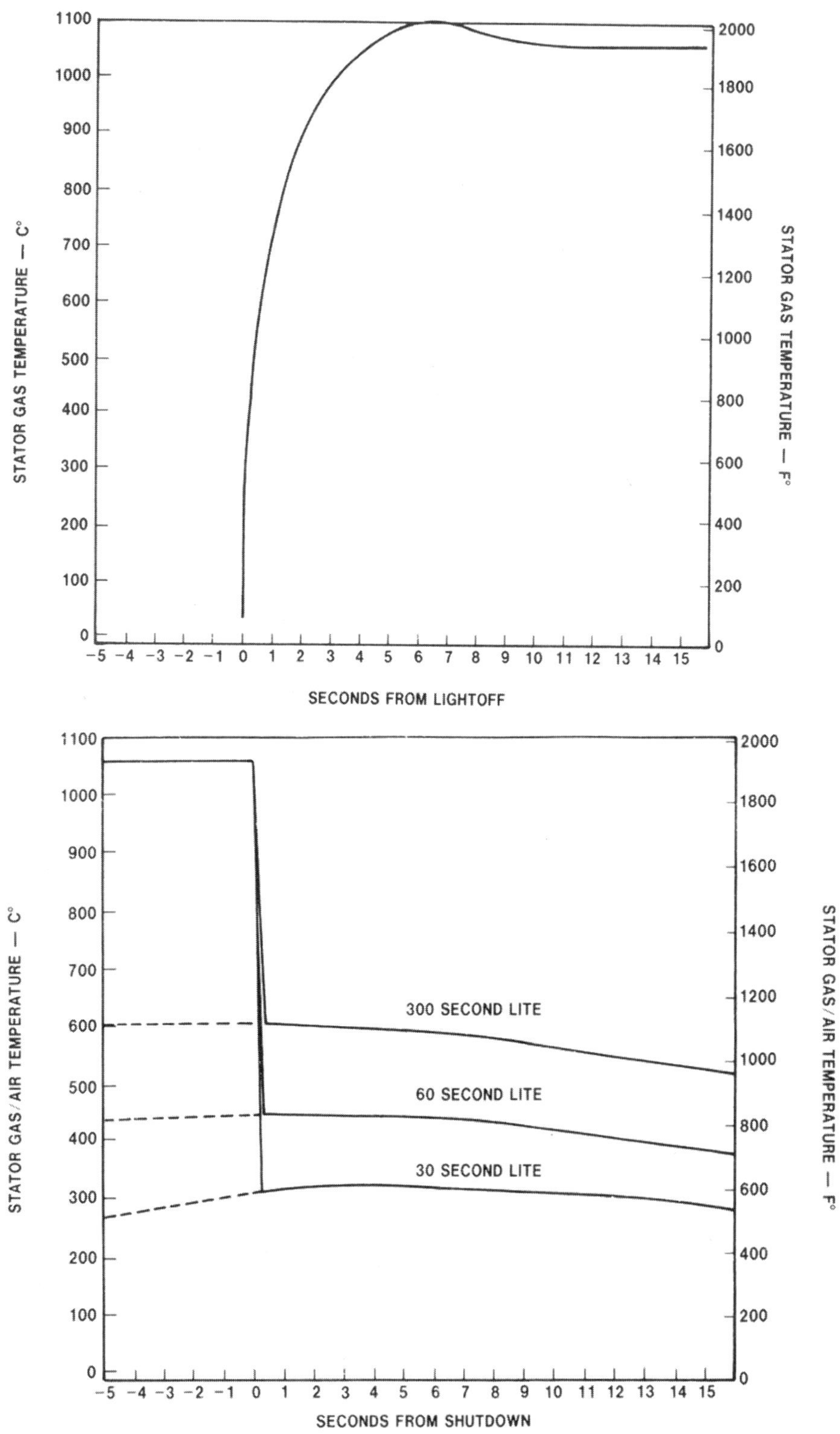

Figure 3.26 Gas Temperature For Light-off Qualification Test Rig Start Up and Shut Down (Reprinted from Ref. [35])

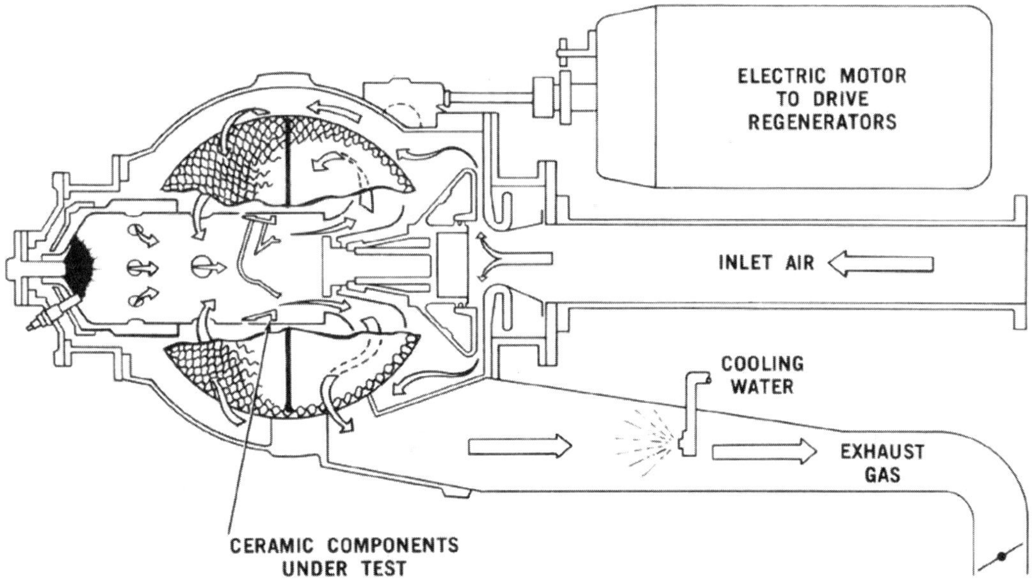

Figure 3.27 Ceramic Structures Test Rig (Reprinted from Ref. [35])

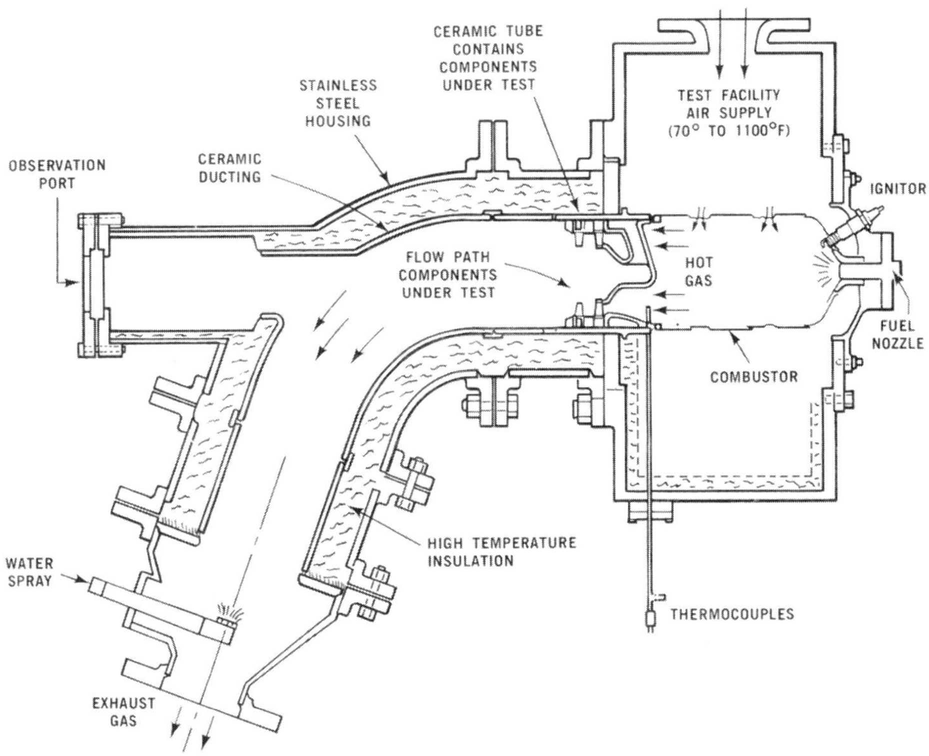

Figure 3.28 1370°C (2500°F) Flow Path Test Rig (Reprinted from Ref. [27])

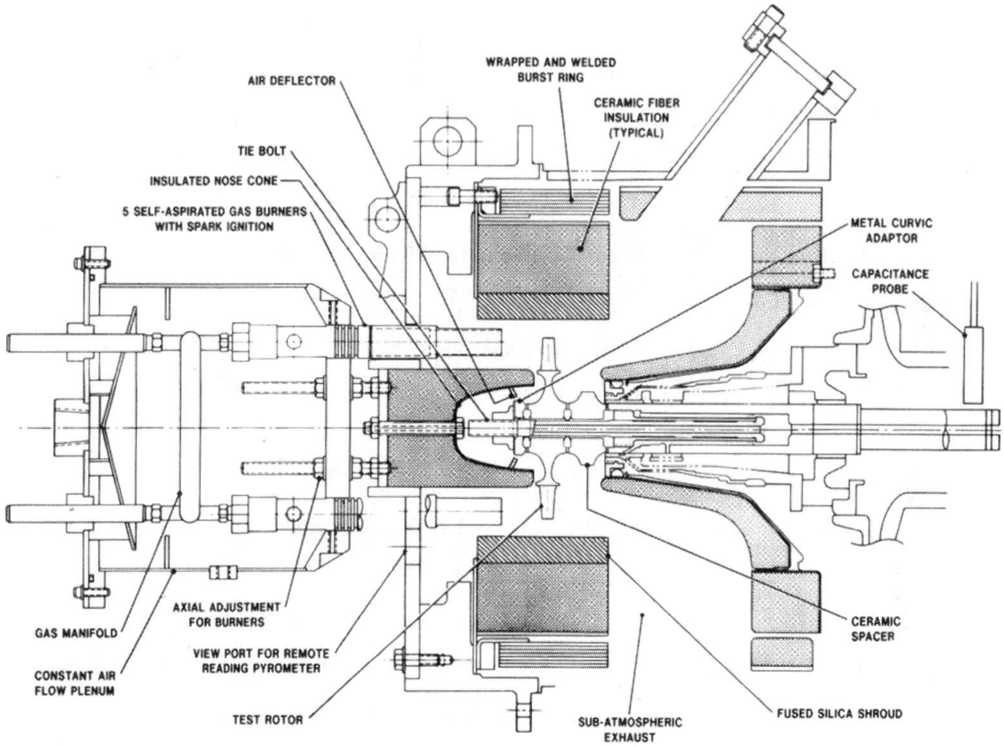

Figure 3.29 Hot Spin Test Rig (Reprinted from Ref. [35])

shows a cross-section of the rig that consisted of an outer stainless steel shell and an inner ceramic flowpath separated by high temperature insulation. The rig utilized a standard engine type combustor assembly operating with preheated, compressed air that was delivered from the test facility at temperatures up to 590°C (1100°F). Airflow, temperatures, and pressures could be manually controlled over the entire operating range via a series of valves. The hot exhaust gas was cooled with a water spray prior to dumping into the cell exhaust system. The combustor exit temperature was monitored by using thermocouples located at the inlet of the nose cone.

The stationary flowpath components could be tested one at a time, all together, or any combination thereof, by utilizing substitute ceramic spacers (except for a nose cone which was required for all testing). A rear observation port provided visual inspection of the components during testing.

This rig was critical in development of the high temperature capabilities of the stationary components prior to incorporation of ceramic rotors since the ceramic structures rig could not go to these high temperatures without over-temperaturing the ceramic regenerators. Interestingly, the components often survived better in this rig than in the ceramic structures rig. Although not fully understood, it was believed that the high temperatures caused a rapid generation of an oxidation layer on the surface of the component that stopped or slowed the weight gain discussed in the previous section. This was not a permanent fix, however, since high airflows could erode this protective surface.

Hot Spin Test Rig

The hot spin test rig was designed to quickly evaluate ceramic turbine rotors at temperature and speed with minimum cost. The principal features of the rig (Figure 3.29) were:

- a single ceramic rotor could be tested
- no stationary ceramic flowpath components were required
- a simplified combustor was used
- containment of all failed rotors was achieved
- rotor hub and blade temperature measurement capability was provided
- automatic failure detection system was provided
- turn-around time after failure was minimized (about 1 week)

The simplified combustion system consisted of up to twelve stainless steel flame tubes attached to a mixing section from a commercially available propane burner and encased in a plenum supported by the air-cooled combustor cover. This configuration eliminated earlier flameout problems encountered during rapid accelerations with bladed rotors.

An Inconel X-750 spiral wound burst ring and high alumina ceramic fiber insulation were used to protect the main housing. A solid fused silica shroud ring eliminated earlier durability problems of erosion of the ceramic fiber insulation above the rotor blade tips.

The engine configuration curvic couplings and folded bolt arrangement were utilized to connect the test rotor to the high-speed shaft. Elimination of the compressor enabled the rotor and shaft to be built up and balanced outside the rig and installed in the rig without disassembly, thereby eliminating any uncertainty present in configurations which first balance the rotor and shaft, then remove the rotor during the assembly of the shaft into the rig. An air deflector was positioned inside the insulated nose cone to direct the rotor bolt cooling air exhaust over the metal curvic adapter keeping it at an acceptable temperature level.

A capacitance probe that measured clearance of the high-speed shaft was utilized as a failure detector in the developed rig. This system replaced an earlier burst detector located above the rotor blade tips consisting of several closely spaced windings of 0.25 mm (0.010 inch) diameter platinum wire encased in insulating material, which exhibited durability problems.

The temperature of the ceramic rotor blades and hub could be monitored by a radiation pyrometer mounted on the test rig and sighted through either a front or rear viewport. Continuous recording on magnetic tape of the outputs of the radiation pyrometer, capacitance probe, speed, cooling airflow, etc., aided in failure analysis.

This rig was used for rotor development and the demonstration of the turbine rotor at design speed and temperatures.

Turbine Rotor Test Rig

The design of the turbine rotor test rig, shown schematically in Figure 3.30, was similar to but less complex than that of the engine, although it used all of the hot flowpath components except for regenerators. The engine centrifugal compressor was utilized to absorb power and provide compressed air to cool the main housing of the test rig. The inlet air to the combustor was supplied by the facility, thereby allowing control of airflow and temperature independent of the rotational speed of the rig. As shown in Figure 3.30, the complete hot flowpath ceramic components, including the ceramic turbine rotors, could be subjected to the aerodynamic, thermodynamic and mechanical loading encountered in a complete engine. However, the time and expense to rebuild this rig after a rotor failure precluded extensive use of the rig; rotor development was conducted in the hot spin test rig.

Modified Engine Test Rig

For rotor testing, the engine was modified so that reduced blade height rotors and/or single turbine rotors could be tested at higher temperatures and lower speeds than planned for the two-stage full-blade-height turbine rotors. The modification reduced the regenerator inlet temperature to an acceptable level by

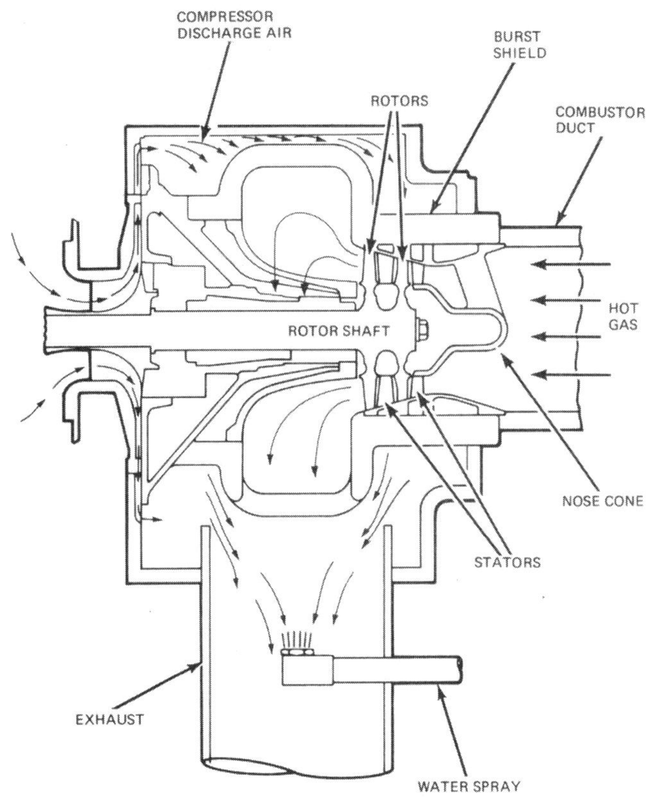

Figure 3.30 Turbine Rotor Test Rig (Reprinted from Ref. [35])

diluting the hot gases exiting from the turbine with combustor inlet and/or compressor discharge air as shown in Figure 3.31.

ENGINE AND RIG TESTING RESULTS OF STATIONARY COMPONENTS

Introduction

The stationary ceramic components shown in Figure 3.1 represent how the combustor, the nose cone, the two stator stages and the two rotor tip shrouds looked at the end of the program. This section will discuss some of the test cycles and why the changes were made to the components.

These stationary ceramic components were subjected to varying environments over the engine operating duty cycle, as shown in Figure 3.32. The engine turbine inlet temperature (TIT) schedule consisted of 175 hours at 1054°C (1930°F) from 55% to 90% speed and 25 hours at 1370°C (2500°F) at 100% speed. Typical turbine inlet temperatures during start-ups and shutdowns are displayed in Figure 3.33.

For convenience in the development of these stationary ceramic flowpath components, a test and evaluation philosophy were established which separated the complex duty cycle into three types of testing. The components were tested and evaluated for the most part, under the following controlled conditions:

(1) 1054°C (1930°F) —Steady state
—Static testing (without turbine rotors)
—Dynamic testing (with turbine rotors)

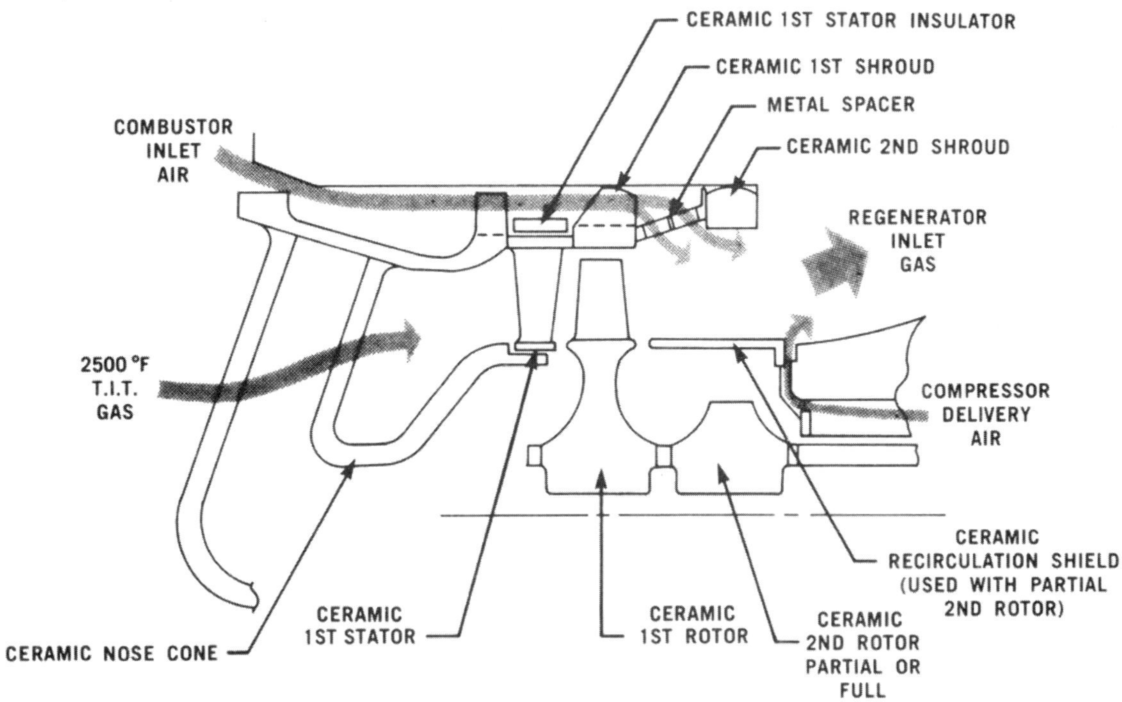

Figure 3.31 Modified Engine Test Rig (Reprinted from Ref. [35])

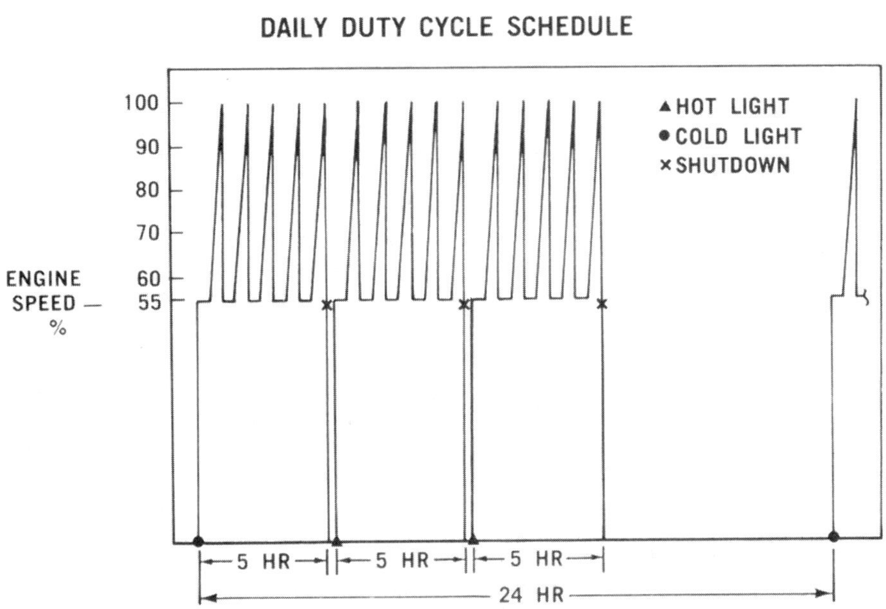

Figure 3.32 Engine Durability Duty Cycle (Reprinted from Ref. [11])

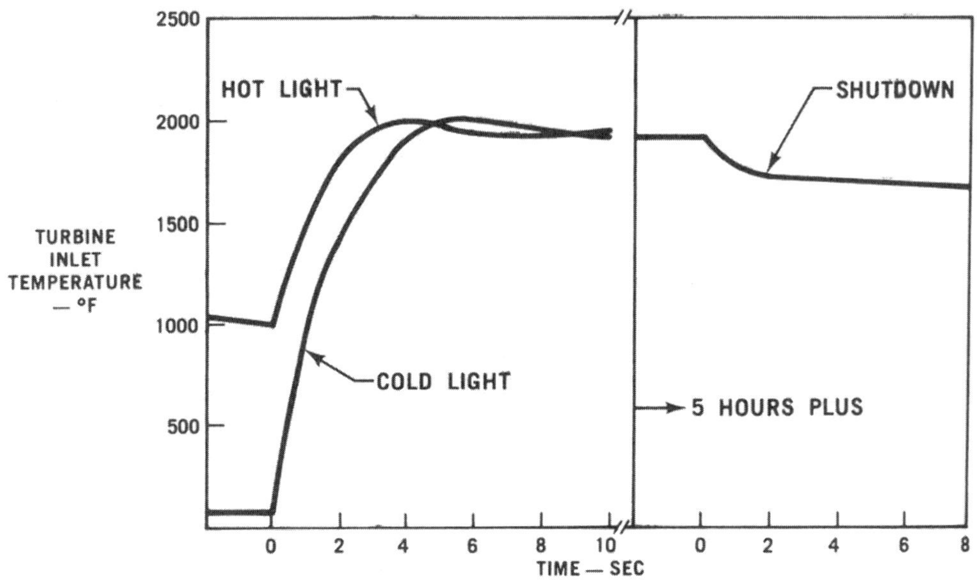

Figure 3.33 Typical Engine Turbine Inlet Turbine Transients (Reprinted from Ref. [11])

(2) 1370°C, 2500°F	—Steady state
	—Static testing
	—Dynamic testing
(3) Cyclic	—Transient
	—A sequence of lightoff, acceleration to 55% speed, and shutdown

Initially all parts were subjected to ten transient cycles. This was followed by static steady state testing of 175 hours at 1054°C (1930°F) and 25 hours at 1370°C (2500°F). This fulfilled the goal of the stationary component development effort.

Ceramic Combustor Testing

The ceramic combustor progressed through two iterations. The first iteration converted it from a test rig to an engine configuration and the second iteration reduced its operating thermal stresses by a reduction in wall thickness. Specific configurations, pertinent design and materials features, and development data used as a basis for progressing to a subsequent configuration are detailed in Figure 3.34, with references which contain more detailed data.

For the original combustor design configuration, a variety of materials were evaluated, including silicon nitride, fused silica, silicon carbide, mullite, silicon carbide-coated graphite, and aluminum nitride. Each material was tested in the Ceramic Combustor Test Rig. During this initial evaluation phase, the most successful material was Refel silicon carbide

The first iteration simply converted the test rig design to the engine design. Although one Refel silicon carbide combustor of this configuration completed 200 hours of duty cycle operation including 25 hours at 1370°C (2500°F), a number of components failed to survive a ten-hour qualification test. A design analysis indicated that the steady state thermal stresses would be reduced by 40°C with a material thickness reduction. The second iteration reflects this change.

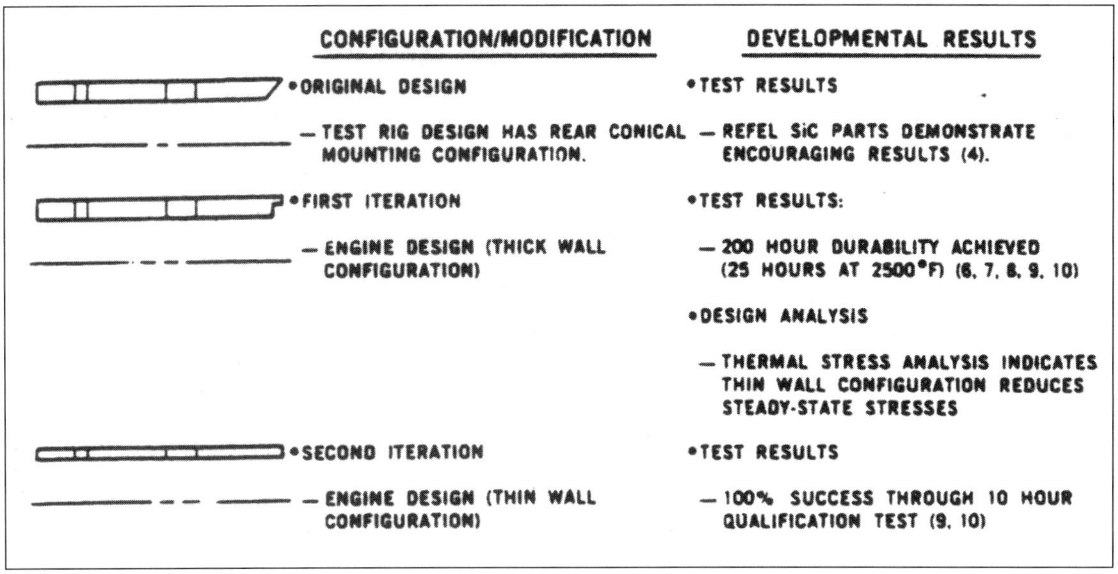

Figure 3.34 Ceramic Combustor Development (Reprinted from Ref. [12])

Test results on the thin wall Refel silicon carbide configuration verified the design analysis. All combustors tested successfully survived a 10-hour screening test. In addition, two combustors completed the 200 hour goal without incident.

Ceramic Nose Cone Testing

The ceramic nose cone progressed through four major iterations during its development. Specific configuration, pertinent design and material features, and resulting development data used as a basis for progressing to a subsequent configuration, are detailed in Figure 3.35.

The original design contained a number of specific features worth noting. The outer shroud was a solid shell with four thermocouple holes located near the front lip. At the rear of the outer shroud was a heavy solid ring called the rear flange. Three radial struts connected the outer shroud to a body called the inner bell. The inner bell was a shell with a tubular section in the center. Initial nose cones were fabricated by injection molding of 2.2 g/cm^3 density RBSN.

Testing of this design produced axial outer shroud cracks, usually through the thermocouple holes. In addition, the inner bell cracked circumferentially, usually at the inner bell tube-to-inner bell junction. Fabrication studies noted that low-density areas were present in the outer shroud front lip and rear flange sections. A design analysis confirmed high stresses were present in the inner bell tubular section junction and also indicated the possibility of adverse thermal gradients in the thicker front lip and rear flange areas.

The first nose cone iteration was designed utilizing this information. It had these modifications: (1) the rear flange was scalloped to reduce its mass, the front lip was cut back, and the four thermocouple holes removed; both these changes resulted in a more uniform thickness outer shroud; (2) the outer shroud was slotted axially in three places, spaced equally between each strut; (3) based on design analysis results, the tubular inner bell section was removed.

	CONFIGURATION/MODIFICATION	DEVELOPMENTAL RESULTS
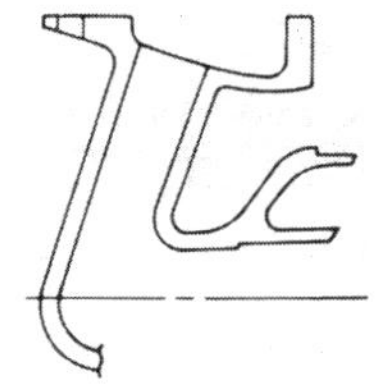	•ORIGINAL DESIGN -CONTINUOUS OUTER SHROUD - 4 THERMOCOUPLE (T/C) HOLES -CONTINUOUS REAR FLANGE -INNER BELL TUBE -2.2 g/cc DENSITY INJECTION MOLDED REACTION BONDED SILICON NITRIDE (RBSN) (20)	• TEST RESULTS: -OUTER SHROUD AXIAL CRACK (21) -INNER BELL CIRCUMFERENTIAL CRACK (21) •FABRICATION RESULTS: -POROSITY IN REAR FLANGE SECTION AND OUTER SHROUD FRONT LIP SECTION (24) •DESIGN ANALYSIS: -CONFIRMS HIGH STRESS AT INNER BELL TUBE TO INNER BELL JUNCTION (20)
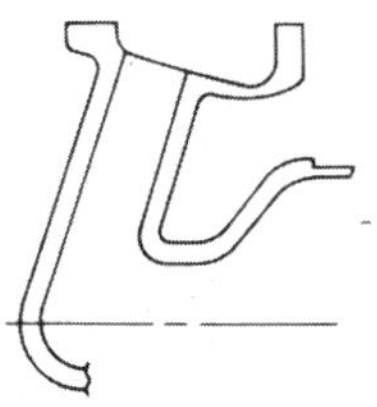	•FIRST ITERATION -CUT BACK CONTINUOUS OUTER SHROUD FRONT LIP AND ELIMINATED T/C HOLES (23) -SCALLOPED REAR FLANGE TO REDUCE MASS AND GRADIENTS (23) -AXIALLY SLOTTED OUTER SHROUD THREE PLACES EQUALLY SPACED (23, 24) -REMOVED INNER BELL TUBE AND MODIFIED INSULATION CONFIGURATION (22) -2.2 g/cc DENSITY INJECTION MOLDED RBSN	•TEST RESULTS: -OCCASIONAL OUTER SHROUD CRACKS (21-23) -AXIAL INER BELL REAR LIP CRACKS •FABRICATION RESULTS: -POOR QUALITY CONTROL (24) -2.2 g/cc DENSITY MATERIAL PRONE TO HIGH OXIDATION (26)
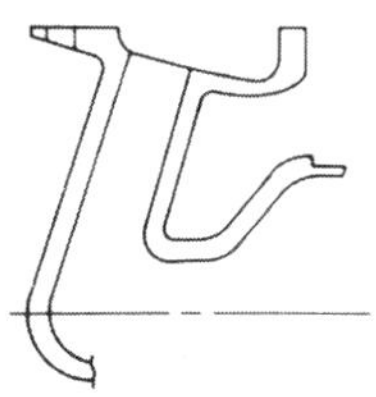	•SECOND ITERATION -2.55 g/cc DENSITY INJECTION MOLDED RBSN (27) -3 T/C HOLES REQUIRED (28) -AXIALLY SLOTTED SHROUD THRU 3 T/C HOLES (28) -MODIFIED INSULATION CONFIGURATION (28)	•TEST RESULTS: -REAR FLANGE SLOT-TO-SLOT CRACKS IN REDUCED AREA SECTION (28) -INSULATION MODIFICATIONS DIMINISHED AXIAL INNER BELL CRACK (28) -BEND TESTS OF REDUCED AREA REAR FLANGE SECTION INDICATE POSSIBLE REDUCTION IN LOAD CARRYING CAPACITY (28) •FABRICATION RESULTS: -TOOLING CONTROLS IMPROVE QUALITY CONTROL
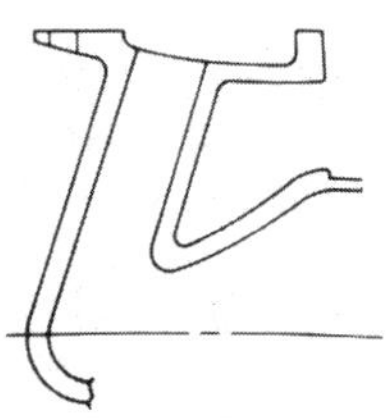	•THIRD ITERATION -2.7 g/cc DENSITY INJECTION MOLDED RBSN (28) -FLOWPATH CONTOUR IMPROVED AERODYNAMICALLY (29) -ANGLED OUTER SHROUD SLOTS TO INCREASE LOAD CARRYING CAPACITY (28)	•TEST RESULTS: -EVALUATION NOT COMPLETE •FABRICATION RESULTS: -MOLDING PARAMETERS NOT OPTIMIZED FOR 2.7 g/cc DENSITY MATERIAL

Figure 3.35 Ceramic Nose Cone Development (Reprinted from Ref. [12])

This nose cone was also fabricated in the 2.2 g/cm^3 density RBSN material. Testing produced the following results: (1) occasional outer shroud cracks occurred; (2) axial inner bell rear lip cracks, a new mode of failure, was observed; (3) data indicated that the 2.2 g/cm^3 density material RBSN was prone to severe loss in strength due to oxidation. An assessment of the inner bell rear lip axial cracks revealed that a change in the insulation configuration had occurred when the tubular section was removed. Additional insulation had been added which changed the thermal conditions in the area of the thin rear lip. This added insulation was subsequently cut back.

The second nose cone iteration reflected major changes, not all due to the prior development data. This configuration, now fabricated in 2.55 g/cm^3 density RBSN, incorporated three thermocouple holes. The reintroduction of these holes was necessary because of problems encountered with upstream hardware adaptation. Axially slotting the outer shroud appeared to be a benefit so this feature was maintained. These slots were oriented to pass through the three thermocouple holes. The insulation configuration was modified as noted earlier.

Testing identified another problem. Rear flange slot-to-slot cracks occurred in the reduced area section. Prior slot orientation was at 120° intervals, so it was recommended that the new slots be angled to approach equal segments at the rear flange. No longer was axial bell lip cracking a problem, since the insulation modification had diminished this mode of failure. In the fabrication area, development of injection molding process controls improved component quality.

The third iteration was the final nose cone configuration. The recommended slot angling was achieved. These components were fabricated in 2.7 g/cm^3 density RBSN. No rear flange failures occurred. Two components of this configuration successfully reached the 200-hour program durability goal.

Ceramic Stator Testing

The first stage stator progressed through four major iterations during its development. Specifics of the first stage iterations are detailed in Figure 3.36. A major effort was expended in developing the stator and demonstrates the required interactions between design, materials, and testing activities.

The original configuration was an assembly of 25 segments, each of which contained a portion of the first stage rotor tip shroud. The initial material used was 2.2 g/cm^3 density injection-molded RBSN. Two major test results were experienced: (1) failures occurred at the segment joints, and; (2) thermal cracks occurred at the vane root junction with the massive outer shroud. Thermal shock rig tests of vane segments confirmed a thermal mismatch at the vane root section.

The first iteration attempted to resolve the vane cracking problem through redesign. The first stage shroud section was removed and the outer shroud redesigned to more closely match the thermal response of the vane. The addition of an inner shroud of segmented configuration was done for aerodynamic reasons. Although bond joint failures were repeated, the vane thermal cracking was altered in that the vane cracking moved to midspan. Decreasing the vane span by removing material from the trailing edge, which also thickened the trailing edge, eliminated this cracking.

The second iteration incorporated this vane modification as a permanent change. Tooling modifications to optimize segment assembly were unsuccessful as failures in the segment bond joint persisted. Thus, one-piece tooling was designed to eliminate the segmented configuration.

The third iteration combined the one-piece design with a higher density 2.55 g/cm^3 RBSN. This combination resulted in a major improvement in part life, but fabrication of such a complex part required considerable development. Outer shroud cracking was diminished but not eliminated. Oxidation of the stators, evidenced by measuring weight gain, appeared to be related to part life, with all parts failing if they exceeded a 2% weight gain. As materials and processes improved, the rate of oxidation (weight gain) decreased. Two stators, 948 and 954, exhibited very little oxidation and survived 175 hours at 1054°C (1930°F) without failure. After this test, stator 948 survived an additional 25 hours at 1370°C (2500°F).

The fourth iteration involved combining the one-piece configuration with a major material improvement, 2.7 g/cm^3 density RBSN. Molding developments and processing developments helped to make these parts more consistent. Fourth iteration stators survived the 200 hour goal. However, the oxidation problem was not fully resolved and experience indicated that stator failures would occur after a weight gain of approximately 2%. Sufficient data were generated to predict the life of a stator, with reasonable accuracy, after 16 hours of testing at 1054°C (1930°F). In most cases, 2% weight gain due to oxidation occurred in less than 200 hours. At the end of the program it was not understood why the 2% weight gain was critical.

The second stage stator development closely paralleled that of the first stage stator. Its four major iterations are detailed in Figure 3.37.

	CONFIGURATION/MODIFICATION	DEVELOPMENTAL RESULTS
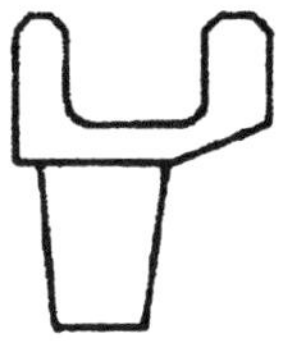	•ORIGINAL DESIGN -25 SEGMENTS OF 2.2 g/cc INJECTION MOLDED RBSN (20) -INTEGRAL FIRST STAGE SHROUD -AERODYNAMIC CLEARANCE AT STATOR ID (TO NOSE CONE)	•TEST RESULTS: -OUTER SHROUD SEGMENT BOND JOINT FAILURES (20) -VANE ROOT THERMAL CRACKS (20) •FABRICATION RESULTS: -SEGMENT BONDING MATERIAL LOW DENSITY AND LOW STRENGTH (20) •DESIGN ANALYSIS: -THERMAL SHOCK RIG TESTS OF VANE SEGMENTS CONFIRM SHROUD MASS-TO-VANE MASS THERMAL MISMATCH (20)
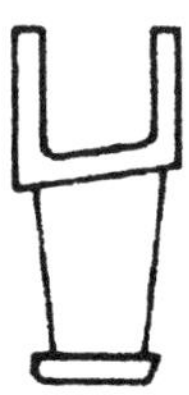	•FIRST ITERATION -SHROUD MASS REDUCED TO MORE CLOSELY MATCH VANE THERMAL RESPONSE (1ST TIP SHROUD SEPARATE PART) (20) -INNER SHROUD ADDED TO DIMINISH AERODYNAMIC LEAKAGE (20)	•TEST RESULTS: -OUTER SHROUD SEGMENT BOND JOINT FAILURES (21) -VANE THERMAL FAILURES MOVED TO MIDSPAN REGION (22) -CUT BACK TRAILING EDGE VIRTUALLY ELIMINATES VANE MIDSPAN CRACKING (22,23)
	•SECOND ITERATION -REDESIGNED VANE TO CUTBACK CONFIGURATION (23) -NEW SEGMENT INJECTION MOLDING TOOLING AND IMPROVED SEGMENT ASSEMBLY FIXTURE (22)	•TEST RESULTS: -OUTER SHROUD SEGMENT BOND JOINT FAILURES (23,24) •FABRICATION RESULTS: -ONE PIECE TOOLING DESIGNED TO ELIMINATE SEGMENTED CONFIGURATION (24)
	•THIRD ITERATION -ONE-PIECE INJECTION MOLDED PART (25) -2.55 g/cc DENSITY RBSN (26)	•TEST RESULTS: -OUTER SHROULD CRACKING POSSIBLY FUNCTION OF RATE OF OXIDATION •FABRICATION RESULTS: -EARLY 2.7 g/cc DENSITY MATERIAL DATA INDICATES LOWER OXIDATION RATE (29)
	•FOURTH ITERATION -2.7 g/cc DENSITY RBSN (28) -EXPANDED PRE-ENGINE TEST QUALIFICATION	•TEST RESULTS: -EVALUATION NOT COMPLETE FABRICATION RESULTS: -MOLDING PARAMETERS NOT OPTIMIZED FOR 2.7 g/cc DENSITY MATERIAL

Figure 3.36 Ceramic First Stage Stator Development (Adapted from Ref. [12])

	CONFIGURATION/MODIFICATION	DEVELOPMENTAL RESULTS
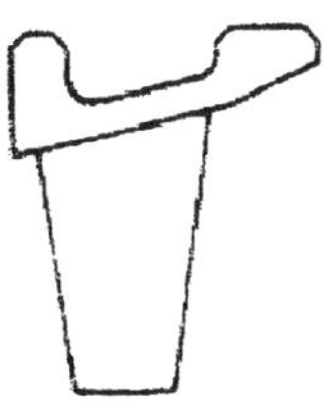	•ORIGNAL DESIGN -33 SEGMENTS OF 2.2 G/CC INJECTION MOLDED RBSN (20) -INTEGRAL SECOND STAGE SHROUD -AERODYNAMIC CLEARANCE AT STATOR ID (TO 2ND STAGE ROTOR PLATFORM)	•TEST RESULTS: -OUTER SHROUD SEGMENT BOND JOINT FAILURES (20) •FABRICATION RESULTS: -SEGMENT BONDING MATERIAL LOW DENSITY AND LOW STRENGTH (20)
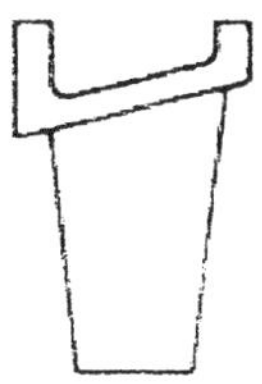	•FIRST ITERATION -2ND TIP SHROUD SEPARATE PART (20) -HIGHER DENSITY, HIGHER STRENGTH SLIP CAST MATERIAL USED AS SEGMENTED ASSEMBLY BACK-UP RINGS (22)	•TEST RESULTS -OUTER SHROUD SEGMENT BOND JOINT FAILURES (21,23) •DESIGN ANALYSIS -NOTCH EFFECT OF SEGMENTED ASSEMBLY INCREASES OUTER SHROUD OUTER SURFACE TENSILE STRESSES DURING STARTUP TRANSIENT (23)
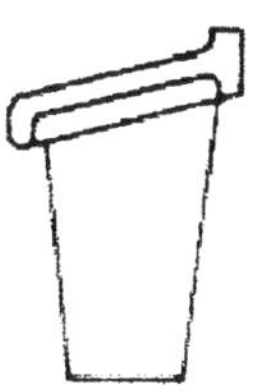	•SECOND ITERATION -OUTER SHROUD REDESIGNED (INVERTED CHANNEL CONFIGURATION ELIMINATES OUTER SHROUD OUTER SURFACE NOTCHES (23) -ONE PIECE SLIP CAST SHROUD (23)	•TEST RESULTS: -OUTER SHROUD CRACKING (SHORT TIME FAILURES) (22-25) •FABRICATION RESULTS: -ONE PIECE TOOLING EVALUATED (NOT POSSIBLE WITH EXISTING BLADE OVERLAP) (26) •DESIGN ANALYSIS: -FLOWPATH REDESIGNED TO ALLOW USE OF COMMON 1ST AND 2ND STAGE STATORS (26, 27)
	•THIRD ITERATION -ONE PIECE INJECTION MOLDED PART (25) -2.55 g/cc DENSITY RBSN (26)	TEST RESULTS: -OUTER SHROUD CRACKING POSSIBLY A FUNCTION OF RATE OF OXIDATION •FABRICATION RESULTS: -EARLY 2.7 g/cc DENSITY MATERIAL DATA INDICATES LOWER OXIDATION RATE (29)
	•FOURTH ITERATION -2.7 g/cc DENSITY RBSN (28) -EXPANDED PRE-ENGINE TEST QUALIFICATION	•TEST RESULTS: -EVALUATION NOT COMPLETE •FABRICATION RESULTS: -MOLDING PARAMETERS NOT OPTIMIZED FOR 2.7 g/cc DENSITY MATERIAL

Figure 3.37 Ceramic Second Stage Stator Development (Adapted from Ref. [12])

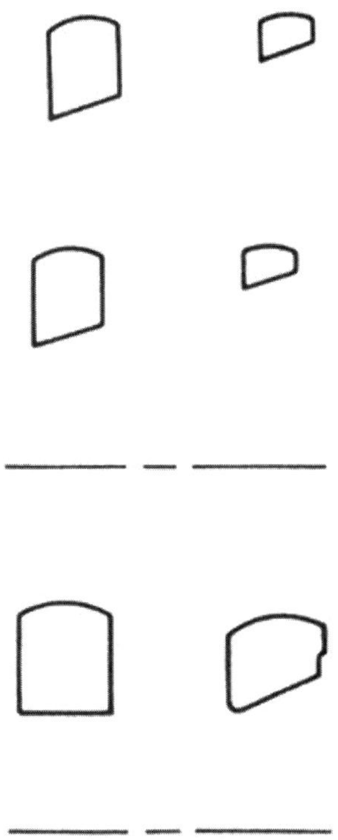

CONFIGURATION/MODIFICATION	DEVELOPMENTAL RESULTS
•ORIGINAL DESIGN - INTEGRAL PART OF STATOR ASSEMBLIES	
• FIRST ITERATION - COLD PRESSED BLANK - RBSN	• TEST RESULTS: - FAILURES AT TEMPERATURE (22)
•SECOND ITERATION - SLIP CAST BLANK - RBSN (2.3-2.7 g/cc DENSITY)	•TEST RESULTS: - DIMENSIONAL INSTABILITY WITH TIME AT TEMPERATURE (23) •FABRICATION RESULTS: -HEAT TREAT BEING INVESTIGATED AS SOLUTION TO DIMENSIONAL INSTABILITY (24,25)
•THIRD ITERATION - HEAT TREATED PRIOR TO TESTING (25) - INCREASED MASS DUE TO FLOWPATH REDESIGN (26,27)	•TEST RESULTS: - DIMENSIONAL STABILITY ESTABLISHED (25) - OCCASIONAL FAILURES AT TEMPERATURE

Figure 3.38 Ceramic Rotor Tip Shroud Development (Adapted from Ref. [12])

The incorporation of material developments was identical with that of the first stage stator. The original stators were fabricated in a segmented configuration from 2.2 g/cm^3 density RBSN. The original second stage stator vane, having a less massive base and a shorter chord vane than the first stage design, did not experience extensive vane root or midspan cracking. However, stator shroud cracking was prevalent as in the segmented first stage stator. One-piece tooling in the original second stage blade design was not possible because of blade overlap. A major ceramic flowpath design modification incorporated common first and second stage stators, thus providing monolithic fabrication capability. This reduced the number of components to be developed and took advantage of the successful first stage stator iterative developments.

Ceramic Tip Shroud Testing

The ceramic turbine rotor tip shrouds were an integral part of the stator assemblies in the original design. With the redesign of the stators to reduce their outer shroud mass, the rotor tip shrouds became separate components. The first iteration design was a simple slotted ring fabricated from cold pressed blanks. Material quality varied widely in the final processed silicon nitride parts and testing at temperature resulted in numerous failures. Figure 3.38 illustrates the progress in tip shroud development.

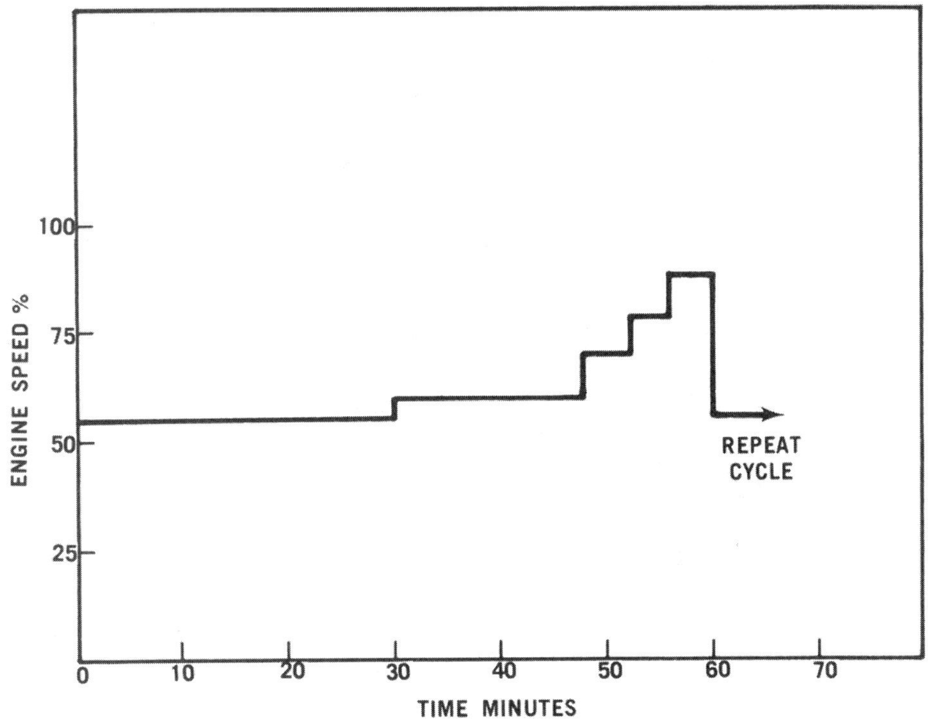

Figure 3.39 Durability Duty Cycle For Ceramic Component Testing with Metal Rotors (Reprinted from Ref. [35])

A material improvement was incorporated through the use of slip cast blanks. Although a majority of short time test failures at temperature were eliminated, a dimensional instability problem remained. A processing heat treatment was developed which pre-oxidized and stabilized the shroud rings. Testing of heat-treated rotor tip shrouds established that this heat treatment results in adequate dimensional stability. Several of these shrouds successfully completed the 200-hour goal.

Ceramic Flowpath Testing

Testing of the ceramic stationary components combined into either a partial or complete ceramic flowpath occurred when sufficient progress had been made on improving individual components to warrant a reasonable likelihood of success. Two such tests of significant importance are described in this section.

100-Hour Durability Test. The first of these tests occurred about midway in the program [7], and comprised a 100-hour duty cycle test in an engine using metal turbine rotors (both first and second stage). The turbine inlet temperature was maintained at 1054°C (1930°F) in order to avoid degradation of the metal rotors, while engine speed was varied over repeated one-hour cycles as shown in Figure 3.39.

The ceramic flowpath consisted of the following components:

- Inlet nose cone of 2.2 g/cm^3 density injection molded RBSN.
- One-piece Design C first-stage stator of 2.55 g/cm^3 density injection molded RBSN.
- First- and second-stage rotor tip shrouds of 2.7 g/cm^3 density slip cast RBSN.

- A spacer of 2.7 g/cm^3 density slip cast RBSN was used in place of the segmented second-stage stator, because of consistent outer shroud cracking problems encountered with this design, as noted in Figure 3.37.

Before the start of the 100-hour test, each component was qualified by Lightoff Qualification Testing. The durability testing was done in 15-hour segments, repeating the basic one-hour cycle noted in Figure 3.39. During each segment, the components were also subjected to two hot shutdowns followed by two hot lightoffs.

Upon completion of the 100-hour test, the nose cone was in good shape and was crack-free. The one-piece first-stage stator had two outer shroud cracks, although the 25 vanes were undamaged and the stator was still serviceable. A complete inspection at 94 hours had shown the stator to be crack-free; therefore, these outer shroud cracks occurred between 94 and 100 hours of testing. The two shroud rings and the spacer were in good condition and free from cracking. The results of this duty cycle testing were very encouraging, and constituted a major milestone of the program.

200-Hour Durability Test. Later in the program a 200-hour durability test of a ceramic stationary flowpath was undertaken [12]. The initial increment of 25 hours at 1370°C (2500°F) was run in the 1370°C (2500°F) Flowpath Qualification Test Rig, shown schematically in Figure 3.28.

The ceramic flowpath consisted of the following components:

- Inlet nose cone of 2.7 g/cm^3 density injection molded RBSN.
- One-piece Design D first-stage stator of 2.7 g/cc^3 density injection molded RBSN.
- First-stage shroud ring of 2.55 g/cm^3 density slip cast RBSN.
- One-piece Design D second-stage stator of 2.7 g/cm^3 density injection molded RBSN.
- A centering ring around the first-stage stator of 2.7 g/cm^3 g/cc density slip cast RBSN.
- Second-stage shroud ring of 2.7 g/cm^3 density slip cast RBSN.

Before the start of testing, each component was qualified by the Lightoff Qualification Test. The stators were further qualified by the Stator Vane Bend Test and the Stator Outer Shroud Pressure Test.

The 1370°C (2500°F) testing was initially conducted in approximately one-hour increments at one pound per second airflow. Parts were inspected after each increment; these increments were eventually lengthened to two, five, and six hours of testing, with inspection after each increment.

The second-stage rotor tip shroud had a partial crack before testing; after five hours at 1370°C (2500°F), this crack had propagated through the shroud in a radial direction, creating in effect a slotted shroud. No further change was noted throughout the testing. The first-stage stator developed one crack through the outer shroud after three hours of testing. No further change was noted in this part throughout the testing. The remaining parts were in good condition and free from cracks after 25 hours at 1370°C (2500°F), and shown in Figure 3.40.

The remaining testing for 175 hours at 1054°C (1930°F) was conducted on these same components in the Ceramic Structures Test Rig, shown schematically in Figure 3.27. This test was conducted at a constant airflow of 0.34 kg/sc (0.75 lbs/s) at time increments varying from 4 to 12 hours, with inspection after each increment.

Upon completion of the 175 hours of testing, the nose cone remained in good shape and crack-free. No further cracking was observed in the first stage stator outer shroud, and all 25 vanes were in good shape and free from cracks. The second-stage stator plus the centering ring and the first-stage rotor tip shroud were also in good shape and crack-free. The second-stage shroud ring had no further cracking. All of these hot flowpath components were still serviceable. The successful completion of 200 hours of durability testing of the entire stationary ceramic hot flowpath at 1054°C (1930°F) and 1370°C (2500°F) constituted attainment of another program milestone, providing further confidence in the feasibility of ceramics for demanding high temperature applications.

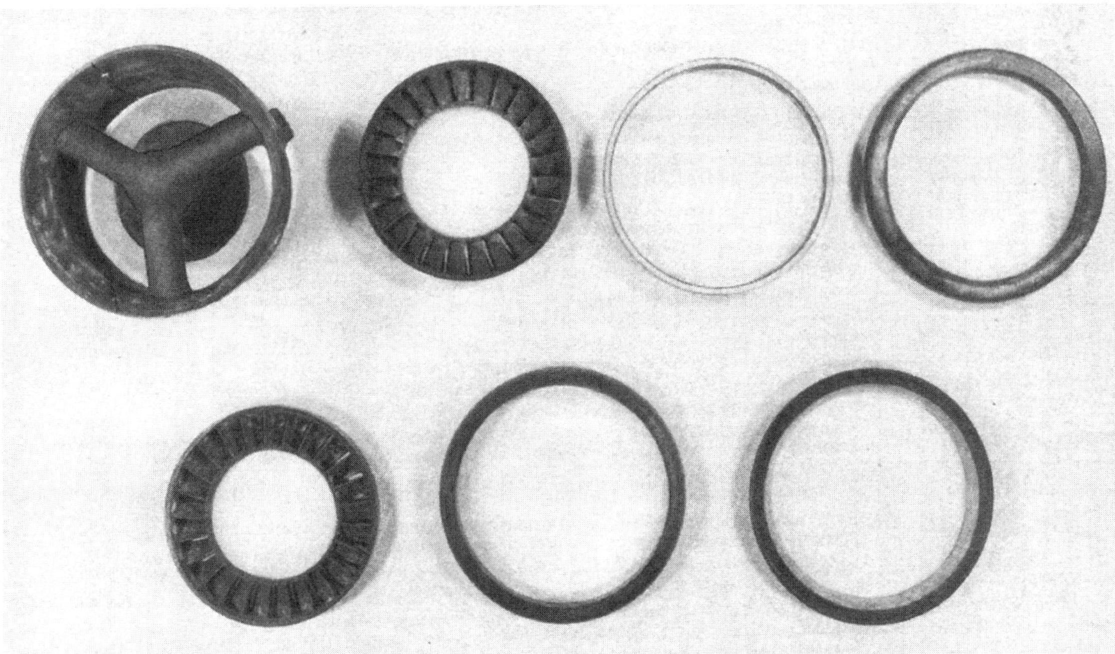

Figure 3.40 Ceramic Flow Path After 25 Hours Testing at 1370°C (2500°F) (Reprinted from Ref. [35])

Stationary Component Testing Conclusions

All of the ceramic flowpath components completed the 200 hour program goal though much work was still needed. This 200 hour demonstration was encouraging and suggested that expanded efforts in ceramic technology R&D would be worthwhile. This led to some of the follow-on ceramic component development programs. Basic material and process development needed to be pursued as evidenced by the stator oxidation problem. Design and analysis techniques needed to be refined along with quality control and NDE to explore and further develop ceramic component reliability.

CERAMIC ROTOR TESTING AND DEVELOPMENT

The design of the rotors for the 820 engine went through several iterations (see Figure 3.41). Design A was the initial design and assumed that metal rotors would be used with stationary ceramic components. This rotor design was based on standard metal design techniques. This rotor included a free vortex (twisted) blade design and a platform on the second rotor extending to the first rotor. The first and second rotors were mated by a press fit.

This same rotor configuration was used for Design B which changed the rotor from metal to ceramic. Design analyses were done assuming fully dense silicon nitride or silicon carbide properties without regard for how the rotors would be fabricated. This step indicated that both materials could be considered for further development. A 15-tooth half-barrel curvic coupling was designed to attach the rotors to each other and to the metal shaft.

The Design C rotor configuration eliminated the platform that extended from the rim of the second stage to the first stage rotor to reduce stresses and ease fabrication. It also refined the rotor attachment, by incor-

CONFIGURATION/MODIFICATION — DEVELOPMENTAL RESULTS

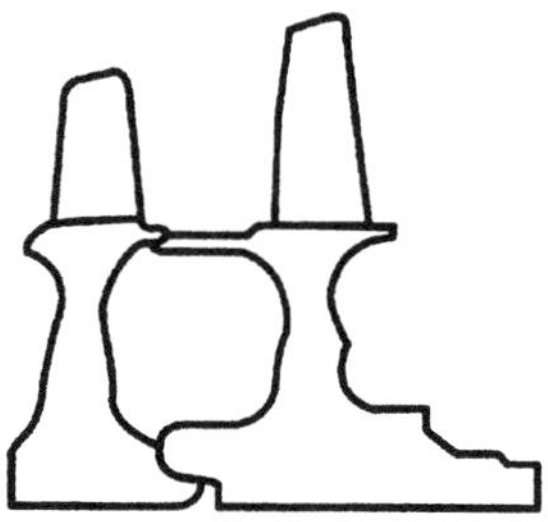

•ORIGINAL DESIGN (A & B)

-DESIGNED USING CONVENTIONAL METAL TECHNIQUES. A=METAL, B=CERAMIC

-METAL DESIGN INCLUDED PRESS FIT FOR ROTORS

-SILICON NITRIDE AND SILICON CARBIDE IDENTIFIED AS POTENTIAL MATERIALS

-MONOLITHIC DESIGN CONCEPT (20)

- CURVIC COUPLING AND FOLDED BOLT ATTACHMENT SCHEMES DESIGNED FOR CERAMIC VERSION (31)

DEVELOPMENTAL RESULTS

•FABRICATION RESULTS:

-MONOLITHIC HOT PRESSING HAS PERSISTANT PROBLEMS (20-22).

•DESIGN ANALYSIS:

- SECOND STATOR REDESIGN FORCES ROTOR REDESIGN (22)

•FIRST ITERATION (DESIGN C)

-ROTOR REDESIGNED TO REDUCE INERTIA AND AERODYNAMIC LEAKAGE (22)

-PLATFORM ELIMINATED BETWEEN FIRST AND SECOND STAGE ROTOR TO REDUCE STRESSES AND EASE FABRICATION.

-DUO DENSITY ROTOR FABRICATION METOD CONCEIVED.

-HOT PRESSED SILICON NITRIDED IDENTIFIED AS HUB MATERIAL

-RBSN USED FOR BLADES

-ATTACHMENT TECHNIQUE IMPROVED TO TAKE TWIST OUT OF THE FOLDED BOLT.

•TEST RESULTS:
-COLD SPIN TESTING OF DUO_DENSITY ROTORS EXPERIENCES BOND FAILURES (23)

-COLD SPIN TESTING OF LOW DENSITY BLADES RESULTS IN FAILURES (23).

-THERMAL SHOCK RIG TESTING PRODUCES BLADE CRACKING (24).

•FABRICATION RESULTS:
-BLADE RING FRACTURES DURING HOT PRESS BONDING FORCES TWO DEVELOPMENTS (23,24)

-A. GRAPHITE WEDGE SYSTEM

-B Si3N4 BLADE FILL TECHNIQUE

-GOOD QUALITY BONDS NOT ACHIEVED (23)

• DESIGN ANALYSIS:
- STRESSES COULD BE REDUCED AND RELIABILITY INCREASED BY ELIMINATING BALANCE RINGS AND USING SECOND STAGE HUB PROFILE IN FIRST STAGE ROTOR (25)

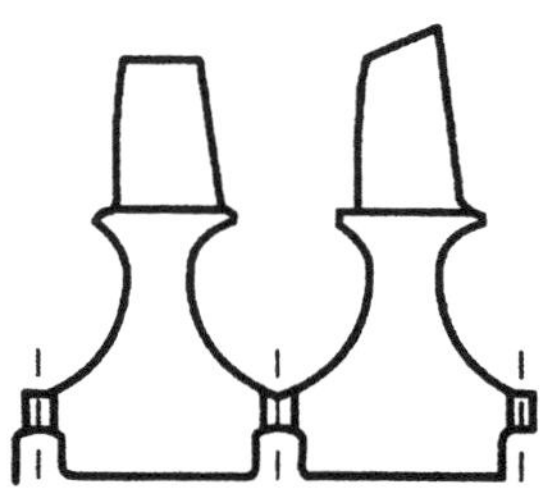

SECOND ITERATION (DESIGN D)

-ROTOR HUBS REDESIGNED TO PROVIDE COMMON PROFILE RESULTING IN REDUCED FIRST STAGE ROTOR STRESSES (25).

- FABRICATION OF BLADE RINGS\ UTILIZING 2.5 G/CC

•TEST RESULTS:

-COLD SPIN TESTING OF HUBS AND ROTORS ACHIEVES 100% N (26,28).

• FABRICATION RESULTS:

-HOT PRESS BONDING OPERATION REQUIRES FURTHER DEVELOPMENT TO PRODUCE FLAW_FREE ROTORS (26-29)

• DESIGN ANALYSIS:

-HELICALLY STACKED BLADE DESIGN CONCEIVED TO ELIMINATE BENDING AND TORSIONAL STRESSES (27).

- 3-D BLADE ANALYSIS CONFIRMS DESIGN GOAL (27).

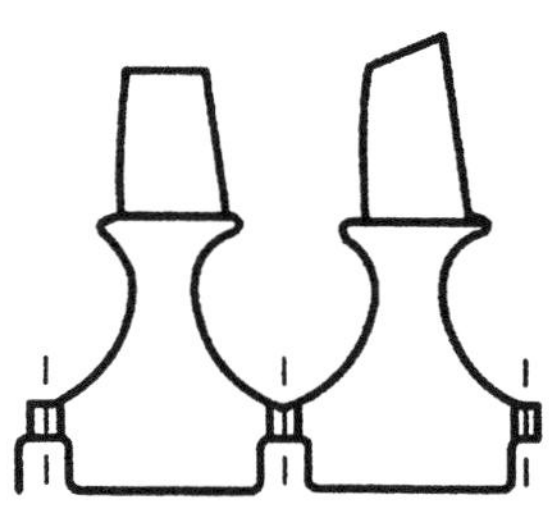

•THIRD ITERATION (DESIGN D')

-ROTOR BLADING REDESIGNED TO REDUCE BLADE STRESS AND IMPROVE FABRICABILITY (27)

-FABRICATION OF BLADE RING UTILIZING 2.7 G/CC DENSITY MATERIAL (28).

- IMPROVEMENTS IN CONTROL OF INJECTION MOLDING, RESULTS IN FEWER FLAWS

•TEST RESULTS:

-ROTOR SUCCESSFULLY COMPLETES 200 DURABILITY OBJECTIVE.

•FABRICATION RESULTS:

-HOT PRESS BONDING OPERATION REQUIRES FURTHER DEVELOPMENT TO PRODUCE CONSISTENT FLAW FREE ROTORS

•DESIGN ANALYSES

- 3 STAGE ROTOR DESIGN WOULD SIGNIFICANTLY REDUCE STRESSES AND INCREASE RELIABILITY

Figure 3.41 Ceramic Rotor Design Configuration (Adapted from Ref. [12])

TABLE 3.4 FAST FRACTURE AND TIME-DEPENDENT RELIABILITY ESTIMATES FOR THE FIRST STAGE TURBINE DISK (EXCLUDING BLADES) AT 1370°C (2500°F) TIT AND 100% SPEED

Throat Thickness Mm (inches)	SCG Threshold Temperature °C (°F)	Reliability (Fast Fracture)	Reliability (Time-Dependent SCG) (25 hours)	Critical Slow Crack Growth (SCG) Disk Region
7.62 (0.300)	770 (1420)	0.99787	0.29186	Throat
10.2 (0.400)	770 (1420)	0.99647	0.82042	Throat
12.2 (0.480)	770 (1420)	0.98986	0.60986	Bore

porating a special tool to take the twist out of the folded bolt during the attachment process. Fast fracture design analyses indicated that hot pressed silicon nitride was a candidate for the rotor hub material. The concept of a duo-density rotor utilizing a hot pressed silicon nitride hub bonded to a reaction bonded silicon nitride rotor blade ring was conceived. This would provide a high strength material for the hub while utilizing a more creep resistant blade material suited to its high temperature, relatively low stress conditions.

Late in the program, the design of the second stage rotor was changed to be identical to the first stage, except for tip length, to take advantage of the development work on the first stage and to minimize excessive tooling costs and timing necessary for a completely different rotor configuration. Also the rotor contour was changed based on advances in design methodology. The rotor contour had previously been based on fast fracture design theories. It was redone based on recently developed time-dependent theories and material properties. As a result, the neck region of the rotor was increased to 12.2 mm (0.48 inches). Table 3.4 shows that the 10.2 mm throat actually had better predicted fast structure and time dependent reliabilities than the 12.2 mm throat. None the less, the 12.2 mm throat was chosen due to (1) the very preliminary nature of the analytical techniques and material properties to predict time-dependent reliability, (2) the ability to start testing with the thick throat and then, if deemed necessary, re-machine rotors to a lesser throat dimension later, and (3) all calculations had assumed purely elastic behavior: any plastic behavior would tend to favor a thicker throat. This design was called Design D.

Design D' was a modification to Design D to reduce blade stress and ease fabrication. This design removed the twist in the rotor blades. The blades were made non-twisted, radially and centroidally stacked airfoils. This reduced the centrifugal stress in the blades from 145 MPa (21 ksi) to 90 MPa (13ksi). The trailing edge thickness of the blades varied from 0.56 mm (0.022 inches) at the tip to 0.86 mm (0.034 inches) at the blade root to facilitate injection molding of the higher strength 2.7 g/cm^3 density silicon nitride material. Design D' was the rotor configuration used in all the final tests referred to in this chapter.

Design E (not shown) was the last iteration, but it was never fabricated or tested. This was a design study based on a three-stage turbine configuration done at the end of the program. It utilized the knowledge gained during the program to predict that the lower stresses present in a three-stage configuration would greatly increase the probability of success. This design would also improve the efficiency (3%) of the engine, but would add cost due to the third stage. This increase in efficiency would occur even though the engine would be operating at a maximum speed (52,452 rpm) that was 81% of the two stage maximum speed (64,240 rpm).

TURBINE ROTOR ENGINE TESTING

Introduction

Engine testing of ceramic rotors was an unprecedented engineering endeavor in which the main difficulty stemmed from the high temperature environment of the ceramic engine and the catastrophic failures characteristic of ceramic parts. High temperatures made instrumentation difficult and expensive. The total failure of ceramic parts left few clues as to initiation of fracture.

Early in the program, tests were conducted on imperfect rotors to gain experience and to quickly identify problem areas. Testing was done at low speeds and temperatures. These parameters were increased as better rotors became available and confidence in the hardware increased.

As mentioned above a major problem was the rotor mounting design. This metal-to-ceramic interface was responsible for one of the two engine rotor failures incurred during the program, was implicated in the other failure, and was a limiting factor in other successful tests.

Rotor Mounting Design

All ceramic rotors tested in engines were mounted to the engine shaft and to each other through face spines called curvic couplings, as shown in Figure 3.9. Rig tests, run before the start of engine testing revealed a problem with the metal-to-ceramic interface designs; e.g., the coupling between the curvic adapter and the first rotor. Different rates of thermal expansion and high coefficients of friction caused failures in the ceramic curvic coupling during thermal cycling.

Commercial lubricants (Nickle Ease. Molykote 321, Electro Film 1000X and Borekote) were tested on sample ceramic curvic couplings against the curvic adapter shown in Figure 3.9. They were cycled through loads and temperatures in an Instron test machine equipped with a furnace. Nickle Ease, with its graphite particles, caused problems in the balancing of rotor assemblies. Molykote 321 worked well in balancing and was used in tests of two rotors. Its low temperature capability limited its useful life in engine tests so it was replaced by Electro Film 1000X. A surface hardening treatment called Borekote was also tried but rejected because of a metallurgical incompatibility with the curvic adapter material. All these lubricants had upper temperature limits of 425°C to 540°C (800°F to 1000°F) and were marginal in that respect because of the high temperatures existing around the curvic adapter during engine tests.

Tests of gold plate (0.0005 cm-0.0002 inches thick) on the metal curvic teeth proved most successful. Gold plate was used in the final three engine tests. Measurements of the gas temperatures around the curvic adaptor in the 1370°C (2500°F) turbine inlet temperature (TIT) tests of rotor 1195 disclosed that the temperature limit of the gold and the adaptor metal were being exceeded. Tests at 1204°C (2200°F) TIT were successful. A modification to the engine reduced the ambient conditions around the adaptor to 593°C (1100°F), so that a follow-up test at 1370°C (2500°F) TIT was totally successful.

Rotor Engine Testing

The most successful rotor tested was number 1195. This rotor had 27.5 blades (36 is maximum). Some of these blades had been removed to balance the rotor after cold spin testing had eliminated the other blades. This rotor was run in two tests. The first was a steady state ten-hour durability test at 45,000 rpm and 1093 to 1260°C (2000 to 2300°F) TIT, and was completed without incidence. This rotor was then further tested at 50,000 rpm and 1204 to 1260°C (2200 to 2300°F) for an additional 25 hours, again without failure. At the end of the 25-hour run the TIT was raised to 1370°C (2500°F) and the rotor run at 50,000 rpm for an additional 2 hours. During this portion of the test, thermocouples in the area of the curvic adaptor indicated an ambient temperature of 1038°C (1900°F). This high temperature had softened the adaptor metal and removed the thin gold plating so that during the shutdown procedure, the rotor failed. It was concluded from examination of the components that sticking of the curvic teeth had started the failure.

A design aimed at ducting bolt-cooling air at 371°C (700°F) over the adaptor was initiated and used for a subsequent engine test. Very limited tests were conducted after this but a short test to 1370°C (2500°F) on another rotor indicated that the curvic adaptor temperature was reduced to 593°C (1100°F).

HIGHLIGHTS

Ceramic design methodology was significantly advanced in the areas of both fast fracture and time-dependent reliability although further work still needed to be done, especially in the area of time-dependent reliability.

Material fabrication and processing techniques were significantly advanced, e.g. the strength of reaction bonded silicon nitride used in stators and rotor vanes more than tripled (97 MPa to 380 MPa) by 1979. The subsequent development at Ford of sintered reaction bonded silicon nitride pushed the strength to 900 MPa by 1986, or more than a factor of 9 improvement in strength.

The advances in design methodology, fabrication and testing technology were demonstrated by hardware successes. By the end of the program, 200 hours of duty cycle durability on turbine test rigs (between 1050°C to 1370°C, 1930°F to 2500°F) was attained on RBSN stationary components including stators, nose cones and turbine rotor tip shrouds. Similarly, 200 hours of duty cycling on test rigs was attained with reaction sintered SiC combustors and stators. A 200-hour test of a duo-density silicon nitride rotor with 980°C (1800°F) rim temperatures, with maximum blade temperatures of approximately 1200°C (2200°F), was run at speeds to 50,000 rpm. Engine tests of up to 100 hours duration were run with an entire uncooled ceramic stationary flow path (only the rotors being metallic) at up to 87% of engine design speed at 1050°C (1930°F). A 37-hour engine rig test was run with an uncooled ceramic rotor at temperatures up to 1370°C (2500°F) and speeds up to 50,000 RPM (100% Design E speed–three stage turbine design).

LESSONS LEARNED

In a program of this duration, many advances in materials, processing, design and testing were developed. Many of these are given in detail in the text. A summary of some of the more important lessons follows:

(1) Differential thermal expansion between metal and ceramics is a major issue as evidenced by the ceramic rotor to metal shaft attachment. Solving the problem of the ceramic to metal interface is critical to success [35].

(2) Large cross section changes in high thermal transient regions should be avoided to minimize thermally induced stresses, e.g. stator vanes to shroud [23].

(3) Mating ceramic components should have edge radiuses to avoid chipping. An example of this is the land on the second stage shroud which moved the contact point away from the lip of the inner liner to prevent edge loading and chipping of the inner liner.

(4) Large ceramic components, such as a nosecone, can be slotted to eliminate cracking due to thermal stresses.

(5) Impurities in some ceramics can cause creep by grain boundary sliding at high temperatures. These impurities may be put in deliberately to aid in forming and/or sintering. The slip cast shrouds demonstrated this phenomenon which seemed to be minimized by a heat treatment.

(6) Failure of ceramics by static fatigue has been demonstrated in the laboratory [61].

(7) Weibull volume fast fracture reliability analysis has been demonstrated to predict early failures [62].

(8) Analytical methods were introduced to predict the time-dependent failure of ceramics subjected to high temperatures and stresses. These methods were later partially validated for turbine components by Baker, Swank and Caverly in 1982 [63].

(9) Measurement of weight gain of silicon nitride parts was shown to be a potential method of predicting imminent failure [13]. This was shown on the silicon nitride stators that typically developed a shroud failure after a 2% weight gain.

(10) Ceramic materials were significantly advanced in density, strength and repeatability during the program. For example, the strength of reaction bonded silicon nitride (RBSN) used in the stator and rotor blades was more than tripled [35] (97 MPa → 380 MPa by 1979 → 900 MPa by 1987).

(11) If a three-stage turbine system was incorporated in the design, the materials that existed at the end of this program may be sufficient for further engine development due to reduced stresses in the turbine rotors [31]. This however would make the system more costly.
(12) Although ceramic materials advanced significantly, their quality in high volume production would have to be proven before a commitment could be made to high volume vehicle applications.

APPLICATION OF THE CERAMIC GAS TURBINE TO PASSENGER CARS IN TODAY'S WORLD

At the beginning of this program, passenger cars were typically large, heavy vehicles that got less than 8.5 kilometers per liter (20 miles per gallon) of fuel. There were emission requirements. Under these conditions the gas turbine had many potential advantages, as sited in the introduction. However as the vehicle got smaller and reciprocating engines adopted electronic controls and got more fuel efficient with reduced emissions, these advantages diminished. If the ceramic gas turbine met all the original targets, it would still have a difficult time competing with today's computer-controlled reciprocating engines for passenger car applications. Further reducing the size of the gas turbine to be more in line with the size of today's vehicles would not solve the problem because the efficiency of turbine engines usually drops as the size goes down since the leakages do not scale well. There is the further business issue that the huge capital investment in building reciprocating engines would have to be replaced with an equally large investment to build gas turbine engines. This last item alone would make conversion to the gas turbine an unlikely event.

FOLLOW ON PROGRAMS

Work on ceramic engine applications at Ford continued after the work on the 820 engine ceased. In 1979, Ford participated in the NASA/DOE project—Improved Gas Turbine Powertrain Conceptual Design Study [64]. The design study led to a program at Garret AiResearch Manufacturing Company to build an advanced gas Turbine called the AGT101 [65]. Ford also participated in other government programs to study the capability of the industry to manufacture ceramic components and to predict time-dependent reliability [63, 66]. Ford internally built and successfully ran a ceramic insulated diesel engine using ceramics for the piston, piston pin, cylinder, head and valves [67–70]. Ceramics were tried with limited success in the reciprocating engine valve train to reduce friction and wear. Ford was also the first in the world to successfully build and run a ceramic turbocharger in a vehicle [71]. This required developing a solution for the ceramic-to-metal attachment [72]. Ford also developed an approximate method for prediction of ceramic component fast fracture reliability [73], and summarized what the designer needs to know to use ceramics in structural applications [74]. Ford contributed to the development of the ceramic reliability program CARES [75] developed by NASA. Ford contributed chapters in two books on how to design with structural ceramics [76, 77]. One significant development was that the knowledge gained in manufacturer of the ceramic regenerators was very beneficial in development of catalytic converters for reciprocating engine emissions improvements.

ACKNOWLEDGEMENTS

The author would like to take this opportunity to thank Mr. Arthur F. McLean and Ford Motor Company for permitting him to work on this exciting project and to thank Mr. McLean for his input and review of this chapter. The author would also like to thank the many members of this close knit team that made coming to work an enjoyable and exciting experience. The work described in this chapter covered many years of work and is covered in much more detail in the references. The author would like to acknowledge that much of the content of this chapter is taken directly from these references of which he was one of many contributors. The author would like to recognize and thank the many contributors who are listed below in alphabetical order:

Ford Motor Company

D. Alexander, R. L. Allor, B. J. Amman, G. W. Andrews, N. Arnon, J. R. Baer, R. R. Baker, P. Beardmore, P. Berry, H. Blair, J. Bomback, J. Borchanian, G. T. Bretz, I. H. Buechel, D. J. Cassidy, K. Casstevens, J. C. Caverly, G. J. Chaundy, H. A. Cikanek, W. D. Compton, W. B. Copple, A. B. Czarnecki, D. A. Davis, G. C. DeBell, E. F. Dore, R. C. Elder, J. Errante, A. Ezis, W. A. Fate, E. A Fisher, J. Gault, M. Goodyear, R. K. Govila, J. W. Grant, D. L. Hartsock, J. Harwood, P. H. Havstad, J. A. Herman, E. R. Herrmann, B. T. Howes, D. W. Huser, R. C. Innes, R. A. Jeryan, C. F. Johnson, R. Kazmarek, K. H. Kinsman, C. A. Knapp, J. T. Kovach, J. G. Lafond, D. MacKenzie, J. A. Mangels, T. W. McLaughlin, A. F. McLean, W. E. Meyer, R. C. McGovern, M. Millberg, J. J. Mittman, R. E. Mogridge, T. G. Mohr, B. J. Moore, J. T. Neil, P. F. Nicholls, J. M. Nicholson, M. D. Noldy, A. Paluszny, J. M. Pieprzak, R. L. Predmesky, L. V. Reatherford, D. E. Ries, J. J. Schuldies, M. Seaman, J. R. Secord, N. L. Smith, E. D. Stiles, K. H. Styhr, M. J. Susko, L. R. Swank, G. Tennenhouse, D. M. Tracy, W. Trela, J. Uy, C. Vassallo, N. F. Waugh, T. J. Whalen, R. M. Williams, W. Wu, S. Zyck

US Government

A. L. Bement, J. Burke, G. E. Gazza, A. E. Gorum, R. N. Katz, E. M. Lenoe, D. R. Messier, H. Priest, M. J. Sinnot, E. C. Van Reuth, E. S. Wright

There were many suppliers, private and government research laboratories that also contributed to these efforts and the author would also like thank them and acknowledge their efforts.

All figures presented in this chapter were originally supplied by Ford Motor company, and a number of them have been published in reports for US government contracts. The authors gratefully acknowledge the permission of Ford Motor Company for the use of these figures in this chapter.

REFERENCES

[1] Arnon, N., Havstad, P. H., Trela, W., "2500°F Testing—Rig Development and Ceramic Gas Turbine Components Test Results," in *Ceramics For High Performance Applications—II*, J.J. Burke, E.N. Lenoe, and R.N. Katz, eds., pp. 317–335, Brook Hill Publishing Company, Chestnut Hill, Mass., 1978.

[2] Baker, R.R., Ezis, A., Hartsock, D.L., Goodyear, M.U., "Developments in Press Bonding of Duo-Density Rotors," in *Ceramics For High Performance Applications—II*, J.J. Burke, E.N. Lenoe, and R.N. Katz, eds., pp. 207–230, Brook Hill Publishing Company, Chestnut Hill, Mass., 1978.

[3] Cassidy, D.J., "NDE Techniques Used for Ceramic Turbine Rotors," in *Ceramics For High Performance Applications—II*, J.J. Burke, E.N. Lenoe, and R.N. Katz, eds., pp. 231–242, Brook Hill Publishing Company, Chestnut Hill, Mass., 1978.

[4] DeBell, G.C., and Swank, L.R., "Testing of Ceramic Rotors Hot Spin Test Rig and Rotor Preparation," ASME paper 80-GT-144, 1980.

[5] DeBell, G.C., LaFond, J.G., Meyer, W.E., Secord, J.R., "Development of a Ceramic Turbine Rotor Hot Spin Test Rig," in *Ceramics For High Performance Applications—II*, J.J. Burke, E.N. Lenoe, and R.N. Katz, eds., pp. 243–258, Brook Hill Publishing Company, Chestnut Hill, Mass., 1978.

[6] Ezis, A., "The Fabrication and Properties of Slip-Cast Silicon Nitride," in *Ceramics For High Performance Applications*, J.J. Burke, A.E. Gorum, and R.N. Katz, eds., pp. 207–222 Brook Hill Publishing Company, Chestnut Hill, Mass., 1974.

[7] Fucinari, C.A., "The Utilization of Data Relating to Fin Geometry and Manufacturing Processes of Ceramic Matrix Systems to the Design of Ceramic Heat Exchangers," in *Ceramics For High Performance Applications—II*, J.J. Burke, E.N. Lenoe, and R.N. Katz, eds. pp. 349–368, Brook Hill Publishing Company, Chestnut Hill, Mass, 1978.

[8] Govila, R.K., "Ceramic Life Prediction Parameters." U.S. Dept. of Energy report AMMRC TR 80-18, 1980.

[9] Grant, J.W. and Davis, D.A., "Aerodynamic Design Considerations for Ceramic Axial Turbines," in *Ceramics For High Performance Applications—II*, J.J. Burke, E.N. Lenoe, and R.N. Katz, eds., pp. 77–99, Brook Hill Publishing Company, Chestnut Hill, Mass., 1978.

[10] Hartsock, D.L., Havstad, P.H., Johnson, C.F., "Fabrication and Testing of Silicon Nitride Turbine Rotors," presented at the ERDA Highway Vehicle Systems Contractors' Coordination Meeting, October 5, 1977, Dearborn, Michigan.

[11] Hartsock, D.L., Baker, R.R., Havstad, P.H., and Buechel, J.J. "Test and Development of Ceramic Components," in *Ceramics For High Performance Applications—II*, J.J. Burke, E.N. Lenoe, and R.N. Katz, eds., pp. 291–316, Brook Hill Publishing Company, Chestnut Hill, Mass., 1978.

[12] Havstad, P.H., Fisher, E.A., "The Development Iteration Process," in *Ceramics For High Performance Applications—II*, J.J. Burke, E.N. Lenoe, and R.N. Katz, eds., pp. 53–76, Brook Hill Publishing Company, Chestnut Hill, Mass., 1978.

[13] Havstad, P.H., "Silicon Nitride Life Prediction Method." U.S. Patent 4,151,740, 1979.

[14] Havstad, P.H., Caverly, J.C., and Baker, R.R. (1978). "Ceramic Turbine Rotors—Engine Test and Development," in *Ceramics For High Performance Applications—II*, J.J. Burke, E.N. Lenoe, and R.N. Katz, eds., pp. 273–290, Brook Hill Publishing Company, Chestnut Hill, Mass., 1978.

[15] Jeryan, R.A., "Use of Statistics in Ceramic Design and Evaluation," in *Ceramics For High Performance Applications—II*, J.J. Burke, E.N. Lenoe, and R.N. Katz, eds., pp. 35–52, Brook Hill Publishing Company, Chestnut, Mass., 1978.

[16] Johnson, C.F., Hartsock, D.L., "Thermal Response of Ceramic Turbine Stators," in *Ceramics For High Performance Applications*, J.J. Burke, A.E. Gorum, and R.N. Katz, eds., pp. 549–562, Brook Hill Publishing Company, Chestnut Hill, Mass., 1974.

[17] Johnson, C.F., Mohr, T.G., "Injection Molding 2.7 g/cc Silicon Nitride Turbine Rotor Blade Rings Utilizing Automatic Control," in *Ceramics For High Performance Applications—II*, J.J. Burke, E.N. Lenoe, and R.N. Katz, eds., pp. 193–206, Brook Hill Publishing Company, Chestnut Hill, Mass., 1978.

[18] Mangels, J.A., "Development of a Creep-Resistant Reaction-Sintered Si_3N_4," in *Ceramics For High Performance Applications*, J.J. Burke, A.E. Gorum, and R.N. Katz, eds., pp. 195–206, Brook Hill Publishing Company, Chestnut Hill, Mass., 1974.

[19] Mangels, J.A., "Development of Injection Molded Reaction Bonded Si_3N_4," in *Ceramics For High Performance Applications—II*, J.J. Burke, E.N. Lenoe, and R.N. Katz, eds., pp. 113–130, Brook Hill Publishing Company, Chestnut Hill, Mass., 1978.

[20] McLean, A.F. et al, "Brittle Materials Design, High Temperature Gas Turbine," AMMRC-CTR 72-3, Interim Report No. 1, March, 1972.

[21] ibid, CTR 72-19, No. 2, September, 1972.

[22] ibid, CTR 73-9, No. 3, March, 1973.

[23] ibid, CTR 73-32, No. 4, September, 1973.

[24] ibid. CTR 74-26. No. 5. April. 1974.

[25] ibid, CTR 74-59, No. 6, September, 1974.

[26] ibid, CTR 75-8, No. 7, April, 1975.

[27] ibid, CTR 75-28, No. 8, October, 1975.

[28] ibid, CTR 76-12, No. 9, April, 1976.

[29] ibid, CTR 76-31, No. 10, October, 1976.

[30] ibid, CTR 77-20, No. 11, August, 1977.

[31] ibid, TR 78-14, No. 12, Volume 1, March, 1978.

[32] ibid, TR 78-14, No. 12, Volume 2, March, 1978.

[33] McLean, A.F. and Baker, R.R., "Brittle Materials Design, High Temperature Gas Turbine," AMMRC TR 79-11, Interim Report No. 13, February, 1979.

[34] McLean, A.F. and Secord, J.R., "Brittle Materials Design, High Temperature Gas Turbine," AMMRC TR 79-12, Interim Report, May, 1979.

[35] McLean, A.F. and Fisher, E.A., "Brittle Materials Design, High Temperature Gas Turbine," AMMRC TR 81-14, Final Report, March, 1981.

[36] McLean, A.F., "Ceramics in Small Vehicular Gas Turbines," in *Ceramics For High Performance Applications*, J.J. Burke, A.E. Gorum, and R.N. Katz, eds., pp. 9–36, Brook Hill Publishing Company, Chestnut Hill, Mass., 1974.

[37] McLean, A.F., "Overview of ARPA/ERDA/FORD Ceramic Turbine Program," in *Ceramics For High Performance Applications—II*, J.J. Burke, E.N. Lenoe, and R.N. Katz, eds., pp. 1–34, Brook Hill Publishing Company, Chestnut Hill, Mass., 1978.

[38] McLean, A.F., and Davis, D.A. "The Ceramic Gas Turbine—A Candidate Powerplant for the Middle- and Long-Term Future," Society of Automotive Engineers Publication 760239, Warrendale, Pa., 1976.

[39] McLean, A.F., "Developments in Ceramic Technology for Gas Turbine Engines," presented at the Joint ASME/I.M.E. International Gas Turbine Conference, London, U.K., April 9–13, 1978.

[40] McLean, A.F., "Ceramic Turbine Housings." ASME paper 82-GT-293, 1982.

[41] McLean, A.F. "Ceramics in Automotive Energy Systems," in *Proceedings of the First International Symposium on Ceramic Components for Engines*, S. Sōmiya, E. Kanai, and K. Ando, eds., October 17–19, 1983 Hakone, Japan, KTK Scientific Publishers, Tokyo, 1983, pp. 21–32.

[42] Nicholls, P.F., Paluszny, A., "Techniques for Design of Small Ceramic Turbine Rotor," in *Ceramics For High Performance Applications*, J.J. Burke, A.E. Gorum, and R.N. Katz, eds., pp. 63–78, Brook Hill Publishing Company, Chestnut Hill, Mass., 1974.

[43] Paluszny, A., "Designing with High Temperature Ceramics." ASME paper 75-DE-28, 1975.

[44] Paluszny, A. and Wu, W., "Probabilistic Aspects of Designing With Ceramics," Transactions of the ASME, Journal of Engineering for Power, 99[4], 623, 1977.

[45] Paluszny, A. and Nicholls, P.F., "Predicting Time-Dependent Reliability of Ceramic Rotors," in *Ceramics For High Performance Applications—II*, J.J. Burke, E.N. Lenoe, and R.N. Katz, eds., pp. 95–112, Brook Hill Publishing Company, Chestnut Hill, Mass., 1978.

[46] Pietsch, A., Styhr, K., "Ceramic Heat Exchanger Applications and Developments," in *Ceramics For High Performance Applications—II*, J.J. Burke, E.N. Lenoe, and R.N. Katz, eds., pp. 385–396, Brook Hill Publishing Company, Chestnut Hill, Mass., 1978.

[47] Styhr, K.H., "Spin Testing of Ceramic Turbine Rotors," in *Ceramics For High Performance Applications*, J.J. Burke, A.E. Gorum, and R.N. Katz, eds., pp. 415–424, Brook Hill Publishing Company, Chestnut Hill, Mass., 1974.

[48] Swank, L.R., Uy, J.C., "Ceramic Turbine Rotor Reliability and Failure Analysis in the Hot Spin Test Rig," in *Ceramics For High Performance Applications—II*, J.J. Burke, E.N. Lenoe, and R.N. Katz, eds., pp. 259–272, Brook Hill Publishing Company, Chestnut Hill, Mass., 1978.

[49] Swank, L.R., "Correlation of Structural Ceramic Time Dependent Failure Data," ASME paper 86-GT-14, 1986.

[50] Uy, J.C., Williams, R.M., Goodyear, M.U., "Study of the Processing and Bond Strength of Hot-Pressed Si_3N_4" in *Ceramics For High Performance Applications—II*, J.J. Burke, E.N. Lenoe, and R.N. Katz, eds., pp. 131–150, Brook Hill Publishing Company, Chestnut Hill, Mass., 1978.

[51] Whalen, T.J., Noakes, J.E., Terner, L.L., "Progress on Injection-Molded, Reaction-Bonded SiC," in *Ceramics For High Performance Applications—II*, J.J. Burke, E.N. Lenoe, and R.N. Katz, eds., pp. 179–192, Brook Hill Publishing Company, Chestnut Hill, Mass., 1978.

[52] Mangels, J.A., "Ceramic Technology Development at Ford Motor Company from 1967 to 1986," Chapter 2 in *Progress in Ceramic Gas Turbine Development. Volume II— Ceramic Gas Turbine Component Development and Characterization*, Mark van Roode, Mattison K. Ferber, David W. Richerson, eds, ASME PRESS, New York, USA, 2003.

[53] Williams, R.M., Uy, J.C., "Ceramic Material Characterization," in *Ceramics For High Performance Applications—II*, J.J. Burke, E.N. Lenoe, and R.N. Katz, eds., pp. 151–178, Brook Hill Publishing Company, Chestnut Hill, Mass., 1978.

[54] Weibull, W., "Statistical Theory of Strength of Materials," Ing. Vetenskaps Akademien Number 151, pp. 1–45, 1939.

[55] Dukes, W.H.. *Handbook of Brittle Materials Design*, AGARD AD719–712, Feb. 1971.

[56] Biasi, V., "Ford Runs Uncooled Ceramic Turbine Engine at 2500°F," Gas Turbine World, July 1977.

[57] Peterson, F.E., "Three-Dimensional Transient Heat Transfer Analysis Using Finite Element Procedure, User Information and Programming Details," Engineering/Analysis Corporation, Redondo Beach, California, August 1971.

[58] Wilson, E.R., "Solid SAP, A Static Analysis Program for Three-Dimensional Solid Structures," Structural Engineering Laboratory, University of California, Berkley, California, September, 1971, Revised March 1972.

[59] Vardar, O. and Finnie, I., "An Analysis of the Brazilian Disk Fracture Test Using the Weibull Probabilistic Treatment of Brittle Strength." International Journal of Fracture, June, 1975.

[60] Manson, S.S., *Thermal Stress and Low-cycle Fatigue*. New York: McGraw Hill, 1966, P. 293.

[61] Evans, A.G. and Wiederhorn, S.M., "Crack Propagation and Failure Prediction in Silicon Nitride at Elevated Temperature," Journal of Material Sciences 9, pp. 270–278, 1974.

[62] Evans, A.G., and Wiederhorn, S.M., "Proof Testing of Ceramic Materials—an Analytical Basis for Failure Prediction," International Journal of Fracture, September 1974, pp. 379–392.

[63] Baker, R.R., Swank, L.R., and Caverly, J.C. "Ceramic Life Prediction Methodology—Hot Spin Disc Life Program." Army Materials and Mechanics Research Center TR 83–44, Interim Report. Contract No. DAAG 46-77-C-0028. Watertown, Ma., 1983.

[64] Howes, B.T., Norman, L.W., Suriano, F., Engineering Staffs of Ford Motor Company and AiResearch Company, "Conceptual Design Study of Improved Automotive Gas Turbine Powertrain—Final report," DOE/NASA/0037-79/1, NASA CR-159580, May 1979.

[65] Schenk, B., Easley, M., and Richerson, D.W., "Evolution of Ceramic Turbine Engine Technology at Honeywell Engines, Systems & Services," Chapter 4, in *Progress in Ceramic Gas Turbine Development, Volume I— Ceramic Gas Turbine Design and Test Experience*, Mark van Roode, Mattison K. Ferber, David W. Richerson, eds, ASME PRESS, New York, USA, 2002.

[66] Caverly, J.C., and Hartsock, D.L. (1980). "Evaluation of Ceramics for Stator Applications—Gas Turbine Engines—Final Report—Reliability Analysis and Material Property Characterization," National Aeronautics and Space Administration Report CR 165270. Cleveland.

[67] Wade, W.R., Havstad, P.H., Ounsted, E.J., Trinker, F.H., and Garwin, I.J., "Fuel Economy Opportunities with an Uncooled DI Diesel Engine." Society of Automotive Engineers publication 841286. Warrendale, Pa., 1984.

[68] Wade, W.R., and Jones, C.M., "Current and Future Light Duty Diesel Engines and Their Fuels," Society of Automotive Engineers publication 840105. Warrendale, Pa., 1984.

[69] Hartsock, D.L., "Partially Stabilized Zirconia Piston Bowl Reliability," Transactions of the ASME, Journal of Engineering for Gas Turbines and Power. Vol 109[4], pp. 367–373, October 1987.

[70] Kabat, D.M., Gawin, I.J., Hartsock, D.L., "Ceramic Valve Analysis, Reliability and Test Results," Society of Automotive Engineers publication 880670. Warrendale Pa., 1988.

[71] DeBell, G.C., and Secord, J.R., "Development and Testing of a Ceramic Turbocharger Rotor," ASME paper 81-GT-195, 1981.

[72] Allor, R.L., and Caverly, J.C., "Method of Attaching a Metal Shaft to a Ceramic Shaft and Product Produced Thereby." U.S. Patent 4,499,646, 1985.

[73] Hartsock, D.L., "A Simplified Structural Ceramic Design Technique." ASME paper 85-GT-100, 1985.

[74] Hartsock, D.L., and McLean, A.F. "What the Designer with Ceramics Needs," Am. Ceram. Soc., Ceramic Bulletin 63[2], 266–270, 1984.

[75] Gyekenyesi, J.P., "Scare—A Post-Processor Program to MSC Nastran for the Reliability Analysis of Structural Ceramic Components." National Aeronautics and Space Administration Technical Memorandum 87188, Cleveland, OH, 1986.

[76] McLean, A.F., Hartsock, D.L., "Design with Structural Ceramics," in *Structural Ceramics*, edited by John B. Wachtman, Jr., Treatise on Material Science and Technology Volume 29, Academic Press, Inc., 1989.

[77] McLean, A.F., Hartsock, D.L., "An Overview of the Ceramic Design Process," in ASM Internationals "Engineered Materials Handbook, Volume 4, Ceramics and Glass," 1991.

CHAPTER

4

EVOLUTION OF CERAMIC TURBINE ENGINE TECHNOLOGY AT HONEYWELL ENGINES, SYSTEMS & SERVICES

Bjoern Schenk and Michael L. Easley
Honeywell Engines, Systems & Services, Phoenix, Arizona, USA

David W. Richerson
University of Utah, Salt Lake City, Utah, USA

INTRODUCTION AND BACKGROUND

Honeywell Engines, Systems & Services (and its prior names AiResearch, Garrett Turbine Engine Company, and AlliedSignal Aerospace) has been working on ceramic components for turbine engines since the early 1970s. Many challenges have been encountered and have required long-term development efforts to resolve. These have included establishing design and life prediction codes, working with ceramic suppliers to generate an extensive design database and also to help them to optimize and refine their materials and fabrication process, extending NDE technologies to detect smaller and smaller material flaws, addressing the critical issues of contact stress, impact, and environmental resistance of ceramic materials, and conducting extensive rig and engine testing of various turbine designs.

The DARPA/Navy **Ceramic Gas Turbine Engine Demonstration Program** (1976 to 1981), and the USAF **Ceramic Components For Turbine Engines Program** (1977 to 1982), provided the foundation for the modern Ceramic Turbine Engine Demonstration Project. These programs utilized a test bed of similar design to the Honeywell production model 331–200 APU. Significant progress was successfully achieved in demonstration of ceramic axial inserted turbine blades, as well as initial exploration of ceramic design technologies, including contact failure, particle impact damage resistance, and pioneering computer-based probabilistic component life prediction.

The DoE/NASA **Advanced Gas Turbine Technology Development Project—AGT 101** (1979 to 1987) pursued several high-risk technologies, including ceramics for future automobile engine applications.

The AGT101 program succeeded in simultaneously demonstrating many related advanced engine technologies, and identified the most critical areas in ceramic engine technology needing further development.

The DoE/NASA **Advanced Turbine Technology Project (ATTAP)** (1987 to 1992) emphasized the development of critical ceramics technologies, using the AGT101 automotive test bed. Significant progress was made in the ATTAP/AGT101 program toward successful impact-resistant ceramic turbine designs and ceramic component processing. However, the AGT101 engine proved to be inadequate as a test bed for the advanced ceramic technologies. This was because the AGT101 was designed as a demonstrator for several high-risk technologies, including the high-temperature ceramic regenerator, low-emission ceramic combustor, low-friction gas bearings, all-ceramic hot section structures, and impact-resistant ceramic turbine. As such, AGT101 reliability as a turbine test bed was not adequate for long-term endurance testing and evaluation.

The DoE **Ceramic Turbine Engine Demonstration Project (CTEDP)** (1993 to 2000) continued refinement of ceramics technologies and design methods, and focused fabrication development on a scale-up of demonstrated technologies to pilot-production levels. To adequately demonstrate ceramic technology, over 6800 hours of endurance testing have been completed, utilizing a reliable test bed based on the Honeywell Model 331–200/250 APU. This APU is a fully-developed gas turbine with current production applications in the Boeing 757 and 767 aircraft, as well as in the Airbus A300 and A310 aircraft. Thus, this APU has higher capability for successfully accomplishing not only long-term endurance testing, but also offers the opportunity for extended field evaluation of ceramic components which is currently progressing in several aircraft applications as well as in a ASE8-1000[CT] industrial engine, a derivative from the very reliable Model 331-turboprop and 331–200/250 APU engines.

The long-term multidisciplinary effort is beginning to pay off. Silicon nitride main shaft seal rings have been in production since 1996 for the Honeywell Model TFE731 aircraft propulsion engine and have solved the number one warrantee problem for that engine. The seal rings have been installed in more than 7000 engines and have accumulated more than six million hours cumulative flight time, with the high time seal runner having accumulated over 3,600 flight hours. Silicon nitride turbine nozzles have received extensive testing in Honeywell's GTCP85 series auxiliary power units used for ground cart power and on board aircraft. The use of ceramic nozzles has been approved by FAA, and nozzles are in service for long-term field tests on eleven MD-80 aircraft. Greater than 40,000 hours operation and 79,000 engine starts have been demonstrated.

The lessons learned from these development programs are currently being applied to the U.S. Army/COSSI Sand Erosion Resistance Kit (SERkit) program for the Honeywell 36–155 Apache Longbow Helicopter APU featuring ceramic turbine nozzles. Another program benefiting from the 30 years of ceramic efforts at Honeywell will be the recently awarded DoE Advanced Microturbines Program. Honeywell will develop and demonstrate the next generation advanced microturbine system featuring ceramic turbine components. Ongoing military as well as internal R&D programs are pushing the ceramic component technology even beyond the capabilities described in this chapter. This is especially true for the continuous improvement of the high temperature ceramic material properties, second generation TBC/EBC development, technologies addressing impact resistance and cooled ceramic component fabrication, low-cost manufacturing technologies, development of next generation compliant layer systems, and the continuous enhancement of design tools and probabilistic life prediction methodologies.

Table 4.1 provides a brief summary of some of the key programs conducted at Honeywell that have led to the evolution of ceramic technology for gas turbine engines. The remainder of the chapter reviews several of the key programs in greater detail.

CERAMIC GAS TURBINE ENGINE DEMONSTRATION PROGRAM

DARPA initiated a program in 1976 with AiResearch Manufacturing Company of Arizona (which later became Garrett Turbine Engine Company during the program) to retrofit ceramics into an existing turboprop

TABLE 4.1 CHRONOLOGY OF KEY CERAMICS PROGRAMS CONDUCTED AT HONEYWELL ENGINES, SYSTEMS & SERVICES

Program	Time	Highlights
In-house Efforts	1970–1975	Design and procurement of combustor liners, segmented and integral nozzles, rotor blades, and integral-manifold plate-fin recuperator. Pioneering efforts on NDE by microfocus X-ray radiography with image enhancement.
TARADCOM Ceramic Reheat Combustor Program [1]	1975–1977	Designed, procured, and tested a reaction-sintered SiC combustor liner.
DARPA/Navy Ceramic Turbine Engine Demonstration Program [2,3]	1976–1981	Retrofitted existing production engine with the goal to demonstrate that ceramics could enable increased power and decreased fuel consumption. Demonstrated 30% increase in power and 7% decrease in fuel consumption with two full stages of ceramic components during 2-h engine runs.
USAF Ceramic Components for Turbine Engines [4]	1977–1982	Reaction bonded silicon nitride (RBSN) material improvement; hot pressed silicon nitride (HPSN) cost reduction; successful redesign of DARPA engine involving elimination of one stage of ceramics by doubling the work of the other stage.
NASA Durability Testing of Commercial Ceramic Materials [5]	1978–1995	Cyclic exposure of test bars of silicon nitride and silicon carbide for up to 3500 hours in an atmospheric burner rig, followed by strength testing to evaluate retained strength and potential degradation mechanisms.
ONR Contact-Stress Analysis of Elastic-Viscoelastic Interface Conditions [6]	1980–1983	Analytical and experimental study that improved understanding of the factors that influence biaxial contact stress.
DoE/NASA Advanced Gas Turbine (AGT) Technology Development Project [7, 8]	1979–1987	Designed, built, and tested a 74.6 kW engine for automotive applications, which was designed specifically for ceramics. Identified most critical areas in ceramic engine technology needing further development.
ATTAP [9]	1987–1992	Used the AGT101 engine test bed. Significant improvements regarding successful impact-resistant ceramic turbine designs and ceramic component processing were achieved.
CTEDP[10]	1993–2000	Continued refinement of ceramics technologies and design methods. Focused fabrication development on a scale-up of demonstrated technologies to pilot-production levels. Over 6800 hours of endurance testing on site, and over 2300 hours of airborne field testing to date utilizing Honeywell Model 331-200/250 APUs.
USAF GTCP85 Series APU Ceramic Components Insertion Program [11]	1992–1995	Extended endurance testing on ceramic turbine nozzles installed in Model GTCP85-180 APUs used in U.S. Air Force A/M32A-60A ground power carts. Accumulated over 17,000 hours of engine operational testing. Several of the individual ceramic nozzle segments accumulated up to 7,542 hours of engine operation time.
GTCP85 Series APU Airborne Field Evaluation	1995–2000	Airborne field evaluation of ceramic radial turbine nozzles in MD-80 aircraft. Over 40,000 hours to date.
ORNL Life Prediction Mcthodology for Ceramic Components of Advanced Heat Engines [12, 13]	1989–1999	Developed leading probabilistic life prediction code for ceramic applications, featuring advanced statistical methods for data analysis as well as enhanced component risk integration schemes.
U.S. Army/COSSI Sand Erosion Resistance Kits for 36-155 Apache Longbow Helicopter APU (SERkit) Program	1999–2001	Design and certification testing of ceramic radial turbine nozzles for APU upgrade kit.
DoE NICE3 Ceramic Turbine Wheel Demonstration	2000–2002	Manufacture ceramic radial turbine wheel and demonstrate reliability in a microturbine using a novel ceramic/metal attachment concept.
DoE Advanced Microturbines Program	2000–2004	Design, fabricate, and test next generation microturbines system achieving 40% thermal efficiency featuring ceramic hot section components.
DoE Hot Section Components for Advanced Microturbines	2000–2002	Develop and evaluate environmental barrier coating (EBC) for silicon nitride microturbine components
DoE High Pressure Burner Rig Development and Operation	2000–2004	Design, build, and operate Mach 1 high pressure burner rig to evaluate ceramic materials and EBCs under realistic gas turbine conditions

TABLE 4.2 PROPERTY CHARACTERIZATION CONDUCTED UNDER THE DARPA/NAVY PROGRAM

4-point flexure strength versus temperature	Tensile strength versus temperature
Elastic modulus versus temperature	Coefficient of thermal expansion
Flexural creep	Tensile creep
Flexural stress rupture	Cyclic fatigue
Thermal shock	Thermal fatigue in fluidized bed
Effect of machining on strength	Effect of static oxidation on strength
Effect of cyclic oxidation on strength	Material response to contact loading
Material response to impact	Fracture analysis of all samples

engine. The primary goal of the program was to demonstrate a 40% increase in power output and a 10% decrease in fuel consumption during a 50 hour engine test.

The ceramics technology at the start of the program was quite limited. The only materials available were reaction bonded and hot pressed versions of silicon nitride and silicon carbide, none of which had a complete design database. Furthermore, AiResearch had not yet fully developed and verified brittle material design and life prediction codes, and silicon nitride and silicon carbide suppliers had very little experience fabricating the size and shape of turbine components that would be required. To address all of these issues, tasks were defined in materials characterization, ceramic component design and life prediction, ceramic component fabrication development, quality assurance development, and rig and engine testing.

Material Characterization

Material characterization linked all other aspects of the program through comparison of material properties versus the anticipated engine conditions, qualification of ceramic components, guidance in material and component improvement, establishment of design criteria, and understanding of component failures. As the program progressed, new characterization needs were identified to gain an understanding of the response of the ceramics to contact stress, impact, and chemical interaction. Table 4.2 identifies key property measurements that were conducted on candidate ceramic materials. Test results determined that RBSN and HPSN were suitable candidate materials to demonstrate the goal of 50 hours of engine testing.

Component Design and Life Prediction

The baseline engine was the AiResearch Model T76-G-10 turboprop aircraft propulsion engine. It contained uncooled metal turbine components, operated at an average turbine inlet temperature (TIT) of about 1000°C (peak of about 1120°C), and was rated at 533 kW. The flow path consisted of a two-stage compressor, a reverse-flow annular combustor, and a three-stage axial turbine.

The goals for the program were to demonstrate a 40% increase in power (to 745 kW) and a 10% decrease in fuel consumption. A parametric engine cycle study was conducted to establish the turbine and combustor section thermodynamic design point that would meet these goals. Some of the key design point parameters are shown in Table 4.3.

A preliminary analysis was conducted to explore various configurations for a ceramic engine. For these analysis iterations, Garrett modified their existing finite element programs to integrate stress fields over both component volume and surface area. The two-parameter Weibull distribution was used to compare the calculated stress field with the material strength and strength distribution parameters. This allowed prediction of failure probability for each finite element and combination of the individual probabilities to

TABLE 4.3 KEY CERAMIC ENGINE (TSE-331C-1) DESIGN POINT PARAMETERS

Combustor efficiency	99.5%
Combustor pressure drop	4.0%
Average Turbine Inlet Temperature (TIT)	1200°C
Maximum hot spot temperature	1370°C

obtain an overall probability for the component. Because planned engine testing was not to exceed 50 hours and since very little time-dependent data were available at the beginning of the program, the analyses were limited to fast fracture probability of failure calculations, including proof testing and transient operating conditions.

Preliminary design indicated that the program goals could be achieved by replacing the combustor liner, transition liner, and first two turbine stages with ceramics. Detailed design was then conducted to define the complete engine layout and to iteratively optimize flow path, sealing, cooling, and component design to minimize stress in the ceramic components.

Hybrid rotors were designed, consisting of 28 individual blades of HPSN inserted into a Waspaloy disk for both the first and second stage rotors. A 60° contact pressure angle with compliant layer was selected due to lower average neck stress, higher neck stiffness and resulting high natural frequencies, and lower overall probability of failure.

Thermal and stress analyses were conducted for 1200°C TIT and 41,730 rpm. The peak stress of the final design was predicted to be 344 MPa. It occurred 125 seconds into the engine acceleration schedule on the suction side of the attachment neck near mid-chord. The peak airfoil stress was predicted to be 107 MPa near mid-span of the trailing edge. Using the lifing tools available at the time, the cumulative probability of survival for each rotor was calculated at 99.983%.

RBSN was selected as the stator material. The stator design consisted of individual vanes each with an airfoil and integral shrouds. For the initial design of both the first and second stage stators, 21 stator vanes were held in position by a RBSN ceramic pilot cylinder and metal structures and seals. Analysis predicted a peak stress of 149 MPa for the final design, which was considered acceptable. However, a stator-passing vibration problem was encountered in the first engine test that required a redesign of the stators, resulting in a new engine cross section with 19 vanes in the first stage and 17 vanes in the second stage.

Analyses determined that ceramics were not required in the combustor and transition liner to achieve the 50 hours engine demonstration, so metal parts were ultimately used. Thus, the engine configuration selected for engine testing consisted of ceramics for the first two turbine stages as shown in Figure 4.1.

Ceramic Component Fabrication Development

At the start of the program, no commercial ceramic suppliers had significant experience with the fabrication of the required turbine components. The only company with proven experience was Ford Motor Company, but they were not interested in being a commercial supplier. However, they did agree to fabricate some initial stator vanes by injection molding. To assure a steady supply, a substantial effort was established at the start of the program to develop an alternate fabrication capability. Norton Company was selected to supply their NC-132 HPSN for rotor blades and to participate in blade machining studies. They also were selected to fabricate stator vanes from their NC-350 RBSN, using a combination of cold isostatic pressing and green machining.

Rotor blades were machined from blanks cut from NC-132 HPSN billets. Slices were also cut between the blanks and machined into test bars to certify that each billet was uniform and within specification. Blade machining was conducted in two stages. The first stage was to fabricate "simulated" blades

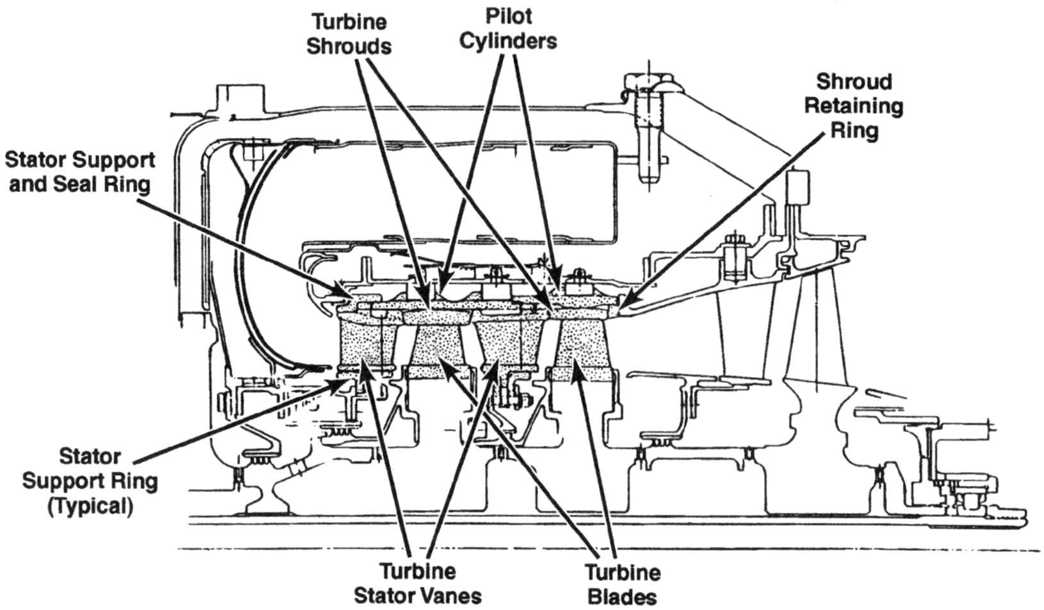

Figure 4.1 Configuration for Engine Testing (Reprinted from Ref. [2])

with the attachment and platform, but a rectangular mass instead of an airfoil. These allowed comparison of various machining techniques and surface finishes and also spin testing to failure to evaluate the design and life prediction codes. Profile grinding and formed-wheel grinding techniques were used to machine the attachment.

Initial complete blades were fabricated by form-wheel grinding the attachment and profile grinding the airfoil. Later, ultrasonic grinding techniques were developed at Garrett that allowed roughing of the airfoil and blade contours and greatly decreased the amount of profile grinding required. Test bars were fabricated by the same machining technique used for finishing the attachment region of the blades. These had average strength of only 473 MPa, which was well below the strength of test bars used to generate data for design. A post-machining oxidation heat treatment of 50 hours at 1000°C was developed that increased the strength to 655 MPa, which analysis determined was acceptable.

Stator vanes were successfully fabricated at Ford, Norton, and AiResearch Casting Company (ACC). The Ford and Norton RBSN materials had 4-point flexure strength in the 250-300 MPa range. The strength of the ACC RBSN was slightly lower. All were suitable for rig testing and later engine testing. Fabrication of other static structure components was conducted at ACC, Ceramics Systems Inc., and Norton. Norton and ACC were selected to fabricate hardware for rig and engine testing. Figures 4.2 and 4.3 show complete engine sets of ceramic components before and after engine testing.

Quality Assurance Development

Three approaches were pursued to assure quality of ceramic components: property measurement of test bars fabricated with each component, nondestructive evaluation (NDE), and proof testing. NDE development involved calculation of the critical flaw size for each component, preparation of NDE standards containing flaws of known size, and evaluation of different NDE techniques to select the one or ones best suited to detect each type of flaw. Critical flaws for the rotor blade attachment peak stress region were calculated to be 35–80

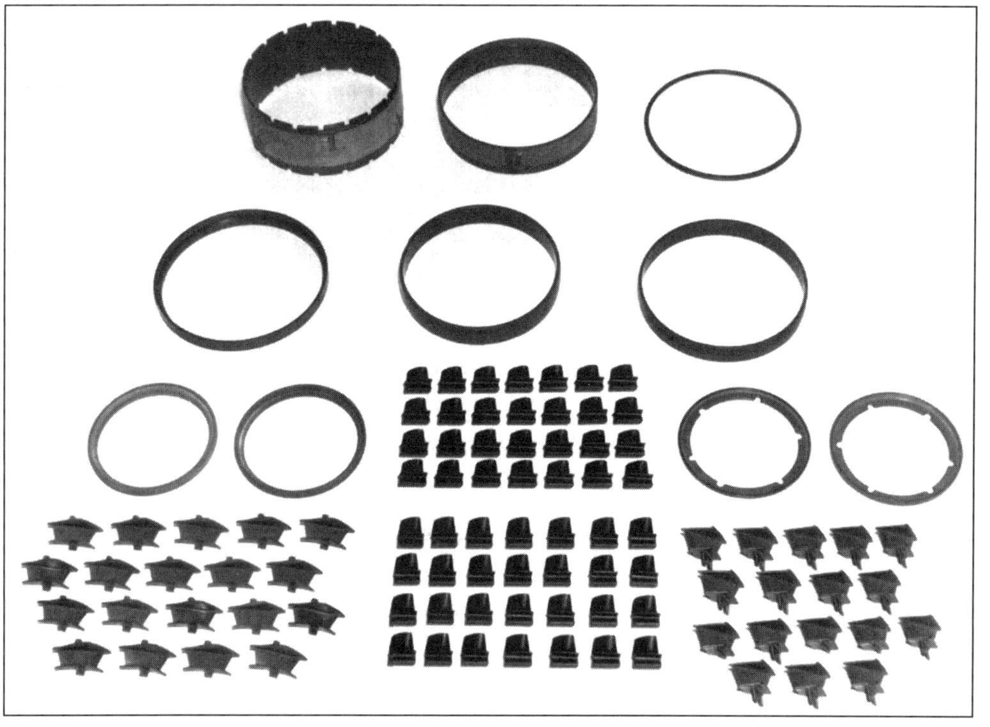

Figure 4.2 Complete Engine Set of Ceramic Components (Reprinted from Ref. [2])

Figure 4.3 Ceramic Components After Engine Test (Reprinted from Ref. [2])

μm for surface cracks and 100 μm for subsurface defects; for the airfoil, 300–700 μm for surface cracks and 1000 μm for subsurface defects. Critical defects for RBSN stator vanes were 60–150 μm along the surface of the trailing edge and 200 μm subsurface along the trailing edge and 200–750 μm for other lower stress regions of the stator vanes.

NDE standards prepared during the program included step-block penetrameters, seeded billets, test bars and rotor blades with Knoop or Vickers indentations, and actual hardware containing known cracks. Conventional NDE methods that were evaluated included x-ray radiography, neutron radiography, fluorescent penetrant, and 10–25 MHz ultrasonic c-scan. Emerging NDE techniques that were evaluated (and in some cases further developed) included microfocus x-ray radiography, image enhancement, >25 MHz ultrasonic c-scan, ultrasonic surface waves, microwave, scanning laser acoustic microscopy, and radioactive gas penetrant. 25-MHz ultrasonic c-scan was able to detect critical defects in flat billets of HPSN and RBSN, but technology did not exist in the mid-1970s to scan contoured or non-uniform thickness parts. The next best approach was microfocus x-ray radiography with image enhancement of the highest stress regions. It was selected along with dye penetrant to be used on all components prior to rig or engine testing.

NDE did not offer complete confidence in component inspection. As a backup, each rotor blade and stator vane was evaluated in a proof test. The blades were inserted individually into a metal hub and spin tested to 130% of design speed. Life prediction analysis estimated a 3% failure rate under this overspeed condition. Of 205 blades tested, six failed, thus qualifying the survivors for engine testing and also verifying the design and life prediction models. Stator vanes were proof tested in a torch thermal shock facility, and were demonstrated to be of suitable strength and quality for engine testing.

Rig and TSE331C-1 Engine Testing

A total of 19 static-structure rig tests were conducted, culminating in a successful 10-hour test to 927°C average TIT and 120% design aerodynamic loading. This test included both stages of ceramic static structure hardware and consisted of a schedule containing two light-offs and 22 transient cycles. None of the ceramic components were damaged.

Initial dynamic rig testing detected a rotor vibration problem due to stator-passing excitation of the natural flexural vibration mode of a second-stage rotor blade. The stators were redesigned from a 21/21 to a 19/17 vane count to solve the problem, and the rotor was successfully operated in the dynamic rig to 45,000 rpm and 927°C.

A total of 27 full engine tests were conducted with mixed results. Most consisted of a successful light-off, smooth operation during a one- or two-hour cycle to design conditions of 1200°C TIT and 41,730 rpm, and successful shutdown. However, disassembly often revealed various levels of chipping or cracking of some of the static structure components. Many tests were completely successful with no sign of damage to the ceramic components.

In spite of these successful tests for individual cycles and for short sequences of cycles, reliable engine operation was not achieved. Some engine tests resulted in complete engine failure even though it contained the same hardware that had operated successfully for several prior cycles. Four complete engine failures occurred, each after the engine had been operating smoothly for some time. Fracture analysis of ceramic components indicated that failure was occurring at contact interfaces by a mechanism that was not understood. Engine testing was stopped, and effort was focused on trying to understand the mechanism of fracture.

The ceramic materials were found to be very resistant to a simple Hertzian load where the only load vector is perpendicular to the ceramic. However, if a biaxial stress is present with a tangential load in addition to the normal load, very high localized tensile stress is present at the trailing edge of the contact, especially if the coefficient of friction is high. Potential sources of biaxial contact stresses in the TSE331C-1 engine included contact surfaces not being normal to the load path, a combination of axial aerodynamic loads and thermally-induced radial relative motion, component misalignment due to tolerance mismatch or cantilevering of the static-structure from the OD, and localized high friction or adhesion at interfaces. To address these sources and solve the problem, the TSE331C-1 engine would require complete redesign, which did not fit within the program funding or schedule. An alternative that would require less time and funds was chosen. This consisted of replacing the first stage rotor of an uprated T76 engine that had a cooled metal static structure that had been demonstrated to operate at 1162°C.

Uprate T76 Engine Tests

Existing ceramic blades and metal disk were modified to replace the metal rotor of the uprated T76 engine. The metal stators were modified to allow increased cooling to allow operation at an average TIT of 1200°C. Two 7.5-hour engine tests were conducted using a severe cruise missile duty cycle. The 15 hours of testing were conducted with no problems. A power output of 882 kW was demonstrated plus a significant reduction in fuel consumption.

The DARPA/Navy/Garrett Ceramic Gas Turbine Engine Demonstration Program demonstrated a substantial increase in engine performance through the use of ceramics and also demonstrated short time durability of ceramic blades. However, the program also demonstrated that contact stress was a critical problem and that substantial further work would be necessary to achieve a durable static structure.

CERAMIC COMPONENTS FOR TURBINE ENGINES PROGRAM

Garrett/AiResearch was awarded a second important program in December 1977. This was Ceramic Components for Turbine Engines, Contract F33615-77-C-5171, funded jointly by The Air Force Materials Laboratory and the Aero Propulsion Laboratory. This program was directed towards a short-life, not man-rated turbine engine such as would be used for a cruise missile. The program was planned to be four phases totaling 41 months of technical effort. Phase I was a preliminary design and manufacturing methods study. Phase II was a component manufacturing effort focused on stator vanes and rotor blades. Phase III was rig testing. Phase IV was planned to be engine testing, but was not conducted because the Air Force decided to switch their efforts towards intermetallics and towards the much higher temperature potential with carbon-carbon composites.

Two important advances were accomplished in this program: (1) improvement in the fabrication of slip cast RBSN vanes at ACC and of HPSN blades at Norton Company and (2) a second generation design of the DARPA/Navy TSE331C-1 engine to solve the contact stress problems.

RBSN development consisted of a first iteration in which a matrix of 29 process variations was conducted and evaluated based on numerical data and fracture analysis. A second iteration involved four variations to achieve further process optimization. The room temperature strength (in 4-point flexure) of the ACC RBSN was improved from a baseline of 207 MPa to 331 MPa with a Weibull modulus >10. In addition, a casting rate of >50 vanes per day with a yield of 80 percent was demonstrated.

HPSN development consisted of forming the concave portion of the airfoil during the hot pressing procedure to decrease the amount of machining required. Effort was also conducted to optimize form-wheel grinding of the attachment dovetail and to improve final spindle grinding of the airfoil. Blades were successfully fabricated and qualified for engine use by overspeed spin testing.

By the time the Air Force program engine design was conducted, there was a basic understanding of the contact stress problems. The challenge was to redesign the engine such that the contact loads were carried uniaxially rather than biaxially. This was achieved by eliminating one whole stage of ceramic components so that each stator vane could be independently supported with no chance of uncontrolled movement due to tolerance mismatch or cantilevering. To accomplish this, the TIT was increased about 55°C and the work of two stages was done by one stage. In spite of the increased work for the single stage, the design was achieved with no increase in stress for the ceramic components, which illustrates the advances gained in design technology.

The static structure was tested successfully in rigs and the rotor blades were validated by spin testing. However, the program did not progress to engine testing because of the Air Force decision to change direction from ceramics towards intermetallics and composites.

ADVANCED GAS TURBINE (AGT) TECHNOLOGY DEVELOPMENT PROJECT-AGT 101

The Advanced Gas Turbine Technology Development Project (AGT) attempted to elevate the state of ceramic gas turbine technology by demonstrating the feasibility of high temperature silicon based materials in small gas turbines, that could eventually propel clean, fuel efficient automobiles. While this federally funded effort fell short of its original, very optimistic goals, it did make several very significant technical accomplishments which provided guidance for maturation of the technology.

The AGT 101 Project was authorized under NASA Contract DEN3-167, and was performed by Garrett Turbine Engine Company with key subcontracts by Ford Motor Company, Standard Oil Engineered Materials Company (Carborundum) and AiResearch Casting Company (ACC). In addition to Carborundum and ACC, Kyocera Industrial Ceramics and NGK Insulators Inc. also provided critical ceramic parts as independent suppliers. This project was conducted under the DOE Gas Turbine Highway Vehicle System Program, which was oriented at providing the United States automotive industry the high-risk, long range technology necessary to produce gas turbine engines for automobiles that will have reduced fuel consumption and environmental impact (principally gaseous emissions).

The project was initiated in October 1979 and was completed in mid 1987, during which it accomplished several ceramic technology milestones:

- Operation of the first complete set of ceramic structures to 1093°C in August 1983
- First engine test with ceramic structures operating to 1149°C turbine inlet temperature in February 1984
- First engine test with ceramic structures and ceramic radial turbine in January 1986
- Completion of an 85 hour long engine test with ceramic structures and ceramic radial turbine in October 1986

AGT 101 Design Characteristics

As noted above, the AGT101 was designed to power a mid-sized automobile. It was compact, so that it could be packaged in the engine compartment of a conventional mid-size automobile and it was designed with features believed to be low cost and mass producible. The engine was originally designed to operate between 50,000 rpm and 100,000 rpm, and was flat rated up to 4267 m (14,000 ft) at 100 hp. The minimum design fuel consumption was a very aggressive 0.18 kg/kWh. This was made possible through the use of very high temperature uncooled components in the engine hot section, and a ceramic rotating regenerator.

The engine cross-section is shown in Figure 4.4. It used a single shaft with an uncooled radial turbine and centrifugal compressor. The rotor was supported by an airfoil bearing near the turbine and an oil lubricated ball bearing near the output pinion gear at the gearbox. Maximum design turbine inlet temperature was 1370°C, and at the regenerator inlet the limit was 1093°C. For maximum efficiency, variable inlet guide vanes on the compressor modulated the compressor flow during off design operation, to keep the cycle temperatures at or near their maximum limits. For design simplicity, most of the static ceramic structures were axisymmetric. The axi-symmetrical design provided more uniform stress distribution, and facilitated component fabrication. The features of the AGT 101 are described more thoroughly in the final report [7].

Suffice it to say that the design goals were extremely ambitious, and they required significant advances in material technology, design technology, and test bed maturity to achieve them. At the inception of the AGT

Figure 4.4 AGT101 Test Bed (Reprinted with permission from Honeywell Engines, Systems & Services)

project in 1979, no usable high strength ceramic parts had been fabricated thicker than 12.7 mm. In addition, no ceramic material existed that would reliably withstand either the operation stresses of the AGT101 rotor operating at 100,000 rpm, or the 1371°C turbine inlet conditions. By the end of the project, the fabrication technology saw breakthrough advances enabling the fabrication, using near net shape processes of the turbine rotor, with its thick hub, and complex structures shown in Figure 4.5. These components were demonstrated in engine testing with turbine inlet temperatures up to 1204°C. Equally important, were advances in "integration technologies," the design approaches for sealing the gas flowpath and managing the structural interactions between the various components in the engine.

Figure 4.5 AGT101 Ceramic Radial Turbine Rotor (Reprinted with permission from Honeywell Engines, Systems & Services)

Through a painful and iterative test and redesign process these components and their integration into an engine system evolved, teaching as much about what didn't work, as what did. Risk mitigation and abatement were key tools employed so that these lessons could be learned as soon as possible and thereby enable needed improvements in the design and fabrication processes. This drove the practices of frequent supplier audits, material property evaluation, careful inspection and screening of the components prior to testing, component proof testing, sub-assembly testing proof testing, and even rig qualification prior to the assembly of a critical component into an "all-ceramic" engine.

Lessons Learned Through Iterations of Design and Rig/Engine Testing

Due to the number of new technologies that needed to be demonstrated and also to the immaturity of design and materials technologies at the start of the program, an aggressive iterative design and test program philosophy was required. Many important lessons were learned as these iterations progressed.

One key lesson was that the design and life prediction tools developed under the prior DARPA and AFWAL programs were not good enough for this new all-ceramic engine, especially associated with the size of the parts and the larger volume of ceramic under tensile stress. Furthermore since the AGT101 engine was not based on a prior-operating engine, the boundary conditions needed for analysis could not be adequately predicted, but instead had to be determined though actual rig testing.

A good example was the turbine shroud. This large axi-symmetric structure performed several key functions while transmitting the highest static loads of any AGT101 component. Critical in its function were to simultaneously support the turbine nozzles and the rest of the hot section ceramic assembly, and it held close alignment with and maintained precise aerodynamic tip clearance for the turbine rotor. The first "engine quality" shrouds were made of reaction bonded silicon nitride (RBSN) and were predicted, using refined, but conventional finite element techniques to have peak transient stresses of 179 MPa.

Since RBSN of that period exhibited average tensile strength of 324 MPa in 4-point bending, the margin was initially assumed to be sufficient. However proof rig testing designed to stress the parts to 125% of the design stress, failed the first parts. Given that subsequent fractography eliminated abnormal material flaws, and that there was correlation between the location of the failure origin and the site of analytical peak stress, it was concluded that the design margin was inadequate. In a redesign that scalloped the flange between attachment points to minimize "hoop stresses" and manipulated part thickness in high heat transfer sections, the peak analytical stress was further reduced to 110 MPa. RBSN turbine shrouds made to this new design were then tested in the proof rigs and survived.

A second important lesson learned regarded solutions to avoiding high contact stresses at interfaces between components. The prior DARPA/Navy program had identified biaxial contact stress due to a sliding movement at ceramic-ceramic and ceramic-metal interfaces to be a source for high tensile stress. An innovative design for transferring contact loads in a rolling contact mode rather than a sliding contact mode was successfully demonstrated for attaching the ceramic turbine shroud to the ceramic diffuser liner and the metal engine case. The rolling contact design assured that the mechanical load was carried uni-axially with a low tensile stress rather than bi-axially with a high tensile stress, yet accommodated thermal expansion mismatch to allow precise control of position of the shroud at all temperatures.

When rolling contact interfaces could not be achieved, crowned surfaces with well-defined points of contact and load path were successfully developed. This approach worked for the interfaces between the combustor baffle and the transition liner and also for attachment of the ceramic rotor to the metal shaft. For some of the other load paths, a compliant element was designed that effectively limited the strains that the ceramic part had to endure. This was accomplished by using mechanical springs and pneumatic pressures to load the parts.

Another important lesson was the benefit of establishing a close partnership with the ceramic material suppliers. The ceramic materials at the start of the project were not reliable enough (low Weibull modulus-large variability in strength) and could not be fabricated to the necessary configurations. Furthermore, every contact interface or surface that could not be formed before final process densification, required diamond grinding that, besides being expensive, had a significant effect on material response to an applied stress. Addressing these issues required continual improvements of material fabrication processing, generation of an extensive property database, and development and refinement of probabilistic life prediction codes that accounted for surface and volume material flaw distributions.

Engine Testing and Development

The first complete set of ceramic structures was demonstrated in the "Ceramic Structures Rig". This development rig had all of the static elements of an engine. It used a functional, rotating regenerator, but did not include the rotating turbomachinery. A watercooled bypass duct enabled concurrent high pressure and low pressure air flow in their respective gas paths. Peak temperature in the rig was 1093°C due to limitations of the alumino silicate regenerator material. During this test, the rig demonstrated that the assembly of ceramic components was capable of operating in conditions typical of the AGT 101 engine. This rig was successfully tested, completing a major project milestone.

The next stage of evolution toward testing of the all-ceramic engine was the "Mod 1, Build 1 Engine." This engine incorporated the majority of the ceramic structures, though it included a metallic diffusion flame combustor and a bore cooled dual alloy radial turbine rotor. Peak operating turbine inlet temperature was limited to 1149°C. In this configuration, the test bed was operated to 100,000 rpm demonstrating the "operating integrity" of the ceramic assembly. Simultaneous with these tests, component development continued with the ceramic turbine rotor, regenerator drive and seal system, and ceramic combustor. A milestone test was performed on a Mod 1, Build 1 engine, operating the engine for 100 (almost continuous) hours at the 1150°C turbine inlet operating condition. All of the ceramic components completed the test successfully.

Testing of the Mod 1, Build 1 engine supported the maturing of several subsystems in preparation for eventual testing of the "all-ceramic" engine. These included refinements in the various static seals, the regenerator seals and drive system, and temperature sensors. Concurrent with the refinement of these engine subsystems, the ceramic turbine fabrication processes were steadily improving.

The AGT rotor development participants included several independent ceramic suppliers, Honeywell, and the US government. Candidate ceramic materials were screened, and tested and evaluated at Honeywell, providing needed feedback to the suppliers for further material development. Initial activity focused on the manufacture of the thick hub, which produced several bladeless "door knob" test specimens. The best of these parts were examined, weighed, X-rayed, and eventually spun to burst at Honeywell. Burst speeds were recorded, photographs taken, and fractography conducted on the remnants, to provide needed clues to the suppliers for further improvements in the process. The sintered silicon nitride materials, by virtue of their high tensile strength appeared to have superior performance in spin testing, and became the focus of further rotor process development. As the integrity of the "door knobs" improved, attention shifted to making bladed parts.

By 1986, ACC (AiResearch Casting Company), Kyocera Ceramics, and NGK Locke, had made prototype turbine rotors with varying degrees of quality. These turbine rotors used materials, which were capable of the stresses induced by the AGT 101 engine, though at lower temperature. Usually these parts held surface defects, or dimensional nonconformance. The best of these parts were carefully examined, defects were removed by hand finishing, and proof spun in a vacuum spin facility to 120,000 rpm.

As the capability to manufacture fully bladed rotors improved, these parts were further evaluated in an AGT 101 test bed with metallic structures. This afforded the opportunity to evaluate progress in ceramic turbine integrity, as well as develop system rotordynamics and designs for turbine attachment with a minimum of collateral risk. The metallic test bed could operate to 100,000 rpm shaft speed at a maximum turbine inlet temperature of 870°C.

The first of these engine tested rotors were from ACC, made from its Code 2 sintered silicon nitride. The failure of two of these rotors in the metallic engine test bed at high speed alerted the test team to the potential for ceramic blade vibration as a failure mode, excited to resonance by the passing frequency of the turbine nozzles. This led to a series of bench tests which used an acoustic method to vibrate strain gauge instrumented blades of two turbine rotors, one made of ACC Code 2 sintered silicon nitride, and the other made of Kyocera's SN220 sintered silicon nitride. The Kyocera rotor had cleaner surfaces and slightly thicker blades than the ACC rotor. During the test, the blades on the ACC rotor consistently failed when excited at resonance, while the Kyocera blades registered higher strains, with no failure.

Later, in January of 1986, an all-ceramic engine was first assembled with a Kyocera SN220M rotor and ran for several cycles at low load. Gradually over the next several builds, of this "all-ceramic" engine, the operating envelope was expanded and improved rotor materials were implemented.

In late 1986, an all-ceramic engine was operated for 85 hours in a single build, with turbine inlet temperatures between 1093°C and 1204°C and engine speeds between 60 and 70 krpm. The rotor was made of Kyocera's latest high temperature sintered silicon nitride, SN250M. Approximately 60 hours into the test, the engine was shut down as a result of fuel starvation. The facility supply of JP-4 fuel was exhausted. The test was continued with Jet-A fuel. At the 85 hour mark the test was terminated as a result of carbon build-up in the combustor, which subsequently came loose and damaged the turbine inducer blades. The remaining ceramic structures shown in Figure 4.6 were in excellent condition. This was the first test during the AGT

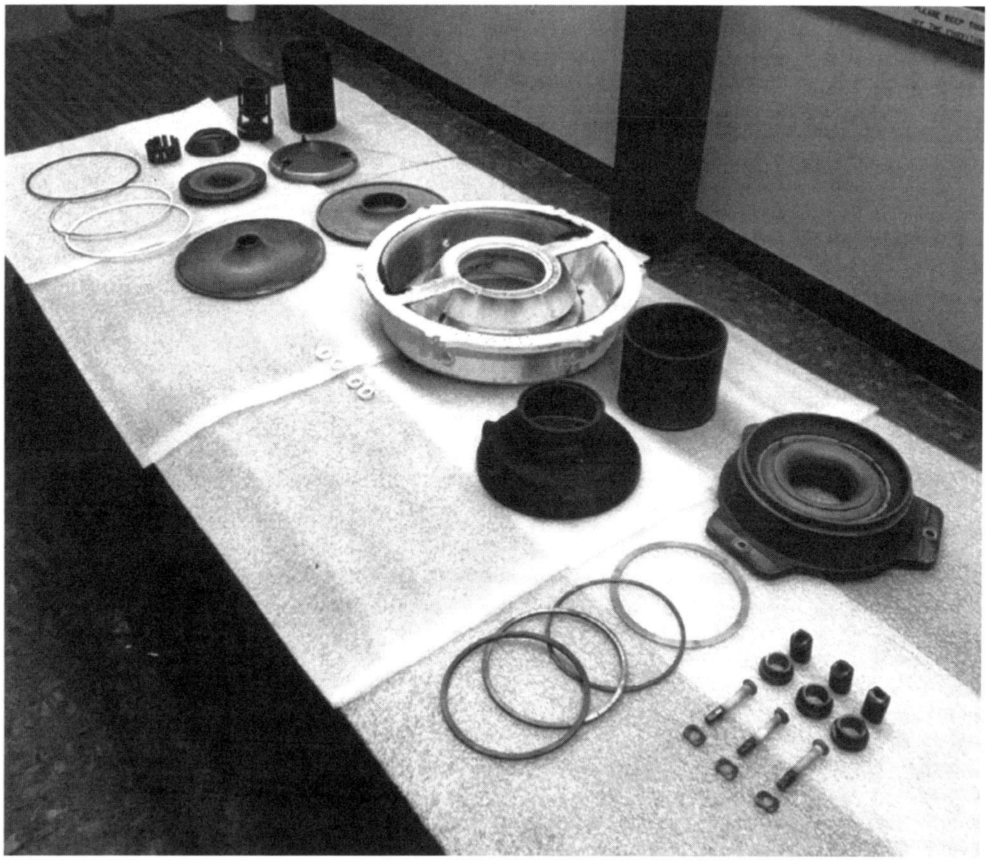

Figure 4.6 Remaining Structural Ceramic Parts After an 85–Hour all–Ceramic Engine Test in AGT101 at 1204°C (Reprinted from Ref. [8])

101 project in which turbine damage was conclusively attributed to combustor generated carbon. The particle impact issue became a driving theme in the development of turbine impact resistance during the follow-on Advanced Turbine Technology Applications Project (ATTAP).

In another test in about the same period, another all-ceramic AGT 101 experienced a turbine failure while operating at 1204°C turbine inlet temperature and 70 krpm, for approximately 12 hours. This failure was attributed to stress rupture of the turbine material in contradiction to the life prediction analysis, which predicted more than adequate life at those conditions. Further analysis and specimen tests of samples from rotors of similar material showed an inconsistency in material properties from rotor to rotor and between fabricated specimens and specimens cut from parts. This fact also drove an emphasis in the follow-on ATTAP and Ceramic Life Prediction Programs to further develop design processes to more accurately predict turbine life and to further improve the processes used to manufacture ceramic parts.

ADVANCED TURBINE TECHNOLOGY APPLICATIONS PROJECT (ATTAP)

The AGT 101 Project used engine testing to explore the boundaries of ceramic turbine technology, and establishing feasibility of the turbine application of the technology. The Advanced Turbine Technologies

Applications Project used engine testing as a validation of the fabrication processes and design methods developed within the project. This project demonstrated significant improvements in test bed operability and reliability as a result of these technology improvements. These technologies included:

- Impact resistant turbines
- Lifing of the ceramic turbine
- High temperature ceramic static seals and springs
- Pressure slip cast sintered silicon nitride components
- Isopressed sintered silicon carbide components
- High temperature, low emission ceramic combustor

As understanding of ceramic technology improved, fewer exploratory engine tests were conducted, and more development was conducted in rigs and in analysis. There were fewer engine failures, with more operating time at higher speeds and temperatures than was the experience during the AGT 101. The test bed was becoming more robust, and its utility as a development tool was improving.

One problem identified late in the AGT 101 project was the susceptibility of the ceramic radial turbine to impact damage from very small carbon particles generated in the combustor during operation. To this end combustor development was conducted to improve the operability of the combustor and reliability of the fuel delivery system and engine controls. In addition a redesign effort was conducted to make the turbine more impact resistant. As an added precaution, subsequent testing of the ceramic radial turbine was conducted using ethanol fuel, which was much less prone to forming combustor carbon than were either Jet A, JP-4, or DF-2 fuels used in previous tests.

Each improvement was fabricated in metal and tested in the metallic engine test bed before committing the design to ceramic. This was especially useful when a redesigned combustor was tested, and found to cause increased combustor exit swirl. The high swirl in the gas flow path produced much higher than design incidence on the radial turbine nozzle, and as a result caused the compressor to surge. The surge event was violent enough to fracture the ceramic regenerator. The problem was corrected in a subsequent test of a redesigned combustor with reduced exit swirl.

The ceramic engine was also tested during ATTAP, also for the purpose of developing and demonstrating improvements to the test bed. During 1991, the ceramic engine with a radial turbine was tested several times, accumulating 13 hours and 78 starts, operating up to 90,000 rpm and up to 1287°C turbine inlet temperature, without ceramic component failure.

In an effort to improve impact resistance, a mixed flow turbine rotor was designed. The prior ceramic radial turbine with its long, unsupported inducer blades was extremely susceptible to impact damage from combustor generated carbon particles as small as 2.5 mm in diameter. The mixed flow turbine (Figure 4.7) sported shorter thicker blades, and slower tip speeds than the radial turbine it replaced. As a result, the impact resistance of the mixed flow turbine was estimated at 4 times that of the radial turbine. The mixed flow turbine also had higher stresses than the radial turbine at the original 100,000 rpm design speed of the AGT. To accommodate this, the AGT Reference Power Train Design was changed to reduce the design speed to 90,000 rpm. The maximum regenerator inlet temperature of 982°C also limited the peak operating temperature at the turbine inlet to approximately 1287°C.

In 1992, the ceramic impact resistant turbine was ready for engine testing. Initial testing of the turbine rotor was performed in the metal structured test bed to assess the operability of the new engine rotor configuration with respect to starting, operability, rotordynamics, and thermal transient stresses. Three mixed flow rotors were qualified for ceramic engine testing, two made of NT154 silicon nitride material from Norton TRW, and one made of GN-10 silicon nitride from Garrett Ceramic Components.

Milestone 4 engine testing in ATTAP demonstrated operation and performance of the all-ceramic mixed flow turbine engine up to maximum AGT 101 design conditions. During this test, the engine was operated up to 90,000 rpm and to a maximum turbine inlet temperature of 1268°C. Cycle state conditions were measured within the engine during this testing to confirm component efficiencies and identify potential sources

Figure 4.7 Impact Resistant Ceramic Mixed Flow Turbine (Reprinted with permission from Honeywell Engines, Systems & Services)

of performance losses. As with the AGT 101 engine, power was again degraded by excessive leakage from the internal engine and regenerator seals.

Milestone 5 testing was intended to demonstrate durability of the engine in a cumulative 100 hour test, 50 hours at 80,000 rpm and 1204°C turbine inlet, and 50 hours at 90,000 rpm and 982°C regenerator inlet. In the first build, an unexplained failure of the NT154 mixed flow rotor halted the first attempt before the engine could be set at the test condition. In the second build, another unexplained failure of a GN-10 mixed flow turbine, occurred after 1.2 hours at the test condition. Each of these failures also damaged the entire set of ceramic static components as well.

In a third attempt at completing the milestone, another NT154 turbine was installed in the test bed with a fresh set of qualified static components and tested. The test bed accumulated 40 hours of operation during the next two builds, interrupted by failures of the LAS Flow Separator Housing. These failures limited the damage to the flow separator housings and the static seals located between that part and the turbine shroud. The test was stopped when the project supply of flow separators was exhausted.

During ATTAP the integrity of the design methods, fabrication processes, and the engine integration technology was significantly enhanced. However, the failures of the two mixed flow turbine rotors during the Milestone 5 event were reminders that the technology was still not completely understood. Shortcomings still existed either in the ability to design highly stressed turbine rotors, or in the ability to fabricate turbines with the consistent material properties required for engineered components. It was also apparent that the AGT 101 test bed, laden with high technical risk, was not a suitable test bed to continue development of ceramic technology. This realization drove the follow-on project (Ceramic Turbine Engine Demonstration Project) to switch to a more reliable gas turbine test bed, based on a fully developed and proven commercial air transport auxiliary power unit design, the Honeywell model 331-200 APU.

GTCP85 SERIES APU DURABILITY AND FIELD TESTING PROGRAMS

During 1992, Honeywell (AlliedSignal Engines at this point in time) initiated work to apply ceramic technology to gas turbine APUs. Using independent research and development (IR&D) internal funding, AE developed ceramic turbine nozzles for the 85-Series APUs, which are used in large numbers in commercial airborne transport aircraft applications and U.S. Air Force ground power carts. This IR&D effort culminated with successful block endurance testing and FAA approval of the ceramic nozzle installation for a commercial air transport field evaluation program. Ceramic nozzles were installed in GTCP85-98DHF APUs and placed in regular commercial revenue service onboard MD-80 aircraft. Between 1994 and the present under this continuing program, accumulated operating time on the ceramic nozzles has exceeded 40,000 hours.

Under a parallel program funded by the DoD (Contract No. F41608-92-C-1420), extended on-site endurance testing was conducted on ceramic turbine nozzles installed in Model GTCP85-180 APUs used in U.S. Air Force A/M32A-60A ground power carts. Four ceramic nozzle equipped APUs accumulated over 17,000 hours of engine operational testing. Several of the individual ceramic nozzle segments accumulated up to 7,542 hours of engine operation time during this endurance testing. These programs were conducted to gain operating experience with ceramic turbine nozzles, and to understand the benefits that ceramics could bring to this engine application.

Two series of extended field trials of ceramic nozzles in 85-Series APUs have been conducted by AlliedSignal Engines. Under the USAF-sponsored Ceramic Nozzle Insertion Program, five Model A/M32-60A ground carts with GTCP85-180 APU engines were refitted with ceramic nozzles and evaluated under normal service conditions at Luke Air Force Base near Phoenix, Arizona. These ground carts were used in regular service on the Luke AFB flightline, providing ground support for starting and servicing of F-16 fighter aircraft. 1,875 operating hours and over 12,000 starts were recorded over the four-year test period.

Concurrent with the U.S. Air Force ground cart endurance and field testing, eleven Model GTCP85-98DHF APUs equipped with ceramic turbine nozzles were installed in MD-80 aircraft and have been flown and operated in regular revenue service. Between 1994 and the present, these engines have accumulated over 40,000 hours of operating time.

The combination of endurance testing and field testing performed by AE has provided invaluable information about the reliability of ceramic turbine nozzles in operating service. One of the early "lessons learned" occurred during the U.S. Air Force-sponsored endurance testing that preceded the military and commercial field evaluations. During the extended operational periods of the GTCP85-180 APU ground cart engines, large sized particles of hard carbon were being generated in the combustors (due to fuel coking) and passing through the engine hot section flowpath. Periodic inspections revealed this material had initiated substantial damage to the ceramic nozzle vane trailing edges. A design change to the ceramic nozzle segments was then accomplished, increasing the thickness of the vane trailing edge from 0.8 mm nominal to 1.27 mm nominal. Further testing with the modified ceramic nozzles indicated this change eliminated potential cracking from combustor carbon, without any performance penalty.

The resulting improvement in ceramic nozzle vane durability was significant, as the trailing edge damage from carbon particles was completely eliminated in subsequent testing. After 2,200 hours of engine operation in a side-by-side "rainbow" test comparison against conventional metallic nozzles, a ceramic nozzle with over 5,000 hours of previous engine testing (total of over 7,200 hours operation)

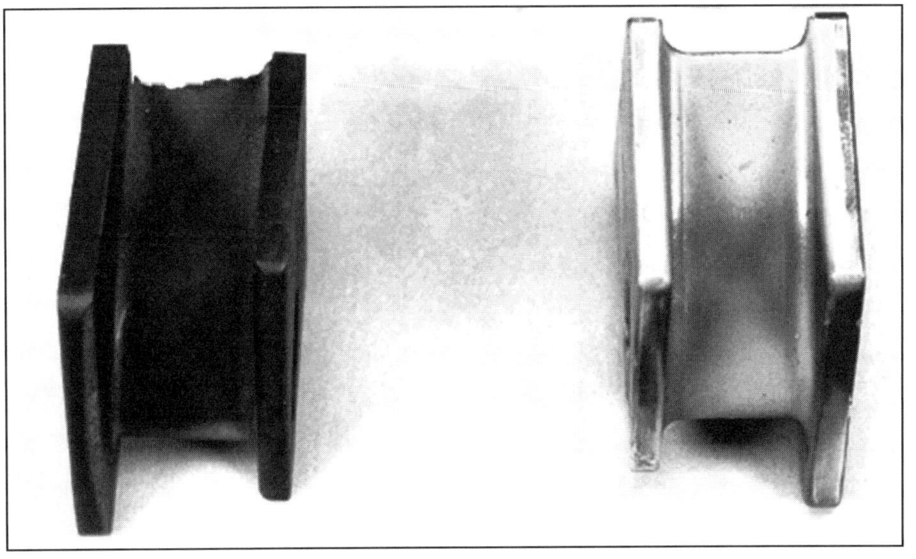

Figure 4.8 Ceramic Nozzles (right) Were More Durable than Conventional Metal Nozzles (Left) in a 2,200-hour Comparative Test (Reprinted with permission from Honeywell Engines, Systems & Services)

exhibited virtually no wear or deterioration (see Figure 4.8). In contrast, the metallic nozzle vanes made of IN713 and MAR-M-247 were visibly thinned and recessed on the trailing edge following this testing in the erosive environment of the Model 85-180 APU hot gas flowpath.

During the field testing discussed above, there were no primary failures attributable to the ceramic nozzles. The only durability issue that has arisen during the 85-Series APU field tests was susceptibility of the ceramic nozzles to foreign object damage (FOD). This FOD typically results from impact to the ceramic nozzles from large, metallic fragments of the compressor impeller blades or turbine blades due to unrelated mechanical failures of the compressor or turbine stages.

CERAMIC TURBINE ENGINE DEMONSTRATION PROJECT

During 1993, the still ongoing Advanced Turbine Technology Applications Project (ATTAP), funded by the U.S. Department of Energy (DoE) and administered by the National Aeronautics and Space Administration (NASA) under Contract No. DEN3-335, was refocused to more effectively address the critical issues slowing the commercialization of structural ceramics in gas turbine engines, i.e. ceramic component reliability demonstration, ceramic design technology refinement, and cost-effective ceramic component fabrication. Honeywell Engines, Systems & Services changed the ATTAP ceramic engine test bed from the AGT101 automotive engine to utilizing the Honeywell Model 331–200[CT] auxiliary power unit (APU) as a test bed/demonstrator, with the first-stage turbine modified to incorporate ceramic nozzles and blades. The program was to provide the supporting experience, design methodologies, and fabrication capability by domestic U.S. ceramics manufacturers necessary for limited commercialization of ceramics in aeronautical applications, and provide a solid technological foundation supporting automotive ceramic heat engine applications. Effective January 1995, the former ATTAP was retitled the Ceramic Turbine Engine Demonstration

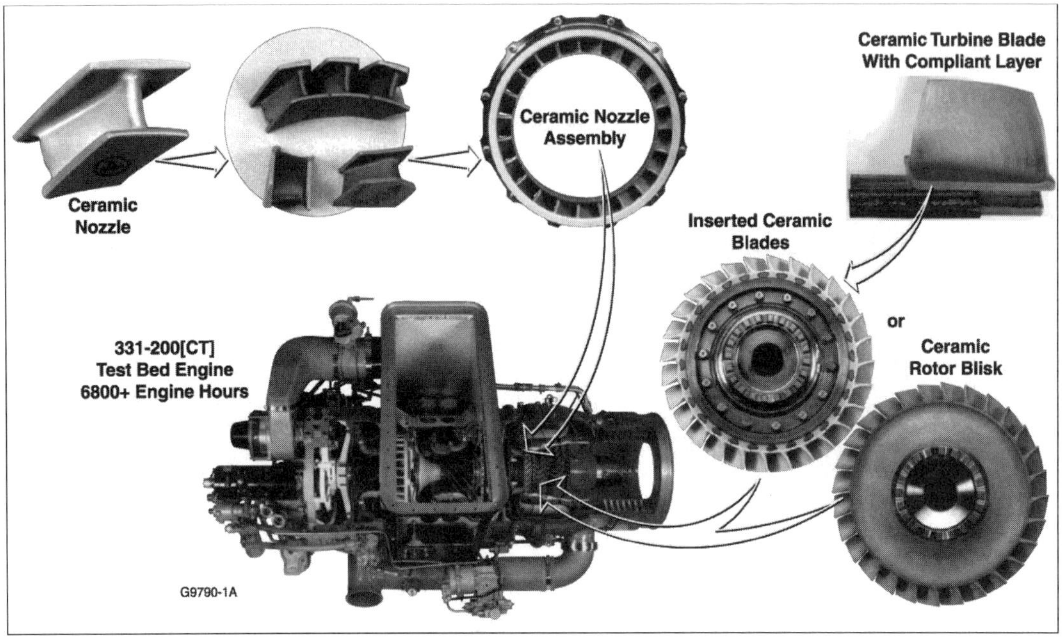

Figure 4.9 Ceramic Turbine Engine Demonstration Test Bed is Based on Proven Model 331-200 Auxiliary Power Unit With Ceramic First-stage Turbine Components (Adapted from Ref. [10])

Project (CTEDP). Since 1997, the CTEDP has been administered by the DoE and authorized under DoE Contract No. DE-AC02-96EE50454.

The CTEDP strategy was developed to augment the maturing ceramics technology by developing the infrastructure and engineering disciplines necessary to overcome those barriers which prevent its commercialization. At the beginning of this program, the principal barriers to the commercialization of ceramics were seen as:

- Immature supporting technologies
- Underdeveloped production capability
- Inadequate demonstration

The overall program plan provided an approach to resolve each of these issues. The following discussions summarize the progress to date in various project activities and outline the lessons learned under this program as well as recommendations to finally achieve commercialization of structural ceramic components in gas turbine engines. This work was initiated in 1993 and scheduled to be complete at the end of 2000.

The Ceramic Turbine Engine Demonstration Project continued refinement of ceramics technologies and design methods, and focused fabrication development on a scale-up of demonstrated technologies to pilot-production levels. To adequately demonstrate ceramic technology, over 6800 hours of endurance testing have been completed, utilizing a reliable test bed based on the Honeywell Model 331-200/250 APU (Figure 4.9). This APU is a fully-developed gas turbine with current production applications in the Boeing 757 and 767 aircraft, as well as in the Airbus A300 and A310 aircraft. Thus, this APU has higher capability for successfully accomplishing not only long-term endurance testing, but also offers the opportunity for extended field evaluation of ceramic components.

The progress of the Ceramic Turbine Engine Demonstration Project towards commercialization of ceramic engine technology may be visualized with the chart in Figure 4.10. This curve, based on cumulative

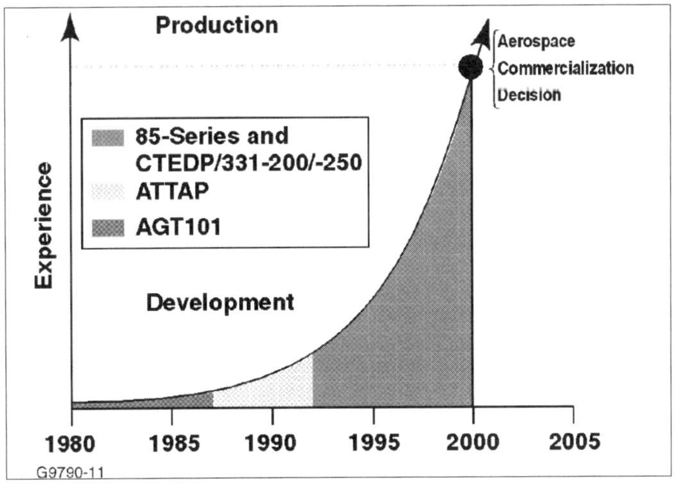

Figure 4.10 CTEDP Targets Solutions for Early Ceramics Commercialization (Reprinted from Ref. [10])

experience in engine demonstration test hours and quantity of ceramic engine components achieved from 1980 to present and projected through 2002, shows the geometric trend in ceramic technology improvement. Figure 4.10 shows that ceramic engine technology is on the threshold of exploitation. The Ceramic Turbine Engine Demonstration Project has enabled very near-term commercialization in low volume, premium gas turbine applications such as airborne APUs, paving the way towards production of ceramic components for industrial microturbine applications for advanced distributed power generation.

Technology Development

All of the technologies described in the following section were identified as critical to the success of ceramics in commercial gas turbine applications

- *Impact Design Methods Refinement.* The end goal of this activity has been the development and validation of a method capable of accurately predicting structural impact damage for any ceramic component from carbon particles (combustor carbon), which resulted in design tools and guidelines for impact-resistant, axial ceramic turbine blades.
- *Contact Design Methods Refinement And Blade Attachment Technology.* The goal of this activity was to develop design tools to predict strength and life for ceramic/ceramic and ceramic/metal interfaces under gas turbine operating conditions, and to generate an improved understanding of the conditions within highly-loaded ceramic interfaces. This knowledge was then used to derive useful tools for designing most cost-effective robust ceramic axial blade attachments for production gas turbines. These design solutions necessarily require a thorough understanding of the blade attachment contact stresses and the environmental and geometric factors that control these stresses. The development activities included:
  - — Design and fabrication of a contact rig for evaluation of ceramic contact interfaces of various configurations under extreme loads and elevated temperatures. Rig operation included the effects of normal contact stress, tangential contact stress, combinations of normal and tangential stresses (sliding motion), cyclic loading, and contact with high background stresses in the ceramic specimens.
  - — Determination of the coefficient of friction for various material combinations.
  - — Determination of the strength degradation which may have occurred as a result of contact damage.

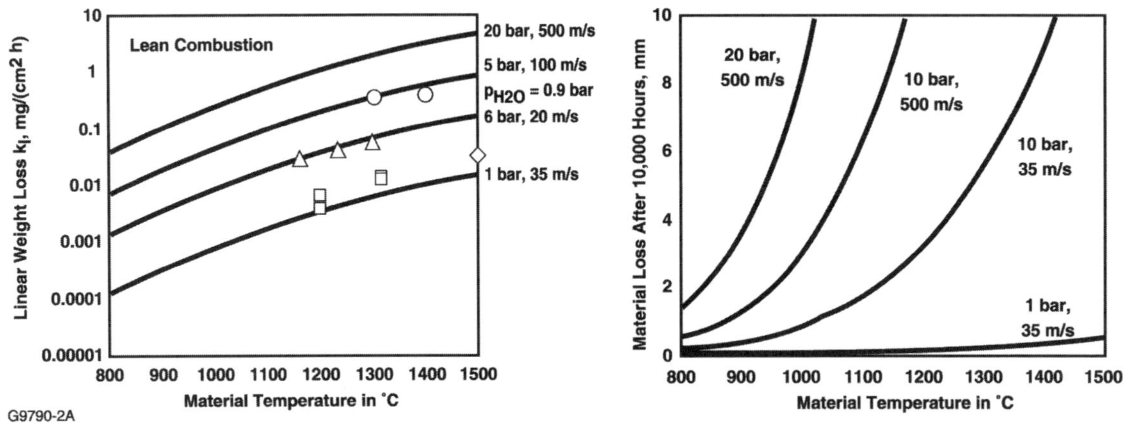

Figure 4.11 Ceramic Airfoil Applications are Extremely Limited by Surface Recession. Extensive Coatings Research and Development Needed for Successful Ceramic Gas Turbine Applications in Aggressive Environments (Adapted from Ref. [10])

— Assessment of compliant layers, surface topography and tolerance effects, and surface strength with respect to machining direction.

— Cyclic thermomechanical ceramic blade attachment subelement rig testing using a standard tensile testing machine which simulated engine disk and ceramic blade attachment deflection and load distribution characteristics.

— Cyclic spin tests at room temperature and at elevated temperature in which various anti-friction treatments were applied to several types of blade compliant layers in order to evaluate wear characteristics and durability of the compliant layer systems.

— Contact stress finite element analysis (FEA) in order to better understand the benefit of the compliant layer in the ceramic blade attachment system, including the effects due to plastic deformation of the compliant layer. The contact rig specimen geometries were modeled, so that a direct link between the analysis and the contact rig data could be established.

This work resulted in a proprietary compliant layer system developed by the Honeywell Research and Technology Center, which demonstrated exceptional life and negligible wear.

- *Oxidation/Corrosion Characteristics of Ceramic Materials.* Auxiliary power units and industrial engines are frequently used in high salt ingestion (e.g., coastal airport) operating environments, which can accelerate turbine component degradation. Surface recession and surface pitting associated with hot corrosion and water vapor attack have the potential to adversely affect the strength of silicon nitride components above 900°C. Environmental component life degradation therefore is of significant concern to the success of ceramic gas turbine applications. Additional experimental activities included burner rig testing of various coated (CVD zircon, CVD tantalum oxide, CVD and plasma-sprayed mullite) and uncoated silicon nitride materials in a salt laden environment, burner rig testing of various uncoated silicon nitride materials in an oxidizing combustion environment at different gas velocities, pressures, and temperatures, and determination of fast fracture strength degradation, weight, and dimensional changes, following burner rig exposure. The experimental results (symbols in Figure 4.11) regarding water vapor accelerated oxidation of silicon nitrides were in close agreement with the empirical surface recession model (solid lines in Figure 4.11) suggested by NASA, [14].
- *Proof Test Techniques.* The focus of this effort has been to identify cost-effective proof testing techniques to ensure the quality of ceramic hardware prior to engine installation. The goals were to develop proof test-

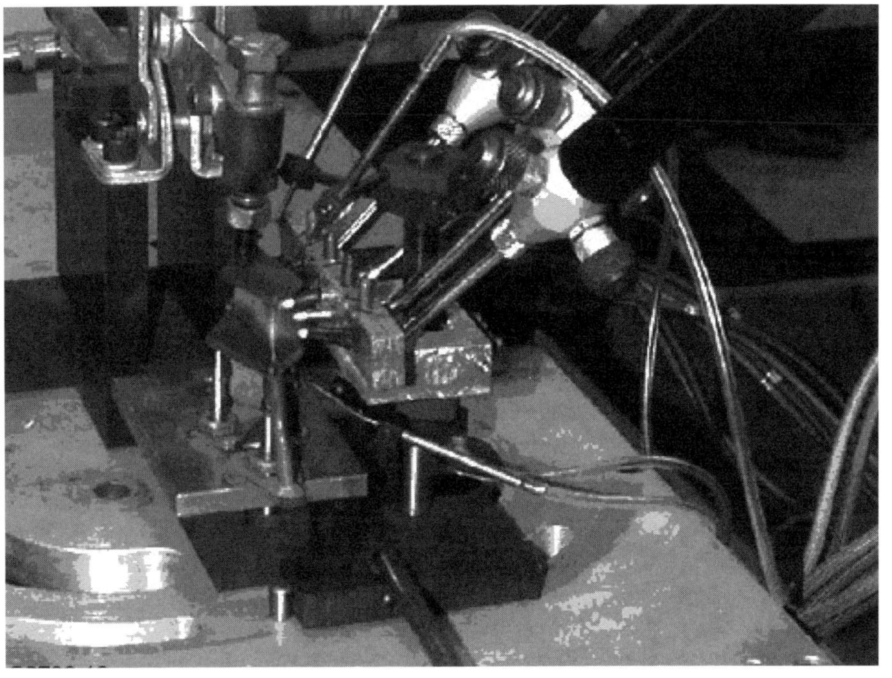

Figure 4.12 Ceramic Nozzle Thermal Proof Test Rig (Reprinted from Ref. [10])

ing methods that are accurate, yet inexpensive, that may potentially be used by ceramic component manufacturers and users. The study concluded that a simple flame impingement test is the preferred nozzle proof test method (Figure 4.12). The flame impingement stresses the vane trailing edges, which typically exhibit the highest stresses in the part during engine operation. The rig creates thermal gradients on the ceramic nozzle by heating and cooling simultaneously. An infrared camera is used to measure the temperature distribution of the ceramic nozzle. Thin-film thermocouples were initially used to evaluate the thermal profile measured by the infrared camera. Repeatability of this proof test process has been assessed systematically and the system was found to be stable and measurement variation was adequate. Additionally, a mechanical test rig was designed and fabricated for proof testing the nozzle attachment features.

- *Thin-Film Instrumentation Development.* Technology for chemical vapor deposition (CVD) methods to apply thin-film strain gages and thermocouples on silicon nitride gas turbine components has been developed and demonstrated. These efforts were the result of work performed in Honeywell's laboratories, subcontract work by the University of Rhode Island (URI, Providence, RI), and collaboration with the NASA Glenn Research Center (Cleveland, OH). The strain gage fabrication efforts used photomasking and sequential deposition methods to apply gage elements to the ceramic turbine blade airfoil surfaces and to attach the required lead films to the blade root area. Under in-house fabrication efforts, Honeywell Engines, Systems & Services has successfully deposited and fabricated thin-film palladium/chrome dynamic strain gages on SN252 and AS800 silicon nitride turbine blades.

Ceramic Components Quantity Fabrication Demonstration

The purpose of this activity has been to develop the capability of the domestic U.S. ceramic engine component suppliers to provide quality components in sufficient quantities to support scale-up to engine

TABLE 4.4 CERAMIC COMPONENT MANUFACTURING SCALE-UP GOALS

Item	Goal
Individual Process Capability	500 Parts/Month
Overall Yield	75 percent
Demonstrated Overall Process Capacity	100 Parts/Month
Cost	<$300 Per Part

production, and to move the component fabrication process out of the laboratory and into an environment where components of consistent high quality can be made economically. Initiated in late 1993, two ceramics manufacturers, Norton Advanced Ceramics (NAC, in East Granby, CT) and Honeywell Ceramic Components (CC, in Torrance, CA), have been challenged to improve their demonstrated fabrication processes and develop methods and procedures to achieve the quantity production goal by building a pilot facility to demonstrate production of over 100 parts per month, on a continuous basis for several months (Table 4.4).

Since NAC announced their business decision to discontinue production of gas turbine engine ceramics and terminated participation in this program by mid 1995, Kyocera Industrial Ceramics Corporation (KICC, in Vancouver, WA) initiated subcontract work during the third quarter of 1995. During their rather short participation in the program, NAC focused their activities on the elimination of iron inclusions and demonstration of closed-loop processing (CLP) of NT164 silicon nitride components.

Already in 1994, CC demonstrated the feasibility of AS800 silicon nitride as a potentially lower cost turbine component material with respect to both material properties and shape capability, which enabled them to replace the glass-encapsulated, hot isostatic pressed (HIPped) GN-10 material with gas-pressure-sintered AS800 for the scale-up demonstration.

KICC initially focused on ceramic nozzle fabrication using their gas-pressure-sintered SN252 silicon nitride material and their proprietary hybrid molding process. By mid of 1996, at the request of Honeywell Engines, Systems & Services, a material change to SN282 silicon nitride was initiated due to the intermediate temperature oxidation problems of SN252, which were discovered during burner rig testing at Honeywell and were confirmed by field test results obtained under the GTCP 85-series APU ceramic nozzle program.

At the end of these subcontract activities in 1999, both suppliers have been able to demonstrate the capability of producing 331–200[CT] ceramic nozzles meeting all blueprint specifications at the rate of 100 parts/month. Both subcontractors also individually developed cost models accurately reflecting the production costs for ceramic nozzles across the supply chain. Overall yield for the complete process was 79.2 percent for CC, and 67 percent for KICC, respectively. Furthermore, it was demonstrated that the cost goal per part could be achieved. In order to further reduce the production cost, design modifications are necessary to reduce the required number of machined surfaces, setups, grinding operations, and/or close tolerance features.

CC also demonstrated their capability to fabricate MOD 2 ceramic blisks for the 331–200[CT] engine (see Ceramic Component Design section).

Engine Demonstrations

The ceramic engine demonstration activities required redesign and modification of the Model 331–200 APU into the ceramic test bed engine. Within the 331–200[CT] test bed, the first axial turbine stage was redesigned to incorporate ceramic turbine nozzle segments and inserted ceramic blades, and finally even a ceramic turbine blisk. This activity included detailed design of ceramic components and modified metallic support structure, test hardware fabrication, component testing to verify the component

TABLE 4.5 331–200[CT] CERAMIC TURBINE DESIGN CYCLE CONDITIONS

Parameters	Units	Value
Corrected Flow	kg/s	0.7058
Corrected Speed	rpm	19,993
Physical Speed	rpm	41,731
Turbine Efficiency	—	85.7
Turbine Inlet Total Pressure	kPa	895
Average Turbine Inlet Total Temperature	°C	982
Combustor Pattern Factor	—	0.18

performance characteristics, and engine demonstration testing. These tasks are summarized in the sections below. The 331-200[CT] turbine design cycle conditions are summarized in Table 4.5.

Ceramic Component Design. The design guidelines recognized that the redesigned components will be used in a later field evaluation program, in which ceramic-equipped engines replace production all-metallic engines; thus the ceramic-equipped engines must be aerodynamically similar to the metallic engines. In addition, to maintain the reliability of the test bed, modifications to the metallic support structures should be minimized. The ceramic turbine nozzle segments and blades were designed with principal consideration given to component reproducibility, while also satisfying structural integrity and aerodynamic performance requirements.

Ceramic Turbine Nozzle Design. The ceramic nozzle assembly (Figure 4.9) has 23 segmented ceramic vanes. For simplicity, the ceramic vane count was reduced to 23, from 29 vanes on the all-metallic engine. Nozzle surface geometry was kept as simply as possible to favor manufacturing, a result of concurrent engineering with the close collaboration of the ceramic component manufacturers.

Over the course of the program the initial nozzle design (MOD 0) was redesigned twice for improved fabrication and reduced stresses of the ceramic nozzle system. The nozzle attachment features were simplified, and the performance and durability of the ceramic nozzle system was improved. Figure 4.13 shows the evolution of the 331-200[CT] ceramic nozzle through the three design iterations.

Analyses were performed to predict the stresses induced in the MOD 2 ceramic nozzle airfoil at steady-state and transient operating thermal conditions. Table 4.6 shows a comparison of predicted stresses in the MOD 0, MOD 1, and MOD 2 ceramic nozzle designs. From these results, thermomechanical fracturing of the MOD 2 airfoil was not expected to present any problems during engine testing. Thermally-induced cracks, found in several MOD 0 ceramic nozzles during proof tests and later during engine testing, indicated that the MOD 0 nozzle design was too highly stressed. The MOD 0 ceramic nozzle stress analysis, performed in prior years, had considered only nominal temperature conditions during the engine start transient and ignored any potential for temperature mal-distribution (hot streaks) from the combustor during the engine start. Additional studies revealed that such hot streaks can significantly increase the transient stress levels in the ceramic nozzles. Due to this finding, the stress analyses performed during the MOD 1 and MOD 2 ceramic nozzle redesign efforts employed worst-case light-off temperatures, to ensure more accurate modeling of actual transient engine operating stresses.

Ceramic Turbine Blade Design. The number of ceramic blades in the redesigned first-stage turbine wheel (Figure 4.9) was reduced to 28 blades from 36 in the original all-metallic engine. This permitted an increase in the size of the ceramic blades, thickening the blade sections for better fabrication, and increasing the spatial availability for a low-stress blade attachment scheme.

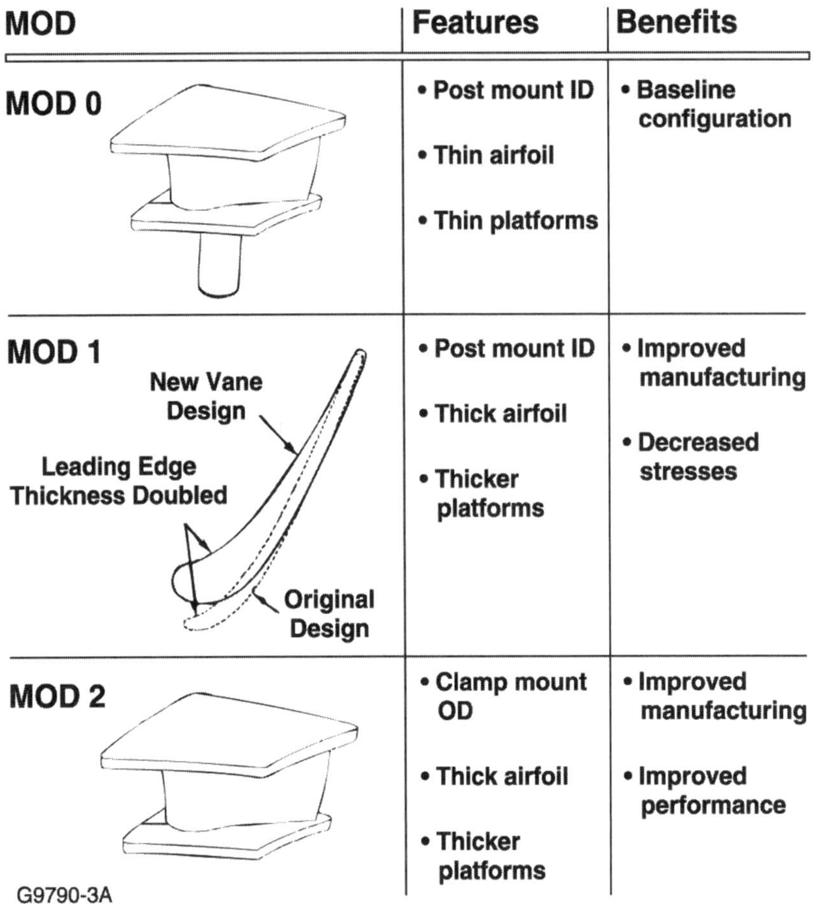

Figure 4.13 331-200 [CT] Ceramic Turbine Nozzle Design Evolution (Adapted from Ref. [10])

TABLE 4.6 PREDICTED 331-200[CT] CERAMIC NOZZLE STRESSES

Nozzle Location	MOD 0 2D Nominal	MOD 1 3D Hot Streak	MOD 2 3D Hot Streak
Peak Transient Stress, MPa			
Airfoil Leading Edge	303	55	55
Airfoil Trailing Edge	283	110	103
Post Fillet	—	69	[N/A]
Peak Steady-State Stress, MPa			
Airfoil Leading Edge	166	35	21
Airfoil Trailing Edge	138	55	45
Post Fillet	—	28	[N/A]

[N/A] = Not Applicable

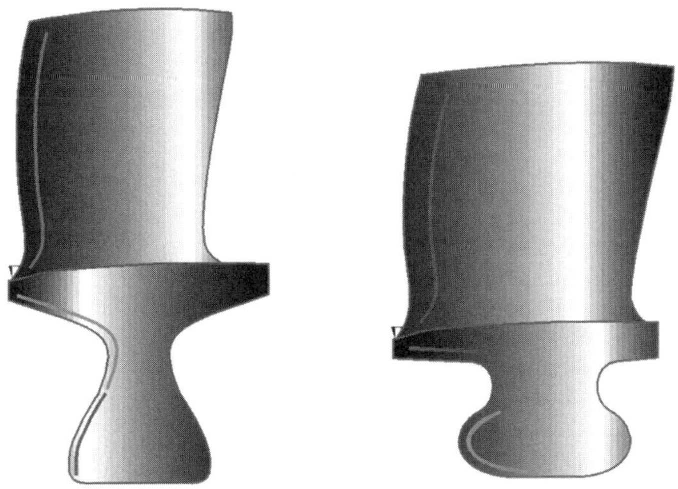

Figure 4.14 331-200 [CT] Ceramic Turbine Blade Attachment was Redesigned to Improve Robustness (Airfoils Shown at Different Angles to Emphasize Changes in Attachment Geometry (Reprinted from Ref. [10])

Derived from previous experience during the T76 project, the initial MOD 0 blade design utilized a single-tang 60 degree attachment contact angle dovetail with a thin foil of compliant metal to distribute the compressive loads between the ceramic blade dovetails and the metallic turbine disk. Each blade assembly was held in place axially with a conventional metallic bend tab. A peak stress of 296 MPa was predicted on the surface in the dovetail region near the edge of the contact zone.

A ceramic blade attachment redesign was initiated when it was learned that the contact stress levels, as a function of blade attachment bearing stress, were strongly affected by variations in the frictional coefficient of the attachment interface, especially during transient operating conditions. It was also considered desirable to move the blade resonant frequencies higher and to make the blade attachment features more robust. The redesign efforts included an appraisal of the shortcomings of the (baseline) MOD 0 ceramic blade design, which were

- the blade first resonant mode fell within the range of engine operating speeds,
- the blade attachment shank was thinner than good design practice suggested, and
- the specified 60 degree contact angle made the blade attachment feature bearing stresses sensitive to interface friction.

For the resulting MOD 1 ceramic blade design (Figure 4.14), the airfoil was not changed from the MOD 0 baseline, and the MOD 0 design issues were addressed with revisions to the attachment geometry. The major features of the MOD 1 design include a thicker, shortened blade shank, which serves to increase the shank section modulus, raising the blade resonant frequencies above the engine operating speed range. A second major change was a reduction in the attachment contact angle from 60 to 45 degrees, which was intended to decrease the sensitivity of the attachment bearing stress to changes in interface friction. Moreover, the broach angle was reduced by 10 degrees resulting in a more favorable attachment stress distribution. A peak stress of 210 MPa is predicted to occur in the blade shank; this is approximately the same as the values for the baseline MOD 0 design. Design considerations for the MOD 1 metallic turbine rotor disk differed from the MOD 0 disk only in the geometry of the blade slots and the method for axially retaining the ceramic blades in the disk. An evaluation of various design options for retaining the ceramic blades

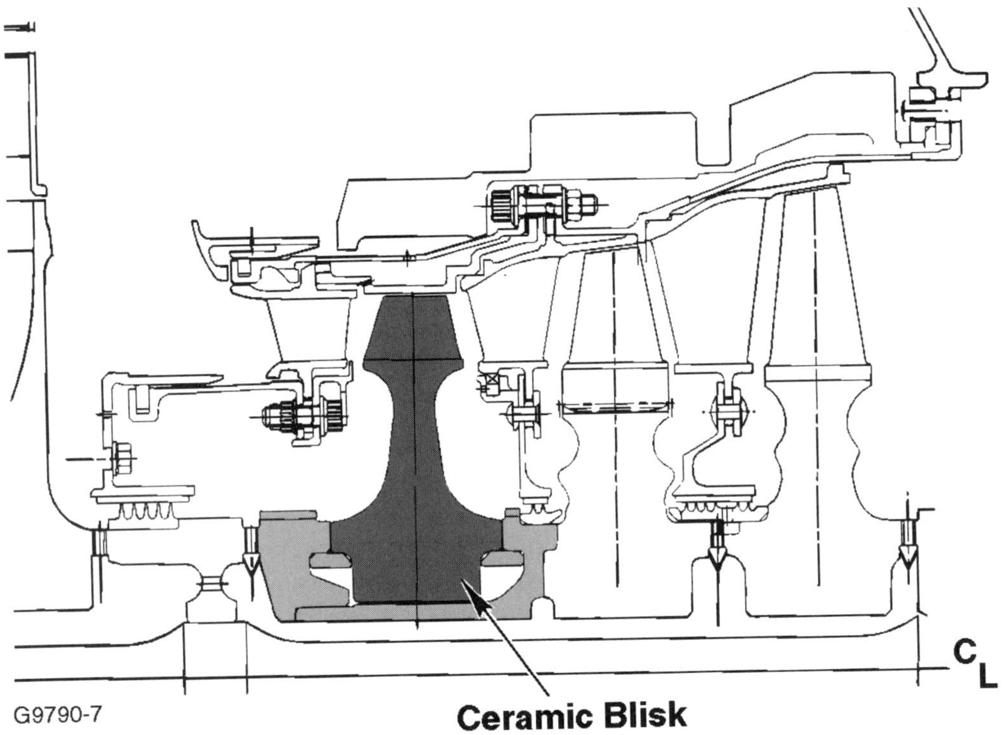

Figure 4.15 Ceramic Turbine Blisk Concept (Reprinted from Ref. [10])

in the metallic disk identified a bolted cover plate design as the favored concept. This design eliminated earlier problems with consistency of the bent tab retainer installation and ceramic machining difficulties encountered with the necessary features on the underside of the blade for alignment of the tab.

Ceramic Turbine Blisk Design. While the program progressed, ceramic turbine design philosophies have been formed which advocate peak design stresses in inserted blades be set much lower than in the present designs. The driver for this approach has been the realization that ceramic materials are more sensitive to the restrained conditions in the blade attachment than for conventional inserted metallic blades. Technology has been developed to improve the compliant layers for inserted ceramic blades; but the sense is that the combined effects of thermal stresses, inertial loads, friction, and contact stresses are not well enough understood at present (especially for transient operating conditions), to achieve successful ceramic turbine designs using the same margins assumed with metallic blades.

To this end, activity had been initiated to identify alternate ceramic turbine blade concepts, and to initiate a MOD 2 ceramic turbine design, based on the most promising candidate. As a result of this effort, an integrally-bladed ceramic rotor ("blisk," see Figure 4.9) version of the turbine rotor including an innovative ceramic/metal attachment scheme (Figure 4.15) was designed. This component retains the airfoil geometry of the MOD 1 inserted ceramic blades, but integrates the 28 blades onto a single cast ceramic rotor disk. This design avoids the cost and durability issues associated with individual inserted blades.

All the blades on two different ceramic blisk rotors were tested in Honeywell's Acoustic Ring Signature Laboratory at room temperature and static conditions. The results from the testing were compared against the analytical predictions, and a favorable comparison was achieved. The test results indicated that the

TABLE 4.7 331-200[CT] CERAMIC TURBINE NOZZLE CUMULATIVE ENGINE TEST EXPERIENCE

Ceramic Nozzle	Hours	Starts
MOD 0	897.6	2481
MOD 1	9.8	22
MOD 2	4232.2	6530
Nozzle Total	5139.6	9033

upstream and downstream vane excitation sources are outside of the operating range for the 331-200[CT] engine. Therefore, potential blade failure due to high cycle fatigue (HCF) was not expected to be an issue during the validation engine testing, and represented a significant improvement compared to the MOD 1 inserted ceramic blade design.

Engine Testing Activities

Successful demonstration of engine operation with ceramic turbine components has served as the validation activity for the ceramic design methods and integrity of the actual ceramic components. Initial development engine testing activities have provided the environment for development and refinement of the ceramic component systems. Subsequent endurance testing and recently initiated field test activities were used to establish the reliability of the ceramic components in both simulated and actual gas turbine operating service conditions.

Two production Model 331-200 APU engines were converted to the 331-200 [CT] configuration for use as test beds for the development and endurance engine test activities. One engine for evaluating the ceramic nozzles with metallic blades, and the other engine for evaluating the ceramic blades with metallic nozzles. Using separate test beds reduced the risks and enhanced data collection for each ceramic component set.

Ceramic Nozzle Engine Testing. Initial MOD 0 ceramic nozzle engine testing revealed cracks in the trailing edge on 2 of 23 ceramic nozzle segments, similar to those seen earlier in the proof testing. Both cracks were thermally induced, and initiated from the trailing edge and propagated halfway through the nozzle vanes. As discussed earlier, two redesign activities of the ceramic nozzles were initiated and resulted in successful engine testing since. Overall, endurance tests of the ceramic nozzles accumulated more than 5,100 engine operating hours (see Table 4.7). None of the MOD 2 ceramic nozzles ever failed during engine testing.

The CTEDP also reached a significant milestone by accumulating more than 2,500 hours of engine testing on a single set of MOD 2 AS800 ceramic turbine nozzles. Post-test inspections revealed no deterioration on the ceramic nozzles.

MOD 2 Ceramic Nozzle Field Testing. A standard Federal Aviation Administration (FAA) 150-hour Engine Block Endurance Test (required for certification of new APUs for commercial air transport service) was completed in 1997 on the ceramic nozzles. The test results, along with supporting data from the development testing and a summary of prior Honeywell ceramic gas turbine development experience was submitted in support of certification of the Honeywell Model 331-200 and 331-250 APUs for operation with ceramic turbine nozzles. FAA certification was granted to Honeywell in July 1999 and enabled the launch of a field demonstration of the ceramic nozzles under actual commercial air transport service conditions. The first two 331-250 APUs equipped with ceramic nozzles were built at Honeywell's Raunheim facility, Germany, in December 1999 and January 2000, and entered service in Lufthansa Airbus A300–600 in early 2000. At the time of this writing, plans included three additional APUs to be subsequently equipped with ceramic nozzles

Figure 4.16 Significant Improvement in Compliant Layer Durability Could be Achieved (Adapted from Ref. [10])

and enter service in early 2001. The five field evaluation APUs will accumulate approximately 20,000 hours of commercial air transport service with ceramic turbine nozzles within the next two years.

Ceramic Turbine Blade Engine Testing. The dual-test-bed strategy paid off during the first engine test on the MOD 0 ceramic blades. In early engine testing of the MOD0 blade, two engine failures occurred. The first was attributed to ineffective compliant layers, which did not adequately distribute the load. The second, which occurred during unloading from max power, was attributed either to blade vibration or excessive contact stress in the attachment as a result of the load transient.

As described earlier, a redesign of the ceramic blades was initiated which resulted in the MOD 1 ceramic blade design. Starting in 1996, development engine testing established "safe" operation of the test bed with the MOD 1 design ceramic turbine blades. Initial testing was performed to resolve the pending issues with respect to blade vibration. An engine test was performed in which MOD 1 ceramic blades were instrumented with thin-film gages for dynamic vibration strain measurement during operation. During this testing, the engine was operated up to 105 percent speed and at each of the design load conditions. The results showed that the ceramic blade strain levels did not exceed 150 μ under any engine operating condition. These strain levels were considered acceptable for further engine testing of the MOD1 blade design without risk of vibration failure.

Subsequent engine testing then focused on evaluating and demonstrating the durability of various compliant layers. Several engine tests were run in blocks approximating 100 hours each, at engine speeds and loads representative of typical APU operation. After each block test, the ceramic blades and compliant layers were evaluated and the configuration adjusted to improve durability. Finally, a compliant layer development breakthrough was demonstrated in the last engine test of 200 hours operation and 50 start-stop cycles, using an improved proprietary Honeywell compliant layer with a stable, high-temperature solid lubricant. Both the ceramic blades and the compliant layers were in excellent condition following the test, as shown in Figure 4.16 (the compliant layers are shown partially pulled out for illustration purposes).

None of the MOD 1 ceramic blades failed during any of the engine testing activities. The CTEDP also reached a significant milestone by accumulating more than 1,000 hours of engine testing using inserted ceramic turbine blades (Table 4.8).

TABLE 4.8 331-200[CT] CERAMIC TURBINE BLADE CUMULATIVE ENGINE TEST EXPERIENCE

Ceramic Blades	Hours	Starts
MOD 0	1.5	4
MOD 1	1043.5	1027
Blades Total	1045.0	1031

Ceramic Turbine Blisk Testing. As discussed earlier, an integrally-bladed ceramic rotor ("blisk") version of the Model 331-200/-250 APU first-stage turbine rotor was designed in 1997, which was designated as the MOD 2 ceramic blade design. A complete 331-200[CT] engine rotating group including a ceramic blisk was assembled and balanced in preparation for planned engine development testing, and in June 1998 testing was initiated. The test bed has since successfully accumulated over 1000 hours of full-speed operation in several test stages, using the same ceramic blisk (Table 4.9). The final 300 hours of blisk engine endurance testing were run for the first time in an all-ceramic first stage (nozzles and blisk) configuration without any difficulties.

The purpose of the blisk endurance testing has been to evaluate the structural integrity of the rotating assembly including the ceramic blisk and its innovative metallic attachment scheme. The test objectives were to demonstrate the durability of the metallic attachment, and to evaluate the cumulative effects of time and cycles on the attachment system. Additionally, blisk cavity temperatures could be evaluated in order to estimate the ceramic/metal attachment temperature and to verify the secondary flow system. The testing successfully proved the feasibility and reliability of the plain friction drive concept without compliant layers (see Figure 4.15) for use in the ceramic blisk field test activity.

During all the testing, the measured engine vibration levels were always normal, and there was no sign of any anomalies. The frictional (ceramic/metal) blisk to shaft attachment never showed any slippage. During two test blocks, the engine also experienced minor foreign object damage (FOD) events from metallic pieces, which had separated from an upstream thermocouple/pressure probe shield. The FOD only resulted in minor chips on the trailing edge of a few blades. However, the FOD damage was deemed to be minor, and did not interfere with continued engine operation during subsequent endurance testing of the very same blisk.

MOD 2 Ceramic Blisk Field Testing. In order to support one of the major objectives of this program, i.e., to provide extensive engine test experience and early field experience, a Honeywell Engines Model IE831-800 industrial engine has been chosen for field testing of the ceramic rotor blisk during 2001. The IE831 engine models are derived from the very reliable AlliedSignal/Garrett Model 331 series turboprop and 331-200/-250 APU engines, with a robust gearbox suited for typical stationary industrial engine installations. Since the flowpaths of the IE831 industrial engine and the 331-200/-250 APU models are identical, the IE831 engine lends itself towards relatively simple introduction of a ceramic blisk in the first-stage turbine. Honeywell Engines, Systems & Services plans to field test the ceramic blisk equipped IE831 engine. The ASE8-800[CT] ceramic blisk field test engine has recently completed over 100 hours of cyclic endurance testing prior to shipment to a field test site in early 2001. The duration of the field test will be a minimum of 10,000 hours or a maximum of two years.

SUMMARY AND CONCLUSIONS

In summary, the long-term multidisciplinary effort of advancing ceramic gas turbine technology development at Honeywell toward commercialization can be characterized by a parabolic learning curve, which is beginning to pay off. Most of this vision has been achieved. The next step will be the commercialization

TABLE 4.9 331-200[CT] CERAMIC TURBINE BLISK CUMULATIVE ENGINE TEST EXPERIENCE		
Ceramic Blisk	**Hours**	**Starts**
MOD 2	1,082 hours	1,072 hours

and production of these technologies in adequate gas turbine engine applications, such as auxiliary power units and microturbines. This might require even better control of the ceramic component fabrication processes, which can only mature if a continuous process throughput can be created by a growing number of applications. Successful large-scale field evaluations of ceramic component technology represent another necessary building block in order to mature the fabrication processes and to help build an increasing level of confidence in ceramic gas turbines. The development efforts over the years have clearly helped to recognize the limitations as well as potential extensions of ceramic gas turbine technology. It has been demonstrated that retrofit applications will only enjoy limited success, since simple ceramic component insertions will always be limited by economics and technical effectiveness when compared to competing technologies. The full benefit of ceramic components in gas turbine applications, however, can only be realized once the application is designed to capitalize on the benefits while considering the limitations, i.e. the most effective applications will be the ones already tailored for ceramic technology in the conceptual stage. Applying the many similar lessons learned in numerous programs will be the key to successful commercialization in the near future. However, long-term commitment from government agencies is still needed to prevent the duplication of effort and to drive and facilitate a sustained, focused, and consistent effort to finally make the large investment in ceramic gas turbine technology development pay off.

REFERENCES

[1] Matuschak, P.E., "Ceramic Reheat Combustor," in *Ceramics for High Performance Applications-II,* Eds. J.J. Burke, E.N. Lenoe, and R.N. Katz, Brook Hill Pub. Co., Chestnut Hill, Mass., 1978.

[2] Richerson, D.W. and Johansen, K.M., "Ceramic Gas Turbine Engine Demonstration Program," Final Report, DARPA/Navy Contract N00024-76-C-5352, May 1982, Garrett Report 21-4410.

[3] Fairbanks, J.W. and Rice, R.W., eds., Proceedings of the 1977 DARPA/NAVSEA Ceramic Gas Turbine Demonstration Engine Program Review, Battelle Columbus Laboratories, Columbus, Ohio, Report No. MCIC-78-36, March 1978.

[4] Wimmer, J.M., Ardans, P., Richerson, D.W., and Lindberg, L.J., "Ceramic Components for Turbine Engines," Final Report, AFWAL Contract F33615-77-C-5171, 1983, Garrett Report 21-2794(13).

[5] Schienle, J.L., "Durability Testing of Commercial Ceramic Materials, Final Report, NASA Contract DEN3-27, January 1996, Report DOE/NASA/0027-1 (NASA CR-198497).

[6] Smyth, J.R., "Contact Stress Analysis of Elastic-Viscoelastic Interface Conditions," Final Report, ONR Contract N00014-80-C-0870, January 1984, Garrett Report 21-4140(2).

[7] Engineering Staff of Garrett Auxiliary Power Divison, "Advanced Gas Turbine (AGT) Technology Development Project," Final Report, NASA CR-180891, 1988.

[8] Boyd, G.L., Easley, M.L., and Kreiner, D.M., "AGT101-Ceramic Gas Turbine Development," ASME paper 87-GT-228, presented at the International Gas Turbine Conference and Exhibition, Anaheim, California, USA, May 31-June 4, 1987.

[9] Engineering Staff of Garrett Auxiliary Power Division, "Advanced Turbine Technology Applications Project (ATTAP), 1992 Annual Report, NASA CR-191088, 1993.

[10] Schenk, B., "Ceramic Turbine Engine Demonstration Project—A Summary Report, ASME paper 2000-GT-0532, presented at the ASME TURBO EXPO 2000, LAND, SEA, & AIR, Munich, Germany, May 8–11, 2000.

[11] Campbell, L., "AlliedSignal/Garrett Model GTCP85-180 APU Ceramic Nozzle Insertion Program for the USAF A/M32A-60A Ground Power Cart—Final Report, USAF Contract No. F41608-92-C-1420, AlliedSignal Report No. 31–13234, Phoenix, AZ, 1996.

[12] Schenk, B., Brehm, P.G., Menon, M.N., Tucker, W.T., Peralta, A.D.: "Status of the CERAMIC/ERICA Probabilistic Life Prediction Codes Development for Structural Ceramic Applications," ASME paper 99-GT-318, presented at the International Gas Turbine and Aeroengine Congress and Exposition, Indianapolis, Indiana, USA, June 7–10, 1999.

[13] Schenk, B., Brehm, P.G., Menon, M.N., Tucker, W.T., Peralta, A.D., "A New Probabilistic Approach for Accurate Fatigue Data Analysis of Ceramic Materials" Transactions of the ASME, Journal of Engineering for Gas Turbines and Power, Vol. 122[4], pp. 637–645, 2000.

[14] Smialek, J.L., Robinson, R.C., Opila, E.J., Fox, D.S., and Jacobson, N.S., "SiC and Si3N4 Recession due to SiO2 Scale Volatility under Combustor Conditions," Adv. Composite Mater., Vol. 8, No. 1, pp. 33–45, 1999.

CHAPTER

5

Ceramic Gas Turbine Development at Rolls-Royce in Indianapolis

Pramod Khandelwal and Peter Heitman

Rolls-Royce, Indianapolis, Indiana, USA

INTRODUCTION

The gas turbine engine is an important powerplant for military, aerospace, and commercial markets. Engine manufacturers have continually strived to increase the efficiency, reduce the size per kW, and reduce the pollution emissions of gas turbines for these applications. Since the early 1970s substantial efforts have been directed towards the use of ceramics to achieve these improvements. Ceramic materials offer the hot strength and corrosion resistance needed to operate at very high rotor inlet temperatures without the need to resort to complex air cooling arrangements. This is particularly important for small engines, where component size and/or cost constraints often make air-cooling impractical. However, to successfully apply advanced ceramics a completely new suite of technologies that focus on material flaw management must be developed. Included in this suite are new methods of component design based on probabilistic analysis, net-shaped component manufacturing methods, and component inspection.

Rolls-Royce has, at its Indianapolis location*, been actively pursuing the development of these technologies since 1972. Table 5.1 provides a timeline and brief highlights of the goals and progress of the key programs. The initial work, which was funded by the Department of Energy through NASA, made use of a heavy-duty truck engine as a development test bed. This engine was a 350 hp two-spool, axial-flow machine with two regenerators. The technology developed under that program was subsequently used as the basis for the design of a new advanced automotive gas turbine engine designated the AGT 100. This engine was a lightweight 100 hp, two-spool machine with only a single regenerator. The Department of Energy funded the

* Formerly the Allison Gas Turbine Division, General Motors Corporation (GMC) and the Allison Engine Company.

TABLE 5.1 TIMELINE AND HIGHLIGHTS OF KEY GOVERNMENT-SUPPORTED PROGRAMS CONDUCTED AT ROLLS-ROYCE IN INDIANAPOLIS (ADAPTED FROM TABLE 2.1 IN CHAPTER 2 OF THIS VOLUME)

Program	Time	Sponsors	Goals	Milestones/Accomplishments Achieved
Heavy-Duty Gas Turbine Engine (HDGTE)	1976–1978	NASA/DOE	Study potential benefits of ceramics for a 270kW GT404–4 truck engine.	Studies predicted substantial benefits.
Ceramic Applications in Turbine Enginers (CATE)	1978–1983	DOE/NASA	Utilize ceramics to increase the axial flow gasifier turbine rotor inlet temperature from 1002°C to 1241°C; demonstrate the engine in a vehicle under realistic on-road driving conditions.	7474 hours of engine testing at 1038°C. Included 231 hours (10,650 km) successful truck demonstration over streets, highways, and the GM proving grounds.
Advanced Gas Turbine Powertrain Systems Development (AGT)	1979–1987	DOE/NASA	Design, build and test a 74.6 kW automotive turbine engine using ceramics to achieve average mileage of ~18km/l (42.8 mpg). This would require TIT of 1288°C and rotor speed of 86,240 rpm.	Designed the two-shaft AGT100 engine using ceramics for the two radial rotors, nozzle guide vanes and shrouds, turbine inlet scroll, combustor liner, regenerator disks, and various transition and seal parts. Subcontracted extensively with ceramic suppliers to improve their materials, especially pressureless sintered SiC and Si_3N_4 that could be fabricated to the required complex shapes. Conducted more than 27 engine builds. Culminated in a series of engine tests that accumulated 99.5 h and 56 start/stop cycles up to a TIT of 1080°C and rotor speed of 83% design, but ended with impact damage to the gasifier inducer tips due to impact of carbon particles. Program determined that improved ceramics, an impact-resistant design, and special design attention to metal/ceramic interfaces were necessary.
Advanced Turbine Technology Applications Project (ATTAP) and **Hybrid Vehicle Turbine Engine Technology Support (HVTE-TS)**	1987–1993 1994–1998	DOE/NASA DOE	Switch to the AGT-5 axial flow design to achieve an impact-resistant ceramic rotor design. Continue support to ceramic suppliers to improve their materials. Demonstrate full engine operation on a chassis dynamometer to measure efficiency and emissions.	Successfully completed a 300-hour cyclic durability test (up to 1371°C) in 1996 of an all-ceramic gasifier turbine stage and demonstrated that emissions met Tier 0 federal standards over the EPA urban driving cycle.
Advanced Turbine Systems (ATS)	1995–2001	DOE	Conduct a long-term field test of uncooled Si_3N_4 first stage nozzle vanes in a 3500–4000 kW Model 501-KB5 industrial gas turbine engine.	Engine with silicon nitride first stage vanes operated for 815 hours, including 793 hours under normal commercial service condition at an Exxon natural gas processing plant. The ceramic vanes survived but had too high a rate of surface recession. Efforts are in progress to engine test the silicon nitride with an environmental barrier coating(EBC).

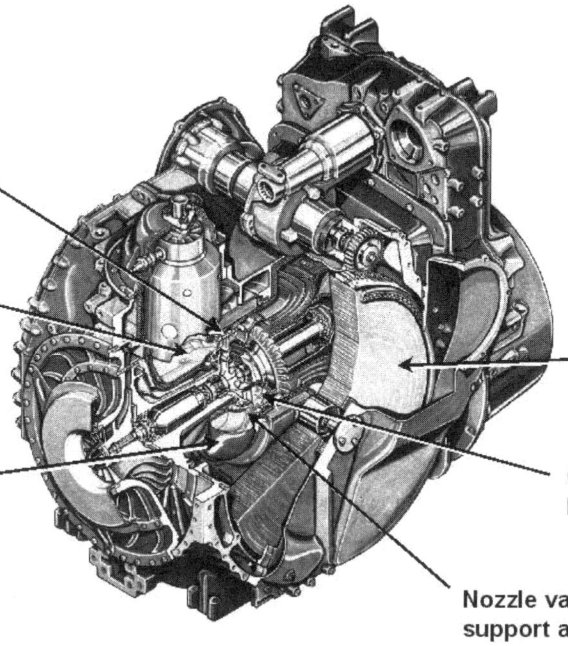

Figure 5.1 Cut Away View of the CATE Ceramic Technology Engine (Adapted from Figure 1 in Ref. [1])

development of this engine with technical guidance again provided by NASA. Following this effort the development of specific ceramic components continued into the 1990s under the ATTAP/HVTE-TS project using a small axial flow gas turbine. Finally, the most recent (1995-1998) ceramic application work addressed the high pressure nozzle of a large industrial machine. This paper will review some of the key ceramic component development activities undertaken in the course of these programs. Included will be consideration of the design, manufacturing development and inspection of ceramic components as well as engine testing.

VEHICULAR GAS TURBINE PROGRAMS: CERAMIC APPLICATIONS IN TURBINE ENGINES (CATE)

Early work (mid 1970s), which was funded by the Department of Energy (DOE) through NASA used a heavy-duty gas turbine truck engine as a development test-bed (IGT 404C). This program was called Ceramic Applications in Turbine Engines (CATE).

The IGT 404C engine was a two spool, regenerative, axial-flow machine as shown in Figure 5.1. The program goal was to demonstrate an entire ceramic hot flow path capable of uncooled operation to an average rotor inlet temperature of 1241°C (2265°F). The all-ceramic gasifier configuration design, shown in Figure 5.2, had 78 individual ceramic components. The nozzle assembly featured 28 individual ceramic vanes captured in slots by inner and outer support rings. The vanes were isolated from mechanical loads by virtue of the inner and outer supports being connected through four, ceramic-shrouded, air-cooled metal support struts ahead of the vane row. The rotor consisted of forty individual ceramic blades that were mounted in a metal disk by dovetail attachments with compliant layers of L605, a cobalt-based superalloy. The dovetail of each blade was coated with boron nitride to serve as a solid film lubricant and diffusion barrier.

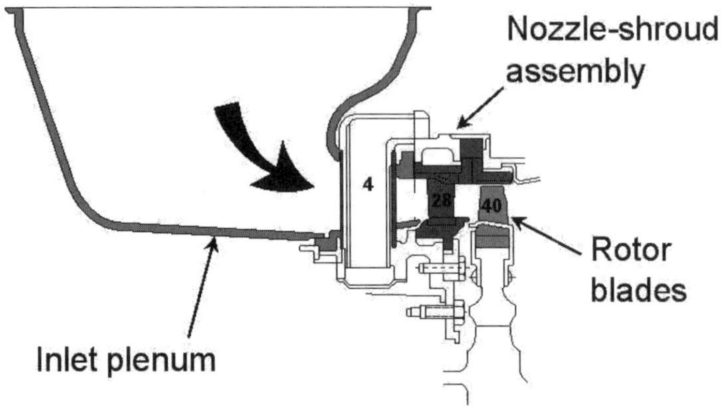

Figure 5.2 Schematic of the CATE Ceramic Gasifier System (Adapted from Figure 5 in Ref. [1])

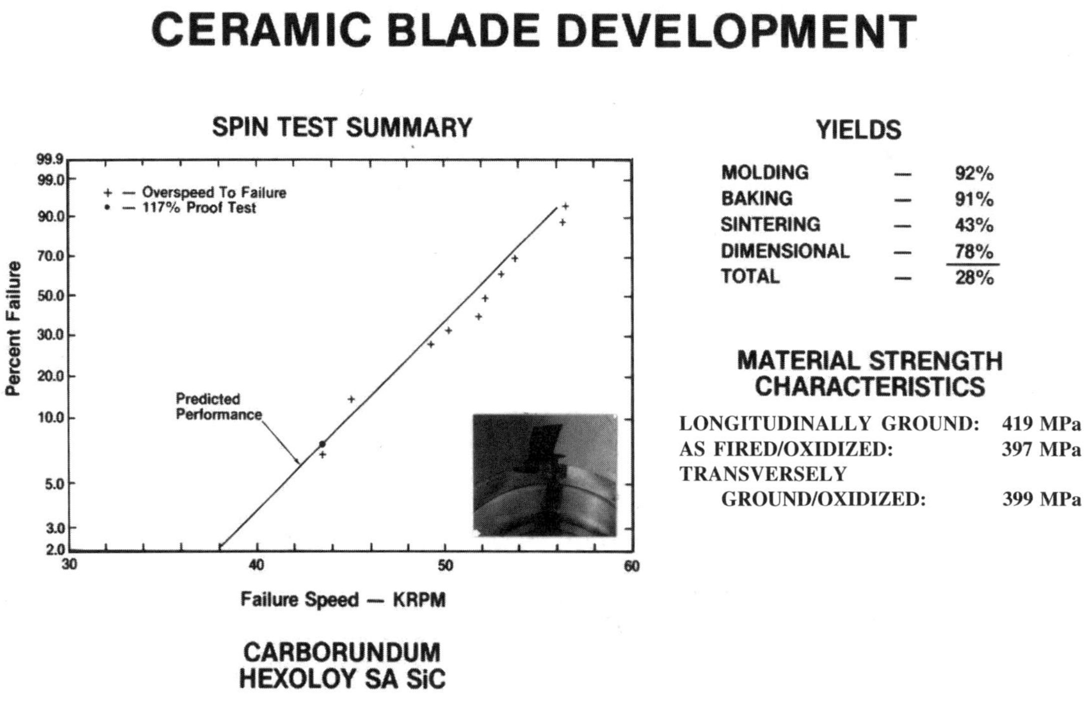

Figure 5.3 Summary of Results of Design Experiments Development Activity for Injection Molded Silicon Carbide Blades (Reprinted with permission from Rolls-Royce)

During 1979 through 1980, the first process development program employing factorial design of experiments was undertaken to address the fabrication of net shaped, injection molded silicon carbide blades. The work was conducted at the Carborundum Company. Upon completion of the program in late 1980 process yield, structural/dimensional quality and strength characteristics were established for a "production run" of over 600 blades. Figure 5.3 summarizes the results. The total process yield of fully

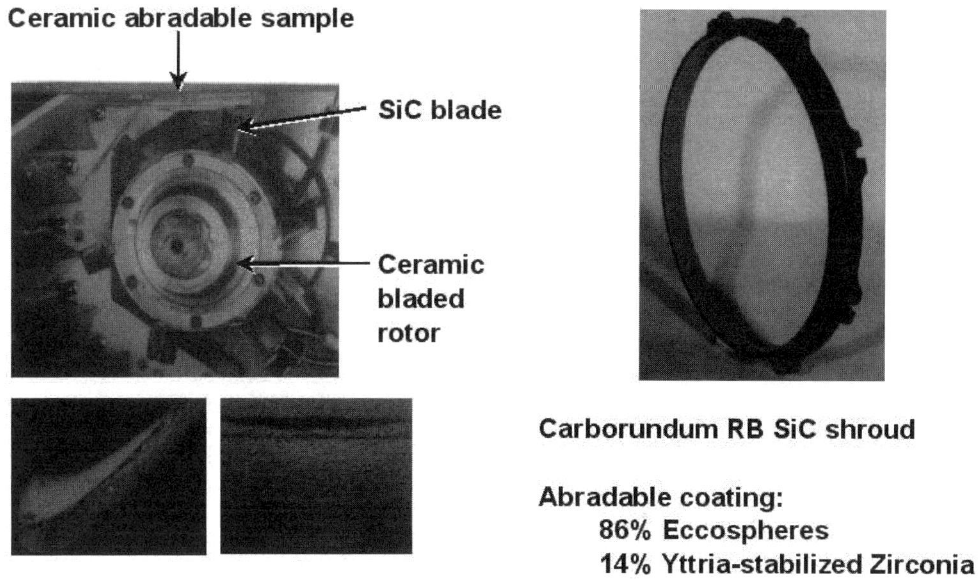

Figure 5.4 Development of an Abradable Seal System Compatible with Silicon Carbide Blading (Reprinted with permission from Rolls-Royce)

machined blades, meeting both structural (nondestructive evaluation—NDE) and dimensional requirements, was 28%. Qualification test material, molded simultaneously with each blade, was used to indirectly assess blade strength characteristics. Various surface finish conditions pertinent to the finished blades were examined. The average fracture strength for the longitudinally ground surface measured 419 MPa (60.7 ksi), while those of the as-fired and transversely ground surface conditions measured 399 MPa (57.5 ksi) and 397 MPa (57.4 ksi), respectively. These strength characteristics were equal to the then state-of-the-art for injection-molded sintered silicon carbide in test bar form. Finally, overspeed spin testing was used to establish actual blade strength characteristics. The correlation between predicted and actual performance was excellent. The types of flaws that precipitated failure were identical to those observed in companion test material. Thus, it was concluded that then-day material characteristics were being achieved in actual rotor blades.

The performance of the very small, low flow turbine engines being contemplated for automotive and truck applications is particularly sensitive to the clearance between the rotor blade tips and the surrounding shroud. To maintain a minimum clearance and at the same time achieve a practical level of stacking tolerance, a rub-tolerant abradable seal system was required. During the CATE program work was conducted to develop an abradable seal system that would be compatible with ceramic blading. The resulting seal system was a porous plasma sprayed zirconia. It contained equal weight percentages of prestabilized zirconia (with 8% yttria) and type FA-A Eccospheres (hollow aluminosilicate spheres). This composition, applied on a metal shroud, successfully ran for 288 hours at 1038°C (1900°F). The abradability of the zirconia-Eccosphere system was evaluated in an intentional rub situation on a test rig. The material was also successfully applied to a structural ceramic substrate.

The Alpha Silicon Carbide Division of Carborundum fabricated the ceramic turbine tip shroud shown in Figure 5.4 from reaction-bonded silicon carbide. This component was produced by green machining a compression-molded ring to near net shape. Following siliconization, the critical dimensions were finish ground. The abradable seal material was applied to the inside diameter of the shroud by plasma spraying.

TABLE 5.2 CATE ENGINE TEST EXPERIENCE (IN HOURS)

Configuration	1038°C (1900°F)	1132°C (2070°F)	Total
Vanes (vane assembly)	3406	900	4306
Tip shroud	1670	900	2570
Regenerator disk	7814	1584	9398
Blades	—	20	20
Engine	—	—	9398

Figure 5.5 Silicon Carbide Nozzle Assembly After 900 Hours of Cyclic Endurance Testing to Temperatures of 1132°C (2070°F) (Reprinted with permission from Rolls-Royce)

Using a sintered alpha silicon carbide blade, twice proof tested to 43,000 rpm (117% design speed), a rub depth of 0.13 mm (0.005 in.) was achieved with no distress to either the blade or the shroud. The abradability demonstrated for this material system was considered suitable for engine application to either metallic or ceramic seal support structures.

Table 5.2 summarizes the engine test experience for the 404C configuration gasifier section. Included are results achieved on individual components tested at lower temperatures. Over 9000 hours of engine testing was completed during the course of the CATE program. The 404C nozzle assembly and tip shroud with an all-metal rotor successfully completed 900 hours of dynamometer testing to a heavy-duty truck engine duty cycle. Figure 5.5 shows the silicon carbide nozzle assembly after test. All components were in excellent condition at the conclusion of the test. Over 9000 hours were accumulated on the regenerator disk and seal assemblies without incident. Only 20 hours of the testing of the gasifier blades were completed, however. This included two engine tests of complete ceramic gasifier sections. Both of these tests

Figure 5.6 Cut Away View of the AGT 100 Automotive Gas Turbine Engine (Reprinted from Ref. [2])

ended in failure after 1 to 2 hours of running. In the first test foreign object damage (FOD) precipitated the failure, while a blade fracture from a large, preexisting flaw caused the second failure. In both instances the complete gasifier section was destroyed. In the latter test the failure-precipitating flaw was in a low stress region beneath the attachment .This flaw was not detected prior to testing because the region below the attachment was not generally examined in great detail. This, of course, dramatically points out a lesson ruefully learned and relearned: the need to consider all parts of a ceramic structure in great detail.

VEHICULAR GAS TURBINE PROGRAMS: ADVANCED GAS TURBINE (AGT) TECHNOLOGY PROJECT

Following the CATE program, the focus of the vehicular turbine program shifted to automotive applications. Rolls-Royce at Indianapolis (then Allison Division of General Motors) formed a team with automotive and ceramic manufacturers to address the goals of the DOE-funded AGT project. The major objective of this initiative was to develop and demonstrate the technology for structural ceramic components that had the potential for both competitive automotive engine life cycle costs and the capability to operate for 3500 hours in a turbine engine environment at turbine inlet temperatures as high as 1371°C (2500°F) without cooling air.

The AGT 100 was a two-shaft design with a single regenerator as shown in Figure 5.6. Both the gasifier and power turbine stages of the AGT 100 were of a radial inflow design. This rotor configuration, which embodies a radical variation in section thickness, represented a very difficult challenge from a manufacturing point of view. Because of the cost constraints that drive the commercial market place, the manufacturing development activity focused on net-shaped processing of both silicon carbide and silicon nitride rotor configurations. Injection molding of a silicon carbide rotor was pursued with the Ceramic Manufacturing Center of the Sohio Engineering Materials Company, while a slip casting approach to develop a silicon nitride rotor was pursued with Kyocera International. These process development activities are summarized in the next two sections.

Silicon Carbide Rotors

In trying to scale the blade processing technology to a monolithic radial inflow rotor, two major issues had to be addressed. First, a uniform green microstructure had to be achieved throughout the rotor, particularly in the highly stressed hub region. This was quite difficult in view of having to fill the tool cavity by injecting and compressing a spaghetti-like mass. Success in achieving a uniform structure in individual blades was accomplished by filling the blade cavity uniformly from the attachment. This approach was not feasible for a radial inflow rotor. The second, and perhaps the more difficult issue, was the removal of the plastic binder from the thick hub section.

Initially, a prototype rotor configuration was used to address these issues. After defining a practical binder that could be readily removed, work focused on investigating the relationship between processing and strength. Fourteen of the 15 rotors tested achieved burst speeds in excess of 100% of design speed (86,240 rpm). The median burst speed was 97,000 rpm. By application of Weibull statistics in conjunction with a finite element analysis, the modulus of rupture (MOR) for the rotors was estimated to be 340 MPa (49.3 ksi). Test material cut from the backface region of actual rotors had a MOR value of 336 MPa (48.7 ksi). Thus, the strength capability realized in the prototype rotor approached that achieved previously in small individual blades.

This processing study served as a starting point for development of the actual engine configuration rotor. This rotor configuration differed from the prototype in that the exducer blade tips were lengthened for aerodynamic reasons. This complicated the tool design that subsequently influenced the molding process. Following a preliminary molding activity, a full-scale process/tool optimization study was initiated. The first phase of this study concentrated on defining a set of injection molding parameters that would produce green (unfired) rotors free of surface imperfections. The major molding parameters considered included temperature, pressure, and injection speed. The second phase of the effort considered a variety of tool modifications also intended to improve green body quality. Of particular interest were gating, venting, and mold temperature control. Over 1100 green rotors were produced and evaluated.

An assessment was conducted at the conclusion of the molding study to establish the capability for meeting dimensional and structural (strength) quality requirements on a consistent basis. Total process yield was 11%. This included a molding yield of 28% and a sintering yield of 85%. Inspection of the sintered rotors, which included visual, dye penetrant, and dimensional examinations, revealed that slightly less than one half of these rotors met quality requirements. Two major discrepancies, one structural and one dimensional, were identified. From a structural point of view, knit lines were randomly dispersed over a backface of many of the rotors. Dimensionally, drooping of exducer portions of the airfoils was evident.

Twelve of the highest quality rotors were selected for further evaluation. Three of the rotors were sectioned into standard sized test bars for strength evaluation. Test material cut from the backface region (note this is the most highly stressed region of the rotor hub) was consistently found to average about 400 MPa (58.0 ksi). The strength of test material, removed from the central portion of the hub, was consistently lower, averaging 286 MPa (41.5 ksi). This lower level of strength resulted from the presence of porosity. The remaining nine rotors were balanced and spun to failure in a vacuum spin pit. An average burst speed of 99,300 rpm (115% design speed) was achieved (note that 100% mechanical speed for the AGT 100 is 86,240 rpm). With but one exception, all of the rotors failed within a range from 95,000 to 105,000 rpm. A single rotor failed below the 100% speed level (82,000 rpm). An examination of the rotor fragments after test revealed that failure occurred from a large pore. The remaining rotors failed from the backface region adjacent to the shaft at or near the surface. If the rotors were completely homogeneous with respect to strength properties and further displayed the strength characteristics achieved in the individual blades noted earlier, the average burst speed for comparison should be 113,000 rpm.

Further studies were conducted to improve quality and yield. The major process modification introduced was the injection of the molding compound into the die cavity through the shaft. Previously, injection was accomplished from the exducer end (nose) of the rotor. This approach allows the initial knitting of the injection strand to occur in the low stress nose area. Further, the nose cavity of the die was extended to provide a removable reservoir for the initial injection stock. To assess the resultant yield and structural capability, a run of 301 rotors was examined. Total process yield increased to 45%. This resulted from an improvement in as-molded quality as well as increased dimensional control of the sintered rotor. Spin test results, however, showed a structural capability similar to earlier rotors. This was attributed to the continuing presence of porosity within the central hub region, as confirmed by strength measurements made on sectioned companion rotors. The lowest strength test bars always fail from large, irregularly shaped pores. However, while the average strength characteristics of material from the backface region is slightly below (10%) that observed for previous rotors, the strength level within the hub has increased by a similar amount. This evening out of strength is consistent with the greater uniformity of the microstructure observed both by destructive and nondestructive means.

While significant progress was made, additional work was required to further improve the microstructure throughout the central portion of the hub. One method of improving microstructure and strength was hot isostatic pressing (HIP). Experimentation conducted at NASA was quite promising. HIP processing complete rotors at 138 MPa (20 ksi) within the temperature range 2050°C to 2172°C (3722°F to 3942°F) resulted in significant increases in component density and strength. Strength increased from 400 to 462 MPa (58.0 to 67 ksi), a 20% improvement. Achieving this level of strength throughout a sintered silicon carbide rotor would have enabled the structural reliability goals established for the AGT 100 reference power train design (RPD) rotor system to be met.

Silicon Nitride Rotors

Kyocera SN220M, at the time a well-developed slip cast silicon nitride material, was selected for the silicon nitride rotors based principally on an assessment of material strength. This assessment considered both machined and fired surface conditions. Test material processing was intended to duplicate that to be used in the fabrication of full-scale rotors. The results of this assessment showed that the average room temperature fracture strength of material tested with an as-fired surface condition measured 462 MPa (67 ksi). The strength-controlling flaws were pores located either on the tensile surface or at the base of surface depressions. For the machined surface condition an average strength of 696 MPa (101 ksi) was observed. The strength controlling defects were internal inclusions, either a glassy phase rich in aluminum, calcium, yttrium, and silicon or a cluster of large grains rich in iron, chromium, titanium, and silicon. Strength was found to decrease markedly at 1000°C (1832°F) and above. The average strength at 1000°C (1832°F) for the machined surface condition measured 607 MPa (88 ksi) while at a test temperature of 1150°C (2102°F), the average strength further decreased to 331 MPa (48 ksi). Fracture always initiated from small surface pores similar to those observed to control strength at lower temperatures. In the case of the as-fired surface condition, average strength decreased to 283 MPa (41 ksi) at 1150°C (2102°F). All materials tested at 1150°C (2102°F) were heavily oxidized, with the as-fired surface oxidized more than the machined surface. The strength-controlling flaws were glassy bubbles near the tensile surface. Based on this data, estimates of the Weibull parameters were made as a function of temperature. These estimates were used in conjunction with a finite element analysis (FEM) to estimate the structural reliability of a SN220M rotor in the AGT 100 environment. The level of reliability predicted met the design goals for the rotor.

At the conclusion of initial process development studies Kyocera International provided eighteen SN220M Si_3N_4 gasifier rotors for analysis and testing. All rotors were proof tested to 100,000 rpm prior to shipment. Nondestructive inspection (visual and fluorescent penetrant) revealed no discontinuities. Dimensional evaluation indicated that all dimensions were within print tolerances with the exception of a minor closing of the exducer throat spacing. This closing was believed to be caused by drooping of the exducer airfoils during sintering and could be corrected by the use of appropriate fixturing.

To evaluate material strength characteristics and component performance, seven SN220M Si_3N_4 rotors were subjected to room temperature spin tests. Of the seven rotors tested, three did not achieve burst due to test facility limitations. All three rotors reached speeds of at least 134,000 rpm. The remaining four achieved burst speeds ranging from 126,800 to 140,700 rpm. A suspended item analysis of the complete set of seven rotors yields on an average burst speed of 141,000 rpm. Assuming the room temperature strength characteristics described above, finite element analysis predicts an average burst speed of 145,800 rpm.

ENGINE DEVELOPMENT TESTING

Two complete engines were been built to support the development and testing of ceramic hot section components under the AGT 100 program. One of these engines is shown in Figure 5.7. More than 27 builds were completed during the course of the program. Each build was intended to examine several specific small engine performance phenomena. Early builds contained a metallic gasifier section and focused on a variety of issues associated with development of a brand new configuration. Having once achieved a consistent running capability and a reasonably well-controlled hot section environment, attention was directed toward the development of ceramic structures. It should be noted that there were no metallic versions of either the combustor assembly or the rear bulkhead. Consequently, all builds contained these ceramic components.

The development of the ceramic gasifier section began with the evaluation of ceramic vanes, both injection molded silicon carbide and silicon nitride. This was followed most recently by rig and engine testing of all of the gasifier static components including the interturbine coupling. The specific components employed are listed in Table 5.3.

The initial testing of this configuration in the S/N 2 engine consisted of six start-stop cycles. The thermal cycle involved was designed to expose the ceramic parts to a moderate level of temperature as a first step. The highest temperature level actually recorded within the scroll was 940°C (1724°F). The average temperature and hold time per cycle were 900°C (1652°F) and one hour, respectively. After the first cycle, the engine was disassembled sufficiently to view the scroll in place, i.e., the regenerator side of the scroll could be viewed but not the support side. All of the ceramic components appeared structurally sound, and no distress of any kind was indicated. Five additional cycles were accumulated giving a total hot running time of 7 hours 51 minutes.

Upon completion of the planned testing, the engine was removed from the test stand and completely disassembled. Teardown inspection showed all ceramic gasifier components unaffected by the test except the outer backplate. This component had developed a crack running radially through a locating slot. A computer simulation of the test identified a need for parts redesign and modification of both the outer backplate and the coupling between the gasifier and power turbine scrolls. Modifications were successfully completed.

Following initial testing of the static structure, attention was directed to the ceramic gasifier rotor. Over 140 hours of total test time were accumulated on eight different ceramic rotors. The first three rotors were silicon carbide and the remaining five were silicon nitride. The second of these tests, completed in August of 1985, consisted of one and one half hours of hot running at moderate speeds and temperature. The specific schedule consisted of a light-off and acceleration to 50% rpm and 540°C (1004°F), a hold for 30 minutes at these conditions, a slow advance to 900°C (1652°F) and 60% rpm and hold for another 60 minutes followed by a controlled shutdown. At the conclusion of the testing, the engine was

Figure 5.7 AGT 100 Ceramic Gasifier Assembly (Reprinted with permission from Rolls-Royce)

TABLE 5.3 CERAMIC COMPONENTS FOR THE ATG 100 PROGRAM

Component	Material	Source
Gasifier turbine scroll	SiC	Carborundum
Gasifier turbine outer backplate	SiC	Carborundum
Gasifier turbine inner backplate	GMAS IISiC	Corning
Gasifier turbine vanes (18)	SiC	Carborundum
Gasifier turbine scroll shim	Zirconia	Feldmühle
Interturbine coupling piston rings (2)	Refel SiC	Pure Carbon
Dilution band (combustor)	SiC	Carborundum
Combustor body	SiC	Carborundum
Pilot flame holder tube	SiC	Carborundum
Interturbine coupling	Refel SiC	Pure Carbon
Power turbine exhaust piston ring	Refel SiC	Pure Carbon

Figure 5.8 Silicon Nitride AGT 100 Gasifier Rotor During Assembly (Left) and After Test (Reprinted with permission from Rolls-Royce)

removed from the test stand, disassembled and the hot section examined. As with the earlier test, significant damage to the inducer blade tips had occurred. Following an examination of the failed rotor, engine test data and analytical studies for potential failure modes, it was concluded that the most likely cause of the observed damage was FOD.

During the ensuing months efforts focused on higher toughness versions of silicon nitride. The first test of a SN220M gasifier rotor occurred in Build 16 of test engine number one. All initial test objectives were met without incident including approximately 20 hours of testing at 60% speed and 1066°C (1950°F) RIT (rotor inlet temperature). The rotor testing continued in Builds 17 through 19 without incident accumulating a total 99.54 hours and 56 start/stop cycles to temperatures of 1080°C (1976°F) and speeds of 83% of N1. During a routine start-up on Build 20 a dramatic increase in rotor shaft whip was observed at 30% N_1 leading to a shutdown. Teardown revealed complete loss of all inducer tips (see Figure 5.8). A post-test failure analysis concluded that failure was caused by impact of large carbon particles produced on the inner wall of the combustor.

Three attempts were made to run an all-ceramic gasifier, the last in May of 1987. In addition to the standard rotor qualifying procedures fluorescent penetrant inspection (FPI), visual, and dimensional) inducer airfoil natural frequencies were experimentally determined and each inducer airfoil was dynamically proof stressed. In each case, catastrophic failure of the hot section occurred early in the testing. In the first case, movement of the static structure into the rotor precipitated failure. In the second case, failure was caused by distortion of the metal portion of the rotor shaft. In the final case, no direct cause could be identified, although airfoil impact resistance remained a major concern.

During the course of the AGT100 program much progress was made in the areas of design, manufacture, non-destructive inspection (NDI) and testing, both of components and engine subsystems. However,

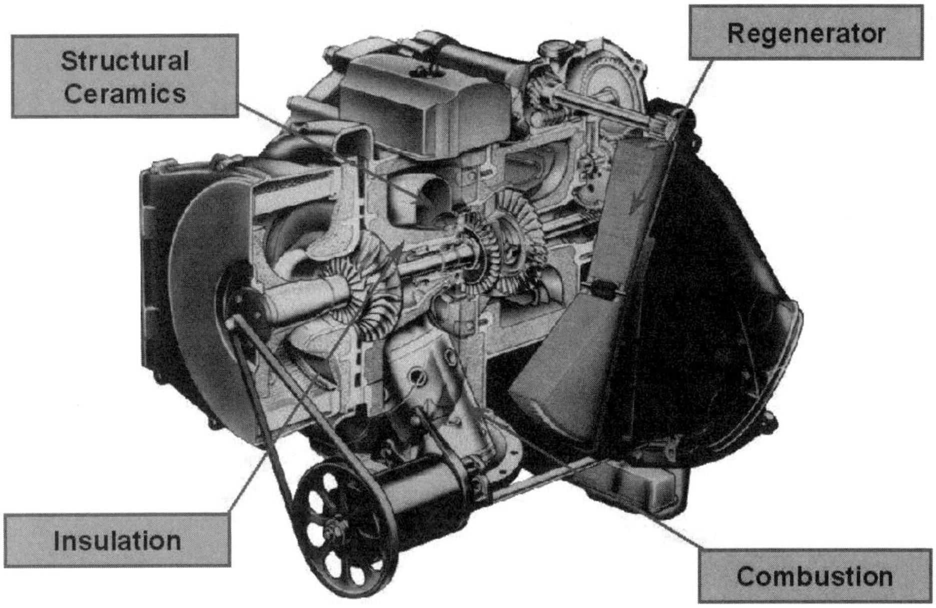

Figure 5.9 Cut Away View of the AGT 5 Automotive Gas Turbine Engine Showing Technology Issues Addressed During the HVTE-TS Project (Reprinted with permission from Rolls-Royce)

two key concerns remained. The first related to the need to more effectively manage the interface between the ceramic static structure and the supporting metal casing to avoid excessive distortion/movement and subsequent interference with the rotating components. The second was the apparent sensitivity of existing ceramic materials/design concepts to random foreign object damage. The sensitivity of rotating structures to impact damage has been experienced by a number of groups working on the development of both gas turbine engines and turbochargers. The problem is particularly severe for radial inflow designs. It was thought that if this sensitivity could not be minimized through innovative design or advancements in material toughness, then a change in configuration would be required if commercialization of small high efficiency and low emissions ceramic gas turbine engines was to be successful.

VEHICULAR GAS TURBINE PROGRAMS: THE ADVANCED TURBINE TECHNOLOGY APPLICATIONS PROJECT (ATTAP)

The AGT program identified challenges that needed to be overcome. The Advanced Turbine Technology Application Project (ATTAP) was initiated under DOE/NASA sponsorship in 1987 to address these challenges. The goals were to solve the FOD problem through a combination of design and materials development, to build and qualify an engine with ceramic components, and to test the full engine on a chassis dynamometer to measure efficiency and emissions.

At the outset of the ATTAP it was decided to move away from the radial inflow design to achieve a robust integral ceramic rotor design more able to resist foreign object damage. The AGT-5 automotive engine developed by General Motors was selected as the test-bed for the ATTAP program. This engine was a two-shaft, regenerative configuration with axial-flow gasifier and power turbines, as shown in Figure 5.9. The engine produced approximately 100 hp at its original metallic full-power turbine inlet temperature of 1038°C (1900°F).

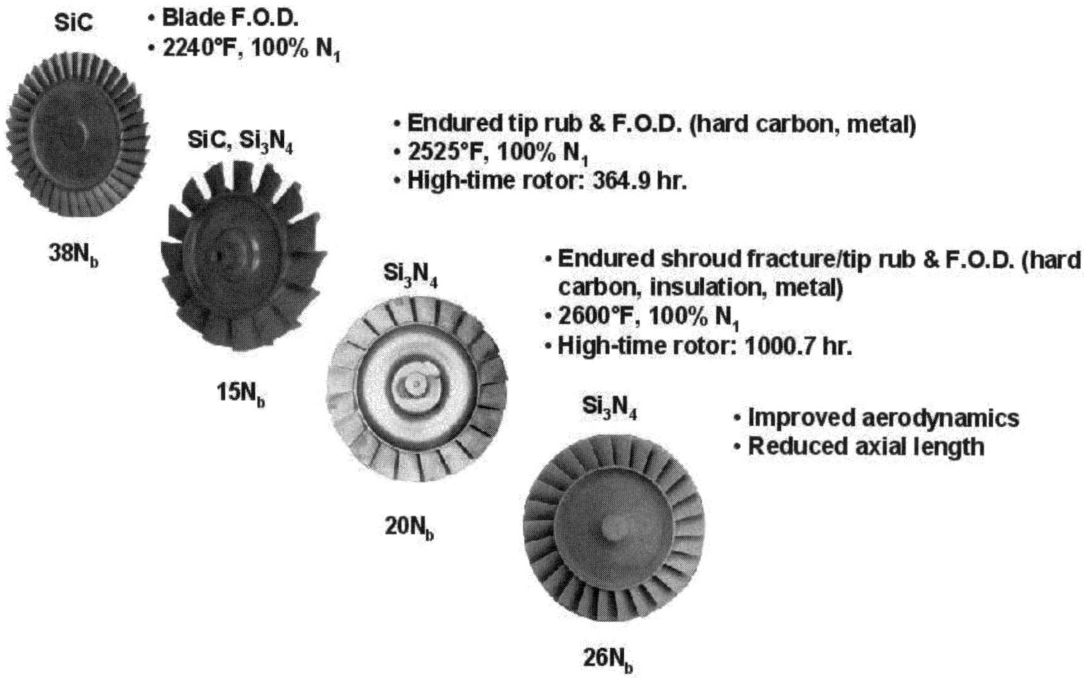

Figure 5.10 Summary of Gasifier Rotor Development Experience Showing Evolution of a Configuration that is both Robust and Aerodynamically Efficient (Reprinted with permission from Rolls-Royce) (Temperature Conversions: 2240°F → 1227°C; 2525°F → 1385°C; 2600°F → 1427°C)

The use of ceramic components in the flow path was projected to enabled the engine to operate at 1371°C (2500°F) without cooling air, which would result in a horsepower increase of 70 to 80%.

The development of an integral axial flow gasifier rotor for the AGT-5 was initiated in 1982 at the General Motors Technical Center. The initial rotor was a high aspect ratio, 38-bladed design produced from silicon carbide as shown in Figure 5.10. Like the AGT 100 hp rotor this early configuration was found to be very sensitive to FOD. During the early stages of the ATTAP, design iterations to improve the fabricability and impact resistance of the rotors were under investigation. The second-generation rotor was a low aspect ratio blade (15) with increased blade thickness and improved impact resistance. The successful engine testing of the 15-bladed silicon nitride rotor configuration enabled the fabrication of designs with improved aerodynamics, while maintaining acceptable impact resistance. A 20-blade gasifier turbine rotor design fabricated by the Kyocera Corporation from SN252 silicon nitride was successfully engine tested for 1000 hours, after which the rotor was retired in engine-ready condition. This rotor endured blade tip rubs into the ceramic shroud ring, plus impacts with particles of insulation, carbon, and molten superalloy combustor parts and vanes, with only minor chipping of the blade tips (see Figure 5.11). The rotor was operated to full design speed and temperatures higher than 1427°C (2600°F), with approximately 50% of the time accumulated on a transient durability cycle. This cycle was developed to simulate the combined Environmental Protection Agency (EPA) federal city/highway driving cycle in an accelerated manner.

The ceramic gas turbine static structure designed for the AGT-5 consisted of a combustor assembly and a scroll/inlet vane assembly. The scroll body, shown in Figure 5.12 along with the rest of gasifier section, was the most geometrically complex ceramic component in the engine. It was the focus of extensive development activities, both from producibility and dimensional tolerance standpoints. An extensive gasifier rig

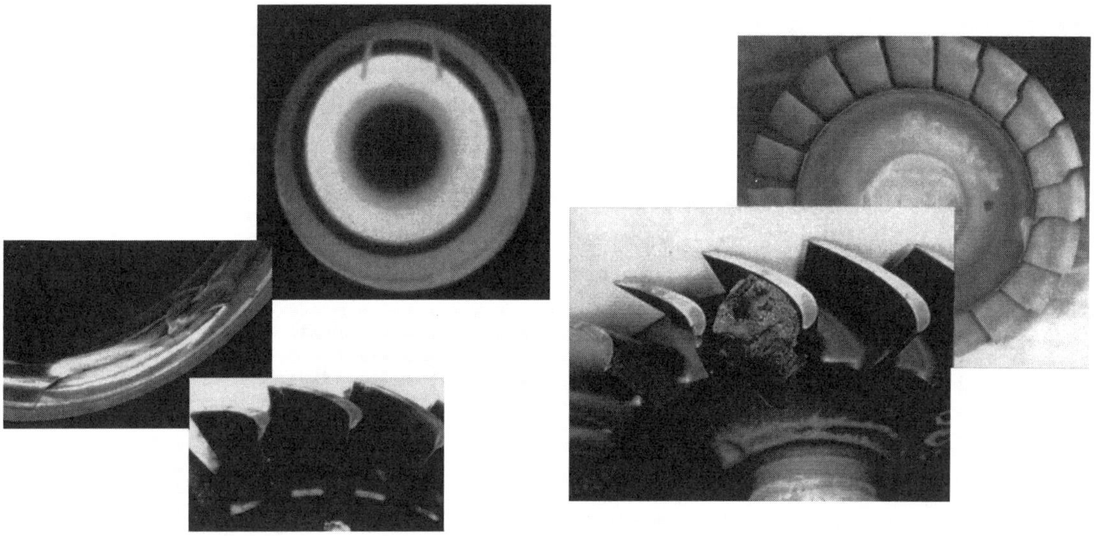

Figure 5.11 Silicon Nitride Gasifier Rotor After 1000 Hours of Durability Testing at RPD Conditions (Reprinted with permission from Rolls-Royce). Rotor Sustained Impact of Hard Carbon at 430 Hours and a Severe Rub at 556 Hours Without Damage

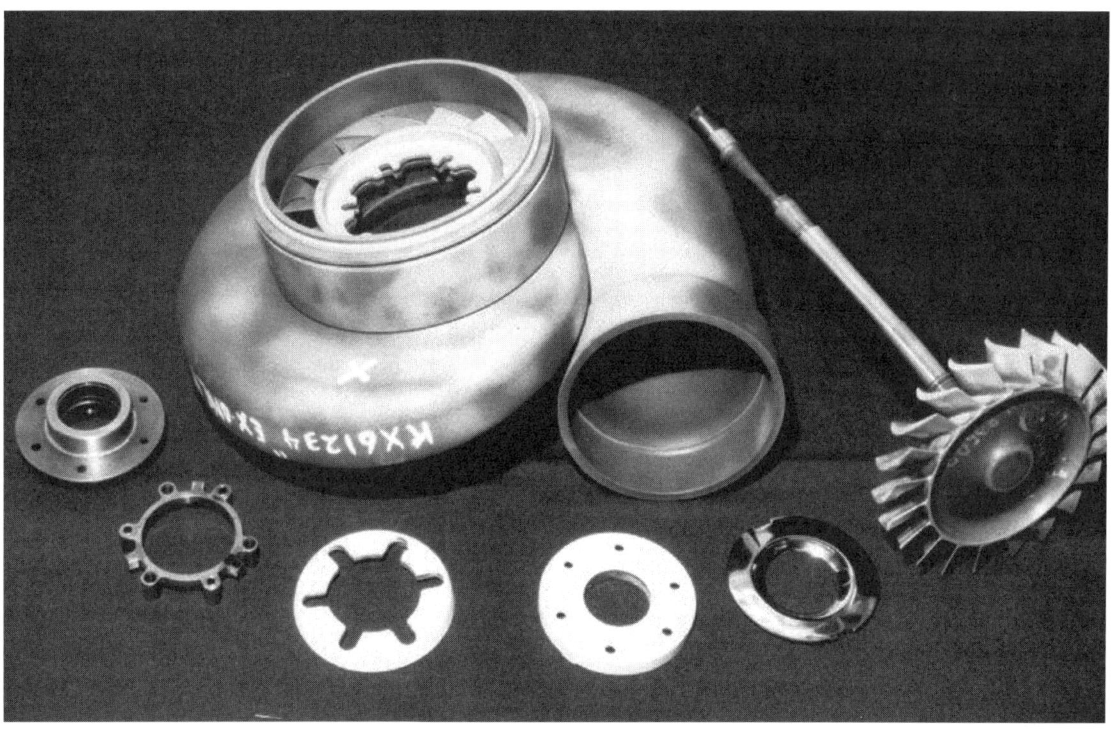

Figure 5.12 Ceramic Hot Section Components for the AGT-5 Engine Including the Scroll and Rotor (Reprinted from Ref. [3])

TABLE 5.4 CERAMIC ENGINE AND GASIFIER RIG TESTING	
Total Ceramic Component Testing	
Engines	**Test Hours**
Durability, system integration	547
Ceramic components running	977
Vehicle running	651
Total Engine Hours	**2175**
Gasifier Turbine Hot Test Rigs	
Ceramic components running	2721
Other testing	210
Total Gasifier Rig Hours	**2931**
Total Hours	**5106**

and engine testing activity as summarized in Table 5.4 supported these development activities. A major milestone was achieved in 1996 with the successful testing of an all-ceramic gasifier section, consisting of a Si_3N_4 gasifier scroll assembly, a Si_3N_4 gasifier turbine rotor, and magnesium-aluminosilicate (MAS) regenerator cores. The automotive durability cycle was utilized for engine testing, with slight modifications to simulate the operating characteristics of an APU for a hybrid vehicle, including cycles run to 100% speed with an RIT of 1371°C (2500°F). Inspections were conducted every 50 hours to assess the condition of the ceramic and metallic hardware. A total of 300 hours of cyclic durability testing (corresponding to 3500 hours of vehicle operation) was successfully completed, with no observable distress to any of the ceramic components (see Figure 5.13).

INDUSTRIAL GAS TURBINE ENGINES: ATS CERAMIC VANE PROJECT

Since 1995 Rolls-Royce at Indianapolis under the Advanced Turbine Systems (ATS) program of the US Dept. of Energy (DOE) has been conducting a project aimed at developing and demonstrating an uncooled ceramic vane in the first-stage of a Model 501-KB5 industrial gas turbine engine. The focus of this program was to develop, fabricate, and test an optimized silicon nitride vane design that is compatible with the current engine structure and space claim.

The approach taken has been to re-design the vane using 2-D thermal and stress analyses developed on earlier programs to achieve a level of aerodynamic performance equivalent to the current all-metal engine. The analyses revealed that such vane airfoils experienced high thermally induced stresses at the thin trailing edges during emergency shutdowns, since they cool much faster than the thick leading edges of the airfoil. The analyses, however, indicated an acceptable probability of survival (POS > 99.9 %) for an optimized new airfoil shape. Further, the aerodynamic performance of this new airfoil shape was actually improved relative to the current Model 501 metal vane profile.

Following the initial 2-D transient optimization studies, more detailed transient and steady state analyses were performed. The most significant source of the vane steady state stresses resulted from the thermal gradients due to the combustor temperature pattern. Three combustor bounding temperature patterns were determined for the purpose of thermal and stress analyses from thermocouple data in Model 501-K engine tests. The 3-D thermal and stress finite element analyses (FEA) were completed for maximum continuous power and emergency shutdown conditions using the three bounding combustor temperature patterns and AS800 (Honeywell Ceramic Components) Si_3N_4 material properties [4]. Maximum stress always occurred at the midspan of the trailing edge of the vane for all transient and steady state power conditions. A maximum thermal stress of 208 MPa (30.1 ksi) occurred at 0.9 s after emergency shut down (see Figure 5.14), where the temperature was only 729°C (1344°F). For steady state (continuous

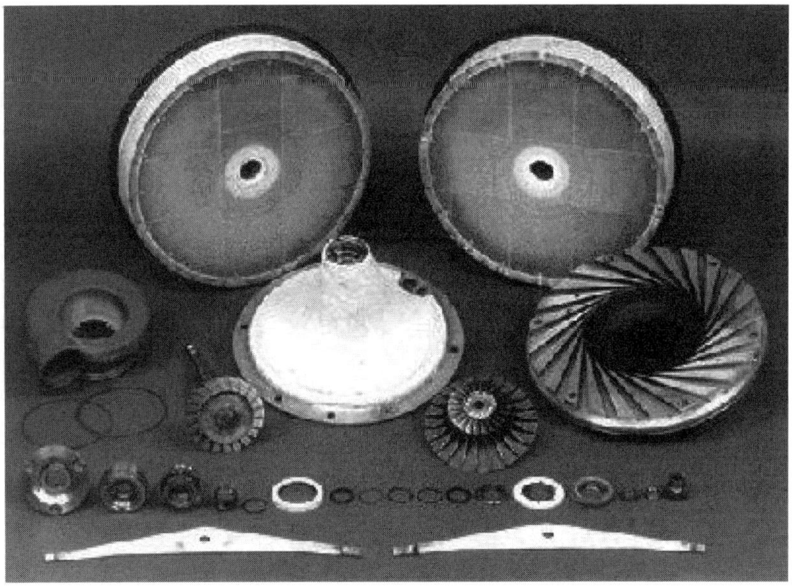

Figure 5.13 AGT-5 Hot Gasifier Rig Components After 300-Hour Durability Test (Reprinted from Ref. [3])

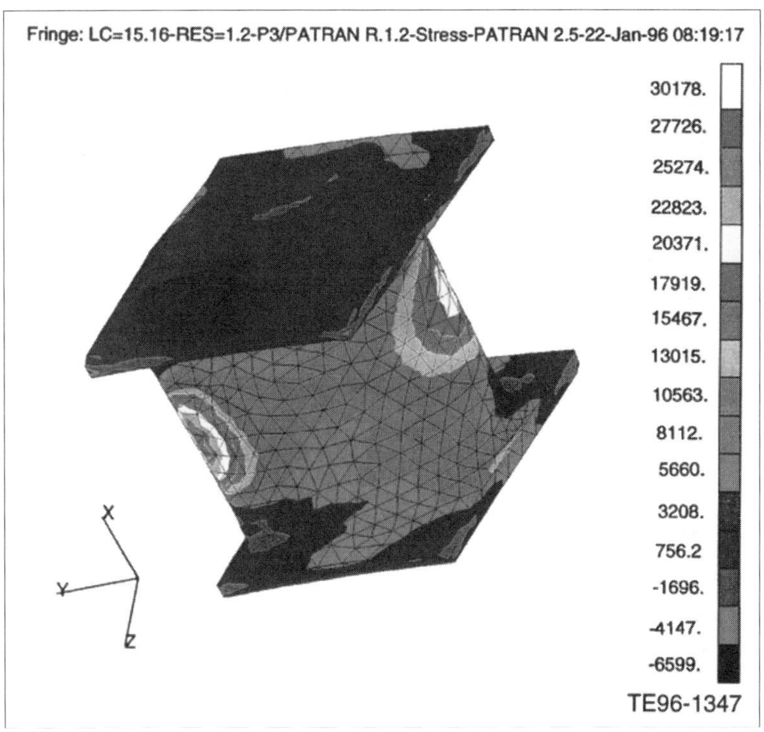

Figure 5.14 Finite Element Stress Analysis of the Ceramic Vane—0.9 s after emergency Shutdown (Reprinted with permission from Rolls-Royce)

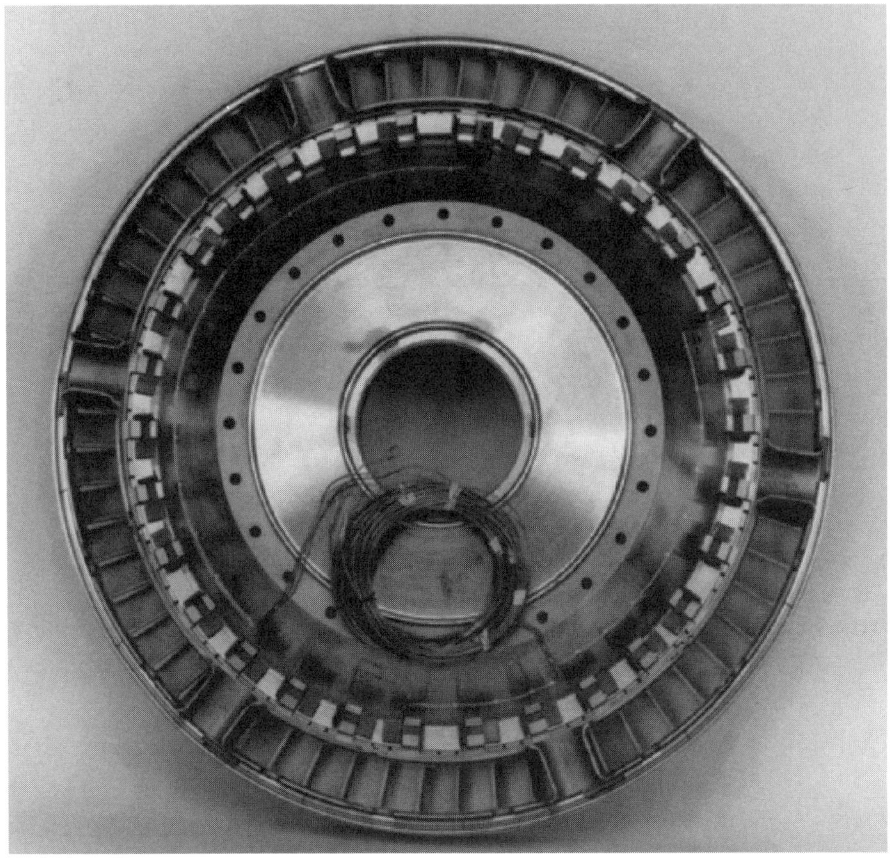

Figure 5.15 Ceramic Vane/Mount Assembly (Reprinted with permission from Rolls-Royce)

power) operation the highest calculated stress was 169 MPa (24.5 ksi) at 1182°C (2160°F). The maximum temperature was calculated to be 1238°C (2260°F) under steady state conditions with a corresponding stress of 150 MPa (21.8 ksi). Analyses revealed that the POS exceeded 99.99% for all 60 vanes under all operating conditions.

Stress rupture tests of the AS800 Si_3N_4 ceramic vane material with durations up to 10,000 hours have been conducted at Oak Ridge National Laboratory (ORNL) and the University of Dayton Research Institute (UDRI). Retained strengths of specimens measured after surviving the 4500 hours stress rupture test exceeded 490 MPa (71 ksi), which is substantially higher than the highest calculated emergency shutdown stress of 208 MPa (30.1 ksi) and the maximum steady-state stress of 169 MPa (24.5 ksi). The results of these tests provided the basis for conducting a long-term field test of the AS800 ceramic vanes.

Figure 5.15 shows the first-stage ceramic vane/mount assembly that replaces a similar metal vane assembly used for conventional 501-K turbines. The inner vane support and an outer sheet metal band that held the assembly together were common to both the ceramic and metal vane assemblies. All of the other parts were designed for retrofit of the first-stage ceramic vanes into Model 501-K turbines. To minimize mount loads and stresses, the vanes are not gripped or otherwise hard mounted but instead, are seated in circumferential grooves of the inner and outer mount segments. Ceramic cloth is used at the interfaces between the metal mount segments and the outer surfaces of the vane platforms. This compliant material reduces

Figure 5.16 Exxon 501-KB5 Turbine at Site (Reprinted with permission from Rolls-Royce)

ceramic contact loads and stresses from the mounts and also insulates the ceramic/metal interface. Insulation allows the mounts to operate at lower temperatures in their alloy working range and reduces the thermal gradient in the ceramic vane, which is the primary source of the steady-state stress.

Rig and Engine Proof Testing

All the AS800 silicon nitride vanes were evaluated using dimensional measurement, FPI, and microfocus radiographic x-ray inspection. The vanes were then proof tested in an atmospheric combustor burner rig for thermal shock testing. Visual, FPI, and radiographic inspection were conducted after burner rig testing.

The ceramic vane/metal mount assembly was proof tested in an actual engine for 22 hours. During disassembly two vanes were found to be slightly chipped and a number of the vanes had developed a thin oxide scale. Oxide scale formation observed on chipped platform surfaces after the other demonstration inspections indicated that the vanes operated many hours after chipping and that cracks had not propagated from the chipped surfaces.

Performance measurements conducted during proof testing indicated that for standard day conditions the turbine with ceramic vanes produced about 1.7% more output power than that for a run of the engine with metal vanes. Some of the improved performance for the ceramic engine probably resulted from the thinner

Figure 5.17 Ceramic Vanes Before (Right) and After (Left) 815 Hours of Operation in the Turbine (Reprinted with Permission from Rolls-Royce)

airfoil of the ceramic vanes compared to the metal vanes, the absence of trailing edge cooling air flow discharge, and the redesigned aerodynamic shape. Thermocouple measurements of the metal vane support temperatures adjacent to the ceramic vanes showed temperatures well below the working limits of the vane support alloy.

Engine Field Demonstration

The first-stage ceramic vane Phase I field demonstration took place in 1999 at the Exxon natural gas processing plant near Mobile, Alabama. Three identical Rolls-Royce Model 501-KB5 turbine skids provided power and process steam at this Exxon plant. Figure 5.16 shows one of the turbines in its skid enclosure at Exxon. The turbine with ceramic first-stage AS800 vanes was installed and operated under normal commercial service conditions at maximum continuous power, except for pre-planned teardowns and ceramic vane assembly inspections with removal of vanes for analyses. These inspections were originally scheduled at 200, 500, and 1000 hours. The Exxon turbine experienced an emergency shutdown (the highest vane stress condition) unrelated to the ceramic vanes at 525 hours and a full power water wash of its compressor with no adverse affect to the ceramic vanes. Analyses of vanes removed during the periodic inspections showed unexpected rates of material loss due to the oxidation of silicon nitride ceramics in the turbine operation environment. The rates of material loss were determined to be excessive for industrial turbine applications, which require part lifetimes of 20,000 hours or longer. Analyses of the vanes showed the need for additional protective measures (e.g. environmental barrier coatings—EBCs). Since the Phase I demonstration had achieved its objective of identifying redesign needs, this demonstration phase was discontinued after a total operation time of 815 hours (22 hours proof testing and 793 hours of field demonstration) for the Exxon turbine with ceramic vanes

Analyses of Field Tested Ceramic Vanes

Visual evidence of ceramic oxidation was apparent during the turbine inspections. As illustrated by Figure 5.17 the color of the AS800 vanes had changed from tan/gray (vane on the right) to white (vane on the left). A white powder covered the ceramic vanes. The white powder was analyzed by ORNL and found to be an oxide of the silicon and lanthanum from the intergranular phase of the ceramic.

The trailing edge thickness was measured for each vane removed during inspections and after demonstration to determine the oxidation loss of material at that location. The data suggest insignificant surface recession for times less than 200 hours. Analyses by Oak Ridge National Laboratory (ORNL)

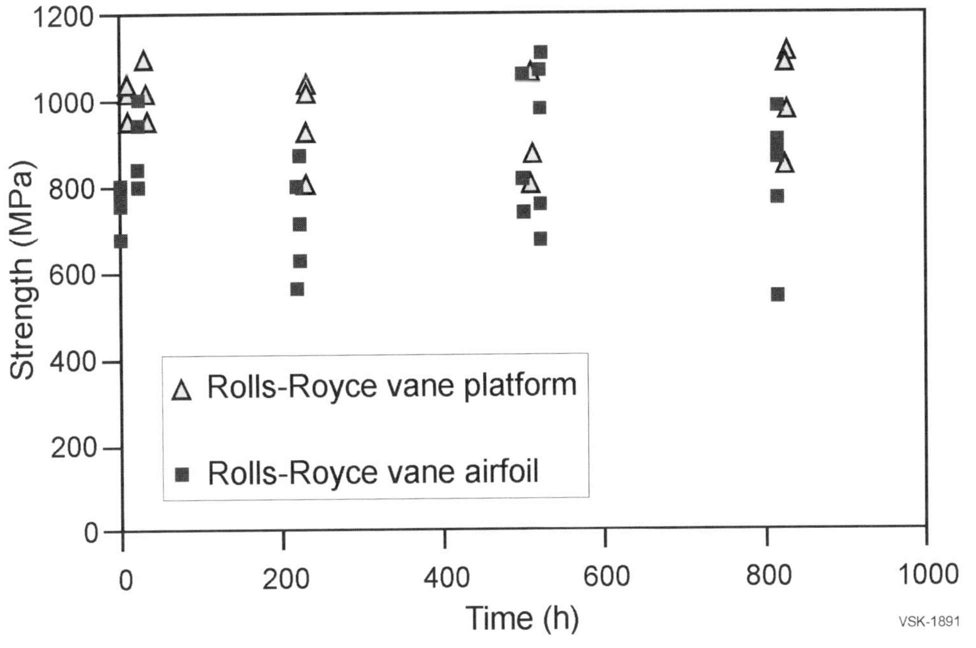

Figure 5.18 Biaxial Strength as a Function of Time for Specimens Dissected From the Engine Tested Vanes (Adapted from Ref. [6])

identified an initial oxide scale on the new ceramic vanes resulting from processing. [6] Apparently, this scale was much more adherent and impermeable to further oxidant penetration to the underlying substrate than the oxide scale formed in the turbine environment. It was observed that the vanes removed after 22 hours had very little measurable trailing edge recession when compared to pre-test measurements. However, the vanes removed after 522 hours had measured trailing edge losses up to 0.32 mm. All 60 vanes removed from the assembly were measured after the engine test was discontinued. Except for vanes with low oxidation levels because of their locations in atypical flow and temperature regions, the decreases in trailing edge mid-span thickness at 815 hours range from 0.13 mm to a maximum loss of 0.61 mm.

The loss of material is apparently due to the inability of the silicon nitride to form a protective silica scale in the high-pressure gas turbine environment. This result is quite different than that experienced during the thousands of hours of testing in low-pressure ratio machines in earlier programs. In the high-pressure environment, the normally protective silica scale reacts with water vapor to form gaseous $Si(OH)_4$, which is swept away by the high velocity gas. The competing processes of scale formation and scale volatilization ultimately lead to a steady-state value of the scale thickness as well as a linear recession of the ceramic substrate. Scanning electron microscopy (SEM) analysis confirmed that during the engine test little or no silica formed.

Several ceramic vanes were removed at the various operating intervals during field engine testing. Biaxial disks (about 6 mm dia. and 0.6 mm thick) were machined from the vane platform and airfoil regions of the engine run vanes and tested at ORNL in fast fracture. Figure 5.18 shows that the average retained strength is similar to the strength of the virgin material. After engine testing, the lowest retained strength of the material was around 500 MPa, which is much higher than the maximum operating stresses of the vanes. Large scatter is indicative of the variation in the combustor patterns resulting in varying oxidation of the AS800 silicon nitride material in the engine.

ROLLS-ROYCE CURRENT AND FUTURE PLANS

Work is ongoing under the ATS program to develop environmental barrier coatings (EBCs) for protection of the vanes. EBC coated vanes of AS800 (Honeywell Ceramic Components) and SN282 (Kyocera Industrial Ceramics Corporation) are being tested in the 501-KB5 engine at the Exxon site in 2001. As of July 2001 over 1800 operating hours had accumulated for this second engine field test. Some of the vanes are periodically removed for evaluation. It is hoped that the EBCs will significantly postpone the attack on the underlying silicon nitride substrate and thereby contribute to ceramic vanes becoming a reliable enabling technology for advanced turbine engines.

ACKNOWLEDGEMENTS

The authors wish to acknowledge the helpful suggestions of Ms. Debbie Haught of the US Department of Energy, Office of Power Technologies, Mr. Tom Sebestyen, formerly with the US Department of Energy Transportation Technologies, who were so kind to review this chapter. Unless otherwise indicated all figures were supplied courtesy of Rolls-Royce.

REFERENCES

[1] Helms, H.E., Heitman, P.W., Lindgren, L.C., Thrasher, S.R., *Ceramic Applications in Turbine Engines*, Noyes Publications, Park Ridge, New Jersey, 1986. From Helms, H.E., Heitman, P.W., Lindgren, L.C., Thrasher, S.R., "Ceramic Applications in Turbine Engines-Final Report," Allison Turbine Divison, for National Aeronautics and Space Administration, and Department of Energy, NTIS Report: DOE/NASA—0017-6, 1984.

[2] "Advanced Gas Turbine Technology Project," Final Report, DOE/NASA 0168-11, NASA CR-182127, August 1988.

[3] "Advanced Turbine Technology Applications Project/Hybrid Vehicle Turbine Engine Technology Support Project (HVTE-TS)," Final Report, DE-AC02-96EE50453 (DEN 3–336), December 1998.

[4] Wenglarz, R., Calcuttawala, S., and Pope, J.E., "Ceramic Vane Demonstration in an Industrial Turbine," ASME paper 97-GT-155, presented at the International Gas Turbine and Aeroengine Congress and Exposition, Orlando, FL, USA, June 2–5, 1997.

[5] Wenglarz, R.A., and Kouns, K, "Ceramic Vanes for a Model 501-K Industrial Turbine Demonstration," ASME paper 2000-GT-0073, presented at the ASME TURBO EXPO 2000, LAND, SEA, & AIR, Munich, Germany, May 8–11, 2000.

[6] Ferber, M.K., Lin, H.T., Parthasarathy, V., and Wenglarz, R.A., "Degradation of Silicon Nitrides in High Pressure, Moisture Rich Environments," ASME paper 2000-GT-0661, presented at the ASME TURBO EXPO 2000, LAND, SEA, & AIR, Munich, Germany, May 8–11, 2000.

CHAPTER

6

Ceramic Gas Turbine Development at Hamilton Sundstrand Power Systems

Tibor Bornemisza PE

Energy Technology Applications Inc., San Diego, California, USA

Anthony Jones

Hamilton Sundstrand Power Systems, San Diego, California, USA

INTRODUCTION

The evaluation of structural ceramic materials for small gas turbine components began in 1972 at Solar Turbines, the predecessor of Hamilton Sundstrand Power Systems, with the design and testing of hot pressed silicon nitride vanes in the 100 HP Titan gas turbine [1]. After Sundstrand corporation acquired the Solar Turbomach division in 1985, which became Sundstrand Power Systems the ceramic turbine wheel demonstration program in the Gemini engine, which started at Solar (see chapter 7 of Volume I) was continued at the division. With the acquisition of Sundstrand Co. by United Technologies the division was renamed to Hamilton Sundstrand Power Systems.

Although the original intent was improved erosion resistance, the favorable high temperature properties of the new structural ceramic materials offered the potential of higher turbine operating temperatures, thus increasing power output and efficiency. The continuing demand for higher power density small gas turbine engines with better fuel economy mandates increased turbine operating temperatures, which exceed the high temperature strength capabilities of the nickel-cobalt super-alloys. Air cooling of the components exposed to the hot gases, although widely used on large gas turbines, has not proved economical on small compact auxiliary power unit (APU) gas turbines.

APU start system weight can be a significant fraction of the total installed APU weight, especially if fast starts are demanded under Arctic environments. An important advantage of ceramics vs. superalloys

Figure 6.1 Ceramic Turbine Wheel Demonstrated in the Gemini Engine (Reprinted with permission from Hamilton Sundstrand)

is the lower density, which may result in up to 60 percent lower ceramic turbine wheel weight and mass moment of inertia. Design studies show that total engine weight is significantly affected by the mass moment of inertia of the rotor assembly due to the secondary effects on the size and weight of the APU start system and wheel burst containment shroud. It is shown that both start system weight, and rotor containment armor weight are proportional to the product of rotational speed squared and rotating assembly inertia. Significant weight savings are therefore feasible with rotating assemblies using lower density materials such as ceramics and composites. On the other hand, for certain applications the fast availability of APU power can be very critical.

CERAMIC COMPONENT DEMONSTRATIONS IN THE T-20G-10C ENGINE

In 1985 a long range Ceramic Technology Development Program was initiated with the objective of developing ceramic turbine components. During the first phase of this company-funded effort a 112 mm (4.4 inch) diameter radial inflow ceramic turbine wheel (Figure 6.1) was designed and tested in the 40 HP Gemini gas turbine. Two T-20G10C silicon nitride (Kyocera SN 220M) rotors were tested at full speed (93,880 RPM) to 1038 °C (1900 °F) turbine inlet temperature (TIT). The first wheel had completed nearly 100 hours of endurance testing at 1010 °C (1850 °F) TIT before it failed due to rubbing of the metal seal plate against the ceramic wheel backface. This problem was corrected by increasing the seal plate clearance before testing the second wheel, which has completed fifty cold starts to full speed and fifty full load-no load cycles. Testing was continued with an endurance test at full load. The second wheel failed after 155 hours of test-

TABLE 6.1 T-20G20C GEMINI CERAMIC TURBINE DESIGN PARAMETERS

Ambient Temperature	°C	15 (59 °F)
Air Flow	kg/sec	0.3 (0.67 lb/sec)
Compressor Pressure Ratio	—	3.58
Turbine Inlet Temperature (TIT)	°C	1204 (2200 °F)
Exhaust Gas Temperature (EGT)	°C	927 (1700 °F)
Estimated Power Output	kW	45
Ceramic Components	Si_3N_4	Silicon Nitride
Combustor Liner	—	HASTELLOY X

ing at full speed and a TIT of 1038 °C (1900 °F). Carbon ingestion (foreign object damage to the blades) was identified as the cause of failure

The test results with the ceramic wheel have confirmed the potential of silicon nitride as a high temperature structural material for non-flight-critical, unmanned turbine applications for limited life expectancy. The need for improved ceramic components and better design integration between the ceramic and the metal hardware were also demonstrated. The objectives of the next program phase were to redesign and test the new ceramic components at 1204°C (2200°F) TIT and to demonstrate faster starts vs. the metallic engine. Additional program tasks included the development of new design and analytical techniques, the introduction of non-destructive and destructive material and component test methods, and the development of acceptance procedures for ceramic components. The design approach and the test results with the silicon nitride turbine wheel and stationary components in the engine hot section are discussed below.

T-20G20C Engine Design

The T-20G10C Gemini engine adapted for the ceramic component development program is a 3.5:1 pressure ratio simple cycle gas turbine with a rated shaft power output of 30 KW. The radial compressor and turbine wheels are installed back-to-back at the end of the shaft in an overhung arrangement. The shaft is supported on two angular contact ball bearings, which are spring loaded axially to assure stable operation at full speed (93,800 RPM) and during engine starts and shutdowns. The location of the turbine rotor at the far end of the shaft assures a relatively cold operating environment for the bearings.

The ceramic engine—designated T-20G20C—was designed to operate at a rated TIT of 1204°C (2200°F), which is 222°C (400°F) higher than the maximum operating temperature of the metallic engine. The design parameters of the T-20G20C engine are shown in Table 6.1. For the high temperature operation the entire hot section of the engine was redesigned. The ceramic (Si_3N_4) components—turbine wheel, turbine nozzle, nozzle shroud and thermal shield—are shown shaded in Figure 6.2.

The radial inflow turbine wheel (Figure 6.2) with thirteen integrally cast blades had a tip diameter of 112 mm (4.4 inches) resulting in a moderate 550 m/sec (1800 ft/sec) tip speed. The boreless wheel was supported at the backface from the cantilevered end of the shaft. The shaft was made of a high temperature alloy with good corrosion resistance and a low coefficient of thermal expansion. The ceramic wheel stub shaft was fitted into the hollow shaft with a combination shrink fit-braze joint, which was designed to transmit the wheel torque, the bending moment and the forces ("g"-loads, thrust, rotor dynamics) acting on the cantilevered wheel.

The ceramic turbine wheel was designed to the same structural criteria, which are nominally used for the design of metal turbine rotors. The rotor had to survive a 25% overspeed condition and the thermal stresses due to the temperature gradients at all operating conditions, especially during rapid starts, load transients and instantaneous shutdowns.

The structural analysis of the turbine wheel was based on finite element techniques with the mass inertia and thermal gradient loads applied at steady state and transient conditions. The thermal loads were based

Figure 6.2 T-20G20C Ceramic Turbine Wheel (Reprinted with permission from Hamilton Sundstrand)

on the temperature distributions calculated with a finite difference computer code for the steady state and the transient operating condition. The accurate evaluation of the thermal stresses required a detailed analysis of the time-dependent temperature and stress distributions during engine starts, shutdowns and rapid load changes. For example, investigation of the stress distribution in a radial inflow turbine wheel during a typical engine start cycle to 100% speed and full load has shown the maximum stress to occur twenty-five seconds after start in the wheel hub area. Seventy seconds after start, the temperatures were practically stabilized and the maximum stress—now at the wheel back face—was 22% lower. The structural analysis work also included the calculation of the blade vibratory frequencies to assure that the blades will not fail due to vibratory excitations.

Since the highest stresses in the static structure are caused by the thermal gradients, the static components were also subjected to time-dependent thermal and stress analysis. The design of the static components provided for the restriction-free movement of the metal and ceramic components to prevent any stresses due to the differential thermal expansion of these parts.

In order to minimize the number of parts and hardware cost, the turbine nozzle and wheel shroud were designed as a single component integral with eleven (11) cantilevered nozzle vanes. To center the nozzle and provide a uniform shroud clearance for the turbine wheel, three radial lugs were placed 120° apart at the outer diameter of the nozzle. The lugs were extended into three matching slots in the nozzle shroud (Figure 6.3). The nozzle shroud itself was centered with a similar radial slot-pin arrangement from the aluminum diffuser shroud.

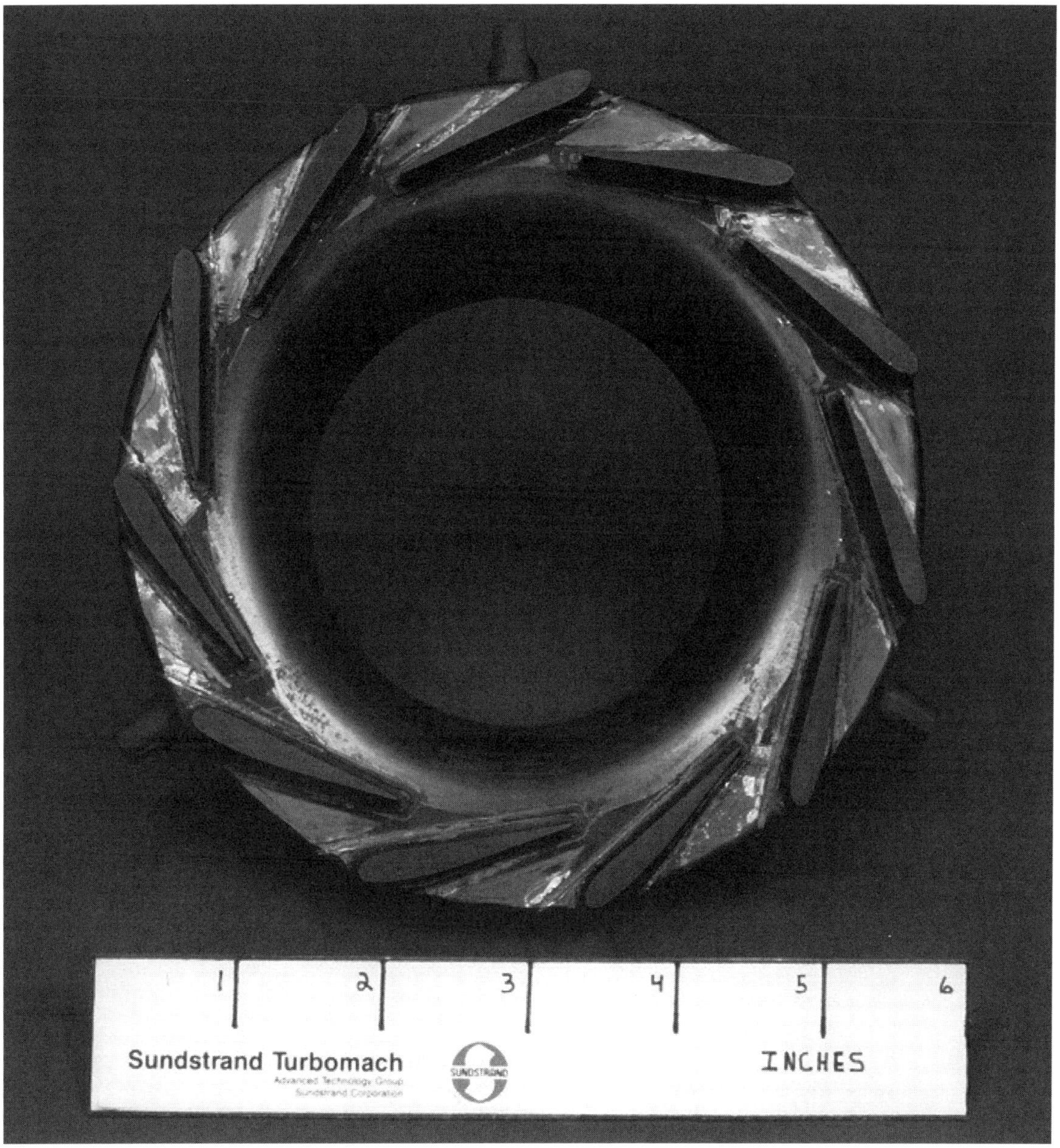

Figure 6.3 T-20G20C Ceramic Turbine Nozzle (Reprinted with permission from Hamilton Sundstrand)

The ceramic thermal shield was made of three segments to facilitate installation and to reduce the thermal stresses. The ceramic components before final assembly are shown in Figure 6.4

An important component of the new engine hot section design was the new combustor. In the simple cycle low pressure ratio engine configuration, the wall of the combustor liner is effectively cooled by the relatively cold compressor discharge air. This consideration led to the material selection of a high temperature alloy (Hastelloy X) instead of a ceramic material for the combustor liner in spite of the high

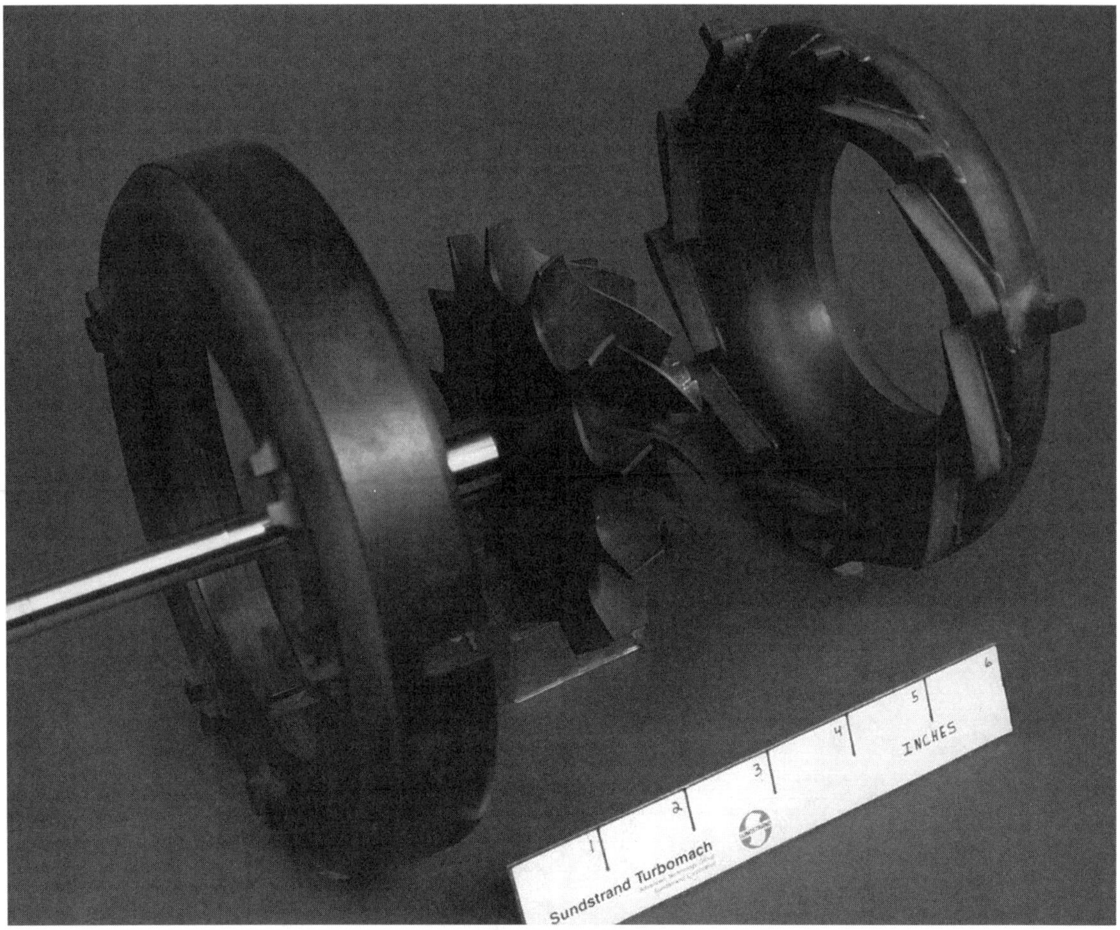

Figure 6.4 T-20G20C Ceramic Turbine Components (Reprinted with permission from Hamilton Sundstrand)

turbine inlet temperature. The interfaces between the combustor liner and the ceramic parts were designed to allow for the relative thermal displacements. The new annular combustor design with four injectors was expected to improve combustion efficiency vs. the old can-type (cylindrical) combustor, thereby reducing carbon formation.

Ceramic Component Design Considerations

The design methodology of ceramic turbine components is strongly influenced by the characteristic material properties of ceramics. For detailed information on specific design issues of ceramics the reader should refer to the references in this book.

Monolithic ceramic materials are brittle and contain randomly distributed flaws. When a ceramic structure is stressed the random flaws result in large stress concentrations. Due to the low ductility, the high stresses around flaws will result in cracking. If the local stress exceeds a threshold level, the crack will grow rapidly and may result in the complete fracture of the part (catastrophic failure). Due to the random distribution of the flaws in the material, the strength of ceramic materials is subject to statistical distribution described by the statistical model attributed to Weibull [3]. Since the failure of the material is caused by the flaws, a higher volume or surface exposed to the maximum stress has a higher number of flaws, i.e., potential sources of cracks. Thus, the probability of failure under uniform stress distribution is strongly volume and surface area dependent.

Reliability analysis of ceramic components involves the rigorous application of Weibull statistics using finite element stress analysis techniques combined with statistical fracture strength data for the material and the selected fracture model [4–6].

The low ductility (brittleness) of ceramic materials mandates designs which avoid stress concentrations within the ceramic components. Special attention has to be paid to the design of the interfaces between components (ceramic/ceramic, ceramic/metal) because the lack of ductility may result in excessive localized contact stresses, leading to cracks and fracture of the ceramic parts. To prevent excessive thermally induced loads, interfaces between the components should not restrict the relative thermal expansion, which may be significant, especially between metal and ceramic parts, due to the low coefficient of thermal expansion of most ceramic materials.

The structural properties of ceramics are affected by many factors such as powder quality (purity, size distribution), component size (volume, surface, area) and shape, and the manufacturing processes used to make the component. For example, improper heat treatment or machining (grinding) may result in excessive residual stresses or "strength-reducing" surface cracks.

Slip cast and injection molded ("green") ceramics may shrink up to 20% during sintering. In theory, the shrinkage should be uniform throughout the part; however, in practice this may not always be the case. Non-symmetrical or complex parts with substantial thickness variation may suffer excessive distortions during sintering, which have to be eliminated by extensive tooling development or component redesign. Because the production processes have a major effect on the geometry, structural strength, and the cost of the finished ceramic component, it is advisable to review the preliminary design with the ceramic manufacturer to assess producibility and cost options.

The structural analysis of ceramic components requires a through knowledge of the loads and the environmental conditions (e.g., temperature) under which the component operates as well as the expected material properties which characterize the various material failure mechanisms.

Fast fracture can be characterized by the fast fracture strength as a function of temperature. The fast fracture strength of ceramic materials is usually characterized by the modulus of rupture (M.O.R.) strength established by four point bending tests of a number of small rectangular test bars. An important measure of material quality is the shape of the Weibull function describing the strength distribution. This is given by the shape parameter (m), generally referred to as the Weibull modulus. For a representative characterization of the strength properties, the test specimens should be made from the actual components with the appropriate surface finish.

Slow crack growth and creep failure are time and temperature dependent material fatigue failure modes and can be characterized by the stress rupture strength, defined for various temperature levels as time to failure. It is important to note that environmental effects (e.g., exposure to corrosive atmosphere) can substantially accelerate this failure mode. Slow crack growth is usually characterized by the crack growth velocity at a given exposure temperature as a function of stress intensity. Creep failure is both temperature and stress dependent. Creep is characterized by the strain rate as a function of time at various temperature and stress levels.

Crack growth failure can also occur due to cyclic fatigue. Rotating test specimens and cyclic tensile loads show that ceramics can fail under cyclic loading. Cyclic failure is characterized by the cyclic fatigue strength defined for a given number of load cycles at various temperature levels [7]. Environmental failure is primarily associated with oxidation, erosion, corrosion, and phase transformation. This failure mode is also temperature and time dependent. Oxidation is characterized by weight gain in air at a given temperature as a function of time.

The structural failure modes of a ceramic turbine wheel are shown summarized in Table 6.2. The check marks represent the potential failure modes induced by the loading mechanisms listed in the left-hand column and the corresponding failure mechanisms of the ceramic material. Addressing these failure modes

TABLE 6.2 CERAMIC TURBINE WHEEL STRUCTURAL FAILURE MODES

	Ceramic Turbine Wheel Structural Failure Modes				
		Slow Crack Growth			
Loading Mechanism	Fast Fracture	Low Cycle	High Cycle	Creep Fracture	Environmental Damage
Centrifugal Force Load	X	X		X	
Rotor Dynamics	X				
Blade Rub	X		X		
Aerodynamic			X		
Foreign Object Damage	X			X	
Thermal Gradient	X				
Thermal Shock	X	X		X	X
High Temperature					X
Combustion Products		X*			X

* Crack growth enhancement due to oxidation of non-oxide ceramics.

during the design and test phases of the ceramic component development program requires a substantial investment in time and resources. The analysis of the long term life expectancy is especially difficult due to the scarcity of the material data.

The assessment of ceramic component life under continuous loading at high temperature requires the use of linear elastic fracture mechanics analysis techniques. The analysis is based on the assumption that total life to fracture depends on the initial stress intensity factor (K), the critical stress intensity, i.e., fracture toughness of the material (K_{IC}), and the velocity of crack propagation in the material, plotted for the corresponding temperature as a function of stress intensity. Life to failure estimates can be made based on integration of the crack growth velocity using the initial stress intensity and the fracture toughness (K_{IC}). Estimate of the initial flaw size can be made based on fast fracture test results which are preferably obtained from hardware load tests, e.g., burst spin testing in case of a turbine wheel [8–10].

The exact calculation of the expected component life has to be based on the stress and temperature distribution in the component as well as the statistical distribution of the material characteristics. Time dependent variations in the component loads and temperatures associated with the duty cycle of the component impose the need to assess possible limitation of the life due to cyclic fatigue [10].

T-20G20C Gemini Ceramic Turbine Engine Test Results

The ceramic components were inspected before engine assembly. The Fluorescent Penetrant Inspection (FPI) microscopic survey revealed no cracks or surface damage. The dimensional inspection revealed no discrepancies.

Engine testing (Figure 6.5) began with engine starts to 100% speed no-load to perform development work on the fuel control system. This involved the optimization of the fuel flow schedule to improve the start capability of the engine. The engine temperature was gradually increased to the design TIT of 1204°C (2200°F). Inspection after twenty minutes of testing at the design conditions did not reveal any problems with the ceramic hardware. Testing was resumed and several fast starts were demonstrated.

Engine starts to 100% speed were obtained in 2.5 seconds using the optimized fuel flow schedule and the standard Gemini electric starter. This compared very favorably against the 8.5 seconds required to start the engine with the metallic turbine components and the standard fuel system.

The engine power output was initially below the expected value due to some leakage problems in the exhaust area. This problem was corrected and the predicted power output of 45 KW was achieved.

Figure 6.5 T-20G20C Engine with Ceramic Components on the Test Stand (Reprinted with permission from Hamilton Sundstrand)

The engine over-temperature capability was also evaluated by running at an average TIT of 1274°C (2325°F) for fifteen minutes without any damage to the ceramic or metal hardware. The engine had completed over one hundred starts to full speed and 15 hours of operation at the maximum design temperature of TIT = 1204°C (2200°F.) Overspeed capability was demonstrated on two occasions at 108% speed with no damage to the hardware. Thermal shock testing included fifty full load/no-load cycles during which the turbine inlet temperature (TIT) was cycled between 677°C (1250°F) and 1204°C (2200°F). The engine had completed a 198 hours endurance test at 1204°C (2200°F) (full power) before the blades of the turbine wheel failed due to carbon ingestion.

Despite the Foreign Object Damage (FOD) suffered by the turbine blades the T-20G20C GEMINI ceramic turbine engine test was considered a success because it demonstrated the technical feasibility of the structural ceramic material at temperatures 177–205°C (350–400°F) higher than the capability of the metallic hardware.

The demonstration resulted in the establishment of procedures for the design, inspection and parts handling specifically tailored for ceramic material properties. The design process revealed the need for the development of a user friendly reliability and life assessment methodology especially for the preliminary design effort. The successful demonstration opened the road for potential applications of ceramic components for non-flight-critical and unmanned turbine applications particularly for non-continuous or limited life operations such as turbine starters and expendable turbojets.

SPSLIFE: A USER FRIENDLY LIFE ASSESSMENT CODE FOR CERAMIC COMPONENTS

Introduction

The design and analysis of the experimental ceramic turbine components at Hamilton Sundstrand Power Systems have highlighted the need for an expeditious approach to the statistical life assessment analysis. The numerous loading mechanisms and failure modes of a ceramic turbine wheel have been discussed earlier and highlighted in Table 6.1 showing the failure modes of an integrally bladed monolithic ceramic turbine rotor. The analytical work to assess the reliability of highly stressed ceramic components in the potential failure modes frequently requires several iterations that are extremely time consuming. The structural design optimization of a turbine wheel involves design changes that may require also revisions of the aerodynamic configuration as well as the finite element model. The designer of a ceramic turbine wheel has to evaluate a considerable number of finite element files containing thousands of numbers describing the stress and temperature distribution within the wheel during the start, shut-down, on-and off-loading transients and at steady state operating conditions. These files are processed through fast fracture and life assessment codes using Weibull statistical analysis. The codes calculate the percent probability of success of the design for the selected failure mode and material. Hamilton Sundstrand Power Systems has been using the structural ceramics reliability codes developed by NASA for the statistical assessment of failure probability [4, 5, 11].

In order to expedite the structural analysis of ceramic components Hamilton Sundstrand Power Systems has developed a proprietary computer code called "SPSLIFE," which can substantially reduce the time spent on the design assessment of ceramic components. The life assessment computations for the various failure modes are performed using the structural analysis and materials files as input data. A number of "menus" are incorporated to request user input data and to guide the user through the problem definition process. The user has the capability to select existing material data files or to create new ones. The computational modules are based on the analytical approaches currently used in industry. The modular construction of the code facilitates making revisions or adding new modules. Graphic display is used to provide a visual summary of the life assessment calculations for the various failure modes. The output files generated provide a detailed review of the results which is helpful for design optimization. The features of the code are illustrated by an application example using a patented design of a ceramic monorotor for the Hamilton Sundstrand Power Systems TJ-50 small gas turbine engine.

Program Objectives

The basic objectives of SPSLIFE can be summarized as follows:

- User friendly. Provide a guided user driven analytical tool for the "non-ceramic expert", but knowledgeable design engineers and stress analysts to perform the life assessment analysis of ceramic components during the preliminary design and structural design optimization phases of a particular project.
- Modularized construction to facilitate expansion of the capabilities and accommodate changes and revisions to the individual modules.
- The visual display of the analysis results should be clear and easy to interpret.
- The output files should be organized to facilitate the detailed review of the results for the purpose of the design optimization.
- Provide for the easy revision of the data files to different levels of users.

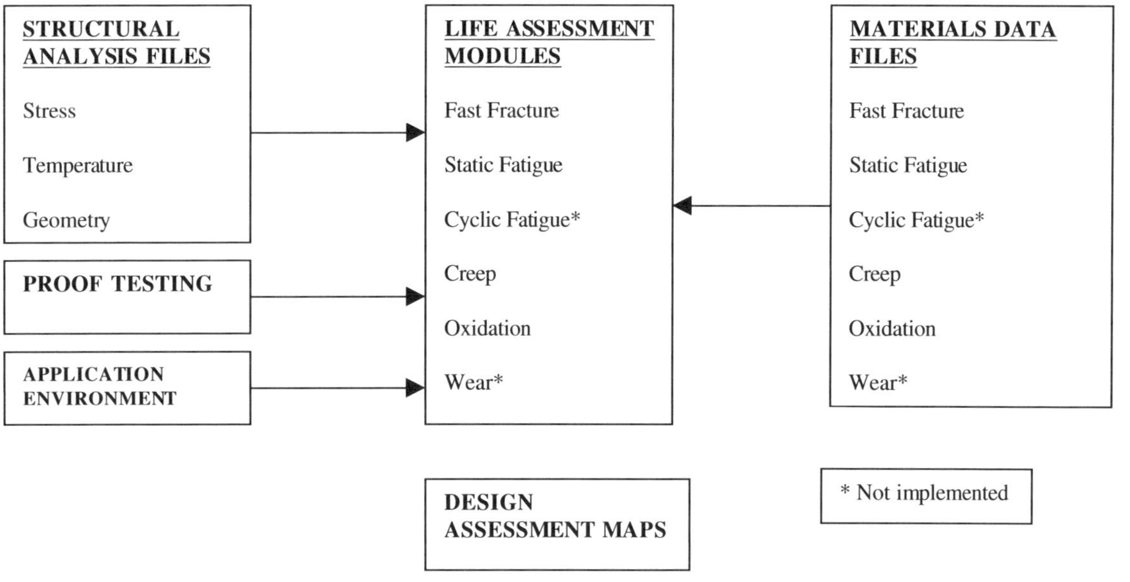

Figure 6.6 SPSLIFE design concept (Reprinted from Ref. [17])

It should be pointed out that it was not the objective of this effort to develop new analytical approaches to ceramic component life assessment. However, the capability was provided to modify, exchange or add new computational modules in a relatively easy way.

Program Structure

The conceptual design of the SPSLIFE modularized computer code for the life assessment of ceramic components is shown in Figure 6.6. The program links ANSYS structural analysis files and materials data files with life assessment modules for different modes of failure to provide quantitative estimates of component life and survival probability. In addition, the program produces presentation quality design assessment maps depicting stress and temperature distributions for the components together with design envelopes for fast fracture, slow crack growth, creep, and oxidation. Cyclic fatigue, proof testing and wear modules are also planned.

SPSLIFE is a user-friendly interactive program. Helpful menus guide the user through the program. The "Welcome to SPSLIFE" menu allows the user to select whether the objective is to define a new problem, to execute an analysis, or to plot analysis results. The "Application Defaults" menu allows the user to specify the problem, select a design life, materials data, and related parameters. The "Applications" menu permits a choice of different analysis types: fast fracture, slow crack growth, creep, oxidation, etc. The "Plot" menu permits graphical superposition of results for different analysis types on the stress-temperature plot, resulting in a Design Assessment Map.

SPSLIFE Design Assessment Maps are component-specific, and provide a graphical overview of the design evaluation results. The maps are based on the "fracture map" concept discussed in the ceramics literature [12] and follow the general layout of Reference [13]. Design envelopes for different failure modes are superimposed on a plot of maximum principal stress versus temperature for all the nodes in the finite element model. Specific equations and procedures used in SPSLIFE are proprietary to Hamilton-Sundstrand Power Systems.

"SPSLIFE" Analysis Modules

The Fast Fracture Model uses enhanced uniaxial Weibull methodology [14], distinguishing between three different categories of flaws: internal or "volume" flaws, as-fired surface flaws and machined surface flaws.

The program requests separate materials data file entries for each category of flaws to ensure that proper attention is given to surface properties. The data files are selected from a materials library that contains background information on each data set. Three fast fracture design envelopes are shown on the Design Assessment Map corresponding to three different survival probabilities; any nodes outside of these contours show that the design goal is not being met. For reference, independently-derived component fast fracture survival probabilities using the NASA CARES and CARES/LIFE codes are shown in the top right hand corner of SPSLIFE Design Assessment Maps.

The slow crack growth or static fatigue module uses Evans-Wiederhorn methodology [15], again distinguishing between internal flaws, as-fired surface flaws and machined surface flaws.

Slow crack growth contours can then be drawn for σ_{sf} for any pre-selected component failure probability F_x. Again, three different design envelopes are shown corresponding to three different survival probabilities; for reference, independently derived component slow crack growth survival probabilities using the NASA CARES/LIFE code are shown below the corresponding fast fracture probabilities (in the top right hand corner of the Design Assessment Map).

The Creep-Fracture module uses Monkman-Grant methodology [16]; a design envelope is shown on the Design Assessment Map, and the calculated creep life corresponding to actual stresses and temperatures is tabulated on the right hand side of the map. The oxidation module uses the fracture mechanics approach of Soma et al. [13]. A design envelope is shown corresponding to the selected design life, and the calculated oxidation life is tabulated on the right hand side of the map. For a detailed description of the analytical approach the reader is referred to Ref. [17].

Applications

SPSLIFE was originally applied to the monorotor of the TJ-50 thrust engine, which was designed with ceramic hot section components. The ceramic components of the TJ-50 are shown on Figure 6.7. The application of SPSLIFE to the TJ-50 monorotor is discussed below in greater detail. The program has been used also for preliminary design evaluation and life assessment of several ceramic components in the DOE/Solar Ceramic Stationary Gas Turbine (CSGT) Program, including two blade designs, one nozzle design and three combustor liner designs. Reference [18] documents the results for the CSGT components.

TJ-50 Monorotor

A 3-D finite element model (Figure 6.8) of the ceramic monorotor was built using brick elements, and triangular shell elements were placed on the surface. Separate files were written in ANSYS for the as-fired and machined surface elements. These files are used by SPSLIFE to allow for property differences between as-fired and machined surfaces.

Figure 6.9 shows the Design Assessment Map for the monorotor for a turbine inlet temperature of 1121°C (2050°F). The temperatures and stresses shown on the node plot are for steady state operation at a speed of 130,000 rpm. The analysis was run for a design life of 4 hours and using a material designated as material A. The rotor was analyzed in the normally machined condition (Condition 1), i.e. the rims and undercut areas were in the machined condition, and the hubs and airfoils were in the as-fired condition. The map shows that creep and oxidation are not a problem for design. However, selected design goals of 90, 95, and 99% survival probability are not met in fast fracture and in slow crack growth. In fast fracture

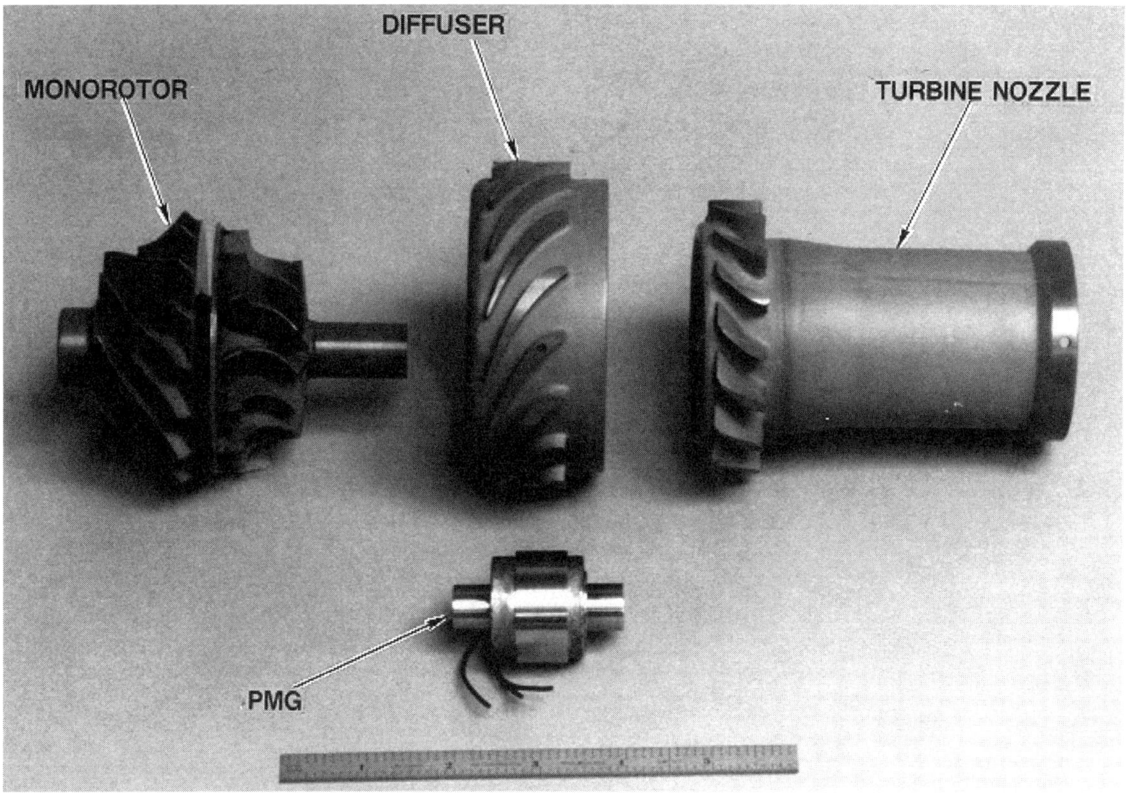

Figure 6.7 TJ-50 Thrust Engine Ceramic Components (Reprinted with permission from Hamilton Sundstrand)

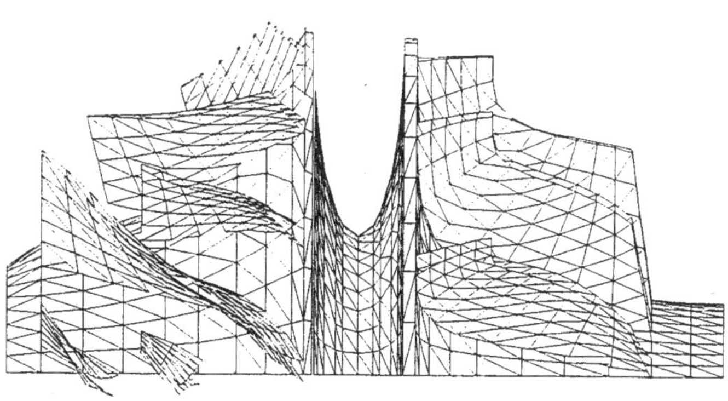

Figure 6.8 Ceramic Monorotor Finite Element Analysis Model (Reprinted from Ref. [17])

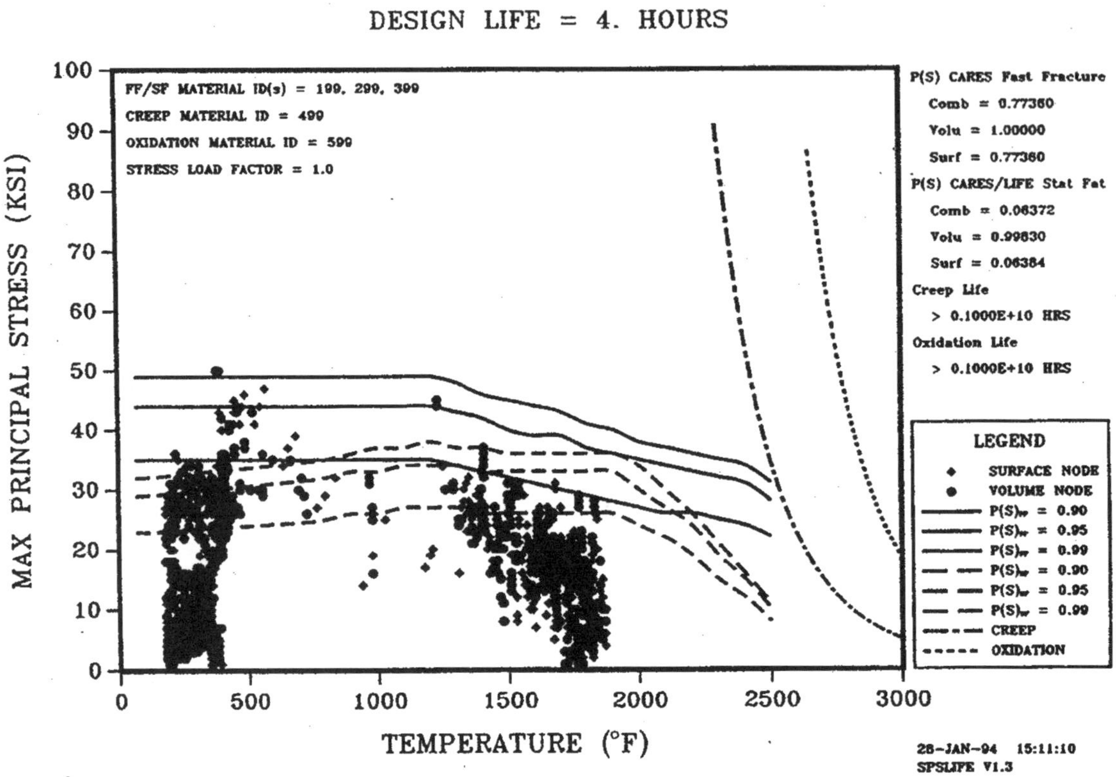

Figure 6.9 TJ-50 Monorotor Design Assessment Map, Material A, Condition 1, Design Life = 4 Hours (Reprinted from Ref. [17]). Temperature conversion: °C = (°F – 32)/1.8.

the three solid lines represent component survival probabilities of 90, 95, and 99%, respectively, and the node plot crosses the highest of these lines, indicating a survival probability of less than 90%. The top right hand corner of the map shows the survival probability determined by the NASA CARES program to be 77%. In slow crack growth, the node plot crosses the three dashed lines, indicating that the lowest design goal of 90% is far from being met. The map shows the slow crack growth survival probability determined by the NASA CARES/LIFE program to be 6%.

Figure 6.10 is a Design Assessment Map for the same component and material, but with better surface properties (Condition 2). For this case, all of the surface elements are assigned to machined surface properties shown in Table 6.2. In fast fracture, all three solid lines are now above the node plot, indicating that the highest selected design goal of 99% fast fracture probability survival has been met. CARES shows a survival probability of 99.52%. In slow crack growth two of the lines are crossed, resulting in an estimated survival probability of 99.52%. In slow crack growth two of the lines are crossed, resulting in an estimated survival probability of 94%. CARES/LIFE shows a survival probability of 88.34%.

To improve the calculated probability of survival the stresses at the highest stress locations on the cast surfaces in the compressor needed to be reduced by re-design, or by improved material properties. Due to the manufacturing complexity of the monorotor combined with the difficulty of reducing the high stresses area it was the decided not to pursue the ceramic monorotor concept and to revert to separate wheels. The ceramic component development for the TJ-50 thrust engine was continued with a metallic compressor and a ceramic turbine wheel.

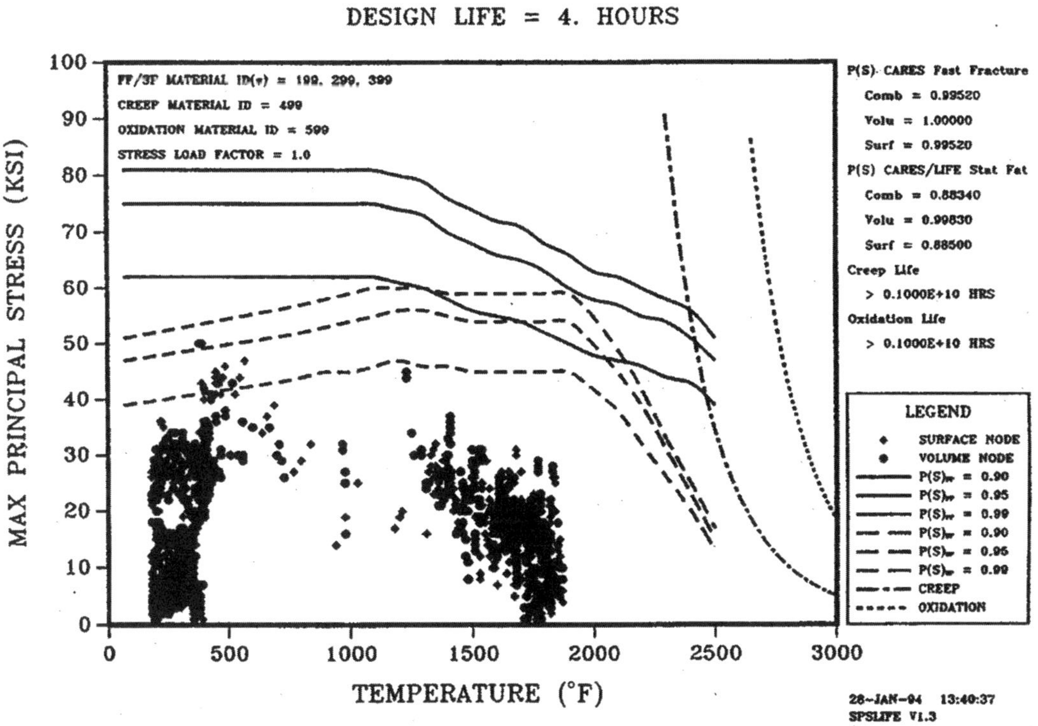

Figure 6.10 TJ-50 Monorotor Design Assessment Map, Material A, Condition 2, Design Life = 4 Hours (Reprinted from Ref. [17]). Temperature conversion: °C = (°F – 32)/1.8.

DESIGN, ANALYSIS AND TESTING OF CERAMIC COMPONENTS FOR AN ADVANCED MILITARY EXPENDABLE TURBOJET ENGINE

Introduction

The objective of the effort was to demonstrate insertion of low-cost, foreign-object-damage (FOD) resistant silicon nitride components in a military application. The application is a growth version of Sundstrand Power System's existing 222 Newton thrust TJ-50 miniature turbojet [19, 20].

In the early '90s Sundstrand designed a miniature turbojet engine for a new generation of small, affordable air vehicles in the 45 Kg weight class. This work was sponsored by DARPA and resulted in the development of a 222 N thrust class engine, the TJ-50 with a nominal diameter of 100 mm.

The thrust of the TJ-50 can be increased to the required level by increasing turbine inlet temperature. The advantage of simply increasing turbine inlet temperature is that the existing turbomachinery can be employed, thereby maintaining a simple, low cost design. However the turbine inlet temperatures being considered exceed the capabilities of simple, uncooled metallic hot section components. Hence the desire to employ ceramic parts.

Figure 6.11 shows the relationship between thrust and turbine inlet temperature for the TJ-50. The baseline engine operates at 1120°C. Based on known material properties and prior experience, a turbine inlet temperature of 1427°C was initially selected for the growth engine with a thrust goal of 356 N in an engine 115 mm in diameter.

A radial inflow ceramic turbine nozzle and rotor were designed to replace the exsisting metal components. The goal was a 305°C increase in turbine inlet temperature and 50% increased thrust. The material

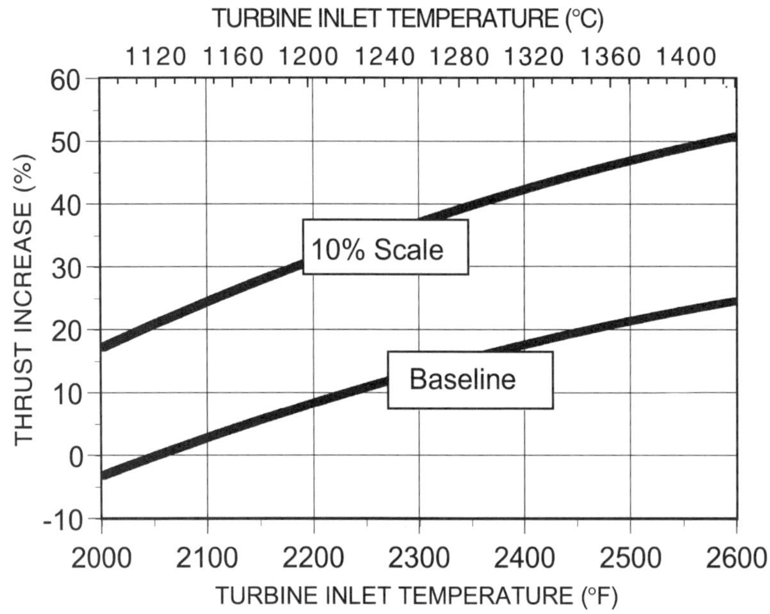

Figure 6.11 Thrust Increase vs. Turbine Inlet Temperature (Adapted from Ref. [20] with Permission from the American Ceramic Society)

selected was Kyocera's SN252. Kyocera fabricated test rotors using the hybrid molding process. They also developed a high temperature brazed rotor/shaft joint and made test nozzles by bisque machining. Under a separate element of the AMP program, Cercom Inc. developed surface strengthening methods aimed at improving the rotor's FOD resistance.

The biggest problems associated with ceramic parts, preventing their adoption into production of military, expendable engines have been F.O.D. sensitivity, high cost, manufacturing lead time, and lack of control over critical cast dimensional features. The primary objective of the AMP program was to address these issues through development of the hybrid molding process for a radial rotor. This development was performed by Kyocera Industrial Ceramics at their Vancouver, WA facilities.

A complete TJ-50C prototype engine was assembled and baseline testing with a metallic turbine successfully completed. Engine quality rotors were spun-tested to a speed of 142000 rpm (20% higher than the engine maximum operating speed) and have been tested in the engine in 1999.

TJ-50 Growth (Ceramic) Engine Design

Configuration. The baseline TJ-50 employs a single stage mixed flow compressor and single stage mixed flow turbine. The two rotors are combined into a single nickel alloy casting known as a monorotor. For the growth engine a more conventional split rotor arrangement was chosen. Silicon nitride was selected for the turbine rotor and a metallic material was used for the compressor. The standard heat shield/seal plate between the rotors was eliminated for lower cost. Gas path sealing is performed at knife-edge seals on the disk rims.

Figure 6.12 compares the configuration of the baseline and growth engines. A metal shaft is attached to a stub shaft on the ceramic rotor and the compressor rotor is an interference fit on the shaft. Ceramic material was also chosen for the turbine nozzle so as to avoid complex, costly cooled metal designs. The interfaces between the ceramic and metal parts of the engine received the most concentrated design effort

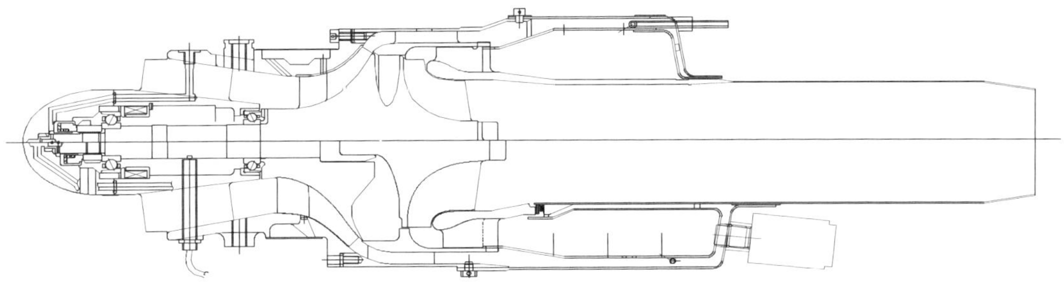

Figure 6.12 Cross Sections of Baseline and Growth TJ-50 (Reprinted from Ref. [20] with Permission from the American Ceramic Society)

since experience has shown these to be the most troublesome areas during actual engine service, once the basic rotor design has been verified.

The basic flowpath was unchanged from the baseline to the growth engine, although the opportunity was taken to scale up by 10% since studies of the installation of the engine in a candidate vehicle showed that this could be accommodated relatively easily.

The combustor design was modified to generate the additional temperature rise, maintaining the same concept as the baseline version. One change considered essential was the reintroduction of an inner liner wall which provides an annulus for the flow of cooling air around the liner and along the turbine exhaust pipe. Conventional metallic materials were employed for the tailpipe for cost reasons, and careful design was needed to tolerate the extremely hot exhaust gases (1200°C+). With its additional heat release rate, in essentially the same volume as the baseline, the TJ-50C combustor is easily the world's most highly loaded design.

Ceramic Materials and Processes

The hybrid molding process was selected for the fabrication of the turbine rotor because it has the potential for producing the lowest cost mass produced rotors. Hybrid molding is in use today for lower temperature ceramic components such as turbocharger wheels. The challenge was to develop the process for Kyocera's high temperature, high strength, SN252 silicon nitride. SN252 was selected for reasons of excellent strength and Weibull modulus at temperatures up to 1400°C (Figure 6.13).

In addition to the development of the basic molding process, a parallel effort was undertaken aimed at eliminating or simplifying as many process steps as possible for further cost reduction. Cost analysis and modeling was performed by Kyocera in conjunction with California Polytechnic College Pomona. Also under the AMP program CamLem Inc. were tasked with investigating means for lowering the cost of the tooling.

After consulting with Kyocera and reviewing prior Sundstrand experience with SN252, a design goal was established for the maximum allowable principal stresses in the ceramic components of 290 MPa for this short-lived application.

Ceramic Component Design

The **turbine rotor design** was performed by iterative finite element analysis. Both centrifugal and steady-state thermal stresses were considered. A steady-state thermal analysis was performed and a correction factor for transient thermal stresses was then applied based on prior experience with several ceramic rotors. Initially, quasi-three-dimensional models (3D blades on 2D axisymmetric hubs) were used to quickly evaluate different designs for the critical back face and blades. Once a reasonable looking design had been established final iterations were performed with fully three dimensional models.

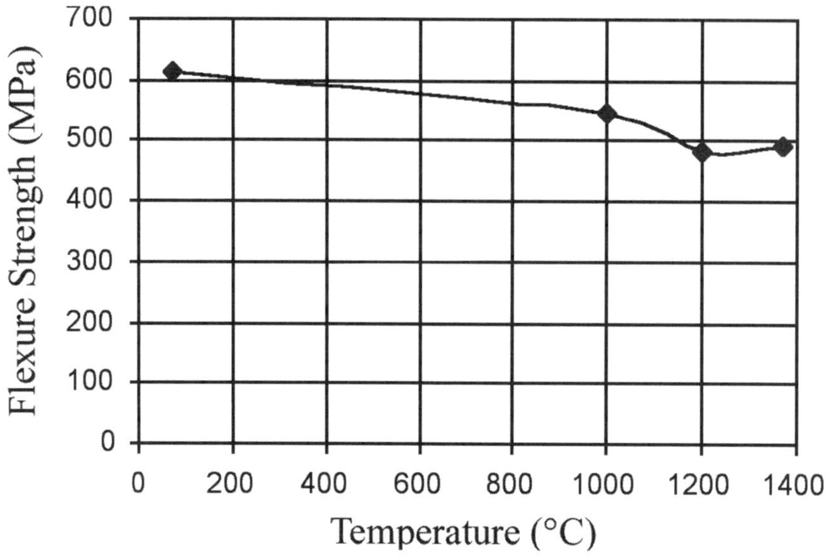

Figure 6.13 Flexural Strength of SN252 (Reprinted from Ref. [20] with Permission from the American Ceramic Society)

The final design achieved a remarkable balance of stress distribution over the entire rotor. Unacceptably high stresses were initially found in the exducer blade root fillet, blade tips in the "knee" region and in the hub between adjacent blade roots. After considerable effort all three regions were reduced to approximately 241 MPa. The design is thus well balanced in the sense that there is no one region of very high stresses as is often the case in metal wheels.

An attempt was made to run a CARES/LIFE failure probability analysis but was unsuccessful due to the size of the model and the incompatibility of CARES/LIFE with the particular version of ANSYS in use at Sundstrand. In light of previous CARES/LIFE and SPSLIFE experience with a similar rotor, and the well established design stress criterion based on prior good Sundstrand experience with the well-characterized SN252 material, this was not felt to be of critical importance.

The **turbine nozzle design** process was comparatively simple in relation to the rotor. Prototype SN252 ceramic nozzles had been tested in the TJ-50 previously with good success therefore a similar configuration was employed for the new nozzle. The design consists of sixteen axial vanes cantilevered from the static shroud (Figure 6.14). The cantilevered vanes are allowed to expand unrestricted by the outer metal shroud. This approach leads to low thermal stresses. The drawback is that some flow "leakage" over the vane tips is inevitable, leading to loss of efficiency. Experience has shown this effect to be small.

The nozzle is located by three "logs" projecting radially from three extended vanes. The logs loosely engage in three slots in the outer metal shroud. A ceramic-to-metal seal at the aft end of the inner shroud applies an axial force maintaining the correct nozzle position prior to engine start. When the engine is operating, pressure loads push the nozzle forward at all times.

Ceramic-To-Metal Joints

Without doubt, successful insertion of high temperature ceramic components into gas turbines is strongly dependent on the details of the ceramic-to-metal interfaces. The two interfaces in the TJ-50C are the rotor to shaft and the turbine nozzle to combustor.

Figure 6.14 TJ-50C Ceramic Turbine Nozzle and Rotor (Reprinted from Ref. [20] with Permission from the American Ceramic Society)

For the shaft joint Kyocera developed a proprietary, high temperature, brazed design which utilized an intermediate low-expansion material Incoloy 909. The 909 ring was brazed to a stub-shaft on the rotor and laser welded to the 15/5 PH steel shaft. Finite element analysis supported the design process. The peak temperature in the joint was predicted to be 700°C which presented a significant challenge due to differential thermal growth.

The nozzle-to-combustor joint utilized a recently-introduced "U-plex" metal seal developed by EG&G. The U-plex seal is similar in principle to previous metal bellows type E-seals but can tolerate much higher crush before permanent deformation. Seal crush provides the necessary force to ensure that the sealing surfaces remain in contact. In principle the seal allows unrestricted radial differential expansion between the ceramic nozzle and the metal exhaust duct. The highest temperature material available for the seal was Waspaloy which is limited to approximately 800°C. Therefore a small amount of cooling air was introduced by drilling very small metering holes through the seal walls. Heat transfer analysis of the complete design showed that the temperature of the seal could be kept within this limit despite a nozzle temperature well in excess of 1100°C.

Figure 6.15 TJ-50C Growth Engine Ready for Test (Reprinted from Ref. [22] with Permission from the American Ceramic Society)

Fabrication

A significant amount of effort was expended in developing the hybrid molding process for the rotor. This was necessary because of the difficulty in hybrid molding an SN252 radial rotor due to the length-to-thickness ratio of the blades.

The development was done in two phases—early risk reduction followed by prototype rotor fabrication. Early in the program, since the TJ-50C rotor was still being designed, tooling could not be manufactured. In order to provide information on the many empirical factors involved in the molding process (e.g. shrink factors), experiments were performed by hybrid molding an existing turbocharger rotor design of very similar size for which a master tool *was* available. The knowledge gained was applied to the design of the tooling for the TJ-50C rotor. The approach proved successful and acceptable engine-quality parts were produced from the first iteration of the tooling. Figure 6.14 shows finished nozzles and rotors.

Prior to delivery of the finished components the turbine rotors were subjected to proof spin test at 142,000 rpm. Apart from the first rotor, which failed at the spin arbor joint due to massive unbalance, all rotors passed spin test without failure. Proof tests were performed on the hot joint design including torque and axial thrust loads. The joint design passed these tests without significant problems also.

Test Results

The prototype engine was assembled in the third quarter on 1998 (Figure 6.15). An all metal rotor and turbine nozzle were used for the first builds to allow the basic mechanical and thermodynamic characteristic to be established. Initial results have been very good with the engine performing slightly better than predicted. Highlights of the test results are:

- Rotor dynamic behavior good up to 120,000rpm
- Turbine inlet temperature of 1150°C achieved
- Thrust in excess of 311 N demonstrated

- E-seal temperatures lower than expected and mechanical condition excellent
- Combustor light-off and stability very good
- Combustor exit temperature distribution too hot on outer wall
- Low carbon build-up

The only problem encountered to date is a skewed combustor exit temperature. Adjustments to the dilution air flow distribution have been made prior to running the ceramic parts.

CONCLUSIONS

The design and analysis of a silicon nitride turbine rotor for an advanced, miniature, military turbojet engine was completed. The design achieved a noteworthy balance of stresses resulting in optimal usage of the SN252 material properties. A unique ceramic-to-metal flexible seal was incorporated in the nozzle design and testing to date has been very successful.

The rotor was fabricated by Kyocera using the hybrid molding process and proof spin tested at a speed 20% above the design goal. High temperature brazed joint development was also successful. Engine quality test rotors and nozzles were delivered.

Shakedown testing of the engine was completed and performance is projected to easily exceed the design goal.

REFERENCES

[1] Bornemisza, T. and Napier, J. "Comparison of Ceramic vs. Advanced Superalloy Options for a Small Gas Turbine Technology Demonstrator." ASME paper 88-GT-228, presented at the Gas Turbine and Aeroengine Congress, Amsterdam, The Netherlands, June 6–9, 1988.

[2] Bornemisza, T. and Rodgers, C. "Fast Start Ceramic Auxiliary Power Unit." SAE 892254, 1989

[3] Kapur, K.C. and Lamberson, C.R. *Reliability in Engineering Design.* John Wiley & Sons, Inc., New York, 1977.

[4] Gyekenyesi, J.P. "SCARE: A Post Processor Program to MSC/NASTRAN for Reliability Analysis of Structural/Ceramic Components." Transactions of the ASME, Journal of Engineering for Gas Turbine Power, Vol. 108, No. 3, (July 1986), pp. 540–546.

[5] Gyekenyesi, J.P. and Nemeth, N. "Surface Flaw Reliability Analysis of Ceramic Components with the SCARE Finite Elements Post Processor Program." Transactions of the ASME, Journal of Engineering for Gas Turbine Power, Vol. 109, No. 3, (July 1987), pp. 274–281.

[6] Pai, S.S. and Gyekenyesi, J.P. "Calculation of Weibull Strength Parameters and Batdorf Flaw Density Constant for Volume and Surface Flaw Induced Fracture in Ceramics." NASA Technical Memorandum No. 100890, (Oct. 1988).

[7] Liu, K.C., Brinkman, C.R. "High Temperature Tensile Strength and Fatigue of Silicon Nitride." Proceedings of the 1989 Automotive Technology Development Contractors Coordination Meeting, Dearborn, Michigan, Oct. 23–26, 1989.

[8] Fuclis, H.D. and Stephens, R.J. *Metal Fatigue in Engineering.* John Wiley & Sons, Inc., New York, USA 1980.

[9] Katavama, K., Sasaki, M. and Itoh, T. "Development of Ceramic Turbine Rotors." ASME paper 88-GT-282, presented at the Gas Turbine and Aeroengine Congress, Amsterdam, The Netherlands, June 6–9, 1988.

[10] Yoshida, M., Kokaji, A. and Koga, K. "Silicon Nitride for Automotive Applications." SAE 890424, 1989. Mayer A., et al. "Extruded Ceramic-A New Technology for the Comprex Rotor," SAE 890425, 1989.

[11] Nemeth, N.N., Powers, L.M., and Janosik, L.A., "Time-dependent Reliability Analysis of Monolithic Ceramic Components Using the CARES/LIFE Integrated Design Program," ASTM Symposium on Life Prediction Methodologies for Ceramic Materials in Advanced Applications—A Basis for Standards, Cocoa Beach, Florida, January 11–13, 1993. Ceramic Engineering & Science Proceedings, 14 [7–8], 318–319, 1993.

[12] Quinn, G.D., and Wiederhorn, S.M., "Structural Reliability of Ceramics at Elevated Temperatures," ASTM Symposium on Life Prediction Methodologies for Ceramic Materials in Advanced Applications—A Basis for Standards, Cocoa Beach, Florida, January 11–13, 1993.

[13] Soma, T., Ishida, Y., Matsui, M, and Oda, I., "Ceramic Component Design for Assuring Long-Term Durability," Advanced Ceramic Materials, 2[4], 809–812, 1987.

[14] Weibull, W., "A Statistical Distribution Function of Wide Applicability," J. Appl. Mech., v. 18, no.3, 293–297, 1951.

[15] Wiederhorn, S.M., "Subcritical Crack Growth in Ceramics," in *Fracture Mechanics of Ceramics,* R.C. Bradt, D.P.H. Hasselman, and F. F. Lange, eds., Plenum Press, New York USA, 1974, pp. 613–646.

[16] Wiederhorn, S.M., et al., "Tensile Creep Testing of Structural Ceramics," Proc. Annual Automotive Technology Development Contractors Coordination Meeting, Dearborn, Michigan, USA, October 28–31, 1991.

[17] Bornemisza T. and Saith, A., "SPSLIFE: A User Friendly Approach to the Structural Design and Life Assessment of Ceramic Components," ASME paper 94-GT-486, presented at the International Gas Turbine and Aeroengine Congress and Exposition, The Hague, The Netherlands, June 13–16, 1994. Journal of Engineering for Gas Turbines and Power. Transactions of the ASME Volume 118, January 1996, pp. 179–183.

[18] Saith, A., Norton, P.F., and Parthasarathy, V.M., "Application of SPSLIFE to Preliminary Design Evaluation and Life Assessment of CSGT Components," ASME paper 94-GT-420, presented at the International Gas Turbine and Aeroengine Congress and Exposition, The Hague, The Netherlands, June 13–16, 1994.

[19] Jones, A.C., "Program Overview: Reduced Cost, Improved F.O.D. Resistance and Improved Attachments For Insertion of Silicon Nitride Turbomachinery Components," Sundstrand Power Systems, San Diego, CA 92123

[20] Jones, A.C., "Design, Analysis and Testing of Ceramic Components for an Advanced Military Expendable Turbojet Engine," 23rd Annual Conference on Composites, Advanced Ceramics, Materials and Structures:B, Cocoa Beach, Florida, USA, January 25–29. 1999, Ceramic Engineering and Science Proceedings, Vol. 20[4], pp. 523–532, 1999.

CHAPTER

7

Ceramic Gas Turbine Development at Solar Turbines Incorporated

William D. Brentnall, Mark van Roode, Paul F. Norton, Steve Gates, Jeffrey R. Price, Oscar Jimenez, and Naren Miriyala

Solar Turbines Incorporated, San Diego, California, USA

INTRODUCTION

The driving force behind the efforts to incorporate ceramic materials into gas turbine engines has been the desire to reduce dependence on strategic metals, improve thermal efficiency, increase output power, enhance component durability, lower emissions, and reduce cost. Ceramic materials can potentially be used for uncooled hot section components at temperatures approaching 1370°C, compared to about 1000°C for cooled advanced superalloy parts. With ceramic component designs, high turbine operating temperatures can be achieved while eliminating, or greatly reducing, component cooling requirements. Solar Turbines Incorporated (Solar) has carried out ceramic development since the 1950s covering a wide range of applications for oxidation, corrosion and wear protection (see Table 7.1).

Ceramic coatings development for oxidation protection of combustor liners and other hot section components started in the 1950s. The first applications involved vitreous phase ceramic coatings (VPCs) for combustor substrates for military aircraft. Numerous combustors were coated with VPCs to extend liner life at a time when strategic alloy elements such as cobalt were in short supply. Subsequently, thermal barrier coatings (TBCs) for oxidation protection of metal components were evaluated [1, 2]. Most of the TBC work until the early 1980s was non-core product-specific contract R&D. The more recent work focused on TBCs for specific gas turbine components (combustor liners, blades and nozzles). TBCs are now the bill-of-material on combustor liners of the Solar recuperated 4.2-MW Mercury 50 engine. During the 1980s and early 1990s, Solar developed oxidation/corrosion-resistant coatings for silicon carbide (SiC) materials for heat exchangers [3]. Since 1998, environmental barrier coatings (EBCs) are being assessed for the protection of ceramic hot section components. Their use will be described further on in the chapter.

Solar developed ceramic recuperators for waste heat recovery in corrosive industrial environments under a number of contracts between 1970 and 1992. A program funded by the Gas Research Institute (GRI)

TABLE 7.1 SUMMARY OF CERAMIC RESEARCH AND DEVELOPMENT AT SOLAR TURBINES INCORPORATED

Time Frame	Ceramic Research and Development
1949–1990	Vitreous Phase Ceramic Coatings
1960–1964	Glass Fiber Technology
1968–Present	Thermal Barrier Coatings
1972–1974	Ceramic Development for 100 hp Titan Engine*
1974–1982	10 kW Gemini Ceramic Gas Turbine Programs (US Army MERADCOM)
1980–1984	10 kW Gemini Ceramic Gas Turbine Program (US Army Ft. Belvoir R&D Center)
1970–1992	Heat Exchanger Technology
1971–1975	Ceramic Ball Bearings
1975–1992	Joining of Ceramics, Ceramic Seals
1981	Rig Test of Ceramic Saturn Nozzle
1986–1994	Ceramic Coatings for Corrosion Protection of Heat Exchangers (GRI)
1986–1994	Coal Fired Turbine Applications (DOE)
1988–1994	High Pressure Heat Exchanger (DOE HiPHES)
1988–1989	Structural Ceramic Composites in Gas Turbines (DOE Study)
1991–2000	Applications of CFCCs to Gas Turbines (DOE)
1992–2001	Ceramic Stationary Gas Turbine Development (DOE CSGT)
1999–2002	CMCs for Advanced Engine Components (NIST)
2000–2003	Advanced Materials for Mercury-50 Combustion System (DOE)

* 100 hp Titan engine is distinct from the Solar 14.5 MW Titan-130 gas turbine

led to the development of a full-scale heat exchanger that operated successfully for three years in an aluminum remelt operation. The system, which incorporated 384 SiC tubes (totaling nearly 610 m in length), reduced energy usage by 40% [4, 5]. Subsequently, ceramic recuperator design for an indirect fired gas turbine was conducted under the High Pressure Heat Exchanger (HiPHES) program funded by the U.S. Dept. of Energy (DOE). Testing was carried out on modules containing SiC/SiC continuous fiber-reinforced ceramic composite (CFCC) headers and monolithic SiC tubes, but a full scale system was never fabricated and tested [6]. The technology did not result in a commercial product because low fuel prices in the U.S. since the mid-1980s provided little incentive for waste heat recuperation. However, Solar developed expertise in joining and sealing of ceramics to metals through these activities.

Solar's ceramic gas turbine R&D started in the 1970s. The initial focus was on improving the erosion resistance of one of Solar's small gas turbines. Subsequently, the main objective became to improve gas turbine performance through increased turbine rotor inlet temperature (TRIT) by replacement of cooled metal components with uncooled ceramic parts. The work stopped at Solar around 1985 when the Turbomach division, which was involved with this activity, was sold to the Sundstrand Corporation. The work continued at Sundstrand Power Systems (SPS), the successor to Turbomach (see Volume I, Chapter 6). Ceramic gas turbine development at Solar resumed when Solar was awarded the Ceramic Stationary Gas Turbine (CSGT) Development contract from the DOE in 1992. This program, which continued through 2001, is the most significant ceramic gas turbine project at Solar to date. The goals of the CSGT program were to improve engine performance (fuel efficiency, output power) and reduce exhaust emissions of NOx and CO. In parallel, Solar participated in a number of smaller programs under the DOE's CFCC Initiative, which benefited the suppliers to the CSGT program. In 1999, under the Advanced Technology Program (ATP) administered by the U.S. National Institute for Standards and Technology (NIST), Solar partnered with Siemens Westinghouse Power Corporation (SWPC) and Composite Optics Inc.-Ceramics (COI-Ceramics) to develop oxide/oxide ceramic matrix composites (CMCs) for gas turbine combustors (see Volume I, Chapter 8).

This chapter focuses on applications involving structural ceramics for gas turbines. The other technologies mentioned (coatings for metal components, heat exchangers, etc.) are not covered here, although a

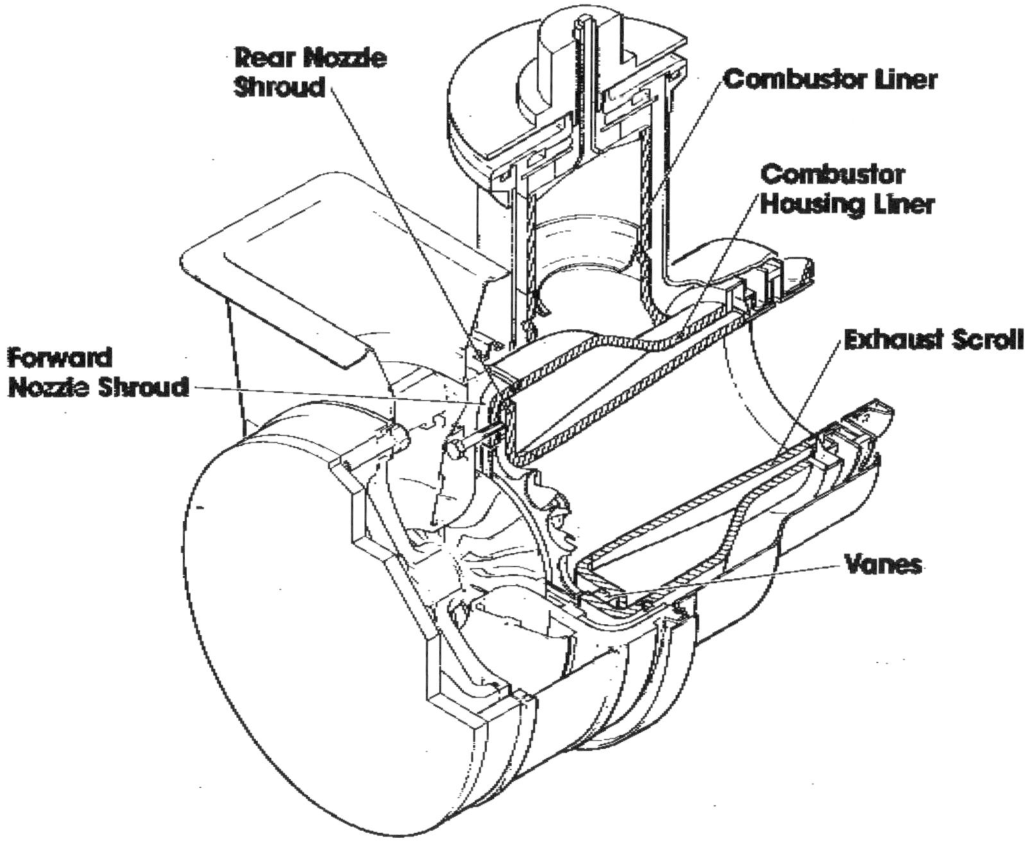

Figure 7.1 Ceramic Components for Evaluation in the Gemini Gas Turbine (Reprinted from Ref. [13])

substantial technology base was generated from these peripheral activities, which contributed to the ceramic gas turbine development efforts.

GEMINI DEVELOPMENT PROGRAM

Ceramics gas turbine development at Solar started in 1972 when hot pressed silicon nitride (HPSN) vanes were evaluated for Solar's 100 hp radial Titan engine. In 1974, a series of programs were initiated for the 10 kW Gemini radial gas turbine engine. Figure 7.1 shows the ceramic components planned for insertion in the Gemini. The all-metal Gemini engine had a turbine inlet temperature (TIT) of approximately 900°C and a pressure ratio of 3.4:1. In the beginning, the Gemini work was funded by Solar. Subsequently, it was sponsored by the U.S. Army Mobility Equipment Research and Development Command (MERADCOM) and its successor, the Fort Belvoir Research and Development Center [7–15]. More than 2300 hours of rig and engine testing were conducted over the 10-year duration of the program.

The initial objective was to improve the high temperature erosion resistance of metal nozzles that suffered severe degradation from ingestion of dust during service of the engine in arid environments. Figure 7.2 shows a hybrid nozzle design with a ceramic insert at the trailing edge of the metal nozzle where erosion

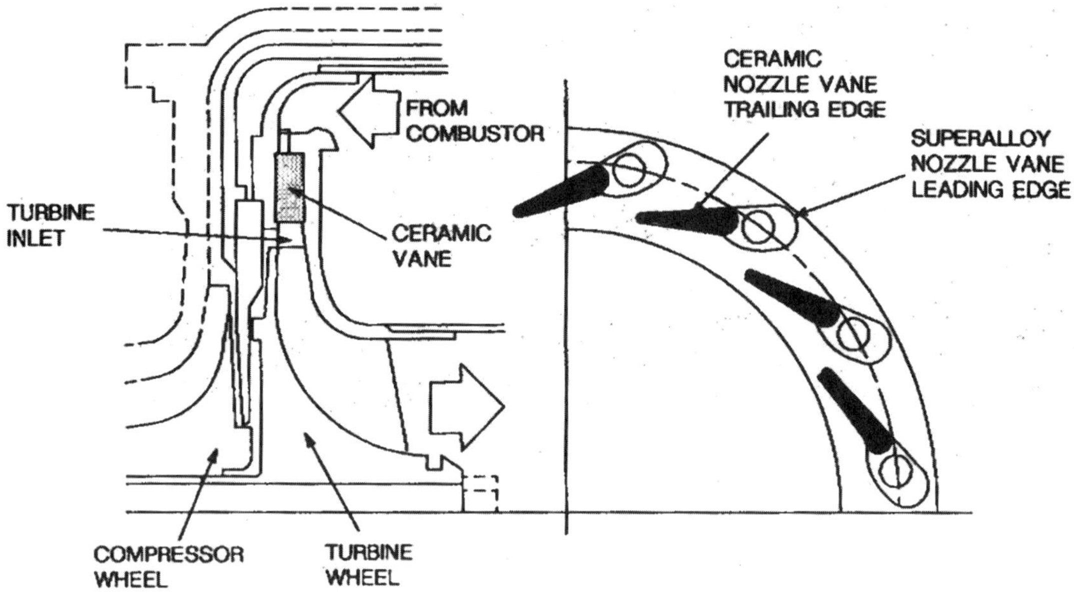

Figure 7.2 Gemini Nozzles with Ceramic Inserts (Reprinted from Ref. [12])

degradation was most severe. The configuration addressed major design obstacles, such as thermal expansion mismatch between ceramics and superalloys, contact stresses in ceramics due to normal surface irregularities or manufacturing misfits, and stresses induced into the assembly due to static and transient thermal distributions. The ceramic vanes were captured in recesses in forward and rearward Inconel 713LC shrouds. The recesses were slightly oversized to avoid thermal expansion mismatch. A specially developed relaxing glass adhesive was applied at the ceramic/metal shroud interfaces to alleviate localized contact stresses. The contact stress problem is a major concern for ceramic component design elsewhere as well. The interested reader is referred to the work in this area by Richerson at Garrett Turbine Engines [16].

The nozzle assembly was evaluated in a cyclic thermal shock test which simulated the most severe transient conditions, and in dust erosion and sea salt corrosion tests at 927°C. Hot pressed silicon nitride (HPSN—Ceradyne Inc.—Ceralloy 147A) vanes had 1/300 of the erosion rate of the metal component. In the sea salt test the corrosion layer on an Inconel 713LC vane was 20 times thicker than that on a HPSN vane. The ceramic vanes performed well in an engine test consisting of 24 hours at steady-state at various loads and 50 start/stop cycles, except for six vanes that suffered trailing edge fractures due to excessive nozzle assembly loads. A redesign of the assembly eliminated the undesirable loads and more than 1000 hours of subsequent engine testing gave positive results. In the next step, various all-ceramic nozzle designs were also evaluated in rigs and in engine testing with a TIT of 927°C. A major milestone was a 200 hour engine test encompassing the full spectrum of engine loads. A total of 50 start/stop cycles was completed without adverse effects to the ceramic components. The shrouds were made of siliconized SiC (Norton NC 430) and the vanes of HPSN (Norton NC 132).

Capitalizing on the high temperature strength properties of the emerging ceramic materials, continued work focused on uprating the MERADCOM synchronous GTED (gas turbine engine driven) generator set to a TIT of 1093°C, increasing the power rating from 10 kW to 15 kW and decreasing the specific fuel consumption by 8%. The ceramic components included the combustor liner and scroll, liner end cap, forward nozzle shroud, rear nozzle shroud with integral vanes, and exhaust scroll. The ceramic parts were held together by pneumatic forces generated by the combustor pressure drop. The scroll assembly, shrouds, and combustor parts were fabricated from SiC (Norton NC-430), the nozzles from HPSN (Norton NC-132). Combustor liners from SiC coated carbon/carbon composite (Refractory Composites Inc.—Recomp 201) and from SiC/SiC CFCC (Norton) were also investigated. The program also included a study of manufacturing methods and production costs. It was shown that the nozzle with ceramic vanes could be fabricated at a cost that was no more than 50% greater than that of the all-metal nozzle.

The initial combustor rig tests resulted in failures from temperature gradients. After redesign, monolithic (Norton NC-430) combustor liners survived up to 40 hours intact, but composite structures suffered severe oxidation damage. In rig testing of full stationary assemblies at TITs up to 1093°C, the combustor liner performed well, but the combustor scroll was damaged due to severe steady-state thermal stresses from local cooling air and combustor gas impingement. Some of the forward nozzle shrouds showed radial through fractures attributable to thermal stress, and the rear nozzle shrouds saw early failures associated with the vanes being integral to the shroud. The exhaust scroll generally showed good durability. After remedial work, all parts, except for the combustor scroll, had at least one successful run.

Application of ceramics for Solar's 1 MW Saturn engine was evaluated in the early 1980s with various ceramic nozzle designs, which included a leading edge nested into cooled metallic airfoils and uncooled ceramic vane inserts in metal shroud rings. The latter design performed well in a 200-hour engine test at a TIT of about 1000°C.

CERAMIC STATIONARY GAS TURBINE DEVELOPMENT (CSGT) PROGRAM

Background

The CSGT program was aimed at improving fuel efficiency and output power, and reducing NOx and CO emissions in land-based gas turbines, by replacement of cooled metallic first-stage blades and nozzles and combustor liner with uncooled ceramic parts. The CSGT program was initiated by the DOE Office of Industrial Technologies (OIT) following a DOE-sponsored feasibility study by Battelle Memorial Institute (Columbus, Ohio), which showed that significant energy gains and emission reductions could be achieved by insertion of ceramics into the hot sections of existing engines in the U.S. fleet of gas turbines, particularly in those engines for power-and-steam cogeneration [17, 18]. OIT-sponsored studies forecasted that ceramic technologies, if successfully applied, would increase the competitiveness and market share of gas turbine based cogeneration, which could result in displacement of utility generating output and produce annual U.S. energy savings on the order of 2×10^{17} Joules (2×10^{14} Btus) [19].

Industrial gas turbines pose design and fabrication challenges different from those of smaller engines. The large component size precludes the use of integral monolithic ceramic rotors, nozzle rings, or combustor liners because no ceramic suppliers have the technology or tooling to fabricate such large parts. The high probability of component failure from processing-induced flaws in large parts poses a high degree of risk as well. Solar, therefore, pursued a design philosophy of individual blades and nozzle vanes integrated into a metal support structure. Consequently, metal/ceramic interface load transfer issues became a critical part of the designs. Component size was also a consideration for combustor liners. Since Solar's engine uses an annular combustor, the outer liner would have to be about 75 cm in diameter, and it was not feasible to fabricate such a large part from monolithic ceramics. However, because of their superior "fracture toughness," CFCCs became candidates for single-piece combustor liners [20]. When the program started in 1992, using CFCCs was considered somewhat of a risk, but Solar was encouraged by the progress made under U.S. government sponsored programs, such as the DOE CFCC initiative and the NASA Enabling Propulsion Materials (EPM) program.

The CSGT program was an integrated team effort involving the sponsor (DOE), the gas turbine manufacturer (Solar), other gas turbine manufacturers, ceramic component suppliers, test laboratories, industrial end users, and consultants. Figure 7.3 shows the program organization. The end user presence was essential

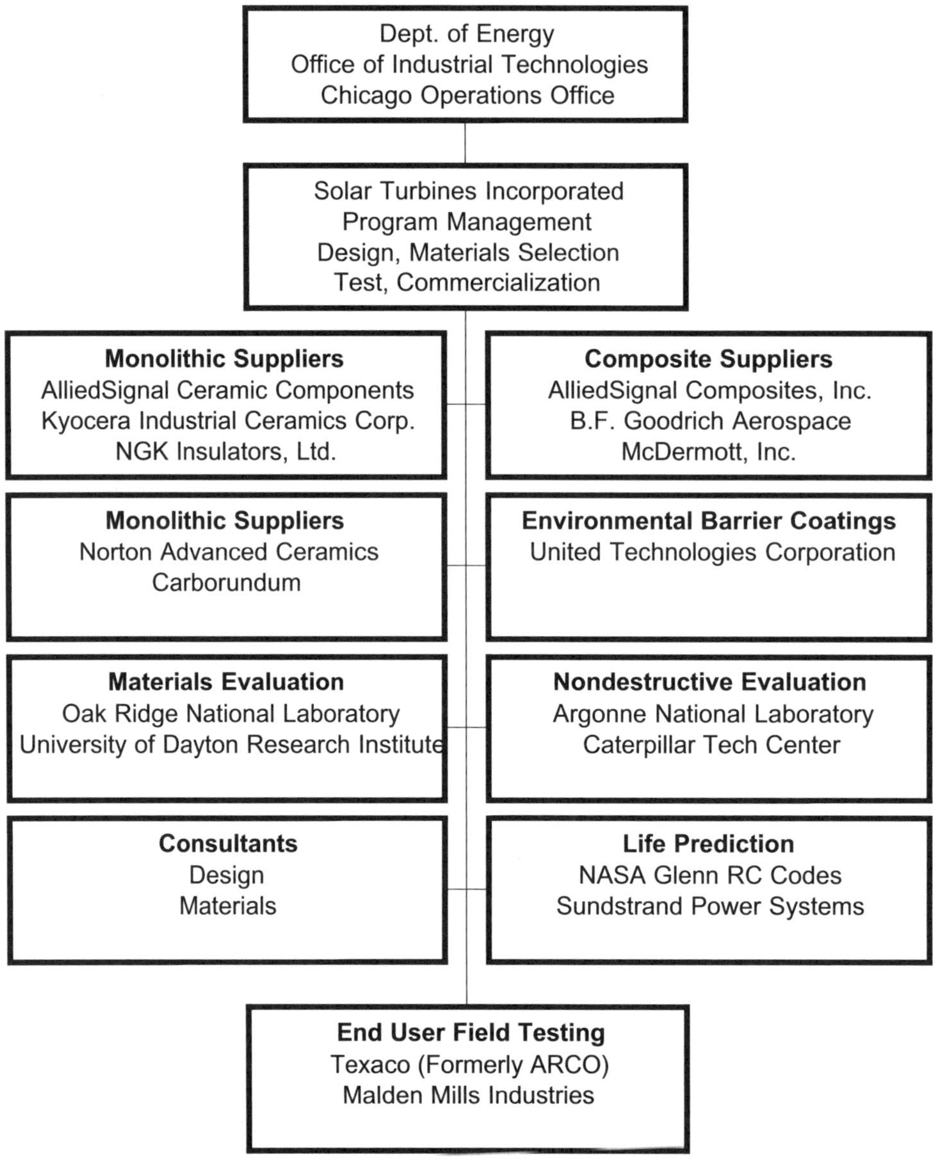

Figure 7.3 CSGT Program Organization (Reprinted with permission from Solar Turbines Incorporated)

for the program since only actual field operation provides adequate proof that the ceramic gas turbine technology has reached the maturity required for commercial application. The end user field tests were invaluable, since it was in those extended tests that the ceramic components revealed their vulnerabilities, insufficient resistance to impact, and surface degradation in the combustion environment of the gas turbine. The field tests also proved the efficacy of protective environmental barrier coatings (EBCs).

The CSGT program started in September 1992. The engine selected for retrofitting ceramic parts was Solar's Centaur 50S. Phase I focused on concept/preliminary engine and component design, ceramic materials selection, technical and economic evaluation, and concept assessment. Phase II, which began in April

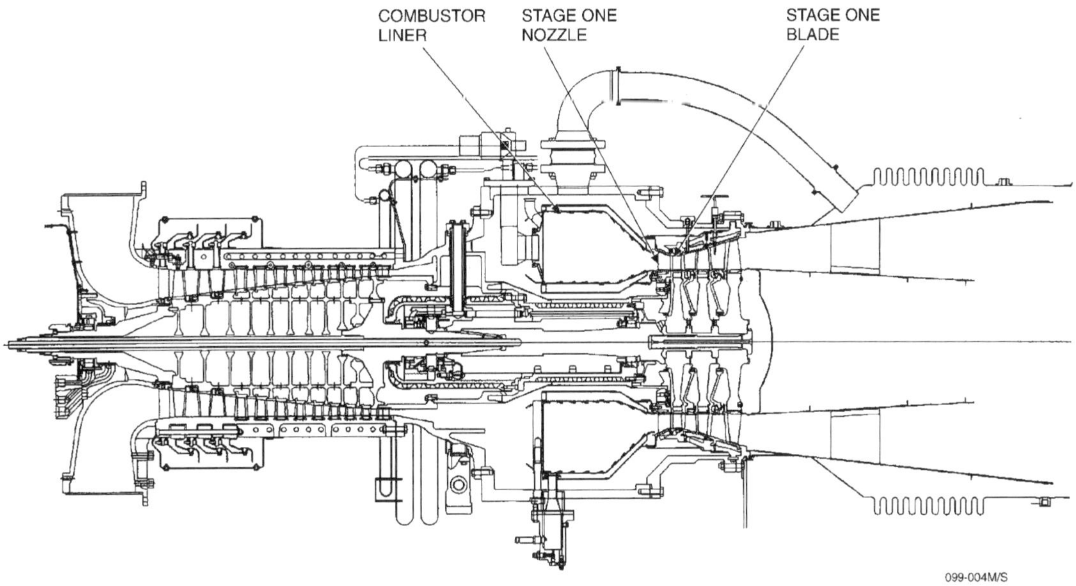

Figure 7.4 Solar Centaur 50S Gas Turbine with Components Targeted for Ceramic Insertion (Reprinted with permission from Refs. [25, 30])

1993, addressed detailed engine and component design, ceramic part procurement, long-term materials property testing, component rig testing, engine testing, and destructive and nondestructive post-test component evaluation. Phase III, which commenced in October 1996, focused on a series of field tests at the cogeneration sites of industrial end users. The field test was originally envisioned to be for 4000 hours at a design TRIT of 1121°C, which is 111°C higher than that of the baseline Centaur 50S. Due to the high risks associated with the monolithic ceramic components, the field test goals were subsequently relaxed to operation at the baseline TRIT of the Centaur 50S engine. Short-term in-house testing at the design target TRIT was conducted toward the end of the program in 2001. By the end of June 2001, more than 33,000 hours of field-testing had been accumulated at two, end user sites. Summaries of the CSGT program progress have been reported in the literature [21–35].

Engine Selection

The Centaur 50S is a single-shaft engine for electrical generation and cogeneration applications. This engine and its uprate, the Taurus 60, are well-established gas turbine products. The Centaur 50S model is equipped with a dry low NOx lean premix (SoLoNOx) combustor. The program plan involved retrofitting the selected engine with uncooled first-stage ceramic blades, uncooled first-stage ceramic nozzles, and ceramic combustor liners. Figure 7.4 shows the Centaur 50S engine layout with the components targeted for ceramic insertion.

The engine has a two-stage gas producer turbine and single-stage power turbine. By uprating the engine from its current TRIT of 1010°C to the design TRIT of 1121°C, it was estimated that the shaft thermal efficiency could be improved from 29.6% to 31.3% in simple-cycle mode. Electrical power was predicted to increase from 4144 kW to 5217 kW (25.6%) in simple-cycle mode. These improvements are typical for a ceramic retrofit of an existing engine design. The CSGT emissions goals were to demonstrate NOx levels of ≤ 25 ppmv and show the potential for NOx levels of ≤10 ppmv at the target TRIT of 1121°C.

CSGT Development Philosophy

The gas turbine development philosophy for the CSGT program involved the following elements:

- Realistic goals for performance and durability of components
- Commercial application targeted
- Use of well characterized ceramic materials or new materials that promise an improvement over existing materials
- CFCCs as well as monolithic ceramics
- Established ceramic design practice including life prediction methodology
- Risk mitigation by limiting the number and type of ceramic components
- Modify interfacing secondary metallic components to accept the ceramic parts
- Laboratory testing to establish a materials database
- Simulated and subscale component testing to screen candidate designs and materials
- Rig and engine testing to evaluate/validate component and engine designs and materials performance
- Post test component evaluation using state-of-the-art metallographic and nondestructive evaluation (NDE) methodologies

Several fundamental design guidelines were established at the onset of the program. The first and most important of these was to establish acceptable stress levels within each component consistent with a sufficiently high reliability for 10,000 hours continuous duty (10,000 hours design life being seen as providing sufficient margin for the required 4000 hour field test) at the target TRIT of 1121°C. Other broad guidelines, such as simplicity of design to minimize stress concentrations, avoidance of planar contacts where possible, predictable mechanical load paths, minimization of relative motion, etc., were used to provide interface integrity under high mechanical and thermal loading during engine operating conditions.

Figure 7.5 illustrates the CSGT program strategy. Design analysis is integrated with life prediction, testing, and post-testing component evaluation. The first step in the design process was to define operating parameters for a specific component, including boundary conditions and material properties. The next step was to predict component stress/temperature distributions through finite element analysis (FEA). Component life was subsequently predicted for monolithic components by reliability analysis using the NASA code CARES/LIFE and the Sundstrand Power Systems SPSLIFE code [36–38]. Only fast fracture and slow crack growth failure modes were considered for predicting the failure modes using the life prediction computer codes. SPSLIFE [37] and the NASA CARES/CREEP [39] computer codes also enable creep life prediction. For the CFCC combustor liner materials, the absence of an adequate life prediction code at the time of the design represented a significant limitation for effective life prediction.

Component performance predictions incorporating fast fracture, slow crack growth, creep, fatigue, and oxidation properties of candidate materials were primarily inferred from specimen and simulated component testing at temperatures similar to those in actual engine operation. Once conceptual design work was completed, subscale components were fabricated and evaluated to validate critical design concepts, such as root configuration and compliant layers for the blade, attachments and cooling schemes for the nozzle and adjacent secondary components, and fixturing to the metallic support structure for the combustor liner. The testing was performed using various rigs, including an attachment tensile rig, a blade spin rig, nozzle thermal gradient and mechanical proof test rigs, and subscale and full-scale combustor rigs. The findings from these tests were assimilated into the design of the component prototypes and simulated parts. The design changes were iterated and next generation hardware fabricated and tested in the rigs and in the Centaur 50S engine until results indicated that the design was satisfactory. A comprehensive interim program evaluation was conducted in 1998 [32] and the results were incorporated in the direction of the program.

Figure 7.6 is a schematic of the CSGT engine hot section showing the three key ceramic components: the CFCC combustor liner, first-stage turbine nozzles and first-stage turbine blades. Detail design information regarding components was presented in prior papers [23–28, 40–42].

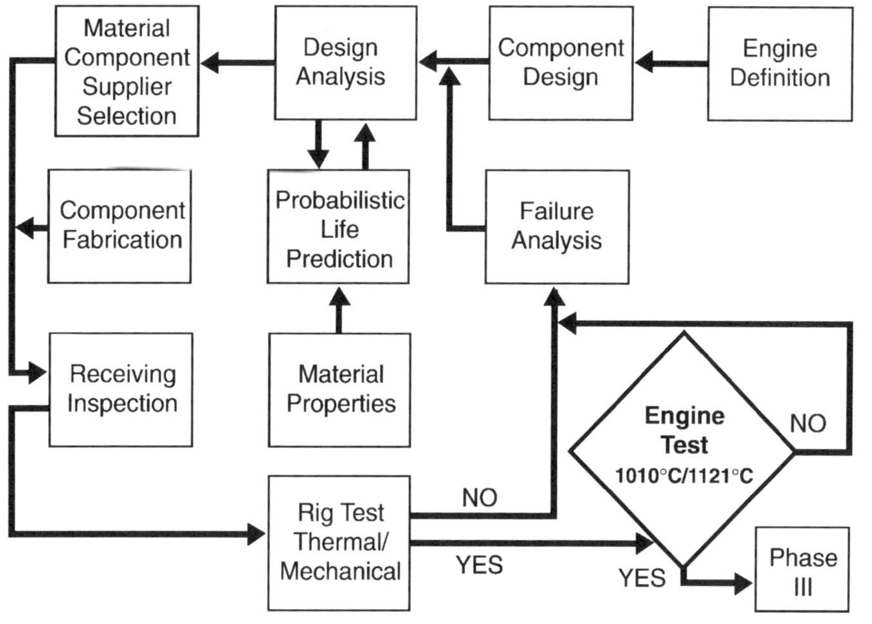

Figure 7.5 Integrated Design and Test Philosophy of the CSGT Program (Adapted from Ref. [23])

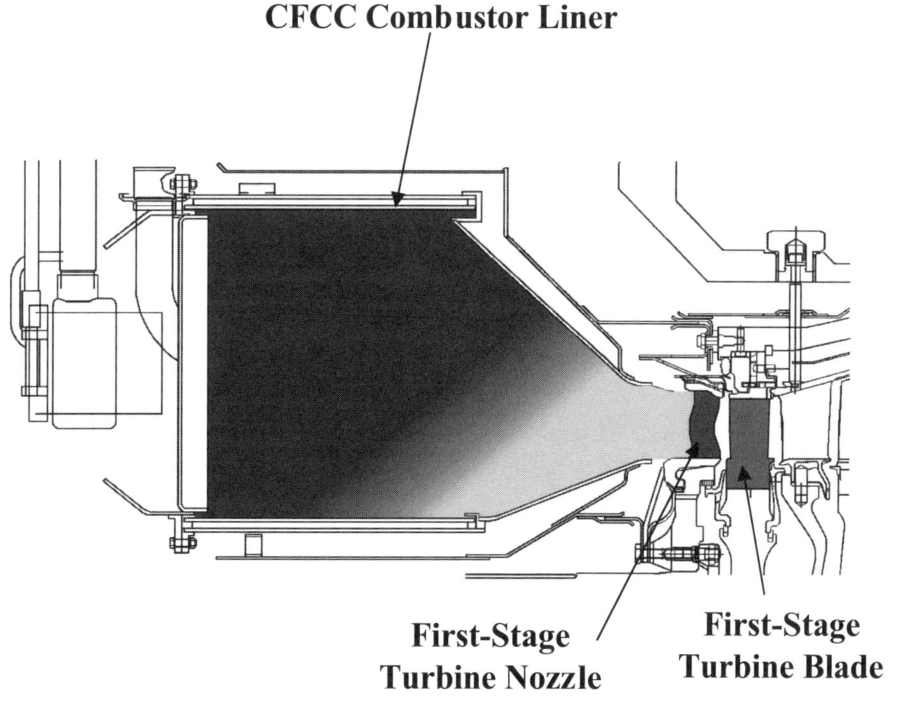

Figure 7.6 Schematic of CSGT Engine Hot Section (Adapted from Ref. [23])

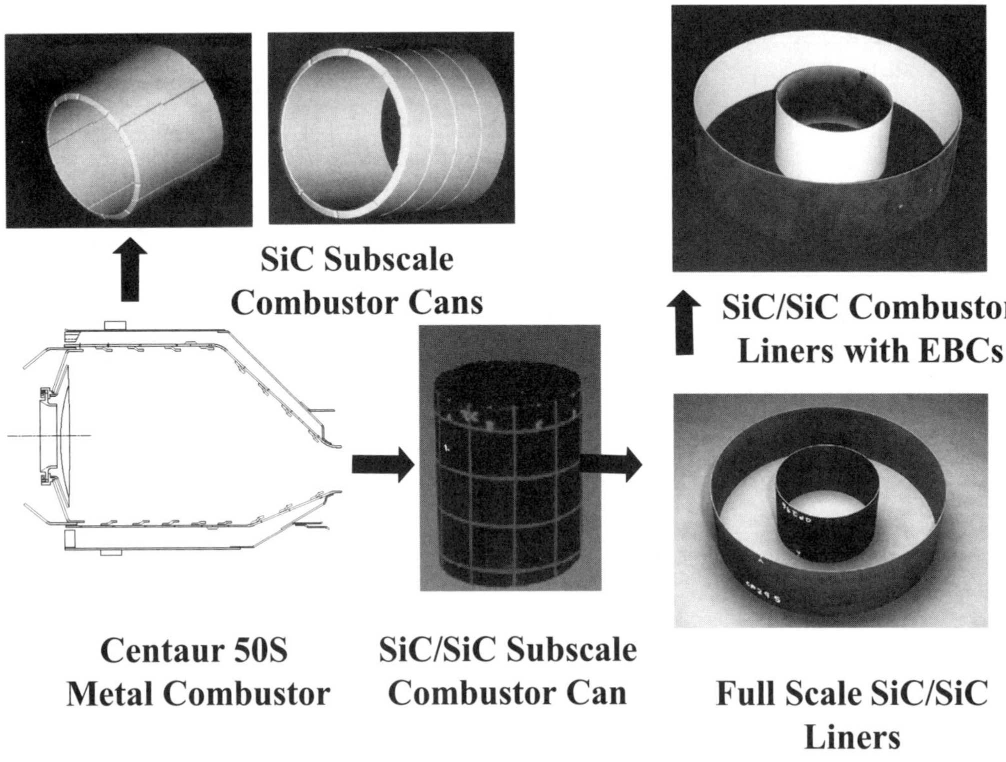

Figure 7.7 Evolution of the CSGT Combustor Liner Design (Adapted from Refs. [45, 46])

Combustor Liner Development

Combustor Liner Design. A ceramic-lined gas turbine combustor has three significant advantages over conventional metal combustor liners. First, given the high temperature capability of ceramics, combustor firing temperature can be increased ("hot wall") without degrading combustor durability. Second, air saved from reduced cooling requirements for the combustor wall can be used to lean out the flame in the primary zone, resulting in lower NOx emissions at comparable conventional firing temperatures. Third, the hot wall combustor reduces near-wall flame quenching that can cause high CO emissions [43, 44].

The ceramic combustor was designed to be interchangeable with the production Centaur 50S (SoLoNOx) combustor which incorporates louver-cooled Hastelloy X combustor liners. To achieve the desired emissions reductions only the cylindrical sections in the primary combustor zone needed to be replaced with ceramic parts. The ceramic combustor liner evolution is shown in Figure 7.7. Initially, both monolithic (silicon carbide) and CFCCs (SiC/SiC, Al_2O_3/Al_2O_3) were considered as combustor liner materials. Due to their relatively low fracture toughness, monolithic liners had to be segmented, and designs based on tiles and rings were conceived. Due to their higher "fracture toughness," CFCC liners can be fabricated as integral parts. The ceramic hot section schematic of Figure 7.6 shows that the combustor consists of a metallic dome section at the fore end, two concentric CFCC cylinders (in metal housings) that form the combustor primary zone, and conical metallic exit sections. The dome and exit sections are film cooled and are essentially identical to their all-metal production engine counterparts. A layer of compliant insulation between the CFCC parts and the metal housing minimizes radial contact stresses. Small axial gaps between the ends of the CFCC parts and the metal housing prevent end loads.

All pressure loads are carried by the metal housing, thereby giving the CFCC cylinders an easy mechanical ride at steady-state conditions.

Based on subscale tests, CFCC materials were selected over monolithic ceramics for the primary zone cylinders [25, 26]. The inner and outer CFCC cylinders are 33 cm and 75 cm in diameter, respectively, and are 20 cm long. Their wall thickness is approximately 0.2–0.3 cm. Figure 7.7 shows a set of enhanced-SiC/SiC CFCC liners of DuPont Lanxide Composites (DLC) fabricated using the chemical vapor infiltration (CVI) process.

Combustor Liner Testing. Candidate combustor materials were first tested in a subscale configuration using 20 cm diameter by 20 cm long cylindrical cans. The goal of the subscale testing was to demonstrate key elements of the full scale combustor design (metal/ceramic interface design, ceramic durability, low emissions potential) in a cost-effective but representative geometry under typical combustion test conditions and mechanical loading. The testing procedure involved sequential tests of 1, 10, and 100 hours. The liners were inspected using NDE (Thermal Imaging and Computed Tomography) methods before and after the test. Intermittent visual inspections were conducted during the longer tests. The subscale combustor testing has proved to be a very effective material screening procedure.

SiC/SiC CFCC materials performed best in the subscale combustor testing, completing successfully the 100-hour test that included 200 cycles (15 minutes between ~480°C and ~1177°C). Following the test visual inspection and NDE at Argonne National Laboratory (ANL) showed no evidence of spallation or delamination. Subscale Al_2O_3/Al_2O_3 CFCC liners fabricated by Babcock and Wilcox (B&W) suffered longitudinal cracks because of matrix shrinkage, and Hexoloy SA (Carborundum) SiC tiles cracked because of contact stresses between the tiles and the adjacent metallic support structure. Also, these tiles were too thick to accommodate the thermal shocks during the test. NT-230 SiC rings fabricated by Norton Advanced Ceramics (NAC) were never tested because of fabrication defects.

Two sets of SiC/SiC CFCC full-cale combustor liners by BFGoodrich Aerospace (BFG) and DLC were proof tested in atmospheric and high pressure rigs. In these full-scale rig tests, liner wall temperatures were monitored with thermocouples. More than 300 hours of full-scale combustor rig testing (TRIT up to 1093°C) were performed under the program. The full-scale tests showed that the SiC/SiC could be used as liner materials in engine testing. The rig testing also presented evidence of the beneficial effect of the "hot wall" ceramic liners. NOx > 15 ppmv was consistently demonstrated in high pressure rig testing at the baseline TRIT of 1010°C. CO levels were also low, typically less than 10 ppmv. A variety of SiC/SiC CFCCs have been tested, including 2-D and 3-D CFCCs fabricated by BFG and 2-D CFCCs fabricated by DLC. The combustor development has been detailed in the literature [45, 46].

Nozzle Development

Nozzle Design. Figure 7.8 schematically represents the nozzle design evolution in the CSGT program. The Centaur 50S metal engine has 15 two-vane cooled first-stage FS-414 metal nozzles, with a Pt/Rh-aluminide coating for oxidation protection. The ceramic nozzle development started with a single vane segment, representing one half of the current metal nozzle with tip shoe removed for ease of ceramic fabrication. The initial nozzle design strategy was to retain the airfoil profiles of the metal nozzle, thereby maintaining the basic aerodynamic performance of the turbine. However, replacing the metal airfoils with a ceramic vane of the same geometric configuration would result in an unacceptably high steady-state stress level (480 MPa) at the hottest temperature location (1288°C), which is incompatible with long service life.

Alternative design concepts considered included airfoils detached from either the inner or the outer shroud, segmented/bolted airfoils (radial and axial), segmented shrouds, various pinned attachments and various clamped attachments, as well as a cooled nozzle. The reasonable alternatives required complete redesign of the nozzle airfoil. The airfoil chord was reduced by 50% and the airfoil was bowed, axially

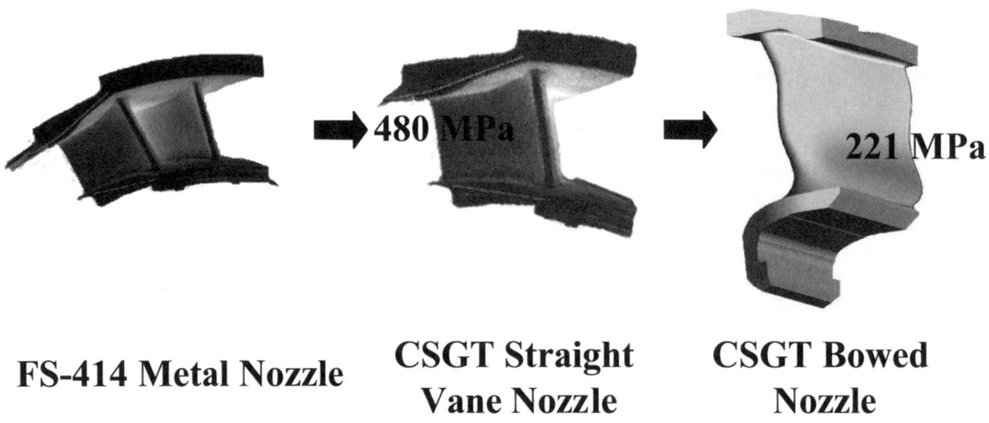

Figure 7.8 Evolution of the CSGT Nozzle Design (Adapted with permission from Ref. [33])

and tangentially, in order to reduce the steady-state stresses to the desired level of about 221 MPa. SN-88 silicon nitride from NGK Insulators, Ltd. (NGK) was selected as the nozzle material based on life prediction analysis, a probability of survival (POS) of 0.9934 for a set of 42 nozzles for 30,000 hours of operation, and creep and slow crack growth test data. A cantilevered "hook" support system accommodates the ceramic-to-metal interface at the first-stage diaphragm. The hook design offers self-assisted clamping from aerodynamic loading without having to apply additional mechanical clamp loads with intrinsic undesirable local contact stress increases. During engine operation, aerodynamic forces act to further load the nozzle in the same direction as the clamping forces. There are 42 single-vane ceramic nozzle segments compared to 15 two-vane metal nozzles to avoid rotor vibratory issues with the nozzle passing frequency excitation at 30 revolutions per second. A peak transient stress of 242 MPa (25% span, 10% chord location) was predicted 3 s after initiation of a fast stop shutdown. CARES/LIFE analysis and thermal gradient proof testing indicated that these stress levels were acceptable for long-term operation. Details of the ceramic nozzle design were documented in the literature [40, 42]. The nozzle material selection process has been documented [45, 47].

Nozzle Testing. The candidate nozzle ceramic materials were Hexoloy SA silicon carbide from Carborundum, and SN-88 (NGK), NT164 (NAC), and SN-253 from Kyocera Industrial Ceramics Corporation (KICC) silicon nitrides. NT164 was eliminated based on initial screening for slow crack growth degradation using dynamic fatigue testing and life assessment (fast fracture and slow crack growth only) using the NASA CARES/LIFE and Sundstrand SPSLIFE codes [36–38]. The high nozzle hot spot temperature (central portion trailing edge) of 1288°C suggested that creep and oxidation could be important contributors to potential component degradation. Long-term creep data were not available at the beginning of the CSGT program. Test programs were therefore initiated at the University of Dayton Research Institute (UDRI), Solar, and Oak Ridge National Laboratory (ORNL) to generate up to 10,000-hour creep rupture data.

SN-88 and SN-253 were selected for 1000-, 5000-, and 10,000-hour tensile creep rupture testing. Limited (1000-hours) testing was also conducted for Hexoloy SA. Figure 7.9 summarizes the results of the creep rupture testing. In the initial 1,000-hour creep test at 1288°C/207MPa in air, all three SN-88 specimens survived the test, two of the SN-253 specimens failed and one Hexoloy SA specimen failed. In creep testing for up to 5000 hours at 1288°C/186 MPa, both SN-253 specimens failed, while both SN-88 specimens survived. In final 10,000-hour creep testing at 1288°C/186 MPa, one SN-88 specimen failed at 7437 hours and

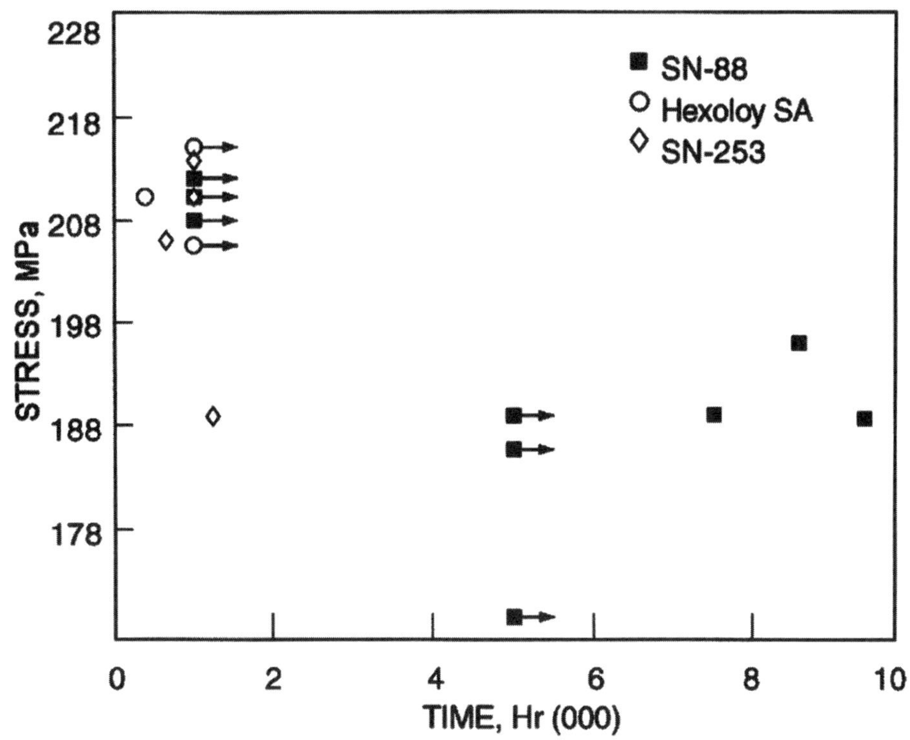

Figure 7.9 Creep Test Data for Ceramic Nozzle Materials (Reprinted from Ref. [26])

one specimen failed at 9497 hours. The temperature/stress combination of 1288°C/186 MPa was selected since at the time (1997) it was believed to be the most severe steady-state condition. Subsequent analysis placed the steady-state stress level at about 221 MPa. The creep rupture test data were the deciding factor in the selection of SN-88 for first-generation nozzle fabrication.

Each nozzle that passed visual and fluorescent penetrant inspection (FPI) was proof tested in the thermal gradient proof test rig and in a mechanical attachment test rig (see Figure 7.10). The thermal gradient proof test generates stress levels equivalent to or greater than those in service and eliminates parts with critical size defects. The basis for the nozzle proof test was provided by a finite element steady-state stress analysis that gave a hot spot steady-state stress at the airfoil trailing edge of 186 MPa when the engine is operated at the design TRIT of 1121°C. This stress level corresponds to a thermal gradient of about 578°C at the location of maximum stress (center of the nozzle trailing edge). Nozzles exposed to temperatures that created stresses in excess of their nominal fast fracture strength failed in the test. The mechanical test rig establishes a proof test load of 125% of the estimated stress in the nozzle attachment area. Nozzles that passed the thermal gradient rig also passed the mechanical proof test.

Blade Development

Blade Design. Figure 7.11 shows the evolution of the blade design. The Centaur 50S has 62 cooled MAR-M247 metal blades with an oxidation resistant platinum (Pt) aluminide coating. The number of ceramic blades was kept at 62. The CSGT blade design has an airfoil shape that is almost identical to that of the metal blade, except for the absence of cooling passages. Two attachment designs were developed. The

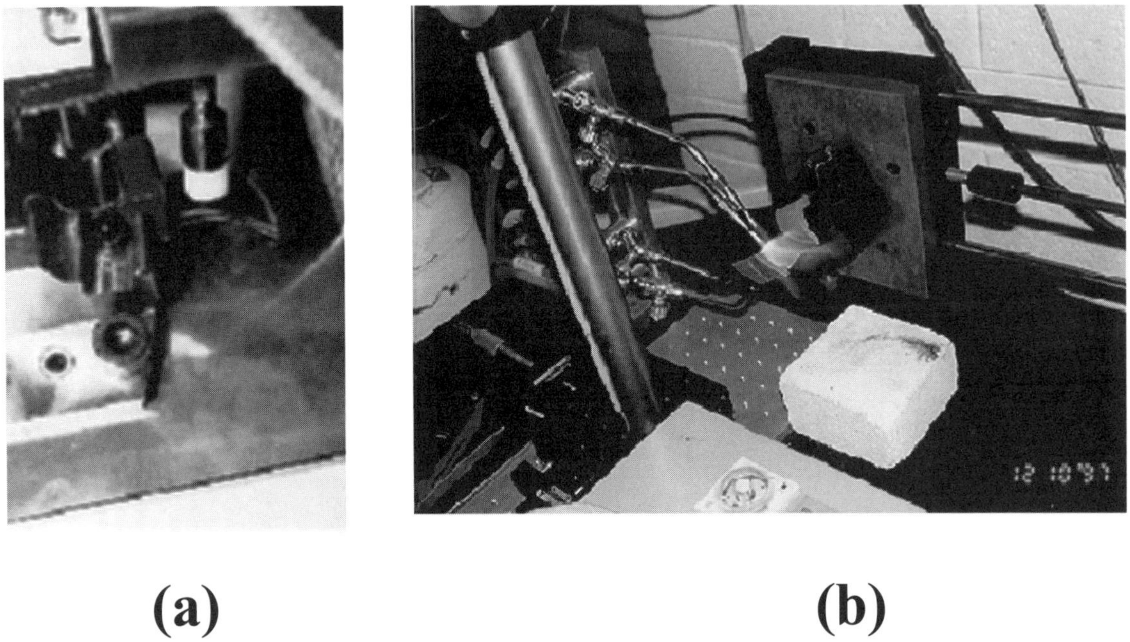

Figure 7.10 Nozzle Proof Test Rigs (a) Attachment Rig, (b) Thermal Gradient Rig (Reprinted with permission from Solar Turbines Incorporated)

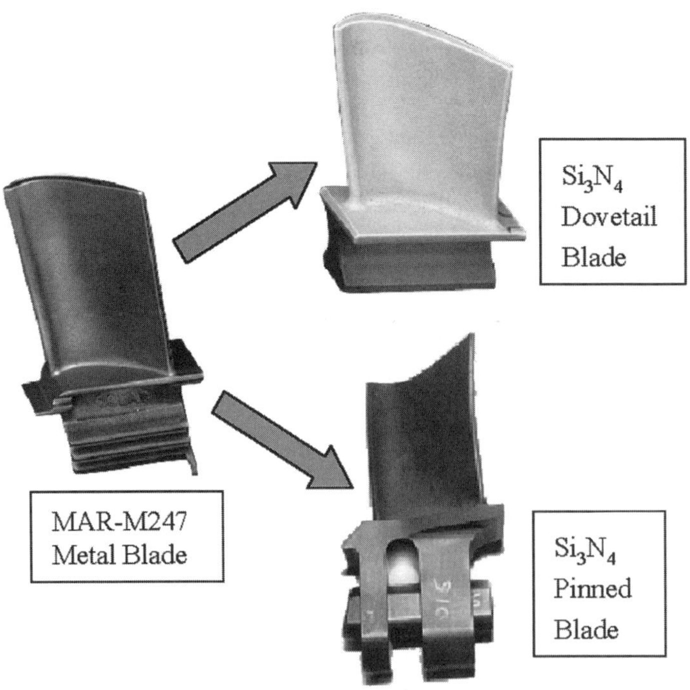

Figure 7.11 Evolution of the CSGT Blade Design (Adapted with permission from Ref. [33])

first design is that of a "dovetail" that requires the use of a compliant layer to act as a buffer for the ceramic/metallic interface. The key to a successful dovetail design is the optimization of the attachment angle. If this is too large, the frictional forces at the interface generated by centrifugal loading become excessive, resulting in potential sticking of the blade in the disk during shutdown. If the blade sticks, the subsequent contraction of the metallic disk as it reaches room temperature will impart a very high "pinching" load onto the blade root, which is capable of initiating failures. If the attachment angle is too small, the blade shank-root fillets become excessive as the bending stresses in the attachment reach critical limits. As a result, various dovetail attachment angles were carefully evaluated. Dovetail blades with a 55° angle were used for early rig and engine testing. Solar defines the dovetail angle as the complement of the angle between the blade median and the root edge. Maximum steady-state stresses in the 55° dovetail design were predicted to be 214 MPa at the blade root neck under the platform at a temperature of 682°C. A later dovetail design with 45° angle was also developed.

Figure 7.12 shows a finite element (FE) stress map for the final 45° dovetail blade configuration and a FE temperature map for an AS-800 silicon nitride 55° dovetail blade. The temperature profile for the 45° dovetail blade is similar to that of the 55° configuration. Dynamic behavior of ceramic blades differs considerably from metal blades due to the much higher elastic modulus of the material. Even though the airfoil did not change, it was important to re-establish the integrity of the blade frequency domain. Dynamic FE analysis of the AS-800 blade correlated very well with the acoustic and holographic techniques (used for metallic blades) and demonstrated that the vibratory modes would not give cause for concern.

An alternative design concept was a blade with a double-lug support attached to the disk with a ceramic pin (see Figure 7.11). A compliant layer is not necessary for this "pinned root" design due to the "round-round" interface geometry, which results in excellent load path predictability. This design was abandoned because (1) the maximum steady-state stress was higher than for the dovetail blade (283 MPa vs. 214 MPa), (2) manufacturing was very challenging because of tight tolerances and intensive machining, and (3) some of the fabricated parts failed in cold spin proof testing at 125% of design stress. Details of the blade design and development were reported in the literature [40, 41].

Blade Testing. All materials selected for blade fabrication and evaluation were silicon nitrides. The initial blade materials were NT164 (NAC) and GN-10 from AlliedSignal Ceramic Components (CC) for the dovetail blade and SN-253 (KICC) for the pinned root blade. While rig and short-term engine testing of first-generation dovetail blades fabricated from GN-10 and NT164 were successful, these first-generation design materials could not provide the required service life based on the CARES/LIFE analysis. For the blades, the maximum steady-state temperature is about 1093°C in the airfoil. At this temperature only the fast fracture and slow crack growth mechanisms were of concern. Table 7.2 shows life prediction data for four silicon nitrides. These data resulted in the elimiantion of NT164 and GN-10 (data not shown but similar to NT164 data) since the POS for a full set of blades of these materials in 30,000 hours was low, about 0.10. These materials had a very low value of the slow crack growth exponent, n, (GN-10: n=24; NT164: n=38) in dynamic fatigue tests [45, 47]. The other materials had POS values of 0.995 or higher for a full set of 62 blades for 30,000 hours.

AS-800 (CC), which became available later in the program, and SN-253 were subsequently selected for second-generation dovetail blade application. AS-800 is a beta-seeded, in situ-toughened Si_3N_4, which retains high strength to temperatures above 1300°C, and is characterized by Weibull modulus values of about 20 and RT fracture toughness values in the 8–10 $MPa.m^{1/2}$ range. SN-253 was later replaced by SN-281, a KICC material that has improved high temperature strength, creep resistance, and oxidation resistance.

Blade root configurations were evaluated in the attachment tensile test set up shown in Figure 7.13. The test specimen consists of two identical blade roots machined on both ends of a rectangular bar, held in a grip

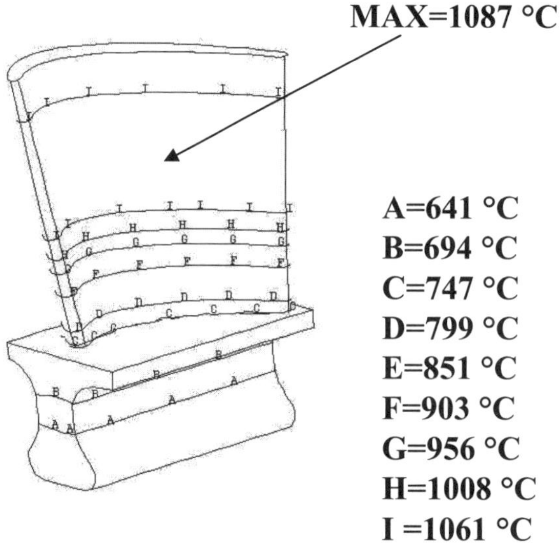

(a) Temperature distribution

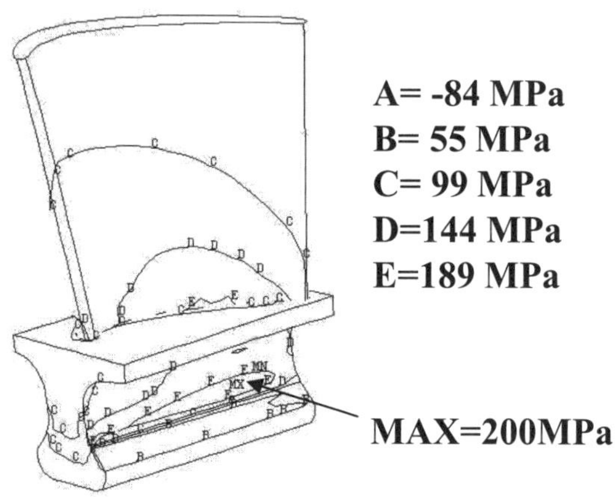

(b) Stress distribution

Figure 7.12 Finite Element Maps for AS-800 Dovetail Blade (a) Temperature Distribution (b) Stress Distribution (Adapted from Ref. [41])

TABLE 7.2 PROBABILITY OF SURVIVAL FOR SILICON NITRIDE DOVETAIL BLADES FROM CARES/LIFE ANALYSIS (REPRINTED FROM REF. [25])

	POS 4000 Hours		POS 10,000 Hours		POS 30,000 Hours	
Si_3N_4	Blade	62 Blades	Blade	62 Blades	Blade	62 Blades
NT164	0.98176	0.31940	0.97502	0.20837	0.96355	0.10004
AS-800	0.99999	0.99976	0.99999	0.99976	0.99999	0.99976
SN-253	0.99996	0.99752	0.99995	0.99690	0.99993	0.99567
SN-88	0.99998	0.99908	0.99998	0.99908	0.99998	0.99908

Figure 7.13 Blade Attachment Tensile Test Set-up (Reprinted from Ref. [26])

that simulates the turbine disk slot. The specimen is subjected to elevated temperature tensile loading to failure and also to stress/temperature cycling at temperatures and stresses representative of design conditions. The test evaluates the load carrying capacity of a specific blade design/material combination, as well as the performance of compliant layer systems over a selected stress/temperature cycle exposure. It was established that a conventional dovetail blade attachment could sustain a load up to about 3–4 times the design load at room temperature and about 2–3 times the design load at the projected operating temperature of the blade root (682°C). Some tendency for "wedging" was observed in the high temperature tests, which led to a redesign of the initial 60° dovetail angle to 55°. This change, along with friction reducing boron nitride (BN) surface layers on the compliant layer, solved this problem. The ruggedness of the compliant layer system was determined in extensive cyclic attachment testing. Several compliant layer systems were evaluated in this manner. The currently preferred nickel-base 0.13 mm thick compliant layer system has withstood 5000 hours of cyclic testing (over 200 24 hour cycles from room temperature to 682°C) indicating that it is likely to perform satisfactorily in an engine operating in continuous duty mode. The alternative pinned root blade configuration was eliminated since attachment testing indicated limited load carrying capacity and/or short life in cyclic testing.

Cold spin testing was conducted to evaluate blade attachment designs under more dynamic conditions and to provide a standard proof test for all blades prior to engine testing. The proof test eliminates blades with life limiting defects that may result in blade failure as a result of fast fracture or slow crack growth. All blades that have been used in engine tests were first evaluated in cold spin testing at 125% of design centrifugal force (CF) load (112% of design speed). Random blades were also tested at 200% of design CF load (141% of design speed). Several hundred dovetail blades of the program materials (GN-10, NT164, AS-800) were evaluated in Solar's standard 125% CF load spin test and none failed the test, indicating the substantial margin of strength of these candidate materials and the ruggedness of the design. Limited hot spin testing (up to 25 load temperature cycles) was also conducted prior to the first engine test in which ceramic blades were evalauted.

Secondary Component Development

Interfacing the metallic components with the ceramic parts is one of the major design challenges for the successful introduction of ceramics into a gas turbine. These challenges arise from two primary sources: thermal mismatch between interfacing metal and ceramic parts and increased heat flows from reductions in cooling air brought about by the use of uncooled ceramics and higher firing temperatures. The first concern was addressed by incorporating sufficient compliance between the metal and ceramic parts to allow effective load distribution, while the second was accommodated by materials uprates for the metal components and redirection of cooling flows, sealing and metal part protection. The two areas that required most attention were the secondary components interacting with the ceramic nozzle and the disk blade environment. The design challenges were detailed in the literature [25, 40–42].

Secondary component redesign for the nozzle environment was made necessary because of the following changes relative to the production Centaur 50S engine: (1) increase of TRIT from 1010°C to 1121°C, (2) first-stage nozzle design change from integral to separate tip shoe to reduce fabrication complexity, (3) uncooled first-stage nozzle, and (4) compliant clamping and sealing for the ceramic nozzle. Materials for both the nozzle support housing and diaphragm were changed to accommodate the higher metal temperatures. The thermal response of the complete diaphragm/nozzle assembly is designed to match the first-stage disk assembly and, thus, to provide superior tip clearance control. A more compliant face seal was selected to minimize all mechanical loads imparted to the ceramic nozzle. The nozzle support housing was completely redesigned and fabricated from two pieces for ease of assembly. The nozzle support structure went through several design iterations to optimize its performance in relation to the ceramic nozzles (see Figure 7.14). Changes from the first to the second generation design included incorporation of a Nextel insulator to reduce the conduction from the nozzle to the metallic support structure and to ease the outer shroud-to-nozzle load distribution, replacement of the segmented outer shroud ring to a single slotted ring to reduce leakage and associated circumferential thermal gradients, and incorporating a double "V" seal configuration to improve cooling effectiveness to the outer shroud and adjacent structure.

The first-stage disk material was uprated from V-57, an iron-based superalloy, which has insufficient creep life at the anticipated rim temperature of 671°C, to Waspaloy, a nickel-base alloy. In order to reduce cooling air requirements, forward and aft boltless rim seals were designed for the first-stage disk to assist in buffering the hot stream from the metallic hardware. The rim seal reduced the cooling air requirements for

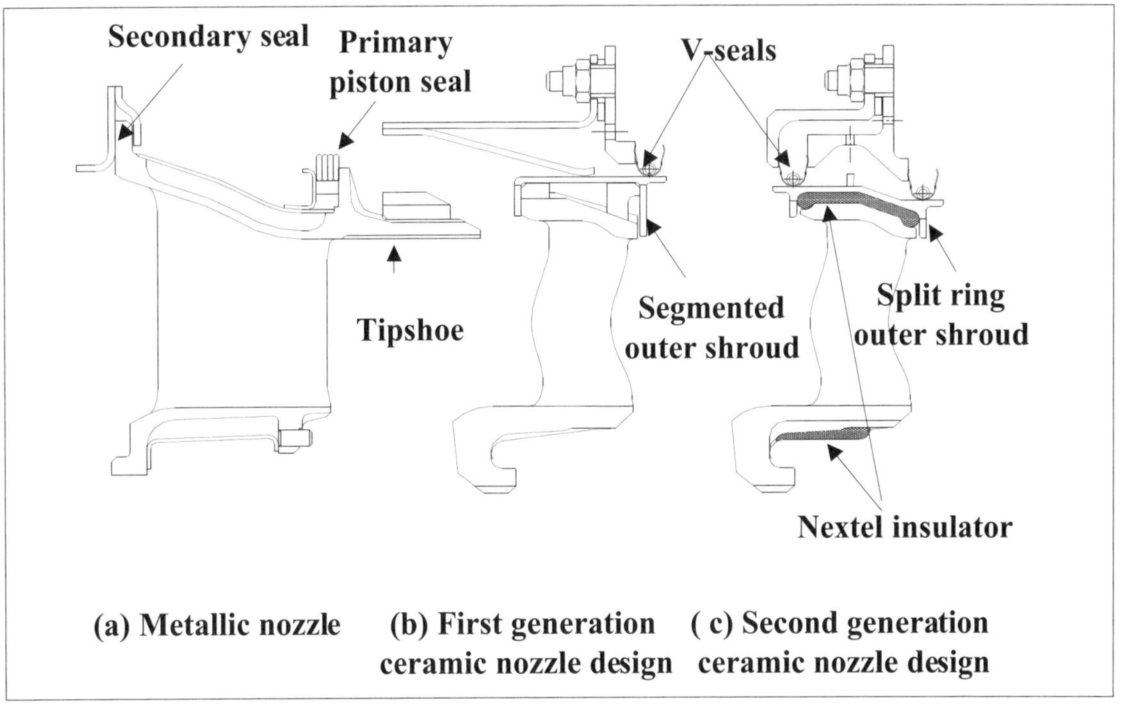

Figure 7.14 Nozzle and Secondary Component Layout for the CSGT Hot Section (Reprinted with permission from Solar Turbines Incorporated)

the dovetail blade/disk combination by approximately 50% over the baseline Centaur 50S design with a 1010°C TRIT.

Engine Testing

While laboratory and rig tests are necessary steps on the road to engine testing, they do not substitute for the "real world." The interplay between the combustion environment and the multitude of dynamic and static stress/temperature interactions between the components of the gas turbine are unique. Engine testing is the ultimate and only true validation of the design, material selection and fabrication of the component. Table 7.3 summarizes five years of Centaur 50S engine test experience with ceramic components for the CSGT program, which include 11 separate builds for in-house Solar engine tests and six field tests. A number of design changes and adaptations were made as a result of the engine tests. Engine test experience has been reported in the literature [24–35].

The engine test philosophy was based on initially evaluating the components separately, before testing them in combination in the Centaur 50S gas turbine. This minimized secondary damage to downstream ceramic components in case of an upstream ceramic component failure. Most of the engine tests were performed at the baseline TRIT of 1010°C of the all-metal Centaur 50S engine. Following separate component qualification, engine testing at the baseline TRIT was repeated with two ceramic components installed.

TABLE 7.3 CSGT ENGINE TEST SUMMARY (AUGUST 1995–JUNE 2001)

Components	Blades	Nozzles	Combustor Liners*
Ceramics	NT164,GN-10, AS-800, SN-281 Si_3N_4	SN-88 Si_3N_4	DLC/H-ACI SiC/SiC (CVI and MI) BFG SiC/SiC (CVI and MI)
Total Test Time	1600 hours	195 hours	> 33,000 hours
Max Time on Single Assembly	1052 hours (AS-800)	104 hours	13,937 hours (DLC/H-ACI SiC/SiC CVI Outer Liner/MI Inner Liner)

* MI: Melt Infiltration Process – H-ACI: Honeywell Advanced Composites, Inc., formerly DLC

Figure 7.15 CSGT Centaur 50S Engine Test Set-up (Adapted from Ref. [25])

Toward the end of the program, a final test at the ultimate design TRIT of 1121°C with all three ceramic components was performed.

In-House Engine Testing. Figure 7.15 shows the engine test rig at Solar on which all in-house testing was conducted. Short (1–2 hours) initial ceramic engine tests started in August 1995. These tests, which were performed with the available NT164 and GN-10 silicon nitride blades, were intended to validate the dovetail blade design. The ceramic blades and interfacing secondary components performed as expected and demonstrated good durability. Two different compliant layers were evaluated in these tests and both appeared to function satisfactorily over the short test duration. The next engine test was conducted with

Figure 7.16 Centaur 50S Rotor with AS-800 Blades During Balancing (Reprinted from Ref. [28])

BFG SiC/SiC CFCC combustor liners for a total of 12 hours at full load over several cycles. This test established that the CFCC material performed adequately. Emission levels were very promising NOx levels < 15 ppmv and CO levels < 10 ppmv were achieved at 3% pilot fuel setting [44].

Subsequently, ceramic blades and CFCC combustor liners were tested simultaneously at the baseline TRIT of 1010°C, but the test time and the number of cycles were increased. Various transient conditions were evaluated based on typical shutdown/restart scenarios common in commercial engine operation. In November 1996, a milestone endurance engine test was conducted in which 62 AS-800 first-stage blades and enhanced SiC/SiC (DLC) combustor liners were evaluated. Figure 7.16 shows the rotor assembly with the first-stage AS-800 blades during balancing prior to insertion in the test engine. In this test, planned for 100 hours, two compliant layer systems (Ni-alloy + BN, and Ni-alloy + Pt) for the blade attachments were used. One test objective was to compare the performance of the two systems and select the best system for incorporation in a field test. The test terminated after 58 hours due to failure of the first-stage blades. There was no evidence of foreign object damage (FOD) as a source of component failure initiation. While most of the blades ruptured above the platforms, only segments of the blade roots remained in alternating slots in one sector of the disk spanning 12 blades (see Figure 7.17). The more heavily damaged blades were associated with the Pt-coated compliant layer system, which caused the dovetail to pinch in the disk following cyclic engine testing.

Interestingly, the condition was subsequently duplicated in a blade tensile attachment test. In the past, the Ni-alloy + Pt compliant layer systems survived many cycles because the test included a step in which ceramic attachment was pinch relieved after each cycle. When this step was eliminated, the failure in the attachment occurred in exactly the way shown in Figure 7.17 after just two cycles. The ceramic attachment incorporating the alternative N-alloy + BN compliant layer system survived 120 thermal and stress cycles over a 5000 hour period after pinch relief had been eliminated. The attachment system was subsequently modified and the failed compliant layer system was eliminated [41]. These modifications to the blade

(c) Attachment test failure
Ni-alloy+Pt compliant layer

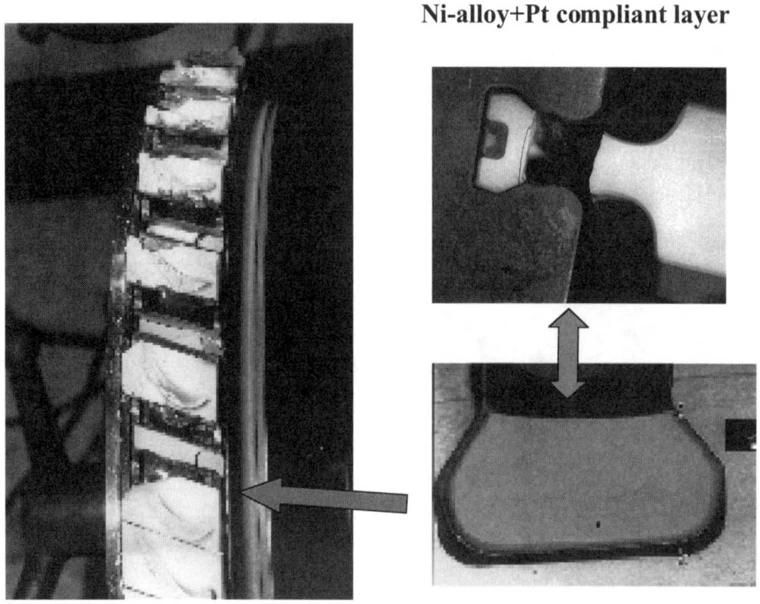

(a) Rotor with AS-800 blade residues

(b) Failed blade root: Ni-alloy+Pt compliant layer

Figure 7.17 Failed Centaur 50S rotor: (a) Rotor with AS-800 Blade Residues, (b) Failed Blade Root—Ni-alloy+Pt Compliant Layer, (c) Attachment Test Failure—Ni-alloy+Pt Compliant Layer (Reprinted with permission from Solar Turbines Incorporated)

attachment system were evalauted in a successful second 100-hour test, which included 12 cold starts and 15 hot restarts. Final inspection revealed the ceramic and metal components to be in excellent condition.

The first engine test of 42 SN-88 silicon nozzles was conducted in January 1997. The nozzles completed a nine-hour test (one hour at full load) in the CSGT Centaur 50S developmental gas turbine at Solar. During tear-down of the nozzle assembly, some minor chipping was found at the inner and outer shroud areas contacting the metal, suggesting a localized excessive contact stress condition as can be seen in Figure 7.18. The ceramic nozzle/metallic attachment was redesigned to eliminate the sources of the localized contact stress [42]. The redesign of the nozzle on the diaphragm is shown in Figure 7.19. By faceting the diaphragm and removing material at the diaphragm contacting edge locations of each ceramic nozzle, edge loading on the nozzle inner shroud was reduced and no chipping was observed in subsequent nozzle engine tests.

A 100-hour nozzle engine test was initiated in September 1998. The test consisted of cold and hot engine restarts and shutdown cycles that progressively increased the severity of stresses and thermal gradients on the nozzle. Borescope inspections were conducted after shutdown cycles. After 68 hours of cyclic engine testing, the borescope inspection revealed cracking in several airfoils and the test was terminated for detailed evaluation. Figure 7.20 shows the nozzle section after the engine test. A detailed failure investigation was conducted at ORNL, Solar and NGK Insulators, Ltd. No chipping was found in the attachment areas of the nozzles. Both scanning electron microscopy (SEM) and x-ray analyses indicated changes in the microstructure and secondary phase due to exposure in the engine environment (temperature, water vapor). Apparently, an extensive damage zone developed in the subsurface region after the 68-hour test and changes

Figure 7.18 SN-88 Nozzle Chipping at the Inner Shroud Occurred During Engine Test (Reprinted with permission from Ref. [30])

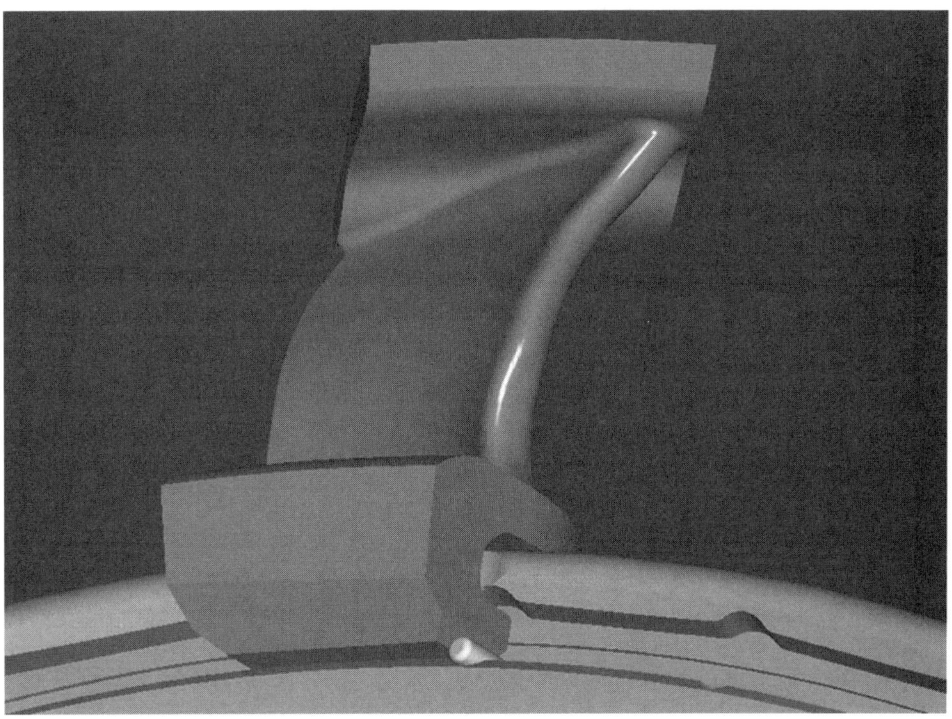

Figure 7.19 Redesigned Diaphragm with Faceted Nozzle Attachment (Reprinted with permission from Ref. [30])

Figure 7.20 Ceramic Nozzle Section After 68-hour Engine Test (Adapted from Ref. [31])

TABLE 7.4 DATA FROM FINAL CSGT IN-HOUSE ENGINE TEST (APRIL 2001)

TRIT (°C)	Shaft Thermal Eff. (%)	Shaft Output Power (hp)	Output Power (kWe)
1010	28.8	5581	3932
1066	29.6	6302	4440
1121	30.1	6910	4868

in the secondary J-phase apparently led to the generation of a large residual tensile stress. It was established that in the affected region of the airfoil, which was at relatively low temperature (about 850°C), the predominant secondary phase (J-phase—$Yb_4Si_2O_7N_2$) underwent a chemical change to $Yb_2Si_2O_7$ plus $Yb_2Si_2O_5$ after the protective silica layer was removed from the Si_3N_4 [48].

In April 2001 the final in-house engine test for the CSGT program was conducted with all three ceramic components, a full complement of 62 first stage SN-281 Si_3N_4 blades (KICC), 42 SN-282 Si_3N_4 nozzles (KICC), and a set of CG SiC/SiC CVI CFCC liners from H-ACI that had been used on a previous engine test. The test was planned for 100 hours at full load. In fact, total test time was 104 hours, most of which was at the baseline TRIT of 1010°C. The engine was also operated for short times (< 1 hour) at TRITs of 1066°C and 1121°C. This time was sufficient to collect performance data which have been summarized in Table 7.4. The engine test was not optimized for emissions performance for the higher TRITs of 1066°C and 1121°C. Liner temperatures increased with TRIT. A maximum temperature of 1284°C was measured on the outer liner at the 1121°C TRIT. Figures 7.21 and 7.22 show the first stage rotor and the nozzle assembly, respectively, following completion of the engine test.

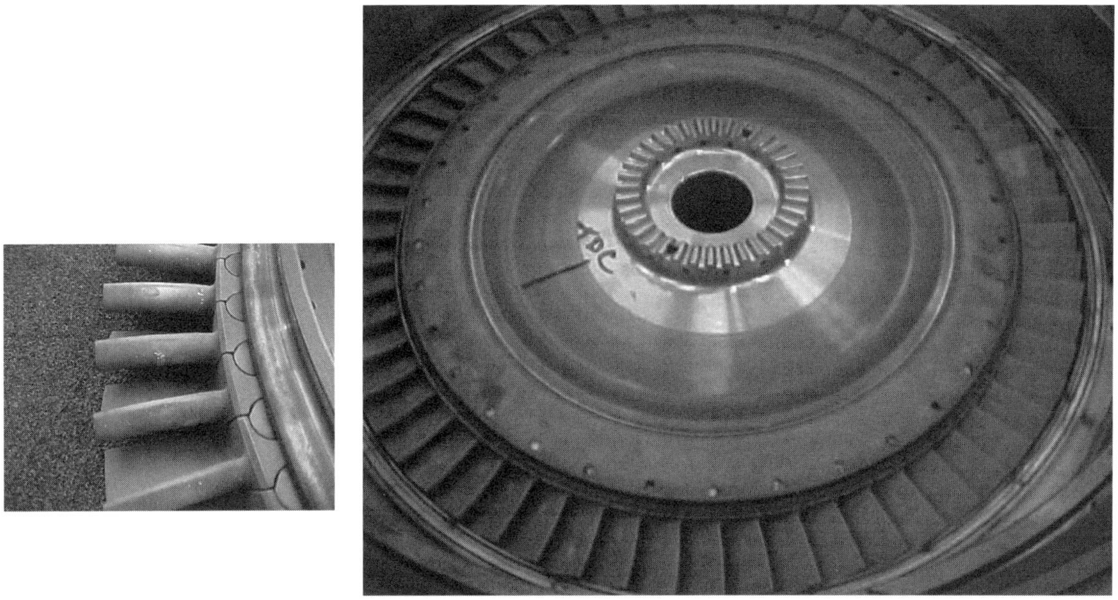

Figure 7.21 First Stage Rotor with SN-281 Blades After the Final 104-hour CSGT In-house Engine Test (Reprinted with permission from Solar Turbines Incorporated)

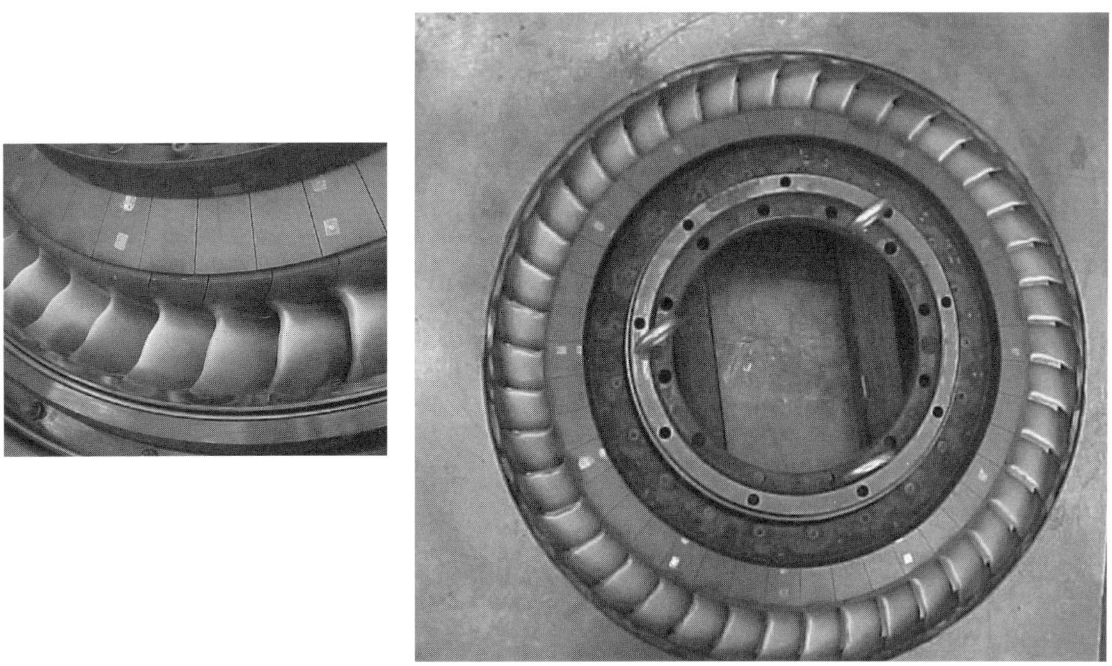

Figure 7.22 First Stage Nozzle Assembly with SN-282 Nozzles After the Final 104-hour CSGT In-house Engine Test (Reprinted with permission from Solar Turbines Incorporated)

To what extent were performance improvements achieved in the engine testing since this was a key objective of the program? The all-metal engine used for in-house engine testing had a thermal efficiency of 28.8% at the nominal baseline TRIT of 1010°C. In the first two engine tests with NT164 and GN-10 blades at the same nominal TRIT, thermal efficiencies were 27.9% and 28.1%, respectively, indicating an efficiency loss of 0.7–0.9%. Metal combustor liners were used for all these tests. For the tests conducted with uncooled nozzles which was conducted with a different test engine, the thermal efficiency was 28.6% versus 30.4% in a similar test with the metal nozzle, a 1.6% efficiency loss. SiC/SiC CFCC combustor liners were used for both of the latter engine tests. In the final engine test the thermal efficiency was 28.8% at the baseline TRIT of 1010°C. The test engine was the same as the one used for the ceramic nozzle engine test.

A major factor that contributed to the lower thermal efficiency with the ceramic parts was the greater first-stage blade cold tip clearance used for the engine tests with ceramic parts. For the tests with ceramic blades the tip cold clearance was increased to avoid a tip rub, while in the engine tests with metal blades, a tip rub is allowed and is accommodated by the presence of a rub-tolerant coating. The increase in tip clearance from about 0.5 mm in the all-metal Centaur 50S to about 1.3 mm is believed to have resulted in a thermal efficiency loss of approximately 0.6%, which accounts for most of the efficiency loss observed. For the tests with ceramic nozzles, the cold tip clearance was deliberately increased over that of the engine with the metal nozzle because of the new non-integral first stage blade tip shoe design. Another factor that might have contributed to the efficiency loss for the engine tests with ceramic parts was the increased amount of cooling air on some of the secondary components. The cooling flows for the first-stage disk were designed for operation at the final program TRIT of 1121°C, but the engine was operated mostly at a TIT of 1010°C. Larger than predicted leakages may also have played a role for the ceramic nozzle tests. Table 7.4 shows that in going from the baseline TRIT of 1010°C to the final TRIT of 1121°C, the thermal efficiency increased by 1.3%, and the output power by 936 kW (i.e. 23.8% over that of the baseline engine). These improvements over the performance at the baseline TRIT are close to the predicted improvements, taking into account variability between engines, and keeping in mind that the changes are from a ceramic baseline at a TRIT of 1010°C to that of an engine with ceramic parts operating at a TRIT of 1121°C. The performance changes from the engine with cooled metal parts at the TRIT of 1010°C to an engine with ceramic parts at the TRIT of 1121°C can be expected to be somewhat greater because of the additional elimination of component cooling in the ceramic components. It should be noted that the predicted performance improvements are largely attributable to the increase in TRIT by 111°C with the gains made because of the elimination of cooling contribute less to the overall performance improvement of the ceramic over the all-metal engine.

The emissions testing to date has been very satisfactory. The program goal of NOx ≤ 25 ppmv was demonstrated in engine testing at the baseline TRIT of 1010°C. As mentioned previously, NOx ≤ 15 ppmv was consistently demonstrated at the baseline TRIT of the metal engines, and NOx ≤ 10 ppmv was occasionally achieved at low pilot settings. CO levels < 10 ppmv were also typically measured.

Field Testing. The CSGT field test is a very significant step on the road to demonstrating commercial viability of the ceramic gas turbine concept. The first CSGT field test was launched in May 1997 when the Centaur 50S engine retrofitted with enhanced SiC/SiC (DLC) CFCC liners and AS-800 (CC) first-stage blades started operation at the ARCO Western Energy (currently Texaco) Bakersfield enhanced oil recovery site. Figure 7.23 shows the CSGT engine being installed in the package at the ARCO site.

The CSGT engine accumulated 948 hours over the duration of the first field test in addition to 4 hours qualification testing and 100 hours endurance testing at Solar, giving a total test time of 1052 hours at full load. The engine met all expectations of performance for a commercial gas turbine. Emission levels for NOx were in the 10–15 ppmv range and CO levels of around 5 ppmv were measured over the 50–100% load range. The CSGT engine was in all aspects fully operational with normal power and steam production. Borescope inspections conducted after 212 and 533 hours of operation showed no change in appearance of the CFCC combustor liners and first-stage ceramic turbine blades compared to the 100-hour acceptance test at Solar.

Figure 7.23 CSGT Centaur 50S Engine Being Installed at the ARCO Bakersfield Site (Reprinted with permission from Ref. [30])

At 948 hours of full load operation, the engine shut down on a fast stop shutdown command alarm. Borescope inspection revealed blade failure during service. The failure cause was in all likelihood impact damage from a metallic locating-pin from the inner combustor liner. Failure analysis indicated that all blade failures occurred above the platform, which excluded failure of the dovetail blade attachment/compliant layer/disk interface. Two other potential causes of blade failure, a rub condition, involving the blade tips and the tipshoes, and airfoil failure, as a result of slow crack growth, were considered, but no evidence was found to support either one.

One concern that surfaced following the completion of the test was the condition of the enhanced SiC/SiC CFCC combustor liners. Both liners, but in particular the inner liner, showed evidence of oxidation degradation. Subsequent analysis revealed that the inner liner was oxidized to a depth of about 0.5 mm. The mechanism of oxidation involves the formation of volatile SiO and Si(OH)x species formed in the interaction between the SiC constituent of the CFCC and combustion water vapor. Critical parameters in the degradation are gas temperature, total pressure, and velocity, and partial pressure of water vapor. The phenomenon of accelerated surface recession of silicon based ceramics under the high pressure combustion environment of the gas turbine was reported in the literature [49, 50], and is also described in detail in Chapter 26 of Volume II of this text.

The oxidation is highly temperature dependent. Design changes were made to the metallic portion of the liners to improve the conduction path from the CFCC liners to the metallic housing and thereby, to lower

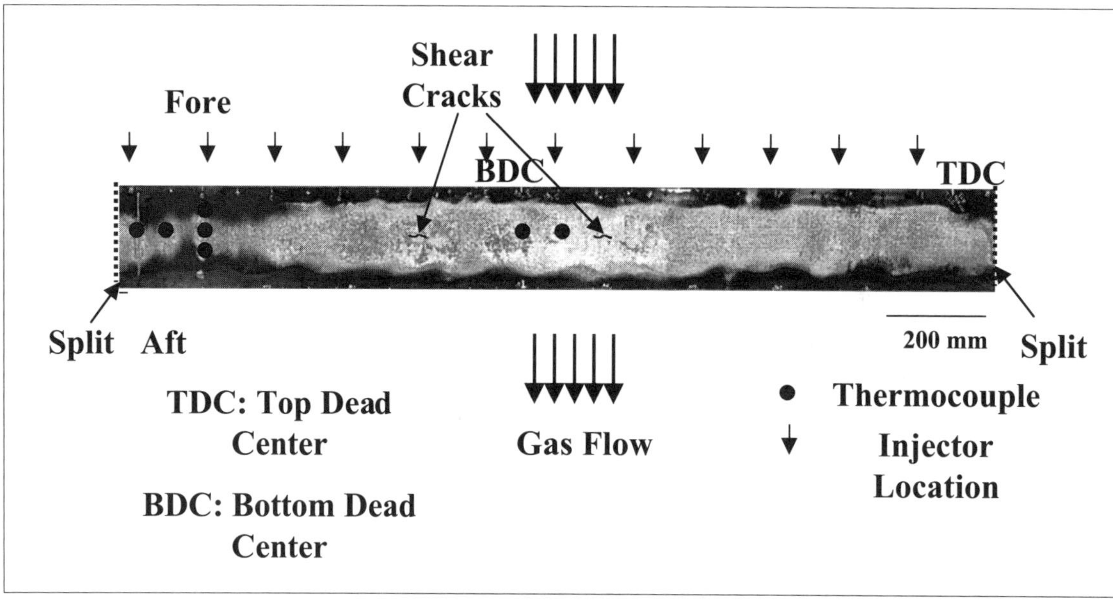

Figure 7.24 The Inner (Hot) Surface of the Hi-Nicalon/SiC Outer Liner After 5028 Hours of Engine Operation (Reprinted from Ref. [53])

the wall temperature for the CFCC liners and slow down the oxidation process. Several material changes were incorporated in later generations of CFCC liners to further mitigate the effects of environmental oxidation. These included replacing the ceramic-grade Nicalon fibers with Hi-Nicalon fibers which are stronger and more stable at higher temperatures and have higher thermal conductivity than the ceramic grade Nicalon fibers, increasing the density of the CFCC liners, and increasing the thickness of the protective seal coat used for the CFCC liners. Also, for later tests the inner liner was fabricated using the MI process, which gives an inherently denser and higher conductivity CFCC than the CVI process used for the earlier liners.

Figure 7.24 shows a composite photograph of an enhanced SiC/SiC outer liner after 5028 hours of cumulative exposure in three subsequent field tests and pre-field qualification tests. The oxidation of the liners (white deposits) is clearly visible. The oxidation is highly correlated with the temperature of the liners monitored with thermocouples. The surface oxidation observed in both the inner and outer liners is correlated with the fuel injector impingement locations. Figure 7.25 shows thermal diffusivity images collected at ANL during NDE of the outer liner in the as-received condition, and after 2266 hours and 5028 hours of engine testing. The NDE scan at 2258 hours identified an area of substantial degradation, which was heavily cracked after 5028 hours. Figure 7.26 shows the microstructure of a degraded liner area. About 80% of the CFCC liner was affected in locations that showed the heaviest degradation. The CFCC material degradation from the field test closely matched the degradation encountered in simulated gas turbine environmental exposure testing at ORNL [51], which enables the lower cost rig test at ORNL to be used to effectively screen various CFCC liner systems. Additional details of the rig and field testing of the CFCC liner materials have been reported [52, 53].

In the most recent field test at the Texaco site, which started in April, 1999, environmental barrier coatings (EBCs) were applied by United Technologies Corporation (UTC) to the SiC/SiC CFCC liners to further increase their life [54]. UTC, under subcontract to the CSGT program, had optimized EBCs, starting with the baseline EBC developed under the NASA Enabling Propulsion Materials (EPM) Program. Figure 7.7

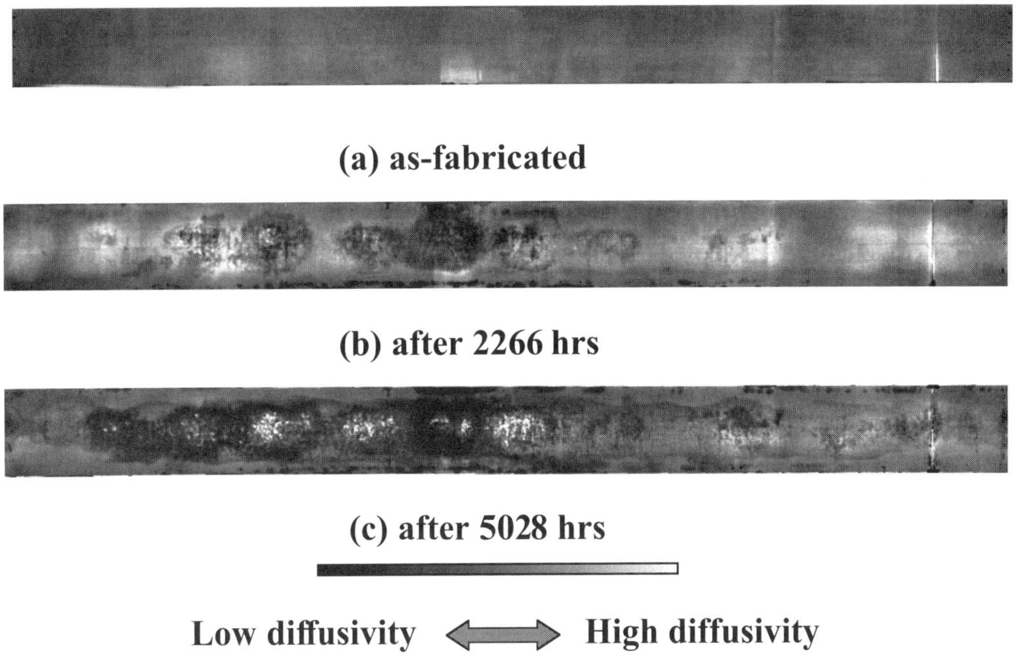

Figure 7.25 Thermal Diffusivity Images of the Hi-Nicalon/SiC Liner: (a) As-fabricated, (b) After 2266 Hours, (c) After 5028 Hours of Engine Operation (Adapted from Ref. [53])

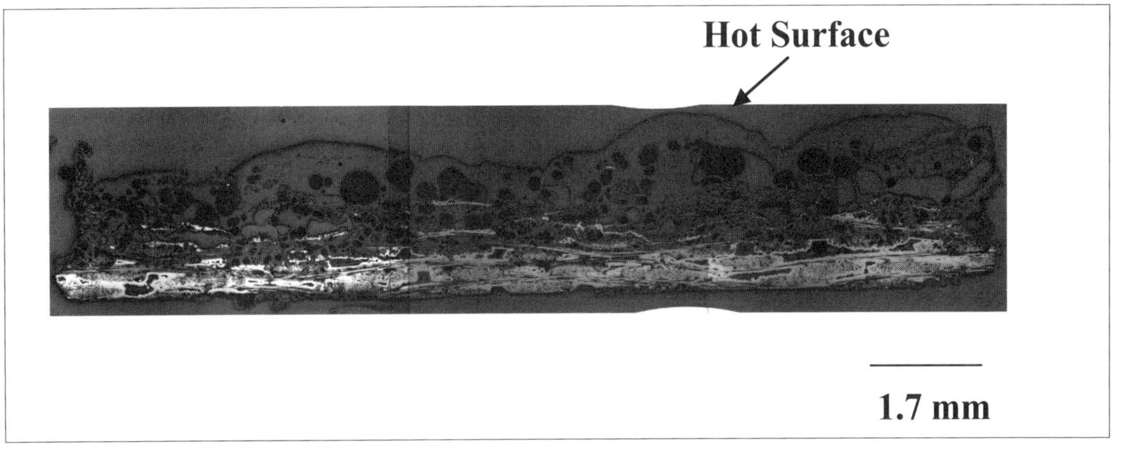

Figure 7.26 Cross-section of Area of Maximum Recession (80% of Wall Thickness) in the Hi-Nicalon/SiC Outer Liner After 5028 Hours of Engine Testing (Reprinted from Ref. [53])

includes a set of EBC coated SiC/SiC CFCC liners. The initial goal of this test was to achieve 8000 hours of production operation prior to removal of the liners. Periodic borescope inspections were conducted throughout the test. It is surmised that an effective EBC will enhance liner life significantly. Additionally, EBCs can be refurbished, thereby further enhancing useful liner life.

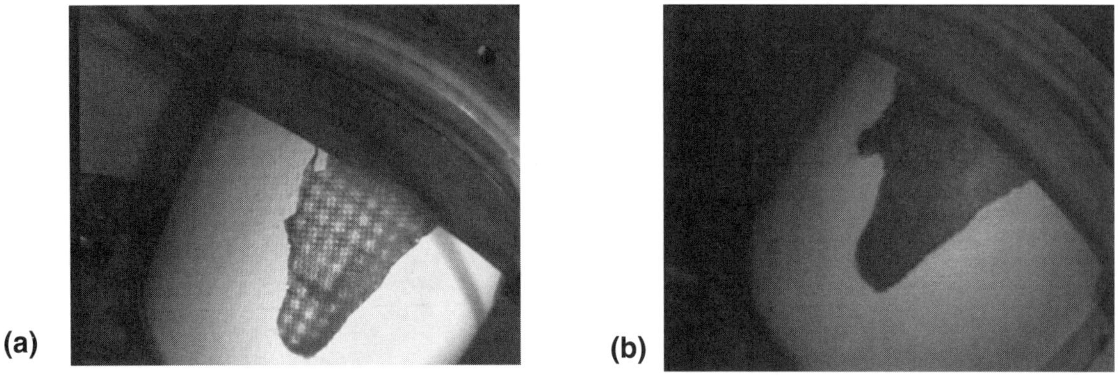

Figure 7.27 Borescope Images of SiC/SiC CFCC MI Inner liner: (a) After 2,897 Hours and (b) After 8,465 Hours of Engine Operation at the Texaco site (Adapted from Ref. [35])

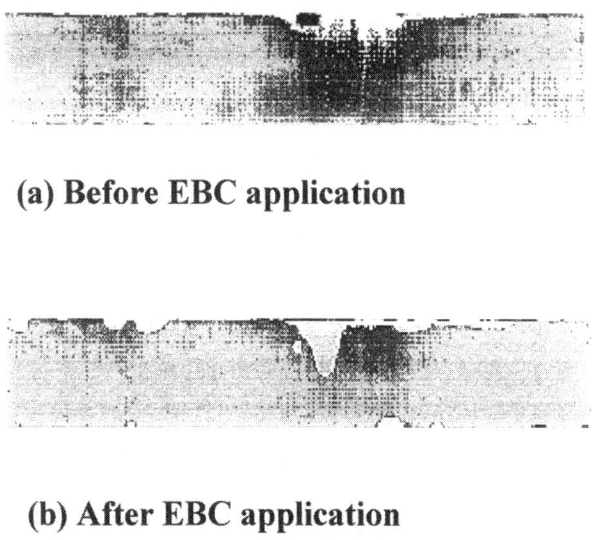

Figure 7.28 ANL NDE Thermal Diffusivity Scans of SiC/SiC CFCC MI Inner Liner: (a) Before EBC Application (b) After EBC Application (Adapted from Ref. [35])

Figure 7.27 (a) shows a borescope view of the MI inner liner after 2897 hours of operation at the Texaco (formerly ARCO) field test site. The EBC had spalled in some areas because of localized delamination near the surface of the CFCC liner that was already observed during a borescope inspection at the 1034 hour mark. The area where the EBC spallation had occurred closely corresponded to an anomaly observed in a thermal diffusivity NDE scan. The anomaly was not observed prior to EBC application (see Figure 7.28). Subsequent borescope inspections indicated that while coating spallation and liner degradation appeared to be gradually increasing over time, the liners remained in functional condition and the decision was made to continue testing beyond the 8000 hour milestone. Figure 7.27 (b) shows the borescope image of the inner liner after 8465 hours of operation. The field test was eventually terminated late November, 2000, after 13,937 hours of production operation and 61 starts, due to a breach of the inner liner observed during borescope inspection, as can be seen in Figure 7.29. Although the liners were still functional, the engine was removed and shipped to Solar for careful teardown and detailed evaluation.

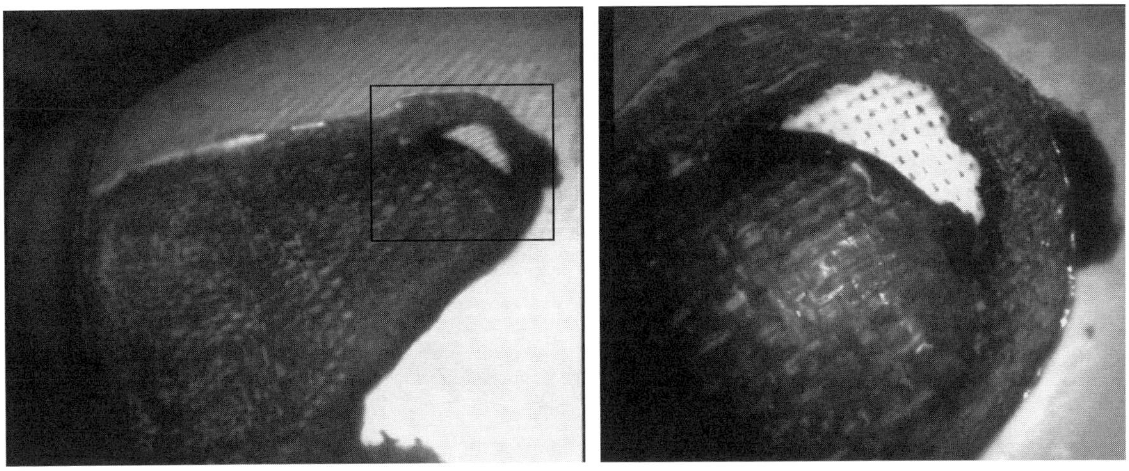

Figure 7.29 Views of SiC/SiC MI Inner Liner After 13,937 Hours of Engine Operation Which Shows that Liner has been Breached (Reprinted with permission from Solar Turbines Incorporated)

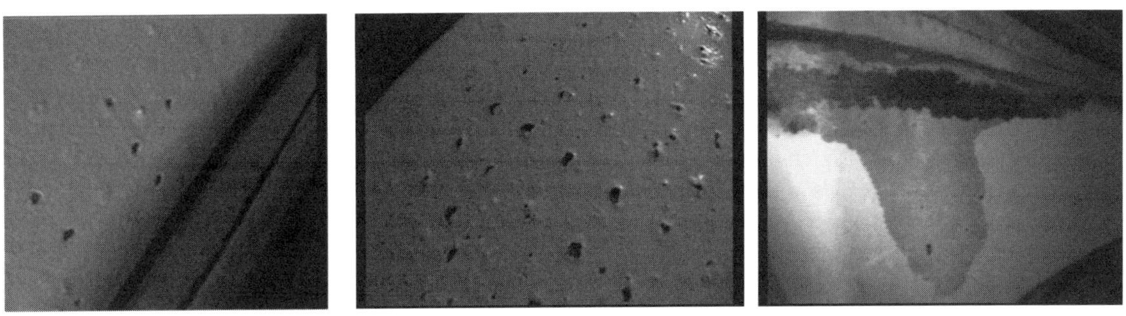

2,897 Hours 8,465 Hours 9,910 Hours

Figure 7.30 Borescope Inspection Views of SiC/SiC CVI Outer Liner After Various Hours of Engine Operation (Reprinted from Ref. [35])

Figure 7.30 shows sequential borescope inspections for the outer liner. In the borescope inspection conducted after 2897 hours of operation, pitting was observed in a grid pattern consistent with small bumps on the CFCC liner surface in the as-received condition. These small bumps were formed during CVI processing of the liner at tooling hole locations. Application of the EBC over these bumps resulted in small cracks forming in the EBC, allowing oxidation of the CFCC through the crack, eventually forming the pits. In future liner fabrication, these bumps can be eliminated through modification of the CVI tooling. The pitted areas at the tooling bump locations gradually increased in size and quantity over time, as shown in Figure 7.30. A large spall between coating layers can bee seen in borescope image at the 9910 hour inspection point.

Figure 7.31 shows the inner and outer liners, respectfully after removal from the engine. While areas where the EBC was missing showed significant degradation, the areas where the EBC remained appeared in good condition. The CFCC liners were sent to ANL for NDE. The program team prepared a detailed

(a) Inner Liner: Hi-Nicalon/SiC-Si MI (H-ACI)

(b) Outer Liner: Hi-Nicalon/Enhanced SiC CVI (H-ACI)

Figure 7.31 SiC/SiC Liners After 13,937 Hours of Field Testing at the Texaco site (Reprinted from Ref. [35])

sectioning and evaluation plan, and portions of the sectioned liners were sent to various team members for detailed evaluation. Solar is planning continued field testing at the Texaco site in the 2001–2003 time frame.

CFCC liners with EBCs were also incorporated into two Centaur 50S engines at the Malden Mills Industries textile factory in Lawrence, Massachusetts. The testing was initiated in August 1999 on the first engine and in August 2000 on the second engine. Both tests were scheduled to continue for at least 8000 hours for each of the two engines. This test differs from the Texaco test in that the engines undergo many more shutdown cycles. The liners for the first installation were a Hi-Nicalon/SiC MI inner liner fabricated by BFG and a Hi-Nicalon/SiC CVI outer liner fabricated by HACI. Both liners contain the optimized EBC deposited by UTC. The EBC coated liners also incorporated a SiC seal cost deposited by chemical vapor deposition between the CFCC surface and the EBC.

The test on the first engine was termianted in October 2000 after 7238 hours and 159 starts (compared to 61 starts at Texaco after 13,937 hours) due to an engine problem not associated with the CFCC liners. The engine was shipped to Solar's overhaul facility in De Soto, Texas, for careful teardown of the engine. The combustor assembly was then sent to San Diego for disassembly and inspection of the CFCC liners. Most of the inner liner appeared in good condition following the test. Some spallation of the coating was discerned at the mid section of the inner liner, while some areas of complete spallation were observed near the top (aft) end of the liner. The outer liner exhibited significantly more spallation following the test. While the coating remained intact in some areas of the liner, spallation between coating layers and complete coating spallation was seen for a large portion of the liner. As was seen in the Texaco field test, pitting, as expected was experienced at the CVI tooling bump locations. Detailed evaluation of the CFCC liners and EBC coating was initiated by the CSGT team, and it was found that the CFCC liners were in sufficiently in good condition to be reused for additional-testing, following stripping of the residual EBC and recoating with a new improved EBC. Details on the recent field testing and coating refurbishment have been reported [35, 46, 55].

The liners for the second engine at the Malden Mills site are a Tyranno/SiC MI inner liner fabricated by BFG and a Hi-Nicalon/SiC CVI outer liner fabricated by HACI. Both liners contain an optimized EBC deposited over a CVD SiC seal coat by UTC. As of June 2001, this test had accumulated over 6,500 hours of operation and 52 starts. This test is scheduled to run for at least 8000 hours prior to removal for evaluation.

SUMMARY OF LESSONS LEARNED

The lessons learned from Solar's ceramic gas turbine programs are summarized below and were also described in more detail in an ASME paper [32]:

- Both ceramic components and adjacent metal parts need to be considered for ceramic gas turbine design. Alleviating thermal expansion mismatch and contact stresses is critical for component performance. Contact stress issues played a major role in failures observed under the Gemini and CSGT programs.
- For the dovetail blade design, the root angle (45–55°), compliant layer material selection and thickness, and the incorporation of an anti-friction surface coating (BN) are essential for the mitigation of contact stresses at the ceramic blade/metal disk interface.
- Even if contact stress problems are solved, impact damage remains a major hurdle for the practical application of monolithic ceramics, particularly for rotating components. This is particularly true for retrofit of existing metallic components where opportunities for a more rugged redesign are limited. This was felt to be a major impediment to a successful monolithic blade design under the CSGT program. Future development of monolithic ceramics with improved fracture toughness will be needed for successful rotating ceramic component designs. This is a major effort that may require substantial government support to be successful.
- CFCCs have superior "fracture toughness" compared to monolithic ceramics, but have considerably less strength. Design stresses for the SiC/SiC CFCC materials are best kept at least 25% below the first matrix cracking stress. This sets an upper limit of about 100 MPa for the best CFCC materials (year 2000 vintage) for most applications. The design operating temperature should not exceed 1200°C for long life.
- Materials selection, environmental conditions in the turbine, and component design need to be considered in designing with ceramics. Computerized life prediction is a useful tool that helps to predict service life by identifying life-limiting factors, but it only gives partial information.
- While rig testing gives useful initial information for the early validation of component design and materials performance, testing in an actual gas turbine for extended time (thousands of hours) is required to fully assess ceramic component designs for commercial applications.
- Since continuous service lives on the order of 30,000+ hours are expected from industrial gas turbines, a data base covering the entire range of temperature, stress, and target service life that the component of interest will experience during all stages of engine operation, including steady-state and transient conditions (start-up and shutdown, trips), must be developed. This database needs to include time independent (fast fracture strength) and time-dependent (oxidation, slow crack growth, creep, fatigue resistance, environmental degradation) properties that are critical for component life prediction.
- Silicon based ceramics—monolithic Si_3N_4 and SiC, as well as SiC/SiC CFCCs—have shown accelerated surface degradation and recession under the conditions of the gas turbine combustion environment. Contributing factors are gas temperature, total pressure, gas velocity, and partial pressure of water vapor. Effective EBCs are critical for meeting component performance and service life targets in gas turbines in commercial operation. State-of-the-art EBCs are showing great promise for CFCCs thanks to vigorous development efforts under sponsorship of U.S. government agencies. EBC development for monolithics is still in its infancy in the year 2000, and will need substantial government support to be successful. Further research will be required to show the temperature limitations and modes of degradation of the EBCs in the gas turbine.

FUTURE POTENTIAL

Substantial progress has been made in the design/development and end user demonstration of ceramics for Solar's gas turbine products. At the beginning of the CSGT program, component designs for industrial-size gas turbines had only been tested for tens of hours and no actual engine testing had ever been conducted in a full 4-MW engine under field conditions. Designs for monolithic blades, nozzles, and CFCC combustor liners have all been demonstrated in an industrial gas turbine, and CFCC combustor liners have operated continuously for almost 14,000 hours.

The two major hurdles preventing commercialization of the ceramic gas turbines are long-term durability and component cost. Of the three components tested, the CFCC combustor liners are most likely to be commercialized first. The driver is the demonstrated reduction in emissions levels of NOx and CO. To be commercially viable a minimum of two years of uninterrupted service life (about 16,000 hours) needs to be demonstrated, and for widespread applications including retrofits to existing gas turbines 30,000 hours of life, representing a full cycle of service life time between overhauls (TBO), is expected. This durability goal seems to be well within reach.

Of equal importance to component durability is cost. The forecasted cost of CFCC combustors is still several times that of the thermal barrier coated metal liners which are a competing option for lower firing engines and which have demonstrated good durability and excellent emissions reduction in engine field testing in Solar's products. Volume scale up, the use of lower cost fibers, and innovative fabrication development will be needed to address the cost hurdle. It is Solar's belief that the cost hurdle can be overcome if the market potential is evident. It is anticipated that future industrial gas turbine product development will result in engines with higher firing temperatures, 1200°C and higher. In such systems the availability of air for component cooling will be at a premium and it is unlikely that conventional TBC-coated liners will meet durability goals while maintaining adequate emissions control.

The development of tailored EBCs for monolithics along the lines currently being devleoped for CFCCs is expected to increase silicon nitride component life substantially. Part life should be no less than 30,000 hours, i.e. through a full TBO cycle. Impact resistance will require rugged designs, higher fracture toughness materials (perhaps some monolithic/composite hybrids), and impact mitigation strategies.

Given the hurdles that need to be overcome, it is expected that stationary monolithic components such as small combustors, nozzles, shrouds, tip shoes, etc., will be commercialized first and rotating components later. In our view commercialization could be viable by about 2004–2005. At that time, new engines will have higher firing temperatures than today and conserving air will be desirable. At these high firing temperatures uncooled or minimally cooled EBC-coated ceramic components may become an integral part of the gas turbine designer's tool kit.

ACKNOWLEDGMENTS

The authors wish to acknowledge the assistance of Jim Napier and from Dr. Arthur Metcalfe with the early history of Solar's work, and the support of personnel from the U.S. Department of Energy, including William P. Parks, Jr., Stephen Waslo, Patricia Hoffman, and Debbie Haught, for the work performed under the Ceramic Stationary Gas Turbine Development program (DOE Contract No. DE-AC02-CE40960).

REFERENCES

[1] Vogan, J.W., and Stetson, A.R., "Advanced Ceramic Coating Development for Industrial/Utility Gas Turbines," DOE/NASA/0109-1, NASA CR 167852, January 1982.

[2] Batakis, A.P., and Vogan, J.W., "Rocket Thrust Chamber Thermal Barrier Coatings," Final Report NASA Contract No. NAS3-23262, NASA CR-175022, July 1985.

[3] van Roode, M., Price, J.R., and Stala, C., "Ceramic Oxide Coatings for the Corrosion Protection of Silicon Carbide," ASME paper 91-GT-38 presented at the International Gas Turbine and Aeroengine Congress and Exposition, Orlando, Florida, June 3–6, 1991. *Transactions of the ASME, Journal of Engineering for Gas Turbines and Power,* 115(1), 1–208, 1993.

[4] Ward, M.E., van Roode, M., and Gildersleeve, R.E., "A New Ceramic Recuperator Technology for Use on High Temperature Processes with Environments Containing Corrosive Agents," AIChE Symposium Series, Heat Transfer-Pittsburgh, 257(83), 264–271, 1987.

[5] van Roode, M., Ward, M.E., and Gildersleeve, R.E., "A New Ceramic Recuperator Technology for Use on High Temperature Processes with Environments Containing Corrosive Agents," paper presented at the ATEE Energie Plus Conference, Paris, France, March 31–April 3, 1987, Actes des Journées Techniques, 1987, pp. 265–287.

[6] Harkins, B.D., and Ward, M.E., "Preliminary Design for an Advanced Ceramic Heat Exchanger Integrated in a Gas Turbine Power Generation Cycle," paper presented at the Annual Meeting of the American Society of Chemical Engineers, December 1991.

[7] Jordan, F.D., Napier, J.C., Metcalfe, A.G., and Duffy, T.E., "Application of Ceramics to a Radial Inflow Gas Turbine," paper presented at the International Automotive Engineering Congress and Exposition, Cobo Hall, Detroit, February 28–March 4, 1977.

[8] Napier, J.C., "Application of All-Ceramic Nozzle to Radial Inflow Turbine," ASME paper 79-GT-96, presented at the Gas Turbine Conference and Exhibit and Solar Energy Conference, March 12–15, 1979, San Diego, California, U.S.A.

[9] Napier, J., "Application of Ceramic Nozzles to 10kW Engine," Final Report, Contract No. DAAK02-75-0138, December 1980.

[10] Napier, J., and Batakis, A.K., "Investigation of Ceramic Uprating of 10kW Gas Turbine (Gemini) Engine," Interim Report, Contract No. DAAK70-79-C-0135, October 1981.

[11] Messier, D.R., Schioler, L.J., Quinn, G.D., and Napier, J.C., "Analysis of Fracture of Reaction-Bonded Si_3N_4 Shrouds," Ceramic Engineering and Science Proceedings, 2[7–8], July 1981.

[12] Arnold, J.P., McGovern, W., Russell, A.D., and Napier, J.C., "Manufacturing Methods for Erosion Resistant Ceramic Nozzles for Small Gas Turbines," ASME paper 82-GT-160, presented at the 27th International Gas Turbine Conference and Exhibit, London, England, April 18–22, 1982.

[13] Arnold, J.P., McGovern, W., Napier, J.C., and Batakis, A.K., "Demonstration of Ceramic Hot-Section Static Components in a Radial Flow Turbine," ASME paper 82-GT-184, presented at the 27th International Gas Turbine Conference and Exhibit, London, England, April 18–22, 1982.

[14] Metcalfe, A.G., and Napier, J.C., "Ceramics for Small Radial Inflow Power Generators", in *Ceramics for High-Performance Applications—III Reliability*, Proceedings of the Sixth Army Materials Conference", Edward M. Lenoe, R. Nathan Katz, and John J. Burke, eds., Plenum Press, New York, 1983, pp. 173–186.

[15] Napier, J.C., and Arnold, J.P., "Advancements in Application of Ceramics to the Gemini Radial-Flow Gas Turbine," ASME paper 85-GT-183, presented at the Gas Turbine Conference and Exhibit, Houston, Texas, March 18–21, 1985.

[16] Richerson, D.W., "Contact Stress at Ceramic Interfaces," *Progress in Nitrogen Ceramics,* ed. by F.L. Riley, Martinus Nijhoff Publishers, The Hague, 1983, pp. 657–666.

[17] Anson, D., Ramesh, K.S., and DeCorso, M., "Application of Ceramics in Industrial Gas Turbines; Task 1, Technology Background," Task 1 Report, U.S. Department of Energy Contract No. DE-AC-2-89CE40878, DOE/CE/40878-2, 1991.

[18] Anson, D., Sheppard, W.J., Norland, D.L., and DeCorso, M., "Application of Ceramics in Industrial Gas Turbines; Economics, Market Analysis, and R&D Plan," Part 2 Report – Tasks 2, 3, and 4, U.S. Department of Energy, Contract No. DE-AC-2-89CE40878, DOE/CE/40878-2, 1991.

[19] Anson, D., Sheppard, W.J., and Parks, Jr., W.P., "Impact of Ceramic Components in Gas Turbines for Industrial Cogeneration," ASME paper 92-GT-393, presented at the International Gas Turbine and Aeroengine Congress and Exposition, Cologne, Germany, June 1–4, 1992.

[20] Parks, Jr., W.P., Ramey, R.R., Rawlins, D.C., Price, J.R., and van Roode, M., "Potential Applications of Structural Ceramic Composites in Gas Turbines," ASME paper 90-GT-251, presented at the Gas Turbine and Aeroengine Congress and Exposition, June 11–14, Brussels, Belgium, 1990.

[21] van Roode, M., Brentnall, W.D., Norton, P.F., and Pytanowski, G.P., "Ceramic Stationary Gas Turbine Development," ASME paper 93-GT-309, presneted at the International Gas Turbine and Aeroengine Congress and Exposition, Cincinnati, Ohio, U.S.A., May 24–27, 1993.

[22] van Roode, M., Brentnall, W.D., Norton, P.F., and Boyd, G.L., "Ceramic Stationary Gas Turbine Development—First Annual Summary," ASME paper 94-GT-313, presented at the International Gas Turbine and Aeroengine Congress and Exposition, The Hague, The Netherlands, June 13–15, 1994.

[23] van Roode, M., Brentnall, W.D., Norton, P.F., and Edwards, B.D., "Ceramic Stationary Gas Turbine Development—Second Annual Summary," ASME paper 95-GT-459, presented a the International Gas Turbine and Aeroengine Congress and Exposition, Houston, Texas, June 5–8, 1995.

[24] van Roode, M., Rohy, D.A., Waslo, S., and Parks, Jr., W.P., "Ceramic Stationary Gas Turbine Development," *Proceedings of the 1995 International Gas Turbine Congress*, Yokohama, Japan, October 23–27, 1995. Vol. I, I-165/I-172, 1995.

[25] van Roode, M., Brentnall, W.D., Smith, K.O., Edwards, B.D., Faulder, L.J., and Norton, P.F., "Ceramic Stationary Gas Turbine Development—Third Annual Summary," ASME paper 96-GT-460, presented at the International Gas Turbine and Aeroengine Congress and Exposition, Birmingham, U.K., June 10–13, 1996.

[26] van Roode, Brentnall, W.D., Smith, K.O., Edwards, B.D., McClain, J., and Price, J.R., "Ceramic Stationary Gas Turbine Development—Fourth Annual Summary," ASME paper 97-GT-317, presented at the International Gas Turbine and Aeroengine Congress and Exposition, Orlando, Florida, June 2–5, 1997.

[27] Parthasarathy, V., van Roode, M., Price, J., Gates, S., Waslo, S., and Hoffman, P., "Review of Solar's Ceramic Stationary Gas Turbine Development Program," *Proceedings of the 6th International Symposium on Ceramic Materials and Components for Engines*, Arita, Japan, October 19–23, 1997. pp. 259–264.

[28] Price, J.R., Jimenez, O., Faulder, L., Edwards, B., and Parthasarathy, V., "Ceramic Stationary Gas Turbine Development—Fifth Annual Summary," ASME paper 98-GT-181, presented at the International Gas Turbine and Aeroengine Congress and Exhibition, Stockholm Sweden, June 2–5, 1998.

[29] Price, J.R., Jimenez, O., Parathasarathy, V., and Miriyala, N., "Ceramic Stationary Gas Turbine Development—Sixth Annual Summary," ASME paper 99-GT-351, presented at the International Gas Turbine and Aeroengine Congress and Exposition, Indianapolis, Indiana, USA, June 7–10, 1999.

[30] van Roode, M., Price, J.R., Gates, S., Waslo, S., and Hoffman, P., "Ceramic Stationary Gas Turbine Development—Opportunities and Challenges" *Proceeding of the 7th International Gas Turbine Congress*, Kobe, Japan, November 14–19, 1999. Vol. I, pp. 65–72.

[31] Price, J.R., Jimenez. O., Parthasarathy, V., Miriyala, N., and Leroux, D., "Ceramic Stationary Gas Turbine Development Program—Seventh Annual Summary," ASME paper 2000-GT-0075, presented at the ASME TURBO EXPO 2000, LAND, SEA, & AIR, Munich, Germany, May 8–11, 2000.

[32] Gates, S., Jr., Price, J., Parthasarathy, V., Miriyala, N., and Jimenez, O., "Recommended Direction of the Solar/DOE Ceramic Stationary Gas Turbine Program," ASME paper 2000-GT-0076, presented at the ASME TURBO EXPO 2000, LAND, SEA, & AIR, Munich, Germany, May 8–11, 2000.

[33] van Roode, M., Price, J.R., Jimenez, O., Miriyala, N., and Gates, Jr., S., "Design and Testing of Ceramic Components for Industrial Gas Turbines," in *Ceramic Materials and Components for Engines*, 7th International Symposium, Goslar, Germany, June 19–21, 2000. Jürgen G. Heinrich and Fritz Aldinger, editors, Wiley-VCH Verlag GmbH, Weinheim, Germany, pp. 261–266, 2001.

[34] van Roode, M., and Price, J.R., "Ceramic Stationary Gas Turbine Program Overview," *Proceedings of the 2000 Advanced Turbine Systems Annual Program Review Meeting*, Washington, D.C., December 4–6, 2000.

[35] Price, J., Jimenez, O., Miriyala, N., Kimmel, J.B., Leroux, D.R., and Fahme, T., "Ceramic Stationary Gas Turbine Development Program—Eighth Annual Summary," ASME paper 2001-GT-517, presented at the 46th ASME TURBO EXPO 2001, LAND, SEA, & AIR, New Orleans, Louisiana, USA, June 4–7, 2001.

[36] Nemeth, N.N., Powers, L.P., Janosik, L.A., and Geykenyesi, J.P., "Durability Evaluation of Ceramic Components Using CARES/LIFE," ASME paper 94-GT-352, presented at the International Gas Turbine and Aeroengine Congress and Exposition, The Hague, The Netherlands, June 13–15, 1994. *Transactions of the ASME, Journal of Engineering for Gas Turbines and Power,* 118[1], 150–158, 1996.

[37] Bornemisza. T., and Saith, A., "SPSLIFE: A User Friendly Approach to the Structural Design and Life Assessment of Ceramic Components," ASME paper 94-GT-486, presented at the International Gas Turbine and Aeroengine Congress and Exposition, The Hague, The Netherlands, June 13–16, 1994. *Transactions of the ASME, Journal of Engineering for Gas Turbines and Power.* Volume 118[1], 179–183, 1996.

[38] Saith, A., Norton, P.F., and Parthasarathy, V.M., "Application of SPSLIFE to Preliminary Design Evaluation and Life Assessment of CSGT Components," ASME paper 94-GT-420, presented at the International Gas Turbine and Aeroengine Congress and Exposition, The Hague, The Netherlands, June 13–16, 1994.

[39] Powers, L.M., Jadaan, O.M., and Gyekenyesi, J.P., "Creep Life of Ceramic Components Using a Finite Element Based Integrated Design Program (CARES/CREEP)," ASME paper 96-GT-369, presented at the International Gas Turbine and Aeroengine Congress and Exposition, Birmingham, U.K., June 10–13, 1996.

[40] Norton, P.F., Frey, G.A., Bagheri, H., Fierstein, A., Twardochleb, C., Jimenez, O., and Saith, A., "Ceramic Stationary Gas Turbine Development Program—Design and Life Assessment of Ceramic Components," ASME paper 95-GT-383, presented at the International Gas Turbine and Aeroengine Congress and Exposition, Houston, June 5–8, 1995.

[41] Jimenez, O., McClain, J., Edwards, B., Parthasarathy, V., Bagheri, H., and Bolander, G., "Ceramic Stationary Gas Turbine Program—Design and Test of a Ceramic Turbine Blade," ASME paper 98-GT-529, presented at the International Gas Turbine and Aeroengine Congress and Exhibition, Stockholm, Sweden June 2–5, 1998.

[42] Faulder, L., McClain, J., Edwards, B., and Parthasarthy, V., "Ceramic Stationary Gas Turbine Development Program—Design and Test of a First Stage Ceramic Nozzle," ASME paper 98-GT-528, presented at the International Gas Turbine and Aeroengine Congress and Exhibition, Stockholm, Sweden, June 2–5, 1998.

[43] Smith, K.O., and Fahme, A., "Experimental Assessment of the Emissions Benefits of a Ceramic Gas Turbine Combustor," ASME paper 96-GT-318, presented at the International Gas Turbine and Aeroegine Congress and Exposition, Birmingham, England, U.K., June 10–13, 1996.

[44] Smith, K.O., and Fahme, A., "Testing of a Full Scale Low Emissions, Ceramic Gas Turbine Combustor," ASME paper 97-GT-156, presented at the International Gas Turbine and Exhibition, Orlando, Florida, June 25, 1997.

[45] van Roode, M., Price, J.R., Richerson, D.W., Parthasarathy, V., and Graves, G.A., "Ceramic Stationary Gas Turbine Development—Monolithic Ceramic Component Development Summary," ASME paper 2001-GT-0457, presented at the 46th ASME TURBO EXPO 2001, LAND, SEA, & AIR, New Orleans, Louisiana, June 4–7, 2001.

[46] Miriyala, N., Fahme, A., and van Roode, M., "Ceramic Stationary Gas Turbine Development—Combustor Liner Development," ASME paper 2001-GT-512, presented at the 46th ASME TURBO EXPO 2001, LAND, SEA, & AIR, New Orleans, LA, June 4–7, 2001.

[47] Parthasarathy, V.M., Price, J.R., Brentnall, W.D., Graves, G., and Goodrich, S., "Material Characterization of Candidate Silicon Based Ceramics for Stationary Gas Turbine Applications," ASME paper 95-GT-249, presented at the International Gas Turbine and Aeroengine Congress and Exposition, Houston, June 5–8, 1995.

[48] Lin, H.T., Ferber, M.K., and van Roode, M., "Evaluation of Mechanical Reliability of Si_3N_4 Nozzles after Exposure in an Industrial Gas Turbine," *Ceramic Materials and Components for Engines*, 7th International Symposium, Goslar, Germany, June 19–21, 2000. Jürgen G. Heinrich and Fritz Aldinger, editors, Wiley–VCH Verlag GmbH, Weinheim, Germany, pp. 97–102, 2001.

[49] Robinson, R.C., "SiC Recession Due to Scale Volatility Under Combustor Conditions," NASA Contractor Report 202331, March 1997.

[50] Smialek, J.L., Robinson, R.C., Opila, E.J., Fox, D.S., and Jacobson, N.S., "SiC and Si_3N_4 recession due to volatility under combustor conditions," Adv. Composite Mater., 8[1], 33–45, 1999.

[51] More, K.L., Tortorelli, P.F., Ferber, M.K., Walker, L.R., Keiser, J.R., Brentnall, W.D., Miriyala, N., Price, J.R., 1999, "Exposure of Ceramics and Ceramic Matrix Composites in Simulated and Actual Combustor Environments," ASME paper 99-GT-292, presented at the International Gas Turbine and Aeroengine Congress and Exposition, Indianapolis, Indiana, USA, June 7–10, 1999.

[52] Miriyala, N., Simpson, J.F., Parthasarathy, V.M., and Brentnall, W.D., "The Evaluation of CFCC Liners after Field-Engine Testing in a Gas Turbine," ASME paper 99-GT-395, presented at the International Gas Turbine and Aeroengine Congress and Exposition, Indianapolis, Indiana, USA, June 7–10, 1999.

[53] Miriyala, N., and Price, J.R., "The Evaluation of CFCC Liners after Field-Engine Testing in a Gas Turbine—II," ASME paper 2000-GT-0648, presented at the International GasTurbine and Aeroengine Congress and Exposition, Munich, Germany, May 8–11, 2000.

[54] Eaton, H.E., Linsey, G.D., More, K.L., Kimmel, J.B., Price, J.R., and Miriyala, N., "EBC Protection of SiC/SiC Composite in Gas Turbine Combustion Environment," ASMEpaper 2000-GT-0631, presented at the International Gas Turbine and Aeroengine Congress and Exposition, Munich, Germany, May 8–11, 2000.

[55] Eaton, H.E., Linsey, G.D., Sun, E.Y., More, K.L., Kimmel, J.B., Price, J.R., and Miriyala, N., "EBC Protection of SiC/SiC Composites in the Gas Turbine Combustion Environment—Continuing Evaluation and Refurbishment Considerations," ASME Paper 2001-GT-0513, presented at the 46th ASME TURBO EXPO 2001, LAND, SEA, & AIR, New Orleans, LA, June 4–7, 2001.

[56] "Recent Evaluation of CFCC Combustor Liners with Environmental Barrier Coatings after ~14,000 Hour Engine Test," More, K.L., Tortorelli, P.F., Lara-Curzio, E., Keiser, J.R., Walker, L.R., Eaton, H.E., Sun, E.Y., Linsey, G.D., Landini, D.J., Craig, P., Ellingson, W.A., Sun J.G., Erdmann, S., Deemer, C., Miriyala, N., Kimmel, J.B., and Price, J.R., paper without publication presented at the 46th ASME TURBO EXPO 2001, LAND, SEA, & AIR, New Orleans, LA, June 4–7, 2001.

CHAPTER

8

Progress in Ceramic Gas Turbine Development Westinghouse/Siemens Westinghouse

Jay Morrison

Siemens Westinghouse Power Corporation, Orlando, Florida, USA

Jay Lane and Michael Burke

Siemens Westinghouse Science & Technology Center, Pittsburgh, Pennsylvania, USA

INTRODUCTION

For over half a century ceramic materials have offered the promise of higher temperature capability for heat engines. For power generation gas turbines the projected benefits of higher temperature materials and, hence, gas temperature capability have always offered improved engine efficiency and increased power output. In addition, as gas turbine combustion temperatures have increased, the use of higher temperature materials for "hot wall" combustion has emerged to reduce Greenhouse gas and NOx emissions. Ceramics thus offer the potential for reduced emissions as well as improved fuel efficiency and increased power output in gas turbine engines. These potential benefits have driven a series of initiatives over the past fifty years for the development and demonstration of ceramic components for gas turbine engines.

Siemens Westinghouse Power Corporation, (SWPC) the erstwhile Power Generation Business Unit of Westinghouse Electric Corporation, has closely followed, and has actively participated in, many of these initiatives. The perspective of Siemens Westinghouse should help to highlight some of the pitfalls experienced in previous ceramic insertion programs. Many lessons can be learned from the technical progress and setbacks of previous programs, but equally important are the trends observed in the forecasting, funding, execution, and packaging of results for technology development and demonstration programs. Despite over 40 years of essentially continuous effort, structural ceramics have still not been implemented commercially in engines. The task of developing ceramic materials to meet engine requirements remains formidable, yet the combination of today's emerging materials together with the experience and knowledge derived from previous programs

TABLE 8.1 GENERALIZED DEVELOPMENT HISTORY OF CERAMIC MATERIALS FOR GAS TURBINES

Time Frame	Materials Technology	Key Advantages / Remaining Technology Issues
1960s and early 1970s	Monolithic Ceramics	Temperature stability / brittleness, properties, contact loading etc.
1980s	Non-Oxide CFCCs	Strain tolerance notch insensitivity / reduced strength, lower temperature limits, long term costs, water vapor attack susceptibility
1990s and 2000s	Oxide-Oxide CFCC and Coatings for Non-Oxide CFCCs	Environmental stability, water vapor stability, Cost effectiveness / Minimal properly database, processing development needed

offers a renewed promise for commercial reality. Through this retrospective look at ceramics development, perhaps some insights will be gained that will assist ongoing and future development efforts.

This chapter will first provide a brief historical overview of the structural monolithic ceramics and CFCCs that have over the years been targets for applications in components of gas turbine engines. Secondly, we review in detail the most significant material development and engine implementation efforts in which Westinghouse has participated. This will provide an engine developer/materials' user's perspective of the technical progress and remaining challenges that resulted from these programs. Thirdly we present the current status of today's material technology and its potential for insertion into advanced power generation engines in the near future. In our review of these past programs we attempt to highlight the key "lessons learned" during the programs and which in our view are still particularly relevant to today's technical efforts.

HISTORICAL PERSPECTIVE AND OVERVIEW

Since the late 1960s, when advanced ceramic compositions and processing technologies first promised that advanced ceramic materials could be employed in turbine engines, several generations of initiatives have been pursued to achieve this goal. While none of these attempts ended in the successful deployment of structural ceramic components in commercial engines, the lessons learned along the way frequently gave rise to material and technology improvements that would facilitate further attempts to develop and use improved materials. The materials progression can be summarized as indicated in Table 8.1.

The technology developments shown in Table 8.1 represent a continuous development of materials and support technology since the early 1960s. The initial monolithic ceramics were, and still are, high temperature capable, strong, high-density single-phase materials [1–5]. These materials, however, fail in brittle catastrophic modes, making them unsuitable for arbitrarily large parts in commercial power generation gas turbine engines. The strain tolerance and distributed failure engendered in continuous fiber ceramic composites (CFCCs) made them more attractive for larger applications. More recently oxide-oxide CFCCs have offered the potential to deliver the improved long term environmental stability that the traditional non-oxide based CFCCs lack.

The drivers for gas turbine manufacturers to incorporate ceramic materials in engines are improved fuel efficiency, increased capacity, and reduced emissions. Reduction in the cooling air required for hot section components results in significant improvements in all three critical areas, thus making high temperature materials of vital importance to gas turbine performance. Ceramics have always offered the greatest potential for reducing or eliminating cooling air while at the same time allowing for higher gas path temperatures. Importantly, the perceived benefits of the technology must be weighed against its cost of implementation. At specific points in history such cost benefit analyses have favored implementation in either larger or smaller engines.

The development of the ceramic materials through the years has targeted a range of gas turbine engine sizes and component sizes and types. Materials-led development programs, in fact, often targeted generic gas turbine applications. Siemens Westinghouse turbines, being at the higher end of the engine power range and consisting of parts that are at the larger end of the size scale have, at times, presented different cost/benefit pay-backs compared to those applications of other engine manufacturers. We recognized early on that some developments that would suit smaller engines would not be suitable for our applications. In the case of ceramics, most emphatically "one size does not fit all"! This review presents the perspective and history of the development of materials for gas turbines at Siemens Westinghouse, a path which has not always directly paralleled the general development path outlined in Table 8.1.

It should also be recognized that the benefits to be derived from new technology insertion change significantly as the established technologies are also improved. Developments in competing materials' technology for large gas turbines have, over the years, meant that the standards to be met by ceramics systems have become a "moving target." Impressively high reliability records achieved by GT manufacturers in recent years have dramatically increased the reliability standards of large industrial GTs since the 1960s. Developments such as improved single crystal superalloys, thermal barrier coatings, 3D-CFD and FEA codes, advanced cooling techniques enabled by both analytical and manufacturing improvements, etc. have advanced the GT state-of-the-art with increased performance engines (higher efficiency and power output). These developments have raised the performance requirements for advanced ceramics.

While the early benefits projected for ceramics in GTs have been realized in recent years (and even surpassed) through these more conventional approaches, it is important to recognize that there still remains significant potential for improved GT performance through ceramic implementation. Siemens Westinghouse continues to evaluate oxide-based CFCCs for GT application. Experience in earlier programs has provided a realistic perspective on the risks and challenges involved. The significant economic (performance, efficiency, power, emissions) benefits from ceramic implementation provide an overwhelming impetus to drive this technology. The following sections detail the experience of Siemens Westinghouse efforts in implementing ceramics materials in advanced gas turbines and present our current viewpoint of the technology needs for future efforts.

Major Ceramics Efforts at Westinghouse/Siemens Westinghouse

For the past 35 years, Westinghouse/SWPC has engaged in ceramic development programs that target the hot gas path components in gas turbines. For these efforts, efficiency benefits have been the main driver, with emissions reductions a secondary driver. The major IR&D and government funded programs at Westinghouse/SWPC are concisely reviewed below in a chronological manner. For more details, in addition to the cited references, the government program contract numbers can be used to acquire the program reports, whereas, for the IR&D efforts the readers are encouraged to contact the authors.

Early Ceramics Implementation Effort. In the early 1960s ceramic lined, hot-wall combustors were developed, fabricated and tested under conditions reaching 2200°C flame temperatures [6]. These combustors were metallic shells lined with castable ceramic bricks that were overlapped and interlocked for mechanical constraint and to prevent radiation directly to the metal shell (Figure 8.1). The tiles were suspended from the metal shell via metallic spring clips, which allowed for differential thermal expansion and generated low attachment stresses. Cooling of the metallic frame was via external convection only. Despite proving the feasibility of the concept, the need for hot-walled combustors was deemed a lower priority (than, e.g., higher temperature vanes) and the program was stopped.

It is worth noting that this method of combustor design has been used successfully for many years by Westinghouse's now parent company Siemens KWU (among others) in both external silo and annular combustors (see chapter 28 in Volume I of this text). While the details of attachments, tile configuration, and material selection vary somewhat from this early Westinghouse concept, basics of the approach used today are similar and the concept of ceramic tiled combustors remains attractive due to the reduction in thermal stresses, simple ceramic construction, and ease of maintenance. Such float-wall or tiled combustor construction is more

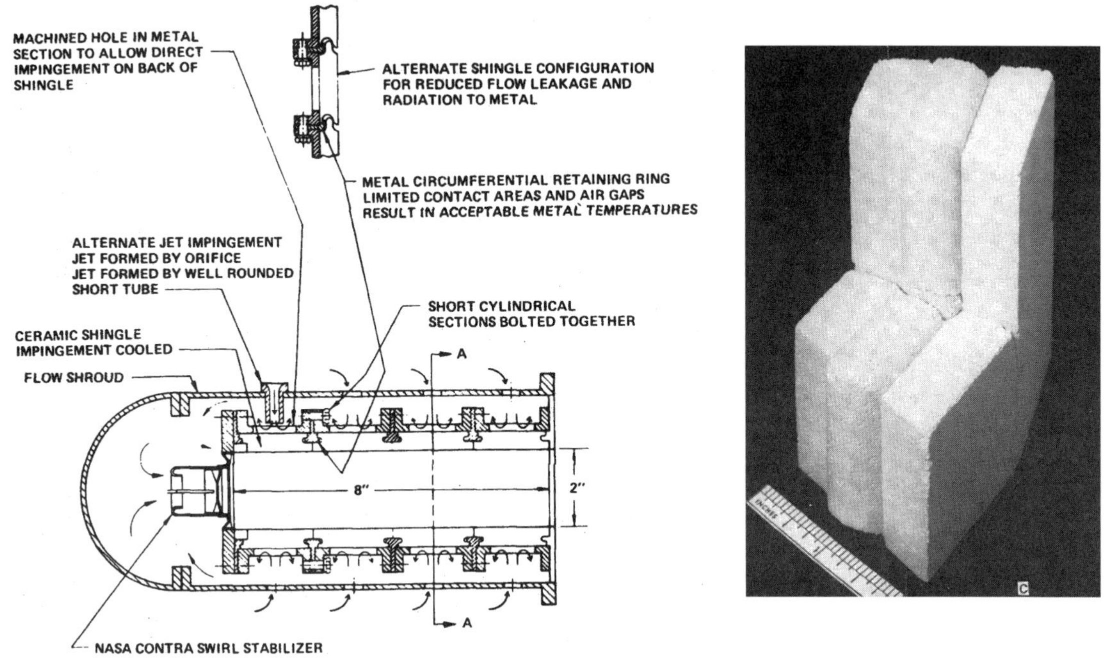

Figure 8.1 Westinghouse Ceramic Tiled Combustor Design Concept and Individual Tile Test article (Reprinted from Ref. [6]) Dimensions in Inches—1 inch = 2.54 cm.

feasible for large, land-based gas turbines since designs are not constrained by size and weight restrictions typical of smaller land-based turbines or aero-engines.

Initial Assessment of Monolithic Ceramic Materials for Gas Turbine Vane Designs. Between 1971 and 1976 Westinghouse participated in the Brittle Materials Design, High Temperature Gas Turbine Program [7]. This program was part of the Ford effort under the same name, but with emphasis on applications for large (30+ MW) industrial gas turbines. Monitored by AMMRC, the program emphasis was on development of design methods for brittle materials and culminated in validation testing of engine components. After initially focusing on both turbine blades and vanes, the Westinghouse effort after 1973 was redirected solely to vanes [7]. The rotor blade effort was subsequently pursued in conjunction with EPRI and continued in parallel with the DARPA program.

At Westinghouse, the Stationary Turbine Project's principal objective was the development of first-stage ceramic stator vanes with associated support hardware and a demonstration engine test. This test was to have been in a 30 MW frame test engine for 100 duty cycles of peaking operation with a vane hot spot temperature of 1371°C. This full engine test proved too formidable for the 5-year program. The objective was revised in 1974 to conclude the effort by demonstrating the vanes in a high temperature static rig. This revised objective was achieved and ceramic vanes were demonstrated successfully.

Engine efficiency improvement was the primary benefit targeted for the ceramic vane design with an increase in combined cycle thermal efficiency of 8% (from 42% to 50%) claimed for the W251 engine. This benefit is the combined result of increasing turbine inlet temperature from 1093°C to 1371°C while concurrently reducing overall cooling airflow by using uncooled ceramic components. [For perspective, Siemens Westinghouse Power Corporation now manufactures land-based gas turbines with greater than

1426°C turbine inlet temperatures at >58% thermal efficiency—without the benefit of uncooled ceramic hot section components.]

A novel three-piece monolithic ceramic vane was designed in three iterations or generations. The vane design was an aerodynamic compromise of the metallic vane it replaced, having half the chord length and half the thickness of the metal counterpart, thus necessitating twice the number of airfoils (Figure 8.2). The attachment design included the use of axial and radial springs to ensure positive retention of the vane platforms and of the vane airfoil within the platforms throughout the full range of engine thermal excursions. Since the vanes were designed to run at 1371°C uncooled, thermal insulator blocks were installed to protect the cooled metal components and minimize thermal gradients in the ceramics (Figure 8.3).

Testing was conducted in a highly instrumented static sector rig using a full-scale production combustor (Figure 8.4). The first generation vane design (Norton HS-130 Si_3N_4 vanes and LAS insulators) was tested in late 1973 for 100+ start/stop cycles but only to 1204°C hot spot temperature. The cycle consisted of: 1) cold start to idle; 2) a 5°C/s ramp to peak temperature; 3) a hold at steady-state for 3–5 minutes; 4) controlled emergency shutdown (instantaneous drop to 1093°C, followed by controlled cool-down to ambient). Although the vanes were functional at the end of the test, cracks were found to emanate from attachment surfaces on some airfoils. The source was traced to machining tolerance deviations that were subsequently fixed and a second set of tests was run in October 1974. This test was run with second generation SiC and Si_3N_4 vanes at 1371°C firing temperature. SiC vanes were found to have cracked almost immediately due to thermal stresses and the test was cut short after 5 cycles due to a metal combustor failure. SiC was discontinued from further testing.

Testing of third generation vanes was performed in three installments starting in October 1975 and completed in March 1976. The vanes accumulated a total of 103 start-up/shut-down cycles (9 hours at peak temperature) with vane hot spot temperatures exceeding 1370°C, but not without problems. The first test (25 cycles) resulted in more vane cracking—this time from an "assembly error" and thermal stresses. The second and third builds of the rig also experienced thermal stress cracking in some of the vanes—albeit less severe—and the assembly was still functional at the end of the test (Figure 8.5). One of the lessons learned from this test was the potential for non-catastrophic failure in "brittle" materials.

Surface recession due to oxidation/corrosion/erosion phenomena was a concern in the early vane rig tests. Therefore, hot corrosion exposure tests were performed on Norton HS130 Si_3N_4 and Norton NC203 SiC test specimens under 3 atmosphere pressure and a flow velocity of 150 m/s using fuel contaminated with sodium, sulfur, and vanadium. After 250 hours of exposure at 1093°C, negligible recession (3 μm) was measured on either material. However, after 125 hours at 1370°C, recessions of 25 μm and 60 μm were measured for SiC and Si_3N_4, respectively [8]. In comparison, Si_3N_4 vanes exposed in the static test rig for 9 hours at 1370°C exhibited recessions of >150 μm at the leading edge.

Lessons Learned While significant contributions were made in this program to the knowledge base for ceramics, it was recognized that much further work was needed in design methods, more reliable and refractory materials, and manufacturing techniques. One of the key observations noted in the final report was, "Nothing has occurred as a result of this program to indicate that ceramics *cannot* be developed to work in electric power generating gas turbines." [7] Specific lessons include:

- "Ceramic design thinking" was vital to developing a 3-piece turbine vane and yielded a configuration dramatically different from original conventional metallic parts. Such brittle material design methods were a direct product of the DARPA Ford/Westinghouse program.
- Contact stress management proved difficult and resulted in several "failures" for vanes and was also identified as a key issue for ceramic substrates for catalytic combustors. This continues to be a major focus of ceramic designs today.
- Probabilistic design methods were necessary for assessing component reliability over many thousands of operating hours. Even short-term analyses require much more intensive computational methods than traditional materials.

(a)

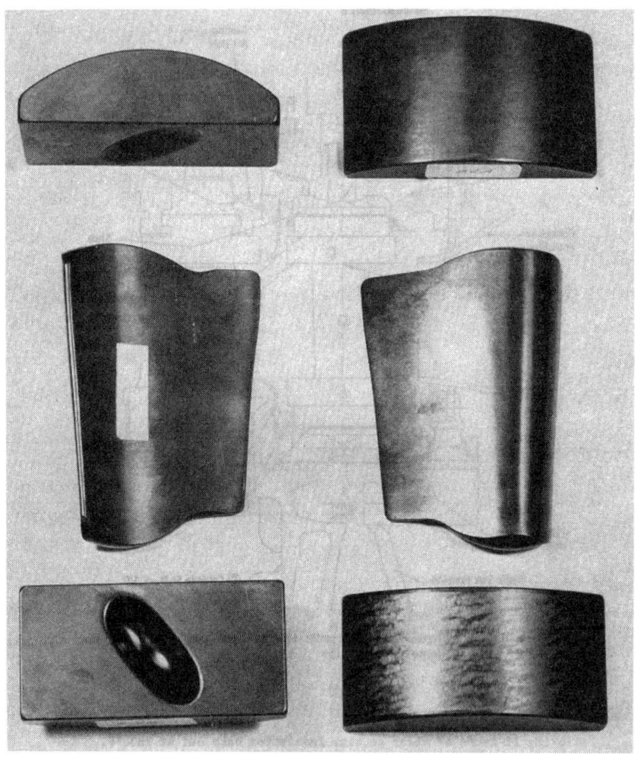

(b)

Figure 8.2 Three-piece Ceramic Vane Design; (a) Compared to Original Metal (Left)—Illustrating Shorter Airfoil Chord Length and Consequent Aerodynamic Performance Loss for the Ceramic Version (Reprinted from Ref. [7]); (b) Third Generation Configuration Showing Airfoil/Platform Engagement Via Torroidal Grooves (Reprinted from Ref. [13])

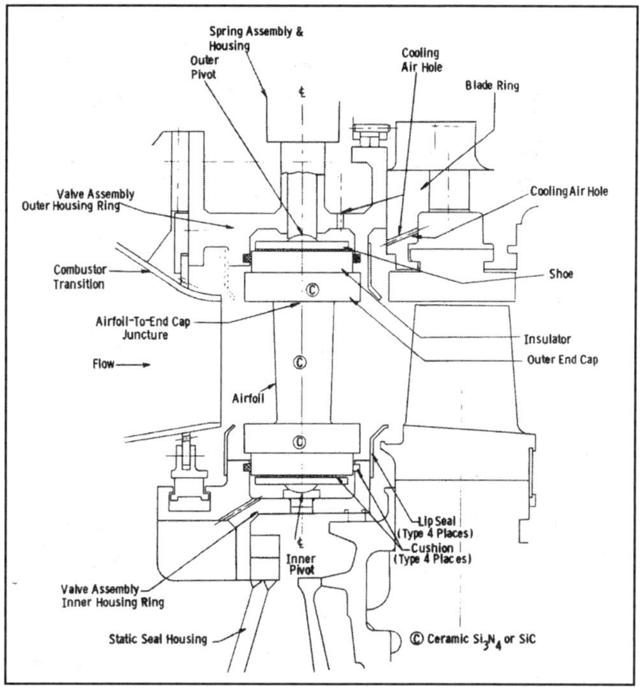

(a)

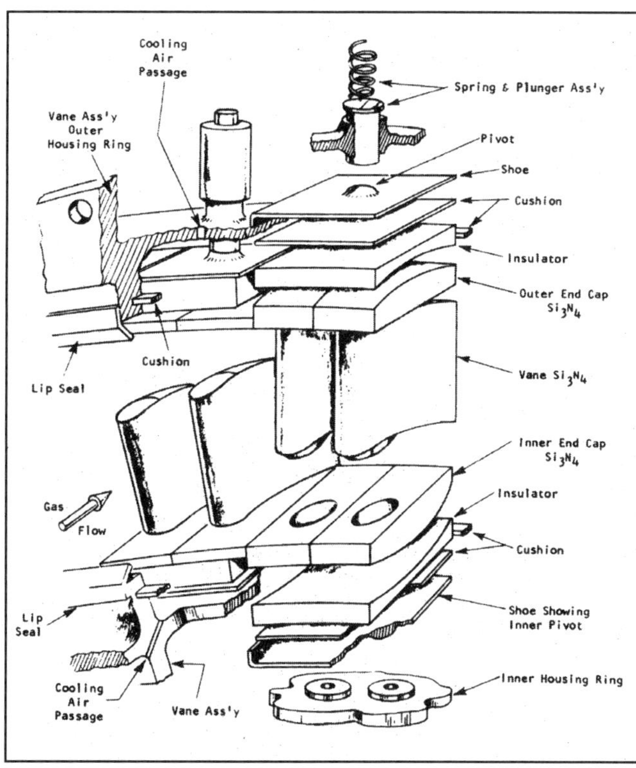

(b)

Figure 8.3 Ceramic Vane Attachment and Engine Installation: (a) Cross Section; (b) Exploded View (Reprinted from Ref. [13])

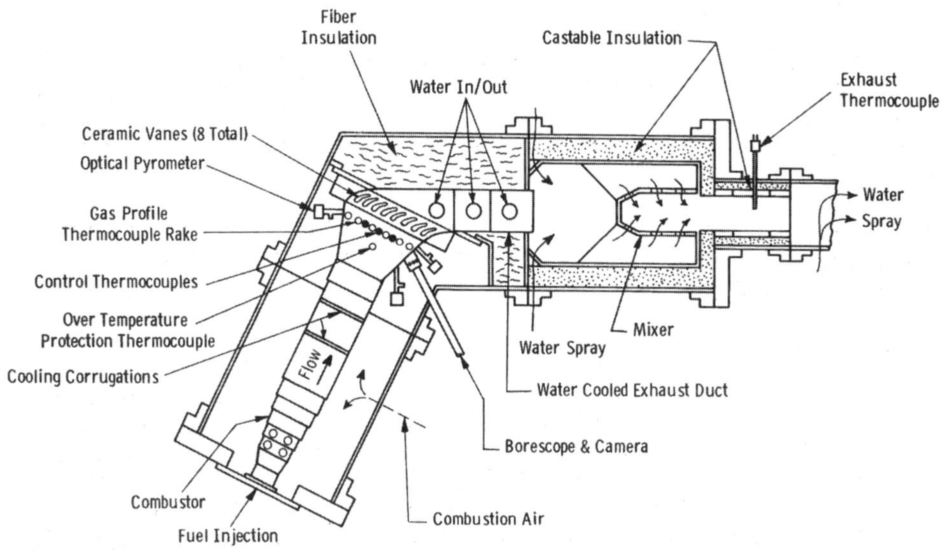

(a)

(b)

Figure 8.4 Vane Sector Rig: (a) Schematic; (b) Si_3N_4 Vane Cascade (8 Vane Assemblies) Arrangement (Reprinted from Ref. [7])

(a)

(b)

Figure 8.5 Sector Rig After 100 Cycles of Testing (9 Hours) at 1371°C (2500°F): (a) General Condition of Ceramic Airfoils Showing Centerline Hot Streaks; (b) Close Up of Thermal Stress Cracks (Reprinted from Ref. [7])

- Non-catastrophic failure of monolithic ceramic parts is possible, as demonstrated with rig testing of ceramic vanes in the DARPA program. This is true primarily for thermal-strain-driven stresses wherein the driving force dissipates with crack propagation.
- Oxidation/corrosion/erosion issues under turbine combustion conditions led to rapid recession and loss of strength for Si_3N_4 materials at temperatures >1093°C. Recession rates of 150 microns were noted for Si_3N_4 vane leading edges after just 9 hours of rig testing at 1371°C, a rate much greater than measured in laboratory tests. This led to the conclusion that the materials were not ready for engine use.
- Due to the oxidation problems and creep concerns of the materials used in this program, the ability to achieve 1371°C turbine inlet temperature was deemed impossible and "blamed" on the lack of adequate material capability.
- Ceramics are not "quick-fix" items for high performance turbomachinery. The initial euphoria based on projected engine performance benefits was tempered by the realization that it would require many years of developmental effort to get there.

Initial Assessment of Ceramic Materials for Rotor Blades. This continuation of efforts to apply ceramic materials to the more difficult rotating gas turbine components was conducted under an EPRI sponsored initiative between 1974 and 1979 [9–12]. This program was designed to complement and build upon the design methods and materials developed under the above DARPA program. The target for application for this program was the same 30 MW nominal baseline engine (W251) with an assumed target TRIT of 1260°C. (Note that this is the averaged rotor temperature as compared to the 1371°C *hot spot* assumed for the vanes in the previous DARPA program). One of the key challenges in this program was to develop analysis methods powerful enough to model a full 3D blade in sufficient detail to capture critical volumes at local stress concentrations while simulating contact friction and sliding phenomena at attachment interfaces. The development of adequate design codes within Westinghouse was thus a major effort in this program.

Twenty conceptual designs were generated in the first year of the program offering many unique methods of attaching ceramic airfoils to metallic disks [9]. These were eventually culled to leave two major concepts which were pursued the remainder of the program – both of which relied on conventional dovetail root attachments (Figure 8.6). The blades were relatively large (>100 mm airfoil height, >175 mm overall height) compared to the vehicular-size turbines which were being developed in parallel [13]. Hot-pressed Si_3N_4 (Norton NC132) was the material of choice primarily because of its superior mechanical properties, the availability of large pieces needed for these parts, and the existence of a materials property database. Design of the blade root was the most critical feature for the ceramic blade. A double curvature root geometry was conceived to effectively manage the contact area and avoid high stresses which result from point contact (Figure 8.7). However, this geometry introduced additional challenges in machining of the complex curvature. Both of the tested blade root configurations utilized this type of contact surface [12].

While no blades were tested in either an engine or rig, there was considerable work on spin testing of "dummy" blades in spin pit rigs. Dummy blades were fabricated for several root design concepts (Figure 8.8) and tested at room temperature both with, and without, compliant layers. For blades tested without compliant layers, results were alarmingly low; failures occurred at predicted nominal stresses of only 63 MPa (not accounting for friction effects). All subsequent spin rig testing was therefore performed using compliant layers (primarily platinum foils), which proved effective for the short term, but longer term durability of the compliant layer remained a concern. Many compliant layer materials (platinum, graphite, stainless steel, etc.) and concepts (perforated foil, felt metals, gauzes, etc.) were tested under fretting wear conditions, with some "composite" constructions of feltmetal and platinum foil working best. Despite these "successes," long term durability of compliant layers remained an issue [11].

Spin test results with compliant layers (0.13 mm platinum foil + 0.5 mm feltmetal) still yielded failures at rotational speeds below predictions. Post-test analysis revealed all fractures emanating from the high stressed root area with cracks initiating at surface, machining-induced flaws. A machining-effects study was

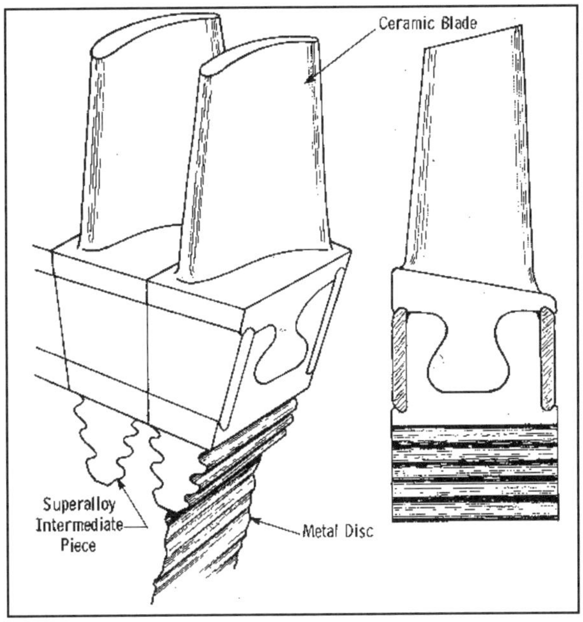

(a)

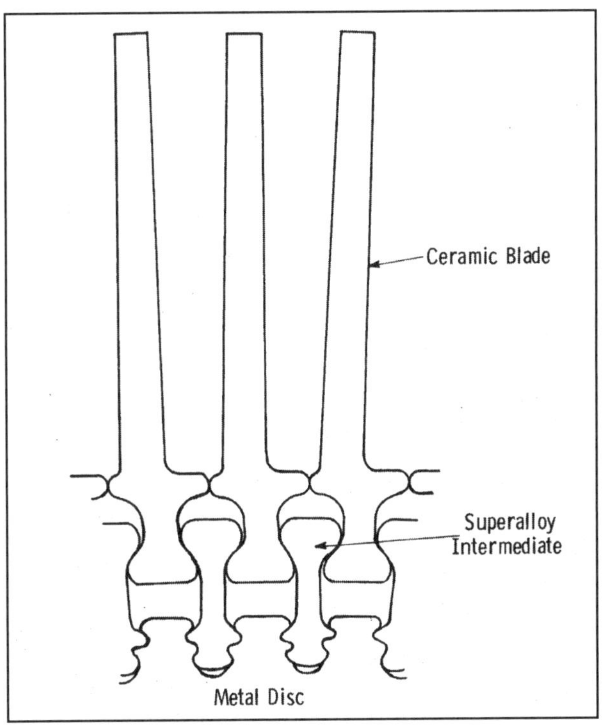

(b)

Figure 8.6 Two of the Twenty Turbine Blade Concepts Generated Under the EPRI Program Which were Down-selected for Further Evaluation: (a) Configuration # 14—Circumferential Dovetail Attachment with Superalloy Intermediate Piece; (b) Configuration # 17—Axial Dovetail with Superalloy Intermediate Pieces (Reprinted with permission from Refs. [9, 10, 12])

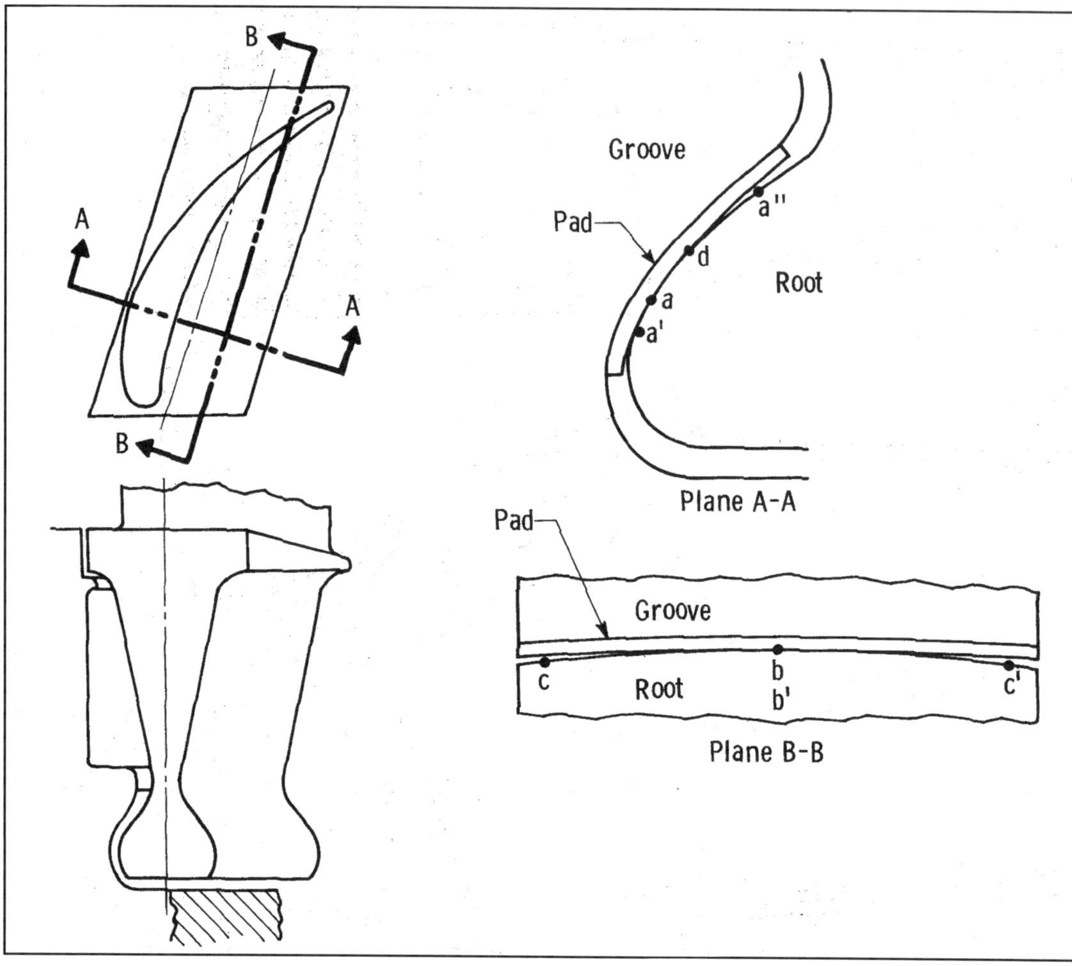

Figure 8.7 Double Curvature Blade Root Design for Controlled Contact Area and Stress (Reprinted with permission from Refs. [10, 12])

therefore conducted to assess the strength effects of various machining parameters. In parallel, a probabilistic analysis was performed on the spin tested blades based on surface area failure assumption. Comparison of the analysis to the Weibull strength distributions measured during the machining-effects study yielded insights into the source of premature failure in the components (Figure 8.9). Improved surface finishing techniques were deemed necessary for future blade manufacture.

Lessons Learned As in the previous DARPA program, this program concluded that a need existed for ceramics with higher strength, higher temperature capability, and improved manufacturing methods. In addition, NDE, failure criteria, and lifetime prediction methods were identified as needed developments. An overlooked conclusion to this program recommended concentration on "smaller combustion turbines for applications where reliability is less important" [11]. Specifically:

- Control of contact stresses in blade root attachments via compliant layers was shown to have significant effect on stresses and failure of the blades. Durability of these layers for long-term service has yet to be demonstrated.

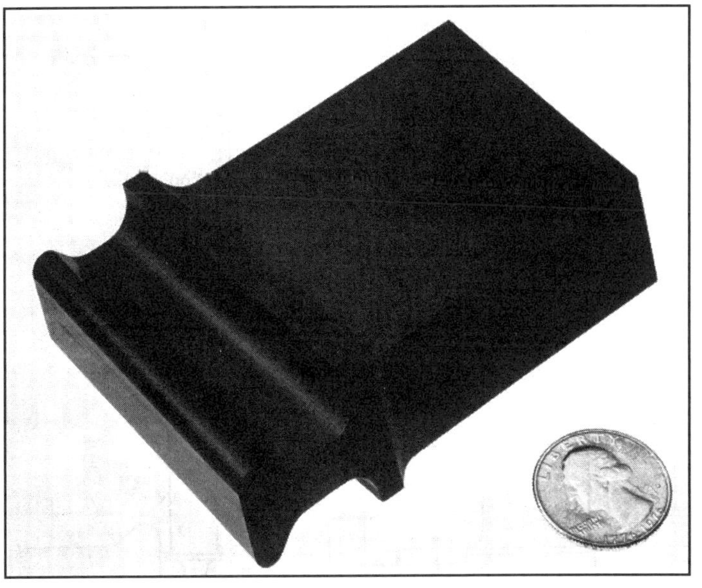

(a)

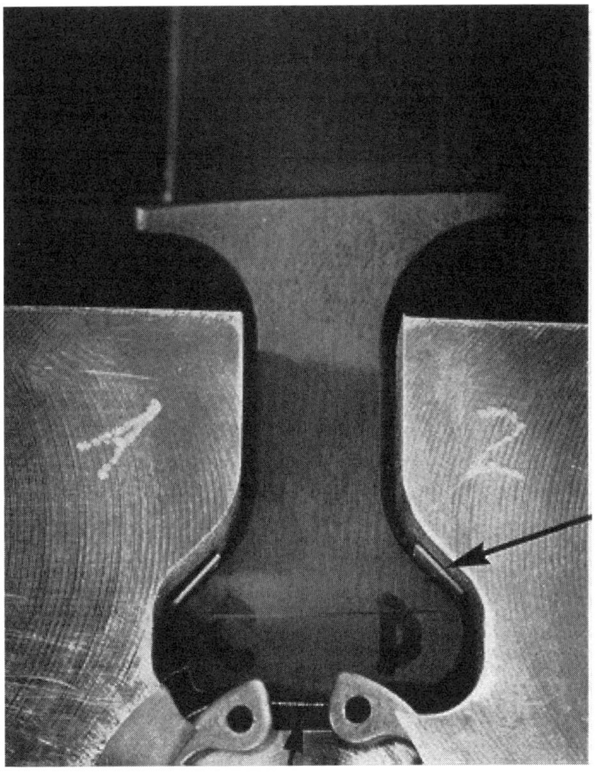

(b)

Figure 8.8 Silicon Nitride "Dummy" Blades for Spin Pit Testing—(a) Blade Blank with Preliminary Design Root Configuration; (b) Blade Installed in Spin Rig with Compliant Layer (Reprinted with permission from Ref. [12])

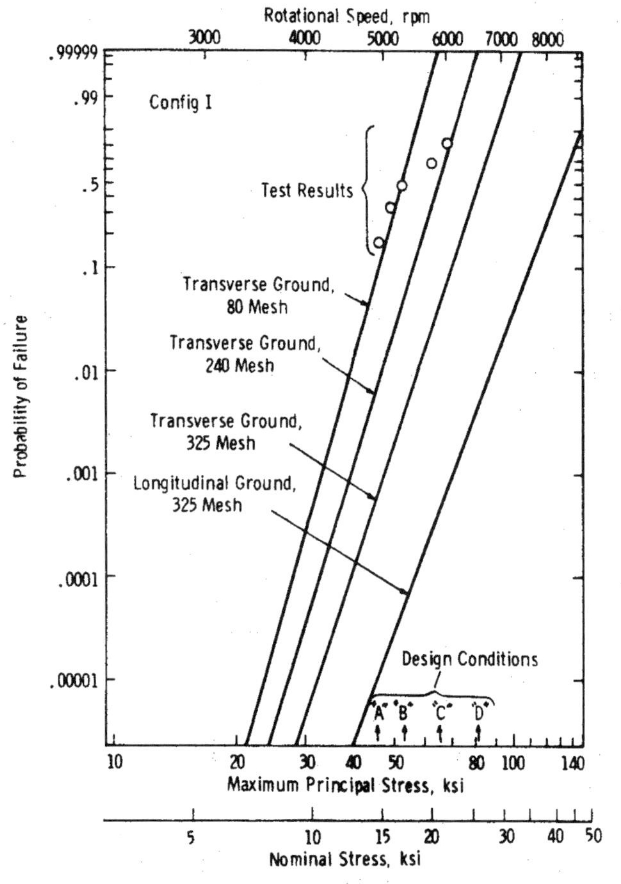

Spin Pit Test Design Conditions - Root Form Configurations I and II					
Root Form Configuration	Design Condition	Description	Speed N, rpm	N/N_{Design}	Maximum Principle Stress (MPa)
I	A	R.T. Design	4947	0.93	320
	B	H.T. Design	5328	1.00	371
	C	10% Overspeed	5861	1.10	449
	D	50% Overspeed	6526	1.22	557
II	A	R.T. Design	4919	0.93	214
	B	H.T. Design	5298	1.00	248
	C	10% Overspeed	5828	1.10	300
	D	50% Overspeed	6489	1.22	372

Figure 8.9 Spin Pit Test Results Compared to Strength Distributions for Various Surface Finishes (Reprinted with permission from Ref. [11]) Stress: 1 ksi = 6.896 MPa

- Transverse direction grinding required for the dovetail blade root (and overly aggressive grinding methods) was shown to be the cause for premature blade failures in spin rig testing.
- Ceramic materials were concluded to be unreliable and better manufacturing practices (higher Weibull, better NDE, machining methods, etc.) needed to be developed to make them viable for turbine blade application.

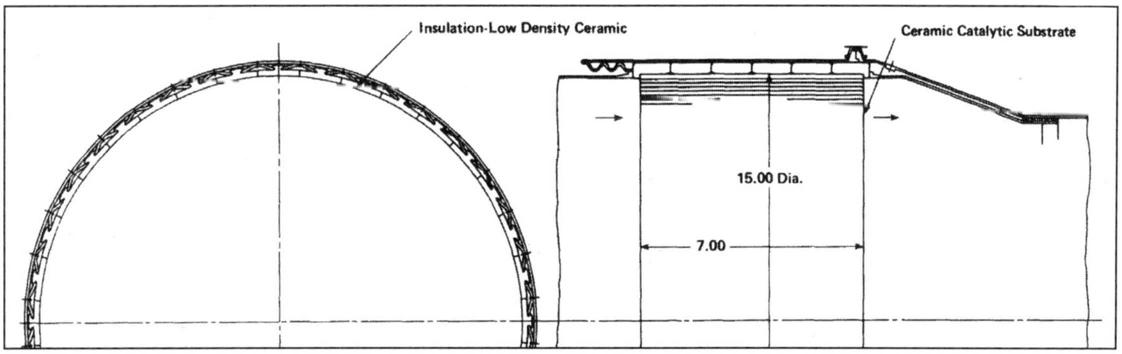

(a)

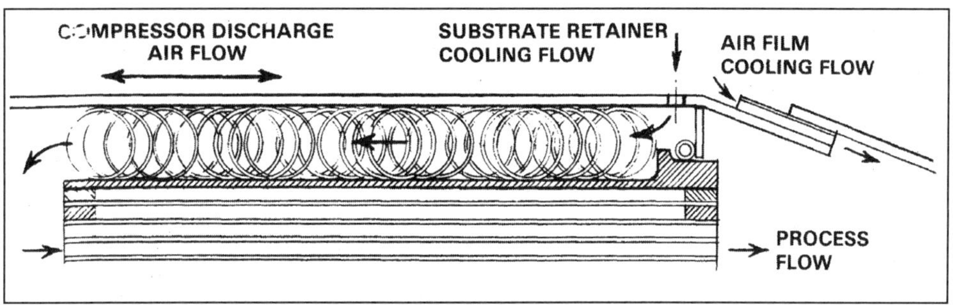

(b)

Figure 8.10 Conceptual Design for Ceramic Catalyst—(a) Ceramic Tile Insulator Concept; (b) Coiled Wire Compliant Layer with Backside Cooling (Reprinted with permission from Ref. [14]) Dimensions in Inches—1 inch = 2.54 cm

Initial Assessment of Ceramic Materials for Catalytic Combustion. Conducted simultaneously with the assessment of the Si_3N_4 rotor blades under the same EPRI program, the catalytic combustor focus was on design, analysis, and short term demonstration testing. Design of a full-scale combustor was based on honeycomb catalyst substrates made of cordierite contained within a Hastelloy-X shell via a compliant insulation layer (Figure 8.10). Such compliant layers were recognized as crucial to the secure retention of low ductility ceramic materials throughout the transient and steady state temperature and stress excursions experienced in turbine operation. Several compliant mounting schemes were developed on a conceptual basis, two of which are shown in Figure 8.10. These and other attachment concepts were identified as potential candidates for subsequent development efforts, but were never directly evaluated in this program.

For testing purposes, both 25 mm diameter and 225 mm diameter catalysts were made by Engelhard from a zircon composite material (DXE-441). The 25 mm tests were run for only ten hours total with incremental increase in adiabatic flame temperatures up to 1482°C. Catalyst effectiveness was found to drop during the higher temperature exposures, but the mechanical design features performed as intended. A 225 mm diameter prototype was also made and tested at 1260°C flame temperature without loss of catalytic activity or evidence of physical damage, but tests were limited in duration to several hours [14]. Despite these "successful" early demonstrations, SWPC acknowledged that much work remained to deploy ceramic substrate catalytic combustors in service.

Lessons Learned

- Catalytic combustors were demonstrated at flame temperatures up to 1482°C. While the ceramic substrates performed well mechanically for these short-term tests, catalyst durability prevented longer term testing.

Initial Assessment of Ceramic Technology Readiness For Gas Turbine Insertion. Between 1977 and 1979 Westinghouse conducted a program funded by DOE to assess the readiness of materials and design technology for ceramic insertion in gas turbine engines [6]. This program was intended to be a multi-phased, long-term development effort, the first phase of which focused on developing a Ceramic Turbine Design Concept and creating a technology readiness plan for future phases [6]. Unfortunately, only the initial design and planning phase was ever pursued. This work, however, was successful in establishing a "clean sheet" ceramic turbine concept that highlighted the potential benefits over conventional metallic designs.

Turbine cycle studies were conducted to assess the impact of uncooled ceramics on overall powerplant performance (Figure 8.11). These studies showed a nearly 3% efficiency benefit for uncooled ceramic components over conventionally cooled metallics for the same turbine inlet temperature (1427°C)—i.e., the benefits were due solely to the reduction of cooling air. Referring to Figure 8.11.b, it is clear that additional performance benefits can also be gained by increasing turbine inlet temperature, but these were not claimed in this study, as it is also clear that alternative technologies (i.e., air-cooled metals) could also account for increasing TRIT.

Three conceptual turbines were designed with various extents of ceramic component insertions in the hot sections. Concept #1 involved complete ceramic vane and blade insertion as well as ceramic hot wall combustors. A cross-section of the turbine hot section (Figure 8.12) illustrates the extent of ceramic application assumed (crosshatched areas), including all four turbine stages of blades (rotating) and vanes (static). A schematic of the key individual ceramic components is also shown in Figure 8.13.

Individual component designs were brought only to the preliminary level (since only the initial phase of the program was funded), so validity of the designs was never challenged—i.e. stress analyses, material selections, and manufacturing issues were not addressed. All four turbine stages were based on the same generic design concept for blades (Figure 8.14) and vanes (Figure 8.15). Several potential ceramic combustor configurations were proffered; all based on the general concept of hot ceramic tiles attached to a cooled metallic shell (Figure 8.16). This program did not develop these design schemes beyond the conceptual phase. Thus, these designs were not subjected to the rigors of detailed analysis, design review, or the realities of manufacturing constraints.

In parallel with the conceptual turbine and ceramic component design studies, an extensive set of design rules and criteria for ceramics was formalized and included in an interim report [15]. These rules covered such issues as statistical design practices, long-term performance effects (creep, stress-rupture, fatigue), fracture mechanics (fracture toughness, slow crack growth, etc.), foreign object damage, and corrosion/erosion. Unfortunately, the opportunity never arose for Westinghouse to exercise these ceramic design rules and criteria.

Lessons Learned

- Tremendous performance benefits were credited to application of ceramic blades and vanes to an otherwise conventional turbine design (3% increase in combined cycle efficiency).
- Without the rigors of detailed design and analysis, manufacturing and testing, the benefits attributed to the concept turbines developed in this program were never challenged or critically assessed.

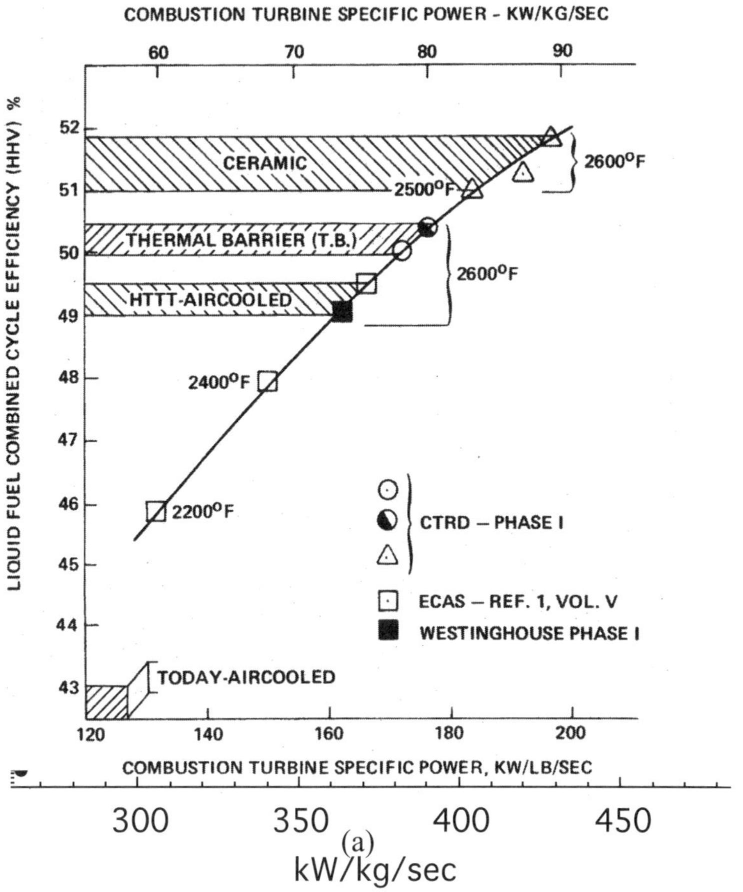

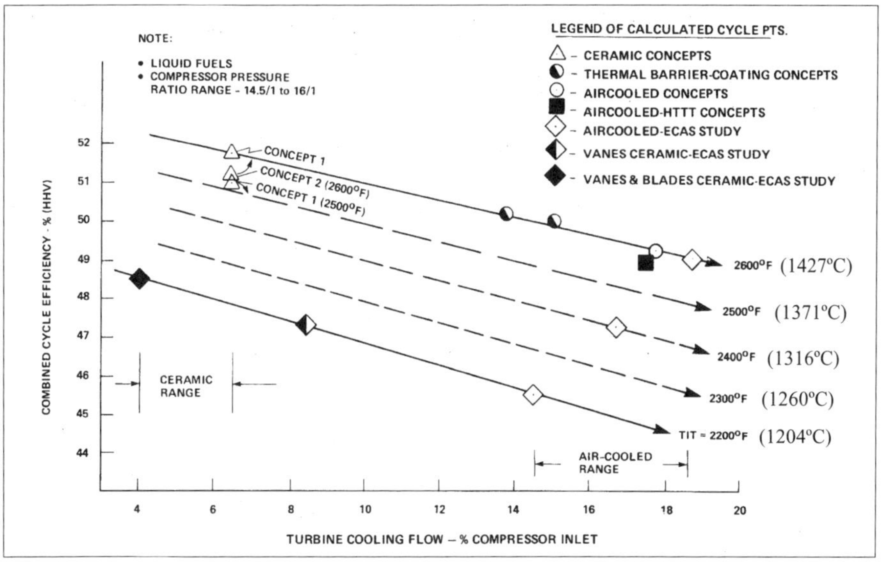

(b)

Figure 8.11 Turbine Cycle Studies Showing Impact of Uncooled Ceramics on Combine Cycle Plant Efficiency—(a) Combined Cycle Plant Efficiency vs. Combustion Turbine Specific Power; (b) Efficiency vs. Turbine Cooling Flow (Reprinted from Ref. [6])

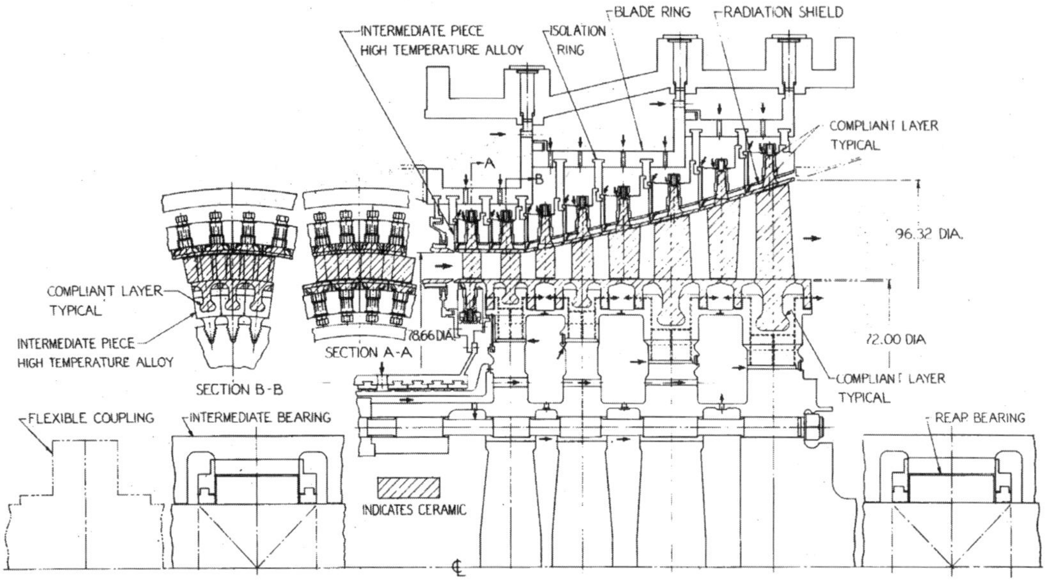

Figure 8.12 Turbine Concept Schematic Featuring Complete Ceramic Hot Section Gas Path Components (Reprinted from Ref. [15]) Dimensions in Inches—1 Inch = 2.54 cm

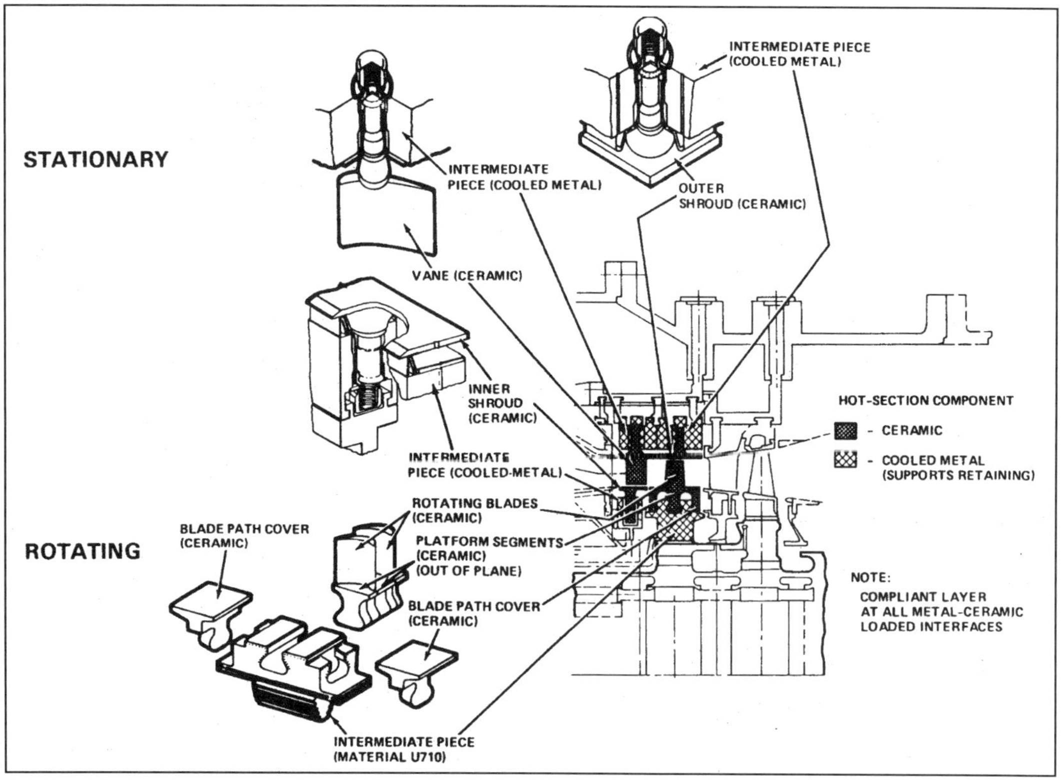

Figure 8.13 Key Ceramic Components (Reprinted from Refs. [6, 15])

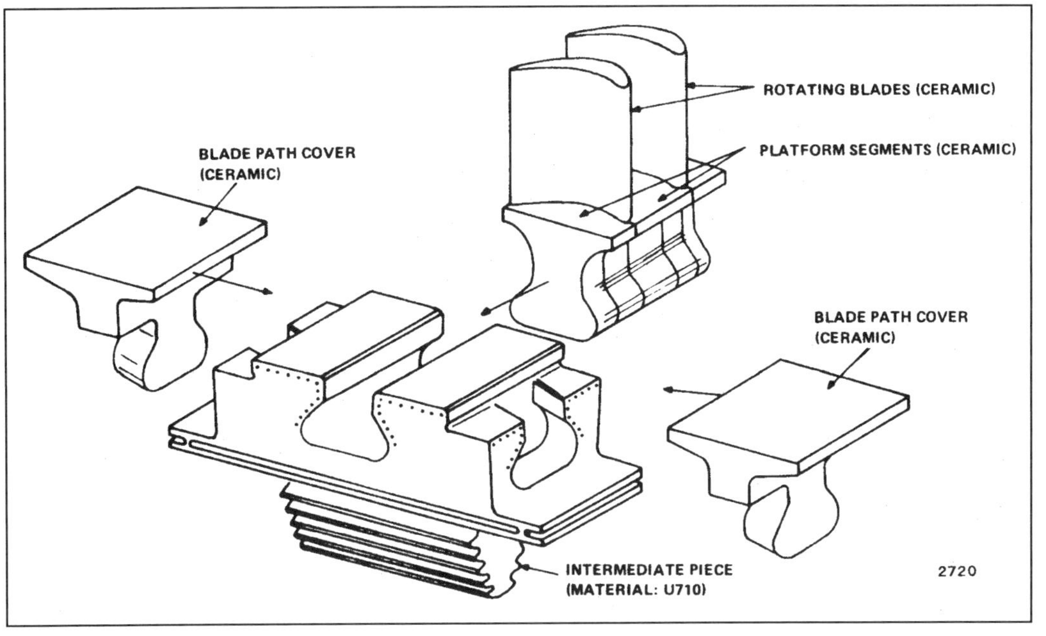

(a)

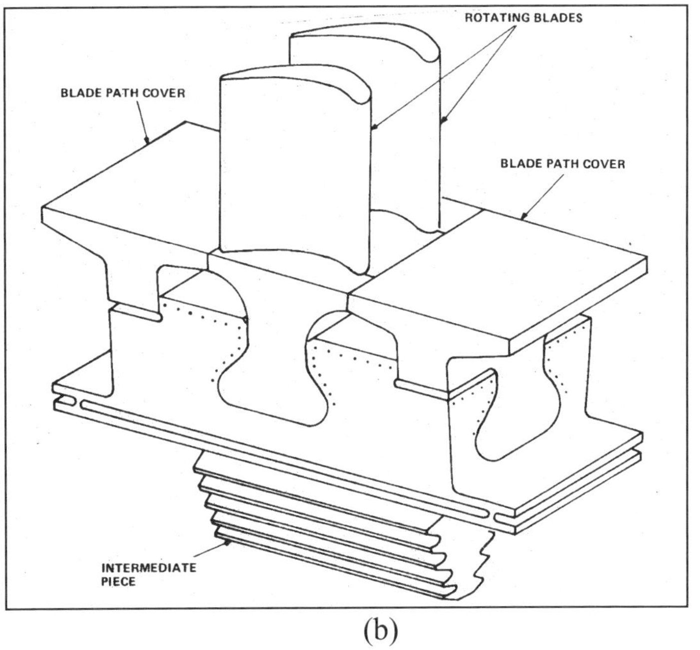

(b)

Figure 8.14 Generic Ceramic Blade Design for Concept Turbine—(a) Exploded View Showing Intermediate Metal piece, Ceramic Airfoils with Separate Platform Pieces, and Blade Path Covers; (b) Assembled Stage (Reprinted from Ref. [15])

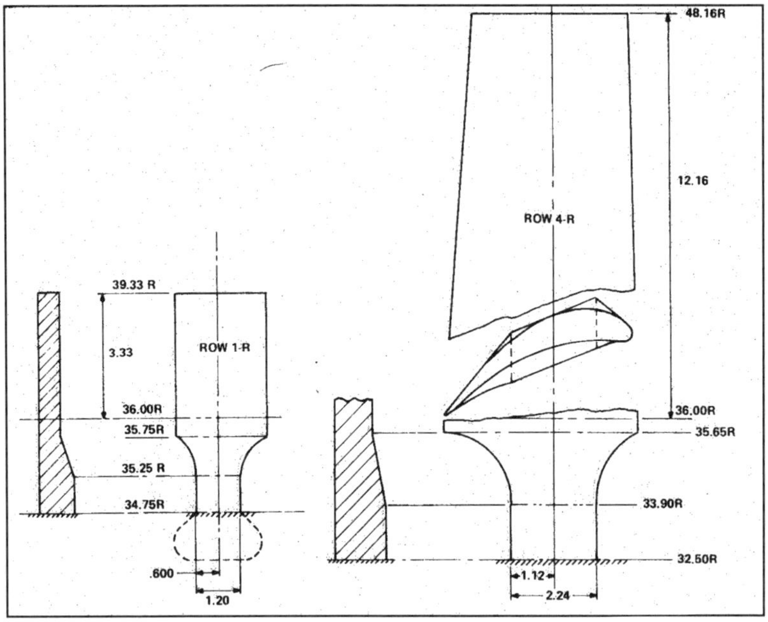

(a)

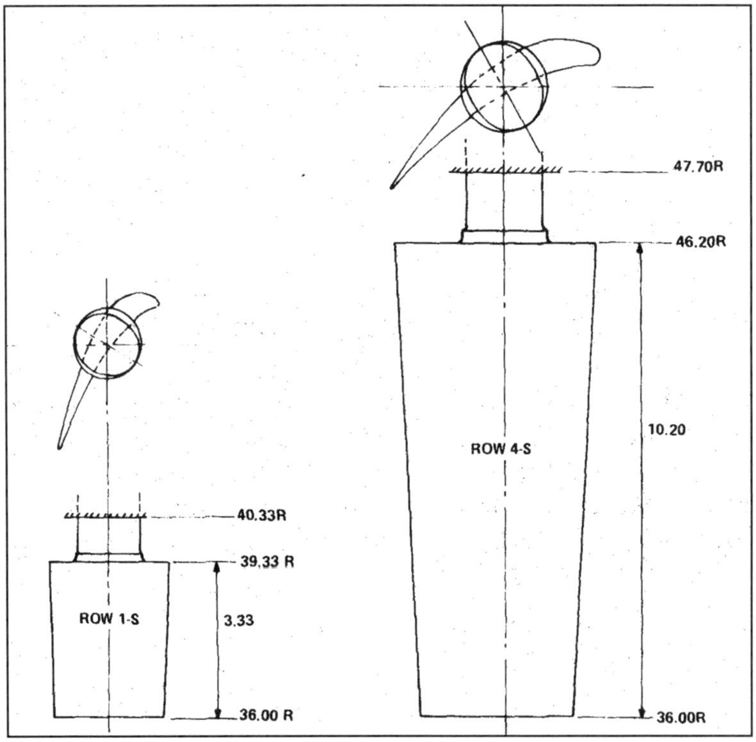

(b)

Figure 8.15 Ceramic Airfoils for Reference Turbine Illustrating Large Sizes Required for Large Frame Land-based Turbines—(a) Blade Sizes for Stages 1 and 4; (b) Vane Design Showing Large (10″) Airfoil Lengths for Rear Stages (Reprinted from Ref. [15]) Dimensions in Inches—1 Inch = 2.54 cm

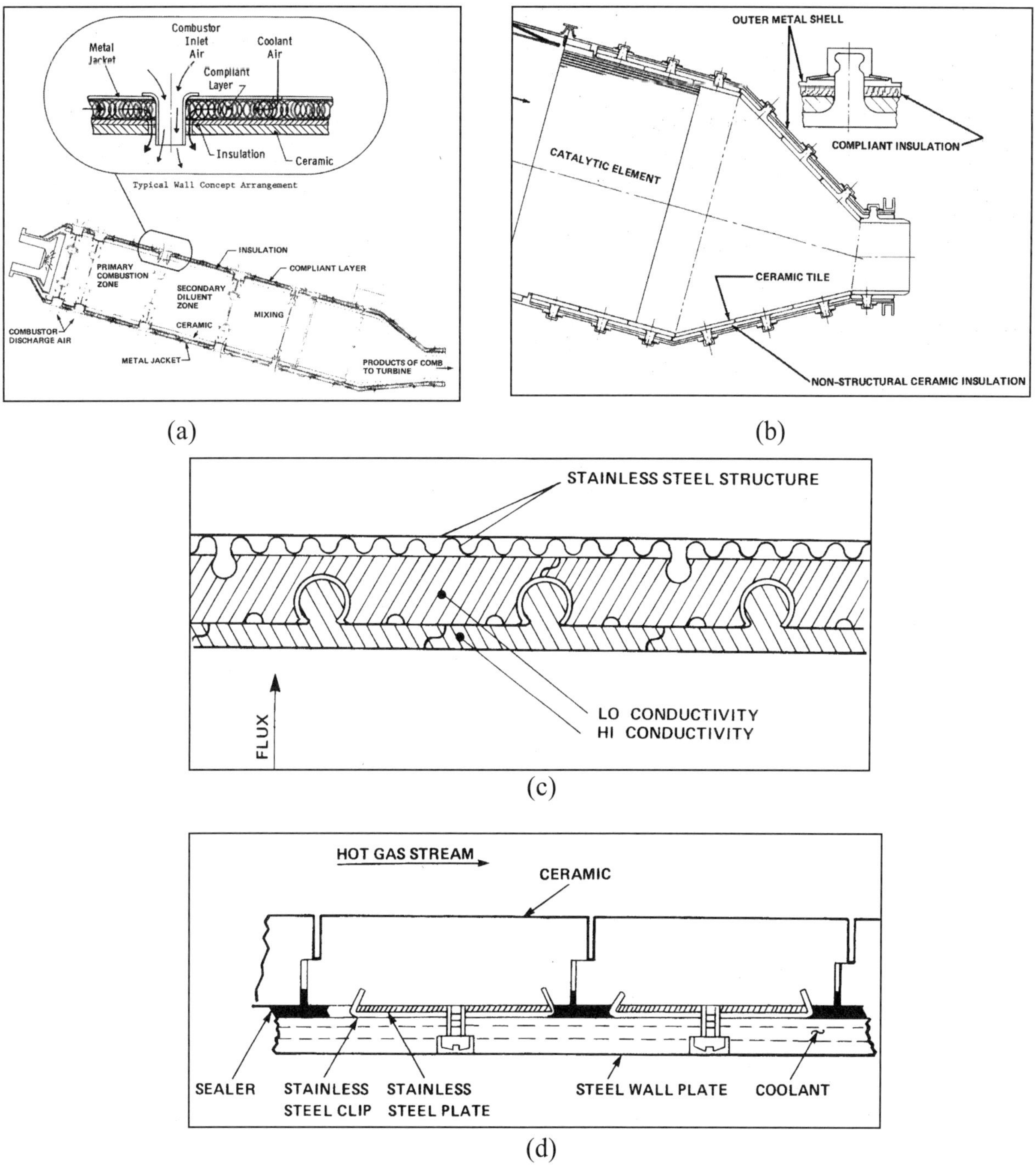

Figure 8.16 Hot-walled Combustor Concepts Based on Ceramic Tiles—(a) Coiled Metal Compliant Layer with High Temperature Ceramic Tiles; (b) Ceramic Tiles with Back-side Post for Through-thickness Mechanical Attachment; (c) Ceramic Thermal Insulation Behind Hot Tiles; (d) Thick Refractory Brick Tiles Retained Using Spring Clips (Reprinted from Ref. [6])

Progress Monitoring of Advanced Monolithics and Early CFCCs. Several factors contributed to the withdrawal of large power generation engine manufacturers' interest in advanced engine concepts and in ceramics materials insertion into engines during the 1980s. Simultaneous with a depressed market for power generation gas turbine sales, developments in thermal barrier coatings, metallic alloys and improvements in incorporating advanced cooling methods combined to reduce the potential advantages of ceramics over metallic systems for gas turbine components applications. With the conclusion of its major ceramic programs in the late 1970s, Westinghouse determined that monolithic ceramics were not viable at the time for the large components and extreme service conditions of its engines. It is important to recognize, however, that in this time frame, great strides were being made along more conventional technology fronts, leading to significant improvements in gas turbine performance without the use of ceramics. However, the promise of ceramics remained as engine operating temperatures continued to rise.

In the 1980s one non-structural application of ceramics materials was pursued for power generation gas turbines at Westinghouse. Continuation of the catalytic combustor work did occur during this time period focusing on more conceptual designs [16] and on short term tests of commercially available ceramic substrates [17]. The program did test a near-full-scale combustor (356 mm diameter element) for over 17 hours at temperatures up to 1430°C without apparent damage to the catalyst substrate. The substrate material used in this test was NGK-X, a magnesium aluminum titanate composition. The investigators recognized that, while these tests demonstrated short time performance capability at that temperature, they did not address the key technology issue of long-term durability of such materials.

Lessons Learned

- Material development and implementation efforts must be continually assessed in comparison to competing technologies accounting for each technology's readiness time frame.
- Viable ceramic catalyst materials and designs are possible, but true validation requires long term durability testing.

Application of Continuous Fiber Ceramic Composites

Evaluation of CFCCs for Non-Structural Turbine Component Applications. The availability of a range of viable CMC materials and potentially cost effective component manufacturing methods led to a resurgence in Siemens Westinghouse' interest in ceramic materials for power generation gas turbine components in the middle 1990s.

In the early-to-mid 1990s, the then Westinghouse Group demonstrated, in principal, the viability of oxide-based CMC systems to withstand the rigors of a combustion environment by fabricating and testing oxide-based combustor liners. A 3D-reinforced Nextel™ 550 fiber mullite/alumina matrix CFCC 150 mm diameter liner was fabricated (Figure 8.17) and installed in a natural gas fired combustor that was used for filter testing. The liner was operated uncooled at 10 atmospheres pressure and temperatures up to 1100°C for a total of 500 hours. The test was considered to be cyclic in that the firing was interrupted to accommodate other testing requirements. The liner successfully survived this test in good condition, without obvious environmental attack or mechanical cracking. This successful test was taken to demonstrate the feasibility of using a CFCC in a gas turbine environment. It was recognized, however, the temperatures to which the materials were exposed in the test rig were lower than those to which non-oxide systems were reported to be capable but were also lower than those that would be required for commercially viable service in a large size power generation engine. The demonstration of surviability was considered sufficiently promising to support further development efforts in the area of applying oxide-based systems to gas turbine components.

Lessons Learned

- CFCC turbine components could be designed with minimal mechanical loads, but thermally-induced strains are unavoidable and represent the most challenging design aspect.

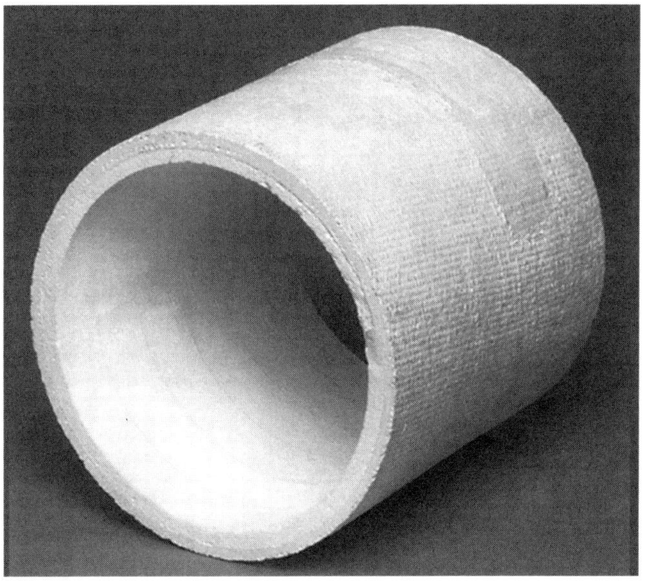

Figure 8.17 Oxide/oxide CMC Subscale Combustor Liner Developed and Fabricated by Westinghouse (Reprinted with permission from Siemens Westinghouse Power Corporation)

Evaluation of CFCCs for Structural Turbine Component Applications. Following on from the demonstration of survivability of the CMC combustion liner a task was undertaken under Westinghouse's ATS program (DOE Contract DE-FC21-95MC32267) to investigate the potential to use CFCCs in a first stage turbine blade tip seal for the ATS engine. In an effort to employ commercially available materials, the material of choice for this task was a DuPont Lanxide Composites (now Honeywell Advanced Ceramic Composites, Inc.) CG Nicalon fiber-reinforced alumina matrix composite made via the DIMOX process. This material was selected based on availability of an extensive database that had been generated to support aeroengine application. Specifically this database presented the high temperature properties of the composite in a very favorable light when the materials long term creep rupture properties were used to define the material performance. A program was therefore initiated to design, validate and fabricate CFCC ring segments (Figure 8.18) based upon this approach.

In our program, early creep-rupture testing of Nicalon/alumina CFCC at 1100° to 1300°C did result in extremely long lifetimes at moderate stress levels (e.g. 8000 hours for 70 MPa at 1200°C), indicating a long-life material at more moderate temperatures (Figure 8.19). In order to support component design criteria, subsequent design-related testing for life assessment (which was to be used to provide input to Virginia Polytechnic Institute's MRLife™ code), however, focused on interrupted creep testing of the CFCC. Residual strengths measured after relatively short creep time showed dramatic reductions due to exposure and/or cooling. Moreover, the specimens were significantly embrittled and the desired strain tolerant composite behavior was lost or greatly reduced. Microstructural studies indicated that oxidation of the silicon-based fibers and of the interface coatings was significant after matrix cracking that the free-sliding interfaces that were the basis of strain tolerant composite behavior were lost.

The unexpected degradation of the DuPont Lanxide material provided implications, not only for it's own intended use but also for that of other non-oxide based systems. The importance of first matrix cracking in environments (i.e. oxidizing and water vapor attack-bearing) which can degrade components of the composite structure was recognized. As Figure 8.20 demonstrates, the importance of limiting non-oxide composites to the regime below first matrix cracking is sufficient to negate most of their traditional strain tolerance and strength advantages. As a result of these findings, Nicalon-alumina was deemed

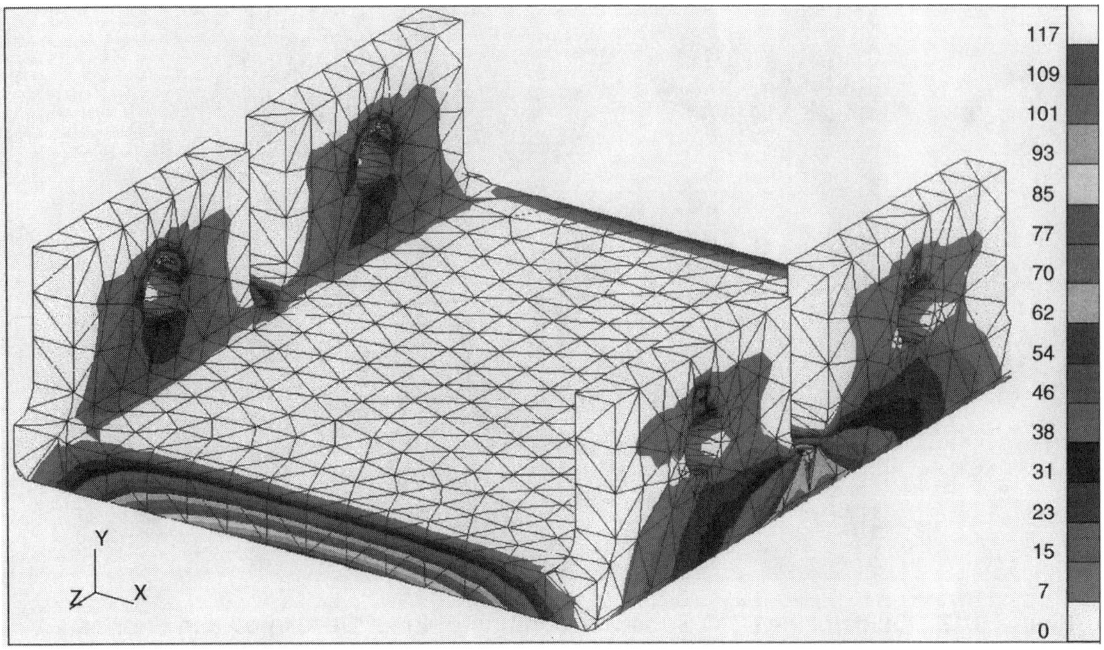

Figure 8.18 CMC Ring Segment Design Showing Results of Thermal and Mechanical Stress Analysis (Maximum Principal Stresses) (Reprinted with permission from Siemens Westinghouse Power Corporation) Stress levels in MPa

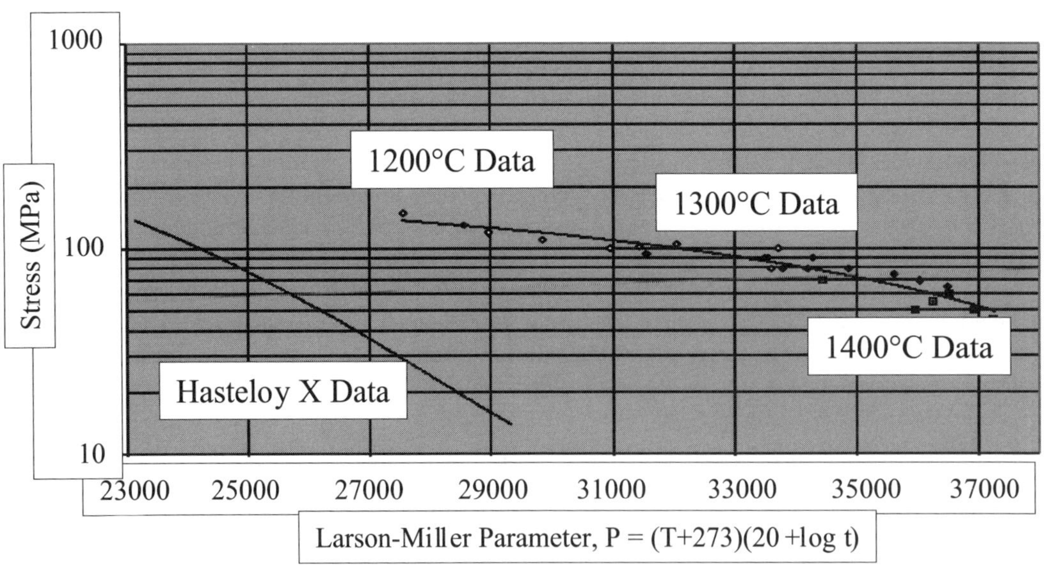

Figure 8.19 Larson-Miller Stress Rupture Data for Nicalon/Alumina Composite Material Made Via the DIMOX Process. Data Indicates Acceptable Life (>24,000 hours) at Reasonable Stresses (>70 MPa) and Temperatures (>1100°C) (Reprinted with permission from Siemens Westinghouse Power Corporation)

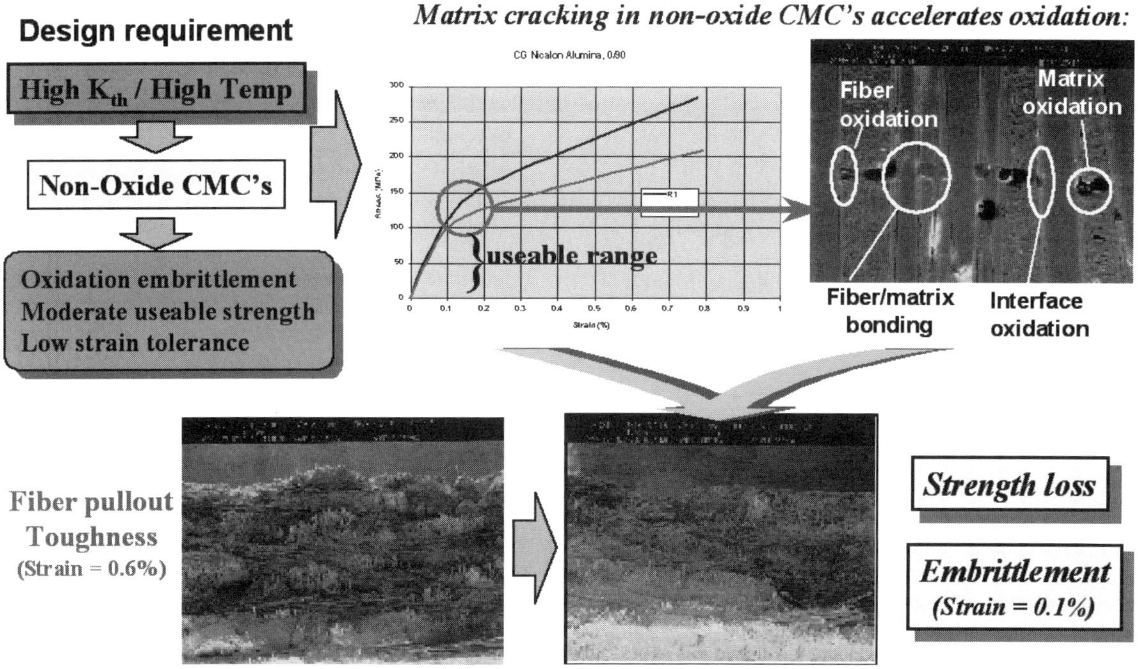

Figure 8.20 Conventional Paradigm for CMC Design and Application—Conclusion: Non-oxides are the Only Viable Candidate Materials. Long Term Environmental Degradation Issues Lead to Embrittlement and Unacceptable Strain-to-failure (Reprinted with permission from Siemens Westinghouse Power Corporation)

tional materials a support program was conducted to evaluate the performance and economic impact (performance benefits and market value vs. component and development costs) of oxide CFCC turbine components. While uncooled CFCC parts are not considered viable for gas path temperatures much above 1200°C, It was determined that minimally cooled CFCCs can function under hot gas temperatures over 1400°C. Performance benefits of CFCCs were calculated based on a reduction of cooling air of at least 90% for the static hot gas path components in a state-of-the-art combined-cycle power plant. This reduction in cooling air leads directly to increased efficiency and power output and reduced emissions. The performance benefits and economic impact that are expected to be derived from the implementation of such components are considerable.

- *Value of Improved Fuel Efficiency.* Fuel efficiency benefits for CFCC components are derived directly from reduction in parasitic cooling air. Based on the 90% reduction, this equates to an efficiency increase (and corresponding fuel savings) of 2%. In a plant of this size, the fuel cost savings would be over $1M annually.
- *Value of Increased Capacity.* In addition to delivering increased fuel efficiency, the reduction in cooling air for the GT components means that more air is available for combustion. For the same total mass flow and constant turbine rotor inlet temperature, more fuel can be added, resulting in increased power for the same engine. For the referenced plant, an 8% increase in power output is predicted (over 30MW). At a conservative value of $0.05/kWh market price for electricity, this 30MW increase in capacity would result in gross revenue increase of $13M annually.

- *Value of Reduced Emissions.* Significant emissions improvements (>50% reduction in NOx) from the use of "hot wall" combustion systems have been reported in smaller, intermediate temperature industrial gas turbines [18]. Similar trends are anticipated in the larger and higher operating temperature engines due to improved temperature patterns enabled by ceramic liners and lower flame temperatures allowed by downstream cooling air reductions. The combination of these two benefits can lead to near-zero (e.g., low single digit ppm NOx) emissions even while using conventional combustion methods. The economic value of this benefit can be compared to the cost avoided for exhaust cleanup. For the reference engine, this amounts to another $1M of annual savings.

This performance and economic assessment concluded that the product enhancements, or market value, offered by oxide CFCC components far outweighed the costs of development and increased cost of ownership for the components themselves. On the basis of these economic and technology assessments, a long-term program was undertaken to develop and demonstrate oxide-based CFCCs for GT engine applications.

Lessons Learned

- Cost-benefit analysis for ceramic vanes, blades, combustors, and transitions retrofitted into existing state-of-the-art engines shows tremendous potential benefits for ceramic insertion. Despite the improvements in engine performance via more conventional technologies, significant room for improved performance still exists through the introduction of ceramics in engines.

Initiative to Implement Oxide-Oxide CFCCs in Advanced Turbine Engine Components. As a result of the cost-benefits study Siemens Westinghouse identified the oxide-oxide materials systems produced by COI Ceramics of San Diego, California. as a potential approach to developing cost effective CFFC components for large, high temperature gas turbine components (Figure 8.21). The Siemens Westinghouse approach is to address and demonstrate the more-simple component manufacturing and less severe operating condition components first of all before proceeding to address more difficult component manufacturing and service demanding applications. Our approach is shown in Figure 8.22 [19]. Our initial program targets a proof of concept demonstration of materials' viability, component manufacturing and systems performance for implementing the oxide-oxide system in a full scale combustor liners.

Our proof of concept program has been underway since November 1999 and is being conducted in partnership with COI Ceramics and Solar Turbines Incorporated under the sponsorship of the Department of Commerce NIST-ATP program. This 3 year program has the objective to develop and demonstrate hybrid oxide CFCC materials for use in advanced gas turbines, with the aim of increasing the power output and reducing engine fuel costs and emissions. The scope of the program encompasses development of CFCC design and lifing tools, database development, component fabrication, NDE, and both rig (Solar and SWPC) and engine (Solar) testing of sub-scale and full scale proprietary CFCC combustors. The first rig test is planned for late 2000. A key component of the program development and on-going technology management is the simultaneous development of materials and component designs. Although the program funding represents a very significant investment by both the Department of Commerce and the industry partners, the program is providing the concerted effort that is needed to simultaneously answer all of the materials, design and manufacturing questions that currently hold back the application of oxide-oxide CFCCs in gas turbine components.

Lessons Learned (to date):

- The conventional view of CFCCs as being "materials" is limiting. Viewing the CFCC as a "structure" and designing components based upon "Composite principles" opens up a wider range of structural implementation for engine applications.

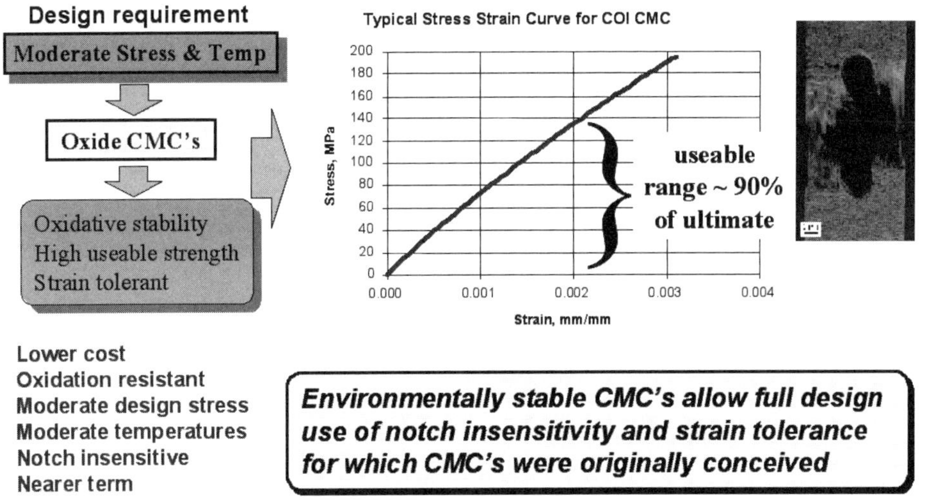

Figure 8.21 Siemens Westinghouse Approach to CMC Design and Application—Conclusion: Oxide-based CMC's are Only Viable Candidate Material for Long-term Operation in Oxidizing Environments.... Find a Way to Make them Work! (Reprinted with permission from Siemens Westinghouse Power Corporation)

Outlook for Future Usage of Ceramics in Industrial Gas Turbines

The benefits to be derived from using ceramic components in gas turbines remain significant and commercially attractive. These benefits not only warrant large scale efforts towards their implementation but also provide the basis for next generation electricity generating systems. Future government plans for power generation systems and propulsion engines universally acknowledge the need for the cross-cutting benefits that the implementation of ceramic technology can provide. However, sustained, consistent government and industrial funding will be required to accomplish widespread use of ceramics in engines. Programs such as the DoE/Solar CSGT program [20] and the high risk NIST-ATP program are examples of the kind of sustained effort the authors believe is required to achieve success.

The development, design and use of CFCCs in gas turbines will require open minds and courageous management support. The advantages of CFCCs should ideally be optimized in clean sheet designs for next generation gas turbines. Successful designs will only be achieved by working within the boundaries of CFCC material properties and manufacturing limits—not the direct-metal-replacement designs which fail to fully utilize the advantages of ceramics and place unattainable targets on material properties. Risk mitigation in the form of extensive component rig and engine bench testing will be necessary prior to field operation.

Hybrid oxide-based CFCCs have the potential to achieve the promise of ceramics in gas turbines while utilizing simple, low cost part designs. Siemens Westinghouse is therefore engaged in evaluating oxide-based CFCCs for gas turbine application. Experience in earlier programs has provided a realistic perspective on the risks and challenges involved. The overwhelming performance and economic benefits (performance, efficiency, power, emissions) from ceramic implementation provide sufficient impetus for continued assessment.

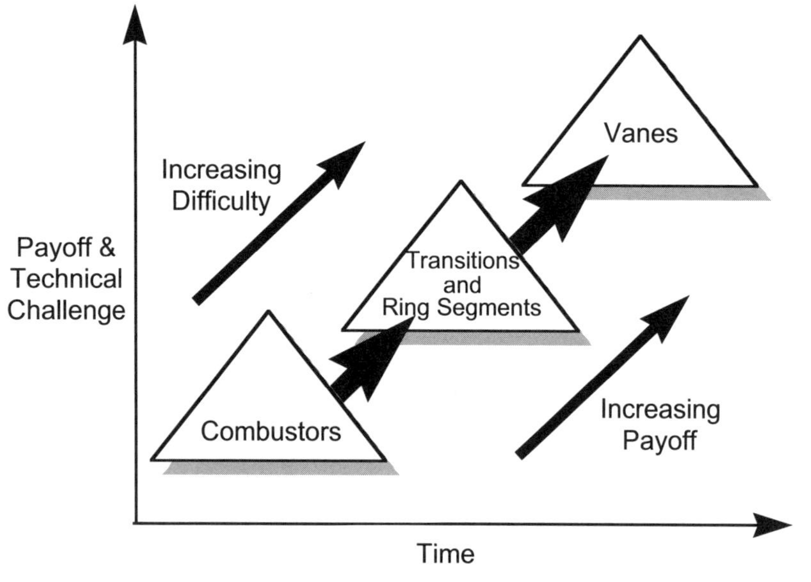

Figure 8.22 Proposed Progression of Development of CMC Gas Turbine Components (Reprinted with permission from Siemens Westinghouse Power Corporation)

SUMMARY & CONCLUSIONS

An early Westinghouse progress report stated, "Nothing has occurred as a result of this program to indicate that ceramics *cannot* be developed to work in electric power generating gas turbines" [7]. This observation is still valid today. In the intervening years, however, many of the barriers to successful commercial introduction have been systematically removed through focused and sustained efforts, making commercial ceramic application a much more real possibility today. The development programs of the past that tied materials developer and engine manufacturers have helped to deliver improved materials and supporting technologies, such as ceramics based design processes and lifing criteria. These programs have also highlighted the critical issues remaining for successful implementation. A key to effective progress on the remaining issues will be to constantly bear in mind the lessons of previous attempts to develop and implement ceramics in gas turbines.

In summary we believe the lessons of our previous programs to be:

Design-Related Enabling Technology

- The implementation of ceramics in gas turbines *has been* accomplished successfully for many years (e.g. by Siemens Westinghouse's now parent company Siemens KWU) for non-structural components but which do endure extreme thermal, erosion, corrosion, and vibration conditions.
- Key issues related to using brittle materials still remain today as the critical challenges to the successful design and commercial implementation of ceramics in engines.
- More rigorous design, lifing and analysis methods based upon the appropriate uses of today's powerful FEA codes are needed to support component design.

Material Capability & Performance

- The promised temperature advantages of ceramic materials have not lived up to early expectations due to the constantly improving performance of conventional materials and the crucially deleterious impact of environmental factors on ceramics that were not understood in initial assessments.

- Long-term durability requirements for large, land-based gas turbine applications preclude the use of environmentally unstable CFCC materials, e.g., the oxidation/corrosion/erosion of Si_3N_4 identified as a problem in early Westinghouse programs and continues to be a critical issue for non-oxide monolithics and CFCCs.
- Oxide CFCCs offer superior strain tolerance (toughness or flaw tolerance), notch insensitivity, non-catastrophic failure, and manufacturing scale advantages compared to non-oxide materials. These advantages offset their slight deficits in temperature capabilities. The design envelope for oxide oxide systems is actually expanded compared to non-oxide systems when environmental effects are taken into consideration.

Material Application & Philosophy

- "Ceramics are not quick-fix items for high performance turbomachinery" but ceramics can be realistically implemented using the appropriate "Ceramic design thinking."
- Designs which view CFCCs as "structures" and/or as components of a broader system open up the possibilities for engine applications. Concurrent engineering of ceramic solutions is essential to success.

Economic Drivers for Ceramics

- The performance and economic drivers to develop ceramics for gas turbines are considerable.
- The engine performance improvements offered by CFCCs, result in tremendous economic benefits for the operator, the environment, and the cost of electricity. For a large, 400 MW-class gas turbine power plant, the projected annual benefits exceed $3 million.
- As conventional (metals based) materials solutions become unable to meet increasing system demands only the development of ceramics based components can fulfill the need for improved efficiency, reduced emissions and low cost of electricity goals.

ACKNOWLEDGEMENT

The authors would like to acknowledge the consistently helpful advice and sustaining support of Dr. Brij Seth of Siemens Westinghouse Power Corporation.

Abbreviations and Acronyms

AMMRC	Army Materials and Mechanics Research Center
ATP	Advanced Technology Program
ATS	Advanced Turbine Systems Program
BN	Boron Nitride
CFCC	Continuous Fiber-reinforced Ceramic matrix Composites
CFD	Computational Fluid Dynamics
CTRD	Ceramic Technology Readiness Development Program
DARPA	Defense Advanced Research Projects Agency
DIMOX	Directed Metal Oxidation
DOE	Department of Energy
EBC	Environmental Barrier Coating
ECAS	Energy Conversion Alternative Study
EPRI	Electric Power Research Institute
FEA	Finite Element Analysis
GT	Gas Turbine
HTTT	High Temperature Turbine Technology Program
IR&D	Internal Research & Development

KWU	Kraftwerks Union (Siemens Power Generation Group)
LAS	Lithium Aluminosilicate
mm	millimeter
m/s	meters per second
MPa	Megapascals
MW	Megawatts
NIST	National Institute of Standards & Technology
NO_x	Nitrogen Oxides
R&D	Research & Development
SiC	Silicon Carbide
Si_3N_4	Silicon Nitride
SWPC	Siemens Westinghouse Power Corporation
TBC	Thermal Barrier Coating
TRIT	Turbine Rotor Inlet Temperature
μm	microns
3D	Three Dimensional

REFERENCES

[1] Beck, R.J. "Evaluation of Ceramics for Small Gas Turbine Engines," Paper No. 740239 presented at the SAE Automotive Engineering Congress, Detroit, Michigan, February 25-March 1, 1974

[2] Johnson, S.M., and Rowcliffe, D.J., "Ceramics for Electric Power-Generating Systems," EPRI AP-4380, Project 2608-1, Final Report, January 1986.

[3] Parks, W.P., Ramey, R.R., Rawlins, D.C., Price, J.R., and van Roode, M., "Potential Applications of Structural Ceramics in Gas Turbines," ORNL/Sub/88-SA798/01, Final Report, September 1989.

[4] Anson, D., Ramesh, K.S., DeCorso, M., "Application of Ceramics to Industrial Gas Turbines—Task 1, Technology Background," Task 1 Report, DOE/CE/40878-1 (DE91013140), March 1991.

[5] Anson, D., Sheppard, W.J., Norland, D.L., DeCorso, M., "Application of Ceramics to Industrial Gas Turbines; Economics, Market Analysis, and R&D Plan," Part 2 Report—Tasks 2, 3 and 4, DOE/CE/40878 (DE91013141), March 20, 1991.

[6] Reike, K.L., "Ceramics Technology Readiness Development Program, Phase I, Conceptual Designs and Material Screening," Final Report, FE-2786-38, November 1979.

[7] Bratton, R.J., Miller, D.G., "Brittle Material Design, High Temperature Gas Turbine—Stationary Turbine Project," Final Report – Volume 1 Program Summary, Contract DAAG 46-71-C-0162, May 1977.

[8] Bratton, R.J., Miller, D.G., "Brittle Material Design, High Temperature Gas Turbine—Stationary Turbine Project Summary," pp. 389–730 in *Ceramics for High performance Applications – II;* J.J. Burke, E.N. Lenoe, and R.N. Katz, eds., Proceedings of the Fifth Army Materials Technology Conference, March 1977, Brook Hill Publishing Company, Chestnut Hill, Mass, 1978.

[9] Bratton, R.J., Miller, D.G., Lange, F.F., Sanday, S.C., "Ceramic Rotor Blade Development (First Semi-Annual technical Report)," EPRI Contract No. RP 421-1, December 3, 1975. Copyright © 1975. Electric Power Research Institute.

[10] Anderson, C.A., Bratton, R.J., Cohn, A., Sanday, S.C., and Lange, F.F., "Progress on Ceramic Rotor Blade Development for Industrial Gas Turbines," ASME paper 77-GT-42, presented at the Gas Turbine Conference, Philadelphia, PA, USA, March 27–31, 1977.

[11] Bratton, R.J., "Ceramic Turbine Components Research & Development: Summary Report," EPRI AP-1539-SY, Research Project 421-1, August 1980. Copyright © 1980. Electric Power Research Institute.

[12] Anderson, C.A., Bratton, R.J., Boorer, C.R., "Ceramic Turbine Components Research and Development, Part 1: Ceramic Rotor-Blade Development," EPRI AP-1539, Part 1, Research Project 421-1, Final Report, August 1980. Copyright © 1980. Electric Power Research Institute.

[13] McLean, A.F., Fisher, E.A., Bratton, R.J., Miller, D.G., "Brittle Materials Design, High Temperature Gas Turbine," AMMRC CTR 75-28, Interim Report Number 8, October 1975.

[14] Lew, H.G., DeCorso, S.M., Dickson, W.H., Hung, W.S.Y., Glenn, R.G., Tobery, E.W., "Ceramic Turbine Components Research & Development: Part 3: Ceramic Combustor Design and Analysis," EPRI AP-1539, Part 3, August 1980. Copyright © 1980. Electric Power Research Institute.

[15] Reike, K.L., "Ceramics Technology Readiness Development Program, Phase I, Conceptual Designs and Material Screening," Interim report, FE-2786-20, July 1978.

[16] Pillsbury, P.W., Reike, K.L., Toof, J.L., Wisniewski, E.J., "Stationary Gas Turbine Catalytic Combustor Development Program, Preliminary Design Report," EPRI AP-2584, Research Project 1657-1, Final Report, October 1982. Copyright © 1982. Electric Power Research Institute.

[17] Pillsbury, P.W., Reike, K.L., "Critical Component Technologies for Stationary Gas Turbine Catalytic Combustors," EPRI AP-4063, Research Project 1657-3, Final Report, August 1985. Copyright © 1985. Electric Power Research Institute.

[18] Smith, K., Fahme, A., "Experimental Assessment of the Emissions Benefits of a Ceramic Gas Turbine Combustor," ASME paper 96-GT-318, presented at the International Gas Turbine and Aeroegine Congress and Exposition, Birmingham, England, U.K., June 10–13, 1996.

[19] Seth, B., Presentation at DOE Workshop—Materials for Next Generation Power Generation, Arlington, VA, August 1999.

[20] Price, J.R., Jimenez, O., Parthasarathy, V., Miriyala, N., "Ceramic Stationary Gas Turbine Development Program—Sixth Annual Summary," ASME paper 99-GT-351, presented at the International Gas Turbine and Aeroengine Congress and Exposition, Indianapolis, Indiana, USA, June 7–10, 1999.

CHAPTER

9

GENERAL ELECTRIC COMPANY

Paul S. DiMascio, Robert M. Orenstein,
Mark S. Schroder, Leonardo Tognarelli
GE Power Systems, Schenectady, New York, USA

Gregory S. Corman, and Anthony J. Dean
GE Corporate Research & Development, Schenectady, New York, USA

INTRODUCTION

GE Power Systems designs and manufactures heavy duty and industrial gas turbines for electric power generation, gas compression, and combined heat and power applications. Ceramic materials will enable design of turbine components that can operate at higher gas temperatures and/or require less cooling air than their metal counterparts. This, in turn, leads to higher turbine output and thermal efficiency, which ultimately leads to lower Cost of Electricity for the customer. In addition, many customers desire increasingly longer intervals between planned maintenance outages, which places further demands upon the turbine materials. Ceramics have great potential for meeting both of these customer driven Critical-To-Quality needs (CTQs). This chapter discusses the philosophy toward the use of ceramic materials in stationary gas turbines that has evolved at GE Power Systems over the past thirty years. In addition, plans for future applications of ceramics and technology development are outlined.

MARKET DRIVERS ... WHY CERAMIC COMPONENTS NOW?

The evolution of gas turbines for electric power generation has been paced by materials capability, with several notable exceptions. Advanced air-cooling technology, originally developed for aircraft engines, was introduced into the F-Class heavy duty gas turbines in the late 1980s. This enabled the turbine firing temperature to be increased by approximately 170°C (300°F) with relatively minor increases in existing hot section material rupture capability. Subsequent advances in both materials and cooling technologies have enabled firing temperature to be further increased by approximately 140°C (250°F), but this has also required implementation of Dry Low NOx combustion technology to retain low emission levels.

The introduction of steam-cooling technology will enable continued increases in firing temperature while maintaining single-digit NOx emissions. Materials requirements for steam-cooled components are significantly more demanding than for air-cooled components, however; leading GE Power Systems to seek the optimum balance between coated metals and ceramic materials for the hottest turbine compo-

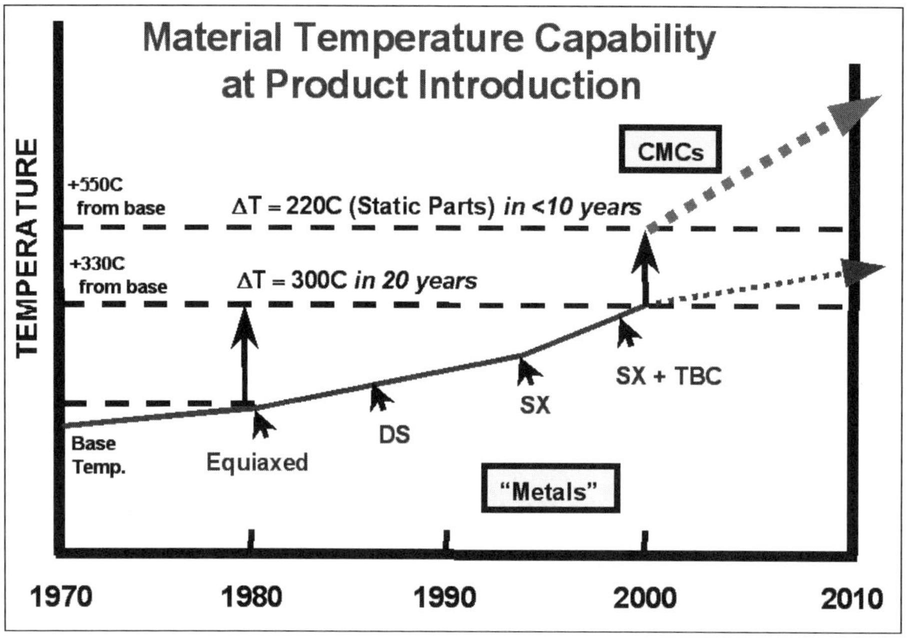

Figure 9.1 Historic and Anticipated Improvements in Gas Turbine Materials Capability (Reprinted with permission from GE Power Systems)

nents. Integration of ceramic components into static structures such as turbine shrouds and nozzles, which require high levels of chargeable cooling in their current configuration, will result in considerable performance payoffs in advanced gas turbines while keeping the risk of introducing new materials to acceptable levels.

Integration of ceramic components is most attractive when cooling flows are less available, as is the case for low emissions combustors; or are chargeable to the cycle, as is the case for turbine shrouds and nozzles. The step improvement in temperature capability afforded by utilizing Ceramic-Matrix Composite (CMC) materials is shown in Figure 9.1. An increment of ~ 220°C (400°F) in the material use temperature can be realized for certain components by using CMCs. It is expected that four years will be required to implement this material replacement on a commercial scale, which compares very favorably with the 20 years it has taken to gain ~ 300°C (550°F) in the material use temperature by progressing from Equiaxed to Directionally Solidified (DS) to Single crystal (SX) cobalt- and nickel-based superalloys, including the very recent application of ceramic Thermal Barrier Coating (TBC). It is expected that ceramics will only be utilized in static structures over the next 15–20 years, however, due to the much higher risk of introducing new materials into rotating structures. Therefore, continued advances in both superalloys and TBCs are also expected, but at a lower rate relative to the expected advances in CMCs, as illustrated in Figure 9.1.

The potential benefits of using CMCs in the 9FA gas turbine were reported by GE in 1998 [1]. The results of the study showed that the use of ceramic materials in stationary components (such as shrouds, nozzles and combustors) could enable increases in combined cycle efficiency of up to 1.4 percentage points and increases in combined cycle output of up to 17%. Although the study was originally conducted using silicon nitride material properties, similar benefits have recently been confirmed for a CMC shroud retrofit application in the 7FA gas turbine. The expected benefit of using a stage one ceramic/metal "hybrid" turbine shroud in a 7FA gas turbine and incorporating this structure into a clearance control system are:

- 0.2 increased combined cycle efficiency and 1.4% increased output by reducing shroud chargeable cooling, plus
- 0.2 increased combined cycle efficiency and 0.2% increased output by reducing bucket tip clearance

GE's fleet of F- and H-class gas turbines (70+ MW output) comprises over 26 GW of installed power, which is projected to grow to over 160 GW by year 2020. Implementation of the hybrid stage one turbine shroud will contribute a combined cycle efficiency gain of 0.4 percentage points in these machines, which corresponds to potential yearly savings in energy cost of ~ $174M and yearly reductions in carbon dioxide emissions of ~ 0.84 MMTCE by 2020. Most of GE's smaller gas turbines operate at significantly lower firing temperatures than in the F- and H-class, and still have large growth potential by using coated metals with advanced cooling technologies, as are used in aircraft engines. Demonstration of ceramic materials in smaller machines will continue, as these provide lower risk opportunities for gaining initial experience. However, GE's strategy for introducing ceramic materials into stationary gas turbines is centered on the largest machines.

ALTERNATIVE TECHNOLOGIES … WHY NOT CERAMIC COMPONENTS?

First and foremost, the general public, including operators of power generation gas turbines, still perceive "ceramic" materials to be highly fragile, like porcelain or glass, which are well known for catastrophic, brittle failures. Thus, there is the widely-held perception by gas turbine users that any and all ceramic materials will fail catastrophically, with unacceptable consequences to their equipment and, more importantly, to their revenue streams. As a result of these concerns, Melt Infiltrated-Ceramic Matrix Composite (MI-CMC) was developed by GE. This material does not fail in a catastrophic manner, as proved by high pressure combustion rig, ballistic impact and gas turbine testing. Nevertheless, gas turbine design and application experience with MI-CMC is very limited in comparison with metals, and considerable "proof-of-concept" development and testing is still needed before the risk of using this material is low enough to satisfy GE's New Product Introduction requirements.

The second reason alternative materials technologies are still needed and being actively developed is component cost. In particular, the cost of composite materials such as MI-CMC is significantly higher than the cost of the materials it is intended to replace, partially because of low initial production volumes. However, projections show that the finished cost of MI-CMC components may reach the levels of TBC-coated, single crystal superalloy components at moderate commercial production levels. In addition, when a cooled component is replaced by an uncooled component, elimination of coolant delivery hardware, seals and other ancillary components can reduce the overall sub-system cost, even after the higher cost of the MI-CMC components is taken into account. Finally, the cost of MI-CMC components must be weighed against the benefit provided by their utilization. Recent studies on advanced (F- and H-class) GE gas turbines show very substantial performance benefits, as discussed previously. GE Power Systems has concluded that the higher cost of MI-CMC components is justified by value of these benefits to the customer.

So what are the alternatives to ceramics? The state-of-the-art materials used for the hottest turbine components are nickel-based superalloys coated with a metal overlay "bond coat" and a ceramic TBC. Substantial backside cooling of these components results in a large temperature drop across the TBC, which is required for acceptable life of both the ceramic coating and the superalloy substrate. Current generation superalloys are limited to bulk temperatures of 850 to 950°C (1600 to 1750°F) for heavy duty and industrial gas turbine duty cycles, with expected improvements of approximately 50°C (100°F) every 5–10years. TBC bond coats are limited to approximately 950°C (1750°F), while the upper temperature limit for TBC is approximately 1200°C (2200°F) due to environmental contaminants. Ongoing development programs aim to increase these limits by 50 to 100°C within the next 5 years. In addition, the effectiveness of the TBC as an insulating layer can be improved by increasing backside cooling effectiveness. Therefore, although coated superalloys do not offer the dramatic increase in temperature capability offered by MI-CMCs and other ceramic materials, further improvements in temperature capability are expected. Furthermore, these improvements can be realized without significant changes in component design.

CERAMIC COMPONENT PROGRAMS AT GE POWER SYSTEMS

Ceramic Technology Readiness Development Program (ERDA)

GE Power Systems has had a long standing interest in ceramic materials. The Ceramic Technology Readiness Development Program, sponsored by the Energy Research and Development Administration, was conducted in the late 1970s to develop ceramic components for high efficiency heavy duty turbines operating on coal-derived fuels. Monolithic silicon carbide and silicon nitride materials were evaluated for combustor and turbine airfoil applications. The available materials were not considered acceptable, however; due to high strength variation (low Weibull modulus), low fracture toughness and poor environmental resistance.

Ceramics for Gas Turbines Program (EPRI/GRI)

The Ceramics for Gas Turbines program, sponsored jointly by the Electric Power Research Institute and the Gas Research Institute, was conducted from 1993–96 to evaluate the feasibility of using monolithic silicon nitride turbine stator components in a MS7001FA gas turbine. The overall feasibility goal was to determine whether life and reliability targets could be satisfied for a conceptual ceramic inner turbine shroud having properties of a commercially available sintered silicon nitride [2, 3]. Reliability and Life models were developed and validated via thermal shock and tensile experiments. The effect of oxidation was empirically evaluated using four-point flexure samples exposed to flowing natural gas combustion products at one atmosphere. The reliability- and life-limiting failure mechanisms were characterized in terms of temperature, stress and probability of component failure, enabling conservative design limits for silicon nitride gas turbine components to be established.

Continuous Fiber Ceramic Composites Program (DOE)

The Continuous Fiber Ceramic Composites (CFCC) program, sponsored by the Department of Energy–Office of Industrial Technology (DOE-OIT), began in 1992 and will continue through 2001. Five different material systems were evaluated by GE Power Systems (GEPS) and GE Corporate R&D (CRD) in the first phase of the program. Both a combustor liner and turbine shroud are candidate components.

The prime material selected for component fabrication was silicon carbide/silicon carbide MI-CMC. GE-CRD is the original inventor of MI-CMC and holds over 15 patents relating to the material composition, processing and use. Process development has been funded for the past eight years by NASA, under the High Speed Civil Transport (HSCT) program, as well as by DOE under the CFCC program. The methods by which MI-CMC components can be fabricated are described in greater detail in Vol. II of this text, Chapter 16.

A material properties database has been created jointly by GEPS and GE-CRD, and initial design limits were established. Shrouds and combustor liners were successfully tested in high pressure combustion rigs and a 2 MW gas turbine in 1998 and 1999 [4]. MI-CMC is used exclusively for both applications for the following reasons:

First, these materials have higher thermal conductivity and lower thermal expansion than other classes of CMCs. The highest steady-state stresses in turbine stator components are caused by temperature gradients. Transient thermal stresses, particularly during turbine trips, are also very high. Considering the fact that the design stress limits for CMCs are lower than those for nickel-based superalloys, the thermal stresses in CMC components must be reduced as much as possible for successful implementation. Use of high thermal conductivity and low thermal expansion materials is critical for achieving low thermal stresses.

Second, the high matrix density results in high matrix cracking strength for MI-CMC. The matrix cracking strength is limiting for component design, because it defines the allowable limits for fatigue, rupture and high temperature environmental stability. Design studies confirmed that MI-CMC gave the largest safety margin compared to CVI SiC/SiC and Dimox™ Al_2O_3 / SiC CMCs. For each material, the in-plane stresses predicted for a conceptual F-class inner turbine shroud were compared to the matrix cracking strength to

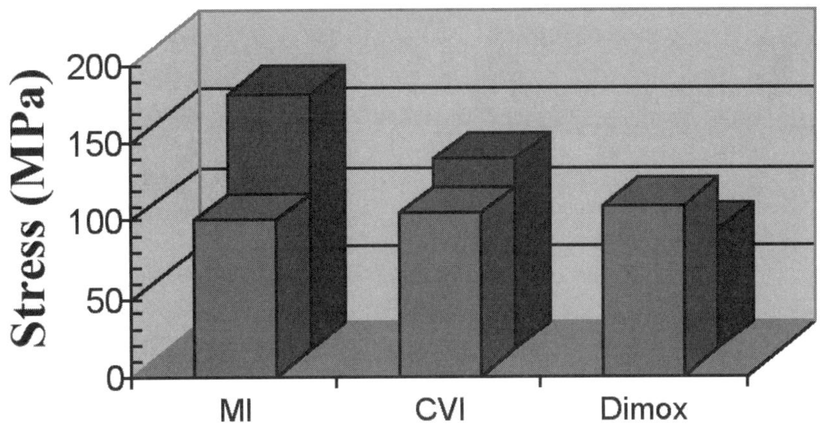

Figure 9.2 Results of Mechanical and Stress Analyses Performed in the CFCC Phase Two Program for an F-Class Turbine Shroud Application. The Predicted Maximum In-plane Stress Levels are Compared to Matrix Cracking stress for three types of CMCs. The MI-CMC had the Largest Safety Margin (Reprinted with permission from GE Corporate Research & Development)

determine the safety margin. MI-CMC gave the largest safety margin, particularly when compared with the Dimox™ (oxide-based) CMC, as shown in Figure 9.2. The unacceptable stress margin for the oxide-based CMC was confirmed by combustion rig testing.

The excellent performance of MI-CMC was further confirmed by impact experiments. Ballistic impact tests were performed using projectile velocities of 311 m/s (1020 ft/s), which is equal to or greater than the velocities of particles in gas turbines that are known to cause Foreign Object Damage (FOD) of metal components. In these tests, holes were blown completely through 0.2 cm (0.08 inch) thick MI-CMC panels; however, impact damage was limited to a region only 2X the diameter of the projectile. One such panel was then exposed in a high pressure combustion rig for over 100 hours, including 100 turbine trip cycles. Destructive evaluations following rig exposure indicated no propagation of the damaged zone.

Conceptual MI-CMC turbine shrouds and combustor liners, manufactured by GE-CRD and two suppliers, were exposed to gas temperatures exceeding 1500°C (2800°F) in combustion rig tests without catastrophic failures. This performance was vastly superior to that of similarly exposed Haynes Superalloy (HS) 188 nickel-based superalloy. Figure 9.3 shows photographs of a MI-CMC shroud and an HS-188 shroud after combustion rig testing. The metal shroud showed severe warping, oxidation and fatigue cracking after only 50 thermal cycles and 2 hours total exposure at 1200°C (2200°F). The MI-CMC shroud showed no dimensional change and only very minor edge cracks after 200 thermal cycles and 100 hours of total exposure.

These very encouraging results have been followed by testing of a one-piece MI-CMC stage two shroud ring, manufactured by GE-CRD, in a GE-2 gas turbine by GE-Nuovo Pignone in Florence, Italy. The GE-2 produces two MW output from a two-stage centrifugal compressor, single can silo combustor and two-stage axial turbine. The current metallic stage two shroud consists of eight segments with leaf seals between the segments. Neither the stage one nor the stage two shrouds are cooled in this machine. A cross section fo the turbine section, showing the positions of the stage one and two shrouds, is shown in Figure 9.4. Bulk average combustion gas temperatures in the vicinity of the stage two shroud are about 780°C (1435°F), with the temperature of the inner casings typically running about 450°C (840°F).

The ceramic shroud design consisted of a thin walled MI-CMC single piece ring measuring 466 mm (18.3 inches) inner diameter by 57.5 mm (2.3 inches) length. The shroud ring was located axially in the engine between two metallic flanges, and additional features were incorporated into the shroud to control its

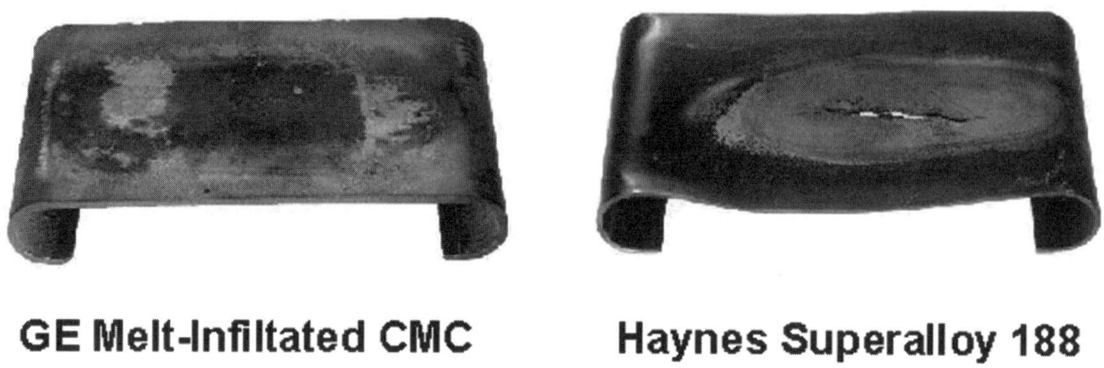

Figure 9.3 Conceptual F-Class Gas Turbine Inner Shroud Components Shown Following Combustion Rig Testing During the CFCC Phase Two Program (Reprinted with permission from GE Corporate Research & Development)

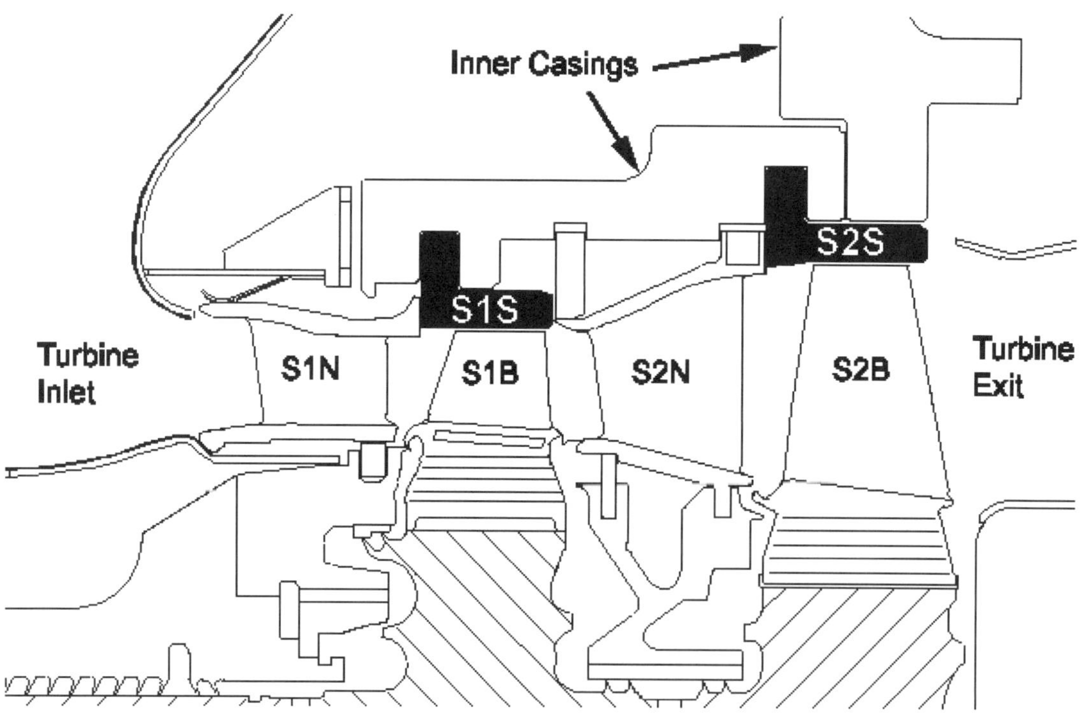

Figure 9.4 Schematic of the Turbine Section of the GE-2 Engine Showing the Positions of the Stage One and Stage Two Nozzles (S1N and S2N), Buckets (S1B and S2B) and Shrouds (S1S and S2S) (Reprinted from Ref. 4)

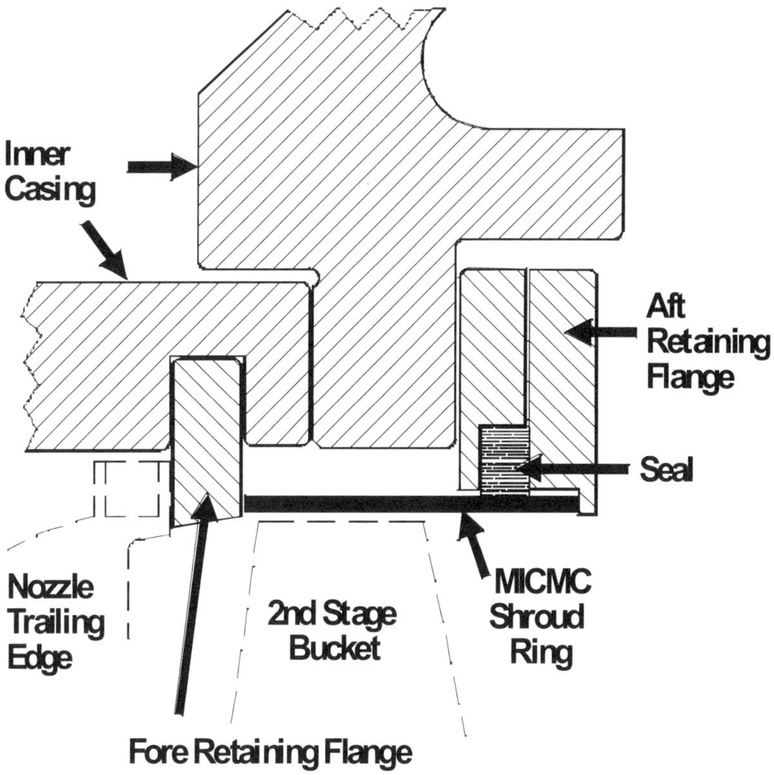

Figure 9.5 MI-CMC Stage Two Shroud Design in the GE-2 Engine (Reprinted from Ref. 4)

radial position. A compliant seal was used to prevent combustion gas leakage around the outer diameter of the shroud. A schematic of the design is shown in Figure 9.5. A photograph of the as-fabricated shroud is shown in Figure 9.6. This shroud has been tested for over 100 hours at full speed/full load conditions, including numerous shutdowns and turbine trips. The shroud also survived multiple blade tip rubs without damage, as shown in Figure 9.7.

LESSONS LEARNED

GE's experience with MI-CMC has been very encouraging, but a number of lessons have been learned which currently limit their application to a small number of components. The most important consideration when using MI-CMC and other ceramic materials in gas turbines is establishing conservative design limits, in terms of both stress and strain, because the absolute strength and strain tolerance of these materials is substantially lower than those of the nickel-based superalloys they are intended to replace. Furthermore, these design limits must account for both the statistical variation in as-fabricated properties, as well as the degradation of properties over time due to the effects of creep, fatigue and environmental attack. Although this has always been GE's practice for metals, it was not considered a priority for ceramics until recently because the material systems were steadily improving and new test methods were needed. Furthermore, and particularly for monolithic ceramic materials, a large number of specimens must be tested and thoroughly evaluated in order to accurately describe the variation in strength, which can be significantly greater than in metals.

Figure 9.6 Prepreg MI-CMC Shroud Shown Prior to Installation in a GE-2 Gas Turbine for Testing by GE-Nuovo Pignone. The Part Measures 466 mm Inner Diameter by 57.5 mm Length (Reprinted from Ref. 4)

Figure 9.7 Photograph of the MI-CMC stage Two Shroud in the GE-2 Engine, Showing the Blade Tip Rub (Reprinted from Ref. 4)

Second, selecting materials with high thermal conductivity is of paramount importance. The highest steady-state stresses in turbine stator components are caused by spatial temperature differences. Transient thermal stresses during turbine trips are also very high. This, and considering the fact that the design stress limits for ceramic materials are lower than those for nickel-based superalloys, dictates that the thermal stresses in ceramic components be reduced to the maximum extent possible.

The ceramic materials with the highest thermal conductivity are non-oxides, typically containing silicon, which form a scale of silica (SiO_2) when exposed to oxidizing environments. This scale is protective against further oxidation in low moisture atmospheres; however, the high partial pressure of water vapor in gas turbine combustion streams causes the silica scale to volatilize by the formation of Si-O-H species, consequently exposing the underlying material to continued attack. This behavior is fundamentally different from the behavior of the nickel-based superalloys and overlay coatings currently used throughout the gas turbine industry, which form a protective layer of either alumina or chromia that are stable in the presence of water vapor.

The surface recession and mass loss of MI-CMCs in the presence of high water vapor pressures have been experimentally verified through furnace tests at GE-CRD and high pressure burner rig tests at NASA. The recession rate is also affected by the temperature, velocity, type of fuel combusted, and fuel/air equivalency ratio. Figure 9.8 shows the results of analytical modeling for typical turbine operating conditions and fuels [5].

The models predict that, in the environment of an F-Class gas turbine operated on natural gas fuel, the recession rate of MI-CMC at 1200°C (2200°F) will be greater than 0.2 cm (0.08 inch) in 24,000 hours, which is unacceptable for reliable gas turbine operation. These results have been confirmed by experimental testing, leading GE to consider protective coatings as well as other means to mitigate this degradation mechanism.

Development of an Environmental Barrier Coating (EBC) for MI-CMC was started under the HSCT program. Several candidate oxide coatings were developed, with the most promising being an Air Plasma Sprayed (APS) Barium-Strontium Alumino-Silicate (BSAS) with a silicon/mullite bond layer [6]. The coating design is schematically shown in Figure 9.9.

SUMMARY AND PLANS FOR FUTURE DEVELOPMENT

Many advanced materials have been developed and/or evaluated by GE in a continuing effort to achieve higher specific power, lower fuel consumption, and lower operating cost of stationary gas turbines. As described above, and in Chapter 16 in Volume 2, GE began evaluation of silicon nitride and silicon carbide materials starting in the early 1970s. Initial efforts involved monolithic materials (i.e., non-composite ceramics with a brittle fracture mode). When GE considered all factors; such as reliability, cost, and performance of the available materials; the decision was made to pursue advanced cooling technologies and ceramic thermal barrier coatings on metal components for the F- and H-class gas turbines.

During the time frame that the F- and H-class gas turbines were being developed, GE continued the evaluation and development of new ceramic materials. By the mid-1990s, Melt Infiltrated-Ceramic Matrix Composite (MI-CMC) materials that had excellent properties and did not fracture in a brittle mode were developed and proved to have high potential for use in gas turbines. Design studies and component testing were therefore resumed, and GE Power Systems is currently committed to explore the full potential of these materials as replacements for coated superalloys in uniquely suited gas turbine hot section components.

GE's patented MI-CMC will be extensively tested in large and small gas turbines over the next five years. Early applications of this material will be turbine nozzle leading and trailing edges, combustor liners and inner turbine shrouds. The latter may be integrated into a Clearance Control System (CCS) developed under the Advanced Turbine Systems program.

The purpose of the CCS is to create the smallest possible gap between the static turbine shroud and the rotating turbine bucket under all operating conditions, while eliminating rotor to stator interference that will result in the bucket tip "rubbing" against the inner shroud. A feedback control system is included, which receives and interprets a capacitance signal from clearance probes mounted within the turbine shroud assembly, and adjusts the temperature of "conditioning" air that circulates within the outer shroud. The temperature

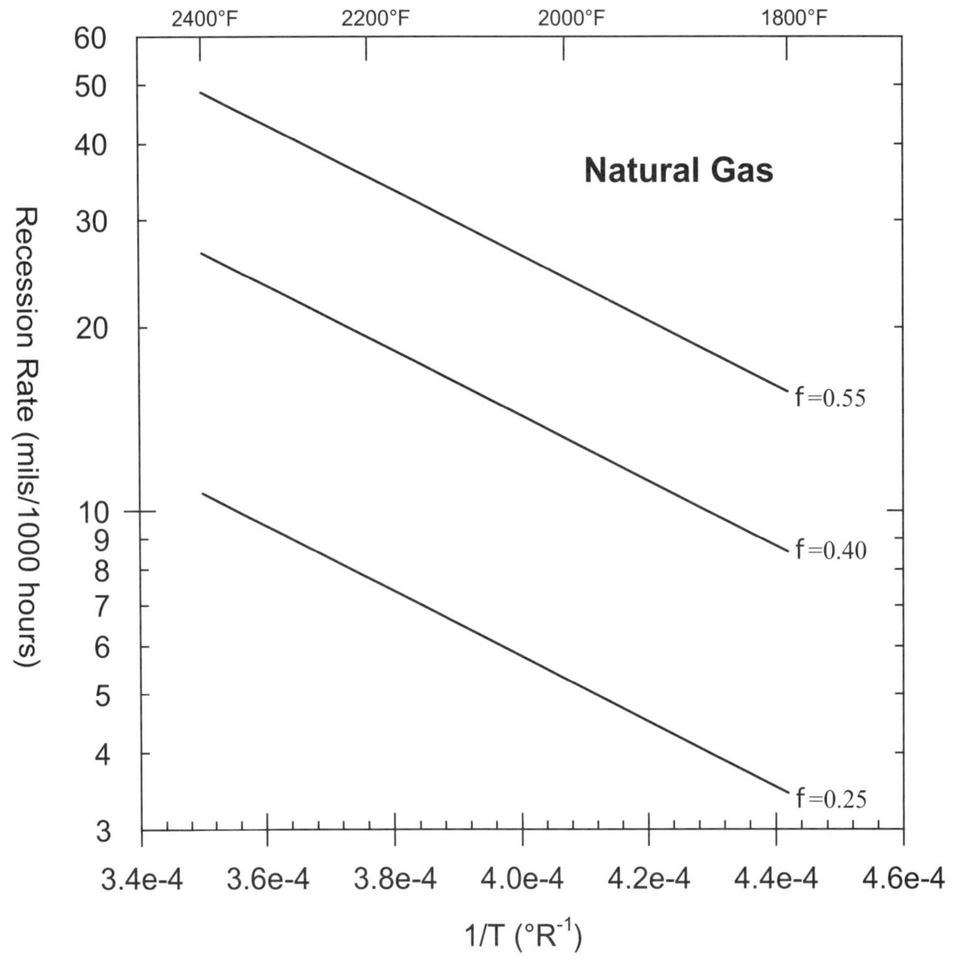

Figure 9.8 Modeling Results of SiC Recession Rate Under Some Representative Turbine Conditions; Φ is the Fuel/Air Equivalency Ratio. A Φ Value of 1 Corresponds to Stoichiometric Combustion (Reprinted with permission from GE Corporate Research & Development) 1 mil = 25 μm. Temperature in °R = °F + 460. Temperature in °C = (°F – 32)/1.8

of the conditioning air controls thermal growth and concentricity of the outer shroud assembly and, consequently, the radial clearance between the inner shroud and the turbine bucket tip.

GE Power Systems, in cooperation with GE Aircraft Engines, will also develop the following supporting technologies for implementing MI-CMC:

Design Integration of MI-CMC Components with Metallic Structures

In addition to design of the MI-CMC components, this will include re-design of the adjoining metal components, design of the ceramic-to-metal attachments, and determining the inspection and acceptance criteria for each component, as well as for the final ceramic-metal assemblies. New design practices must be developed in order to accomplish this.

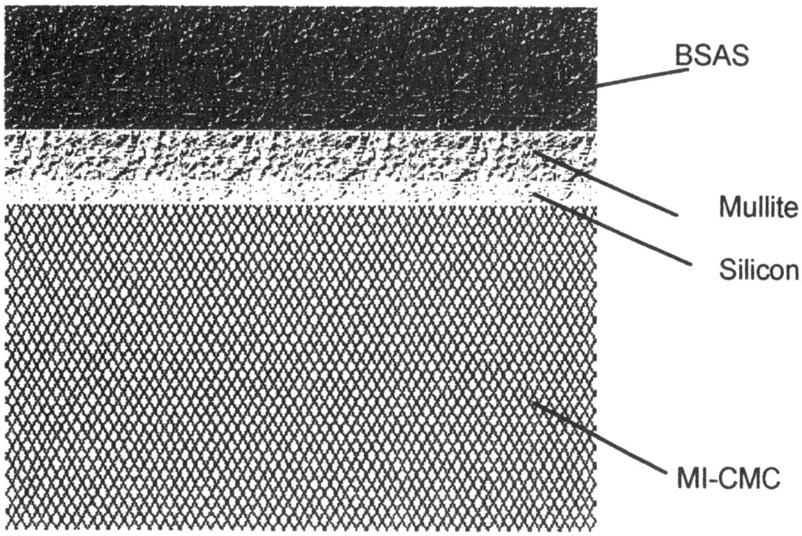

Figure 9.9 Schematic of Multi-layer Environmental Barrier Coating (Reprinted with permission from GE Corporate Research & Development)

Non-Destructive Evaluation (NDE) Methods

Infra-Red (IR) thermal imaging and Ultrasound (UT) are promising methods for inspection of MI-CMC; however, both materials-specific method development and calibration standards are needed before either technique can be used with confidence. In addition, linkage between NDE and probabilistic design methods is needed, both to establish inspection limits (i.e. define the size and nature of a "critical flaw"), and to ensure that designs are "flaw-tolerant."

Cooled Ceramic Components

Development of cooled MI-CMC components will extend the range of applications, particularly for first stage turbine nozzles. Cooling technology may include hot-side film cooling or increasing the cold-side heat transfer coefficient by increasing the MI-CMC surface roughness. This technology is of more critical importance in aircraft engines, however, due to higher peak turbine inlet temperatures.

Ceramic-Metal Joining

The capability to join MI-CMC to metal is desired, as it will improve flexibility in design and may enhance system performance because air leakage through mechanical joints will be reduced. However, the large mis-match in thermal expansion coefficient between MI-CMC and most metals requires the development of "compliant" joints. The demonstration of joint reliability at temperatures above approximately 1000°C (1830°F), as well as process robustness in large-scale manufacturing, remain significant challenges.

REFERENCES

[1] Grondahl, C.M., and Tsuchiya, T., "Performance Benefit Assessment of Ceramic Components in an MS9001FA Gas Turbine," ASME Publication 98-GT-186 (1998).

[2] DiMascio, P.S., Orenstein, R.M., and Rajiyah, H., "Reliability of a Conceptual Ceramic Gas Turbine Component Subjected to Static and Transient Thermomechanical Loading," Transactions of the ASME, Journal of Engineering for Gas Turbines and Power, 120[2], 263–270, 1998.

[3] EPRI TR-108260, "Evaluation of Ceramic Materials for Use in Gas Turbine Components," Electric Power Research Institute, Pleasant Hill, CA (1997).

[4] Corman, G.S. et al., "Rig and Engine Testing of Melt Infiltrated Ceramic Composites for Combustor and Shroud Applications," ASME Publication 2000-GT-638 (2000).

[5] Luthra, K.L. Private communication.

[6] U.S. Patent 5,985,470 "Thermal/environmental barrier coating system for silicon-based materials," Spitsberg, I and Wang, H.; Nov. 16, 1999.

CHAPTER

10

United Technologies Pratt & Whitney/ United Technologies Research Center Experiences with Ceramics and CMCs in Turbine Engines

John Holowczak and David C. Jarmon
United Technologies Research Center, East Hartford, Connecticut, USA

Mel I. Mendelson
Mechanical Engineering Dept., Loyola Marymount University, Los Angeles, California, USA

Gary D. Linsey
United Technologies Research Center, Pratt & Whitney Programs Office, East Hartford, Connecticut, USA

INTRODUCTION

The development and introduction of ceramic components into gas turbine engines at United Technologies Pratt & Whitney (P&W) began in the 1970s. This effort continues today at two operating divisions of United Technologies Corporation (UTC), United Technologies Research Center (UTRC) and Pratt & Whitney (P&W).

Recent efforts have been concentrated in transitioning silicon carbide fiber composite, and monolithic silicon nitride into such applications as combustors and cooled turbine vanes, respectively. These efforts are driven by the need to reduce cooling air requirements for turbine and combustor hot section components (in both aircraft, and aeroderivative stationary engines). In addition, weight reduction for improved thrust to weight ratios is important in aircraft engines.

Work began in the early 1970s with silicon nitride ceramics for turbine vanes and vane end walls. These applications were expanded in the late 1970s and early 1980s with a focus on individual turbine blades from silicon nitride. In the mid 1980s, the focus shifted to engine applications testing of ceramic matrix composites (including glass ceramic, CVI SiC, and sol-gel oxide matrices). Engine tested components included nozzle liners and flaps.

The early 1990s saw extensive engine testing of glass-ceramic matrix composites (COMPGLAS™) in compressor static structure components. The early to mid-1990s saw development of CVI SiC/SiC and MI-SiC/SiC composites for combustor applications in future hypersonic transport aircraft engines. In addition, cooled silicon nitride vanes were developed for ground based power generating engines. The late 1990s saw successful development and field testing of Environmental Barrier Coated SiC/SiC composite combustors in a cooperative effort between P&W/UTRC and Solar Turbines Div. of Caterpillar, Inc

FIRST P&W TRIALS OF MONOLITHIC CERAMIC TURBINE COMPONENTS

While tungsten carbide/cobalt was studied as a candidate turbine airfoil material in the mid-1950s, the first efforts in monolithic ceramic turbine engine components at Pratt & Whitney did not begin until the early 1970s. Efforts focused on the use of hot pressed silicon nitride in turbine vane endwalls (platforms) [1], and in turbine vanes [2]. Preliminary designs were developed for a first stage vane of an industrial gas turbine (combustor exit temperature of 1260°C) and a second stage vane for an experimental aircraft engine (combustor exit temperature of 1370°C). Note that second stage vanes would see considerably cooler gas temperatures than those at the combustor exit/turbine entrance.

First stage vanes were seated in pocketed cooled superalloy platforms with support on both the inside diameter and outside diameter of the engine. The ceramic platform area was minimized to improve tolerance to thermal shock. The second stage vanes used a cantilevered attachment with a span of about 5 cm (see Figure 10.1). The vanes were completely diamond ground from hot pressed billets.

The vane prototypes were subjected to a variety of impact, thermal cycling, burner rig, and combined testing. Crack growth was observed in vanes that were thermal cycled after precracking in either impact (carbon particle @ 250 m/s) or thermal shock (fluidized bed). Investigators were encouraged that while vane cracking was extensive in some cases, the vanes themselves remained intact with minimal lost material.

Turbine vane end walls (platforms) were designed to fit in between individual turbine vanes as depicted in Figure 10.2. The end walls gave the potential for a 1% of total core flow reduction in cooling airflow for the vanes, which were designed to be fabricated from NC-132 hot pressed silicon nitride from Norton Company. The thrust specific fuel consumption was expected to increase by approximately 1% as a result of this development. The initial plans included building prototypes and testing them in a cascade rig, including thermal cycling.

USE OF CERAMICS IN P&W'S TURBINE ENGINES (LATE-1970S TO LATE-1980S)

During the late 1970s to mid-1980s monolithic ceramics were tested in P&W's demonstrator engines in an attempt to improve the thrust-to-weight ratio. Engine performance was enhanced principally by increasing the operating temperatures of the turbines. The most popular ceramics were Norton Company's

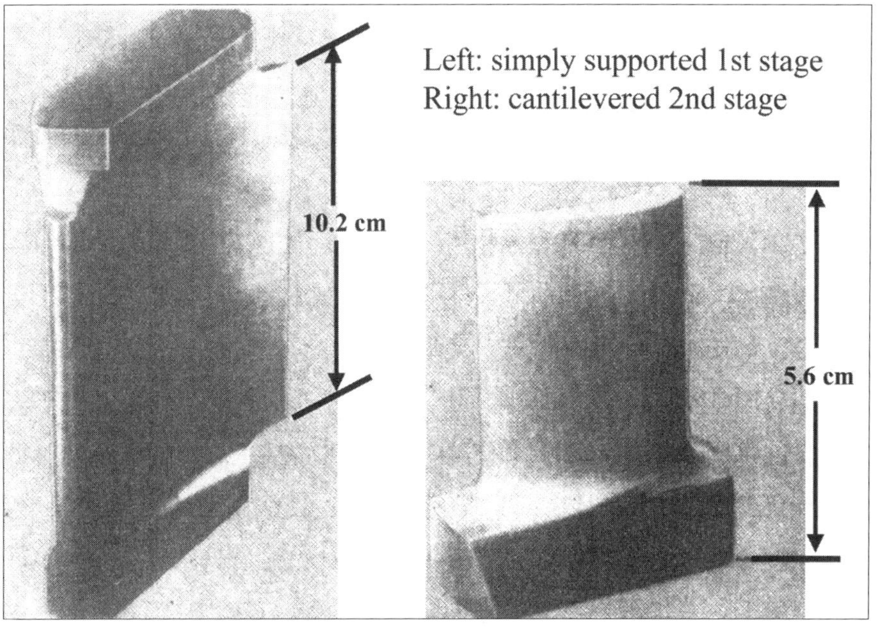

Figure 10.1 Hot-pressed Silicon Nitride Prototypes, First Stage Simply Supported Vane (Left), and Second Stage Cantilevered Vane (Right). (Reprinted with permission from United Technologies Pratt & Whitney)

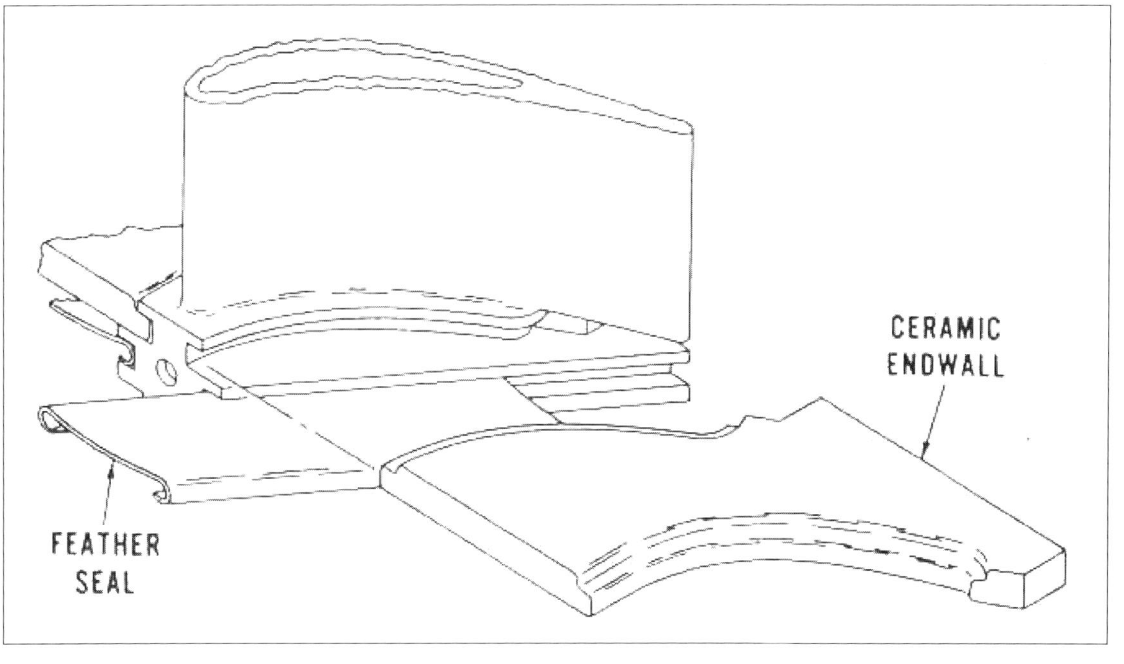

Figure 10.2 Final Design Concept for Ceramic Vane Endwall (Platform). Ceramic Endwalls Reduce the Need for Cooling Air (Reprinted with permission from United Technologies Pratt & Whitney)

Figure 10.3 Hybrid Rotor Consisting of Hot Pressed Norton NC 132 Silicon Nitride Blades and Endwalls, in a Forged Nickel Alloy Disk (Reprinted with permission from United Technologies Pratt & Whitney)

hot-pressed NC-132 Si_3N_4 and reaction bonded Si_3N_4 (RBSN) materials developed at Norton, Ford, and Allied-Signal. Norton's reaction bonded SiC and Carborundum's sintered SiC were also materials that were evaluated. By the early to mid-1980s hot isostatically pressed (HIP) Si_3N_4 emerged as the highest strength ceramic that could be fabricated into complex shapes, such as integrally bladed turbine rotors. Also, by the mid 1980s, ceramic matrix composites reached a stage in development where they began to see limited engine testing.

Through the mid-1980s Si_3N_4 ceramics received the most attention within P&W and UTRC. The advantage of Norton NC-132 was its high fast fracture strength of approximately 700 MPa. It exhibited one of the highest strengths of all commercially available ceramics at that time; however, its strength degraded with temperature and time during use. It addition, NC-132 was expensive and difficult to fabricate into parts, and in many cases it had to be machined and ground to the final shape. Reaction bonded Si_3N_4 (RBSN) ceramics had approximately half the fast fracture strength of NC-132. However, its strength also degraded with temperature and time. The advantage of RBSN was its low cost and ease of fabrication into net-shape turbine parts.

During 1974–79, a DARPA sponsored program at P&W was conducted to develop a ceramic blade/metal disk (hybrid) rotor. The objective was to design, fabricate and spin test the turbine rotors in a limited-life (expendable) engine. NC-132 ceramic blades and endwalls were machined and bonded into a circumferential Ni-based superalloy (AF2-1DA material) disk by isothermal forging the disk halves around the ceramic blade roots with a Pt interfacial layer [3]. The hybrid rotor was spin tested at 45,000 rpm and 1230°C. One spin-tested rotor exceeded both the predicted life of 88 to 260 hours [4], and the 50 hour design life for the engine. The test rotor is depicted in Figure 10.3.

Figure 10.4 Burner Rig Testing of Hollow RBSN Turbine Vanes (Reprinted with permission from United Technologies Pratt & Whitney)

The blade roots in the hybrid rotor were polished successively down to 1 μm diamond to reduce the surface flaw size developed during rough grinding. When the ceramic bladed rotors were spin tested to failure, the blades failed at the root of the blade's trailing edge, which had been predicted analytically to incur the highest stresses. The failure origin occurred on the surface from critical flaws that originated during grinding, which could not be eliminated by diamond polishing [5]. The contact stress was highest at the blade root attachment on the trailing edge of the blade. The blade root temperature was ~ 1230°C, and the stress at the blade root was predicted by finite element analysis to be about 520 MPa.

The hybrid rotors were also manufactured for United Turbine's automotive gas turbine engine [6]. In addition, a superalloy shaft was joined to a NC-132 turbine rotor for use as a Porsche turbocharger [7]. During the 1970s, NC-132 ceramic ball bearings (35 mm diameter) with a metal race were successfully run in demonstrator engines at 2.7 million DN with reduced heat generation (typical steel bearings ran at 2.3 million DN).

During 1977–79, surfaces of reaction bonded Si_3N_4 parts were coated with sputtered Si_3N_4 coatings [8], CVD Si_3N_4 coatings [9], and high temperature oxynitride glasses [10] in order to seal the open porosity and improve its strength. However, the coatings negatively affected the parts by producing lower strengths [11].

Due to the high expense of engine testing, most of the monolithic ceramics were rig tested in the late 1970s and early 1980s. In 1979, RBSN Si_3N_4 vanes were slip cast and reaction bonded at P&W. Typical room temperature flexural strengths of these materials were measured in the 210 to 280 MPa range. The vanes were rig tested at P&W to simulate an advanced turbine engine gas generator (ATEGG) environment (see Figure 10.4). Testing was done with JP-4 fuel in a burner rig. The vanes survived 12 hours of testing and 350 cycles from room temperature to 1200°C. The parts also survived thermal shock testing by water quenching from 1090°C to room temperature [12].

In the early 1980s, P&W analytically predicted that hollow ceramic vane structures had the lowest stresses. In addition, air cooling was shown to reduce the thermal stress in the outer wall of hollow vanes [13]. Based on this work, P&W designed and fabricated its ceramic vanes to have hollow structures. In a joint project with UTRC, hollow Si_3N_4 blades were fabricated by chemical vapor deposition (CVD) [14]. However, these parts were never tested.

In 1981, P&W reaction bonded Si_3N_4 hollow vanes were rig-tested at the Air Force's (Wright-Patterson AFB, Dayton, OH) heat transfer test facility. Holes were drilled in the trailing edge for gas discharge. The vanes were successfully thermally cycled from 425-~1760°C for 30 minutes under forced gas flow (138 kPa pressure using 0.81 kg/s of cooling air at 200–300°C) [13].

In 1983–85, Norton's NC-132 hot-pressed Si_3N_4 and NC-203 reaction sintered SiC were machined into catalytic igniter supports for the flame holders (in the afterburner section of an engine). NC-132 igniters were successfully tested for 1000 hours in a JT9D turbine engine. To our knowledge, this was the first ceramic part that was qualification tested for a production aircraft turbine engine [14, 15]. Eventually NC-132 became the bill of materials for this engine application.

In the mid-1980s, ceramic matrix composites (CMCs) began to work their way into low stress engine applications. These materials exhibited much greater impact toughness than monolithic ceramics. Matrices included glass-ceramics (principally COMPGLAS®, see below) and matrices made by chemical vapor infiltration (CVI), the later principally from silicon carbide (SiC). For example, COMPGLAS® (Nicalon® SiC/lithium aluminosilicate glass matrix) composites from UTRC and Nicalon® SiC/CVI Si_3N_4 matrix composites were fabricated into divergent seals and tested in the nozzle section of ATEGG engines (see next section). In these composites, the 2-D fiber lay-ups were prone to delaminate with time during engine cycling. Nicalon® SiC/CVI SiC matrix composites were also formed into convergent and divergent flaps for 2-D nozzle designs for advanced engines. Some 2-D divergent flaps were tested in a NASA Lewis engine for 23 hours at temperatures up to ~1150°C. Nicalon® SiC fibers were also impregnated with CVI SiC matrices, polymer precursor SiC matrices and sol-gel oxide matrices. Parts were constructed and rig tested under engine-simulated conditions.

In the early to late 1980s, coated C/C composites were tested in simple, low stress engine applications. The applications tested included combustor liners, tail cones, heat shields, augmentor liners and nozzle flaps and seals. The benefits were reduced weight, reduced cooling flow and reduced dependence on strategic materials. Some C/C composites were CVI/CVD-coated, while other C/C composites were impregnated with resins and glass coated. Most of these structures suffered from oxidation and reproducibility problems.

In an attempt to overcome these problems, in the mid-to-late 1980s, Nextel® 312 and 440 fibers were used with sol-gel processed silica and mullite matrices. Nextel® 312/silica composites were fabricated into hollow rib-core and truss-core structures. These structures were light weight, but the fibers often lost their strength during processing and testing. The 2-D fiber lay-ups were easy to fabricate into structures; however, the 3-D woven structures exhibited greater cross-ply strength and irregular void size distribution [16]. These composites were used in a variety of low stress, moderate temperature (~1100°C) applications in the augmentor and nozzle sections of advanced demonstrator engines.

THE COMPGLAS YEARS (GLASS-CERAMIC MATRIX COMPOSITES)

COMPGLAS® is the trade name chosen for a family of ceramic matrix composites (CMC) developed and tested from 1974 to 1995 by UTRC and P&W, divisions of United Technologies Corporation (UTC). The family was developed as an extension of polymer matrix composites and is composed of several classes of fibers reinforcing a broad range of glass and glass-ceramic compositions. However, the three major types of COMPGLAS® developed at UTC were:

- Silicon carbide fiber reinforced glass-ceramics for structural gas turbine engine applications up to 1300°C
- Carbon fiber reinforced glass for gas turbine applications up to 400°C and tailored to meet the high stiffness and dimensional stability requirements for space satellite applications
- Oxide fiber reinforced glass and glass-ceramics for gas turbine engine applications up to 1100°C

The SiC fiber reinforced glass ceramics were found to be the best for gas turbine applications based on both their combination of strength and toughness, as well as environmental durability. During the many years of development, numerous fiber/matrix combinations were developed. This was due to environmental durability requirements and also to changes in the fiber composition by the manufacturer, Nippon Carbon Company, of Japan. The two most successful SiC fiber reinforced glass-ceramic systems that were developed and tested were UT-22 and UTRC-200. UT-22 is a Ceramic Grade (CG) Nicalon® SiC fiber reinforced barium magnesium aluminosilicate with a boron nitride fiber/matrix interface with a use temperature of up to

1200°C [17, 18]. UTRC-200 is a Nicalon® SiC fiber (carbon coated) reinforced lithium aluminosilicate with 3% boron in the matrix and a use temperature of up to 900°C [19].

One of the key advantages of the COMPGLAS® approach is that gas turbine components of complex geometry can be fabricated. The processing of COMPGLAS® components is similar to the processing of polymer matrix composites, except that it is done at higher temperatures (1000°C to 1400°C) in an inert atmosphere and generally with graphite tooling. UTC fabricated components by hot pressing, composite transfer molding, hot isostatic pressing, and glass transfer molding. Fiber reinforcements used included: 2-dimensional ply and cloth lay-ups, chopped fibers, and various 3-dimensional woven preforms.

It was recognized early on that the successful application of COMPGLAS® to gas turbines would require significant rig and engine testing to uncover the key factors controlling durability. As had been previously learned in the case of polymer-based composite gas turbine applications, it was not possible to fully simulate material behavior in the laboratory. This is due to the important effects of component geometry, fiber architecture, multiaxial stress state, temperature distribution, and gas composition and flow. Thus, from 1984 to 1996, UTC fabricated, rig and engine tested a significant number of COMPGLAS® gas turbine engine components. Much of this component testing was based on material compositional and manufacturing development work at UTRC during the 1974 to 1985 time period. These efforts had been funded by the Office of Naval Research, Air Force Office of Scientific Research (AFOSR), and UTC. The range of components tested was related to a desire to explore a full range of stress states and temperature/environmental regions of the engine. By this approach, it was hoped to gain confidence and establish the limits that would govern component useful life. Examples of the types of components tested are as follows:

- *Compressor Inner Shrouds.* During the 1980s, COMPGLAS® high pressure compressor (HPC) inner shrouds were developed for positioning the 3rd, 4th and 5th stage inlet guide vanes (IGV) in the F119 engine. Due to the lower density of the COMPGLAS® relative to traditionally used superalloys, a total weight savings of nearly 10 kg was obtained. Also, since the COMPGLAS® exhibited exceptional wear resistance, it eliminated the need for nonmetallic bushings, resulting in a significant reduction in part count when compared with metal designs. The COMPGLAS® inner shrouds were initially produced using graphite fiber reinforced glass fabricated by composite transfer molding. This was later changed to UTRC-200 and fabricated by hot pressing to improve wear characteristics and increase oxidative stability. A single COMPGLAS® segment is shown in Figure 10.5, and the segments positioned together to form a ring are shown in Figure 10.6.

A total of 1600 hours of engine test time at a maximum temperature of 315°C was accumulated on the graphite/glass shrouds, and 260 hours on the UTRC-200 shrouds in F119 engines. In addition, 180 hours of test time were accumulated in the PW5000 HPC rig. Despite significant unanticipated structural loading during engine testing, the COMPGLAS® shrouds exhibited only minor localized chipping (which was addressed via fiber architecture modification) and never failed in test. In addition, a fully functional pilot plant was successfully established within P&W Manufacturing. This plant produced the hardware for all of the engine tests and demonstrated the ability to produce in a true manufacturing setting. Ultimately, the COMPGLAS® shrouds achieved "Bill of Materials" status in the F-119 engine development program, but were finally replaced by metal when cost considerations became more important than the weight savings.

- *Bulkhead Heat Shields.* During the 1980s, UTC ran burner rig tests on COMPGLAS® JT9D-7R4 bulkhead heat shields and COMPGLAS® panels, as shown in Figure 10.7. The 5 cm diameter nozzle impingement burner rig delivered a high gas velocity flame at 0.7 Mach. Specimens were thermally cycled by shuttling them vertically into and out of position in front of the flame. A typical cycle was fifty seconds of flame exposure followed by ten seconds of forced air cooling out of the flame.
- *Combustor Liners.* The segmented combustor was selected as an important application area for COMPGLAS®. Significant weight savings and improvement in durability were anticipated, and several test programs were conducted using various material compositions and segment designs. During 1983–85, UTC ran a sector rig test on a COMPGLAS® segmented combustor liner assembly under the NASA-

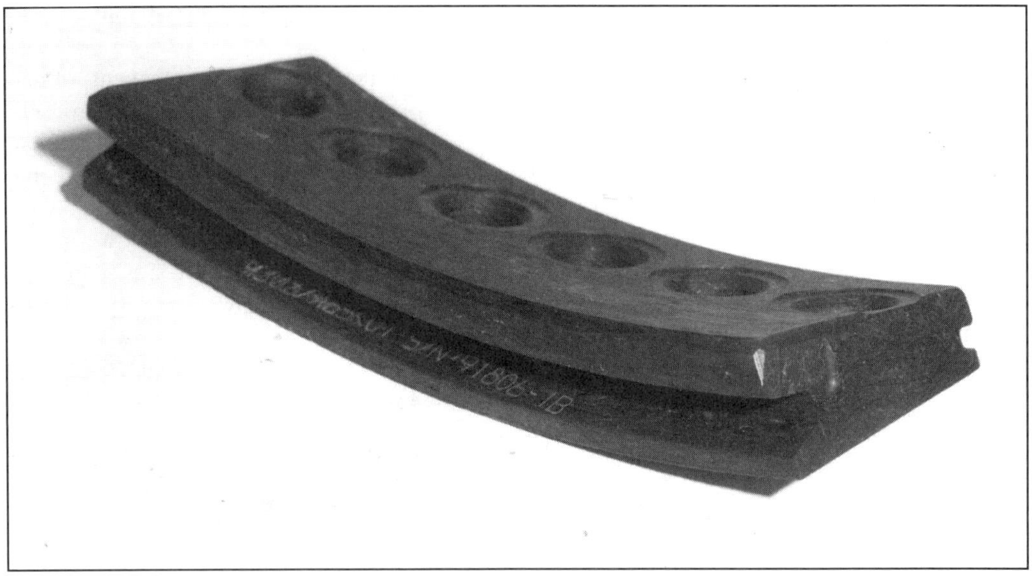

Figure 10.5 COMPGLAS® Inner Shroud Segment 13 cm Long (Reprinted with permission from United Technologies Research Center)

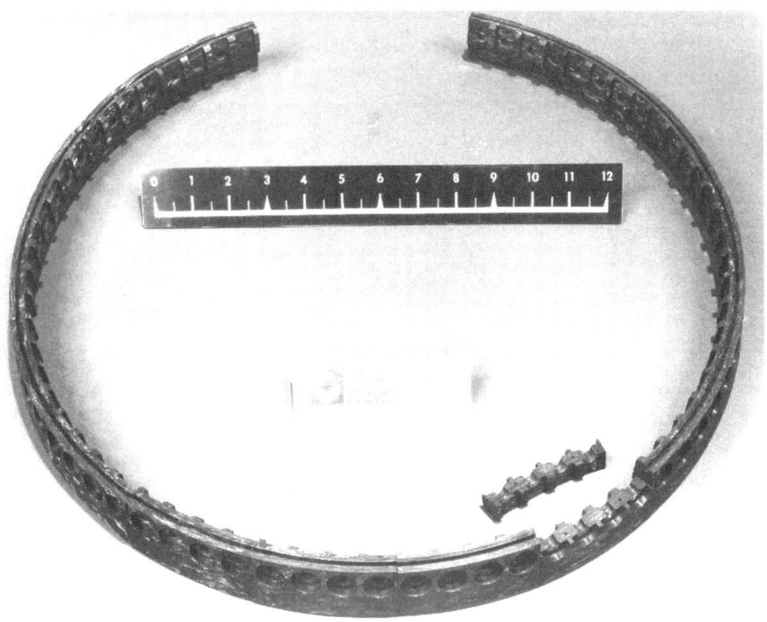

Figure 10.6 COMPGLAS® Inner Shroud Ring 46 cm in Diameter (Reprinted with permission from United Technologies Pratt & Whitney)

Figure 10.7 COMPGLAS® Bulkhead Heat Shield Burner Rig Test (Reprinted with permission from United Technologies Research Center)

funded Energy Efficient Engine (E^3) Program [20]. P&W analyzed the potential operating conditions, and estimated the thermally and mechanically induced component stresses versus expected overall engine operating life. The combustor panels were fabricated by hot pressing and mounted to a metal shell for testing shown in Figure 10.8. Figure 10.9 shows the COMPGLAS® combustor liner sector after the 18 hour sector rig test with temperatures up to 1150°C. The pattern on the left segments in Figure 10.9 is due to thermal paint used to measure the maximum surface temperature. No structural failure was noted during the test. Post-mortem examination and testing of the COMPGLAS® segments confirmed that oxidative degradation had occurred in a manner anticipated, based on laboratory testing of multiaxial reinforced composite specimens.

During 1983–89, UTC ran a second sector rig test on COMPGLAS® segmented combustor liners under the Naval Air Propulsion Center's ATEGG 685 Combustor Panel Program. The test was run for 39 hours, with 6 hours at or above the goal temperature. The test was stopped when a broken fuel nozzle (standard metal component) melted and subsequently burned through a first stage metal vane. Despite the failure of these nearby metal components, the COMPGLAS® combustor liners showed only minor distress.

Due to the combustor's high operating temperature, a hybrid concept that had COMPGLAS® functioning as the support structure for monolithic ceramic tiles was investigated under Air Force funding (Contract F33615-86-C-2617) [21]. Tiles composed of SiC segments supported by COMPGLAS® were tested in the XTC-65 combustor at the P&W X-960 facility. This facility provided preheated inlet air at sufficient quantities, temperature and pressure to represent design operating conditions of the XTC-65 core. The entire test

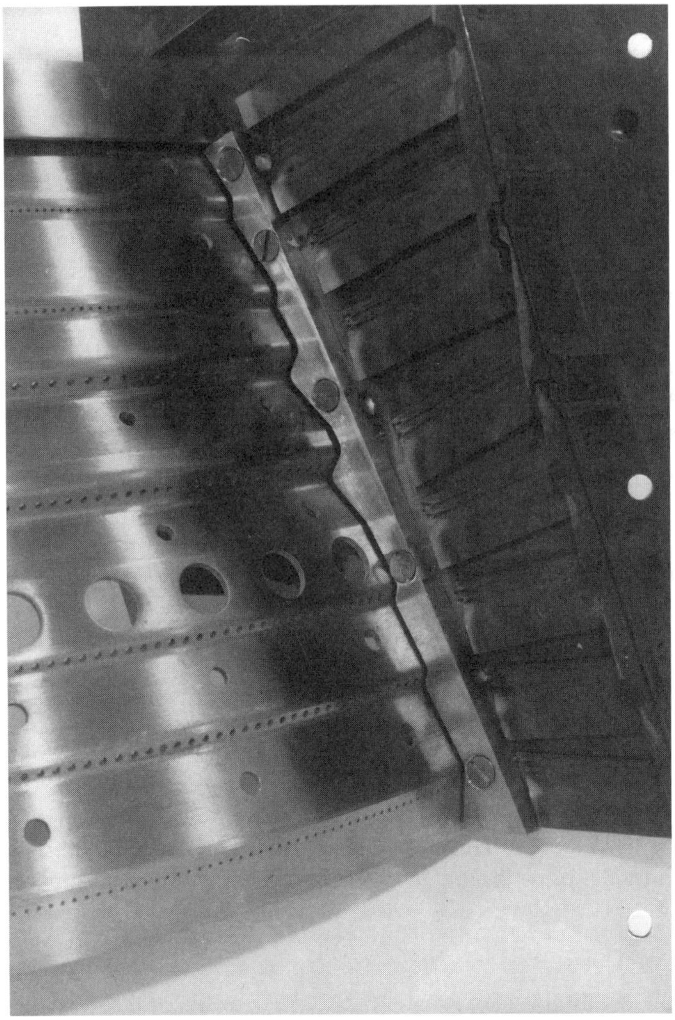

Figure 10.8 E³ Metal Shell Which Supported COMPGLAS® Combustor Panels (Reprinted with permission from United Technologies Pratt & Whitney)

rig program included 8 hours and 840 cycles of test exposure. Post-test analysis revealed that tile cracking was a serious problem; however, fracture occurred primarily because of correctable deficiencies in materials properties and alignment of constituents during fabrication. However, as in previous test programs, no structural failure of the COMPGLAS® components took place.

- *Turbine Support Ring.* During 1986–92, UTC developed and tested COMPGLAS® turbine support rings under the Air Force funded Advanced Turbine Rotor Design (ATRD) Program (Contract AFWAL-TR-88-2129) [22]. The ATRD support rings are located in a hub to which the turbine blades are diffusion bonded and function as a primary load carrying member for turbine stresses. An ATRD ring is shown in Figure 10.10; these rings had several innovative features. The rings were designed and manufactured to have a graded modulus, with the modulus increasing from inner diameter to outer diameter. This graded modulus enabled the very high rotationally induced stresses to be distributed evenly over the ring cross-section, thus eliminating the inner diameter stress concentration experienced in traditional metal, uniform modulus, thick wall ring structures. The modulus was graded by chang-

Figure 10.9 E^3 COMPGLAS® Combustor Panels After Sector Rig Testing (27 cm Top to Bottom) (Reprinted with permission from United Technologies Pratt & Whitney)

Figure 10.10 COMPGLAS® "Advanced Turbine Rotor Design" Ring (21 cm O.D.) (Reprinted with permission from United Technologies Research Center)

Figure 10.11 COMPGLAS® Divergent Seals (12 cm × 30 cm) (Reprinted with permission from United Technologies Research Center)

ing the ratio of intermediate and high modulus fibers. The ATRD rings were 3-dimensionally reinforced and fabricated by glass transfer molding.

The ATRD rings were never engine tested; however, the theory of the graded modulus concept was validated in spin and static rig testing. A 480 MPa hoop strength was measured in spin testing of a graded modulus, carbon fiber reinforced glass ring in 1988. A 760 MPa hoop strength was measured in static burst testing of a dual SiC fiber reinforced glass ring in 1990. The ATRD ring work was stopped in 1993 because increased operating temperatures resulted in excessive thermo-mechanical mismatch between the COMPGLAS® ring and the nickel hub.

- *Nozzle Divergent Seals.* In 1985, COMPGLAS® PW1120 divergent seals were fabricated with a 0/90 layup of SiC fiber reinforced lithium aluminosilicate by hot pressing (Figure 10.11). One of the COMPGLAS® divergent seals was tested in the FX-413 engine for 69 hours total test time with maximum temperatures up to 900°C. During the test, gas velocities ranged from 1.58 to 1.76 mach numbers. Post-test analysis revealed delaminations and localized blistering. This important finding pointed clearly to the significant role of fiber orientation in composite material oxidative degradation. The cross 0/90 cross ply created paths for rapid oxidation and delamination into the structure that had not been observed in unidirectionally reinforced laboratory specimens. In addition, the divergent seals were fabricated from one of the early COMPGLAS® systems. If these seals had been UT-22, burner rig data indicates that they would have performed well.
- *EPM Acoustic Tiles.* In the early 1990s, UTC fabricated COMPGLAS® acoustic tiles for noise suppression in the nozzle of the High Speed Civil Transport (HSCT) under the NASA funded Enabling Propulsion Materials (EPM) Program [23]. The tiles were 30 cm × 30 cm × 0.25 cm and had either 10,000 or 20,000 holes with a diameter of 0.76 mm laser machined through the thickness. A new attachment method, referred to as Miller Fasteners [24], was developed to attach the tiles to the metal support structure. The Miller fasteners minimized interlaminar shear and tensile stresses in the composite fastener and composite panel. In 1995, a COMPGLAS® acoustic tile was tested in Pratt & Whitney's hot acoustic rig along with tiles fabricated from other CMC systems. The total test time was 23 hours with sound pressure levels up to 142 db/hz (11 hour) and temperatures up to 488°C (9 hour). The COMPGLAS® tile showed no evidence of damage except for some cracking at the Miller Fastener slots due to incorrect machining of fastener angles. The panel is depicted in Figure 10.12.

Figure 10.12 COMPGLAS® Acoustic Tile (30 cm × 30 cm) (Reprinted from Ref. [23])

COOLED SILICON NITRIDE VANES FOR AERODERIVATIVE STATIONARY GAS TURBINES

During the early 1990s the corporation recognized that significant advancements had been made in the quality and complex shape fabrication capability of gas turbine grade silicon nitrides. An application in the high pressure turbine of the FT8 aeroderivative engine was identified as potentially benefiting from monolithic ceramics principally through reduction in cooling air requirements. Under funding provided by DARPA, U.S. Dept. of Energy-OIT, and Office of Naval Research, UTRC and P&W have been developing a cooled silicon nitride turbine vane. The vane is designed to fit the FT8 production engine, but the technology has applicability to any other aeroderivative engine the corporation should choose to develop.

The FT8 is a 25.5 MW stationary engine derived from the JT8D flight engine. While its average turbine inlet temperature, as listed in Ward's Gas Turbine World, is a moderate 1190°C, hot streak temperatures are considerably higher. Uncooled ceramic vanes operating in hot spot locations would have unacceptably high operating temperatures for acceptable long term durability. In order to harness the higher material operating temperatures of silicon based ceramics, it will be a requirement for these types of components to have internal cooling systems. Goals for the ceramic vane program include reducing the cooling air required by the first stage vane by approximately 5% of total core flow, while limiting surface material temperatures of the silicon nitride vanes to a level allowing extended operational intervals in field use. It is important to note that having several percent of total core flow coming through a Bill of Materials metal first stage vane reduces T4.1 (rotor inlet temperature). The reduced cooling air requirement for lightly cooled ceramic vanes raises T4.1. The saved cooling air can be burned with additional fuel for increased power density, or used to dilute the combustion flame and reduce NOx emissions.

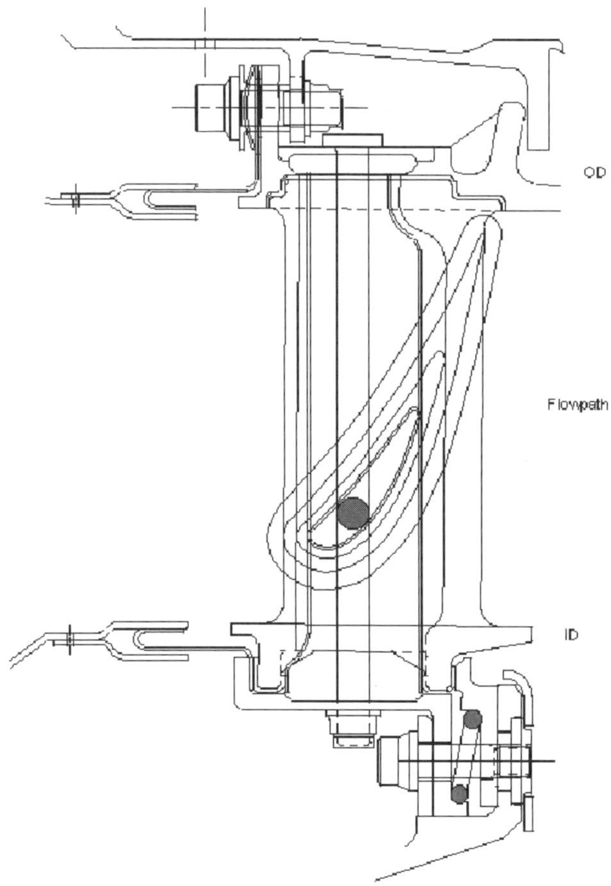

Figure 10.13 Spanwise Section and Chord Section Superimposed Views of the Cooled Silicon Nitride Vane for the High Pressure Turbine of the FT8 Aeroderivative engine (Reprinted from Ref. [25])

UTRC and P&W cooperatively developed a design for a cooled silicon nitride turbine vane, and details of this design and testing can be found in papers by Day et al. [25], and Holowczak et al. [26]. The basic vane design is shown in both cross section and top views in Figure 10.13. An Incoloy 909 bolt keeps the vane suspended vertically between 2 spring plates, while a second spring plate assembly holds the ceramic ID platform. The spring systems help to soak up thermal expansion mismatch. The outer platform of the vane is a superalloy that has a pocket to receive the silicon nitride vane airfoil. A metallic cooling baffle is used to direct impingement cooling onto the inner surface of the vane. Cooling air is vented by either tapered slots or circular cross section holes, at the vane trailing edge. Both Honeywell Ceramic Components and Kyocera Industrial Ceramics Corporation have successfully developed methods for fabricating these large (nominal 7.5 cm span), hollow vanes. The fabricated vanes have circular holes exiting truncated trailing edges, and are produced from gas turbine grade silicon nitride (AS800, and SN282, respectively). Efforts to develop tapered slots with conventional trailing edge geometries are ongoing. A Kyocera SN282 silicon nitride FT8 vane is depicted in Figure 10.14.

Vane surface temperatures were allowed to reach 1300°C in the design at maximum steady state power condition. Although judged to be excessive for long term use, this was deemed adequate for the 50 to 200 hour long high pressure combustor sector/vane cascade testing that is planned. The high surface temperature

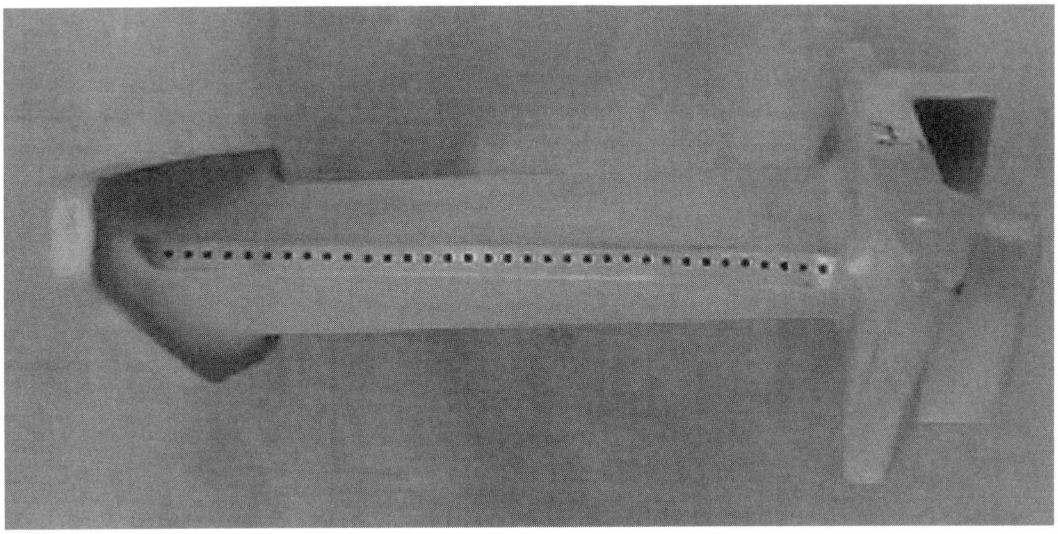

Figure 10.14 Trailing Edge of Engine Sector Test Candidate Vane Manufactured by Kyocera Industrial Ceramics Corp., Showing Trailing Edge Discharge Holes (Reprinted from Ref. [26])

was desirable from the standpoint of accelerated life prediction and degradation mechanisms that were anticipated. At this temperature, peak tensile stresses of approximately 250 MPa exist in the relatively cool (1090°C) interior of the vane, while the outer surface of the airfoil has appreciable compressive stress.

For the vane demonstration test, an existing FT8-1 sector rig was adapted for use. The rig utilizes a single can of the 9-can FT8 engine, along with a 40 degree are slice of the combustor to HPT transition duct. The rig is fed compressed air from a mechanical compressor and is then heated by a natural gas fired vitiated heater to simulate T3 air coming from a compressor in an actual turbine. The rig was previously built for evaluation of the emissions performance of several different injector and combustor configurations for the FT8-1 engine. The rig provides for pressure, temperature and temperature distribution conditions nearly identical to the FT8-1 engine. Pressures within the combustor are approximately 20 atmospheres due to the high compression ratio of the FT8 engine, and this is known to affect surface degradation rates for silicon based ceramics such as silicon nitride. The rig utilized FT8 production engine hardware, including arc segments of both the Tangential On Board Air Injection (TOBI, inner support) and the HPT vane outer assembly support. This was done in an attempt to simulate the same mounting stresses that the vanes will be subjected to in an actual engine [26]. The rig is shown in cross section in Figure 10.15.

As of the end of 2000, the cooled silicon nitride FT8 vanes have survived one trip shutdown event from 70% of full power engine condition. The vanes themselves appeared to suffer no effects from the trip shutdown, as shown in Figure 10.16. Continuation of the 50-hour demonstration test is expected during the first quarter of 2001. Efforts have also been initiated at UTRC in development of environmental barrier coatings for silicon nitride turbine vanes.

SILICON CARBIDE FIBER REINFORCED SILICON CARBIDE CERAMIC MATRIX COMPOSITES—1990S EXPERIENCE

Much of the development during the last decade at UTRC and P&W in ceramics and CMCs has been focused on silicon carbide fiber reinforced silicon carbide matrix composites (SiC/SiC) with an

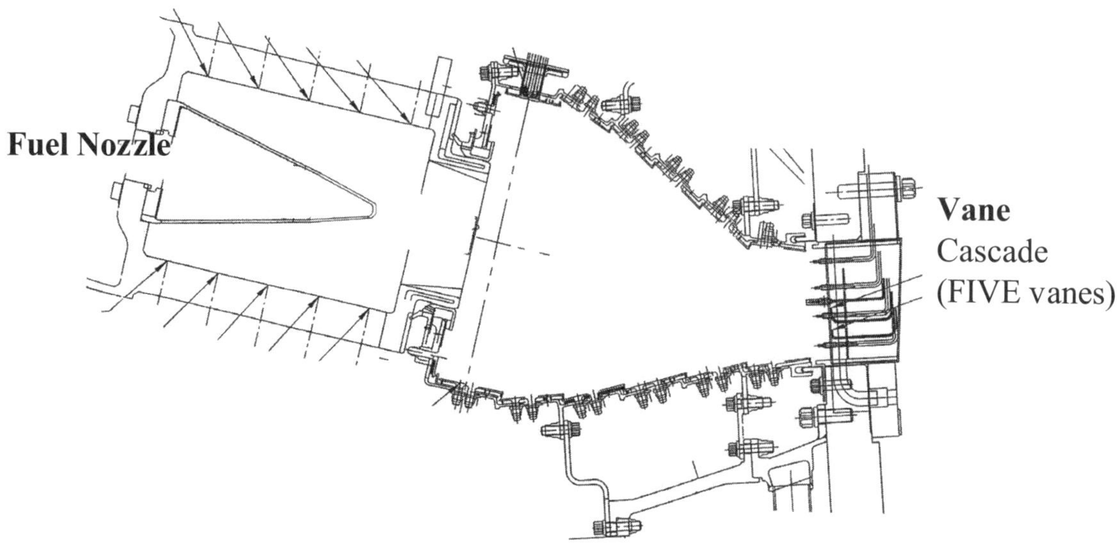

Figure 10.15 Cross section of FT8 Vane Cascade Sector Rig (Reprinted from Ref. [25])

Figure 10.16 Silicon Nitride Vane Cascade Following Trip Shutdown from 70% of Full Power Condition (Reprinted from Ref. [26])

Environmental Barrier Coating (EBC). P&W/UTRC, working with the National Aeronautics and Space Administration (NASA), the Department of Energy (DOE), the Department of Defense (DOD), and many Universities and other corporations, has demonstrated the potential for this class of advanced materials. The efforts have also revealed some of the issues that remain prior to complete acceptance and qualification testing in production engines. This section will focus on key contributions UTC has been involved with during the NASA High Speed Civil Transport Enabling Materials Program (HSCT-EPM) and the DOE Ceramic Stationary Gas Turbine (CSGT) program.

NASA High Speed Civil Transport Enabling Propulsion Materials Program

The goals of the High Speed Civil Transport (HSCT) aircraft were; (1) 8000 to 9500 kilometer range, (2) 250–300 passengers, (3) 340,000 kg gross weight, (4) Meet FAR 36 – Stage III noise goals, (5) provide equivalent standard passenger fares, and (6) 3–8 g/kg NOx emissions. Two engine components were initially identified as critical to the environmental acceptability of the HSCT. Therefore, research on those two components (combustor and nozzle) was initiated first as the Enabling Propulsion Materials (EPM) Program. The NASA High Speed Civil Transport Enabling Materials Program focused on developing the materials required to enable the HSCT aircraft. One of the enabling materials was a CMC combustor liner that was able to meet the HSCT combustor environmental, thermal, structural, economic and durability requirements. The CMC combustor liner task included combustor design, CMC development, environmental barrier coating (EBC) development, design database, analytical modeling and rig testing.

Two combustor concepts were studied (1) Rich-burn/Quick-quench/Lean burn (RQL) and (2) Lean Premixed Prevaporized (LPP). The RQL design concept was successfully demonstrated in a NASA Sector Rig for 200 hours [27]. The CMC selected for the HSCT combustor liner was SiC/SiC reinforced with a BN coated stoichiometric SiC fiber. An EBC was developed to minimize silicon carbide recession during long term exposure to oxygen plus water vapor and demonstrated in the 200 hour NASA sector rig test. A detailed design database was generated to provide data to the combustor liner designers and the life prediction group. Analytical models were developed and preliminary validation was successful. The designs, material, and models were brought together at NASA for the sector rig test to demonstrate the materials and designs and validate the models.

HSCT Ceramic Matrix Composite Combustor Liner Design

United Technologies initially conducted a conceptual design and later a preliminary design of the RQL combustor concept that was the concept selected for Sector Rig evaluation at NASA. The design effort was critical in that the designers were required to optimize the design to take full advantage of the CMC material. The design effort helped guide the material development effort to insure that the material would be able to handle the harsh operating environments. During 1999, two hundred hours of sector rig testing were completed at NASA and demonstrated both the CMC and the EBC materials. Six different RQL sector components were fabricated (Figure 10.17), evaluated via nondestructive evaluation (NDE) techniques, then rig tested. Post-test analysis was conducted to determine if there were any detrimental effects of the environmental exposure.

HSCT Ceramic Matrix Composite Material Development

Material development was conducted on three processes to fabricate SiC/SiC components: Polymer Infiltration Pyrolysis (PIP), Chemical Vapor Infiltration (CVI) and Melt Infiltration (MI). Each process has advantages and disadvantages based on the specific requirements of the engine component designs. Due to the relatively low through-thickness thermal conductivity at that time of the PIP matrix material, most of the effort was concentrated on the CVI and MI processes.

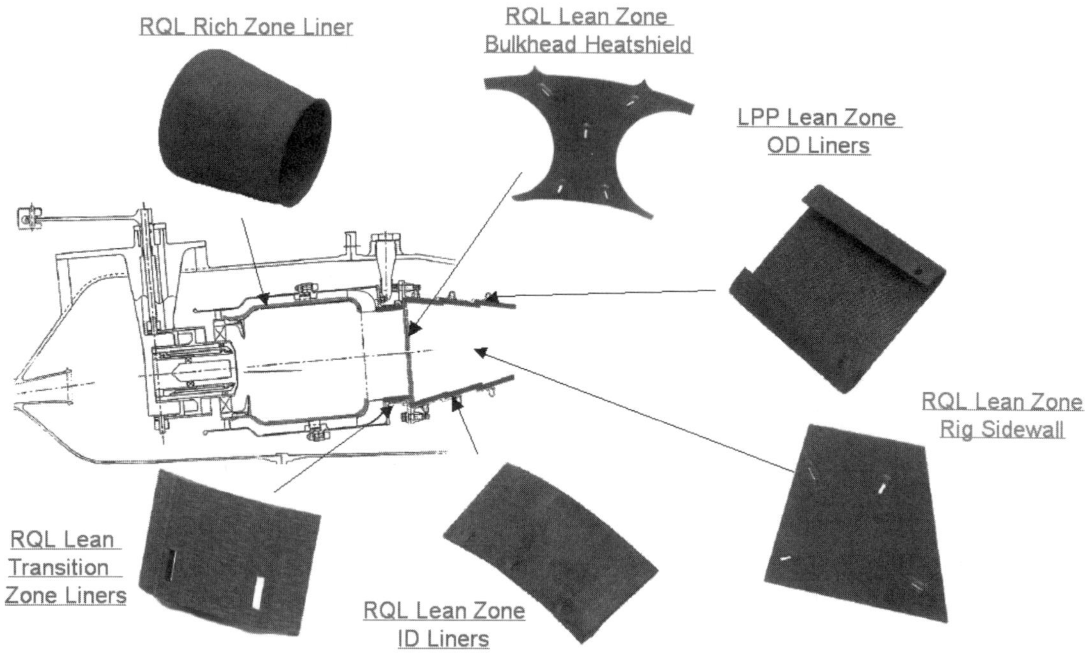

Figure 10.17 High Speed Civil Transport Rich-Burn Quick-Quench Lean-Burn (RQL) Combustor Liner Sector Test SiC/SiC Components (Certain Components have United Technologies Environmental Barrier Coating) (Reprinted from Ref. [27])

The advantages of PIP processing (practiced by Dow Corning) include complex-shape forming capability due to the use of standard Polymer Matrix Composite (PMC) fabrication methods as the first step in the manufacturing process. The PIP process consists essentially of fabricating a PMC using a silicon based preceramic polymer, such as polycarbosilane or polysilazane, followed by heat treatment to convert the cured polymer matrix to a silicon base ceramic material such as silicon carbide or silicon nitride. Since mass loss and subsequent volume shrinkage are inherent to the polymer conversion process, multiple polymer infiltration and pyrolysis cycles are required to achieve a CMC material with high density.

The advantages of CVI processing (practiced by BFGoodrich and Honeywell Advanced Ceramic Composites) include both a relatively mature fabrication process and a material that does not contain residual metal. Since the through-thickness thermal conductivity of CVI material was marginal, emphasis was placed on material and process modifications to increase conductivity. These included attributes such as CVI process optimization, matrix chemistry/crystallinity enhancements and the reduction of residual porosity and microcracking through the use of fillers and hybrid processing. While fillers such as silicon carbide whiskers or platelets did show promise as a concept, densification of the CVI matrix was hindered by addition of fillers and/or hybrid processing. The most promising route to increase through-thickness thermal conductivity was determined to be the use of three dimensional fiber reinforcements combined with a stoichiometric silicon carbide fiber.

The advantages of MI processing (practiced by BFGoodrich and Honeywell Advanced Ceramic Composites) include both higher through-thickness thermal conductivity and a lower residual porosity in the material. Since the resultant material contains residual silicon, significant effort was expended to react the silicon to form silicon carbide. Although complete conversion was never obtained, processing improvements were accomplished to reduce the residual free silicon, fabricate more uniform boron nitride fiber coatings,

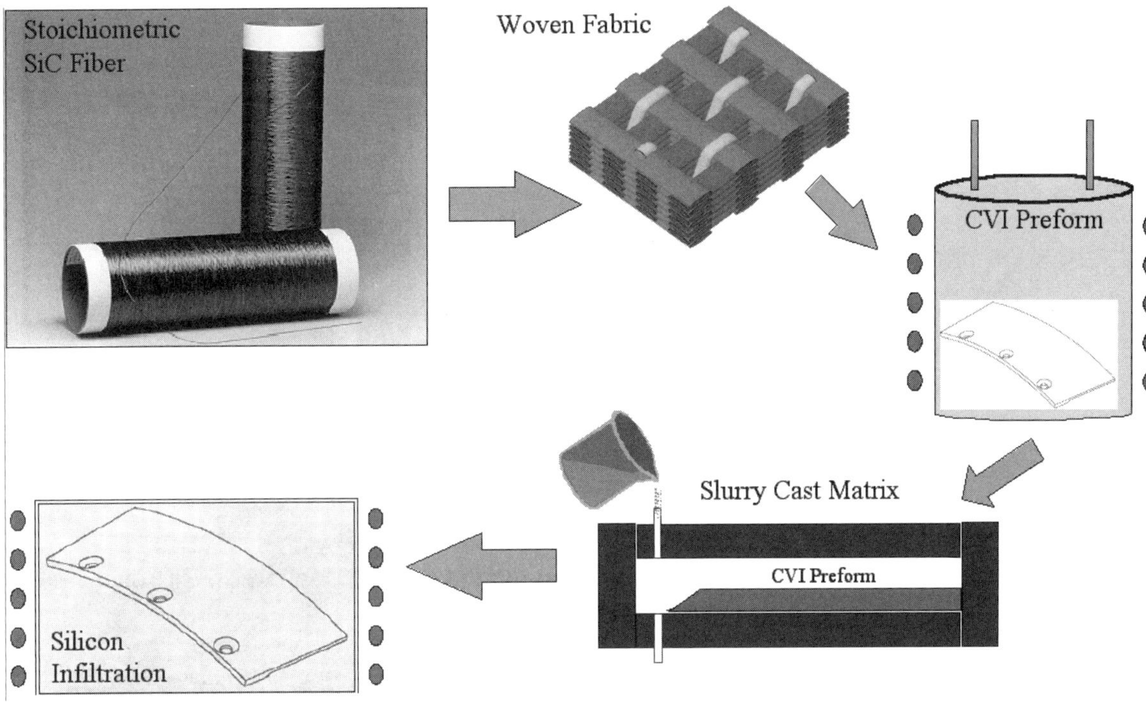

Figure 10.18 SiC/SiC Manufacturing Process Developed on the NASA HSCT Program (Reprinted with permission from United Technologies Pratt & Whitney)

of magnitude improvement in recession resistance compared to that of uncoated components. This performance improvement met the goals for the HSCT program.

HSCT Ceramic Matrix Composite Design Database

In order to generate design data early in the program and to have a final database for the final design, two databases were developed for the SiC/SiC system. These databases included physical, mechanical, mission life and environmental testing data. The initial design database, conducted in 1996, consisted of data from approximately 150 tests using 16 CMC panels. Within this study, both Sylramic and Hi-Nicalon® silicon carbide fiber reinforced panels were included. The final design database, compiled in 1998, consisted of data from approximately 250 tests using 23 CMC panels all with Sylramic fibers. Fatigue specimens from these evaluations were tested for almost 10,000 hours to define the long-term durability properties of the CMC material. During the final two years of the program, material characterization was focused on reproducibility of the CMC material. In addition to the design databases, specialized testing was conducted throughout the program to evaluate the CMC material. This testing included wear testing, ring-burst testing, atmospheric burner rig testing and quartz lamp testing.

HSCT Nondestructive Evaluation Methods

Nondestructive evaluation (NDE) methods were developed and utilized as an aid to determine whether panels and components were uniform for material screening efforts and as an aid in understanding physical and

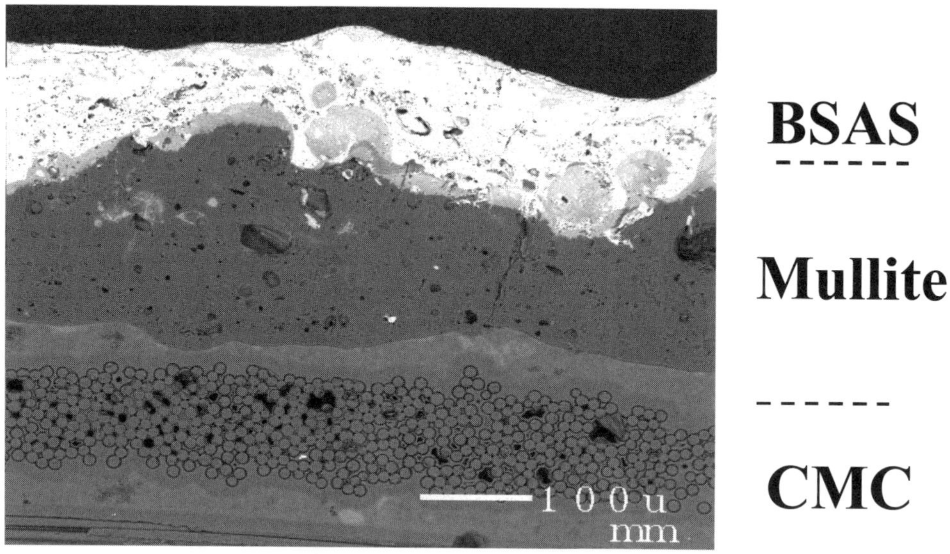

Figure 10.19 United Technologies Environmental Barrier Coating on SiC/SiC (Reprinted from Ref. [28])

mechanical property test results. Three techniques were chosen for this effort: ultrasound, thermography and x-ray. Ultrasound and x-ray were proven techniques that United Technologies had access to and for which development was not needed. Thermography development was conducted to create a non-contact inspection technique that could be used in field inspections. This capability was demonstrated during the sector rig test program.

HSCT Ceramic Matrix Composite Life Prediction

Analytical methodology and procedures required to predict structural performance and component life were developed. Failure modes and damage mechanisms were identified, extreme environmental conditions anticipated in the HSCT combustor were understood and then analytical models were developed and verified through benchmark testing and analysis. The primary analytical code developed to predict the life of a specimen, subcomponent or component was "NASALife." Many improvements were made to predict the life of a CMC in a combustor environment (such as routines to predict mean stress effects, mission cycles and overload cycles). This code was used extensively in the prediction of the lives of the sector components and the preliminary design of the combustor.

Other codes were developed to enhance the predictions of the components and to better understand the material. They included the codes CSTEM, pcGINA and NESTEM/CEMCAN. CSTEM was the computer code selected to perform coupled structural and thermal analysis in the inelastic regime. Specific enhancements included an improved gaseous radiation model, creep and plasticity models for orthotropic materials, integration of the structural material properties, a simplified linear load-ramping routine and an automated pre- and post-processing link. The code pcGINA is a lamina property prediction code enhanced with drop-down and online help menus, graphical feedback screens and a knowledge-based optimization algorithm to optimize the fabric for user-prescribed boundary conditions. Also, a probabilistic heat transfer/structural analysis code (NESTEM) and micromechanics-based CMC property-prediction code W-CEMCAN were developed. These codes helped in optimization of material and quantification of uncertainties, scatter and effects on overall behavior of liner components. They provided better understanding of the material and supported material development activities.

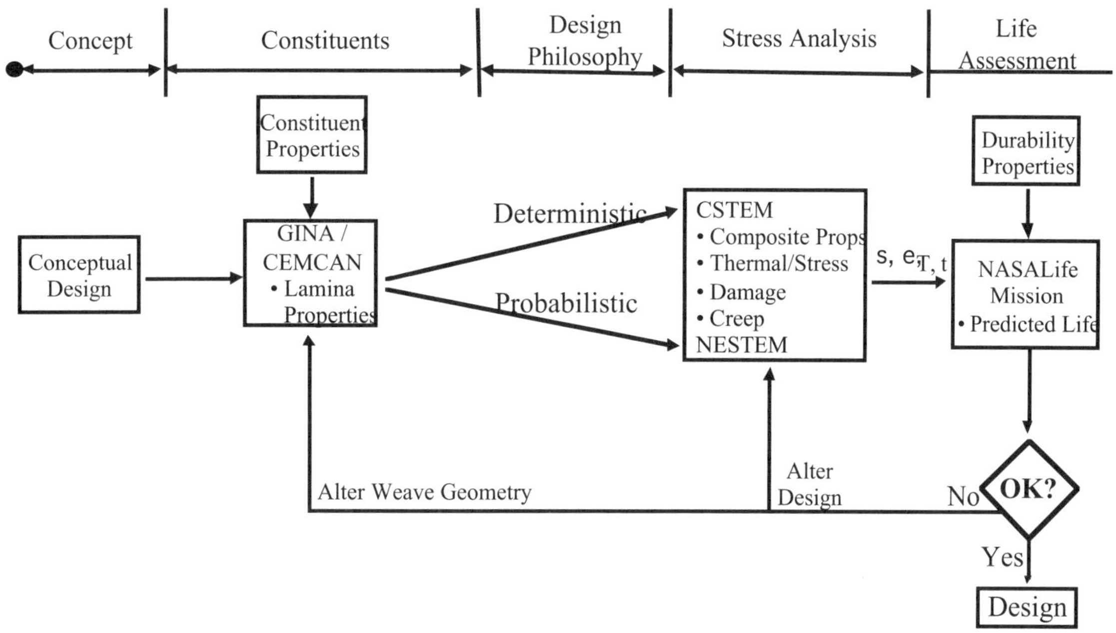

Figure 10.20 Analytical Model Based CMC Component Design Philosophy (Reprinted with permission from United Technologies Pratt & Whitney)

Model validation was initially conducted using simple flexural and open-hole specimens. It was demonstrated that linear-elastic analysis provided extremely conservative predicted lives. Both CSTEM with damage mechanics and the pseudostress approach improved the predictions, however, they were still conservative for conditions with notches. More complex benchmark testing was conducted with four-inch diameter cylinders with axial thermal gradients. The CMC component design philosophy including the models that were utilized is shown in Figure 10.20.

HSCT ROL Sector Rig Component Evaluation

The goal of the RQL sector rig component evaluation task was to demonstrate CMC capability for 200 hours of HSCT conditions to validate the stress analysis but not to demonstrate full life durability (due to the high cost of rig operation). This rig, run by NASA and built by Pratt & Whitney, provided all of the challenges to the material that would be encountered in a HSCT combustor environment. While the liner configurations were slightly different that those finalized in the preliminary design of the RQL combustor, the stresses and temperatures of the components were similar. Five different liner components (shown in Figure 10.17) were tested with two components coated with the EBC. Although some components cracked during the testing (for reasons unrelated to the CMC itself) most of the components successfully completed the 200 hour test. Post-test NDE and destructive evaluation showed little or no degradation.

Solar Turbines Ceramic Stationary Gas Turbine Program

The Solar Turbines Ceramic Stationary Gas Turbine program is discussed in detail in Chapter 7. United Technologies provided the Environmental Barrier Coatings (EBC) required to meet the program goals. The

Figure 10.21 Solar Turbines SiC/SiC Combustor Liner with United Technologies Environmental Barrier Coating (Reprinted from Ref. [28])

EBC system was derived from the HSCT-EPM EBC system described above. The EBC system was scaled-up from 5 cm by 5 cm coupons to a 75 cm diameter, 20 cm long cylinder and optimized to provide long-term (14,000 hour) protection during engine operation [28]. The Solar Turbines CMC combustor liner with the United Technologies EBC is shown in Figure 10.21.

ACKNOWLEDGMENTS

The authors would like to thank all of their many partners and colleagues at Solar Turbines Incorporated, Rolls-Royce Indianapolis, U.S. Dept. of Energy, NASA-Glenn Research Center, U.S. Air Force, Office of Naval Research, and Defense Advanced Research Projects Agency.

REFERENCES

[1] Hayes, C., Nessler, C., and Zabierek, D. "Turbine Vane Ceramic Endwall," in *Ceramics for High Performance Applications,* Proc. 2nd Army Materials Tech. Conf., Hyannis, MA Nov. 13–16, 1973, Eds J.J. Burke, A.E. Gorum, and R.N. Katz., Brook Hill Publishing Chestnut Hill, MA.

[2] Nessler, C., "Prototype Ceramic Vanes," in *Ceramics for High Performance Applications,* Proc. 2nd Army Materials Tech. Conf., Hyannis, MA Nov. 13–16, 1973, Eds J.J. Burke, A.E. Gorum, and R.N. Katz., Brook Hill Publishing Chestnut Hill, MA.

[3] Carruthers, D. and Walker, B.H., "Design, Fabrication and Evaluation of GATORIZED^{R} Ceramic-Wrought Alloy Attachment Concepts," DARPA Contract N00019-74-C-0484, P&W Aircraft Group, Final Report FR-9787, January 1979.

[4] Mendelson, M.I., McLeod, S.A. and Walker, B.H., "Life Prediction of Ceramic Blade/Metal Disk Turbine Rotor," *Ceramic Engineering and Science,* Vol. 1, pp. 540–548, (1980).

[5] Mendelson, M.I., "Fracture of Hot-Pressed Si_3N_4 Turbine Blades," Basic Sci Div, 82nd Amer. Ceram. Soc. Meeting, Chicago, IL, April 1980. Published in Pratt & Whitney Proprietary Report, 1981

[6] McLeod, S.A., and Mendelson, M.I., "Hybrid Blade/Attachment Design," published in P&W Tech Report FR-14633, United Turbine Contract #22.884, May 15, 1981.

[7] Mendelson, M.I., "Ceramic-Metal Joining for Turbine Applications," SAE Technical Paper 810206, *Soc for Automotive Engineers Conference,* Detroit, MI, Feb. 23–27, 1981. Published in Vol. 90, *SAE Transactions* (1982).

[8] Mullaly, J. (P&W), personal communication, 1977–79.

[9] Galasso, F. (UTRC), personal communication, 1977–79.

[10] Layden, G. (UTRC), personal communication, 1977–79.

[11] Mendelson, M.I., P&W IR&D, 1977–1979.

[12] Walker, B.H. and Mendelson, M.I., "Development of Integral Ceramic Blade-Metal Disk with Circumferential Blade Attachment," *Proceedings of AGARD Specialists Meeting:* Ceramics for Turbine Engine Applications, Cologne, Germany, Oct. 8–10, 1979.

[13] Mendelson, M.I., and Zabierek, D., "Evaluation of Air-Cooled Si_3N_4 Vanes," *Ceramic Engineering and Science,* Vol. 3 (9–10), pp. 630–641 (1982).

[14] Galasso, F., Veltri, R., and Mendelson, M.I., "Manufacture of Hollow CVD Silicon Nitride Articles," *US Patent 4,500,483,* Feb 19, 1985.

[15] Mendelson, M.I., "Hot-Pressed Si_3N_4 Afterburner Supports," Presented at 32nd ASME Gas Turbine Conference, Anaheim, CA, June 1–4, 1987. Published in Proc. Technical News Brief: American Metal Market, June 5, 1987.

[16] Mendelson, M.I., "Characterization of Nextel™ 3-D Woven Fiber Structures," *Ceramic Engineering and Science,* Vol. 9 (9–10), pp. 834–842, (Sept–Oct, 1988).

[17] Brennan, J.J., Chyung, K., Taylor, M.P., "High Strength Thermally Stable Magnesium Aluminosilicate Glass-Ceramic Matrix SiC Fiber Composite," U.S. Patent No. 4,589,900 May 20, 1986.

[18] Brennan, J.J., Chyung, K., Taylor, M.P., "High Strength Thermally Stable Magnesium Aluminosilicate Glass-Ceramic Matrix SiC Fiber Composite," U.S. Patent No. 4,588,699 May 13, 1986.

[19] Brennan, J.J., Layden, G.K., Prewo, K.M., "Oxidation Resistant Fiber Reinforced Composite Article," U.S. Patent No. 4,857,485 Aug. 15, 1989.

[20] Dubiel, D.J., et al., "Pin Fin and Ceramic Composite Segmented Combustor Sector Rig Test Report," Pratt & Whitney for NASA-LRC. Report No. NAS CR-179534, Sept. 1986.

[21] Lohmann, R.P. and Able, E.C., "Nonmetallic Liner Development Program," Pratt & Whitney for Aero Propulsion and Power Directorate—Air Force Systems Command, Final Report. WL-TR-91-2072, Nov. 1991.

[22] Cairo, R.R., "Advanced Turbine Rotor Design Program," Pratt & Whitney for Aero Propulsion and Power Directorate—Air Force Systems Command, Final Report. No. WL-TR-92-2009, Apr. 1992.

[23] Pratt & Whitney/General Electric Co., "Enabling Propulsion Materials Program Annual Technical Progress Report, Volume 3: Task B—Exhaust Nozzle Materials," Rept. FR-22596-3, Nov. 30, 1996.

[24] Miller, R.J., Palusis, M.E., and Jarmon, D.C., "Composite Fastener For Use In High Temperature Environment," U.S. Patent 6,045,310, Apr. 4, 2000.

[25] Day, W.H., Kouris, K., Holowczak, J.E., Blair, M.F., Hale, S.E., and Bornemisza, T., "Development of Cooled, Ceramic First Stage Vanes for the FT8 Aeroderivative Gas Turbine," ASME paper 98-GT-133, presented at the ASME International Gas Turbine and Aeroengine Congress and Exhibition, Stockholm Sweden, June 2–5, 1998.

[26] Holowczak, J.E., Bornemisza, T.G., and Day, W.H., "Sector Testing of Cooled Silicon Nitride Ceramic HPT Vanes for the FT8 Aeroderivative Engine," ASME paper 2000-GT-0662, presented at the ASME TURBO EXPO 2000, LAND, SEA, & AIR, Munich, Germany, May 8–11, 2000.

[27] Brewer, D., Ojard, G., and Gibler, M., "Ceramic Matrix Composite Combustor Liner Rig Testing," ASME paper 2000-GT-0670, presented at the ASME TURBO EXPO 2000, LAND, SEA, & AIR, Munich, Germany, May 8–11, 2000.

[28] Eaton, H.E., Linsey, G.D., More, K.L., Kimmel, J.B., Price, J.R., and Miriyala, N., "EBC Protection of SiC/SiC Composites in the Gas Turbine Combustion Environment," ASME paper 2000-GT-0631, presented at the ASME TURBO EXPO 2000, LAND, SEA, & AIR, Munich, Germany, May 8–11, 2000.

CHAPTER

11

Ceramic Composite Turbine Engine Test Experience at Williams International

William Fohey, Dr. Timothy Nielsen, and Scott Cruzen

Williams International, Walled Lake, Michigan, USA

WILLIAMS INTERNATIONAL CORPORATE BACKGROUND

Williams International, located in Walled Lake, Michigan, was established in 1955. Williams proved that it was feasible and practical to produce miniature turbofan engines that performed well and could be built economically. This made it possible to develop advanced long range strategic cruise missiles, which would not have been possible without Williams International's F107 engine. The F107 had unprecedented performance of 2700 N of thrust from a 30.5 cm diameter engine with very low fuel consumption. Because of their low cost, small size, and amenability to modification, the F107 and other Williams engines became the test beds for a variety of advanced materials, including ceramics and ceramic matrix composites. The F107 and its derivative F112 engine provided propulsion for the U.S. Air Force Air-Launched Cruise Missile (ALCM) and Advanced Cruise Missile (ACM), and the U.S. Navy Tomahawk Cruise Missile. Over 6,500 cruise missile engines have been produced.

DESIGN PHILOSOPHY

New technologies such as ceramic composites are always assessed based on performance and cost impact to a product. Use of ceramics has offered potential performance improvements that have been and will be weighed against the cost of the material. By increasing the turbine inlet temperature (TIT), it is possible to increase the thrust within the same engine envelope. Williams International has been a leader in evaluating and assessing the capability and applicability of new material systems due to the smaller sized components and cost effective testing with limited life engines. When ceramics are considered for short life applications, the material temperatures, stresses and environmental conditions are usually more aggressive than those possible for long life or man rated applications with the same material system. Cost is a very important parameter in an expendable application, since one of the primary goals is low cost.

Each engine and component in the engine has unique environmental and structural requirements. The first task is to define the requirements of the engine component that is being considered for transition to ceramic material systems. Fabrication limitations of the selected material system usually have a significant influence on the design of the component.

Designing for a ceramic component requires a careful analysis of the structural and thermal stresses. Since ceramic materials have very high elastic moduli, thermal gradients cause thermal strains that can result in significant stresses. Experience has shown that thermal stress and not mechanical loading causes the highest stress levels in some ceramic components.

Attachments between the ceramic component and metal hardware are critical. The thermal expansion mismatch between the two materials must be accommodated and the attachment hardware temperature must be held to a manageable level. Ceramics usually do well with a clamp group arrangement or in a trapped assembly, and perform relatively poorly with splined assemblies or conventional bolted fasteners.

The advantage of incorporating ceramics into turbine engines is increased performance with potential for lower weight and cost. With the use of ceramic engine components in a reusable turbojet drone engine the TIT could be increased from 982°C to about 1371°C. For an increase in specific fuel consumption of 11%, specific thrust would increase about 47%. To run at a TIT of 1371°C would require the replacement of the hot section components with suitable ceramic or refractory materials.

Primary candidates for ceramic turbine engine components are the combustor, turbine nozzle and the turbine rotor. Each of these components has different geometries, structural constraints, and environmental requirements.

Combustor material property requirements include high temperature capability, low thermal expansion, oxidation resistance, and ability to withstand severe thermal gradients and thermal shock with moderate mechanical stresses. Williams International's ceramic composite and carbon-carbon combustor components have ranged in thickness from 1.5 mm to approximately 3 mm thick with a compound curved surface with a diameter range of 22 cm to about 30 cm.

Turbine nozzles require high temperature capability, ability to withstand severe thermal gradients, good oxidation resistance, and compatibility with the combustor liner material. The nozzle must have a thermal expansion match with the combustor liner or be designed to account for the difference. In addition the chemical compatibility with surrounding structures is important such that no chemical reactions occur between individual ceramic components or the ceramics and neighboring metals. Nozzles have a complicated geometry that comprise many individual airfoils and usually inner and outer shrouds. To perform efficiently, the nozzle airfoils must have relatively thin trailing edges, on the order of one-half millimeter in thickness. This imposes a requirement that the material must be fabricable in thin sections while retaining sufficient strength and oxidation resistance. The nozzles tested by Williams are assembled into a ring structure that fits inside a metal support structure.

The turbine rotor is a very demanding structural component. A material system needs high specific tensile strength (strength divided by density), high fracture toughness, oxidation resistance at both high and intermediate temperatures, low thermal expansion and good thermal shock resistance. Large variations in cross sectional thickness can be found in the turbine rotor. The hubs of small turbines can be 30 mm thick and the trailing edges of blades typically thin down to 0.5 mm or less. These variations in thickness need to be considered in the design as wells as fabrication and coating processes, if required.

The first high temperature composite material evaluated at Williams was carbon/carbon (C/C), tested in a combustor in an F107 engine under contract to the USAF. This work was conducted from May 1982 through April 1989. IRAD programs and other contracted efforts have evaluated C/SiC, SiC/SiC, and C/C combustors, turbine nozzles, and turbine rotors in a WR24-7 turbojet engine from 1982 through to the present (see Figure 11.1).

RIG TESTING

Before committing hardware to an engine test, ceramic components and subelements undergo extensive rig testing. The rig tests are designed to simulate a turbine engine environment without risking an entire

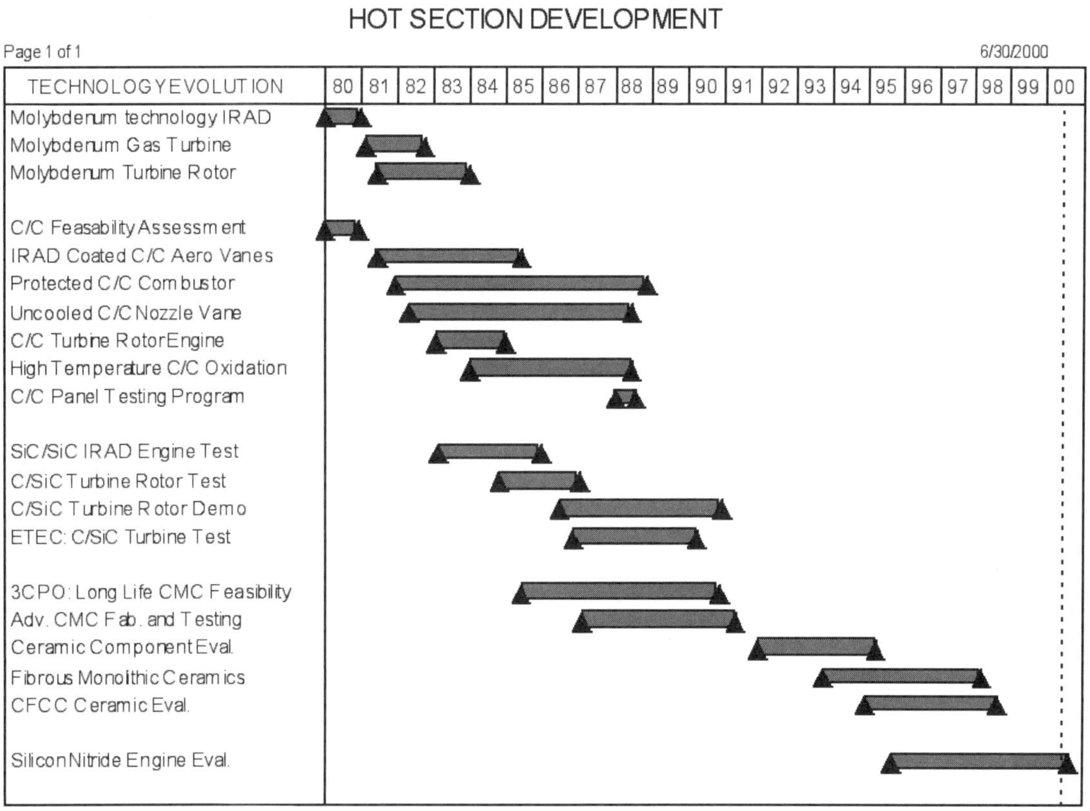

Figure 11.1 Timeline of Ceramic Composite Development at Williams International (Reprinted with permission from Williams International)

engine. Rig testing provides critical information about the material limiations and confidence in the material system prior to engine testing.

Rig testing includes burner cascade exposure, stressed oxidation, thermal shock, as well as cold and hot spin. Burner rigs are used for basic material evaluation in a high velocity combustion environment using jet fuels. The stressed oxidation rig applies a tensile load to a test specimen while it is subjected to a high velocity combustor environment. Thermal shock rig testing is used to evaluate materials subjected to extremely high thermal transients. Vacuum spin pit testing structurally loads a component by spinning at high speeds. Hot spin testing subjects a material to a combustion environment while rotating at high speeds.

Burner Rig Testing

The burner rig was developed to evaluate materials in a combustion environment. This provides the opportunity to assess material temperature capabilities and limitations, erosion and chemical stability in an actual combustion environment. Based upon the results of the rig testing along with other test results, engine applications for selected materials were recommended.

The burner rig is designed to use various types of jet fuel. The test rig is used to evaluate the thermal and chemical stability of materials and to evaluate protective coating systems. Testing has been conducted at temperatures from 815°C to 1870°C and pressures from 100 kPa to 310 kPa. The velocity can be controlled from Mach 0.05 to Mach 1 with a mass flow range of 0.0045 to 0.45 kilograms per second.

TABLE 11.1 BURNER RIG MATERIAL EVALUATIONS

Material	Company	Date	Description
C/C 2D cloth layup	Hitco	1983 to 1988	Testing at 1232 to 1370°C – 2 hours at 1371°C <0.1% wt loss – 4 hours at 1371°C <1% wt loss – 5 hours at 1371°C 1.63% wt loss – 7 hours at 1371°C <1% on one with 2 vanes broken and lost
SiC/SiC 2D cloth layup	Amercom	1985 to 1988	Initial screen test for 30 min from 1090 to 1760°C – No significant weight/dimension changes 10 hours of testing at 1090, 1230°C – Slight weight gain from oxide at both temperatures 5 hours of testing at 1370, 1510 and 1650°C – Weight gain up to 0.5% due to oxidation at all 3 temperatures
SiC/CAS-IIB	Corning	1985 to 1988	Initial screen test for 30 min from 1090 to 1430°C – 1260°C white layer formed on the convex surface – 1370°C ablation of the LE exposing fibers – 1430°C LE ablated and start of delamination 10 hours of testing at 1090, 1200, 1260, 1320 & 1370°C – 1090°C and 1200°C no significant change – 1260°C glassy surface layer formed – 1320°C leading edge ablation after 5 hours – 1370°C heavily damaged after 3 hours
SiCp/Al_2O_3 Dimox	Lanxide	1992	Tested at 1204, 1370, 1593 and 1760°C – 10 hours to testing conducted up to 1593°C – Degradation of the material at 1760°C – Material usable limite appears to be 1370°C
SCS-6/RBSN Unidirectional	NASA LeRC	1993	Tested for 10 hours at 1121, 1232 and 1343°C with RT42 coating – Cracks developed parallel to the fibers
Si_3N_4/BN Fibrous Monolithic	ACR	1995	2, 4 and 10 hours tested at 1370 and 1540°C – 10 hours at 1370°C specimens in excellent condition – 10 hours at 1540°C oxidation only 1–2 cells deep Testing at 1760°C caused severe oxidation as expected since it is above the oxidation temperature of boron nitride

Most of the burner rig testing has been targeted for short-life expendable engine technology. Testing is typically limited to 10 to 50 hours, although material temperatures may be higher than required for longer life applications. Materials for combustors, nozzles, turbines and ducts were evaluated in the burner rig test (see Table 11.1).

The burner rig nozzle and test-section and specimen holders are fabricated from a castable zirconia, and the airfoil vanes rest in grooves within the zirconia test section. The configuration for most of the burner rig testing has been a cascade arrangement of three nozzle vanes that simulate the environment of a turbine inlet nozzle (see Figure 11.2). Vane temperature is monitored using a Type B platinum/platinum-rhodium thermocouple located in the gas stream approximately 8 mm upstream of the vanes.

Two types of static vanes have been evaluated. The most basic were rectangular or wedge shaped specimens. Typical engine airfoils have also been evaluated.

The fibrous monolithic ceramic material developed by the University of Michigan and Advanced Ceramic Research, ACR, is a promising new material. The fibrous structure gives the material the toughness of a fiber-reinforced system without the high cost of the reinforcing fiber. This material used silicon nitride with a weaker second phase of boron nitride. The oxidation of the boron nitride is the limiting factor in the use of this material system. The material performed well at 1370°C for short periods of time. ACR is developing new material systems based on the fibrous monolithic principle.

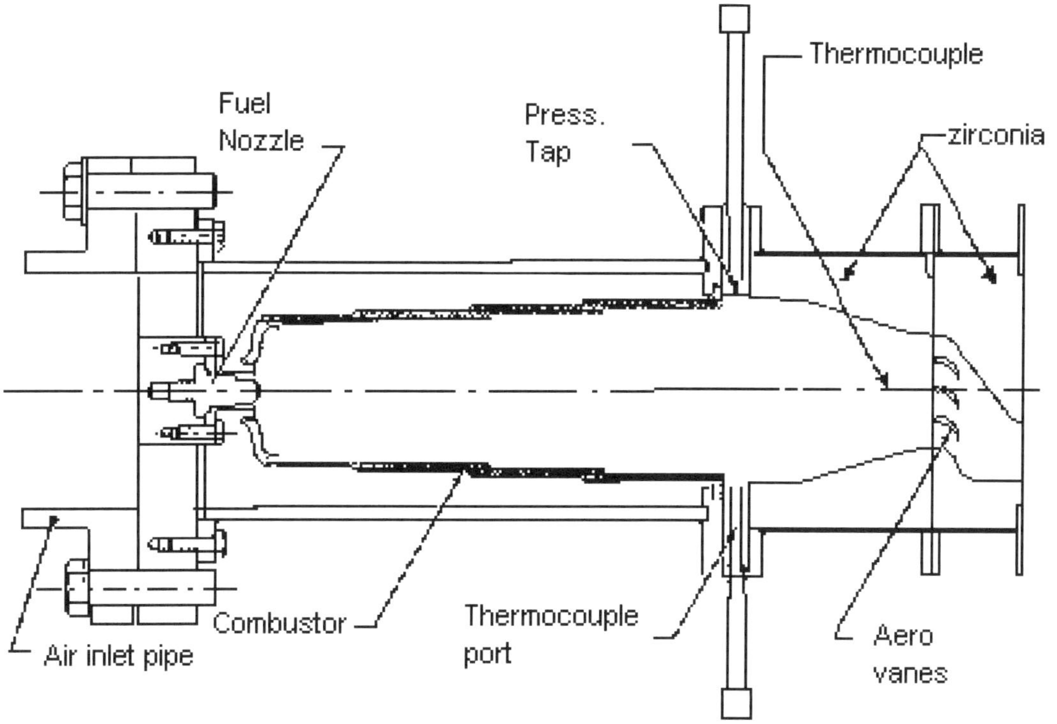

Figure 11.2 High Velocity Burner Rig

Stressed Oxidation Rig

In an effort to evaluate materials for critical structural applications the stressed oxidation rig was developed. Unlike the burner rig that exposes samples without mechanical loading, the stressed oxidation rig applies a mechanical load to the materials while in a combustion environment. This simulates the conditions such as would be seen in the turbine blades and turbine rotors. Stressed oxidation testing can expose weakness and strengths in material systems without putting engine hardware at risk.

The stressed oxidation test rig applies a mechanical load to a test specimen while it is mounted in the burner rig. A load frame applies a monotonic tensile load to a standard edge loaded test specimen. The load is applied to the stressed oxidation test specimen with an air cylinder attached to a load cell. The load frame is mounted on the end of the burner rig (See Figure 11.3).

This provides the capability to stress materials and/or coating system while they are exposed to an actual combustion environment. Results from this rig testing were used to define fiber architectures, oxidation protection systems and maximum stress levels for highly loaded ceramic components such as turbine rotors. A summary of the testing can be found in Table 11.2.

The initial C/SiC testing in 1992 was to evaluate the material for high-pressure turbine rotor engine applications near 1600°C.

The results of the stressed oxidation testing has shown that some material systems have dramatically lower stress capability than what would be expected from high temperature tensile and creep tests. Some of the difference observed in the testing of the C/SiC and NBSiC material systems may be attributed to internal oxidation after matrix cracking.

In tests with C/SiC, it was found that 0°/±60° rotated fabric fiber architecture provided better results than the 0°/90° fiber architecture. The difference may be due to the fact that carbon fibers have a very low

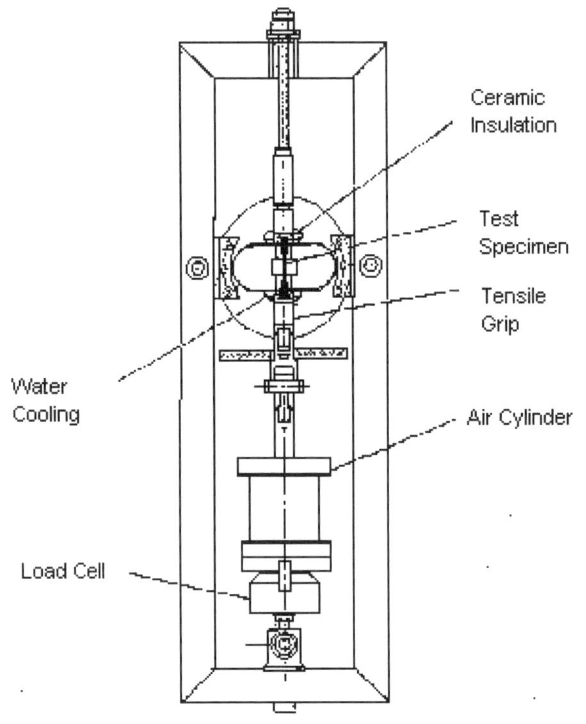

Figure 11.3 Stressed Oxidation Test Rig

TABLE 11.2 STRESSED OXIDATION TESTING

Material	Company	Date	Description
C/SiC	DuPont	1992	0/90 and 0/±60 architecture with UTRC & Chromalloy coatings – 0/±60 specimens survived longer than the 0/90 specimens
SCS-9A/RBSN	NASA-LeRC	1994	Material fabricated by NASA – Reaction Bonded Silicon Nitride silicon carbide fibers
SCS-9A/NBSiC	Textron Systems Division	1998	– 0° test survived 1200°C for 10 hours at 55 and 69 MPa – 0° test survived 1315°C for 10 hours at 83 MPa – ±45° test survived 1200 and 1315°C for 10 hours at 28 MPa – Extensive fiber pullout after thermal exposure
Si_3N_4/BN	ACR	1998	Fibrous Monolithic Ceramic – Hot pressed unidirectional and quasisotropic layup

coefficient of thermal expansion along the length of the fiber, but a large positive expansion across the diameter of the fibers. Cracking of the oxidation protective coating appeared to be associated with the 90° tows on the sides of the test specimens. A combination of mechanical strain and thermal expansion mismatches may have caused the cracks in the coating and reduced the composite capability.

Textron SCS-9A reinforced nitride bonded silicon carbide was evaluated in the stressed oxidation test rig. Two fiber architectures were evaluated, 0° and ±45°. The fiber content was low to simulate the architecture of a burner can that was fabricated via filament winding. Test specimens that survived the stress oxidation testing were tensile tested at room temperature. Specimens tensile tested after stressed oxida-

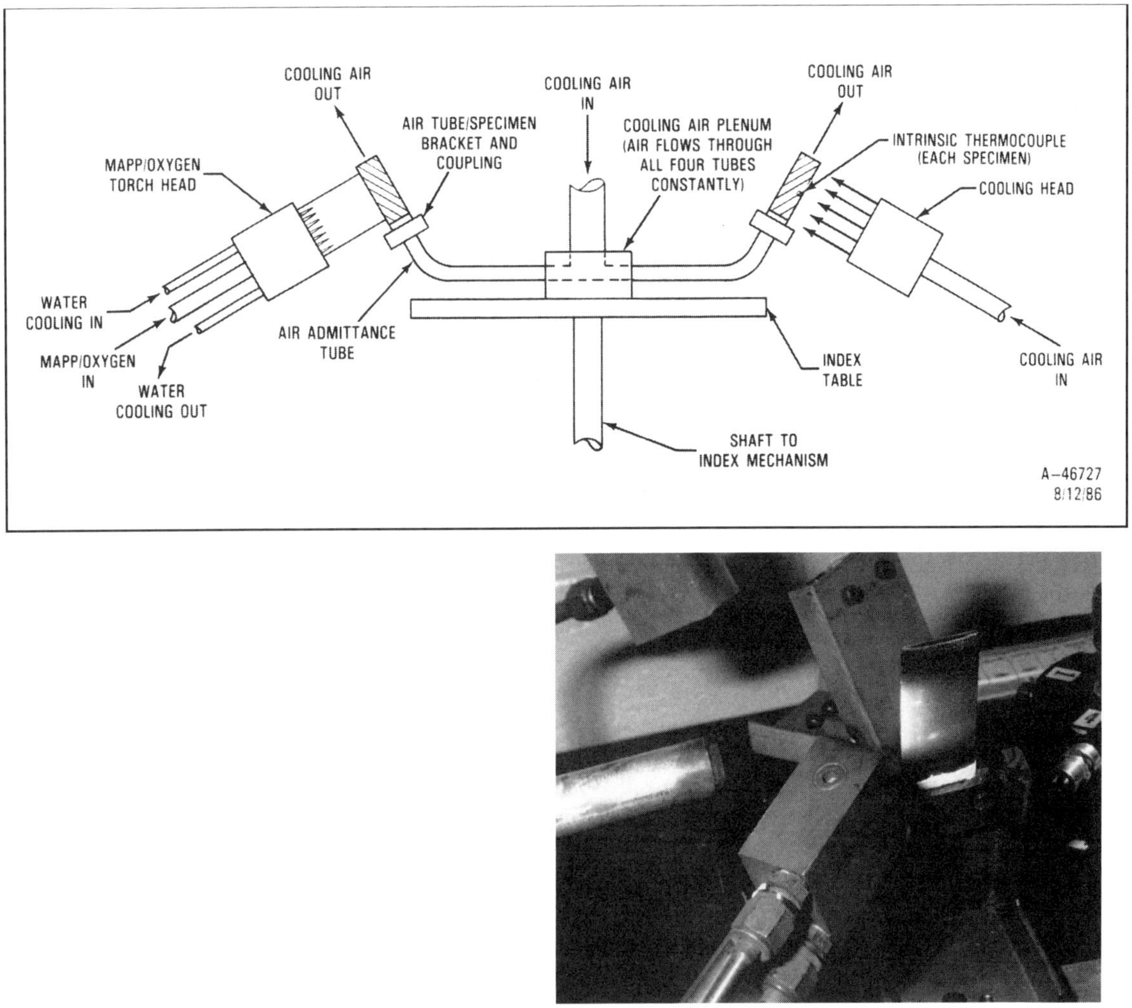

Figure 11.4 Thermal Shock Rig (Reprinted with permission from Williams International)

tion testing had substantially more fiber pullout than specimens fractured in the monotonic tensile test without any conditioning.

Thermal Shock Rig

Development of the thermal shock rig was initially to evaluate thermal barrier coatings. This then expanded into evaluating thermal shock properties of ceramic materials. Components such as nozzle vanes are exposed to high thermal shocks on engine startup and shut down. This rig evaluated materials for turbine and rocket applications that require high thermal transients.

The thermal shock rig can hold up to four test specimens on a rotary table, as shown in Figure 11.4. There are two MAPP gas/oxygen torches, positioned to heat two samples at once. A programmable controller is used to index the rotary table and move the specimens from the heating station to the cooling station. The cycle times can be controlled from 1 second to 1 hour. The rig is capable of very high temperatures and has been used to test specimens from 540°C to 2200°C.

TABLE 11.3 VACUUM SPIN TEST MATERIALS

Material	Company	Date	Description
C/SiC & SiC/SiC	SEP	1985	– SiC/SiC rotor spun to 60,000 RPM and held for 1 minute – C/SiC rotor spun to 72,000 RPM successfully
C/C	Hitco	1987	– Spun to burst for proof testing
SiC/SiC	Amercom	1988	– Burst speed 50,660 to 52,560 RPM corresponds to a stress of 150 to 160 MPa – Panel in-plane strength was 85 MPa considerably lower than the burst stress
C/SiC	DuPont	1988	– Spin to burst tests verified quasisotropic layup properties for turbine disk application
SiC/CAS-II	Corning	1988	– Stress of 170 MPa at burst – This material had lower fiber content and a higher density than Nicalon/CAS-IIB disks
SiC/CAS-IIB	Corning	1988	– Burst speed 43,800 to 49,820 RPM corresponds to a stress of 130 to 170 Mpa – Panel in-plane strength was similar to the stress at burst
Si_3N_4	Norton Technical Ceramic Lab.	1992	– Slip cast disk obtained expected silicon nitride properties
Si_3N_4		1993	– Low pressure injection molded disk were warped and did not obtain the expected stress levels at burst
SiC_w/Si_3N_4	Cercom	1992–1999	– All turbine blades proof spun prior to engine testing – Burst testing of subelement blades conducted which verified the tensile properties and correlated data from test specimens

Since this rig was originally designed and built to test and evaluate thermal barrier coating systems, it has the capability to supply cooling air to one side of a test specimen while the opposite side is exposed to the torches. Some of the test specimens were cooled internally.

Two CMC material systems were evaluated for thermal shock capabilities using the rig, DuPont-Lanxide SiC particulate/Al_2O_3 Dimox material and DuPont-Lanxide C/SiC CVI material. The samples were rotated into the torch and held for 1 minute then rotated in front of the air cooling jet. The C/SiC material used a Chromalloy CVD SiC, RT42, oxidation protective coating. The temperatures were monitored using an optical pyrometer and thermocouples and measured at 1760°C. The alumina based material with the SiC particulate reinforcement cracked after 1 cycle. The C/SiC samples retained 100% of their flexural strength after 50 thermal cycles and had 75% of their flexure strength after 100 cycles.

Vacuum Spin Pit Testing (Room Temperature)

Spin pit testing is performed to proof test a rotating component or determine the ultimate strength of a material system and/or fiber architecture. The test article is spun in a vacuum. This rig is capable of spinning a part at 156,000 RPM and has an inside diameter of approximately 35 cm. Spin disks were fabricated from various material systems and fiber architectures (see Table 11.3). Quasi-isotropic layups of plain or satin weave cloth and radial spiral weave were tested. A modified spin disk design, called a paddle-wheel disk, is frequently used to evaluate the strength of rotor blades. Thin flat paddles machined on the periphery of the disk are designed to simulate blade root stresses at speeds below the disk burst speed.

SEP C/SiC polar woven rotors were cold spun to 53,000 RPM and 60,000 RPM. In both cases the rotor lost the blades. Failure of the blades was attributed to radial fiber misalignment that caused the blades to fail by interlaminar shear.

Most of the ceramic spin disks were approximate 15 cm diameter with a thickness of 3.8 mm. Spin disks were spun to burst, and then the stresses were calculated based on the material densities. Engine test rotors were proof spun to 110% stress condition in this pit prior to engine testing. Inserted ceramic blades were also proof test spun and burst tested to verify mechanical properties prior to engine testing.

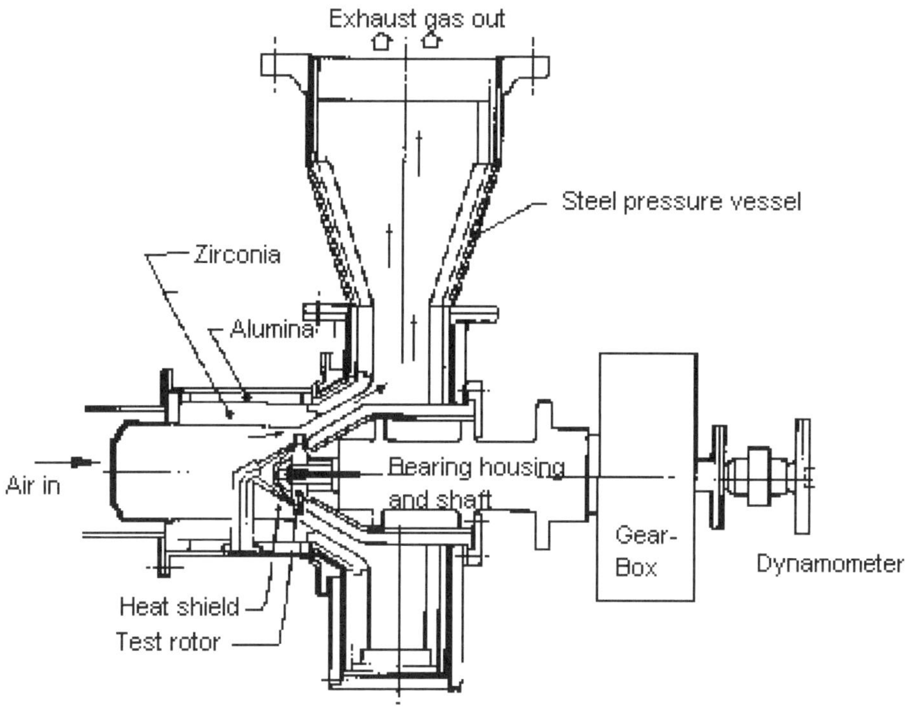

Figure 11.5 Hot Spin Combustor Rig

Hot Spin Rig Testing

The hot spin rig (Figure 11.5) is capable of testing full size, ceramic turbine components in a combustion environment. The temperature capability of the rig is 815°C to 2090°C with hot gas pressures ranging from 140 to 1,030 MPa. The rig is capable of spinning a part up to 60,000 RPM. Full size ceramic turbines have been tested in the hot spin rig. Several C/SiC rotors were made at DuPont using the CVI process and tested in this rig.

It was found through the hot spin rig testing that the C/SiC turbine rotors had a problem with moderate temperature oxidation in the bore of the turbine rotor. As a result, additional development was conducted to develop an additional coating that worked with the high temperature coating system.

ENGINE TESTING

Extensive engine testing of ceramic components has been conducted using several different engines including the F107, WR24–7, WJ24–8 and FJ44. Ceramic components including combustor walls, turbine nozzles, turbines, and bearings have been engine tested. The selection of materials and design of the engine components is dependent upon the fabrication limitations and material properties.

Combustor designs used in Williams International engines have a relatively uniform cross section, are lightly mechanically loaded, and are exposed to high thermal stresses in an oxidizing environment. Mechanical stresses are usually less than 70 MPa.

Turbine nozzles require lower tensile strengths than a rotor. Tensile strengths of greater than 70 MPa are usually required, depending upon the part size and design. As with the combustor, they need to be oxidation resistant and be able to withstand thermal cycling.

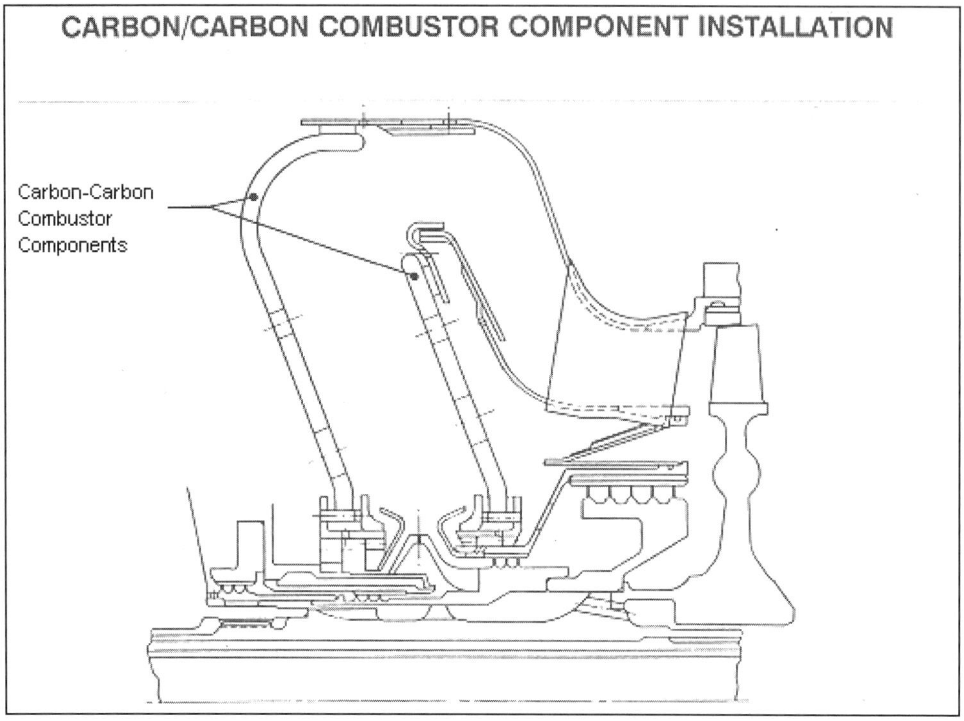

Figure 11.6 Carbon/Carbon Combustor Design

Turbine rotors require high hoop and radial strength. They also must be able to handle hot streaks, thermal cycling and be oxidation resistant. For the turbine rotor, structural, geometric, environmental, and mechanical loading is the most demanding of the components tested. Two types of ceramic rotors have been tested, inserted-blade turbines and integral blisks (bladed disks).

F107 Engine Test Experience

The first high temperature composite engine test experience was with the F107 "Cruise Missile" engine. Coated C/C combustor components were designed, fabricated and tested in this limited life gas turbine engine (see Figure 11.6). Early C/C combustors would delaminate because the silicon carbide conversion coating was too thick and strong for the shear strength of the C/C. It was important to control the thickness of the conversion oxidation protective coating to produce a successful component.

The first successful C/C combustor component was fabricated from a two-dimensional laminate using the Vought ACC-4 process and coated with Vought's Type-I—Mod I oxidation protection coating.

Two sets of combustors were fabricated and tested for a cumulative test time of 13 hours and 44 minutes. The first set of hardware was tested for 2 hours and 35 minutes in 5 start cycles. Maximum component surface temperatures were approximately 1315°C. Despite a loosening of the clamp group attachment on the forward combustor cover during engine operation, both C/C components remained in excellent condition after testing.

A second set of C/C combustors components were tested in the next series of engine tests. Two start cycles were performed with a total run time of 1 hour 3 minutes. Combustor surface temperatures reached 1340°C. A turbine rotor failure halted testing at this point, though no damage occurred to the

TABLE 11.4 COMBUSTOR TEST SUMMARY FOR THE WR24-7 AND -8 ENGINES

Material	Company	Date	Description
Nicalon/SiC	Amercom	1993	– Engine run time 3 hours 42 min, surface temperature 1090–1150°C – Forward combustor plate cracked in 2 engine tests – High thermal stress probable cause of the cracking
Nicalon/SiNC	Kaiser Aerotech	1993	– Engine run time 3 hours 5 min, surface temperature 1090–1150°C – Combustor components limited by the pyrolysis temperature – Exceeding the pyrolysis temperature affects the component
Nicalon/Al_2O_3	DuPont-Lanxide	1993	– Engine run time 20 hours 34 min, surface temperature 1090–1370°C – Cracking noted between combustor air inlet holes – Residual metal leaked and oxidized from the component with no apparent adverse affects
Nicalon & SiCp/Al_2O_3	DuPont-Lanxide	1994	– Engine run time 10 hours 54 min surface temperature 1260°C max – Addition of SiCp eliminated cracking around holes – Thermal conductivity increased 50% with SiCp
C/Si_3N_4	General Atomics	1994	– Engine run time 20 hours 23 min surface temperature 1260°C max – The Si_3N_4 may not have been stoichiometric – Material performed well in the engine testing
Nicalon/SiNC	Kaiser Aerotech Dow Corning	1994	– Engine run time 35 hours 10 min surface temperature <1370°C – Material performed very well and complex shapes easily fabricated
HPZ/SiC	DuPont-Lanxide	1994	– Engine run time 33 hours 37 min surface temperature 1370°C – Material performed similar to Nicalon/SiC
Si_3N_4/BN Fibrous Monolithic	Advanced Ceramics Research	1997	– Engine time 11 hours 10 min surface temperature 1315–1370°C – Oxidation of the BN phase is the limiting factor – Material performed very well

C/C components. The engine was rebuilt, and this second set of components was used for further testing. Ten hours and 6 minutes of additional engine testing were accumulated on this hardware, and combustor temperatures reached approximately 1540°C.

The coating on the forward combustor hardware sustained local damage in three places, but the component was still intact and structurally sound. The aft combustor plate was still in excellent condition after testing. Both components were well within the weight loss criterion goals for this program.

A problem encountered with coated C/C combustors was coating failure around sharp edged holes. Large thermal gradients from compressor discharge air flowing into the hot combustor can cause substantial thermal stresses. Potential moderate temperature oxidation can also occur around the hole where silicon carbide coating systems can have open cracks and fail.

WR24-7 and WJ24-8 Turbojet Engine

The WR24-7 engine and the upgraded thrust version WJ24-8 engines are used on Navy target drones. These small engines were extensively modified to incorporate many ceramic components. Modifications were made to the combustor, turbine nozzle assembly and the turbine and shafts assemblies.

Combustor Component Testing. The combustor design of the WR24-7 and WJ24-8 engine are almost identical and use the same basic design as the F107 engine. The combustor is composed of two shells that, when assembled, define a toroidal area. The fuel is delivered through the shaft into the toroidal area through rotating shaft-mounted fuel nozzles. The combustor is similar in configuration to the F107 combustor as seen Figure 11.6.

Many material systems have been fabricated into the two combustor components for this engine. Following in Table 11.4 is a list of some of the CMC material systems evaluated in these engines.

Figure 11.7 WR24-7 Turbine Nozzle Vane

A limiting factor for some of the material systems is the temperature limit of the ceramic fibers, such as Nicalon or HPZ. The pyrolysis or processing temperatures were other limiting factors in some systems due to chemical and physical changes that occur when they are exceeded. CVI systems that use a carbon interface layer required an oxidation protective coating to retain mechanical properties. Coating cracks such as seen around air inlet holes can cause oxidation of the carbon fiber interface down the length of the fibers. All components with oxidation concerns were seal-coated after machining, usually by the manufacturers of the ceramic components.

Each of the material systems has its attributes and combustor components may not be the best or only application of the material systems. Improvements to fibers and interface coatings will improve the performance of some of the material systems.

Nozzle Testing. After the combustor, the next component the hot combustion gas encounters is the turbine nozzle. The WR24-7 engine has been used to test ceramic nozzle vanes. Figure 11.7 shows the tiny size of the vanes.

The vanes were attached using metallic inner and outer shroud rings. The removable nozzle assembly consisted of bolt-on inner and outer retaining rings and shrouds with vane-shaped slots. The trapped assembly provided clearance for the thermal expansion difference between the metal and ceramic. The ceramic nozzle assembly held 26 vanes, while the standard engine nozzle has 29 vanes.

Separate engine builds and engine tests were conducted on the Amercom SiC/SiC and Corning SiC/CAS-IIB vanes. A rainbow test was also performed after the initial engine test using a combination of three different ceramic nozzle vanes assembled in the same engine test (see Figure 11.8). For the rainbow test, Textron Specialty Materials SCS-6/Si_3N_4, was included into the engine test. The rainbow engine test ran for 16 hours 46 minutes. Following in Table 11.5, is a list of materials along with engine run times and temperatures achieved during the testing.

The intent of the ceramic turbine nozzle design was to run the entire ceramic vane hot with very little thermal gradient. Many of the small thin vanes were found to have cracked at mid-span or chipped at the edges. Cracks and chips were due to handling and inserting the vanes into the retaining rings. The vanes were difficult to assemble and allowed vibration of the vanes in the assembly.

Turbine Rotor Experience. The first composite rotor engine tested at Williams International was a C/C rotor fabricated by Hitco, which used a CVD oxidation protective coating applied by Midland Materials.

Figure 11.8 WR24-7 Ceramic Turbine Nozzle Vanes (Rainbow) Before Engine Testing

TABLE 11.5 SUMMARY OF TURBINE NOZZLE VANE TESTING

Material	Company	Description
C/C	Hitco	– Angle interlock material – Successfully tested for 6 hours at 1370°C
Nicalon/SiC	SEP	– 2D cloth layup
Nicalon/SiC	Amercom	– Engine tested 21 hours 13 min, avg. temperature 1000°C – Max temp 1070°C for 1 hour 3 min
Nicalon/CAS-IIB	Corning	– Engine tested 22 hours 9 min, avg. temperature 1000°C – Max temp 1070°C for 1 hour 3 min
SCS-6/Si_3N_4	Textron Specialty Mat'l	– Engine tested 16 hours 46 min, avg. temperature 1000°C – Max temp 1070°C for 1 hour 3 min
SCS-9A/RBSN	NASA LeRC	– Engine tested in a WJ24-8 turbojet engine

Two rotors were engine tested up to 45,000 RPM with a turbine inlet temperature of 1290°C for a total run time of 45 minutes. At about 10 minutes after engine start, progressive blade tip oxidation and blades were lost due to full SiC conversion of the carbon fiber. This test demonstrated that C/C rotors were structurally feasible and that oxidation protective coatings could protect the C/C for short periods of time.

This first rotor was fabricated from rotated fabric that is dependent upon the shear strength of the matrix in the turbine blade portion of the rotor. After this test, radial blisk designs were made to provide continuous fiber from the hub to the blade tips.

The next engine test was conducted with an SEP SiC/SiC rotor. This rotor ran for 1 hour with 4 start cycles. The turbine speed was between 40,000 and 50,000 RPM. The exhaust gas temperature was between 780°C and 1090°C. The test ended with a rotor burst at 50,000 RPM with a turbine inlet temperature of 1260°C. The bore stress at failure was 200 MPa. The rotor burst into four quadrants and the bore temperature was approximately 430°C. The failure analysis concluded that burst occurred due to progressive matrix cracking under sustained loading. Thermal stresses were not analyzed for this rotor.

The following SEP SiC/SiC rotor test ran for 2 hours with 2 engine starts. The shaft speed was varied between 40,000 and 50,000 RPM with the exhaust gas temperature between 715°C and 830°C. Two adjacent blades failed at mid-span after 2 hours of engine run time. The failure of the blades may have been due to low shear properties and the lack of continuous fibers from the hub to the tip as a result of blade twist in the design.

An SEP C/SiC quasi-isotropic rotor was successfully proof spun at room temperature to 72,000 RPM. This rotor was subsequently run in the WR24-7 turbojet engine for 25 minutes with two start cycles. The shaft speed was between 41,000 and 55,000 RPM with a turbine inlet temperature of 775°C to 1140°C. Blade loss occurred at 55,000 RPM with a turbine inlet temperature of 1050°C. The blade densities were significantly higher than the hub density. No significant amount of weight loss was detected due to oxidation.

Critical issues for ceramic composite rotors are fiber volume content, fiber architecture, fiber tow size and tow spacing, and oxidation behavior. These early engine tests with ceramic composites illustrate some of the problems and lessons learned. It was determined that the fiber architecture between the hub and blades must be continuous. The fiber tow size must be small to effectively reinforce the blade, due to the thin trailing edges of the turbine blades. Oxidation is always a concern with carbon fibers or interface coatings. Two temperature regimes have to be considered for oxidation protection. The first is the high temperature regime (>1000°C) in the turbine blades. The second is the intermediate temperature regime, 540°C in the bore of the turbine rotor where the high temperature coatings have open craze cracks.

LESSONS LEARNED

Metal to Ceramic Attachment Designs

Attachment of ceramic components to the metal engine components is critical to the success of the component. Most fiber reinforced ceramic composites have relatively low interlaminar shear and tensile strengths. Turbine rotor attachments must take this into account when the attachment is designed. In general, ceramics can accommodate higher compressive loads than tensile or shear loads. This is one of the reasons that clamp group type attachment concepts are used. Ceramics can also be damaged by point loading. In most clamp group attachments, a compliant layer is usually needed to distribute the load over a larger area.

Designing for High Modulus Materials

Ceramic components are stressed both mechanically and thermally. Thermal stresses in some instances can be the larger stress contributor. A small strain can result in a large stress due to the fact that ceramics usually have very high elastic moduli. A high thermal gradient across a ceramic part can cause high thermal stresses due to the thermal straining of the part.

Oxidation Protection

Most ceramic composite systems require oxidation protection. Carbon/carbon and carbon/silicon carbide ceramic composite systems require the most oxidation protection. Silicon carbide CVD coatings have a thermal expansion mismatch relative to carbon/carbon that can cause the coating to crack. These craze cracks are open at low temperature and do not close until the composite is reheated to near the original processing temperature. Therefore, intermediate oxidation is a problem with silicon carbide coated carbon/carbon. From approximately 370°C to 700°C, the silicon carbide cracks are open, leaving the substrate open to oxidation.

Thin effective coatings for rotors or thin nozzle vanes is a major problem. On rotors, the coating is a parasitic load on the blade. Therefore you want it to be as thin as possible. However, the coating that provides good oxidation protection does not have the same CTE as the substrate, resulting in chipping and spalling. Any defect in the coating is catastrophic to the composite given enough time.

Fiber Architectures

The fiber architecture for each part is dependent upon design and loading of a component. Turbine rotor disks require strength in the hoop direction while the turbine blades require strength along the axis of the blade. Fiber alignment in the load direction is very important. Misalignment of fibers can degrade the capability of a material to carry the required loading.

FUTURE WORK

There are several additional improvements that need to be made to ceramic matrix composites:

- High strength oxidation resistant fibers that are stable above 1650°C.
- Oxidation resistant fiber interface coatings, which provide toughness to the composite.
- Material property data base sufficient to design engine components
- Rapid densification techniques that would lower the processing time and component cost.
- Improved fiber architectures that tailor the mechanical properties to the component requirements.
- Improved coatings to protect the ceramic matrix composites from intermediate and high temperature oxidation.

ACKNOWLEDGEMENTS

The authors would like to thank the many people and organizations that made the ceramic work described in this chapter possible. The following is a partial list: Dr. Ben Wilcox, Steve Wax and Dr. William Coblenz of DARPA; Dr. Steven Fishman of ONR; Steve Szaruga, Don Schmidt, Dr. Ronald Kerans, Dr. Allan Katz, and Larry Zawada of the Air force Materials Laboratory; Don Zabrieck, Elmer Buchanan, William Troha, Dr. Charles MacArthur, Dr. Charles Stevens, Carlos Arana, Mark Dale and others from the Air Force Propulsion Laboratory, Jill Jonkouski of DOE; Dr. Ramakrishna T. Bhatt of NASA Glen; Bill Barker of the Navy; and Dr. Sam Williams and Gregg Williams of Williams International.

Except for when otherwise indicated the figures in this chapter have been reprinted with permission from reports of US Government programs that are not in the public domain.

NOMENCLATURE

ACRONYM	*DESCRIPTION*
ACM	Advanced Cruise Missile
ALCM	Air Launched Cruise Missile
CAS	Calcium Aluminosilicate
C-CAT	Carbon-Carbon Advanced Technologies
C/C	Carbon/Carbon
C/SiC	Carbon/Silicon Carbide

CMC	Ceramic Matrix Composite
CVD	Chemical Vapor Deposition
CVI	Chemical Vapor Infiltration
DIMOX®	Directed Metal Oxidation
HPZ	Dow Corning silicon carbide based fiber
TIT	Turbine Inlet Temperature
JETEC	Joint Expendable Turbine Engine Concept
MAPP	Methylacetylene Propadiene
NASA	National Aeronautics and Space Administration
NBSiC	Nitride Bonded Silicon Carbide
RBSN	Reaction Bonded Silicon Nitride
RPM	Revolutions per Minute
SCS-9A/BN	Silicon Nitride/Boron Nitride
SEP	Sociéte Européene De Propulsion of Bordeaux, France
SiC	Silicon Carbide
$SiCp/Al_2O_3$	Silicon Carbide particulate/Alumina
SiNC	Silicon Nitride and Silicon Carbide
UTRC	United Technologies Research Center

CHAPTER

12

Evolution of Ceramic Gas Turbine Development Programs at Engine Manufacturers in Japan

Takane Itoh

Tokai University, Yokohama, Kanagawa, Japan

Akira Sugawara

New Energy and Industrial Technology Development Organization, Tokyo, Japan

INTRODUCTION

In Japan, almost all of the primary energy has to be imported from foreign countries, which poses very high risks for the nation's energy supply. In 1973 more than seventy seven percent of the energy consumption depended on pertroleum. Therefore, Japan faced the necessity of a drastic reduction in energy consumption, especially in petroleum, in many fields after the energy crisis of the 1970s.

One area of high energy consumption was the large power generating plants which were heavily dependent on petroleum. These large power plants were primarily operated in the combined cycle mode and improvement of the gas turbine efficiency was required. Another field was distributed electric supply systems or cogeneration systems for public use where small efficient engines with low pollutant emissions were needed. Then there was the transportation sector where automobiles were the biggest petroleum consumer while their energy efficiency was less than 25%. Against this background,several research and development programs aimed at upgrading the gas turbine efficiency were planned and carried out, as shown in Figure 12.1. In this chapter, some brief summaries of ceramic gas turbine (CGT) development activities conducted in Japan will be described for each of the key applications.

CGT PROGRAMS FOR LARGE POWER PLANTS

The Ministry of International Trade and Industry (MITI) started the development project of an advanced 100 MW reheat cycle gas turbine with 55% thermal efficiency in 1978. Although the TIT (turbine

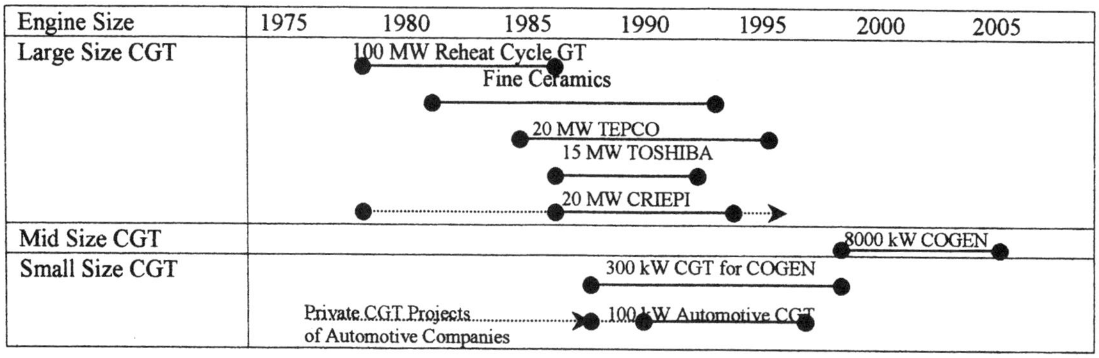

Figure 12.1 Timeline of Major Ceramic Gas Turbine Programs in Japan

inlet temperature) target was 1500°C and there were no plans to use ceramic materials for the engine components research and development of ceramic combustors and ceramic turbine nozzle vanes, etc., as future attractive technologies, was started in the latter half of the project. That was the first national program concerning application of ceramics into a gas turbine, and Ishikawajima Harima Heavy Industries (IHI), Kawasaki Heavy Industries (KHI), and other engine manufacturers gained a lot of experience with ceramics and related application technologies. In 1981, three years after the start of the advanced gas turbine program, another national program entitled 'Fine Ceramics' was initiated. This was a generic research program for understanding the characteristics of ceramic materials and advancing their properties. It was not a program for developing ceramic components for engines but a materials development oriented program. Most of the ceramic companies and some engine manufacturers contributed to the program. However, in the third phase of the program, it was recognized that the application technologies were also essential if ceramics were to be widely used for many applications in the near future. One of the most attractive applications was components for a gas turbine for a coal gasification combined cycle plant, and the focus of the program was slightly changed, and the development of fabricating technologies of turbine stator vanes and turbine blades was added to the objectives. That program was continued to the end of Fiscal Year 1992, and Toshiba, IHI, Toyota and other engine manufacturers learned about ceramics and their design criteria in the program.

The necessity of the development of a highly efficient combined cycle for the coal gasification system had been also recognized in electric power companies, to lessen the dependence on petroleum for power generating plants. Tokyo Electric Power Co., Inc. (TEPCO) conducted cooperative research programs for the application of ceramics to a 20 MW power generating gas turbine with Toshiba Corporation, Hitachi Ltd., and Mitsubishi Heavy Industries, Ltd. during the period from 1984 to 1995. The programs were aimed at reducing cooling air drastically and upgrading thermal efficiency. The initial TIT goal was 1300°C and a combustor liner and transition piece, first and second stage turbine nozzle vanes, and the first stage rotor blades were selected for ceramic fabrication. A similar research and development program was also carried out at the Central Research Institute of Electric Power Industry (CRIEPI), where the conceptual design of ceramic components for a 20MW coal gasification combined cycle power plant had already started around 1978, and research and development had been advanced in cooperation with Hitachi Ltd.. Parallel to these programs, Toshiba started another in-house development program in 1987, and they developed a first stage ceramic turbine blade and nozzle vane of a 15MW power generating gas turbine with a TIT of 1300°C for a future high efficiency ceramic gas turbine.

During the development period of these programs, the remarkable advancement of gas turbine technology led to the achievement of 49% thermal efficiency in Liquified Natural Gas (LNG) combined cycle power plants using 1300°C class metal gas turbines without ceramics. This realization resulted around 1990 in both of the above 20MW programs increasing their TIT target from 1300°C to 1500°C. Recently, LNG combined

TABLE 12.1 COMMERCIALIZED CERAMIC PARTS FOR RECIPROCATING ENGINES IN JAPAN

		Manufacturers		
Components	**Materials**	**Automotive**	**Ceramic**	**Year Introduced**
Glow Plugs		Isuzu	Kyocera	1981
		Mitsubishi	Kyocera	1983
		Mazuda	NTK	1985
Swirl Chamber in Diesel Engine		Isuzu	Kyocera	1983
		Toyota	Toyota	1984
	Si_3N_4	Mazuda	NGK	1985
Rocker Arm Pads		Mitsubishi	NGK	1984
		Nissan	NTK	1987
Turbocharger Rotor		Nissan	NTK	1985
		Nissan	NGK	1985
		Isuzu	Kyocera	1988
		Toyota	Toyota	1989

cycle power plants using a 1500°C class metallic gas turbine with 53% of gross thermal efficiency have just started service operation in Japan. This means that the time of the introduction of ceramics into the large power generating plants needs to be extended further and these recent circumstances have had a rather negative effect on research and development activities for large ceramic gas turbine power plants.

CGT PROGRAMS FOR AUTOMOBILES AND SMALL COGENERATION SYSTEMS

On the other hand, for small gas turbines, for which the adoption of sophisticated air cooled turbine blades is difficult, there is no effective way to increase the thermal efficiency other than increasing the TIT while introducing ceramics into the turbine hot section, and adopting a heat recovery system. Therefore, research and development of the ceramic gas turbine has been carried out actively by automobile manufacturers beginning in the latter half of the 1970s. Another reason why automobile manufactures intensified the development of the ceramic application technology at that time was the great expectation that ceramics were also effective for the performance improvement of reciprocating engines. Actually, some ceramic parts for gasoline and diesel engines had been developed and some of them have been used in the market. Examples are shown in Table 12.1. Progressive experiences with these parts in the market provided very useful data for the ceramic gas turbine development. Especially, the success of the ceramic turbocharger rotors demonstrated that the reliability and cost, which are the most critical issues for commercialization of ceramic components, were reasonable for practical use in the real market. In the beginning, research and development of the CGT was conducted privately on a rather small scale by several automobile manufacturers (Toyota, Nissan and Mitsubishi Motors).

It was eventually recognized that there were many challenges to be addressed and that each program by itself was too small, both in terms of manpower and financial resources, to develop and complete each ceramic gas turbine within the expected short development period. Also, the scope of these development programs was compared with the programs conducted by R&D organizations overseas at that time. Those considerations led to the conclusion that it would be advantageous to integrate all of the technologies that had been developed in each private program into one common national project.

In the mean time, the realization of small high-efficient 300 kW class gas turbines for a distributed electric power supply system or a cogeneration system had also been recommended. Again, it was argued that ceramic components and a heat exchanger were as indispensable for those efficient small gas turbines as they were for the automotive CGT. Initially, there was a plan to start a new ceramic gas turbine (CGT) development program, where research and development of each ceramic gas turbine for automobiles and cogeneration systems would be conducted in parallel under the same management because of the great deal of commonality of technologies between both CGT systems. Finally, however, two separate programs were proposed, mainly because of a complicated budget problem.

The automotive 100kW CGT program was initiated in FY 1990 after a two-year feasibility study. The program was a 7-year, 50% cost sharing program managed by the Petroleum Energy Center (PEC). The objectives were to develop a 100kW automotive CGT with a thermal efficiency of 40% while meeting the exhaust emissions regulations for gasoline fueled passenger cars. Three automobile manufacturing companies (Toyota Motor Corp., Nissan Motor Co. Ltd. and Mitsubishi Motor Corp.) and seven petroleum refinery companies collaborated on the program. In that program, only one engine configuration (regenerative, single-shaft CGT) was adopted, in contrast to the three for the 300kW program described below. The headquarters of the CGT development team was established at the Japan Automobile Research Institute (JARI) and the principal gas turbine engineers of each automobile manufacturer moved to JARI. The engine was designed and tested at JARI, while the components were developed by the individual engine manufacturers according to the initial agreements between JARI and them. At the end of the program, in March of 1997, the experimentally obtained thermal efficiency of the 100kW CGT at the TIT of 1350°C was 35.6%, only ~ 4% below the target of 40%. This was the world's first all-ceramic gas turbine engine which demonstrated experimentally that a well designed CGT had high potential for excellent thermal efficiency and also for low exhaust emissions. In order to reach the thermal efficiency goal of 40%, at least two more years of work would be needed. However, a program extension in time was not allowed because of problems in obtaining an unexpected supplementary budget at that time. The program was closed out and the consortium was dissolved at the end of March 1997.

As for the 300kW CGT program for cogeneration and mobile power generation, it was initiated in Fiscal Year 1988 as a 9-year, 100% funded program. The program was sponsored by MITI and managed by the New Energy and Industrial Technology Development Organization (NEDO). Three engine manufacturing companies (IHI, KHI, Yanmar Diesel Engine Co., Ltd. (YDE)) and the main ceramic manufacturers participated. Three different configurations (CGT-301,302 and 303) were selected as the reference gas turbine engine and were developed by each company. CGT301 was a regenerative single-shaft ceramic gas turbine for cogeneration developed by IHI with the emphasis on the applicability of the design technology to larger sized gas turbines. CGT302 was a regenerative two-shaft ceramic gas turbine also for cogeneration use developed by KHI. CGT303, developed by Yanmar Diesel Engine Co., Ltd., was a regenerative two-shaft engine for mobile power generation and was equipped with rotary regenerators instead of a recuperator. At the intermediate program review in 1993, the review committee advised that the development period should be extended to the end of Fiscal Year 1998. At the end of the program, in March of 1999, successful data were reported from the engine test of the CGT302 which had achieved the program thermal efficiency goal of 42%. That was the highest efficiency ever achieved with an uncooled gas turbine and the program was completed with a great success.

As a follow on program to the 300kW CGT program, a new 5-year program, "Research and Development on Practical Co-generation Technology," was started by MITI and NEDO in 1999. The main purpose of the program is to evaluate long-term durability and reliability of ceramic components in a "hybrid gas turbine" (using both metal and ceramic components in the turbine hot section). The reference engine selected was a medium size gas turbine (8000kW-class) for cogeneration, the M7A-02 of KHI. The engine will be retrofitted with ceramic combustor liners and transition pieces, and turbine nozzle vanes. KHI, Kyocera and three gas companies are involved in the program.

As seen from the above history of the CGT in Japan, development activities can be classified into two main categories, 15-20 MW CGT programs for large electric power plants (TEPCO, CRIEPI, Toshiba) and small CGT programs, including the 300 kW CGT programs for small cogeneration systems (KHI,IHI,YDE) and the 100 kW CGT program for automobiles (JARI Team). Although the details of each program will be described in several of the following chapters, the major objectives, plateaus reached and key challenges remaining are simply summarized in Table 12.2.

TABLE 12.2 OUTLINE OF KEY PROGRAMS RELATED TO CERAMIC GAS TURBINES IN JAPAN

Programs	Major Participants	Major Objectives	Targets	Plateaus Reached	Key Challenges Remaining
CRIEPI 20 MW Program	CRIEPI Hitachi	To examine the effects of applying ceramics to gas turbine for power plant. To study structural design concepts for ceramics to improve strength, reliability	To develop ceramic combustor, turbine stator vanes for 20 MW, 1300°C class gas turbine	Pattern factor of Si_3N_4 combustor was about half of that of a conventional metallic combustor. Ceramic components survived after shut-down test from rated condition	Not shown clearly, but it is desired to demonstrate the durability and long term reliability of ceramic components in engine environments
		To examine possibility of applying ceramics to higher temperature gas turbine with TIT of 1500°C	To develop ceramic combustor, turbine stator vanes for 20 MW, 1500°C class gas turbine	SiC combustor survived after shut-down test from rated condition. SiC stator vanes were subjected to the thermal loading cascade test under rated load, and no cracks were observed	
TEPCO 20 MW Program	TEPCO Toshiba MHI Hitachi	To confirm the adaptability of ceramics for a 20 MW-class gas turbine with TIT of 1300°C	To develop ceramic combustor, 1st and 2nd stage nozzles and 1st stage rotor blades of 1300°C class gas turbine	Hybrid rotor blade survived in hot spin test at the design speed	To improve the reliability of ceramic components by means of progress in material, design, quality assurance and technology
		To study the effects of utilization of small amount of cooling air in 1500°C-class gas turbine		Assembly test of ceramic combustor and ceramic hybrid nozzle vanes was successfully conducted at rated condition	To confirm the long term endurance and reliability of ceramics under actual engine conditions
Toshiba 15 MW Program	Toshiba	To upgrade 15 MW, 1300°C-class gas turbine	To evaluate ceramic 1st stage turbine stator and rotor blades in hot gas of 1300°C	Ceramic stator vanes survived a 1300°C steady state test with trip tests. Hybrid sleeve type rotor blades survived after 72 hours of 110% overspeed hot spin test at 1350°C	Not shown clearly, but it is desired to assure long term life and reliability in actual engine environment
300 kW CGT Program	IHI	To develop and demonstrate 300 kW-class highly effient ceramic gas turbines for cogeneration and mobile power generation	Thermal eff.: 42% TIT: 1350°C Emission: less than regulatory values	Thermal eff.: 35% at 312°C Accumulated operating hours: 280 hours NOx emissions: 5.6 ppmv at full load	To investigate much more time-dependent properties of ceramics. To control the hot section metallic structure and rotor tip clearance
	KHI			Thermal eff.: 42% at 1396°C Accumulated operating hours: 334 hours NOx emissions: 31.7 ppmv at full load	To idendtify the long term reliability of ceramic components in actual engine. To investigate and overcome FOD
	YDE			Thermal eff.: 12% w/o regenerators Accumulated operating hours: 21 hours GG rotor operated at 1350°C for 3 hours	To minimize deflection of metallic engine housing and air leakage from regenerators
100kW CGT Program	JARI Toyota Nissan Mitsubishi	To develop and demonstrate 100 kW-class highly efficient ceramic gas turbine for automobiles	Thermal eff.: 40% TIT: 1350°C Exhaust emissions: less than regulatory values	Thermal eff.: 35.6% at 1350°C Accumulated operating hours: 162 hours NOx emissions: 0.15 g/km Max.output: 92.3 kW	To minimize thermal deformation of an engine housing at elevated temperature. To identify the long term reliability of ceramics and ceramic components. To minimize the rotor tip clearance

CHAPTER

13

JAPANESE AUTOMOTIVE CERAMIC GAS TURBINE DEVELOPMENT

Takane Itoh
Tokai University, Yokohama, Kanagawa, Japan

Yusaku Yoshida
Japan Automobile Research Institute, Tsukuba, Ibaraki, Japan

Shoji Sasaki
Toyota Motor Corporation, Toyota, Aichi, Japan

Masafumi Sasaki
Nissan Motor Corporation, Ltd., Kitami, Hokkaido, Japan

Hiroshi Ogita
Mitsubishi Motors Corporation, Tokyo, Japan

PRIVATE PROGRAMS IN AUTOMOBILE MANUFACTURING COMPANIES

Japanese automotive ceramic gas turbine (CGT) development activities were initiated by main automobile manufacturing companies such as Toyota Motor Corp., Nissan Motor Co., Ltd., and Mitsubishi Motor Corp. in the latter half of the 1970s as their own internally funded projects, after 10–15 years experience with automotive "metal" gas turbine engines. The main reason for switching their development activities from metal to ceramic gas turbines was the requirement to raise the turbine inlet temperature (TIT) over 1350°C without any complicated cooling system in order to compete with highly efficient diesel engines.

At Toyota which had some experience with ceramic turbocharger rotors, the ceramic gas turbine project was conducted in cooperation with an in-house ceramic material team. A two-shaft, regenerative 110 kW GT41 engine, shown in Figure 13.1, was selected as the base engine for the development of several ceramic components.

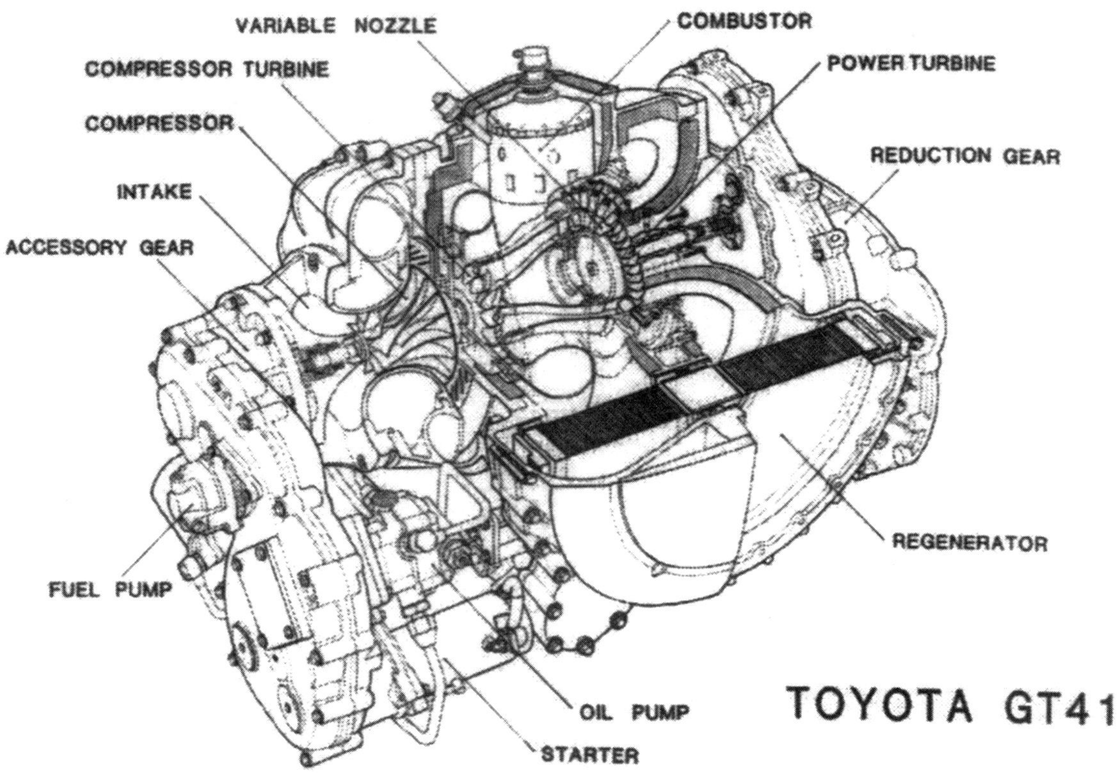

Figure 13.1 GT41 Engine Toyota (Reprinted with permission from Ref. [1])

The TIT was 1350°C and the turbine rotating speed was 68,000 rpm for the gas generator turbine and 53,000 rpm for the power turbine. A scroll duct, a radial gasifier turbine rotor, a transition duct and an axial power turbine rotor were manufactured from silicon nitride, and the combustor liner was made from silicon carbide because of its higher oxidation resistance. The component tests were conducted for each ceramic component under various conditions. Those tests indicated that the ceramics had a high potential as materials for an automotive gas turbine engine and that there were still a large number of technically unexplored areas in ceramic materials. Those ceramic components were installed into the GT41 exhibited at the 27th Tokyo Motor Show in 1987.

Nissan commenced R&D of automotive gas turbine engines in 1963, concentrating on the development of gas turbine powered buses. In 1976 the third generation bus equipped with the 2-shaft, regenerative YTP12 (220 kW) gas turbine engine was the first gas turbine powered vehicle in Japan, approved by the Ministry of Transport for driving on public roads. After much experience from road tests with metal gas turbine powered buses, it was concluded that the fuel economy needed to be substantially improved in order to make gas turbines suitable for practical use. This required a major increase in TIT and the introduction of new materials such as ceramics for high temperature components. Nissan started R&D of ceramic gas turbines in 1978, changing the output range of its base engine from 220 kW for buses to 74kW for passenger cars, after consideration of the difficulty of manufacturing larger ceramic components. All hot section components of the new 2-shaft regenerative gas turbine, such as the combustor, turbine rotors, variable nozzles, turbine shrouds, and other gas path components were made from silicon nitride or silicon carbide. They were tested on dedicated component test benches and in assembly test rigs. The first ceramic gas turbine engine was built in 1984. Since a complex shaped and highly stressed radial turbine rotor was believed to be the most critical

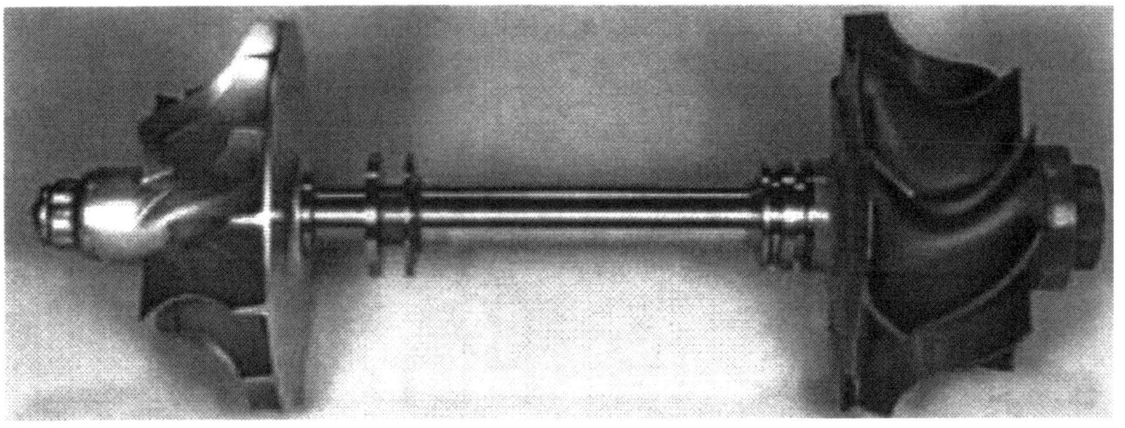

Figure 13.2 The World's First Ceramic Turbocharger Rotor (Nissan) (Adapted with permission from Ref. [1])

component for developing and incorporating ceramic materials, efforts were focused on the development of small radial inflow turbine rotors. The ceramic application technologies developed for the gas turbine engine program were also applied to the development of ceramic turbocharger rotors for passenger cars, and the world's first passenger car equipped with a ceramic turbocharger, shown in Figure 13.2 was released in 1985 [1]. Since then, many ceramic turbocharger rotors have been introduced into the market and their high reliability as a ceramic component has been proven.

In 1969 Mitsubishi Motors in collaboration with Mitsubishi Heavy Industries started the development of a 180 kW 2-shaft regenerative metal gas turbine for truck use, which promised to have excellent noise, vibration and pollutant emissions profiles. The gas generator turbine was of the single-stage radial type and the power turbine was of the single-stage axial type with variable geometry nozzles. Engine and vehicle tests were conducted successfully. However, the thermal efficiency was only 28.6%, and this was the single most critical problem for a small gas turbine engine compared with a comparable diesel engine. Since 1979, the development of ceramic hot section parts such as the combustor, the scroll, segmented turbine nozzles, shown in Figure 13.3, had been given priority and the emphasis was placed on the development of a silicon nitride radial turbine rotor. In 1991, after several improvements in design, fabrication, and nondestructive testing methods, a tip speed of 700m/s at a TIT of 1410°C was achieved in the hot spin tests.

One of the most important conclusions in this R&D program was that a ceramic gas turbine engine was attractive for the future automotive engine although additional efforts would be required, especially in the development of ceramic application technologies. Another conclusion was that a cooperative program carried out by a consortium of concerned automotive manufactures would in all likelihood be more effective than competitive R&D by the individual companies, in overcoming the many challenges that needed to be solved in key technologies.

THE 100 KW AUTOMOTIVE CERAMIC GAS TURBINE PROGRAM

In consideration of the aforementioned circumstances the Japanese Ministry of International Trade and Industry (MITI) undertook a feasibility study for the development of a ceramic gas turbine engine (CGT) in the spring of 1987. Based on the results of this study, the Petroleum Energy Center (PEC) was commissioned to undertake seed development research to identify technological challenges involved in the development of a CGT for automotive use, for a two year program starting in FY (fiscal year) 1988. This preliminary study evolved into the seven-year "Development of a 100 kW Automotive Ceramic Gas Turbine (100 kW CGT)

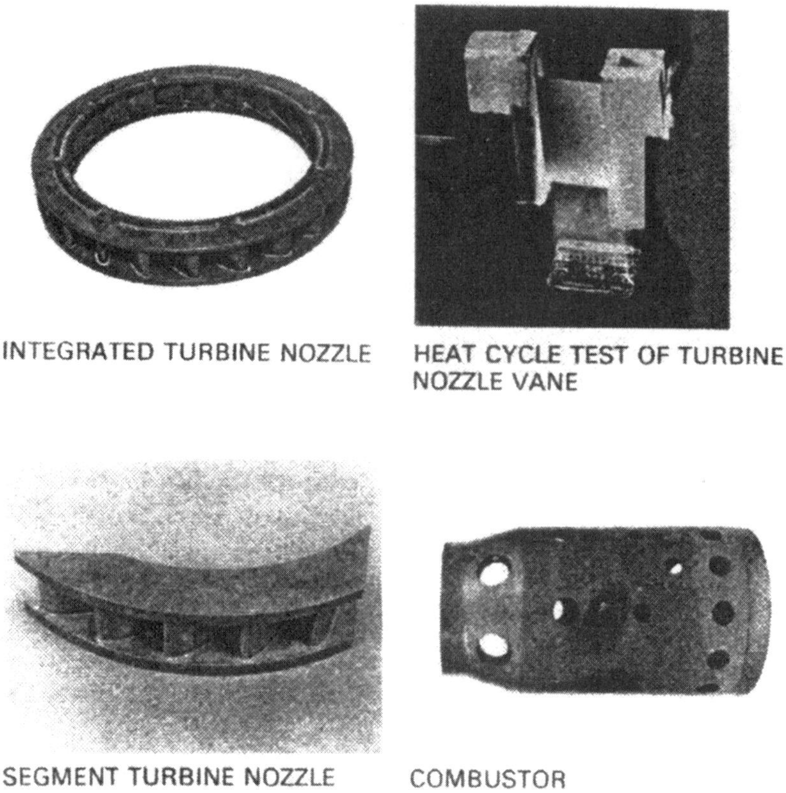

Figure 13.3 Ceramic Components (Mitsubishi) (Adapted with permission from Ref. [1])

Program," initiated in 1990 with the objectives of demonstrating the potential and technical feasibility of the CGT, with engine performance goals as shown below [3, 4]:

- Turbine Inlet Temperature: 1350°C
- Engine Output Power: 100 kW
- Thermal efficiency at the design point: >40%
- Exhaust emissions: to meet the Japanese regulations for gasoline fueled passenger cars

The project was managed by PEC and was a 50/50 cost-shared program financially supported by MITI. The design and development of the CGT engine system was carried out under the leadership of the Japan Automobile Research Institute (JARI) in Tsukuba, in cooperation with the participating companies (Toyota Motor Corp., Toyota Central Research & Development Laboratories, Inc., Nissan Motor Co., Ltd., and Mitsubishi Motor Corp.). Major Japanese ceramic suppliers: Kyocera Corp., NGK Insulators, Ltd. (NGK), NGK Spark Plug, Ltd. (NTK), and Asahi Glass Co., Ltd. (ACG), supplied ceramic components for the engine. Engine component development activities were carried out by the responsible engine companies themselves, and the engine tests were carried out by JARI. In this program ceramic matrix composites (CMC) were also tested and evaluated for four stationary components (a back plate, an orifice liner, an extension liner and an inner scroll) and rotors by petroleum companies (Nippon Oil Co., Ltd., Tonen Corporation, and Idemitsu Kosan Co., Ltd.) [4]. They were evaluated only in an individual stand-alone test rig, and many advantages were confirmed, including thermal shock resistance and so forth [24]. However, they were not scheduled to be assembled as engine components, and the development of the CMCs was not

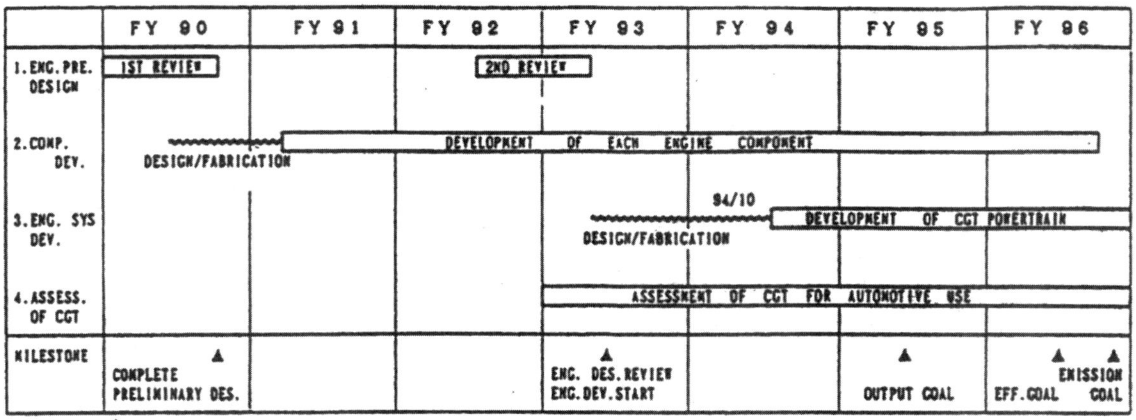

Figure 13.4 Project Schedule (Reprinted from Refs. [3, 4])

the responsibility of the automotive manufacturers. Therefore, the CMCs will not be described in this chapter. The program was concluded in March 1997, conform to the program schedule. The project schedule is shown in Figure 13.4.

The design, fabrication and test results for each ceramic component (stationary components and turbine rotors), sub-assemblies and the whole CGT engine will be described in each of the following sections.

ENGINE CONFIGURATIONS AND SPECIFICATIONS

While both twin-shaft and single-shaft type automotive gas turbines had been developed, and while both types had their respective advantages and disadvantages, a single-shaft type was selected for the base engine design. This choice was based on a preference for a simpler structure with fewer ceramic components, key issues in a program that only had a limited 7-year development period. The regenerative single shaft engine configuration had a centrifugal compressor, a radial inflow turbine, a pre-vaporized pre-mixing lean combustion system and two disks in the rotating regenerator. The detailed structures were investigated with emphasis on the following key items:

- Minimizing air leakages through the clearances between ceramic components, which greatly affects engine performance
- Supporting static ceramic components within the metal housing without excess stresses caused by the deformations from elevated temperatures
- Minimizing the thermal deformation of the regenerator seal platform to minimize air leakages
- Maintaining the concentric alignment between the high-speed rotating parts and the static parts in order to retain the minimum tip clearances

Figure 13.5 shows a cutaway of the engine. The engine specifications are listed in Table 13.1 The TIT of 1350°C was selected by considering the higher thermal efficiency of the engine and expected characteristics of the candidate ceramic materials. The compressor pressure ratio was set at 5:1 by considering its effect on the thermal efficiency under part load condition, and on the maximum stress on the ceramic rotor. The initial maximum speed was set at 110,000 rpm and was revised to 100,000 rpm at the interim review in 1993.

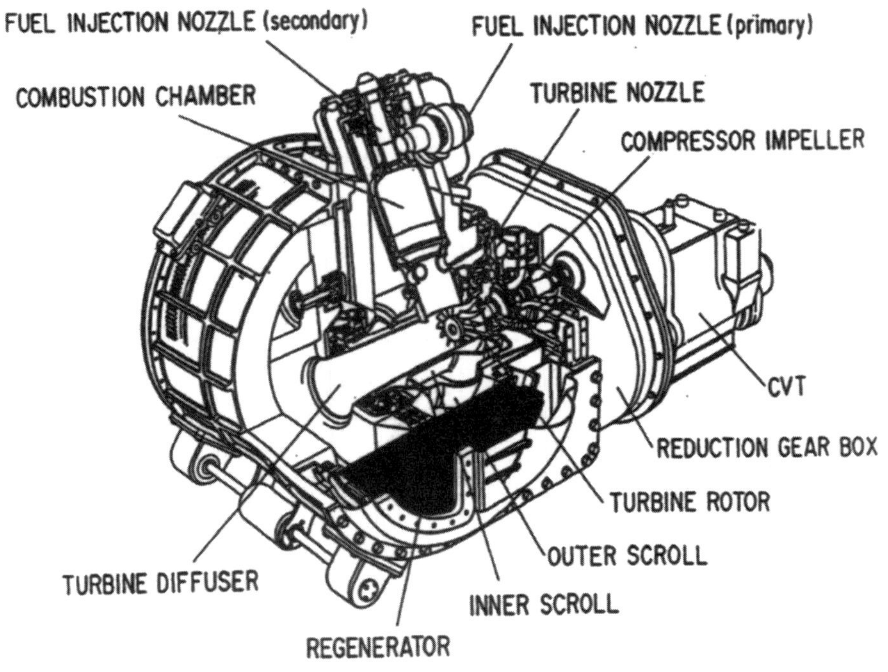

Figure 13.5 Cutaway of the Engine (Reprinted from Ref. [5])

TABLE 13.1 ENGINE AND COMPONENT DESIGN SPECIFICATIONS [3, 4]

Components	Type	Specifications	
Engine	Single shaft +CVT	Output Power	100 kW
		Thermal Efficiency	40%
		Pressure Ratio	5.0
		Turbine Inlet Temperature	1350°C
		Air Flow Rate	0.445 kg/s
		Air Leakage	2%(not including HE)
		Pressure Loss	6.7%(not including HE)
		Mechanical Loss	9.1%
Compressor	Centrifugal	VIGV	Radial blades
		Outlet Flow Path	2-Way scroll type
		Adiabatic efficiency	81.0%
Turbine	Radial inflow	Inlet Flow Path	Scroll type
		Outlet Flow Path	Diffuser type
		Adiabatic Efficiency	87.5%
Combustor	Single can	Combustion	PPL+diffusion flame
		Combustion Efficiency	99.5%
		Pressure Loss	3.0%
Heat Exchanger	Rotary regenerator	Number of Cores	2
		Mounting and Driving	Outer Drive, Outer Support
		Effectiveness	93.0%
		Air Leakage	5.0%
		Pressure Loss	6.0% + 0.3%

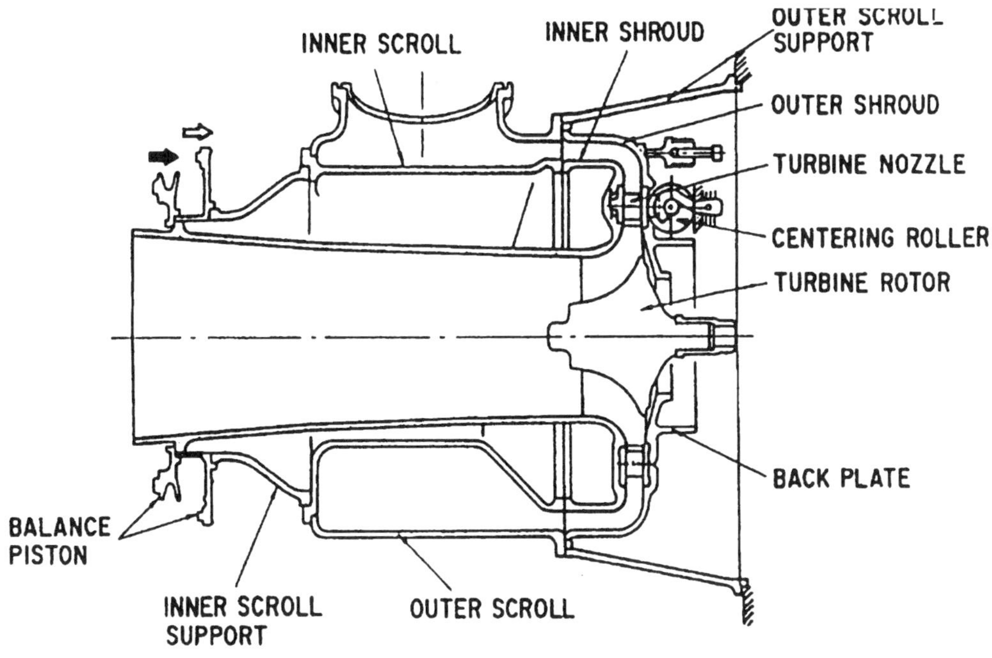

Figure 13.6 Assembly of the Major Ceramic Components (Adapted from Ref. [4])

DESIGN APPROACHES FOR THE CERAMIC COMPONENTS

Engine Structure and Elastically Supporting System of Ceramic Assemblies

Figure 13.6 shows an assembly of the static ceramic components and the turbine rotor of the engine without a combustor assembly. As shown in this figure, each stationary ceramic component had flat surfaces on both ends and they were stacked up successively and elastically supported by thrust forces of the seal and support structure. Air leakages from the joints of the ceramic components were minimized by a face-to-face contact seal concept. The seal and support structure (elastic support system), consisting of bellows and springs, shown in Figure 13.7, elastically supported the stacked ceramic assembly to the metal engine housing and absorbed the differences in thermal expansions between the ceramic assembly and the metal housing.

Figure 13.8 shows the combustor assembly in which the stack-up technique was also applied [19]. The segments were pulled up to the upper side and held together by the pull-up holder with elastic loading by springs. The combustor dilution liner was placed tangential to the turbine outer scroll and was independently supported to allow relative movement between the combustor assembly and the other stationary ceramic component assembly [8].

General Design Concept for the Ceramic Components

Based on prior experience, the following guidelines were thought to be very important for designing more reliable ceramic components:

- Avoid stress concentrations
- The shape of the ceramic component should be as simple as possible for ease in manufacturing, in defect inspection, and also for stress analysis
- Account for volume effects of ceramics (the larger, the weaker)

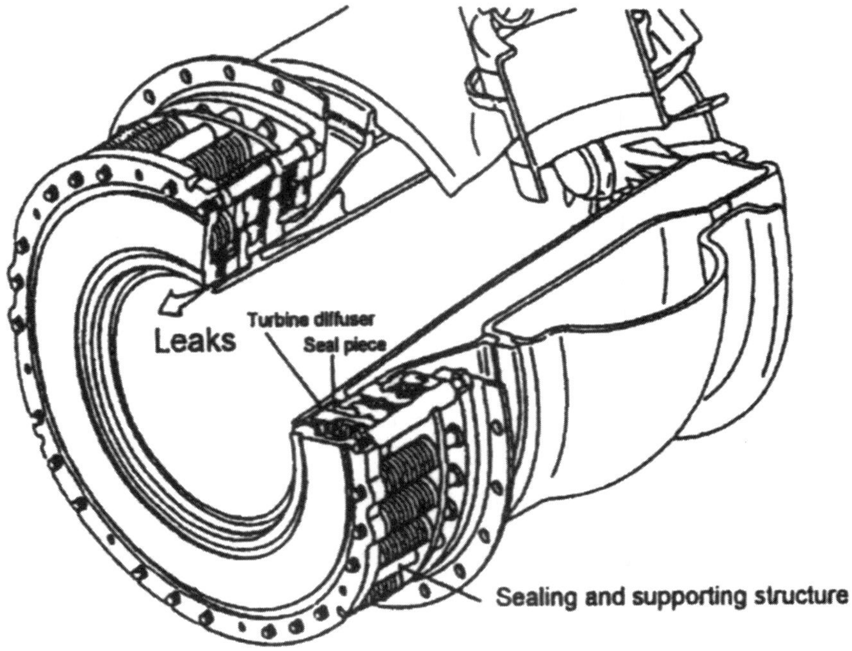

Figure 13.7 Seal and Support Structure (Adapted with permission from Ref. [24])

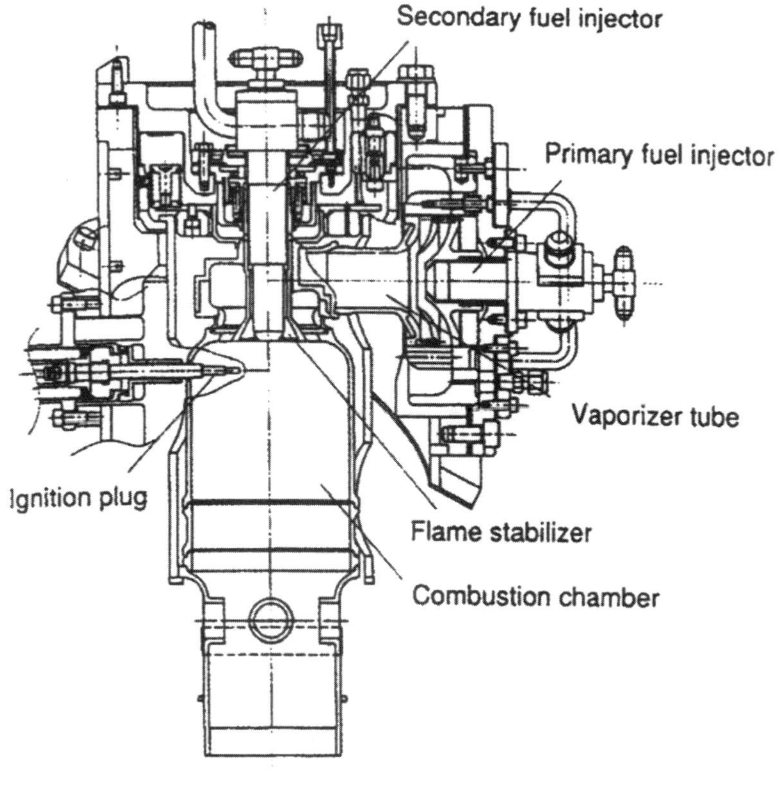

Figure 13.8 Combustor Assembly (Adapted from Ref. [19])

Figure 13.9 Ceramic Stationary Components (Reprinted with permission from Dr. Itoh)

Figure 13.9 shows the fabricated ceramic stationary components. Before fabricating each component, its strength and reliability were checked by stress analysis. The reliability goal, material properties, boundary conditions, analysis code, etc. were all included in the investigation of the strength and reliability of the ceramic components.

As for the reliability goal, the time-dependent failure probability of each component was established at 10^{-5} or less after 100,000 km of a vehicle driving cycle over a period of 10 years. Thermal boundary conditions for each component were assumed by using the temperature, pressure and gas flow rates for each stage of the engine obtained from engine cycle analysis. Temperature and stress distributions for each component were calculated by using finite element modeling (FEM) codes under the rated operating and cold starting conditions [7]. Figure 13.10 shows an example of stress analysis for the turbine outer scroll. The maximum stress of 209 MPa occurred after about 100 seconds after the cold start at a location just down from the inlet where the combustor dilution liner was inserted.

Table 13.2 shows the summary of calculated maximum stresses in the main ceramic components in the first design phase. In most cases transient thermal stresses were dominant for stationary ceramic components.

The combined mechanical and thermal stresses were used in the reliability and life prediction analysis based on Weibull's theory, and the fast and slow crack growth (SCG) mechanisms. The analysis code used for each component was slightly different because the automobile manufactures each had their own analysis code that was very similar to the NASA CARES/Life Code [9]. Reliability analysis showed that all ceramic components met the above mentioned reliability goal.

MATERIAL SELECTION AND COMPONENT FABRICATION

The ceramic material for each component was selected by considering material properties based on the flexural strength and Weibull modulus of test specimens at the assumed operating temperature and fabricating processes. The summaries of the materials, suppliers and fabricating processes for all ceramic components are shown in Table 13.3 [7].

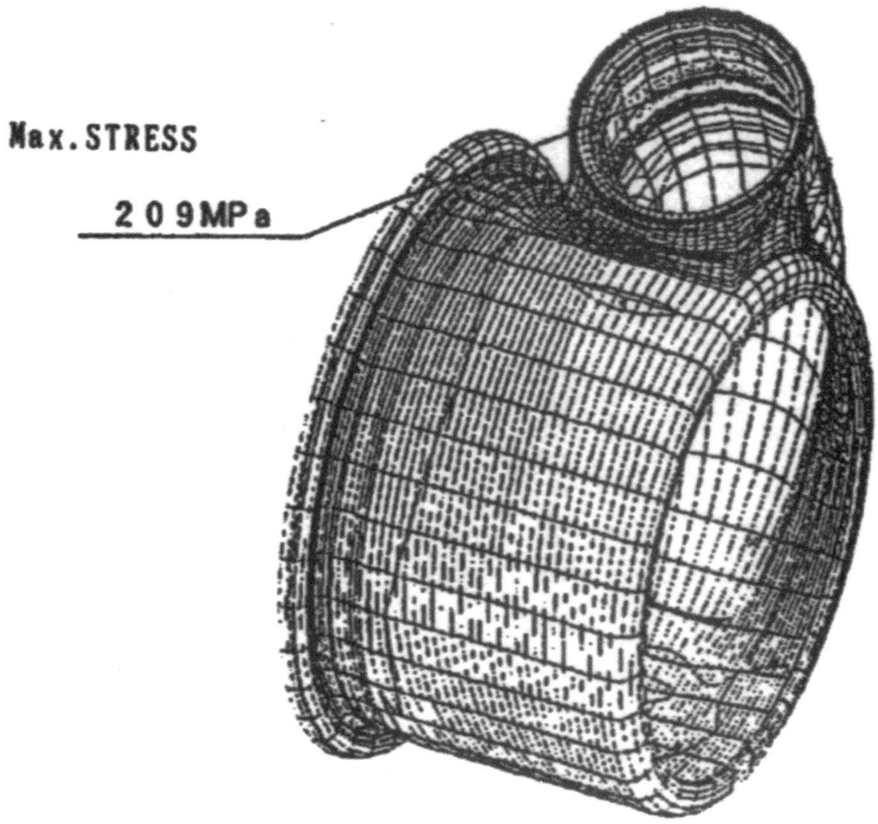

Figure 13.10 Maximum Stress in the Turbine Outer Scroll (Reprinted from Ref. [7])

TABLE 13.2 SUMMARY OF THE MAXIMUM STRESSES IN THE MAIN CERAMIC COMPONENTS

Ceramic Components	Maximum Stress (MPa)	Conditions
Turbine Rotor	293 (Blade Root Fillet)	At rated power
Turbine Nozzle (Stator)	140 (Trailing Edge)	10 s after cold starting
Inner Shroud	414 (Curved Portion)	19 s after cold starting
Inner Scroll	106 (Flange Peripheral)	80 s after cold starting
Outer Scroll	209 (Near the Inlet)	100 s after cold starting
Back Plate	145 (Rounded Corner)	50 s after cold starting
Inner Scroll Support	85 (Inner Surface)	100 s after cold starting
Outer Scroll Support	0.6 (Rear Flange)	At rated power
Roller	324 (Contract Stress)	At just cold starting
Orifice Liner	93 (Contact Portion)	30 s after cold starting
Dilution Liner	47 (Near around Holes)	30 s after cold starting
Combustion Liner	28 (Upper Flange)	70 s after cold starting

TABLE 13.3 LIST OF MATERIALS FOR CERAMIC COMPONENTS

Component	Supplier	Material	Fabrication Process
Combustor			
PP-Swirler	NTK	EC-152 Si_3N_4	Injection Molded
PP-Scroll	NTK	EC-152 Si_3N_4	Injection Molded
VG-Nozzle	NTK	EC-152 Si_3N_4	Injection Molded
Combustor Liner	Asahi Glass	C-800A SiC-Al2O3	Cold Isopressed
Orifice Liner	Asahi Glass	C-600R SiC-Al2O3	Cold Isopressed
Dilution Liner	Kyocera	SN-252 Si_3N_4	Slip Cast
Holder	NGK	SN-88 Si_3N_4	Cold Isopressed
Outer-Scroll	Kyocera	SN-252 Si_3N_4	Slip Cast
Inner-Scroll	Kyocera	SN-252 Si_3N_4	Slip Cast
Inner-Shroud	NGK	SN-88 Si_3N_4	Cold Isopressed
Outer-Shroud	NGK	SN-88 Si_3N_4	Cold Isopressed
Turbine Stator	NGK	SN-88 Si_3N_4	Injection Molded
Turbine Rotor	Kyocera	SN-252 Si_3N_4	Slip Cast
	Kyocera	SN-253 Si_3N_4	Slip Cast
	Kyocera	SN-281 Si_3N_4	Slip Cast
	NGK	SN-90 Si_3N_4	Injection Molded + CIP
	NGK	SN-91 Si_3N_4	Injection Molded + CIP
	NGK	SN-88M Si_3N_4	Injection Molded + CIP
	NTK	EC-152 Si_3N_4	Injection Molded + CIP
	NTK	EC-155 Si_3N_4	Injection Molded + CIP
	NTK	EC-157 Si_3N_4	Injection Molded + CIP
Turbine Backplate	NTK	EC-152 Si_3N_4	Cold Isopressed
Outer-Scroll Support	Kyocera	SN-252 Si_3N_4	Slip Cast
Inner-Scroll Support	NTK	EC-152 Si_3N_4	Cold Isopressed
Seal Piece	Hitachi K.C.	HXL	Cold Isopressed
	NTK	EC-152 Si_3N_4	Cold Isopressed
Regenerator	Kyocera	SN-220 Si_3N_4	Slip Cast
Seal Platform	Hitachi K.C.	HXL	Cold Isopressed
Regenerator Core	NGK	MAS	Extrusion
	Corning Inc.	LAS	Wrapped

TESTING OF CERAMIC COMPONENTS

Stationary Components

Each stationary ceramic component was evaluated according to the several steps shown in Figure 13.11. First, specimens cut from one of the fabricated components were subjected to 4-point flexural strength tests, and the strength was compared with that of Japan Industrial Standards (JIS) test specimens of the same materials to check the problems involved in its fabricating process. Test results showed that the strength of specimens cut from fabricated components was about the same as that of the JIS specimens.

Secondly, each component was subjected to a mechanical proof test. Two types of mechanical tests were available. One was an internal pressure test performed by applying hydraulic pressure to verify the overall strength and to reject components with large size defects generated during the fabricating processes. The other was a concentrated mechanical load test to evaluate whether a critical region in a component had sufficient localized strength, by loading this critical area to a mechanical stress of about 120% of the predicted maximum stress experienced during engine operation. Finally, all ceramic stationary components which had survived the mechanical proof tests were assembled as in an engine, and subjected to a stationary components assembly test with the ceramic combustor assembly. In the early stage

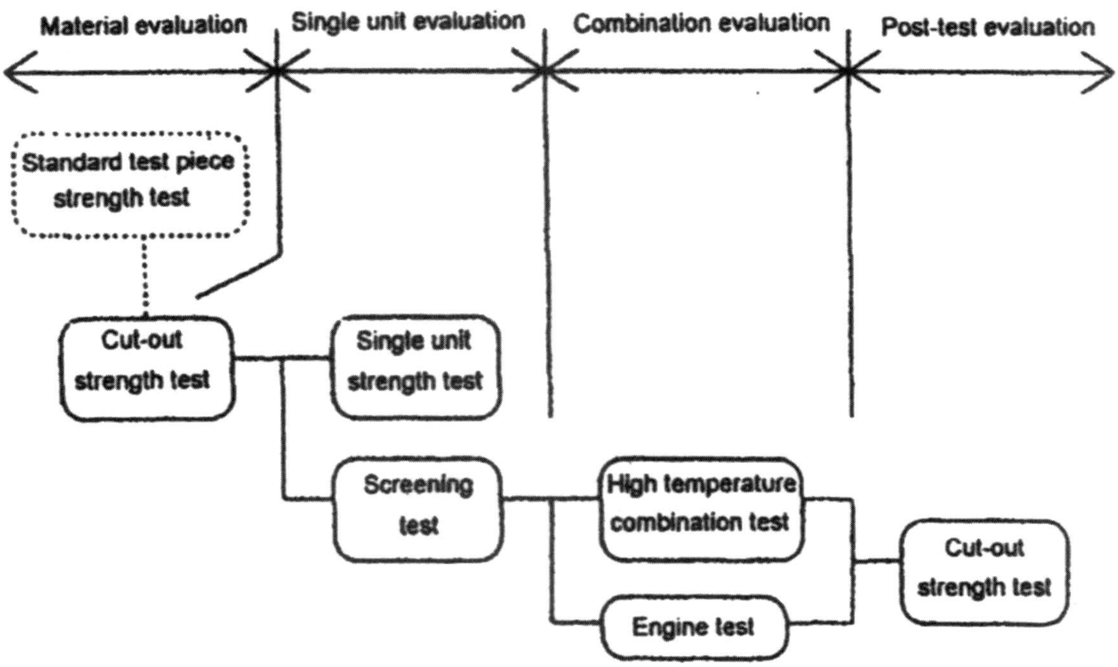

Figure 13.11 Evaluation Flow Chart for Stationary Ceramic Components (Reprinted with permission from Ref. [24])

of the testing, the components were subjected to almost the same conditions (TIT of 1,200°C) as in the engine during a 100 hour test which included 200 cold starts [7, 12]. In those assembly tests, various types of damage (oxidation, cracking and chipping) were experienced in some components [12]. Most of them were caused by underestimating the thermal stresses and stress concentrations, and by strength degradation from oxidation. Some of the damage observed is shown in Figure 13.12. Cracks in the outer-scroll are shown in (a) and cracks in the trailing edge of the turbine nozzle in (b).

After improvements in the design and processing of some components based on an analysis of the test data, all ceramic stationary components were installed into an actual engine housing, and a 100-hour high temperature endurance test at a TIT of 1350°C with 200 cold start cycles was carried out. In that test, an improved seal and support structure was also installed and its function was evaluated. The test results were satisfactory [24].

Combustor

A pre-vaporized-premixing lean (PPL) combustion system had been developed for the CGT engine to meet the more stringent target for NOx emissions. Therefore, a rather complicated combustor configuration, as shown in Figure 13.8, consisting of several ceramic parts, had to be designed. Durability tests were conducted for the ceramic components on combustor test benches, including a 100 hour durability test at a combustor outlet temperature of 1350°C. No cracks or other evidence of damage were observed on any ceramic and metallic parts of the combustor. After the endurance test, the ceramic liner was cut into several rings and the rupture strength of those rings was measured. No degradation in strength was observed [24].

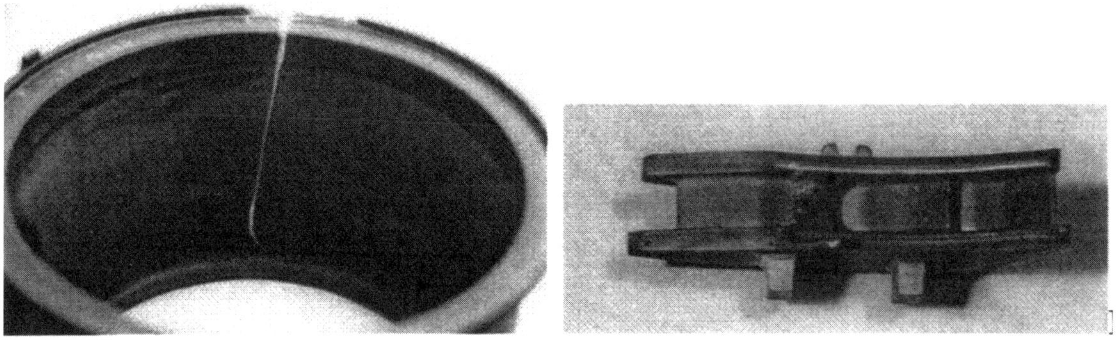

Figure 13.12 Typical Damage Experienced in High Temperature Assembly Tests (Adapted with permission from Ref. [24])

Turbine Rotors

The turbine rotor was one of the most critical ceramic components and also it was a very critical aerodynamic component. Therefore, it was very important to maintain compatibility between aerodynamic performance and high reliability of the ceramic component. The first turbine rotor design had an inlet outer diameter of 127mm and an exit diameter of 92mm. The number of blades was 14 and the rated rotating speed was 110,000 rpm. Three types of silicon nitride materials supplied by Kyocera, NGK and NTK were selected, because the turbine rotor was the most critical component and it was too risky to select only one rotor material at that time. Figure 13.13 shows a flowchart for the rotor development process.

Several types of tests were conducted, such as flexure tests on specimens sectioned from a fabricated rotor, cold spin tests, hot spin tests (at 1200°C in the earlier stage of the project, and later at 1350°C) [7, 11], and an endurance test with turbine stators at an elevated temperature of 1200°C. In the endurance tests some of the rotors broke at an unexpected low speed. It was concluded that blades were forced to vibrate at about their resonant frequency by the aerodynamic exiting force from wakes generated by the turbine nozzle vanes [12]. Therefore, strain gauges were attached to the turbine blades and the mechanical stress of the blades at resonance was measured in an aerodynamic performance test rig, and internal damping characteristics of ceramic materials were also measured [24]. The results from those tests were analyzed and the results were used to improve the rotor geometry, the number of nozzle vanes, and the manufacturing processes. Figure 13.14 shows the temperature and stress distributions of the 6th generation rotor at the rated conditions. The maximum stress was 290 MPa, and the maximum material temperature was 1090°C at the blade inlet tip, and the temperature of the higher stress region was 900–1000°C [24]. Figure 13.15 shows the final rotor design.

The high temperature endurance test, as well as cold and hot spin tests, also served as a proof test to select reliable rotors for the final evaluation in engine tests. Figure 13.16 shows the endurance test rig. One of the rotor designs was subjected to the endurance test and completed 100 hours of continuous operation at a TIT of 1350°C without failure. Two of the 6h generation rotors passed a hot spin proof test, in which the rotor was continuously run at 108,000 rpm and a TIT of 1,350°C for 6 minutes [24]. These conditions were equivalent to the endurance test goals of 100,000 rpm for 100 hours.

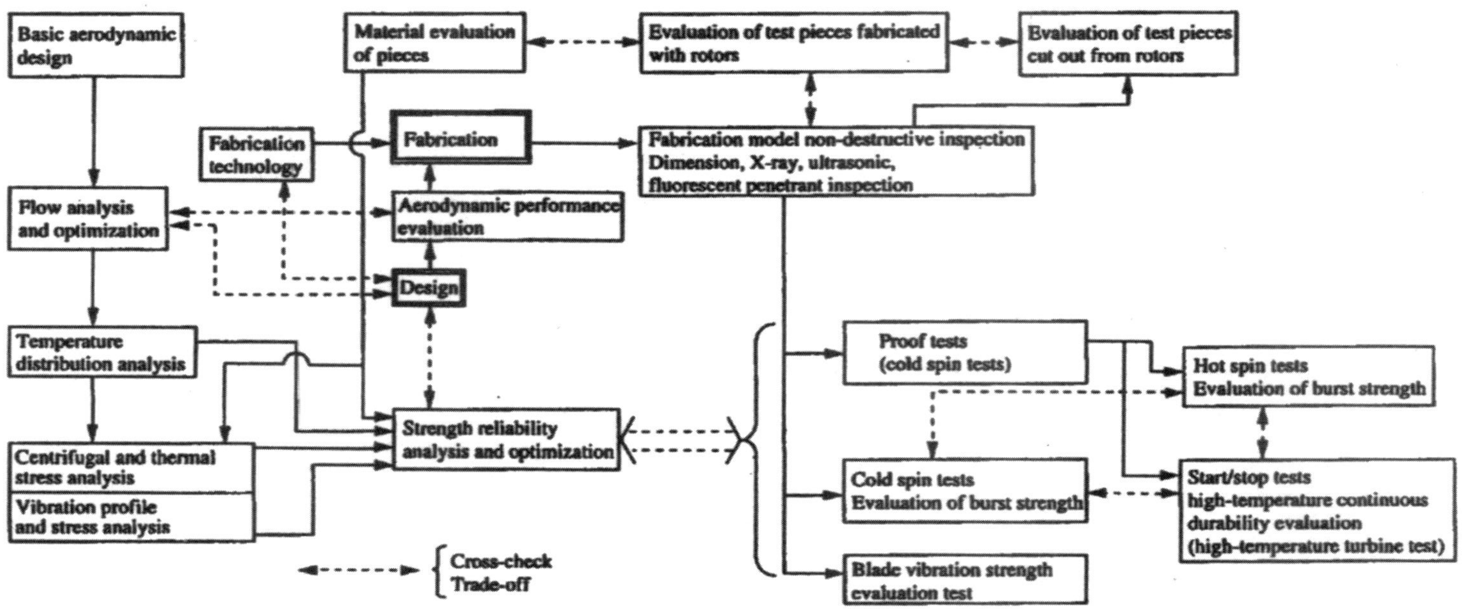

Figure 13.13 Flow Chart for the Rotor Development Process (Reprinted with permission from Ref. [24])

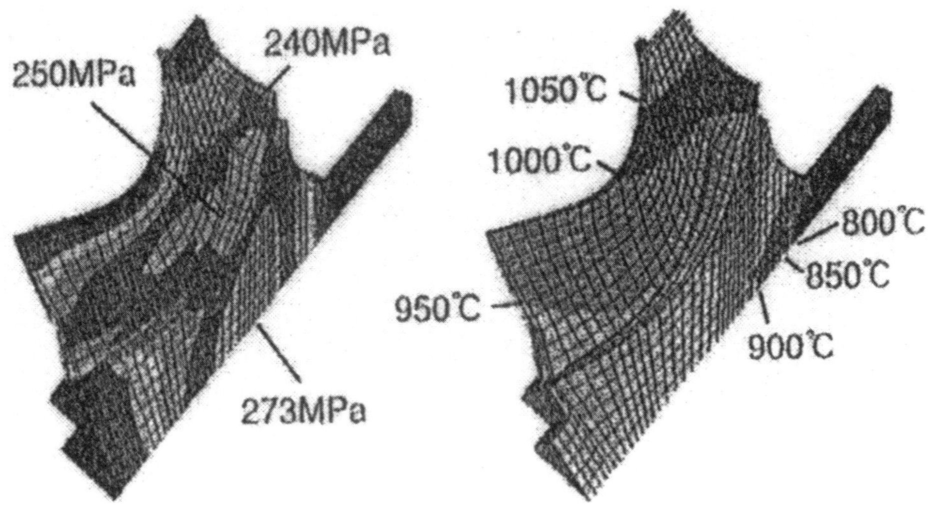

Figure 13.14 The Temperature and Centrifugal Stress Distribution of the 6th Rotor (Adapted with permission from Ref. [24])

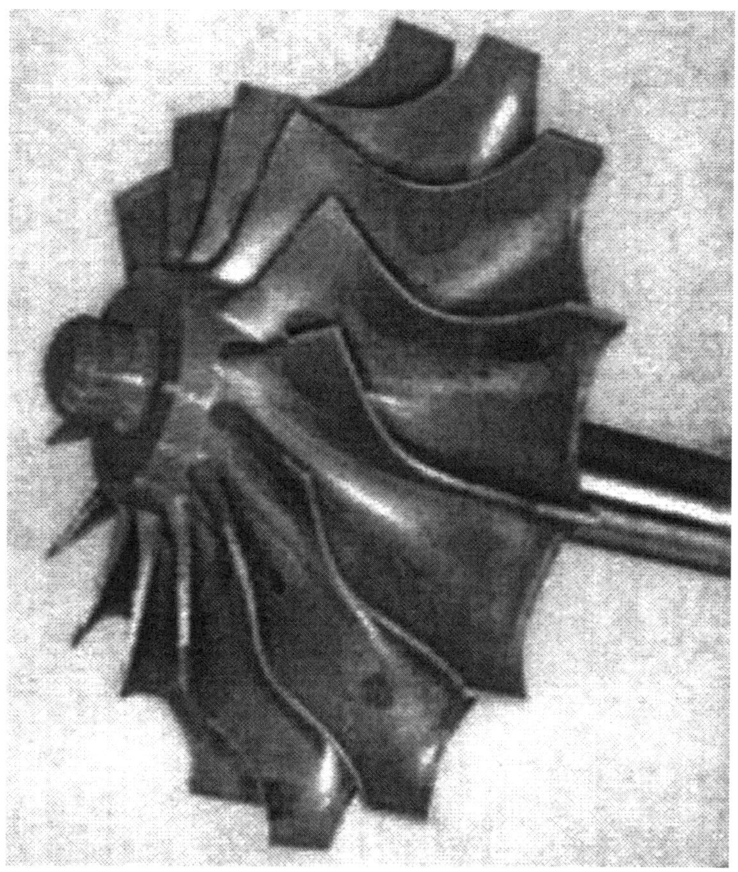

Figure 13.15 The Final Generation Rotor (Adapted with permission from Ref. [24])

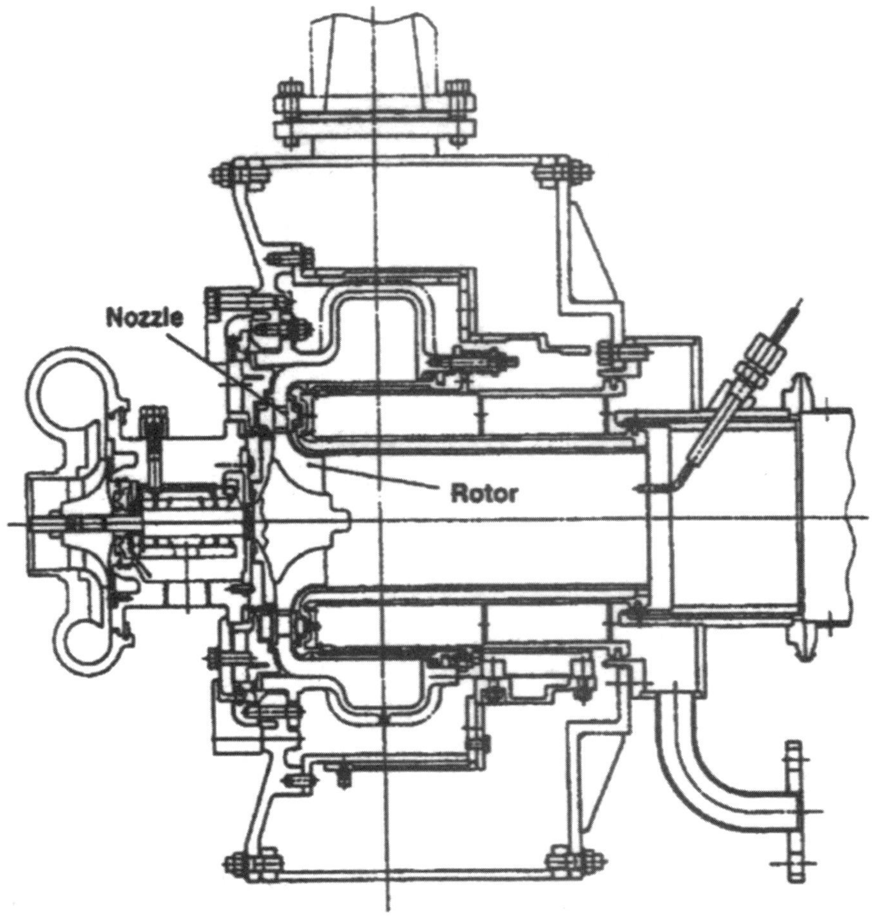

Figure 13.16 Rotor Endurance Test Rig (Reprinted with permission from Ref. [24])

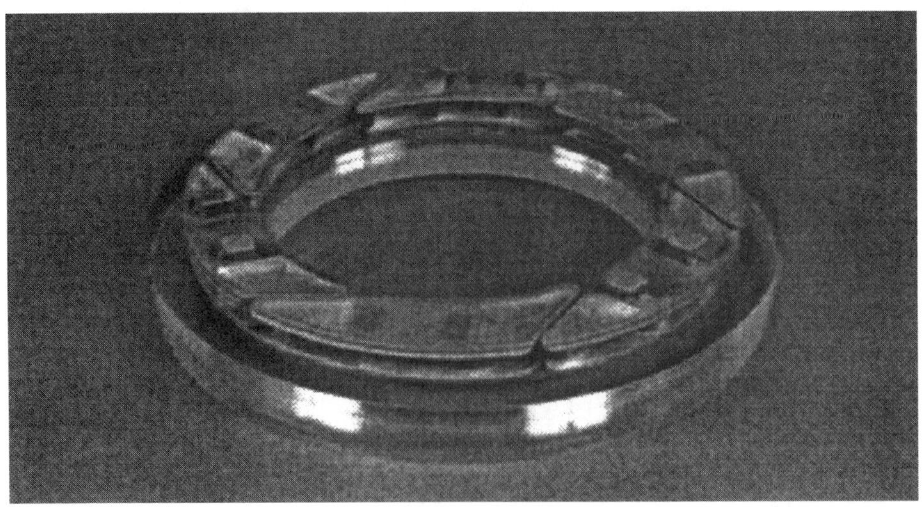

Figure 13.17 Turbine Nozzle Assembly (Reprinted with permission from Ref. [24])

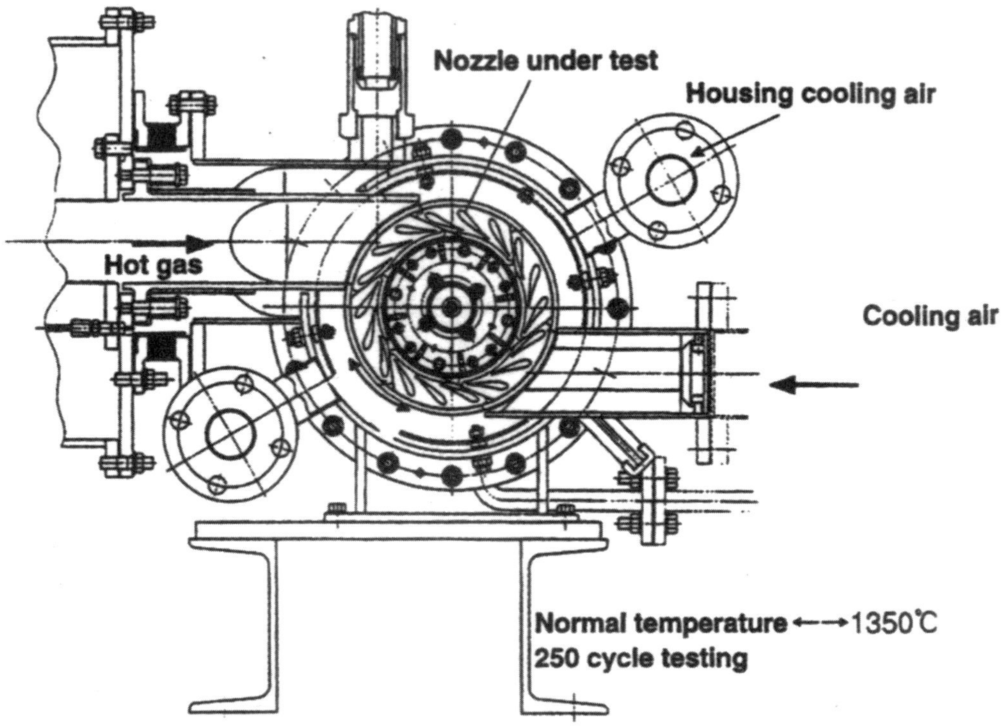

Figure 13.18 Thermal Shock Rig for Turbine Nozzle (Reprinted with permission from Ref. [24])

Turbine Nozzle (Stator Vanes)

Figure 13.17 shows the turbine nozzle, which was divided into six segments, because of ease of fabrication. The FEM analysis showed that high thermal stresses in excess of 140 MPa were generated at the trailing edge during a cold start and shut-down of the engine. Therefore, a turbine nozzle thermal shock endurance test was performed. Figure 13.18 shows the test rig where the nozzle segments were exposed alternatively to hot gas of 1350°C and to cooling air of 35°C to simulate the transient thermal stresses encountered in the engine. The temperatures at the leading and trailing edge of a stator measured by thermocouples showed that the temperature gradients in the vane were 1.2 to 1.4 times larger than the calculated values. In the initial design, the total number of stator vanes was 21, but a nozzle configuration with 18 vanes was adopted in the final design to avoid resonant vibration of rotor blades at their third critical mode. After some correction to the shape of the trailing edge and corner radius at the root of the vane, all tested nozzles survived in a 250-cycle endurance test.

The temperatures at the leading and trailing edge of a stator measured by thermcouples showed that the temperature gradients in the vane were 1.2 to 1.4 times larger than the calculated values. In the initial design, the total number of stator vanes was 21, but a nozzle configuration with 18 vanes was adopted in the final design to avoid resonant vibration of rotor blades at their third critical mode. After some correction to the shape of the trailing edge and corner radius at the root of the vane, all tested nozzles survived in a 250-cycle endurance test.

Engine Testing

The development processes were carried out in a stepwise manner as shown in Figure 13.19. The final stage of the process was engine testing. All engine components including subsystems such as the regenerator,

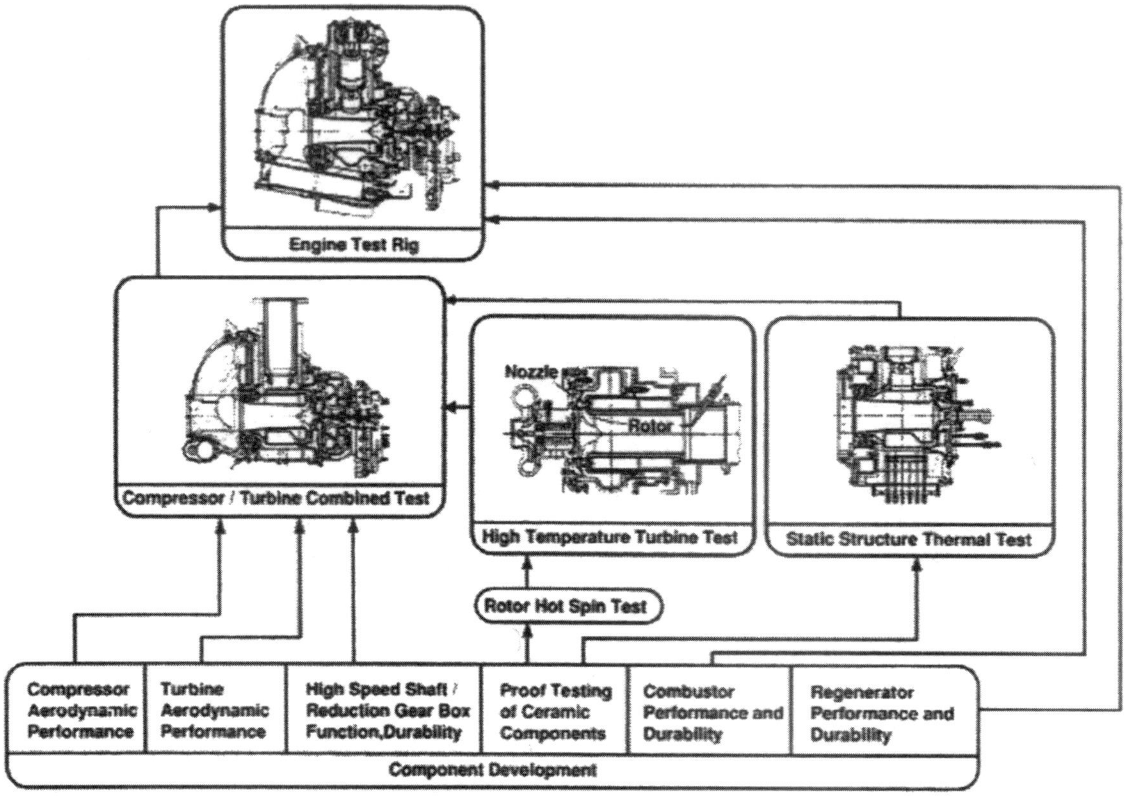

Figure 13.19 Engine Development Flows (Reprinted with permission from Ref. [24])

the compressor, the combustor and all the ceramic components were assembled into an engine, and the engine was installed on a test bench. In that engine test, the following items were evaluated: output power, thermal efficiency, engine exhaust gas emissions, performance and functions of each main component and the engine housing, etc.. In the early stage, the engine tests were conducted at a TIT of 1200°C, and some problems were experienced such as excessive heat losses and air leakages, cracks and chipping in thermally insulating liners, and deformation of the engine housing at elevated temperature. After making adequate improvements, the engine test at the target design TIT of 1350°C was initiated with modified components in February of 1997. The data obtained at the end of the scheduled period of the 100 kW CGT program showed that the maximum output power was 92.3 kW and the highest thermal efficiency was 35.6% at a TIT of 1,350°C, and the accumulated engine run time was 162.3 hours. The measured thermal efficiency was somewhat lower than the efficiency of 39.4% predicted from the test data for each installed component. Data analysis showed that the deficiency was caused by unexpected heat loss and air leakage coming from the engine housing structure, and it was concluded that it was not so serious and that corrective action would be possible by redesigning the engine housing.

SUMMARY AND CONCLUSIONS

Key Lessons Learned

- It was demonstrated that the ceramic gas turbine is a potentially highly efficient engine with low exhaust emissions.

- The internal damping coefficient of silicon nitride materials was too small, and a ceramic turbine rotor and nozzle assembly should be carefully designed to avoid resonant vibrations.
- The structural design of a metal engine housing was also critical for a high performance ceramic gas turbine.
- Current silicon nitride materials are not yet sufficiently reliable for parts used at temperatures above 1250°C and stresses in excess of 250 MPa.
- Underestimation of thermal stresses in stationary ceramic components on occasion was the primary cause of their failure.

Deficiencies Identified

- Long term durability of ceramic components in an engine environment.
- The reliability of ceramic rotors at 1350°C.
- Oxidation resistance of silicon nitride materials at 1350°C.

Additional Development Needed

- To minimize thermal deformation of an engine housing at elevated temperature.
- To develop adequate insulating materials to minimize heat losses of an engine by thermal radiation.
- To identify the long life reliability of ceramic components in the actual engine environment.
- To conquer FOD (foreign object damage) problems.
- To minimize the operating tip clearance of a ceramic rotor.
- To develop low cost fabricating processes and proof test methods for ceramic components.

Plateaus Reached

All ceramic stationary components of the engine had survived in the 100 hour endurance test at the design TIT of 1,350°C with 200 cold starting cycles. The final engine performance is shown below:

- Output power: 92.3 kW at a TIT of 1350°C, and 100,000 rpm
- Thermal efficiency: 35.6% at a TIT of 1350°C, and 90,000 rpm
- Exhaust emissions: NOx: – 0.15 g/km, CO – 0.14g/km, UHC – less than background level
- Accumulated operating time: 162.3 hours

REFERENCES

[1] Special Issue R&D of Automotive Ceramic Gas Turbines in Japan, New Technology Japan, Japan External Trade Organization, 1988, (ISSN-0385-6542).

[2] Sasaki, M., Kumakura, H., and Suzuki, D., "Low NOx Combustor for Automotive Ceramic Gas Turbine-Conceptual Design," ASME paper 91-GT-369, presented at the International Gas Turbine and Aeroengine Congress and Exposition, Orlando, FL, USA, June 3–6, 1991.

[3] Itoh, T. and Kimura, H., "Status of the Automotive Ceramic Gas Turbine Development Program," ASME paper 92-GT-2, presented at the International Gas Turbine and Aeroengine Congress and Exposition, Cologne, Germany, June 1–4, 1992.

[4] Itoh, T. and Kimura, H., "Status of the Automotive Ceramic Gas Turbine Development Program," Trans. of the ASME, J. Eng. for Gas Turbines and Power, 115[1], 42–50, 1993.

[5] Nakazawa, N., Inaba, S., Izumi, T., Munekiyo, M., Sasaki, S. and Fujimori, K., "The Turbine Components Development for the 100 kW Automotive Ceramic Gas Turbine," ASME paper 93-GT-64, presented at the International Gas Turbine and Aeroengine Congress and Exposition, Cincinnati, Ohio, USA, May 24–27, 1993.

[6] Sasaki, M., Kumakura, H., and Suzuki, D., "Development of a Low Emission Combustor for a 100 kW Automotive Ceramic Gas Turbine (I)," ASME paper 93-GT-63, presented at the International Gas Turbine and Aeroengine Congress and Exposition, Cincinnati, Ohio, USA, May 24–27, 1993.

[7] Itoh, T. and Kimura, H., "Status of the Automotive Ceramic Gas Turbine Development Program-Year 3 Progress," ASME paper 94-GT-11, presented at the International Gas Turbine and Aeroengine Congress and Exposition, The Hague, The Netherlands, June 13–16, 1994.

[8] Kumakura, H., Sasaki, M., Suzuki, D. and Ichikawa, H., "Development of a Low Emission Combustor for a 100 kW Automotive Ceramic Gas Turbine (II)," ASME paper 94-GT-33, presented at the International Gas Turbine and Aeroengine Congress and Exposition, The Hague, The Netherlands, June 13–16, 1994.

[9] Nemeth, N, Powers, L., Janosik, L. and Gekenyesi, J., 1994, "Durability Evaluation of Ceramic Components Using CARES/Life," ASME paper 94-GT-362, presented at the International Gas Turbine and Aeroengine Congress and Exposition, The Hague, The Netherlands, June 13–16, 1994.

[10] Yoshida, M., Tsuruzone, S., Ono, T. and Gejima, H., "Development of Silicon Nitride Rotors for Gas Turbine Applications," ASME paper 94-GT-332, presented at the International Gas Turbine and Aeroengine Congress and Exposition, The Hague, The Netherlands, June 13–16, 1994.

[11] Watanabe, M. and Ogita, H., "Evaluation of Ceramic Rotor Strength by Cold and Hot Spin Tests," ASME paper 94-GT-460, presented at the International Gas Turbine and Aeroengine Congress and Exposition, The Hague, The Netherlands, June 13–16, 1994.

[12] Nishiyama, T., Iwai, M., Nakazawa, N., Sasaki, M. and Katagiri, H., "Status of the Automotive Ceramic Gas Turbine Development Program-Year Four Progress," ASME paper 95-GT-447, presented at the International Gas Turbine and Aeroengine Congress and Exposition, Houston, Texas, USA, June 5–8, 1995.

[13] Kato, J., Otsuka, M. and Sugiyama, K, "Performance Evaluation of 100 kW Automotive Ceramic Gas Turbine at TIT 1200°C Operation in a Rotating Sub-Assembly Test," ASME paper 95-GT-448, presented at the International Gas Turbine and Aeroengine Congress and Exposition, Houston, Texas, USA, June 5–8, 1995.

[14] Nishiyama, T., Iwai, M., Nakazawa, N., Sasaki, M. and Katagiri, H., "Status of the Automotive Ceramic Gas Turbine Development Program-Year Five Progress-," ASME paper 96-GT-36, presented at the international Gas Turbine and Aeroengine Congress and Exposition, Birmingham, England, U.K., June 10–13, 1996.

[15] Nakazawa, N., Ogita, H., Takahashi, M., Yoshizawa, T., and Mori, Y., "Radial Turbine Development for 100 kW Automotive Ceramic Gas Turbine," ASME paper 96-GT-366, presented at the International Gas Turbine and Aeroengine Congress and Exposition, Birmingham, England, U.K., June 10–13, 1996.

[16] Sasaki, M., Kumakura, H., Suzuki, D., Ichikawa, H., Ohkubo, Y., and Yoshida, Y., "Development of a Low Emission Combustor for a 100 kW Automotive Ceramic Gas Turbine (III)," ASME paper 96-GT-119, presented at the International Gas Turbine and Aeroengine Congress and Exposition, Birmingham, England, U.K., June 10–13, 1996.

[17] Nakazawa, N., Ogita, H., Takahashi, M. and Kawaguchi, Y., 1997, "Development in the Full Assembly Test Rig of the 100 kW Automotive Ceramic Gas Turbine," ASME paper 97-GT-210, presented at the International Gas Turbine and Aeroengine Congress and Exposition, Orlando, FL., USA, June 2–5, 1997.

[18] Nakazawa, N., Sasaki, M., Nishiyama, T., Iwai, M. and Katagiri, H., "Status of the Automotive Ceramic Gas Turbine Development Program-Seven Years' Progress," ASME paper 97-GT-383, presented at the international Gas Turbine and Aeroengine Congress and Exposition, Orlando, FL., USA, June 2–5, 1997.

[19] Sasaki, M., Kumakura, H., Ichikawa, I., Ohkubo, Y. and Yoshida, Y., "Development of a Low Emission Combustor for a 100 kW Automotive Ceramic Gas Turbine (IV)," ASME paper 97-GT-462, presented at the International Gas Turbine and Aeroengine Congress and Exposition, Orlando, FL., USA, June 2–5, 1997.

[20] Nishiyama, T., Iwai, M., Nakazawa, N., Sasaki, M. and Katagiri, H., "100 kW Ceramic Gas Turbine for Automobiles—Status of Engine Development," J. of the Gas Turbine Society of Japan, Vol. 22, No. 87, p. 38–57, December, 1994 (in Japanese).

[21] Nakazawa, N., Ogita, H., Takahashi, M., Yoshizawa, T. and Mori, S., "Development of the Turbine Rotor for the 100 kW Automotive Ceramic Gas Turbine," J. of the Gas Turbine Society of Japan, Vol. 24, No. 93, pp. 84–91, June, 1996 (in Japanese).

[22] Nakazawa, N., Ogita, H., Takahashi, M. and Kawaguchi, Y., "Development of the Full Assembly Test Rig of the 100 kW Automotive Ceramic Gas Turbine," J. of the Gas Turbine Society of Japan, Vol. 24, No. 93, pp. 60–66, March, 1997 (in Japanese).

[23] Special Issue "100 kW Automotive Ceramic Gas Turbine Development Project," J. of the Gas Turbine Society of Japan, Vol. 25, No. 98, pp. 3–70, September, 1997 (in Japanese).

[24] "100 kW Automotive Ceramic Gas Turbine," the Petroleum Energy Center, March, 1998. (This reference was translated into English from the reference [23]: J. of the Gas Turbine Society of Japan, Vol. 25, No. 98, p. 3–70, September, 1997.)

CHAPTER

14

DEVELOPMENT OF 300kW-CLASS REGENERATED SINGLE-SHAFT CERAMIC GAS TURBINE CGT301

Tadashi Sasa, Shinya Tanaka, and Takao Mikami

Ishikawajima-Harima Heavy Industries Co., Ltd., Tokyo, Japan

INTRODUCTION

It is expected that the application of ceramics to gas turbine engines will increase their durability and fuel efficiency. A number of R & D programs have been carried out in the past decades in several countries. Especially, continued improvements in the mechanical properties of sintered silicon nitride ceramics have enabled the utilization of ceramic components under more severe operating conditions than with all-metal components. However, due to the brittle nature of the materials, the ceramic components, and consequently, the ceramic engines, are highly sensitive to defects. This defect-sensitivity is the most crucial problem in the development of the ceramic engines.

In the present article, the participation of Ishikawajima-Harima Heavy Industries Co., Ltd. in the technology development of ceramic gas turbines, especially the development of a 300kW-class ceramic gas turbine, known as CGT301, was reviewed with the emphasis on the design technology of ceramic gas turbine components.

DEVELOPMENTS OF APPLICATION TECHNOLOGY OF CERAMICS TO ENGINES

Ishikawajima-Harima Heavy Industries Co., Ltd. (IHI for short), one of the manufacturers of gas turbines for aircraft and power generation, has participated in the research and development of ceramic gas turbine technologies since the 1970s. IHI started work on ceramic gas turbines in the Japanese national "advanced gas turbine" project during the period from FY1978 to FY1987. In this project the technology development for the application of ceramics to gas turbines was initiated, especially the evaluation of the mechanical properties of ceramics, e.g. the effect of stress concentration on the strength, the effect of component volume on the

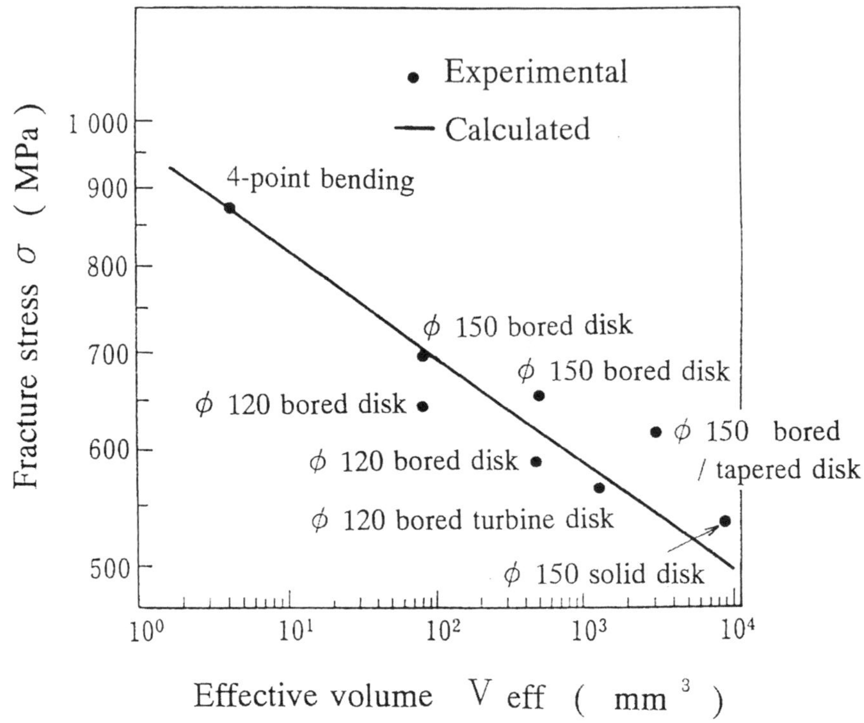

Figure 14.1 Correlation of the Fracture Data of Ceramic Components with Those of Test Specimens (Adapted from Ref. [3])

strength and the biaxial strength of ceramics. A feasibility study on the application of silicon-based ceramics to the combustion chamber or the gas-path was also carried out, and the advantages of using silicon nitride over silicon carbide for such applications with high thermal stress was pointed out [1]. Those projects would be recognized as some of the earliest ones in this field.

The efforts in IHI were continued under the subsequent national "Fine Ceramics" project, which was promoted by the Japanese government during the period from FY1981 to FY1992. In that project, the fracture behavior of ceramic materials and components was fairly well characterized with reference to both the fast and the time-dependent fracture of ceramics with a statistical approach. The statistical approach was then applied to the structural design of ceramic components. Besides the development of computer programs to evaluate the reliability and the life of ceramic components, a design guide for structural ceramic components was developed [2]. The design guide was developed to help designers determine allowable stresses for components under various loading conditions by considering the statistical distribution of the material properties, the volume of the components, the mode and the distribution of the stress experienced by the components, uncertainty in the loading conditions, etc. It had been widely recognized that fairly large discrepancies existed between the mechanical properties of ceramic components and those obtained from ceramic specimens. The methodologies for fabrication, evaluation or application of ceramic components, especially those for sintered silicon nitride components, were substantially further developed in that project, and a level of reasonably good correspondence was obtained between the mechanical property data measured from standard test specimens and those from actual components Figure 14.1 shows an example of how the fracture data of ceramic components and test specimens correspond with the predictions based on Weibull statistics [3].

Based on the design technologies developed in these studies, primary models of a blisk type rotor with a blade height of 35 mm and an outer diameter of 180 mm, and segmented nozzle vanes were designed and fabricated from silicon nitride and silicon carbide. A turbine test rig was installed with these ceramic components together with additional ceramic components such as a combustor liner, gas path, and was operated at a TIT (Turbine Inlet Temperature) of 1400°C to confirm the structural reliability of the turbine rotor and turbine nozzles [3].

It should be noted that the technologies developed for the design and application of ceramics in these national projects were successfully applied to commercial products such as ceramic turbochargers [4], and ceramic heat exchangers for industrial furnace waste heat recovery [5].

It is widely recognized that monolithic ceramics are limited in their tolerance to fracture energy even though they have outstanding high temperature strength and durability. The emergence of high strength silicon carbide fibers in the 1980s accelerated the development of ceramic matrix composites (CMC) as an alternative material to the monolithic ceramics. IHI has been also involved in the research and development of CMCs through programs supported by the Japanese government, e.g.. the "High Performance Materials for Severe Environments Program," started in FY1989 and continued until FY1996, and the Advanced Material Gas Generator Program started in FY1993. Although the CMCs showed considerable improvement in the mechanical properties and the fabrication technologies, it was decided that the material was still at too early a stage of development to be applied to the 300kW-class ceramic gas turbine. Therefore, the activities of IHI in the field of CMCs will not be included in the present chapter.

300KW-CLASS CERAMIC GAS TURBINE PROGRAM AND BASIC DESIGN OF THE CGT301

The 300 kW-class ceramic gas turbine program was planned and initiated by the Ministry of International Trade and Industry of the Japanese Governement in FY1988 and continued until FY1998. IHI joined that program to develop one of the 300 kW-class regenerated ceramic gas turbine engines for cogeneration applications, known as the CGT301, together with the ceramic component manufacturers, NGK Insulators, Ltd. and NGK Spark Plug Co., Ltd.. The development of the CGT301 engine emphasized the applicability of the design technology to larger (MW) size gas turbine systems. It was decided in the initial feasibility study that the multistage single-shaft axial-flow type turbine was the only practical development strategy for the above-mentioned design philosophy. The multistage turbine rotors with a single metallic shaft also led to the selection of the turbine rotor composed of a metallic disk with inserted ceramic blades, known as the "hybrid turbine" concept.

The initial basic studies led to the design of the CGT301 as a single-shaft engine system with a 2-stage axial-flow turbine, a compressor which combined an axial stage and a centrifugal stage, a single-can type combustor, and a shell-tube type heat exchanger. The basic structure and the system layout of the CGT301 engine are shown in Figures 14.2 and 14.3, respectively [6].

The gas turbine system has been developed in 3 phases; the Phase 1 engine with a TIT of 900°C, installed with all-metal components as the primary metallic gas turbine was developed in the period between FY1989 and FY1991, the Phase 2 engine with a TIT of 1200°C installed with ceramic components as the primary ceramic gas turbine was developed in the period between FY1991 and FY1995, and finally, the Phase 3 engine with a TIT of 1350°C as the pilot ceramic gas turbine was developed in the period between FY1996 and FY1998 [7]. Table 14.1 shows the design goals for the CGT301 engine and those for each component to attain the target of the engine thermal efficiency for Phases 2 and 3.

TURBINE DEVELOPMENT

Turbine Aerodynamic Performance Development

The aerodynamic designs of the turbine flow-path and the nozzle/blade airfoils were iterated for both the Phase 2 and Phase 3 engines to attain the design goal shown in Table 14.1. The ceramic component

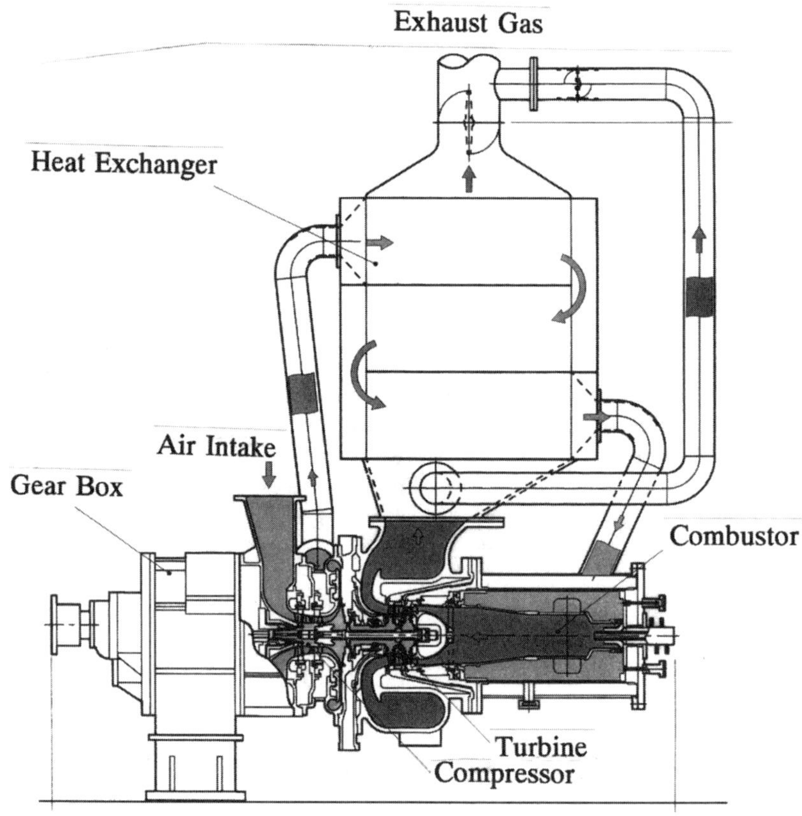

Figure 14.2 Basic Structure of the CGT301 Engine

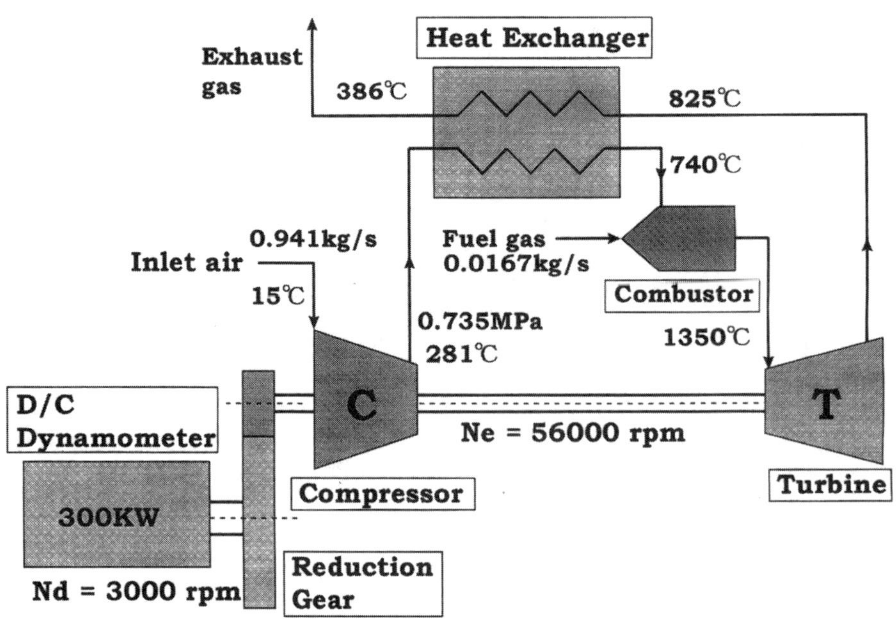

Figure 14.3 System Layout of the CGT301 Engine

TABLE 14.1. DESIGN GOALS FOR ENGINE AND EACH COMPONENT OF THE CGT301 CERAMIC GAS TURBINE [15]

Item			Intermediate Target (Primary CGT)	Final Target (Pilot CGT)
Engine	Turbine Inlet Temperature		1200°C	1350°C
	Rated Speed		56,000 rpm	56,000 rpm
	Thermal Efficiency		30%	42%
Component	Compressor	Pressure Ratio	7.3	7.3
		Efficiency	79%	81.5%
	Combustor	Efficiency	99%	99%
		NOx Emission	70 ppm	35 ppm
	Turbine	Efficiency	83%	87.5%
	Heat Exchanger	Thermal Effectiveness	84%	84.5%
		Pressure Loss	4.5%	4.5%

design must be performed considering not only the aerodynamic performance but also the structural reliability and the forming capability for the components.

The testing of the aerodynamic characteristics of the turbine was carried out for each stage through a rig-test at ambient temperature, and the value for the rated condition of the turbine was evaluated. Figure 14.4 shows the aerodynamic characteristics of the first stage turbine of the Phase 3 engine in the finalized design. The improvements in the efficiency through the design iterations resulted in a turbine efficiency of 83% for the Phase 2 engine, and 87.3% for the final Phase 3 engine. The latter value was very close to the final goal shown in Table 14.1.

Ceramic Materials

The design of the ceramic turbine blades inserted into a cooled metallic disk required the selection of ceramic materials which retain high strength up to moderately high temperatures (1200°C) even for the Phase 3 engine (TIT of 1350°C). Based on this criterion, EC-152 of NGK Spark Plug Co., Ltd., and SN-84HD of NGK Insulators, Ltd. were selected for the Phase 2 engine. Later, ST-1 of NGK Spark Plug Co., Ltd., which showed improved high-temperature strength compared with EC-152, was selected for the Phase 3 engine. On the other hand, much higher temperature performance was expected for the stationary components such as the turbine nozzles or the combustor liner, especially considering the non-uniformity in the combustor outlet temperature. The application of ceramic materials with high temperature durability up to at least 1350°C was necessary for the stationary components of the Phase 3 engine, which resulted in the selection of SN-88 of NGK Insulators, Ltd.

The materials properties of EC-152 and ST-1 supplied by NGK Spark Plug Co. are shown in Table 14.2 [8]. For the properties of ceramic materials supplied by NGK Insulators, Ltd., the interested reader is referred to Volume II, Chapter 7 of this publication.

Ceramic materials generally show a fairly large statistical distribution in their fracture behavior due to their high sensitivity to internal flaws. The 2-parameter Weibull distribution function, shown below, was used for the analyses.

$$P_f = 1 - \exp[-\int V(\sigma/\sigma_0)^m dV] \quad (1)$$

where P_f: failure probability of the component, V: volume of the component, σ: stress distribution in the component, σ_0: specific strength of the material, m: Weibull modulus.

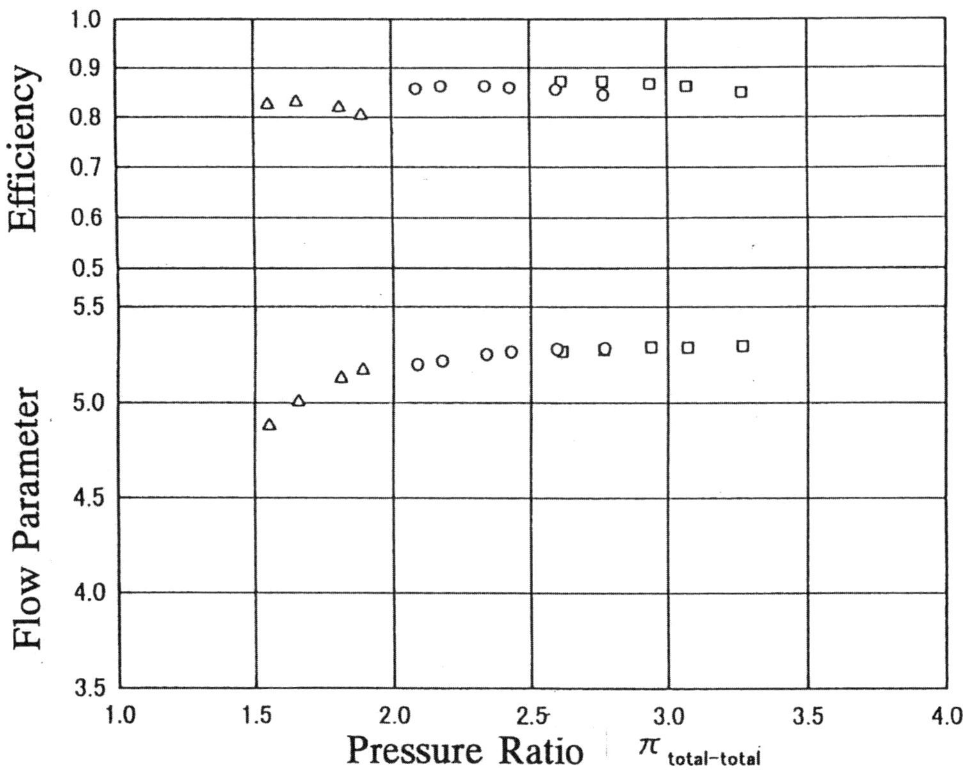

Figure 14.4 Aerodynamic Characteristics of the Turbine of the Phase 3 Engine (First Stage Turbine)

TABLE 14.2 PROPERTIES OF SILICON NITRIDE MATERIALS FOR TURBINE BLADES (NGT SPARK PLUG CO., LTD.)

Material code	EC-152	ST-1
Density (g/cm^3)	3.26	3.37
Bending strength (MPa, 4-point, JIS R1601, JIS R1604)		
RT	1020	1090
1000°C	830	990
1100°C	810	940
1200°C	760	880
1300°C	600	680
Elastic modulus (GPa, JIS R 1602)	318	317
Poisson ratio	0.26	0.28
Fracture toughness (MPa.m$^{1/2}$, SEPB, JIS R1607)	6.0	6.6
Hardness (GPa, Vickers, JIS R 1610)	14.7	16.7
Oxidation weight gain (mg/cm^2)		
1000°C, 1000 hours	0.10	0.16
1100°C, 1000 hours	0.25	0.26
1200°C, 1000 hours	0.49	0.46

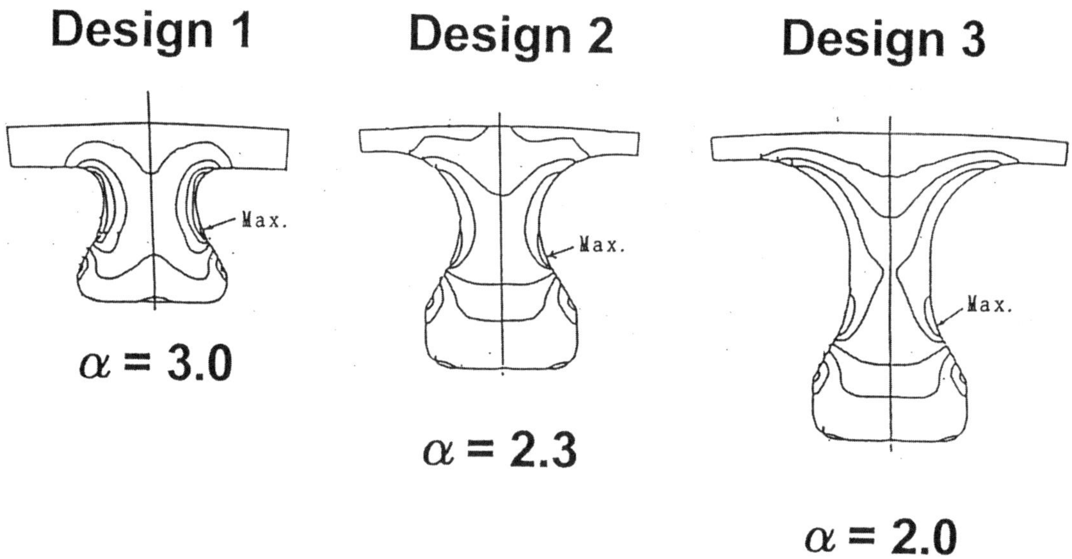

Figure 14.5 Blade Root Geometry and Stress Concentration of Ceramic Turbine Blade: α = Stress Concentration Factor

All the ceramic components were inspected nondestructively, including by fluorescent dye penetrant inspection, ultrasonics, and micro-focus X-ray. Overload cold spin proof testing was also carried out for the ceramic blades prior to their insertion into a rig or an engine at a rotational speed to give a proof stress twice as high as the centrifugal stress at the rated speed.

Structural Design and Reliability of the Turbine Rotor

The design guideline described by the equation (2) [2] was applied to the evalution of the reliability of the ceramic components.

$$\sigma_a = S_t / (K_0 K_1 \beta \alpha) \quad (2)$$

where σ_a: allowable stress in the design, S_t: characteristic strength of the material with consideration of the component service condition, K_0: general safety factor, K_1: statistical design factor, β: a factor correlated with the stress distribution, α: a factor correlated with the volume under load.

In the case of the ceramic turbine blades inserted into the metallic disk, the stress concentration at the root zone at high rotational speeds is most crucial. By modifying and optimizing the blade root geometry the peak stress values were lowered to within the allowable level as shown in Figure 14.5. The blade-root geometry determined above was applied to the engines of both Phases 2 and 3 [9]. The stress distribution of the blades was further computed with a 3-dimensional finite element method (3D-FEM). High efficiency and high reliability often need to be traded off, and the initial geometry of the blade for the Phase 3 engine with higher efficiency was found to give a higher risk of failure than that for the Phase 2 engine. The design of the blade was iterated, and it was found that the optimized displacement between the center of gravity of the blade root and that of the airfoil along the direction of rotation gave the least

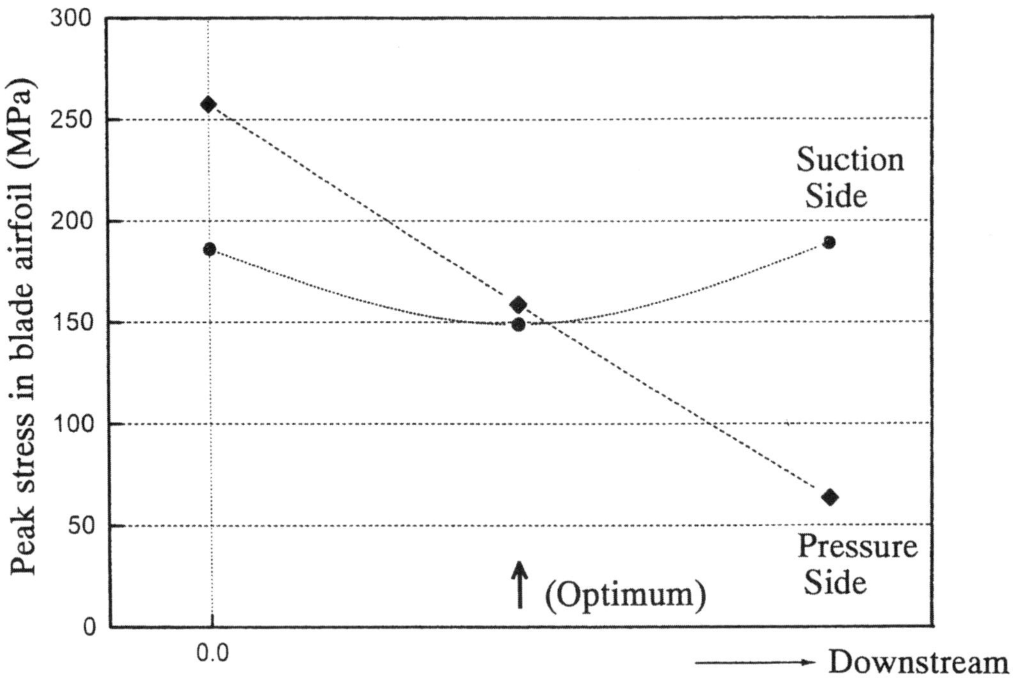

Figure 14.6 Effect of Displacement of the Center of Gravity on the Peak Stress in Blade Airfoil (First Stage Turbine Blade of the Phase 3 Engine) (Adapted from Ref. [19])

risk of failure without deteriorating the efficiency. Figure 14.6 shows the effect of the displacement of the center of gravity on the peak stress in the airfoil for the first stage turbine blade of the Phase 3 engine, where the optimum value is indicated by the arrow. Figure 14.7 shows the stress analyses for the ceramic turbine blades of the final design for the Phase 3 engine in consideration of both the centrifugal force and the thermal stress, where the stress level of the entire blade was kept within the allowable value at the rated condition. The vibration modes of the blades were also evaluated by both the impulse method and the holography method to avoid resonance vibration.

Substantial effort was expended in the quality assurance of the fabrication process of the ceramic components considering the high sensitivity of these components to internal defects. In addition, it should be recognized that ceramic components are highly sensitive to "defects" in the design/analysis process, the assembling process or during operation of the components and the engine [10]. An example of this kind of "defect," which could result in premature failure of the ceramic blades is given in Figure 14.8, which shows the effect of improper contact at the ceramic blade root/metallic disk interface on the peak stress. The introduction of additional compliance into the components or the system is also expected to minimize the sensitivity to defects. A compliant layer was therefore used at the interface between the ceramic blade root and the metallic disk. Figure 14.9 shows a tensile test fixture for simulating the ceramic blade root inserted into the metallic disk, in which the strain non-uniformity was measured by strain gauges attached to the ceramic specimens [11]. Through these evaluations, nickel alloy foil with a thickness of 0.15 mm was selected as the compliant layer. Testing showed that the compliant layer was effective in decreasing the strain non-uniformity at the interface within 10%.

Reliable operation of the turbine rotor with the optimized design and the blade root compliant layer was confirmed in the turbine rig test at a TIT of 1200°C for the Phase 2 engine, and subsequently at a TIT

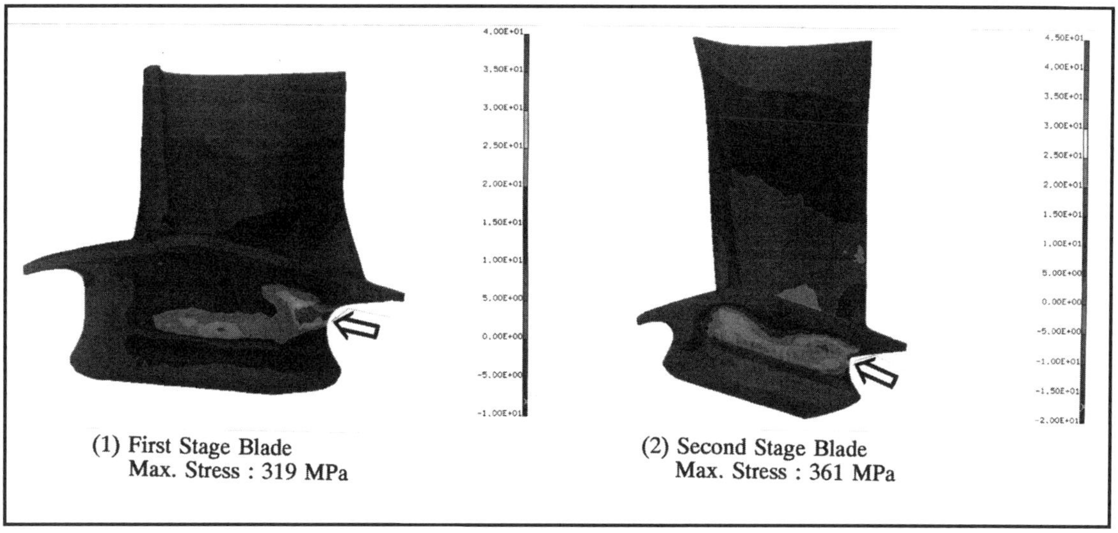

Figure 14.7 Stress Analyses of Ceramic Turbine Blades of the Phase 3 Engine

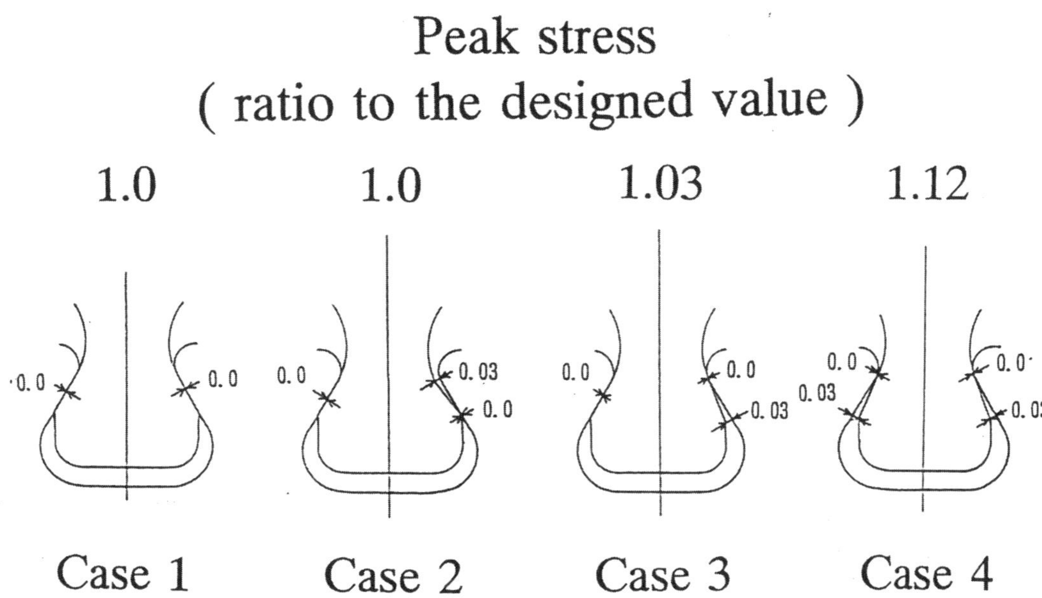

Figure 14.8 Effect of Improper Contact at the Ceramic Blade Root/Metallic Disk Interface on the Peak Stress (Interface gap: mm)

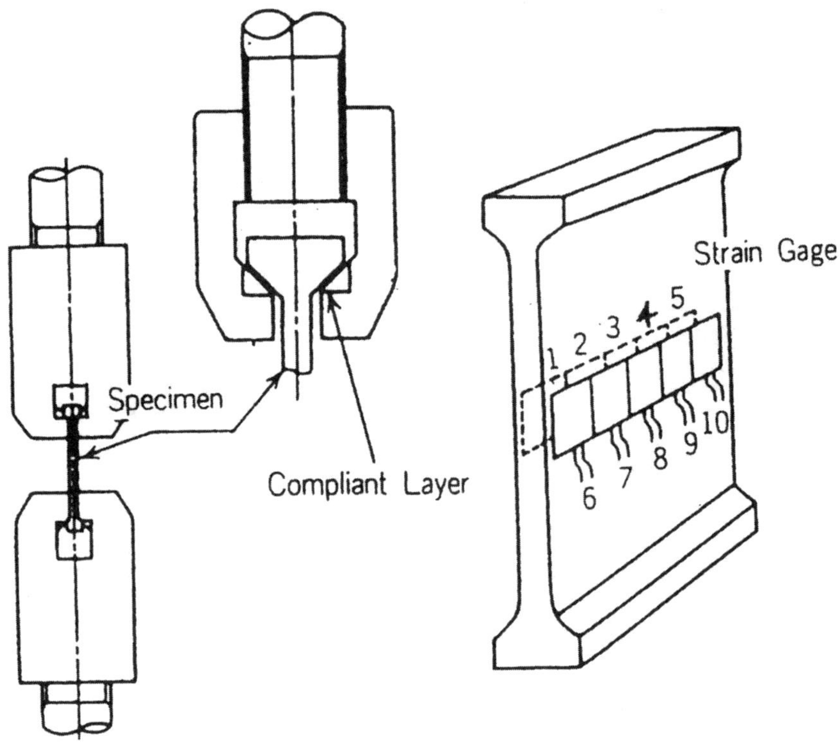

Figure 14.9 Tensile Test Fixture to Evaluate Compliant Layer at Ceramic Blade Root/Metallic Disk Interface

of 1350°C for the Phase 3 engine. Figure 14.10 shows the first stage turbine rotor after the rig test at a TIT of 1350°C.

Apart from aerodynamic factors, the performance of the turbine was also largely affected by the amount of cooling air, leakage of high pressure gas or air, etc. In order to solve one of these problems, a new ceramic component design concept was applied to seal the space under the ceramic blade platform and to minimize the leakage of high pressure gas at that location. Figure 14.11 shows the "seal-damper," which was composed of porous fiber-reinforced ceramic composites with high deformation capability. This component was also effective in minimizing blade vibration at the resonant frequencies [12]. The seal-damper was first tested in the turbine test rig, and then incorporated into the actual engine.

Structural Design and Reliability of the Turbine Stationary Components

The stationary part of the turbine was designed using the concept described below. The frame structure of the turbine was formed by metallic components. Ceramics were only utilized for the most crucial high temperature components, which were designed as small as possible and with the simplest geometry.

The turbine nozzles were designed as segmented vanes with integral inner and outer platforms, which were placed circumferentially and were fixed in the inside of a ceramic retainer ring. Ceramics were also applied to the hot section gas path components, which were stacked as individually separate simple-shaped components. The ceramic retainer rings and gas path components were supported by metallic support casings.

Figure 14.10 First Stage Turbine Rotor After the Rig Test at a TIT of 1350°C

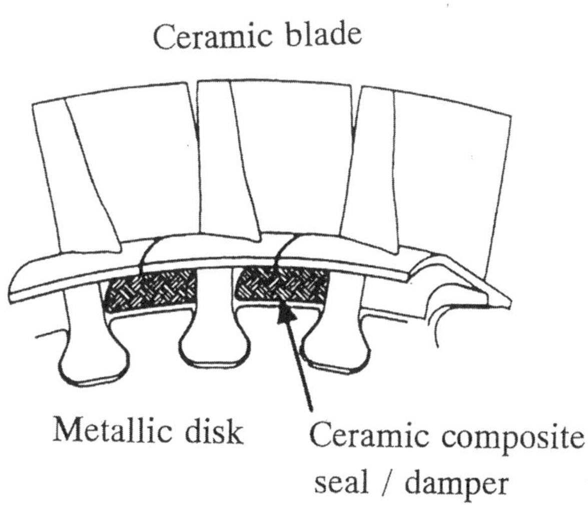

Figure 14.11 Seal-damper Applied to Ceramic Turbine Blades

The risk of failure of the ceramic stationary components is considered to be highest during engine shutdown. Figure 14.12 shows the FEM analysis of the transient thermal stress for the first stage nozzle vane of the Phase 3 engine, where the failure probability was found to be reasonably low.

All the stationary ceramic turbine components were installed in a turbine test rig with basically the same configuration as the engine, and thermal cycle tests were performed for up to 100 cycles which were identical to the start/stop condition of the engine. The tests proved the reliability of all the stationary ceramic components under the operating conditions of the engine.

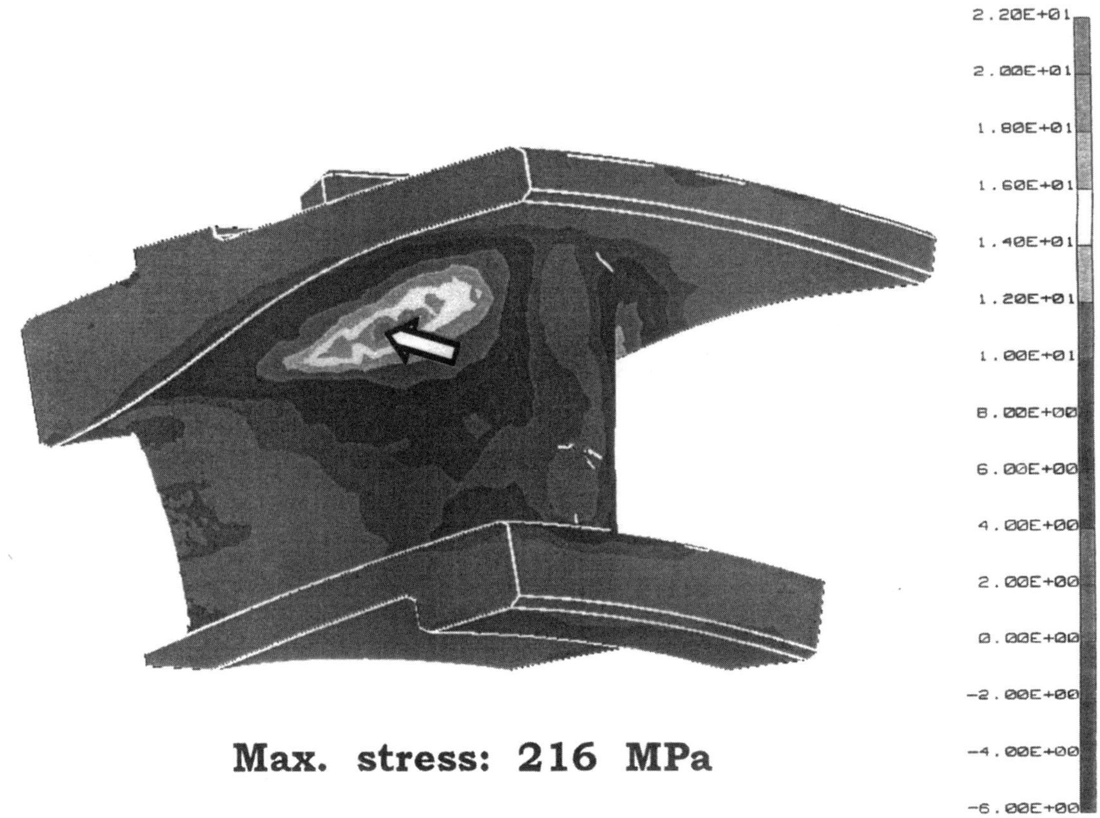

Figure 14.12 Transient Thermal Stress Analysis of First Stage Nozzle Vane of the Phase 3 Engine (4 seconds After Shut Down of the Engine)

COMBUSTOR DEVELOPMENT

The combustor liner was composed of stacked ceramic rings supported by springs. This design was expected to release the transient thermal stress of the chamber and also the thermal expansion mismatch with the metallic outer casing. The combustor design was based on the lean premixed concept to attain the target NOx emission level shown in Table 14.1. The air was supplied primarily through the swirler adjacent to the fuel injection nozzle, and secondarily through the dilution holes in the combustion chamber. No cooling air was supplied to the combustor liner. The optimum geometry of the combustion chamber was determined through CFD (Computational Fluid Dynamics) analyses of the mixing characteristics of the fuel and the air.

Figure 14.13 shows the final design of the combustor for the Phase 3 engine. A mechanism to control the opening of the dilution holes by a slide valve was developed and applied. A supplemental fuel injection mechanism was applied there to improve the premixing of the fuel and the air [13]. The geometry of the chamber was modified and the diameter was enlarged in the final design to improve the combustion efficiency under lean combustion part load operation. An air swirler with wider area was also applied to decrease the pressure loss and to promote the premixing.

Figure 14.14 shows the NOx emission levels and the combustion efficiency obtained in the rig test under the rated condition of the engine. By optimizing control of the ratio of primary/secondary air and that of primary/secondary fuel injection a combustion efficiency of 99.9% was achieved above 70% load, and the NOx emission level was measured at 5.6 ppm at full load, which was 1/6 of the target shown in Table 14.1. These

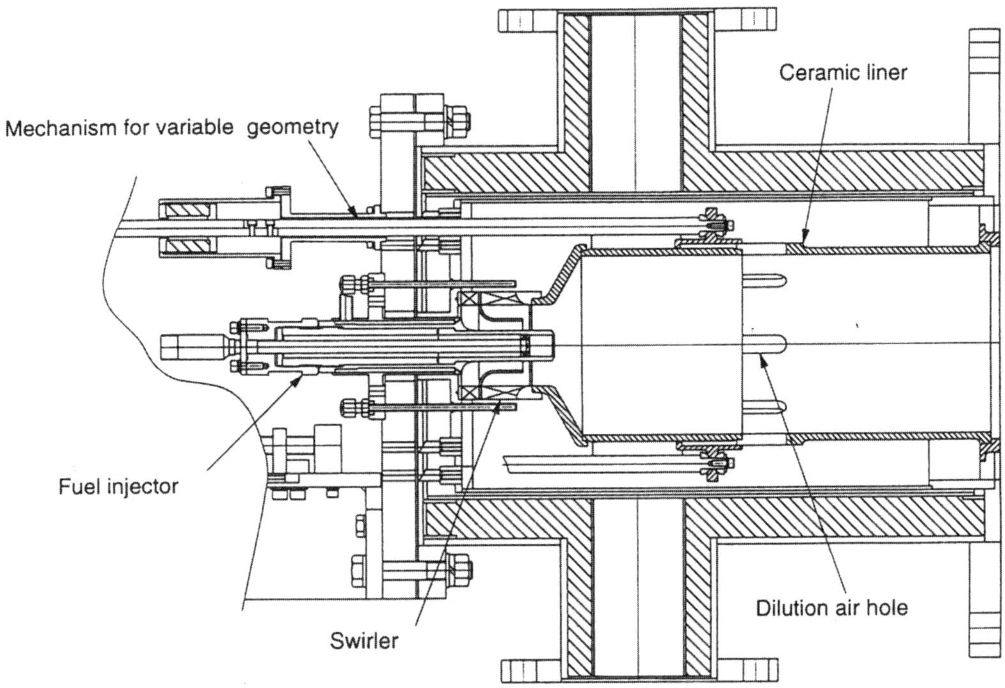

Figure 14.13 Combustor for the Phase 3 Engine (Adapted from Ref. [13])

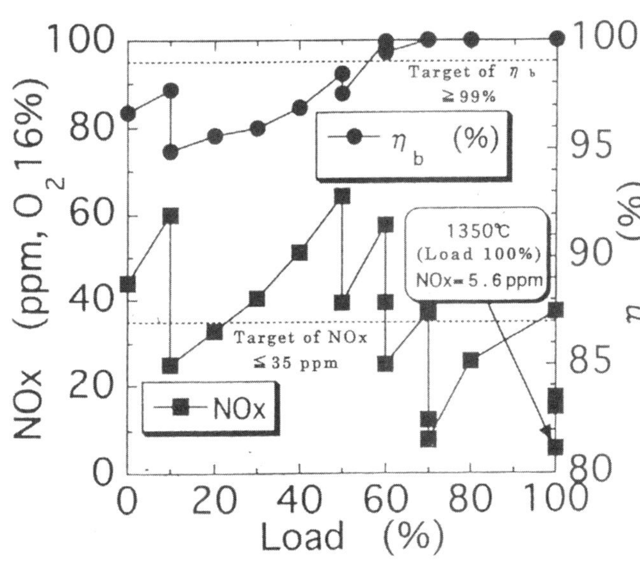

Figure 14.14 NOx Emission and Combustion Efficiency of the Combustor for the Phase 3 Engine

Figure 14.15 Ceramic Tube Assembly of the Heat Exchanger for the Phase 3 Engine (Adapted from Ref. [14])

results show the noticeable superiority of the combustor with the uncooled ceramic liner, in which extremely homogeneous premixing of air with fuel can be realized.

HEAT EXCHANGER DEVELOPMENT

The shell-tube type CGT301 heat exchanger consisted of a metallic U-tube assembly for the lower temperature part, and a straight tube assembly for the higher temperature part. The straight tube assembly for the Phase 2 engine was designed and fabricated as an all-metal welded structure. The plan was to replace the all-metal high temperature section by a ceramic tube assembly later for the Phase 3 engine.

The ceramic tube assembly design consisted of tubes with inner fins, which were bonded tightly with ceramic end plates on both sides (Figure 14.15). The geometry of the inner fins, the number and the configuration of the tubes were determined by heat transfer analysis in conjunction with a consideration of the

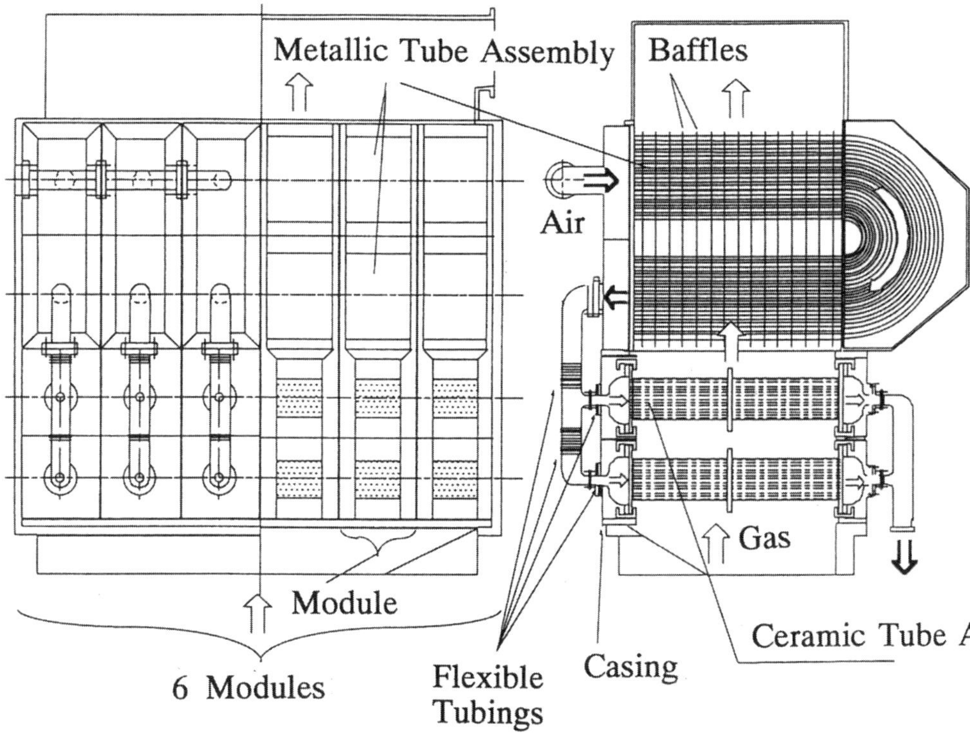

Figure 14.16 Design of the Heat Exchanger for the Phase 3 Engine (Adapted from Ref. [14])

ceramic component fabrication capability. Based on the optimum ceramic tube assembly design, the heat exchanger for the Phase 3 engine was completely redesigned as a module-type structure. The whole structure of the heat exchanger consisted of six vertically divided modules. Figure 14.16 shows the final design of the heat exchanger for the Phase 3 engine.

The ceramic tube assembly was fabricated from SN-84 of NGK Insulators, Ltd., and its structural reliability was evaluated through repeated thermal shock testing by radiation heating which simulated the start/stop condition of the engine. The assembly was tested for up to 200 cycles and the air leakage throughout the testing remained reasonably low, less than 0.1% of the total flow.

The ceramic plates of the tube assembly were mechanically joined to metallic headers by using metallic clamps and metallic bolts with seals and gaskets. A combination of materials of different thermal expansion coefficients were selected to minimize the leakage from the joined parts, and to assure reliable operation of the heat exchanger for over 100 cycles.

A partial model of the heat exchanger, where 2 modules out of 6 full modules were assembled and installed with ceramic tube assemblies, was fabricated and the characteristics were evaluated in a rig test (see Figure 14.17) [14]. The thermal effectiveness of 86% and the pressure loss of 2.5% of the model obtained under the rated condition were superior to the target values shown in Table 14.1.

COMPRESSOR DEVELOPMENT

A combination of an axial-flow stage and a radial-flow stage was selected to achieve simultaneously high efficiency and wide range of operating conditions. Figure 14.18 shows the structure of the compressor.

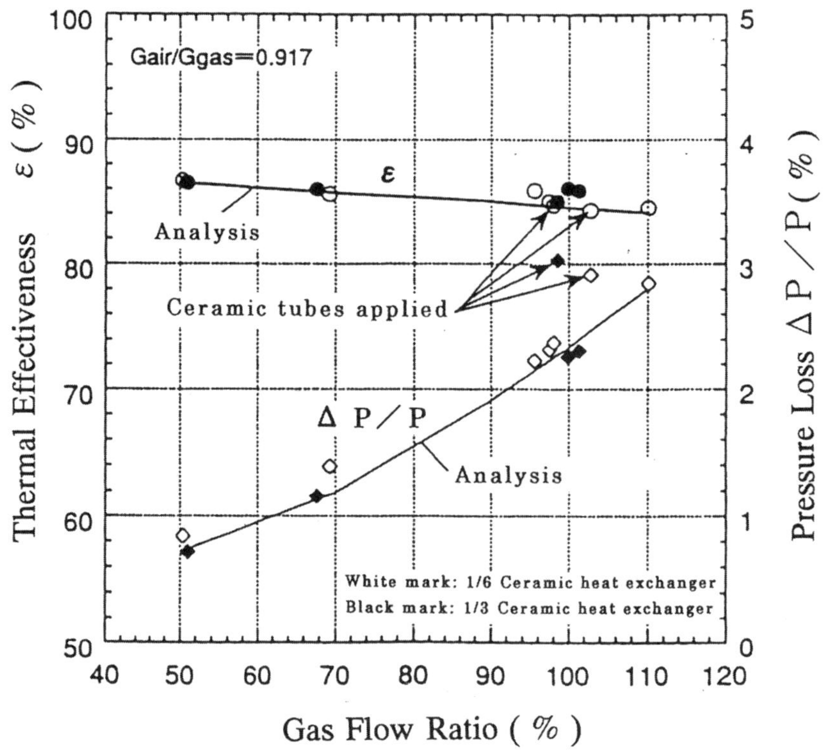

Figure 14.17 Thermal Effectiveness and Pressure Loss of the Heat Exchanger for the Phase 3 Engine

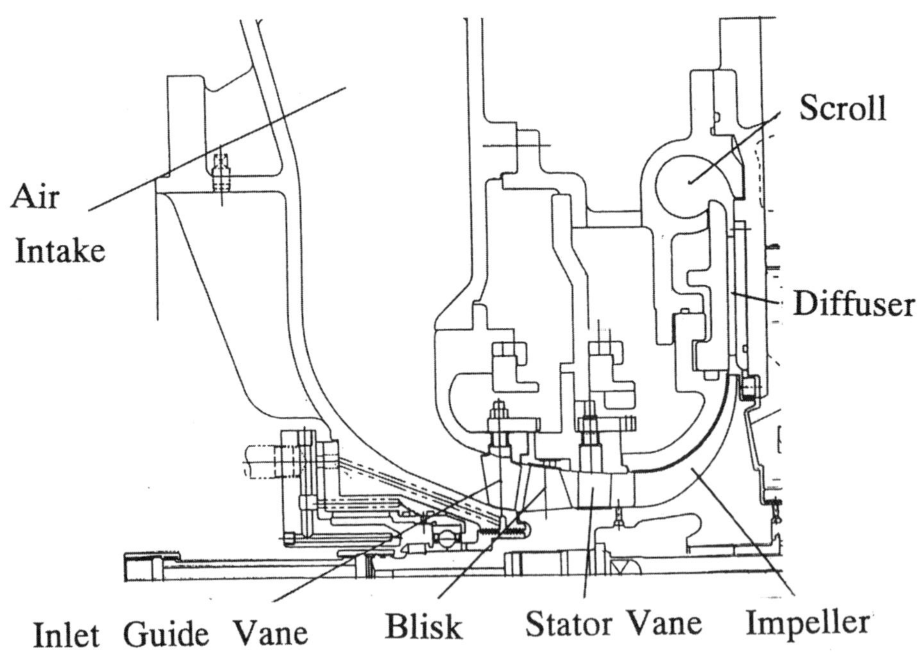

Figure 14.18 Compressor with an Axial-flow Stage and a Radial-flow Stage

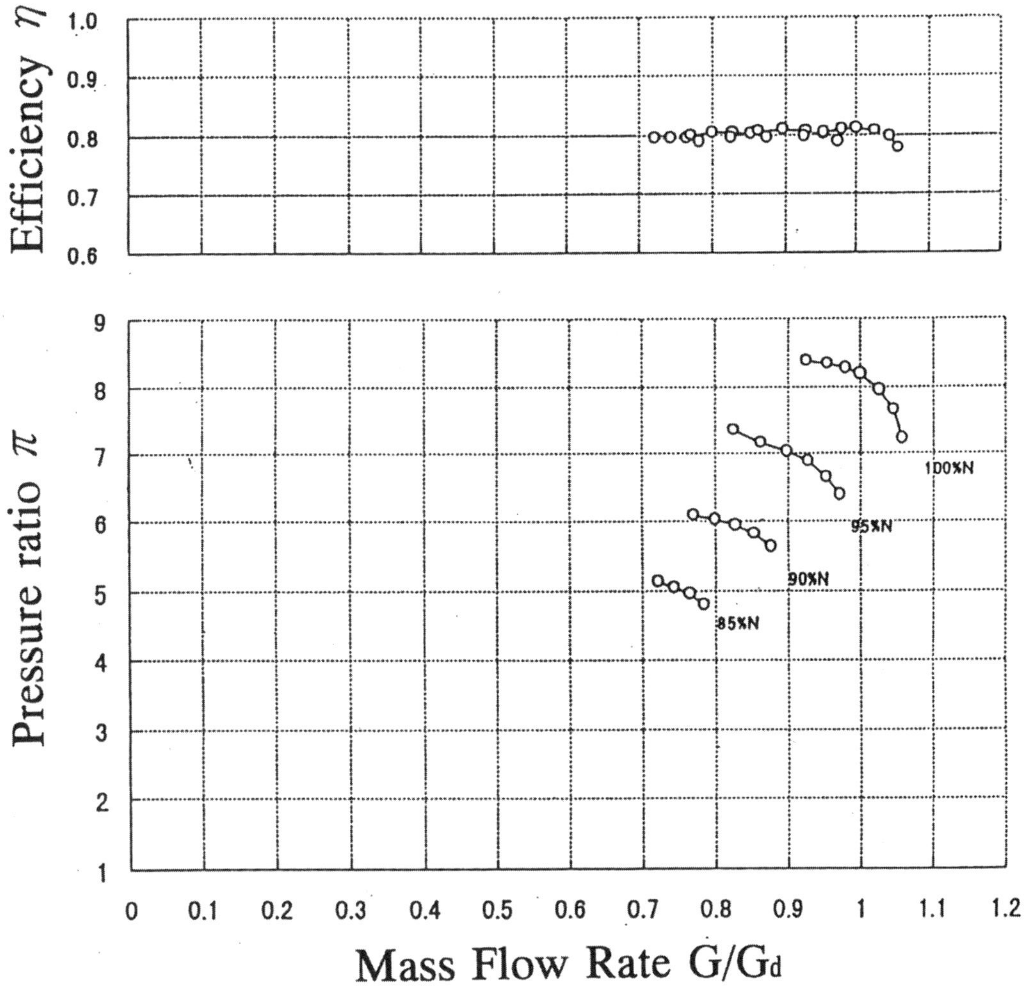

Figure 14.19 Characteristics of the Compressor for the Phase 3 Engine

The optimization of the flow-path of both the axial stage and the radial stage, the development of advanced airfoil profiles for the inlet guide vane, the blade, the stator-vane, the impeller, and the diffuser, etc., were carried out through CFD and structural analyses, and the characteristics were evaluated by rig tests. Abradable shrouds were applied to the blades and the impeller. The characteristics of the compressor for the Phase 3 engine evaluated in rig testing are shown in Figure 14.19. The efficiency of 81.1% was fairly close to the target shown in Table 14.1.

ENGINE DEVELOPMENT

Rotor Dynamics Evaluation Through the Phase 1 Primary Metallic Gas Turbine Testing

In the CGT301 engine design the combustion chamber was placed along the same axis as the turbine and the compressor. This configuration had the advantage that a large ceramic turbine scroll between the

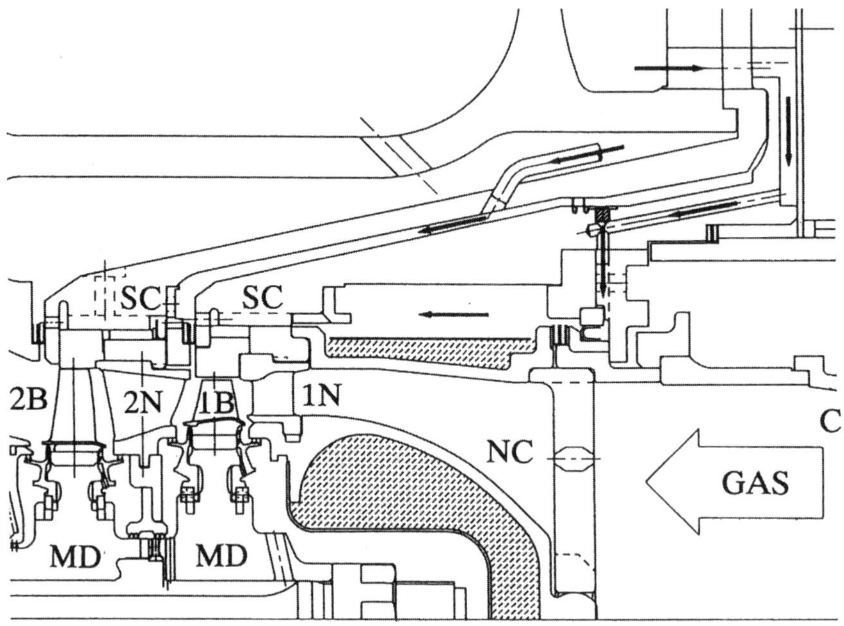

Figure 14.20 Turbine Hot Section of the Phase 2 Engine at the Initial Stage. C; Combustor, NC; Nose Cone, 1N; First Stage Nozzle, 1B; First Stage Blade, 2N; Second Stage Nozzle, 2B; Second Stage Blade, MD; Metallic Disk, SC; Support Casing

combustor outlet and the turbine inlet could be eliminated, but on the other hand, it gave a high thrust to the compressor side of the rotor system. A balance piston was introduced to cancel the thrust by using the compressed air pressure. For the rotor system the turbine shaft was supported by bearings only on the compressor side, giving an overhang structure. Oil film dampers were applied to the bearings in order to stabilize the rotor dynamics. The dynamic characteristics of the single-shaft turbine/compressor of this structure were evaluated through testing of the Phase 1 metallic engine at a TIT of 900°C and reliable operation of the rotor system in the engine was confirmed [7].

Development of the Phase 2 Engine: Primary Ceramic Gas Turbine

Figure 14.20 shows the initial turbine hot section design of the Phase 2 engine. Ceramic stationary components were supported by cone shaped metallic support casings. The turbine nose cone and the strut were designed as separate ceramic components after consideration of the fabrication capability and the risk of failure. The thrust on the nose cone was accommodated by the metallic support casing via the ceramic nozzle vanes and the retainer ring.

The development and testing of the Phase 2 engine were started by initially installing all-metal components in the turbine hot section. Ceramic components were introduced successively as the developments proceeded. The operating temperature of the engine was gradually increased as ceramic components were inserted in the hot section. Finally, all the ceramic components were installed in the turbine and the combustor of the engine, while the heat exchanger remained metallic [15].

The first-stage ceramic nozzles fractured in early testing of the Phase 2 engine. The fractures originated in the section of the nozzle inserted in the nose cone groove. It was determined that the thrust on the nose cone was non-uniformly supported only by a limited number of segmented nozzles due to heterogeneous

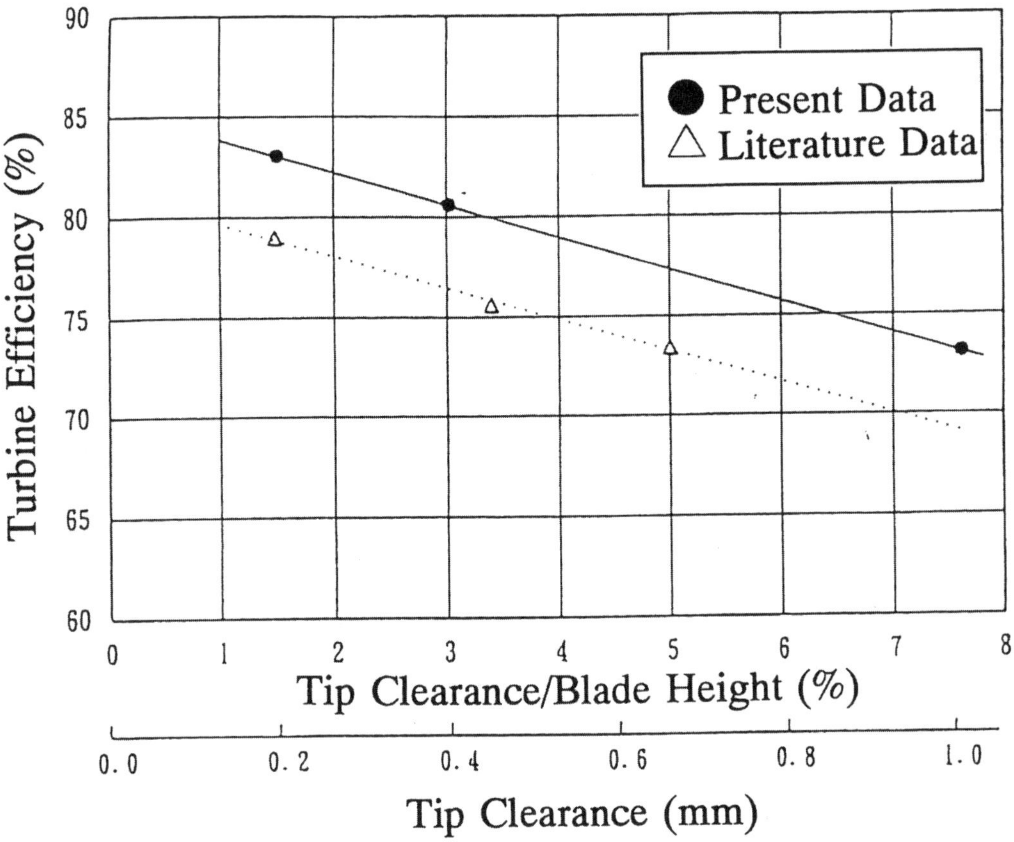

Figure 14.21 Effect of Turbine Tip Clearance on the Turbine Efficiency for the Phase 2 Engine (Adapted from Ref. [19])

contact. By improving the dimensional accuracy and by applying compliant materials at the critical nozzle section, the engine operated without a problem up to the target TIT of 1200°C [15]. The engine was tested at the TIT of 1200°C for more than 20 hours of cumulative operating time to confirm the reliability. The ceramic components after the test were inspected nondestructively, including by fluorescent dye penetrant inspection, and no degradation was found, except for a slight surface color change on the first stage blades. The material properties data also suggested no degradation after these operating conditions.

The effect of the turbine tip clearance on the turbine efficiency for the Phase 2 engine was evaluated by a turbine rig test, where the design value of the clearance was 0.2 mm (see Figure 14.21). Attempts to decrease the tip clearance towards the design value in the Phase 2 engine testing, however, resulted in failure of the blades, which consequently led to failure of most of the ceramic components in the hot section. Through measurements of the temperature distribution of the turbine metallic components and their deformation analyses, it was concluded that the cone shaped metallic support casings showed fairly large deformations during high temperature operation of the engine, which then made it difficult to accurately position the turbine stationary components relative to the blades. The improvements in the efficiency of the Phase 2 engine evidently necessitated the modification of the structure of the support casing. On this basis, the support casing was completely redesigned as a block type and was installed in the engine to improve the positioning of the turbine stationary components. After these improvements in the structure of the engine,

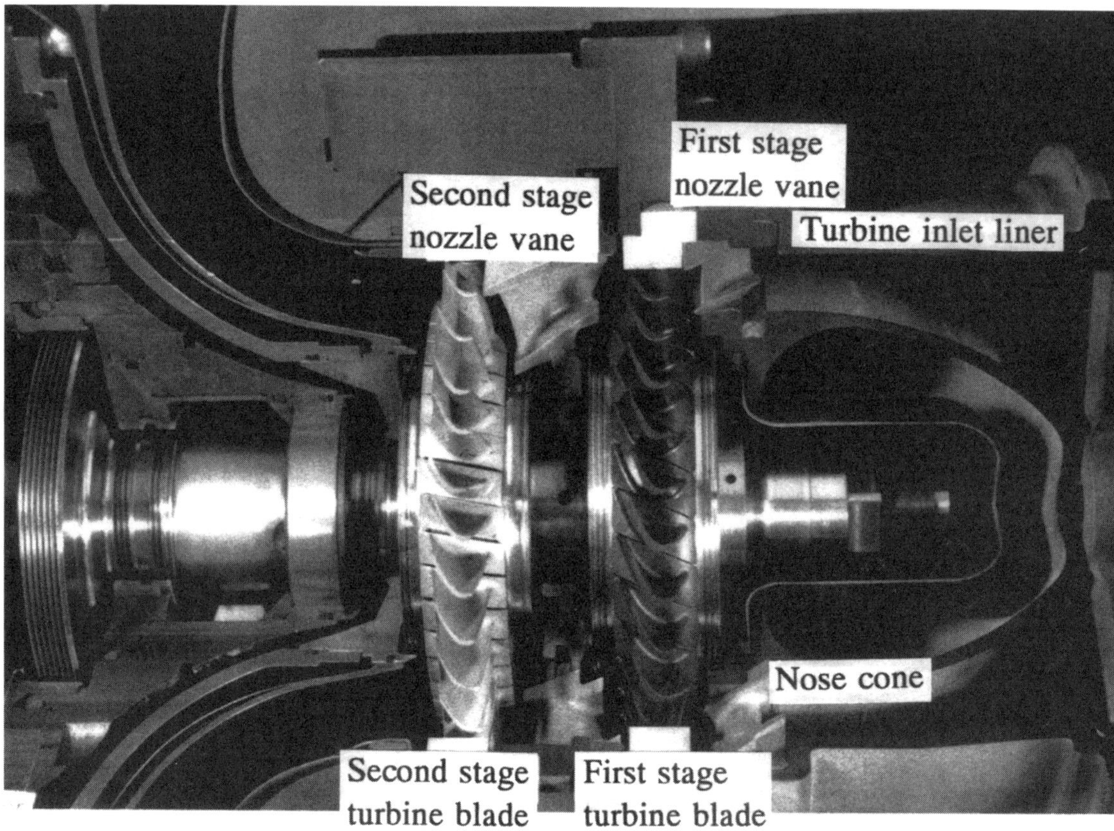

Figure 14.22 Inside View of the Turbine Hot Section of the Phase 2 Engine

including other modifications of the turbine stationary parts, the engine gave a thermal efficiency of 27% at the TIT of 1200°C. Figure 14.22 shows a view of the turbine hot section after the tests.

Development of the Phase 3 Engine: Pilot Ceramic Gas Turbine

The Phase 3 engine was developed by modifying the Phase 2 engine. The engine components, such as the turbine, the combustor, and the compressor, were replaced by more advanced designs with higher efficiency and improved reliability developed based on the results of component testing. Modifications were also made to improve the high temperature durability of the engine hot section up to the design TIT of 1350°C.

The CGT301 design was based on the stacking of a number of relatively small ceramic components. This design concept had the advantage of minimizing the risk of failure of each component. At the same time a disadvantage was the increased number of joints between the parts, with the increased potential for inaccurate positioning of adjacent components and also a higher possibility of leakage of high pressure gas or air. The development of the high performance Phase 3 engine thus necessitated technologies for estimating and controlling the temperature and the deformation of the hot section structure and also for minimizing leakage [16].

Figure 14.23 shows the cross section of the main part of the Phase 3 engine with the ceramic components. All the ceramic components excepts the heat exchanger tube assembly were actually installed in the

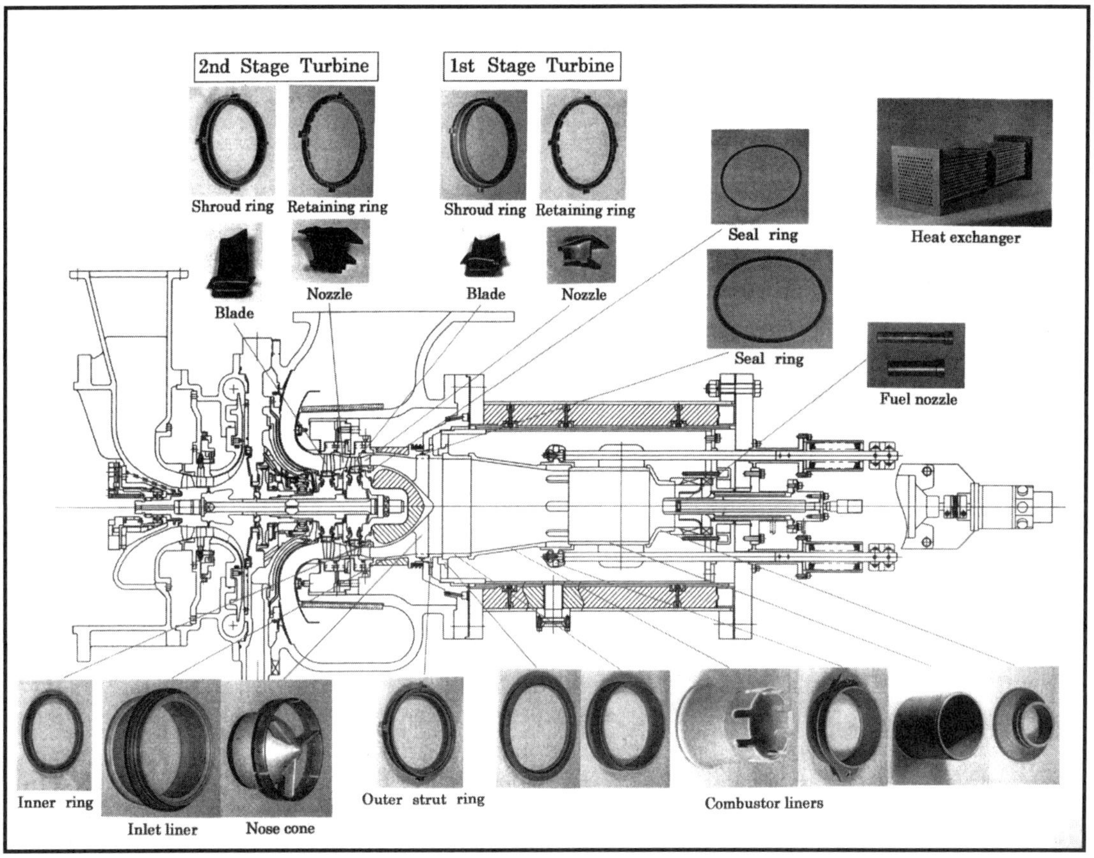

Figure 14.23 Phase 3 Engine and Ceramic Components (Reprinted from Ref. [18])

engine. The turbine blades were fabricated from ST-1 of NGK Spark Plug Co., Ltd. The combustor components, the turbine nozzles and the gas path components were fabricated from SN-88 of NGK Insulators, Ltd. The metallic heat exchanger used in the Phase 2 engine continued to be used in the Phase 3 engine with the replacement of the straight tube assembly in the higher temperature part by more refractory metallic materials. Figure 14.24 shows the hot section of the turbine in detail. The design of the turbine nose cone/strut was changed to give an integral structure, which had become feasible because of the development in the fabrication technology for the ceramic components. In this design the nose cone by itself was able to support the thrust applied to it while avoiding the high stress on the nozzles [17].

A ceramic seal ring was applied to the joint between the combustor outlet and the turbine inlet, which minimized leakage of the hot gas. The cooling of the metallic disk and the metallic casings was improved to minimize the amount of cooling air, and thermal insulation was applied to the hot parts to minimize heat losses [18]. These improvements enabled successful operation of the Phase 3 engine at the design TIT of 1350°C, and the improvements in the engine thermal efficiency [19]. The cumulative operating time of the Phase 2 and 3 engines reached 280 hours with 350 cumulative start/stop cycles. Especially, the cumulative operating time of the Phase 3 engine at full load was 10 hours, and the nondestructive inspection of the ceramic components after the test by fluorescent dye penetration confirmed no degradation. Figure 14.25 shows the engine thermal efficiency up to the TIT of 1350°C obtained during testing of the Phase 3 engine under reliable operating conditions. For comparison with the values obtained during engine operation, the

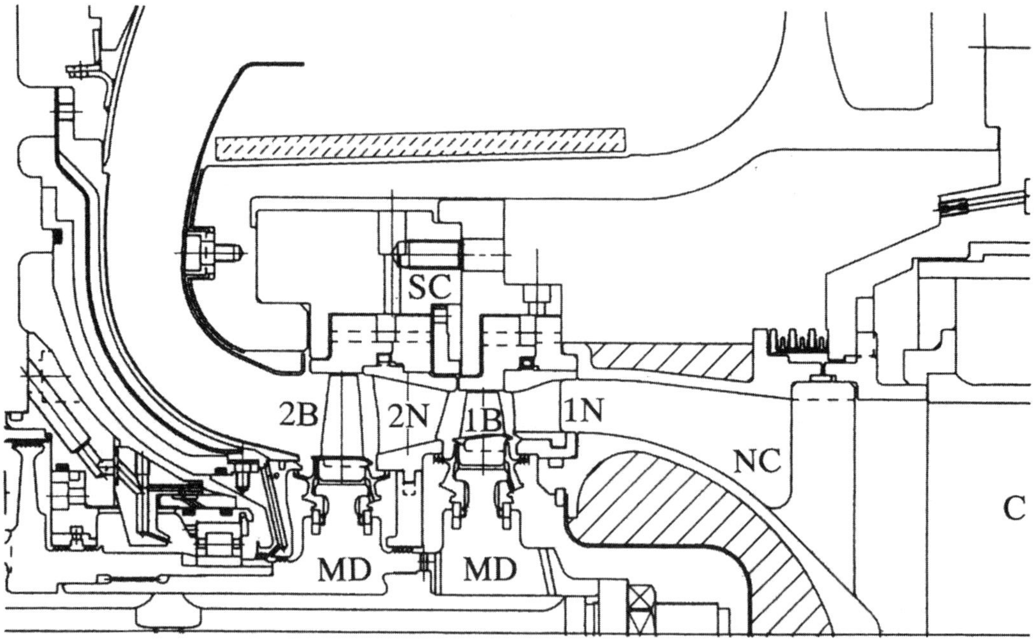

Figure 14.24 Turbine Hot Section of the Phase 3 Engine. (Notation: Same as in Figure 14.20)

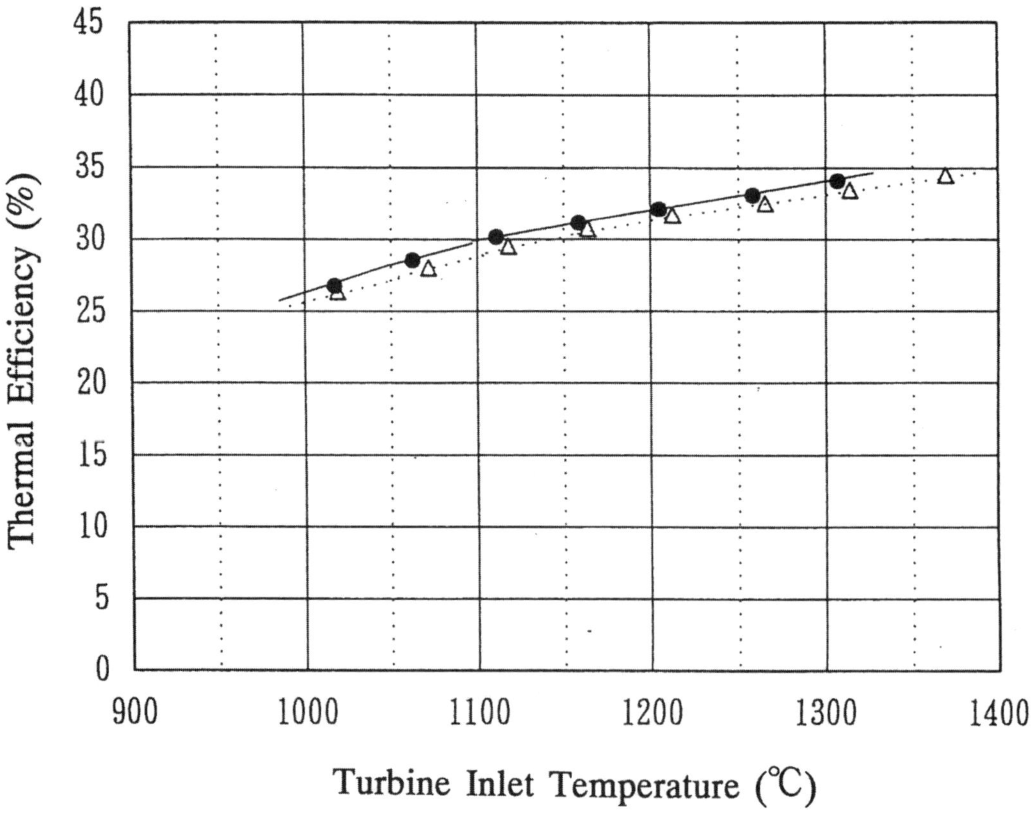

Figure 14.25 Thermal Efficiency of the Phase 3 Engine

TABLE 14.3 FINAL ACCOMPLISHMENTS IN CHARACTERISTICS OF ENGINE AND COMPONENTS OF THE CGT301

Item			Accomplishments	Target
Engine	Turbine Inlet Temperature (TIT)		1350°C	1350°C
	Thermal Efficiency		35%	42%
Component	Compressor	Efficiency	81.1%	81.5%
	Combustor	Efficiency	99.9%	99%
		NOx Emission	5.6 ppm	35 ppm
	Turbine	Efficiency	87.3%	87.5%
	Heat Exchanger	Thermal Effectiveness	86%	84.5%
		Pressure Loss	2.5%	4.5%

value of the engine thermal efficiency calculated from the efficiencies of the individual components (compressor, combustor, turbine, heat exchanger) obtained from rig testing was approximately 41% (Table 14.3). The discrepancy with the overall engine efficiency (35%) was mainly attributable to the turbine tip clearance in the engine. The clearance for the blades was deliberately increased to 0.8 mm at the initial installation into the engine at ambient temperature to avoid their unexpected failure. The clearance in the actual engine operation was roughly estimated at 0.5 mm due to the thermal expansion and the centrifugal deformation of the metallic disk, while its design value was 0.2 mm. Other factors also contributed to the engine efficiency discrepancy, i.e., improper matching of the turbine/compressor combination due to the lowered turbine efficiency, a lower performance of the all-metallic heat exchanger compared to that installed with ceramic components, inaccuracies in the geometry, and the positioning of the segmented ceramic components.

SUMMARY

The 300kW-class regenerated single-shaft ceramic gas turbine CGT301 was developed with emphasis on the applicability of the design technology to larger size gas turbines. The design philosophy led to the selection of a 2-stage axial-flow "hybrid turbine" where ceramic components were utilized in combination with metallic strutures. The most challenging application of ceramics to the hybrid turbine was undoubtedly the turbine blades. The optimization between the aerodynamic design and the structural design together with the application of optimized compliant materials, as well as the development of fabrication technologies and NDE/proof-testing, resulted in the successful development of ceramic blades inserted into a metallic disk, and the reliable operation of the engine with the hybrid turbine at the design TIT of 1350°C.

In the hybrid turbine, ceramic stationary components were basically designed as small sized components with simple geometry, and were also used in combination with metallic support structures. In the present study it was found that the problems encountered with the stationary components were more profound than those with the blades/rotor, especially those with the design and the evaluation of the performance of the metallic structures supporting the ceramic components in the hot-section. Using ceramics for these parts might solve the problem to a certain degree. However, this would be impractical for larger size gas turbines. While these metallic structures were modified, some uncertainties still remained with the prediction and the control of the performance of these hot section metallic structures in actual engine operation. This kind of limitation affected the control of the turbine tip clearance of the engine. It became the major cause of the discrepancy between the actual thermal efficiency of 35% at the design TIT of 1350°C and the value of 41% calculated from rig testing of individual components. Further design changes would be needed to improve the performance of the hot-section metallic structures and the resulting engine performance and thermal efficiency. The application of abradable shrouds would be an alternative method to decrease the turbine tip clearance, which was however not introduced in the CGT301 engine due to a lack of reliability at the present stage of the technology. Another remarkable achievement of the CGT301 program was the successful development of the ceramic combustor with high reliability, high combustion efficiency and extremely low NOx emission levels. This shows the superiority of the combustor with an uncooled ceramic liner. The development of the module-type shell-tube heat exchanger composed of metallic U-tubes and ceramic tube assemblies is also a noteworthy outcome of the present program, which revealed the high performance and high reliability at full load conditions.

Reliability of ceramic components against fast fracture can now be reasonably well managed and correlated with test specimen data. Life prediction of ceramic components necessitates knowledge of time-dependent properties of the ceramic materials. A well developed time-dependent property data base was however lacking during the performance of the CGT301 program. The establishment of such a data base remains a future task. Apart from the performance of the ceramic materials and components, the performance of other materials and components in the engine must also be taken into account. Factors such as the dimensional instability of metallic components, especially the deformation of stationary components at high temperatures, creep, mechanical and thermal fatigue, oxidation/corrosion, erosion, etc. of all the materials/components including compliant and seal materials, changes in the performance of the components after extended service must be considered carefully. It should be pointed out that ceramic components are sensitive not only to the internal defects originating from the ceramic material itself, but also to the "external deficiencies and defects" originating from immature mechanical designs and analyses, assembly or operation of the components and the system. Therefore assuring the reliability and the life of the ceramic gas turbine requires not only highly advanced quality control in the process of component fabrication, but also in the processes of design, analysis, assembly or operation. Technology needs to be developed not only to minimize the internal and external defects, but also to minimize the sensitivity to defects. The efforts to pursue high reliability also have to be optimized in a trade off with efforts to pursue high efficiency.

ACKNOWLEDGEMENTS

The authors would like to express deep appreciation to the Agency of Industrial Science and Technology of the Ministry of International Trade and Industry, and to the New Energy and Industrial Technology Development Organization for the administration and the financial support of the national research programs for the development of ceramic gas turbines. Unless otherwise indicated the figures were provided courtesy of IHI.

REFERENCES

[1] Sasa, T., Koga, S., and Kurita, S., "Reliability of Ceramic Materials for Advanced Gas-Turbine Combustor Associated with Thermal Stress Fracture," Yogyo-Kyokai-Shi, Vol.90[8], pp. 448–458, 1982 (in Japanese).

[2] Suzuki, A., Hamanaka, J., Sakai, K., Nakamura, Y., and Hayashi, S., "A Tentative Design Guide for Fast Fracture of Fine Ceramic Components," Japan Society Mechanical Engineering, Vol. 53 [495A], pp. 2134–2138, 1987.

[3] Okamura, J., Tagashira, K., Katoono, K., Mikami, T., and Sasa, T., "High Temperature Spin Testing and Thermal Shock Testing," Final Report of Fine Ceramics for Future Industries Research Program, Fine Ceramics Research Association, pp. 2089–2131, 1993 (in Japanese).

[4] Sasa, T., "Application of Ceramics to Turbomachinery," Turbomachinery, Vol. 18 [5], pp. 9–34, 1990 (in Japanese).

[5] Yoshimura, Y., Itoh, K., Ohhori, K., and Hori, M., "Research and Development of Ceramic Heat Exchanger," Ishikawajima-Harima Engineering Review, Vol. 34 [6], pp. 409–412, 1994 (in Japanese).

[6] Saito, M., Sasa, T., Sakakida, M., and Kurata, K., "Development of 300 kW Class Ceramic Gas Turbine (CGT301)," ASME paper 93-GT-49, presented at the International Gas Turbine and Aeroengine Congress and Exposition, Cincinnati, OH, USA, May 24–27, 1993.

[7] Sakakida, M., Sasa, T., Akiyama, K., and Tanaka, S., "300 kW Class Ceramic Gas Turbine Development (CGT 301)," ASME paper 94-GT-125, presented at the International Gas Turbine and Aeroengine Congress and Exposition, The Hague, The Netherlands, June 13–15, 1994.

[8] Makino, K., Mizuno, K., and Shimamori, T., "Development and Fabrication of Silicon Nitride Components for Ceramic Gas Turbine," ASME paper 97-GT-67, presented at the International Gas Turbine and Aeroengine Congress and Exposition, Orlando, FL USA, June 25, 1997.

[9] Mikami, T., Tanaka, S., Tatsuzawa, M., and Sakida, T., "Status of the Development of the CGT301, a 300kw Class Ceramic Gas Turbine," ASME paper 96-GT-252, presented at the International Gas Turbine and Aeroengine Congress and Exposition, Birmingham, UK, June 10–13, 1996.

[10] Sasa, T., "Performance of Ceramic Gas Turbine Engine and Parts," *Proceedings of the 6th International Symposium on Ceramic Materials and Components for Engines*, pp. 52–57, Arita, Japan, October 19–23, 1997.

[11] Tagashira, K., Okamura, J., Mikami. T., and Nishi, M., "Evaluation of Reliability of Inserted Ceramic Blades by Hot-Spin Testing," 21st CIMAC, G01, pp. 1–4, 1995.

[12] Iwaki, F., Tagashira. K., and Nishi, M., "Application and Strength Evaluation of Ceramic Blade with Platform Friction Damper," J. Gas Turbine Society Japan, Vol. 25 [98], pp. 106–109, 1997 (in Japanese).

[13] Saitou, T., and Sato, J., 1997, "Development of Ceramic Gas Turbine Combustor for 300kW-Class (CGT301)," ASME paper 97-GT-440, presented at the International Gas Turbine and Aeroengine Congress and Exposition, Orlando, FL, USA, June 25, 1997.

[14] Yoshimura, Y., Itoh, K., Ohhori, K., Hori, M., Hattori, M., Yoshida, T., and Watanabe, K., "Development of Shell-and-Tube Type Ceramic Heat Exchanger for CGT301," ASME paper 95-GT-208, presented a the International Gas Turbine and Aeroengine Congress and Exposition, Houston, TX, USA, June 5–8, 1995.

[15] Tatsuzawa, M., Taoka, T., Sakida, T., and Tanaka, S., "Development of 300kW-Class Ceramic Gas Turbine (CGT301) Engine System," ASME paper 95-GT-201, presented at the International Gas Turbine and Aeroengine Congress and Exposition, Houston, TX, USA, June 5–8, 1995.

[16] Tagashira, K., Mikami. T., and Obata, S., "Cooling Characteristics of an Air-cooled Turbine Disc for Small Ceramic Gas Turbines," 22nd CIMAC, G01, pp. 1–9, 1998.

[17] Mikami, T., Tanaka, S., Tagashira, K., and Koga, S., "Application of Ceramics to the CGT301, a 300kW Class Ceramic Gas Turbine," 9th CIMTEC—World Ceramic Congress, Part D, pp. 241–252, 1998.

[18] Taoka, T., Tanaka, S., Tatsuzawa, M., and Sakida, T., 1998, "Current Status of the CGT301, Ceramic Gas Turbine," ASME paper 98-GT-288, presented at the International Gas Turbine and Aeroengine Congress and Exposition, Stockholm, Sweden, June 2–5, 1998.

[19] Sakida, T., Tanaka, S., Mikami, T., Tatsuzawa, M., and Taoka, T., "Development of the Ceramic Gas Turbine Engine System (CGT301)" ASME paper 99-GT-104, presented at the International Gas Turbine and Aeroengine Congress and Exposition, Indianapolis, IN, USA June 7–10, 1999.

CHAPTER

15

DEVELOPMENT OF THE 300kW CERAMIC GAS TURBINE CGT302

Tetsuo Tatsumi

Kawasaki Heavy Industries, Ltd., Akashi City, Hyogo, Japan

INTRODUCTION

Energy saving technologies are being developed in many industrial fields to minimize environmental impacts by reduction of CO_2 and NOx emissions. The "New Sunshine Project" is a Japanese national project for developing new energy saving and environmental protection technologies. The research and development of a Ceramic Gas Turbine (CGT), which is a part of this project, is expected to be one of the most effective energy saving technologies.

The research and development on the ceramics and ceramic gas turbine in Kawasaki Heavy Industries, Ltd. (KHI) may be traced back to the "Moonlight Project" prompted by MITI from 1978 to 1988. This is the forerunner of the "New Sunshine Project" and the high temperature inter-cooled re-heat cycle advanced gas turbines, AGTJ-100A and -100B rated at 100MW, were developed aiming to prove 55% thermal efficiency [1]. KHI's role in this program involved the development of the combustion system and the evaluation of the ceramic combustor liner and transition duct with temperatures up to 1500°C [2].

Separately from this project, in 1984 KHI started its own research on a recuperated cycle ceramic gas turbine by retrofitting an existing S5 24kW small gas turbine [3] with a ceramic radial turbine wheel, a turbine nozzle and a combustor, and with a metallic recuperator [4]. This research was carried out in cooperation with Kyocera Corporation (KC) and Sumitomo Precision Products (SPP). The ceramic components were made of Kyocera's SN220M silicon nitride. We succeeded in operating this ceramic gas turbine for approximately 100 hours safely at the rated speed of 96,000 rpm and 1100°C TIT. The efficiency of the baseline simple cycle engine and the recuperated ceramic version were 14% and 17%, respectively. The ceramic version was the first prototype design and the efficiency was not satisfactory. We could have improved it but we stopped this research to concentrate to the CGT project which started in 1988. The time line of these early research activities is shown in Figure 15.1.

In 1988, following the "Moon Light" project, the New Energy and Industrial Technology Development Organization (NEDO) started a CGT project [4, 5]. It began as a 9 year program, but was later extended by 2 years, to develop three different types of ceramic gas turbines, namely CGT301, CGT302 and CGT303

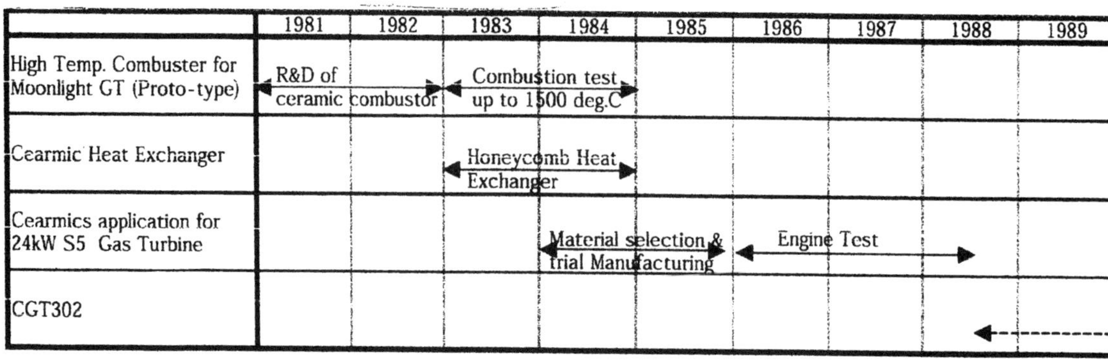

Figure 15.1 Time Line of Ceramics R&D at KHI Before CGT302 (Reprinted with permission from KHI)

under a contract with the Agency of Industrial Science and Technology of MITI [6–11]. The development objective was to achieve 42% thermal efficiency at a 1350°C turbine inlet temperature (TIT). KHI was awarded a contract for the development of the CGT302, a recuperated two-shaft CGT, in cooperation with KC and SPP. After ten years of effort, we achieved 42.1% thermal efficiency and 31.7 ppm NOx emission, and also accumulated 2,100 operating hours at a 1200°C TIT [12–23].

Focus of the Program (Why Ceramics?)

In the history of gas turbine efficiency improvement, the TIT has been steadily increased. For this purpose, sophisticated air cooled turbine blade designs have been adopted for large size gas turbines such as jet engines for aircraft and heavy duty gas turbines for grid power generation. However, it is very difficult to manufacture air-cooled blades and nozzles for small gas turbines because of size limitations. Consequently, the TIT and efficiency of small gas turbines has been limited to approx. 950°C and 20%, respectively. Moreover, increased cooling air leads to a loss of efficiency, as well as a reduction in combustion air, which, in turn, increases the NOx emissions. Therefore ceramics are used to increase TIT without the need for additional air cooling and the associated penalties in efficiency and emissions.

Development Targets

The development targets are shown in Table 15.1. The ambitious efficiency target of 42% was chosen because it is comparable to that of gas and diesel engines.

Development Organization

The development organization is shown in Figure 15.2. Kawasaki Heavy Industries, Kyocera Corporation and Sumitomo Precision Products are responsible for system development and evaluation of the gas turbine, high temperature ceramic components, and the high temperature heat exchanger, respectively.

Development Schedule

The development schedule is shown in Figure 15.3. This development consisted of the following four phases.

TABLE 15.1 DEVELOPMENT TARGETS

	Target
Thermal Efficiency	42%
Turbine Inlet Temperature	1350°C
Output Power	300 kW class
Exhaust Emission	Satisfy the national regulations

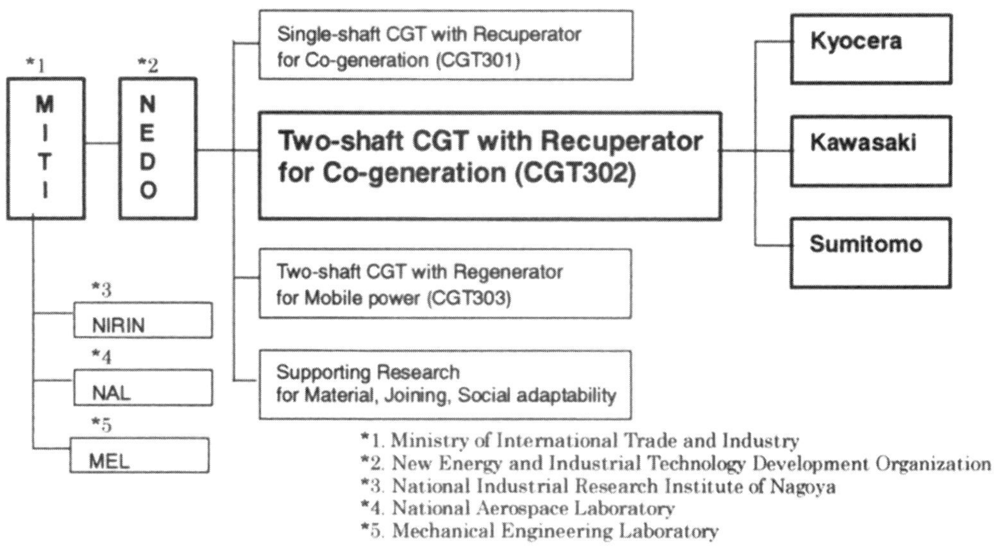

Figure 15.2 Development Organization (Reprinted with permission from KHI)

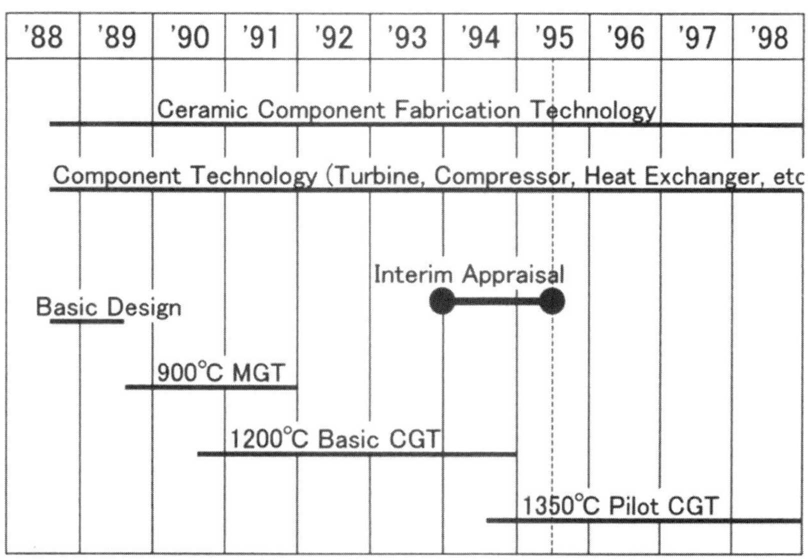

Figure 15.3 Development Schedule (Reprinted with permission from KHI)

Phase 1: Basic design (Ceramic gas turbine)
Phase 2: Basic MGT (900°C Metal gas turbine)
Phase 3: Basic CGT (1200°C Ceramic gas turbine)
Phase 4: Pilot CGT (1350°C Ceramic gas turbine)

The performance target for each phase corresponds to a part load condition of the final target of Phase 4, because the basic designs for all phases are the same.

GAS TURBINE DESIGN

Parametric Study of the Gas Turbine Cycle

A detailed parametric study of the gas turbine cycle was carried out before engine design. Figure 15.4 shows the results. The optimum pressure ratio is affected by the combination of TIT and heat exchanger efficiency while the gas inlet temperature at the heat exchanger is mainly affected by the combination of TIT and pressure ratio. We chose a pressure ratio of 8:1, a TIT of 1350°C, and 82% as the heat exchanger thermal effectiveness. This pressure ratio, which is higher than that of former or existing automobile CGTs, resulted in a lower exhaust gas temperature. It also allowed the use of a metallic recuperator, which was considered more reliable than a ceramic recuperator. The selected design target efficiencies of the compressor, gas generator turbine, and power turbine were 82%, 84%, and 88% respectively.

Candidate Ceramic Materials and Characteristics

Table 15.2 shows physical characteristics of the ceramics represented by silicon nitride and corresponding design consideration required for the ceramic components. For the high temperature structural gas turbine components silicon carbide (SiC) and silicon nitride (Si_3N_4) were the candidate materials because of their high temperature strength and heat resistance at the project start. Si_3N_4 was subsequently selected because of the better fracture toughness and more extensive manufacturing experience compared to SiC, while recognizing that it had lower oxidation resistance than SiC. SN252 was selected as the material for the 1200°C BASIC CGT.

Engine Design Strategy

The design strategy for the CGT302 was the following:

(1) Adopt a conventional component layout to concentrate the development effort on ceramic utilization while avoiding other difficulties.
(2) Adopt the metallic plate-fin type recuperator as a heat exchanger instead of a ceramic rotary regenerator which has been commonly used in the CGT for automobile use.
(3) Adopt a modular configuration to facilitate assembly and disassembly for the numerous program tests.

Conventional Components Layout

One of the main design philosophies for the CGT302 was to simplify its structure and minimize the anticipated development difficulties because of the utilization of the ceramics. Accordingly, we adopted conventional component design methods and layout, and selected a metallic recuperator, single-can combustor, and a turbine scroll, all of which had previously been demonstrated, and which were expected to reduce risks derived from unknown structural factors.

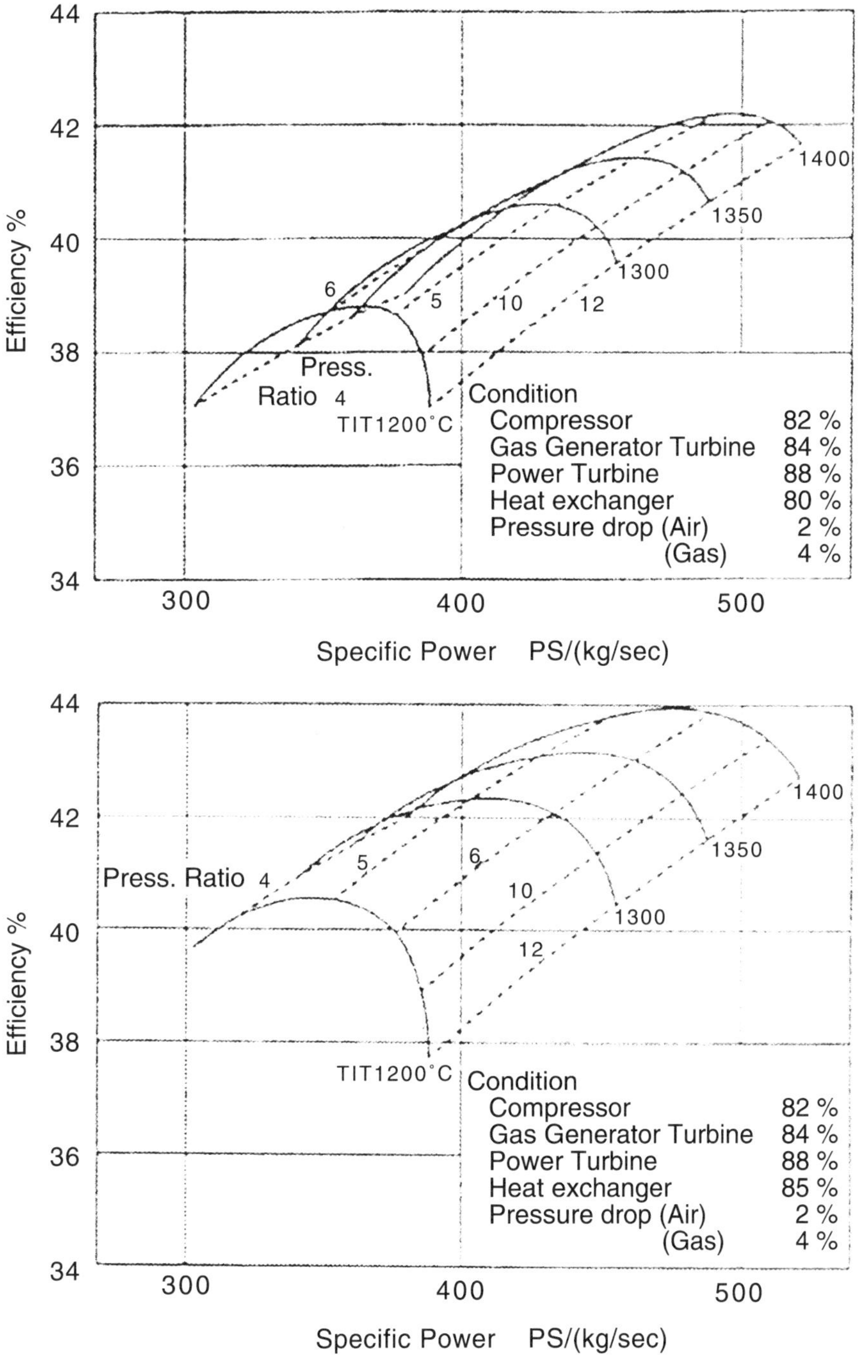

Figure 15.4 Parametric Study of the Gas Turbine Cycle (Reprinted from Ref. [17])

TABLE 15.2 DESIGN CONSIDERATIONS FOR CERAMIC MATERIALS

	Physical Characteristics	Design Consideration
Advantages	* Retains high strength at high temperature	* Suitable for high temperature components
	* Superior resistance against oxidation	
	* High specific strength	* Suitable for high speed rotation components
	* Superior abrasion resistance	* Suitable for components subjected to particulate velocities and high temperature gas flow
	* Low coefficient of thermal expansion	* Turbine tip clearance small because clearance change is small
Disadvantages	* Brittle	* Stress concentration and impulsive load should be avoided
	* Uncertainties in strength	* Difficult to define the design standard
	* Few design data due to recent developed materials	* Special consideration is required for design, machining and inspection
	* Large difference of thermal expansion compared with metal, and less elasticity	* Special consideration is required for joining, binding and assembling

Engine Configuration

The CGT302 consists of a single stage centrifugal compressor, single stage axial gas generator turbine, single stage axial power turbine, single can combustor, a plate fin recuperator, speed reduction gearbox for output shaft, and other auxiliary equipment. The cut away view of CGT302 is shown in Figure 15.5.

Design Features

The main design features of the CGT302 are the following:

- Advanced utilization of ceramics with conventional components layout
- Blisk type turbine rotor
- Metallic recuperator
- Threefold casing to minimize heat radiation loss

ADVANCED UTILIZATION OF CERAMICS

Silicon nitride, the primary candidate ceramic material, is much more brittle than metals and its thermal expansion coefficient is almost 1/4 of that of typical heat resistant alloys. Considering these characteristics, we developed advanced utilization technologies for ceramics contrary to the conventional component layout. These technologies are:

- Composite-component structure (Monolithic—FRC (Fiber-Reinforced Composite) hybrid component
- Stress-free elastic supporting structure
- Stress-free independent supporting structure

Composite-Component Structure (Turbine Nozzle)

The turbine nozzle is a large component consisting of complicated shaped nozzle vanes and an inner and an outer shroud. Among the turbine components, the nozzle experiences the highest temperatures and temperature gradients. Accordingly, a segmented configuration [24, 25] was adopted to reduce thermal stress. This method has the advantage of easy manufacturing but the disadvantage of stress concentration at

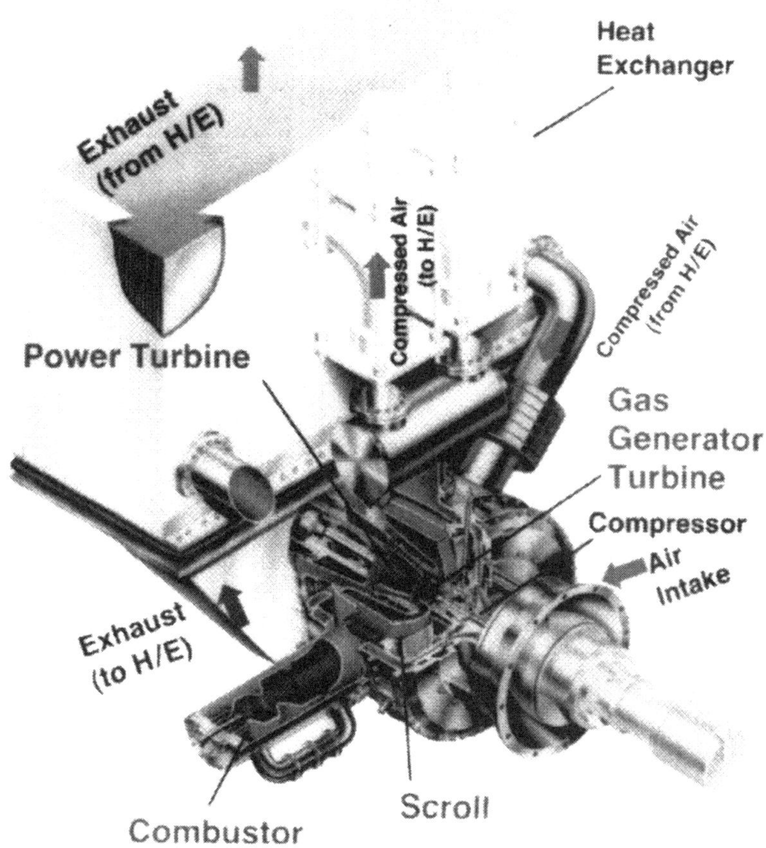

Figure 15.5 Cut-away View of the CGT302 (Adapted from Ref. [13])

the joining position and gas leakage between the segments. Consequently, the concept of composite components (see Figure 15.6) was used to overcome these disadvantages.

The fabricating process for the composite components consists of the following steps:

(1) Join the ceramic segments with ceramic fibers
(2) Impregnate ceramic fibers with an organosilicon polymer
(3) Convert the polymer into SiC
(4) Repeat steps (2) and (3) for sufficient densification of the matrix
(5) Apply a SiC-CVD coating to prevent fiber separation

This process is schematically shown in Figure 15.7.

The resulting composite component has a monolithic ceramic gas path with high heat resistance and a FRC backup ring with high fracture toughness. We applied this technology to the gas generator turbine (GGT) nozzle and the power turbine (PT) nozzle. Utilizing this technology, GGT and PT nozzle segments were assembled tightly enough to seal gas leakage between the nozzle segments. The component centering was precise enough to minimize the turbine tip clearance. These characteristics of little gas leakage and minimum turbine tip clearance greatly contributed to the good turbine efficiency. An example of this composite component technology is shown in Figure 15.8 for the GGT turbine nozzle.

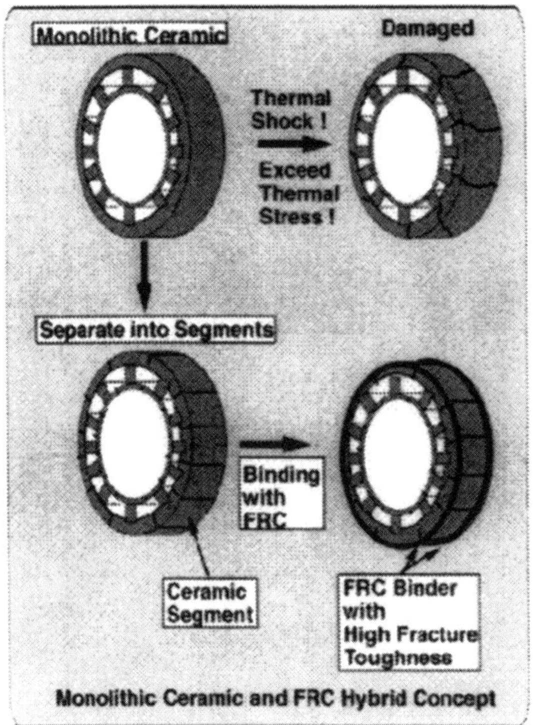

Figure 15.6 Turbine Nozzle Concept (Reprinted from Ref. [17])

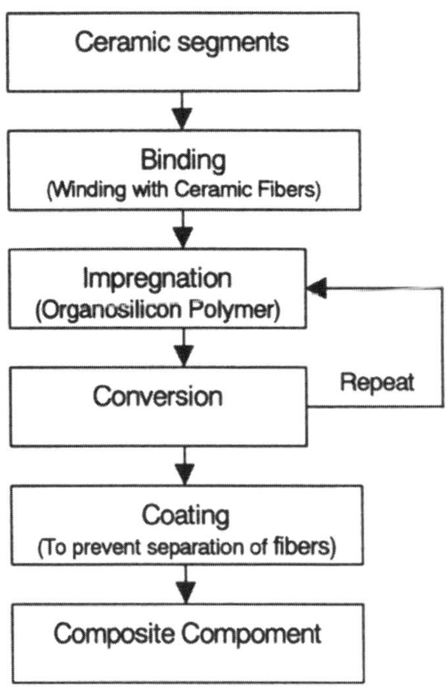

Figure 15.7 Fabrication Process (Adapted from Ref. [24])

Figure 15.8 GGT Turbine Nozzle (Composite Component) (Reprinted from Ref. [17])

Stress-free Structure

A stress-free elastic supporting structure and independent supporting structure were developed to protect ceramic components from excessive force derived from collision of components and loss of tightening force due to the thermal expansion difference between ceramic and metallic components and asymmetric deformation or expansion of metallic supporting structures.

In this approach ceramic components were independently supported while maintaining appropriate clearance between them. Accordingly, ceramic components could be displaced freely, avoiding interference when the metallic support was deformed asymmetrically.

Ceramic coil springs and wave rings made of Si_3N_4 were utilized in the CGT302 to maintain clearance between the ceramic components and to accommodate differences in thermal expansion between metallic support and ceramic components. Figure 15.9 shows the configuration of the stress-free supporting structure.

Blisk Type Turbine Rotor

Two types of ceramic turbine were considered. The first was a "BLISK" type, for which the blades and disk were made in one-piece, and the other was the hybrid type, which was made by using a slotted metal disk and inserted ceramic blades. The BLISK type is difficult to fabricate for a large size turbine and has to be joined with the metal shaft, but it has better properties than the hybrid type as far as heat resistance,

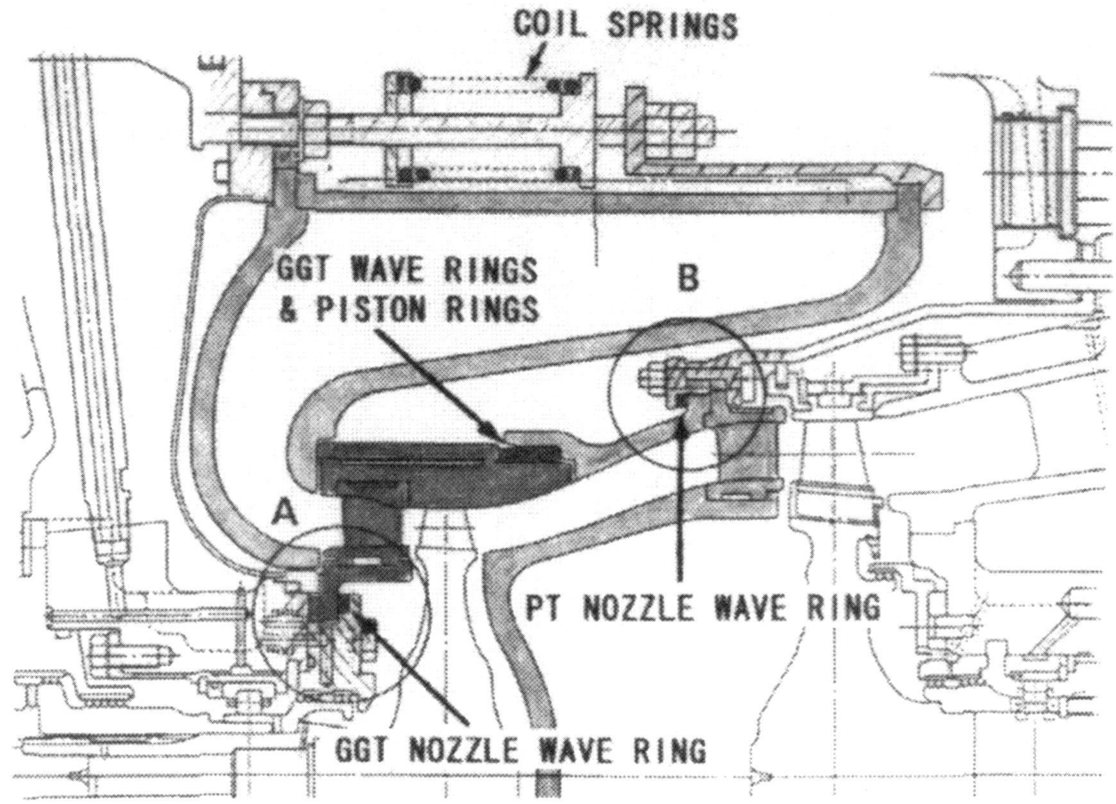

Figure 15.9 Configuration of Stress-free Supporting Structure (Adapted from Ref. [13])

oxidation resistance, strength-weight ratio and manufacturing cost are concerned The manufactured GGT and PT rotors are shown in Figure 15.10. The outer diameter of the PT rotor is 192 mm and this has to be the biggest BLISK type ceramic rotor in the world. The turbine tip clearance greatly affects turbine efficiency especially in these small size gas turbines. The ceramic BLISK turbine with the ceramic shroud is more suitable than the hybrid or conventional metal turbine for minimizing the tip clearance under various operating conditions, because the coefficient of thermal expansion of ceramics is low. The BLISK turbine also helps to obtain better performance with no gas leakage between the blades.

A relatively small BLISK type turbine rotor was selected for a GGT. Initially, a hybrid type was selected for the PT of the program. However as the BLISK manufacturing technology was improved, a BLISK PT rotor was eventually selected for the final design.

Turbine Scroll

The turbine scroll is the biggest ceramic component in the CGT302. It is approximately 360 mm in diameter. Manufacturing challenges such as deformation at sintering, and thermal stresses originating in temperature gradients or external forces generated by joining need to be considered. This component was designed in three parts which were supported elastically and independently from other components by using Si_3N_4 coil springs. These springs were compressed to absorb the thermal expansion difference between the ceramics and metals. At the early stage of the development, the outer scroll was designed and fabricated as

Figure 15.10 BLISK Type GGT and PT Rotors (Reprinted from Ref. [17])

a composite component consisting of 4 wound segments, but later it was changed to one piece based on the engine test experience. Figure 15.11 shows the supporting configuration of the scroll.

Combustor

Compared to the other ceramic components for the CGT the combustor has a rather simple shape. However, it is exposed directly to the combustion flame and has an extremely large temperature distribution in both the radial and axial directions. The single can configuration adopted inevitably causes asymmetry of the casing. Therefore, careful design was required to support it freely and thus minimize excessive force caused by the interference of asymmetrical deformation of the combustor casing. This component was fabricated as a simple one-piece cylindrical combustor liner to facilitate a reliable support system and manufacturing. It was supported by coil springs arranged to push it against the scroll gas inlet port [26–28]. This configuration is shown in Figure 15.12.

Excellent low NOx emission characteristics of the dry low emission (DLE) combustion system can be expected by adopting the ceramic liner. Since the ceramic liner does not need cooling air, the latter can be redirected through the fuel injection system into the combustor primary zone where it reduces the maximum gas temperature.

Metal Recuperator

To obtain the maximum thermal efficiency by fully utilizing a high TIT, the CGT302 adopted a heat recovery system consisting of a plate-fin type counter-flow recuperator made of heat resistant metal. There

Figure 15.11 Supporting Configuration of Turbine Scroll (Reprinted from Ref. [17])

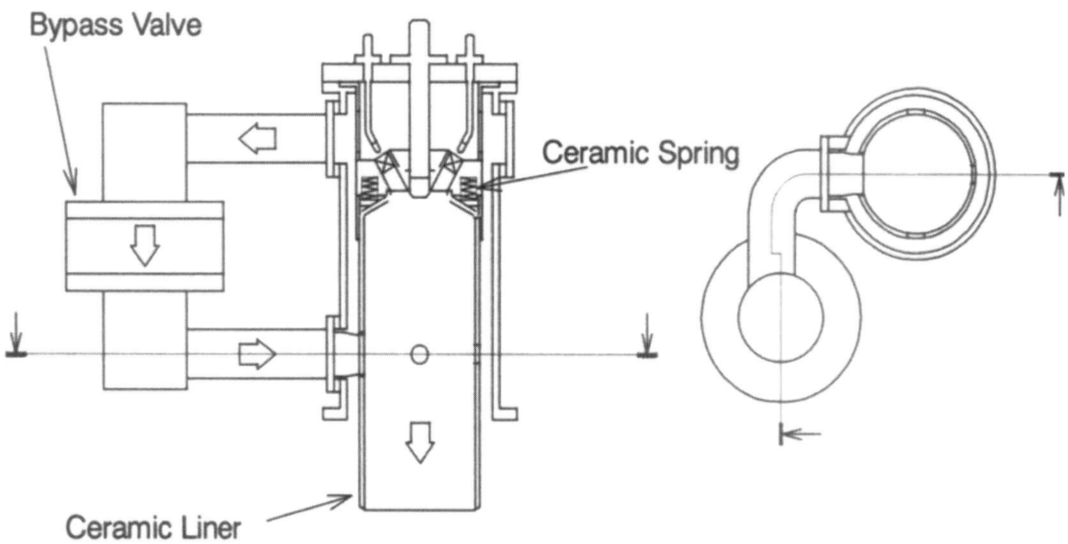

Figure 15.12 Schematic Drawing of the DLE Ceramic Combustor (Reprinted from Ref. [17])

TABLE 15.3 DESIGN PARAMETERS OF THE RECUPERATOR

	Gas	Air
Inlet Temp.	823°C	296°C
Outlet Temp.	390°C	717°C
Pressure Loss	2%	4%

are two very important considerations in the design of the heat exchanger for the CGT. One is the type of heat exchanger and the other is material. The feasible candidate types are rotary regenerators and recuperators, with candidate materials being heat resistant metal and ceramics. We have to choose a suitable configuration carefully from these candidates according to the design specification and the concept of the gas turbine, because the choice of these parameters may greatly affect performance, reliability and size of the gas turbine. We adopted a recuperator which does not have the disadvantages of carryover loss of compressed air and a high temperature moving contact seal. For the material, we chose metal because there were many difficulties such as manufacturing, thermal stress, high temperature seal, etc. if ceramics were selected for the recuperator.

On the other hand, recuperator efficiency tends to be lower than that of a rotary regenerator due to its size limitation. The parametric cycle-performance study defined an optimum pressure ratio of 8:1 which is higher than that of the former CGT for automobile application. This high pressure ratio resulted in a low exhaust gas temperature (820°C), so that we could use a heat resistant metal instead of ceramics. The design parameters of the recuperator are listed in Table 15.3.

Adoption of a metallic plate-fin recuperator seems to be one of the most important factors that led to successful operation of the CGT302, because we could avoid the above mentioned disadvantages and difficulties of the ceramic regenerator.

Threefold Casing to Reduce Heat Radiation Loss

We designed a threefold gas path casing structure for the CGT302, i.e. from inside to outside there is a turbine scroll containing firing gas of 1350°C, the inner chamber of the main casing contains 720°C recuperated compressed air, and the outer chamber of the main casing contains 300°C compressor discharge air. The aim of this configuration was to minimize the heat radiation loss to the atmosphere by avoiding the large temperature variation between the casings. The threefold casing schematic is shown in Figure 15.13.

RELIABILITY EVALUATION OF CERAMIC COMPONENTS

Because the materials used for the CGT302 were relatively new, the database required for life prediction was lacking. Throughout the program, data related to time-dependent behavior (especially cyclic load fatigue) were collected for SN252, SN281, SN282 and interim candidate materials [29, 30].

At the initial design stage, due to the lack of data for time-dependent failure, the ceramic component's failure probability was evaluated according to the fast fracture criterion. We calculated failure probability from 4-point bending strength, Weibull modulus and effective volume. For the actual design, we took the following stress values as design limits assuming that the maximum temperature was below 1200°C for highly stressed components:

For machined surface: 294 MPa (30 kgf/mm^2)
For as-fired surface: 245 MPa (25 kgf/mm^2)

This value was derived from the ceramics manufacturer's experience.

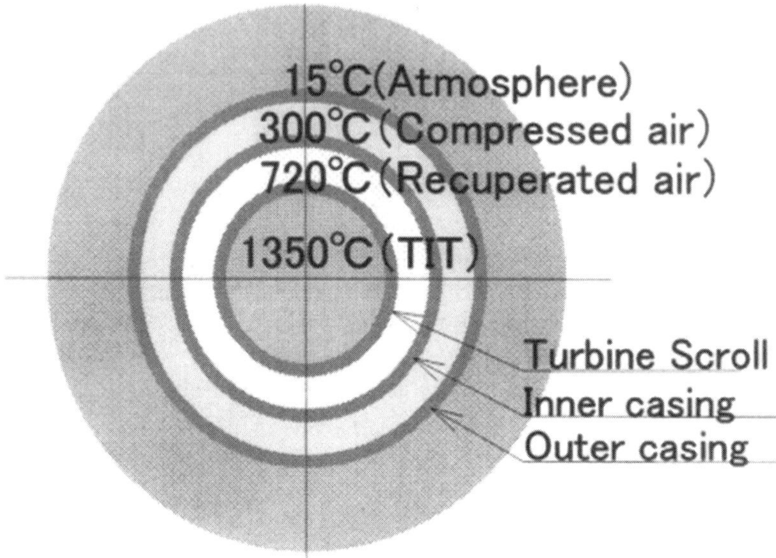

Figure 15.13 Threefold Casing Schematic (Reprinted from Ref. [17])

Stress Analysis of BLISK Type Rotors

Results of stress analysis for the GGT rotor with 76,000 rpm rated speed (initial design) are shown in Figure 15.14.

Based upon this analysis, there was a major concern with the fillet radius because of a stress concentration of 1.5 to 2. The maximum thickness around the disk center was eventually decreased to make forming easier, and the rated rotational speed was decreased and blades were thickened to increase FOD (Foreign Object Damage) resistance. This changed the stress distribution of the final rotor design. A similar stress analysis was also performed for the PT rotor. The steady state thermal stress (incorporating centrifugal stress) is shown in Figure 15.15.

We carried out a reliability evaluation using CARES according to the fast fracture criterion. The resulting failure probability was 10^{-4} to 10^{-5}, which was acceptable for the demonstration phase of the program. We also carried out the steady state and transient thermal stress analyses and consolidated it with the centrifugal stresses for the GGT and PT rotors.

Stress Analysis for the PT Blade

For the hybrid type PT rotor, we carried out elastic-plastic stress analysis to evaluate the effects of the stress concentration and relaxation accurately at the contact portion of the ceramic blade and metal disk. Figure 15.16 shows an example of the calculation.

Stress Analysis for the GGT Nozzle

For this component steady-state and transient thermal stresses were estimated neglecting any loading contribution from the surrounding structure. This assumption was used because the nozzle was designed to be stress-free. A thermal conduction analysis was initially performed to estimate temperature distributions, which were then used to calculate thermal stresses. The results indicated a maximum thermal stress of 217

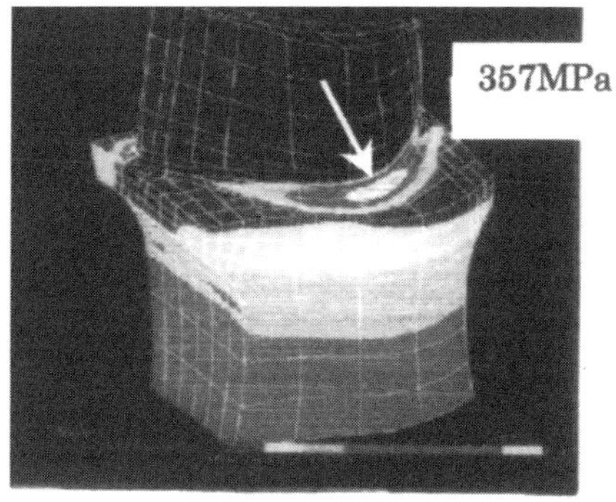

Stress concentration on the blade root (Fillet R=1mm)

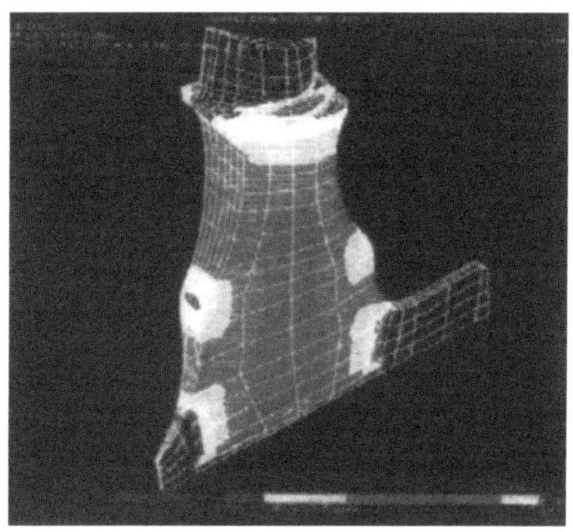

Stress distribution of the GGT rotor (Fillet R=2mm)

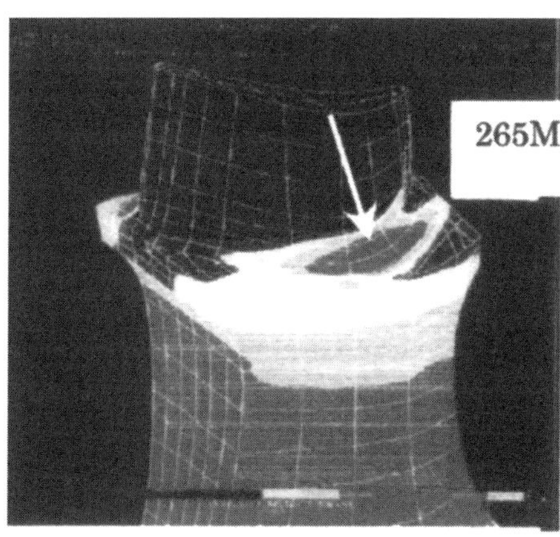

Stress concentration on the blade root (Fillet R=2mm)

Figure 15.14 Stress Distribution in the GGT Rotor (Reprinted with permission from KHI)

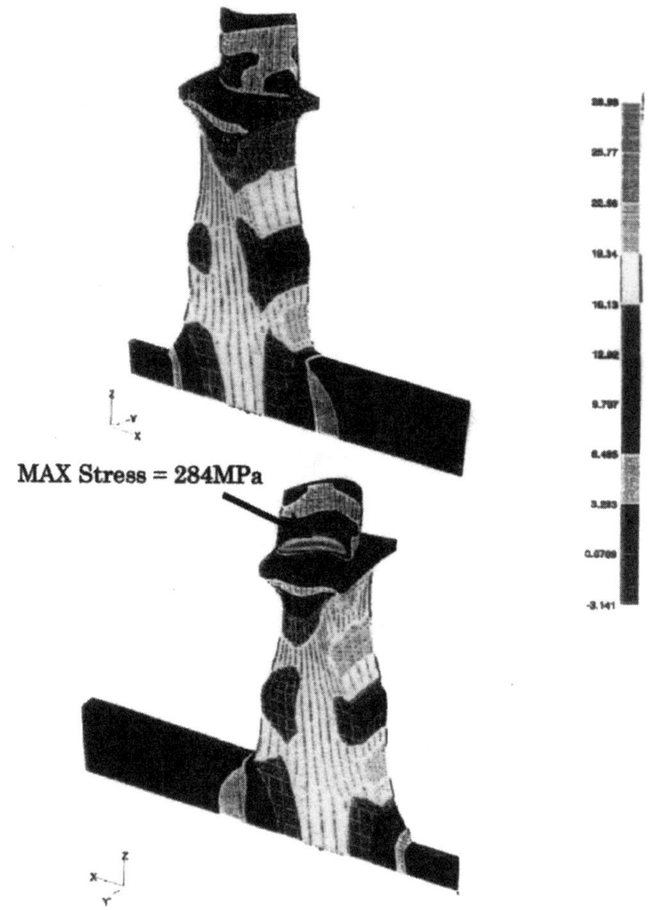

Figure 15.15 Stress Distribution in the PT Rotor (Reprinted with permission from KHI)

MPa on the outer surface of the nozzle segment after 5.9 seconds from the start of the engine. This stress level was acceptable. A result of the stress analysis is shown in Figure 15.17.

ENGINE TEST RESULTS

900°C Metal GT

The MGT was manufactured as a test bed for the CGT and was tested to confirm the performance ability and appropriateness of the design concept of the CGT. In the MGT design, metal parts having the same design as the ceramic components were used. Engine testing of the MGT was started at the end of 1990. We confirmed stable operation and improved the basic performance through the MGT testing. The efficiency and output power at the initial stage were only 17.3% and 42 kW, respectively, at the 913°C TIT, but by improving compressor performance [31, 32], modification of the compressor matching point to that of the turbine, and heat insulation etc., the final stage performance of the MGT was improved to 23% efficiency and 59 kW output at a 899°C TIT. Although the 900°C TIT only represented a 1/5 part load condition, this efficiency

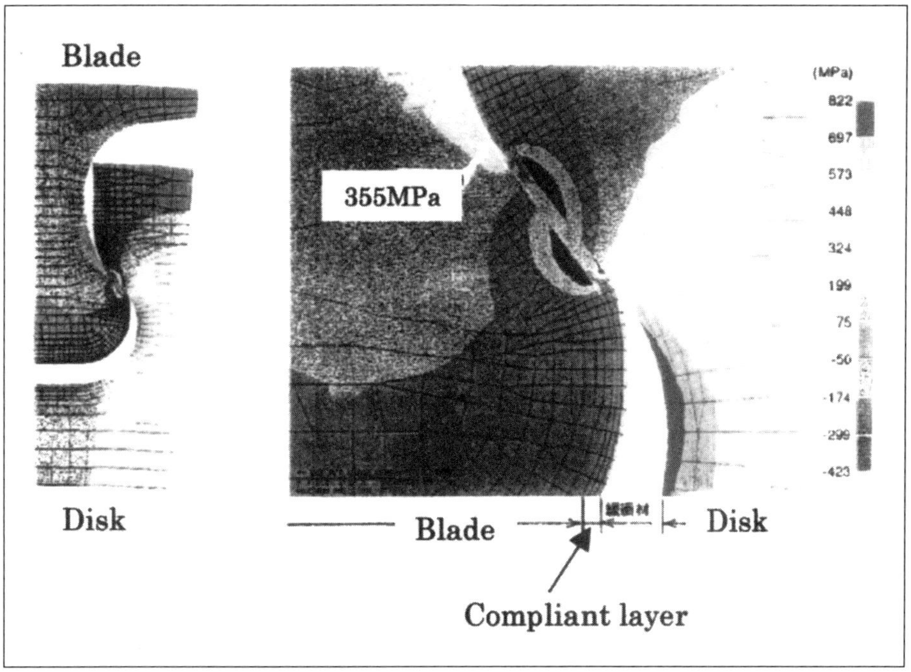

Figure 15.16 Stress Distribution Near the Contact Point of the PT Blade and Disk (Reprinted with permission from KHI)

corresponds to the efficiency of the rated condition of the conventional 1 MW class gas turbine and cleared the expected target value of 23% and 56 kW for the MGT. The performance improvement at the early stage is shown in Figure 15.18 and Figure 15.22 compared with the later development progress of the CGT.

Through these experiences, the design concept was substantiated and the superior performance potential of the CGT was demonstrated. The cumulated operating time of the MGT was 97 hours and the number of start-stops was 282.

1200°C Basic CGT

The high temperature metallic components in the MGT were replaced with ceramic components for the 1200°C Basic CGT. The ceramic material for the Basic CGT was SN252 Si_3N_4. The development of Monolithic-FRC hybrid components, elastic supporting, independent supporting, metal-ceramic joining, etc. had progressed and these results were applied to the Basic CGT design in 1991.

In 1992, before installing ceramic components in the Basic CGT, we conducted component tests such as a cyclic thermal-shock test for stationary components and cold and hot spin tests for turbine rotors. A thermal-shock test cycle consisted of start up till the temperature reached 1200°C, maintaining 1200°C for 10 minutes until the components' temperature became stable, shutdown till the temperature dropped to 150°C, keep cooling for 6 minutes until the components are fully cooled. The temperature logging chart for the 93rd to the 95th cycle is shown in Figure 15.19.

From 1992 on, the Basic CGT was repeatedly tested. Before the all-ceramic configuration tests, we conducted a confirmation test by installing a single ceramic component or the minimum number of ceramic components in a step by step manner. The test step sequence is shown in Figure 15.20.

In 1993 and 1994, we experienced unexpected obstacles in these tests, but most of them were not attributable to the ceramic components. For example, as summarized in Table 15.4, problems arose with gears,

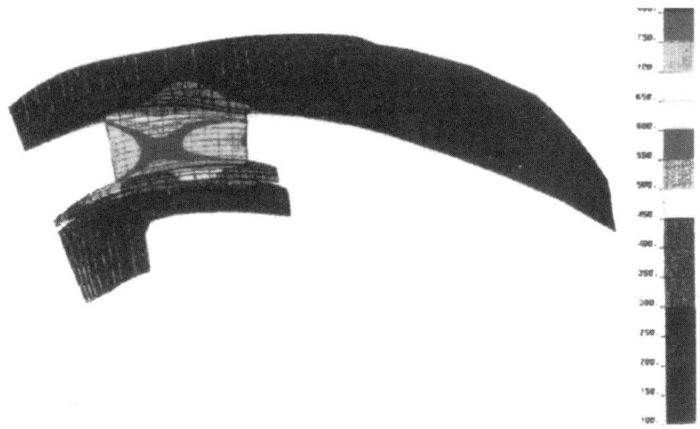

Temperature distribution of GGT nozzle
after 5.9 second from start

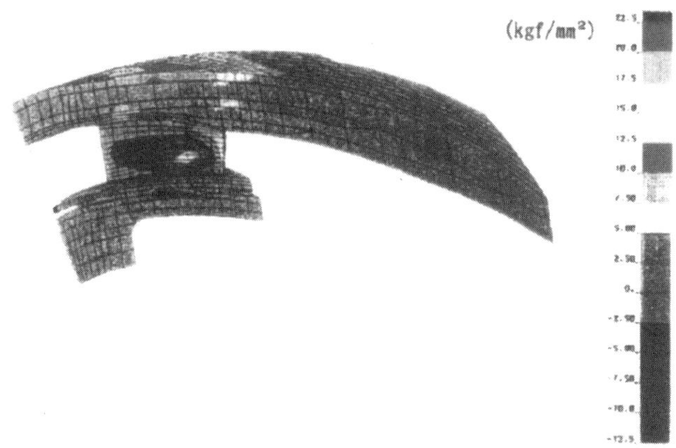

Stress distribution of GGT nozzle
after 5.9 second from the engine start

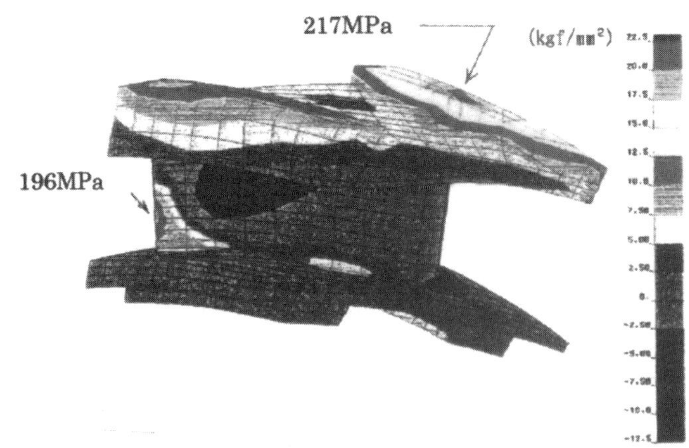

Stress distribution on the nozzle segment
after 5.9 second from the engine start

Figure 15.17 Stress Distribution in the GGT Nozzle Under Transient Conditions (Reprinted with permission from KHI)

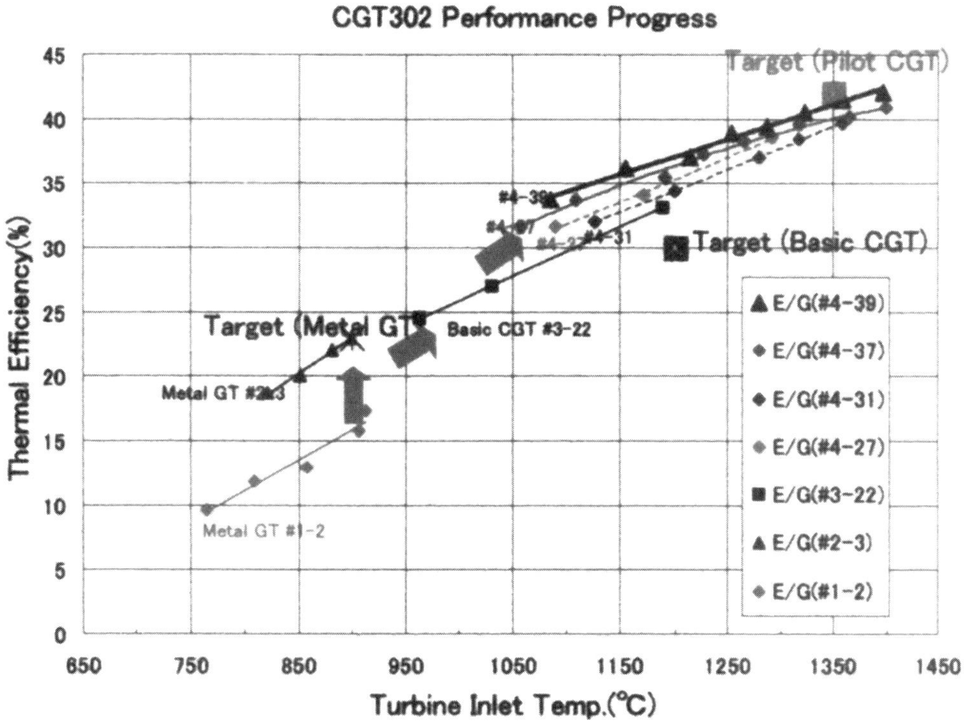

Figure 15.18 CGT Performance Progress (Adapted from Ref. [17])

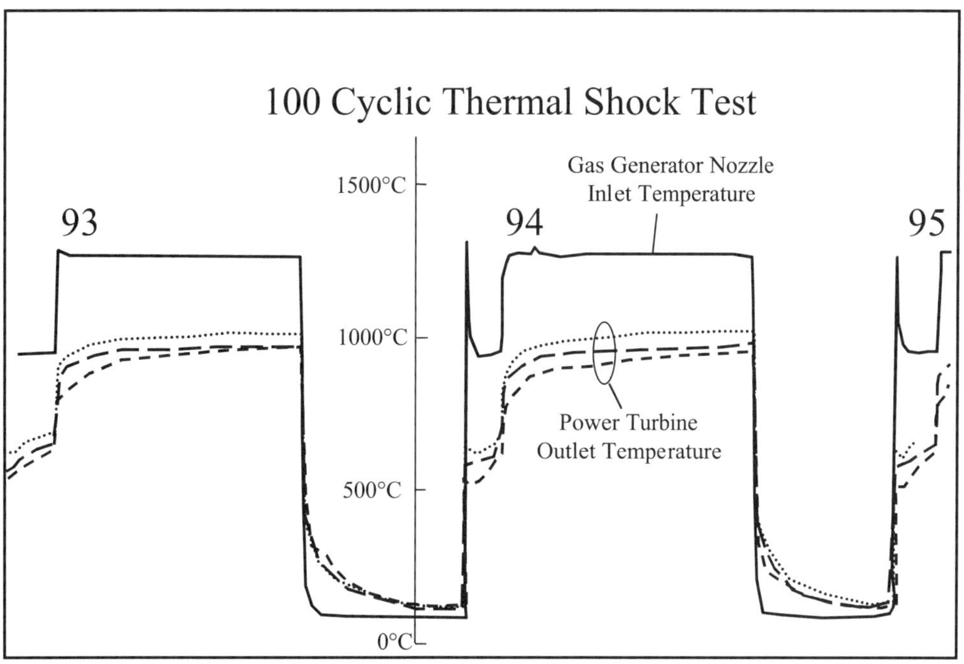

Figure 15.19 Temperature Logging Chart for the 93rd to the 95th Cycle (Reprinted with permission from Ref. [21])

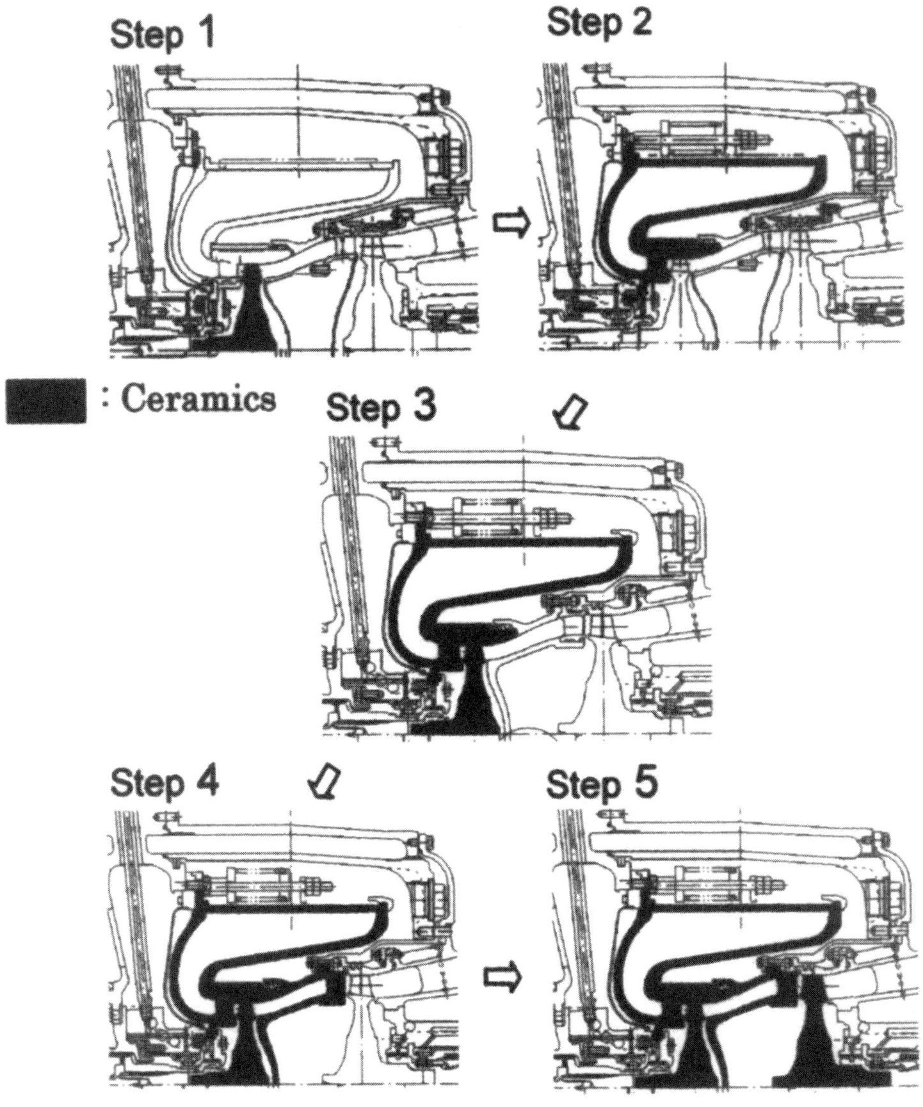

Figure 15.20 Test Step Sequence for the CGT (Reprinted from Ref. [17])

bearings, oil leaks, etc., and they were serious enough to delay the test schedule. After appropriate remedial action, we demonstrated the durability of the CGT through 40 hours of operation at 1200°C TIT in 1994. We have also obtained 33% thermal efficiency and 164kW at 1190°C TIT in 1995 as shown in Figure 15.18 and Figure 15.22. This efficiency is comparable with that of 100 MW class gas turbines and cleared the expected target of 30% and 140 kW for the Basic CGT.

These R&D results for the Metal GT and the Basic CGT were evaluated by the neutral committee for industrial technology assessment in 1994 and 1995, and progression to the Pilot CGT stage was approved [9, 10].

1350°C Pilot CGT

Fortunately, the Basic CGT development was successfully demonstrated and we could move forward to the Pilot CGT stage without any basic design change. The differences between the Basic CGT and the Pilot

TABLE 15.4 PROBLEMS EXPERIENCED AND REMEDIAL ACTION

Problems	Cause	Countermeasure
(1) Lube oil leakage	Seal air imbalance	Improve seal design
(2) Excessive rotor vibration at start acceleration	Inadequate rotor balance	Improve balance method and stiffened the metal shaft
(3) Pull out of ceramic turbine from metal shaft	Overheating of joining section	Improve cooling and joining method
(4) Bearing seizure from overtemperature	Excessive reduction of residual radial clearance of bearing at high temperature operation	Optimize the clearance by trial and error through engine testing
(5) Auxiliary gear teeth damage	Lack of strength	Change material and surface hardening heat treatment
(6) FOD (Foreign Object Damage) for GGT blades	Unavoidable ceramic or metal debris (insufficient cleaning) Poor toughness and strength of ceramics	Reduce turbine tip speed

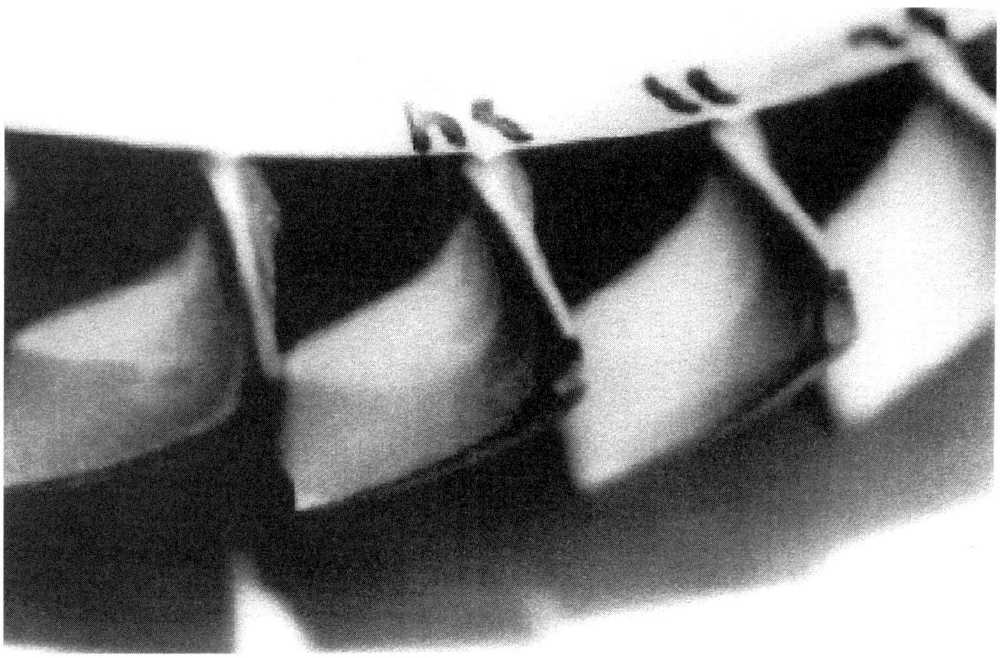

Figure 15.21 An Example of the Chipped Blades (Reprinted from Ref. [14])

CGT were the ceramic material, the replacement of the hybrid PT with a BLISK design, and a reduction in turbine tip speed to improve FOD resistance. Specifically, the ceramic material was changed from SN252 to SN281 for the GGT rotor, and to SN282 for all other ceramic components. The difference between SN281 and SN282 is that a HIP process is applied for SN281. The turbine tip speed at rated condition was reduced to 480 m/s from 570m/s to improve FOD resistance; 480 m/s was selected because of the successful 20 hour engine test operation at this tip speed while FOD had been often observed at tip speeds over 500 m/s. Accordingly, the rotational speed of the gas generator and power turbine were reduced to 64,000 rpm from 76,000 rpm, and to 47,800 rpm from 57,000 rpm, respectively. An example of chipped blades as a result of FOD is shown in Figure 15.21.

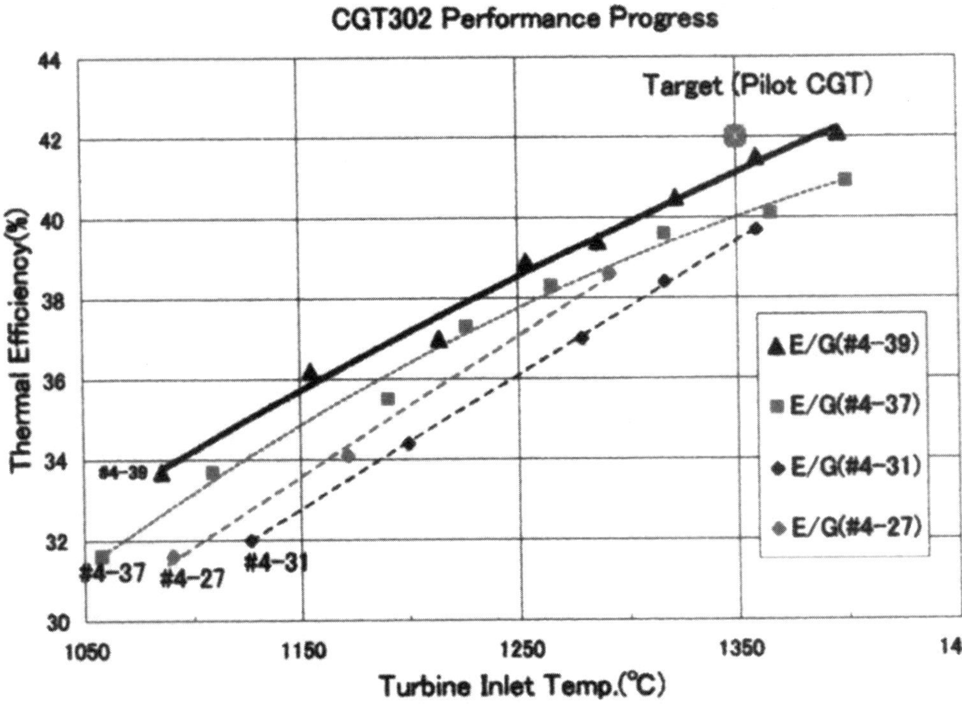

Figure 15.22 Pilot CGT Performance Improvement (Reprinted from Ref. [19])

We tested the Pilot CGT by increasing the TIT stepwise by 50°C starting at 1200°C, the TIT at which successful operation had been confirmed by the Basic CGT, and in each step we confirmed stable operation and performance. To operate the engine safely at the stepwise increased TIT, we needed some modifications such as cooling of the joint of the GGT rotor, and clearance of roller bearings in the hot section. After these steps, we succeeded in obtaining 30 hours of 1350°C TIT operation in 1997.

To obtain further performance improvement, we redesigned the compressor to increase the air flow rate to get better matching to the turbine [33], because we found that the turbine nozzle throat area was slightly larger than the optimum value. We also improved the manufacturing accuracy of the turbine nozzle throat area because it had become evident from the engine test and dimensional inspection that the throat area fluctuated in each product and in each throat. Also, we redesigned the GGT rotor blade shape to match the new reduced rotational speed. After these improvements, in 1998, we measured an efficiency of 38.6% at 1290°C TIT with 254 kW output power by engine test #4–27 in Figure 15.22.

To reach 40% efficiency at 1350°C TIT further improvements in the nozzles were required as well as confirmatory engine tests. We investigated the best combination of the throat area of the GGT and PT nozzles and got 40.5% at 1412°C TIT with 285 kW and 39.7% at 1359°C TIT with 277 kW by test #4–31. To achieve the 42% target efficiency, improvements such as a larger heat exchanger with increased stack layer, an abradable PT shroud to minimize the tip clearance, and a PT rotor with thinner blades for improved turbine efficiency, and an improved compressor, were implemented. After several attempts, we achieved 42.1% at 1396°C TIT with 322 kW and 41.5% at 1359°C TIT with 302 kW in test #4–39 that was carried out in the early morning of March 15, 1999. The achieved efficiencies are comparable or superior to those of the largest gas turbines for power generation that have almost 1000 times the output power of the CGT. Comparisons of the performance improvement history for power output and efficiency are shown in Figure 15.23 and Figure 15.24, respectively.

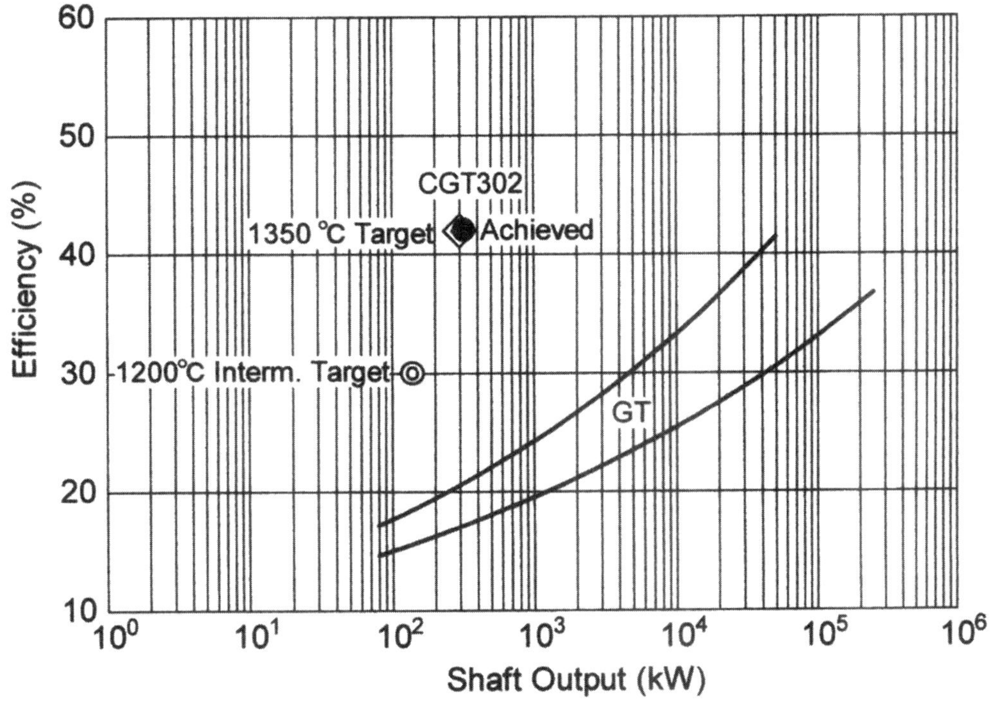

Figure 15.23 CGT302 Efficiency Compared with that of a Conventional GT (Reprinted from Ref. [17])

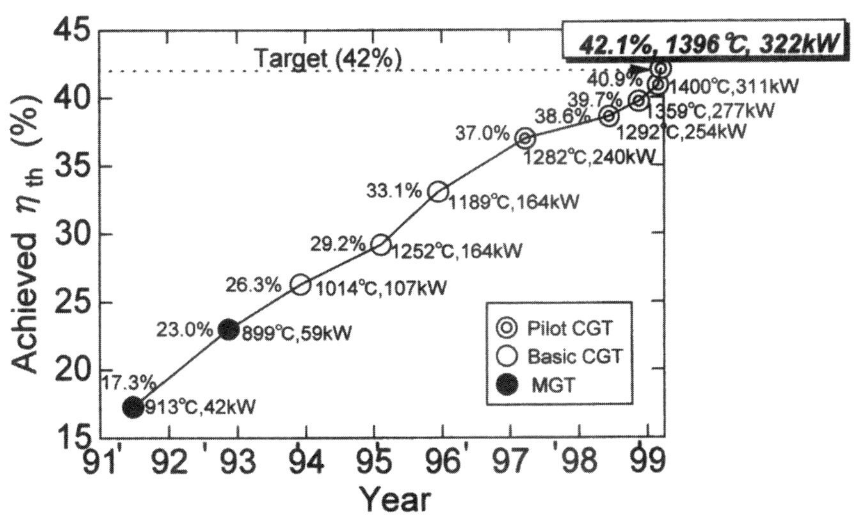

Figure 15.24 Efficiency Improvement History (Reprinted from Ref. [17])

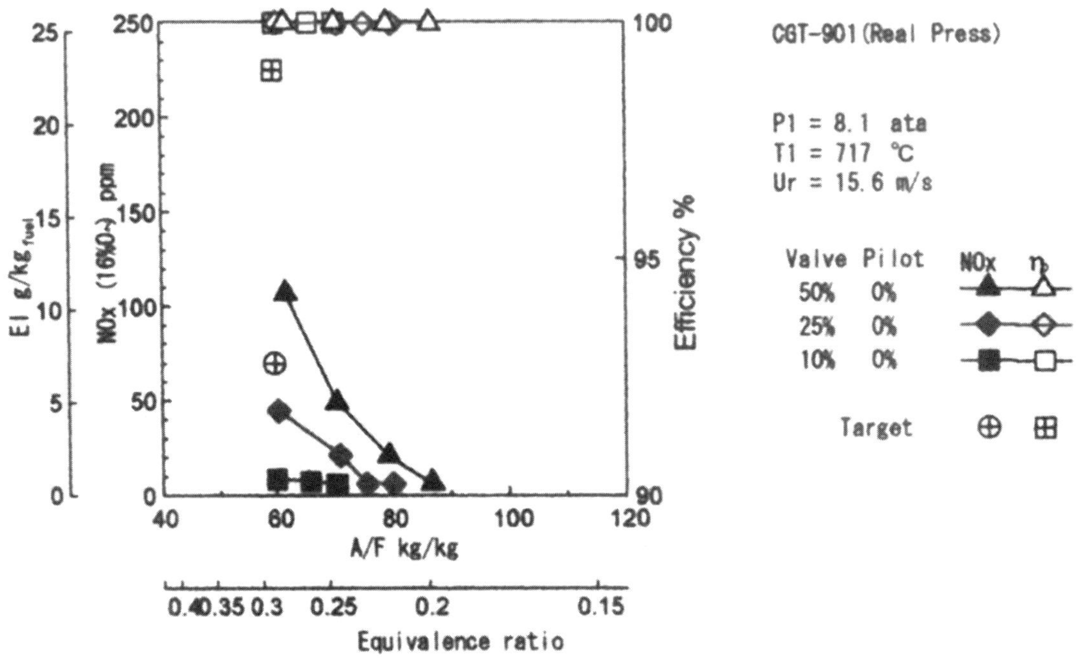

Figure 15.25 NOx Emission in Rig Test (Reprinted from Ref. [17])

As far as NOx emissions, we confirmed 9 ppm by the rig test and 31.7 ppm (O_2 =16%) by the engine test at 1350°C TIT. These values are lower than the target of "<70ppm". Measured NOx emission data are shown in Figure 15.25 and Figure 15.26 [26–28].

The cumulative operating time for the Basic CGT and the Pilot CGT was 334 hours and the number of start-stops was 645.

1200°C Endurance Test

At the final stage of this project a new research task involving 1,000 hours of 1200°C TIT engine endurance testing was added. The 1200°C was considered to be a reasonable TIT for future commercial use of ceramics. The package type test facility was prepared in 1997 and weekly start-stop continuous operation was conducted from the beginning of 1998 till March 1999. We conducted over 2,100 cumulative operating hours with a total of three engines. The longest endurance operation for the same engine was 782 hours and the high time component was the combustor liner, which was operated for up to 1,000 hours. We experienced three unscheduled terminations of the test because of damage to the engine. We concluded in each case that it was initiated by GGT rotor failure. Possible causes of the failure included ceramic material deterioration, turbine blade resonance and foreign or domestic object damage. However the precise cause of the GGT rotor damage could not be definitively established. Investigation of the combustor liner after the 1,000 hours of 1,200°C operation showed a little more surface roughness at the inside surface of the downstream area than the upstream area, but the material strength was not changed. The endurance test results are listed in Table 15.5.

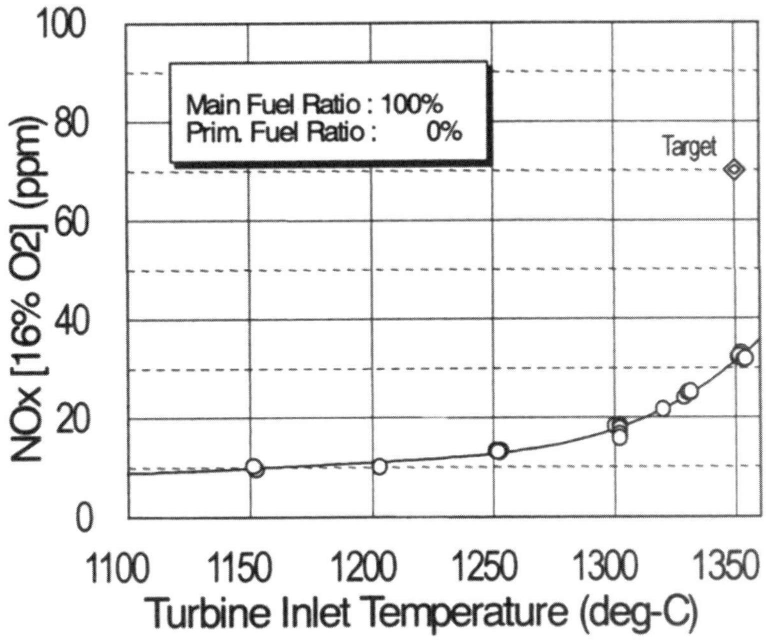

Figure 15.26 NOx Emission in Engine Test (Reprinted from Ref. [17])

TABLE 15.5 ENDURANCE TEST RESULTS

Duration	Time	Observations
Target	1,000 hours	Weekly start-stop
Cumulative operating time for 1200°C	2,117 hours 57 min	
1st attempt	592 hours	GGT rotor damaged
2nd attempt	519 hours	GGT rotor damaged, combustor support ring survived
3rd attempt	782 hours	Combustor liner and support ring survived
4th attempt	218 hours	Combustor liner reached 1,000 hours, the test was halted
Other	7 hours	Checking and adjustment
Total cumulative operating time	2,146 hours 10 min	
Longest time for the same assembly	782 hours	
Longest time for the same component (combustor liner)	1,000 hours	
Operation time of heat exchanger	2,118 hours	

RESEARCH AND DEVELOPMENT ON PRACTICAL INDUSTRIAL CO-GENERATION TECHNOLOGY

Based on the successful results of 300 kW ceramic gas turbine development, MITI and NEDO started the "Research and Development on Practical Industrial Co-generation Technology" project in 1999. The purpose of the project is to encourage prompt industrial applications of co-generation technology that employ

TABLE 15.6 DEVELOPMENT TARGETS FOR 8,000 KW CGT PROGRAM

Items	Targets
Output Power	8,000 kW class
Thermal Efficiency	34% or more
Turbine Inlet Temperature	1,250°C
Operation period	4,000 hours
Exhaust gas properties	Standard regulation values or less

Note 1: Thermal efficiency is the ratio of heat input based on the lower heating value of the fuel to the gross shaft horsepower of the engine under typical ambient conditions (air pressure of 101.3 kPa and air temperature of 15°C)

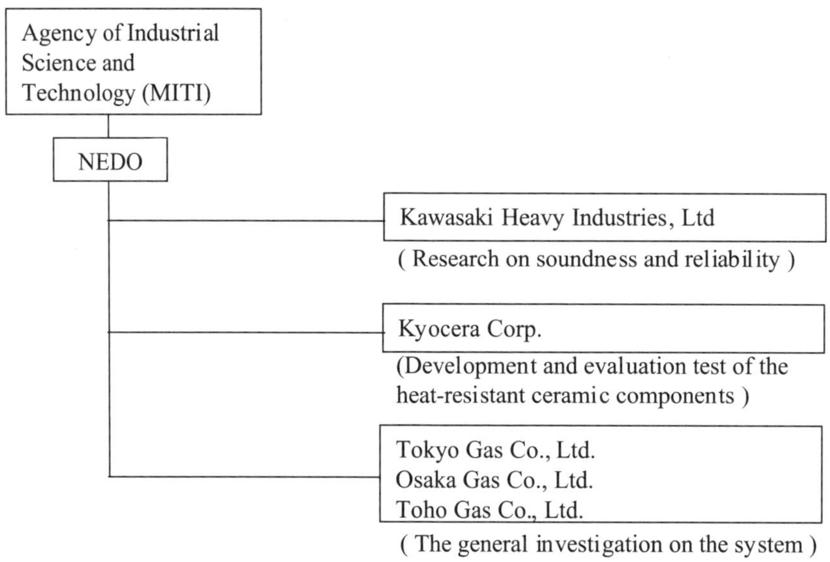

Figure 15.27 Development Organization for the NEDO 8,000 kW Program (Reprinted from Ref. [35])

hybrid gas turbines (using both metal and ceramic parts in its high-temperature section) by confirming its reliability and soundness. The development activities are performed through material evaluation tests and long-term operation tests for a medium size hybrid gas turbine (8,000 kW class). It is expected that the development will result in lower pollution by reducing the emission of CO_2 with highly efficient use of energy. An initial report on the project has been presented [34]. The Development targets are listed in Table 15.6. The development organization and schedule are shown in Figures 15.27 and 15.28, respectively [35].

SUMMARY

The 300 kW CGT development project promoted by MITI and NEDO was successfully finished in March 1999. Through this project, we confirmed that the uncooled 1350°C TIT ceramic gas turbine can be operated in a stable mode and we also achieved the project target with 42.1% efficiency and 31.7 ppm NOx emission. Key technologies such as monolithic-FRC hybrid components, stress free supporting, independent

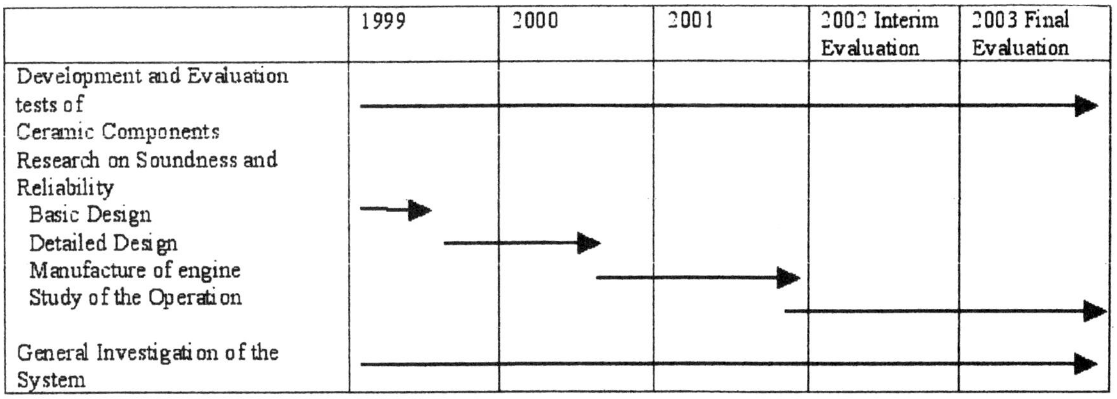

Figure 15.28 Schedule for the NEDO 8000 kW Program (Reprinted from Ref. [35])

supporting, ceramic-metal joining, etc. for the application of ceramics to gas turbines were successfully developed and verified. Engine endurance testing with 2,100 cumulative operating hours, demonstrating the potential for future commercial use, was accomplished.

However, before commercial use of the ceramic gas turbine can be realized, there still remain several difficulties to be solved such as:

(1) The soundness and reliability of the ceramic components after long-term operation with the actual condition of the gas turbine environment has still not been adequately demonstrated.

(2) Degradation of ceramic materials under the high temperature, high pressure and high water vapor pressure in the gas turbine hot section has recently been reported [36, 37]. This phenomenon should be clarified further as to what factor is the main cause of the degradation, and how the various factors affect ceramics.

(3) Moving ceramic blades could not clear long-term (1,000 hours) operation and Foreign or Domestic Object Damage. Further improvements or research will be required to address and overcome these problems.

(4) The popular consensus for structural ceramics still seems to be: "Ceramics are very brittle and they very risky for application for structural components". Convincing demonstrations will be required in order for ceramics to be acknowledged as valid structural materials.

Long-term operation of the actual gas turbine will be required to clear these difficulties. The "Research and Development on Practical Industrial Co-generation Technology" was initiated to meet this requirement and to encourage accelerated commercialization.

REFERENCES

[1] Arai, M., Imai, T., Teshima, K., Koga, A., "Research and Development on the HPT of the AGTJ-100B," ASME paper GT-87-261, presented at the International Gas Turbine Conference and Exhibition, Anaheim, California, USA, May 31-June 4, 1987. *Transactions of the ASME, Journal of Engineering for Gas Turbine and Power*, 110[2], 259–264, 1988.

[2] Mori, K., Kitajima, J., Kimura, T., Nakamura, T., "Research on high Temperature Combustor for Advanced Reheat Gas Turbine," ASME paper 86-GT-281, presented at the International Gas Turbine Conference and Exhibit, Düsseldorf, West Germany, June 8–12, 1986.

[3] Hoshino, A., Sugimoto, T., Tatsumi, T., Nakagawa, Y., "Development of a 30PS Class Small Gas Turbine and It's Power-Up Version," ASME paper 88-GT-2, presented at the International Gas Turbine and Aeroengine Congress, Amsterdam, The Netherlands, June 6–9, 1988, and Transactions of the ASME, Journal of Engineering for Gas Turbine and Power, 111[2], 225–231, 1989.

[4] JETRO, "R&D OF AUTOMOTIVE CERAMIC GAS TURBINES IN JAPAN," New Technology Japan Special Issue, Japan External Trade Organization, 1988.

[5] Yamagishi, K., Yamada, Y., Echizenya, Y., Ishiwata, S., "Current Status of Ceramic Gas Turbine R&D in Japan," ASME paper 89-GT-114, presented at the International Gas Turbine and Aeroengine Congress and Exposition, Toronto, Canada, June 4–8, 1989.

[6] Nagamatu, S., Mizuhara, K., Matsuda, Y., Iwanaga, A., Ishiwata, S., "Current Status of Industrial and Automotive Ceramic Gas Turbine R&D in Japan," ASME paper 91-GT-10, presented at the International Gas Turbine and Aeroengine Congress and Exposition, Orlando, Florida, USA, June 3–6, 1991.

[7] Honjo, K., Hashimoto, R., Ogiyama, H., "Current Status of 300 kW Industrial Ceramic Gas turbine R&D in Japan," ASME paper 92-GT-3, presented at the International Gas Turbine and Aeroengine Congress and Exposition, Cologne, Germany, June 1–4, 1992.

[8] Shimada, K., Ushijima, H., Yabe, A., Ogiyama, H., Tsutsui Y. "Advanced Ceramic Technology Developed For Industrial 300 kW CGT (Ceramic Gas Turbine) Research And Development Project In Japan," ASME paper 93-GT-188, presented at the International Gas Turbine and Aeroengine Congress and Exposition, Cincinnati, Ohio, USA, May 24–27, 1993.

[9] Murayama, T., Nagata K., Taki M., Ogiyama H., "Current Status of 300kW Class Industrial Ceramic Gas Turbine R&D in Japan," ASME paper 94-GT-482, presented at the International Gas Turbine and Aeroengine Congress and Exposition, The Hague, The Netherlands, June 13–16, 1994.

[10] Murayama, T., Nagata, K., Abe, H., "Current Status of 300 kW Industrial Gas Turbine," ASME paper 95-GT-445, presented at the International Gas Turbine and Aeroengine Congress and Exposition, Houston, Texas, USA, June 5–8, 1995.

[11] Arakawa, H, Suzuki, T, Saito, K, Tamura, S, Kishi, S, "Research and Development of 300kW class Ceramic Gas Turbine Project in Japan," ASME paper 97-GT-87, presented at the International Gas Turbine and Aeroengine Congress and Exposition, Orlando, Florida, USA, June 2–5, 1997.

[12] Watanabe, A., Tatsumi, T, Tanimura, K., Takehara, I., Fujii, T., "Development of 300kW Class Ceramic Gas Turbine (CGT302)," ASME paper 94-GT-19, presented at the International Gas Turbine and Aeroengine Congress and Exposition, The Hague, The Netherlands, June 13–16, 1994.

[13] Tatsumi, T, Inobe, I, Takehara, I, Ichikawa, Y, "Current Status of CGT302 Ceramic Gas Turbine," GTSJ, 95-Yokohama-IGTC-125, *Proceedings of the 1995 International Gas Turbine Congress*, Yokohama, October 23–27, 1995. Vol. I, I-153/I-156, 1995.

[14] Takehara, I., Inobe, I., Tatsumi, T., Ichikawa, Y., Kobayashi, H., "Research and Development of Ceramic Gas Turbine (CGT302)," ASME paper 96-GT-477, presented at the International Gas Turbine and Aeroengine Congress and Exhibition, Birmingham, U.K., June 10–16, 1996. *Transactions of the ASME, Journal of Engineering for Gas Turbines and Power*, 120[1], 186–190, 1998.

[15] Ichikawa, Y., Tatsumi, T., Nakashima, T., Takehara, I., Kobayashi, H., "Current Status of CGT302 (progressing to Final Phase)," ASME paper 97-GT-461, presented at the International Gas Turbine and Aeroengine Congress and Exhibition, Orlando, Florida, USA, June 2–5, 1997.

[16] Kobayashi, H, Tatsumi, Nakashima, T, T, Takehara, I, Ichikawa, Y, "Current Status of Ceramic Gas Turbine (CGT302)," ASME paper 98-GT-501, presented at the International Gas Turbine and Aeroengine Congress and Exhibition, Stockholm, Sweden, June 2–5, 1998.

[17] Tatsumi, T., Takehara, I., Ichikawa, Y., "Development Summary of the 300kW Ceramic Gas Turbine CGT302," ASME paper 99-GT-105, Presented at the International Gas Turbine and Aeroengine Congress and Exhibition, Indianapolis, Indiana, USA, June 7–10, 1999.

[18] Takehara, I., Tatsumi, T., Ichikawa, Y., "Development Summary of CGT302 Ceramic Gas Turbine," IGTC'99 Kobe OS-101, "Proceeding of the 7th International Gas Turbine Congress, Kobe, Japan, November 14–19, 1999. Vol. I, pp. 57–64.

[19] Takehara, I., Tatsumi, T., and Ichikawa, Y., "Summary of CGT302 Ceramic Gas Turbine Research and Development Program," ASME paper 2000-GT-0644, presented at the International Gas Turbine and Aeroengine Congress and Exposition, and Users Symposium, Munich, Germany, May 8–11, 2000.

[20] Hoshino, A., "Recuperated Two Shaft Ceramic Gas Turbine for Co-generation," Journal of the Gas Turbine Society of Japan, 17[6], 13–20, 1989 (in Japanese).

[21] Tatsumi, T., "CGT302 Two Shaft Ceramic Gas Turbine for Co-generation," Journal of the Gas Turbine Society of Japan, 22[87], 18–23, 1994 (in Japanese).

[22] NEDO and CGT Technology Development Committee, "R&D Report of the 300kW Class Ceramic Gas Turbine," 61–121, 1999 (in Japanese).

[23] Takehara, I., "R&D of the Ceramic Gas Turbine CGT302," Journal of the Gas Turbine Society of Japan, 27[5], 308–309, 1999 (in Japanese).

[24] Nishio, K., Fujioka, J., Tatsumi, T., Takehara, I., "Development of 300 kW Class Ceramic Gas Turbine (CGT302)," ASME paper 95-GT-264, presented at the International Gas Turbine and Aeroengine Congress and Exposition, Houston, Texas, USA, June 5–8, 1995.

[25] Nishio, K., Igashira, K., and Okazaki, S., "R&D of the Composite Material of Ceramics for Engines and Combustors," *Proceedings of the International Symposium on Ultra-high Temperature Materials '96* Tajimi, Japan, pp. 103–110, 1996.

[26] Okuto, A., Kimura, T. Takehara, I., Nakashima, T., Ichikawa, Y., and Tatsumi, T., "Development of a Low NOx Combustor for 300kw-Class Ceramic Gas Turbine (CGT302)," ASME paper 98-GT-272, presented at the International Gas Turbine and Aeroengine Congress and Exhibition, Stockholm, Sweden, June 2–5, 1998.

[27] Kimura, T., "Research and Development of the Low-NOx Combustor for Ceramic Gas Turbine," Proceeding of the 251st Lecture Meeting in Kansai branch of The Japan Society of Mechanical Engineers, pp. 120–122, 1992 (in Japanese).

[28] Kimura, T., Okuto, A., and Kitajima, J., "Research and Development of the Pre-mixed Lean Burn Low-NOx Combustor for CGT," Proceeding of the 9th Autumn Gas Turbine Lecture Meeting of the Gas Turbine Society of Japan, pp. 87–92, 1994 (in Japanese).

[29] Fujishita, K., and Ono, H., "Fatigue Strength of Si_3N_4 for CGT302 at High Temperature," 6th International Symposium on Ceramic Materials & Components for Engines, Arita, Japan, October 19–23, 1997, pp. 228–232, 1997.

[30] Fujishita, K., and Ono, H., "High Temperature Strength Characteristics of the Si_3N_4 for Gas Turbine Use, Fatigue Test and Fast Fracture Strength Test under the Bi-axial Stress Condition," *Proceedings of the 73rd National Convention Lecture Meeting of The Japan Society of Mechanical Engineers,* Vol. I, 461–462, 1995 (in Japanese).

[31] Tohbe, Y., Fujii, T., Hashimoto, K., Suga, S., Sugimoto, K., Isozaki, T., Sugimoto, T., and Tatsumi, T. "The Application of Three Dimensional Flow Analysis to Centrifugal Impeller Design," The Kawasaki Technical Review 115, 29–37, 1992 (in Japanese).

[32] Suga, S., Hashimoto, K., Tohbe, Y., Fujii, T., "Aerodynamic Design of Centrifugal Compressor with Computational Fluid Dynamics," Journal of the Gas Turbine Society of Japan, Vol. 23[89], 101–107, 1995 (in Japanese).

[33] Sakai, T., Tohbe, Y., Fujii, T., and Tatsumi, T., "Development of a High Pressure Ratio Centrifugal Compressor for 300kW-Class Ceramic Gas Turbine," ASME paper 97-GT-480, presented at the International Gas Turbine and Aeroengine Congress and Exhibition, Orlando, Florida, USA, June 2–5, 1997.

[34] Abe, T., Sugiura, T., Tsutsui, Y., Matsunuma, T., Okunaga, S., and Nojima, K., "Research and Development of Practical Industrial Cogeneration Technology in Japan. ASME paper 2000-GT-0655, presented at the International Gas Turbine and Aeroengine Congress and Exposition, and Users Symposium, Munich, Germany, May 8–11, 2000.

[35] NEDO, "Research and Development on Practical Industrial Co-generation Technology," FY 2000 Annual Report, March 2001.

[36] Ferber, M.K., Lin, H.T., Parthasarathy, and Brentnall, "Degradation of Silicon Nitrides In High Pressure, Moisture Rich Environments," ASME paper 99-GT-265, presented at the International Gas Turbine and Aeroengine Congress and Exhibition, Indianapolis, Indiana, USA, June 7–10, 1999.

[37] More, K.L., Tortorelli, P.F., Ferber, M.K., Walker, L.R., Keiser, J.R., Miriyala, N., Brentnall, W.D., and Price, J.R., "Exposure of Ceramics and Ceramic Matrix Composites in Simulated and Actual Combustor Environments," ASME paper 99-GT-292 presented at the International Gas Turbine and Aeroengine Congress and Exhibition, Indianapolis, Indiana, USA, June 7–10, 1999.

CHAPTER

16

DEVELOPMENT OF 300KW CLASS CERAMIC GAS TURBINE CGT303

S. Arakawa and I. Ohashi

Technical Research Division, Yanmar Diesel Engine Co., Ltd., Amagasaki, Hyogo, Japan

INTRODUCTION

A ceramic gas turbine is expected to have excellent characteristics such as low fuel consumption, low exhaust emissions, multiple fuel adaptability, etc. The CGT303, a two-shaft regenerative ceramic gas turbine for a mobile generator set, was researched and developed from 1988 to 1997, as part of the "Moonlight Project," conducted by the Agency of Industrial Science and Technology of the Japanese Ministry of International Trade and Industry (AIST/MITI), which was later renamed as the "New Sunshine Program." The CGT303 development program was funded by the New Energy and Industrial Technology Development Organization (NEDO). The program was carried out by Yanmar Diesel Engine Co., Ltd. (YDE), Niigata Engineering Co., Ltd., Kyocera Corporation and NGK Spark Plug Co., Ltd. The time chart for the development program is shown in Figure 16.1 [1]. The maximum output power of 300kW and thermal efficiency of 42 percent at turbine inlet temperature (TIT) of 1350°C (1623K) were the objectives of this development. To achieve a thermal efficiency of 42 percent or better, which is as high as the thermal efficiency obtainable with diesel engines, two means are available. The one is to rise the TIT and the other is to recover the exhaust heat. To meet the former objective, ceramic components of excellent high-temperature strength were used to eliminate the need for cooling, while for the latter, energy possessed by the exhaust gas, which has formerly been discharged to the atmosphere, is recovered and used to raise the temperature to the inlet of the combustor by way of the heat exchanger.

To meet the requirement of compact design for the engine intended for this mobile generator set application, a rotary regenerator was adopted for the heat exchanger. To attain the maximum thermal efficiency by making full use of those features, the pressure ratio of the compressor selected was 4.5:1. As the result of adopting the low pressure ratio, the size of the engine components had to be increased. Keeping in mind the disadvantage of using ceramics with their relatively low fracture toughness, making sure that the engine components had sufficient strength became our primary concern. Ceramic components were designed appropriately ensuring adequate strength by using sophisticated computer analysis. Foreign object impact testing and hot spin testing were carried out to certify the strength of ceramic turbine rotor. Thermal cycling testing

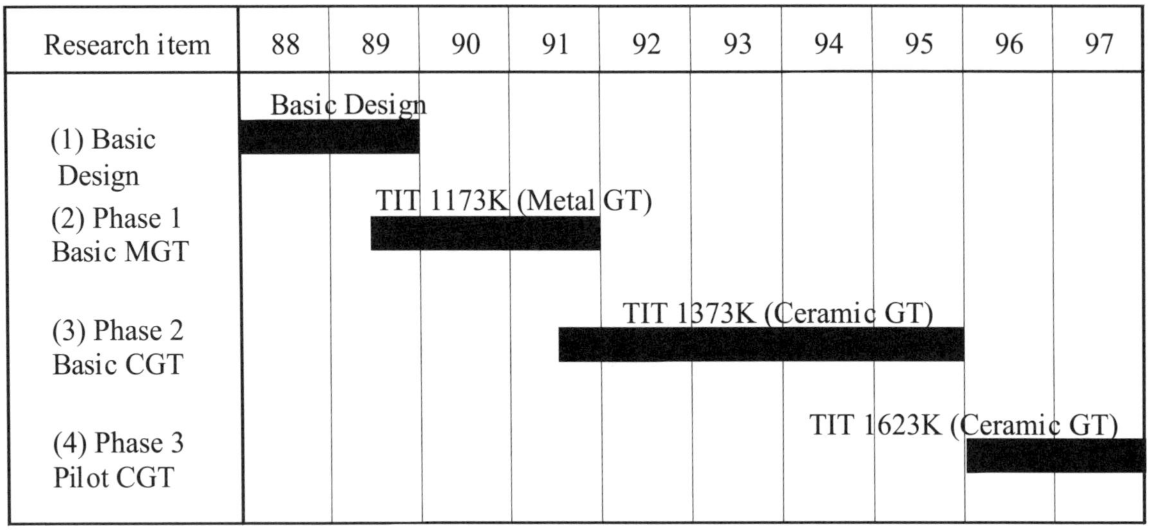

Figure 16.1 The Time Chart of the Development Program (Adapted from Ref. [1])

was also carried out to prove the design concept for static ceramic components. The paper covers the design concept of the CGT303 and the results of the ceramic component development as follows:

(1) Impact testing simulating foreign object damage for the ceramic turbine nozzle and the rotor blade.
(2) A study for the process and geometry for the fit between the ceramic rotor and the metal shaft.
(3) A study for the supporting systems and geometry for large ceramic static components, which were installed and verified in a thermal cycling test rig.

Engine cycle heat flow and engine specifications are shown in Figure 16.2 and Table 16.1 and the material, and forming process and manufacturer of each ceramic component are shown in Table 16.2. Various aspects of the CGT303 development program have been discussed in published papers and in program reports [1–11].

DESIGN CONCEPT

The cross section of the CGT303 is illustrated in Figure 16.3 [2, 3]. For ceramic components which have a low fracture toughness, flexibility of support to reduce stresses from interference between mutually adjacent components was of great concern. In order to improve concentric alignment in assembly, ceramic components were divided into five functional units. They were going from upstream to downstream: combustor, gas generator turbine, intermediate duct, power turbine, and heat exchanger. These units were supported independently.

The gas generator unit consisted of a single-stage centrifugal compressor, the combustor, and a radial-inflow turbine. For the centrifugal compressor, 26 impeller blades (including 13 short blades) and outer diameter of 183 mm were selected. The tip speed of the impeller at the design point (55,000 rpm) was 527 m/s. The combustor was a single can type, mounted tangentially to the scroll swirling direction.

The reasons for adopting a radial-inflow design for the gas generator turbine rather than an axial-flow design were the following:

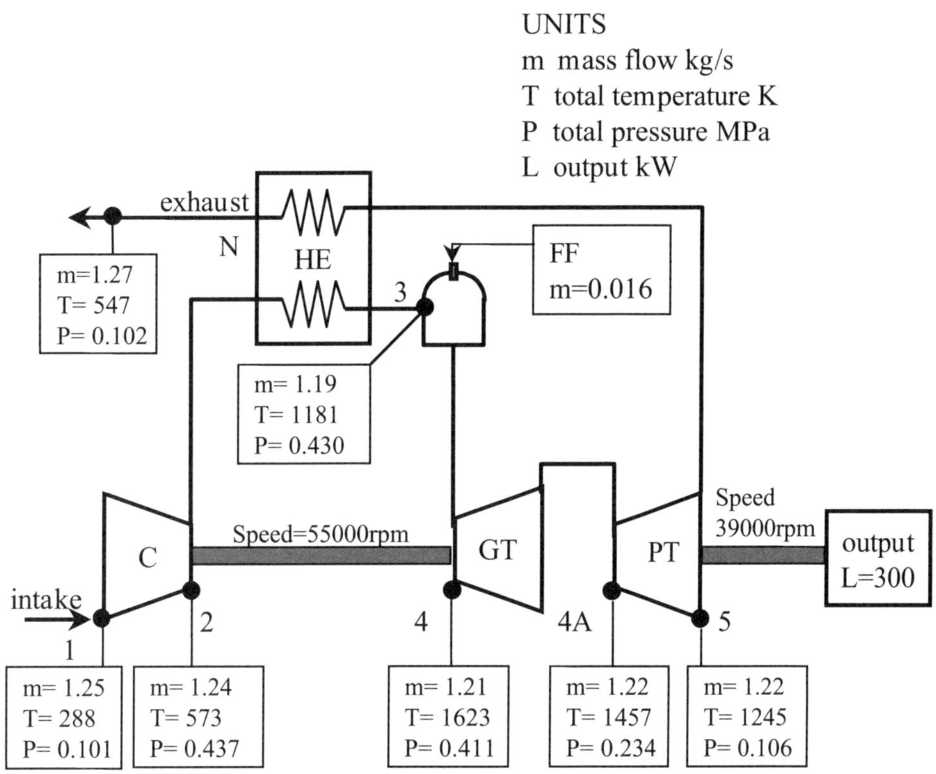

Figure 16.2 Engine Cycle Heat Flow (CGT303) (Reprinted with permission from Yanmar Diesel Engine Company, Limited)

TABLE 16.1 ENGINE SPECIFICATIONS (ADAPTED FROM REF. [2])

Items		Units	Specification
Engine configuration		–	Regenerative (1/LP/R)
Output power		kW	300
Output speed		rpm	3600
Outer dimensions		mm	1420 × 865 × 1155
Dry mass		kg	300
Fuel		–	Gas oil, gaseous fuel
Compressor	Type		Single, centrifugal
	Speed	rpm	55,000
Gas	Type		Single, radial
Generator	Speed	rpm	55,000
Turbine			
Power	Type		Single, axial
Turbine	Speed	rpm	39,000
Combustor		–	Single, can
Heat exchanger		–	Rotary regenerative
Control system		–	Electric

TABLE 16.2 CERAMIC COMPONENT SPECIFICATIONS

Parts name	Material	Forming process	Manufacturer
Combustor flame tube	Si_3N_4	Drain casting	Kyocera Corporation
Gas generator turbine—scroll A	Si_3N_4	CIP	Kyocera Corporation
Gas generator turbine—scroll B	Si_3N_4	Drain casting	Kyocera Corporation
Gas generator turbine—scroll C	Si_3N_4	Slip casting	Kyocera Corporation
Gas generator turbine nozzles	Si_3N_4	CIP	Kyocera Corporation
Gas generator turbine rotor	Si_3N_4	Slip casting	Kyocera Corporation
Gas generator turbine shroud	Si_3N_4	CIP	Kyocera Corporation
Gas generator turbine back shroud	Si_3N_4	CIP	Kyocera Corporation
Intermediate ducts	Si_3N_4	CIP	Kyocera Corporation
Power turbine nozzles	Si_3N_4	Injection molding	NGK Spark Plug Co., Ltd.
Power turbine blades	Si_3N_4	Injection molding	NGK Spark Plug Co., Ltd.
Exhaust diffuser	Si_3N_4	Slip casting + CIP	Kyocera Corporation
Regenerator cores	Cordierite	Projection molding	NGK Insulators, Ltd.
Support springs	Si_3N_4		NHK

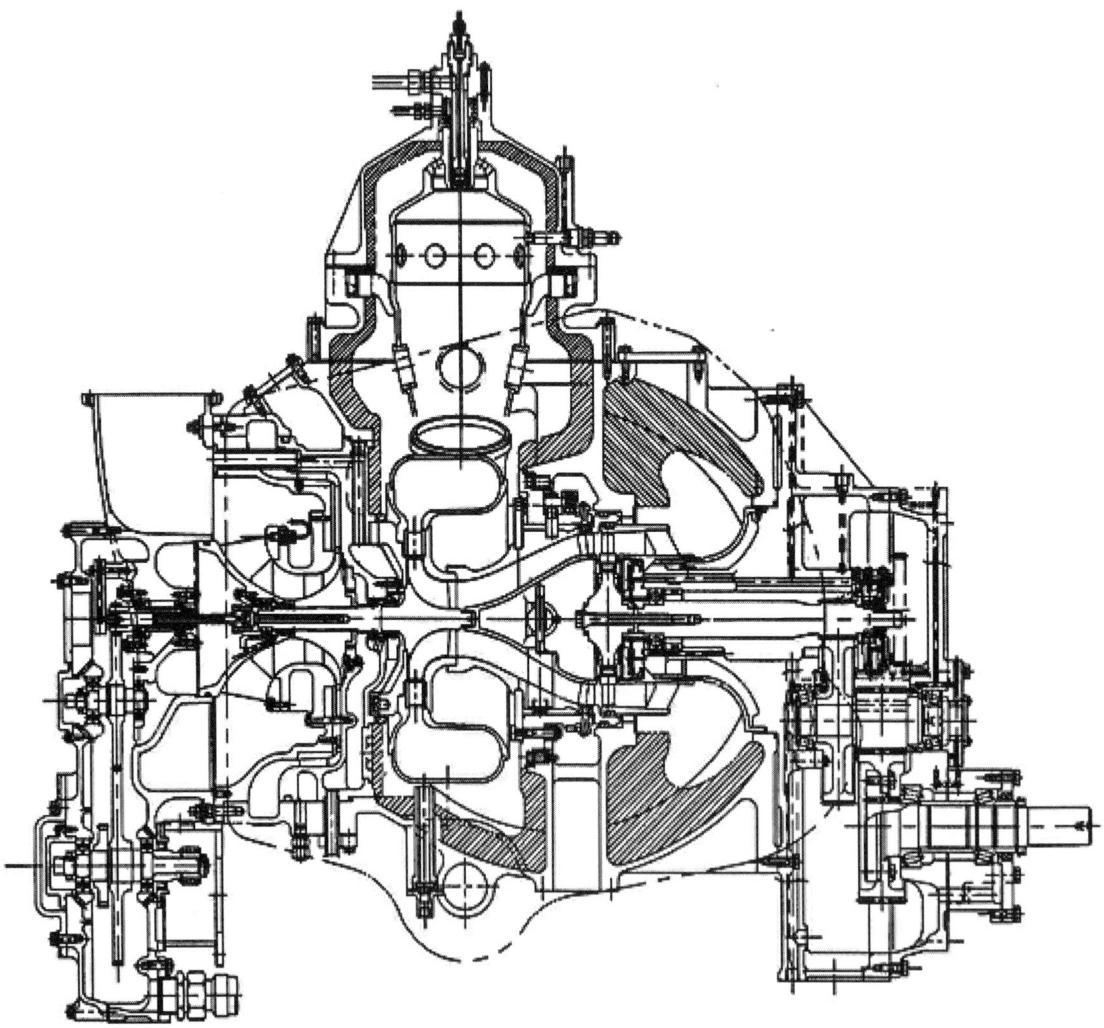

Figure 16.3 Cross Section View (CGT303) (Reprinted from Ref. [2])

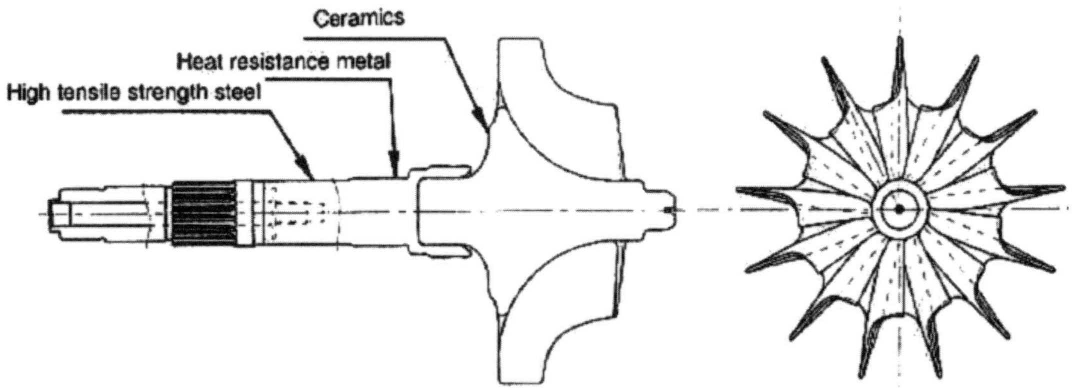

Figure 16.4 Gas Generator Turbine Configuration (Reprinted with permission from the Gas Turbine Society of Japan, Ref. [4])

(1) The material temperature decreases with decreasing radius ratio and so the temperature at the blade root, where the highest stress occurs, could be kept low. This was advantageous for ensuring sufficient strength for the components.
(2) The rotor had a simple shape with a smaller number of blades than for the axial type, and the blades could be thicker.
(3) The ceramic radial-inflow turbine had been successfully produced for turbochargers.

Therefore, this technology and experience were applicable to the fabrication of this radial-inflow turbine rotor.

For the power turbine, an axial turbine allowed a compact and simple gas passage from the gas generator turbine through to the power turbine. A hybrid concept with ceramic blades set in the metal disk was selected to minimize the size of the ceramic components. For the nozzles of the power turbine, a variable geometry system was adopted so that the engine could retain a high thermal efficiency over a wide load range. The ceramic structural components comprising the gas passage were placed on the central axis of the engine and were supported by loading in the direction of the engine front, with ceramic springs (40 coil springs) mounted on the central main support.

GAS GENERATOR TURBINE DEVELOPMENT

The radial gas generator turbine (GGT) had a 162.6 millimeter diameter, and its rated speed was fixed at 55,000 rpm (tip speed 468 m/s). It was assumed that under the start up transient conditions, the maximum stress might reach about 230 MPa in the center portion of the hub, and the fracture probability might be 0.3% at the design TIT of 1350°C (1623K). As shown in Figure 16.4, the ceramic rotor and INCOLOY 909 shaft were coupled by means of brazing and shrink fit. Cold spin testing was carried out by the ceramic supplier (Kyocera Corporation) and those ceramic rotors that were qualified by the spin test at 120% of design speed were delivered. To confirm the strength in the hot rotating state, the delivered rotor was installed in a hot spin test rig. The hot spin test, carried out to simulate turbine operation in an engine up to a TIT of 1200°C (1473K), was conducted by Niigata Engineering Co. Ltd.

After these spin tests, the ceramic GGT was assembled into the metal gas turbine and a problem was detected in the engine. An abnormal vibration occurred at 87% speed. It was revealed that the GGT rotor had a slight tilt at the fitting portion resulting in increased unbalance. Supposedly this was due to the high temperature at the shaft-rotor fitting which exceeded the design temperature and softened the brazing metal.

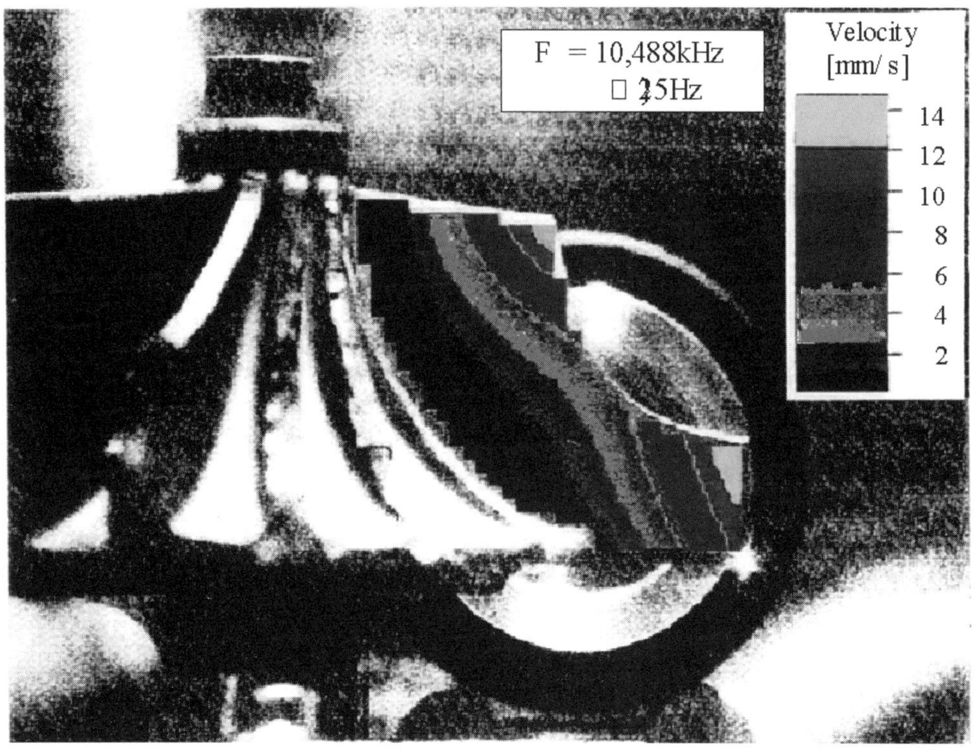

Figure 16.5 Measured GGT Blade Vibration Amplitude (Adapted with permission from the Gas Turbine Society of Japan, Ref. [4])

The flow rate of the cooling air was increased from 1 g/s to 3 g/s and at the same time, the ratio of the outside diameter of the joint (d/D) was changed from 0.8 to 0.75 to increase contact force between the shrinking surfaces. Those countermeasures resulted in stable operation of the turbine up to 100% speed at a TIT of 1200°C (1473K).

Vibrational characteristics of the blade were investigated to establish the critical speed of the GGT. It was expected from the result of calculations that the natural frequencies of the first, second and third mode were in the range of 11kHz to 13kHz. This was in good agreement with the laser beam measurement. Figure 16.5 shows an example of the GGT blade vibration amplitude. When the resonance of the nozzle with the wake was taken into consideration, the critical speed would be between 60 and 70% of the design speed. Though any abnormal vibration did not occur throughout the entire operation of the engine it was recommended to avoid operation within this range (below 70% speed).

During testing in the hot spin rig (Figure 16.6), chipping occurred on the blade inlet portion. The observation of the crack revealed that a foreign object had hit the blade inlet. To evaluate the blade strength against foreign object damage (FOD), a foreign object impact test was introduced which had been developed and conducted by the Mechanical Engineering Laboratory of the Agency of Industrial Science and Technology of MITI. The test rig consisted of a particle launcher, a velocity measuring unit, and a test piece holding unit (Figure 16.7). As a result of the impact tests, the GGT blade strength against foreign object damage was evaluated to have an impact energy of 0.6J, which was equivalent to 6 mg FO impact to the blade tip. In order to increase the foreign matter collision strength, an improved material (Kyocera/SN281) was introduced, and a new blade (GGT/Phase2 blade) profile was examined. From these studies, a bowed inlet shape was selected which is shown together with the first generation blade design in Figure 16.8. The new profile blade was fabricated and installed in the FO impact test rig to evaluate the toughness against FOD. The evaluation showed

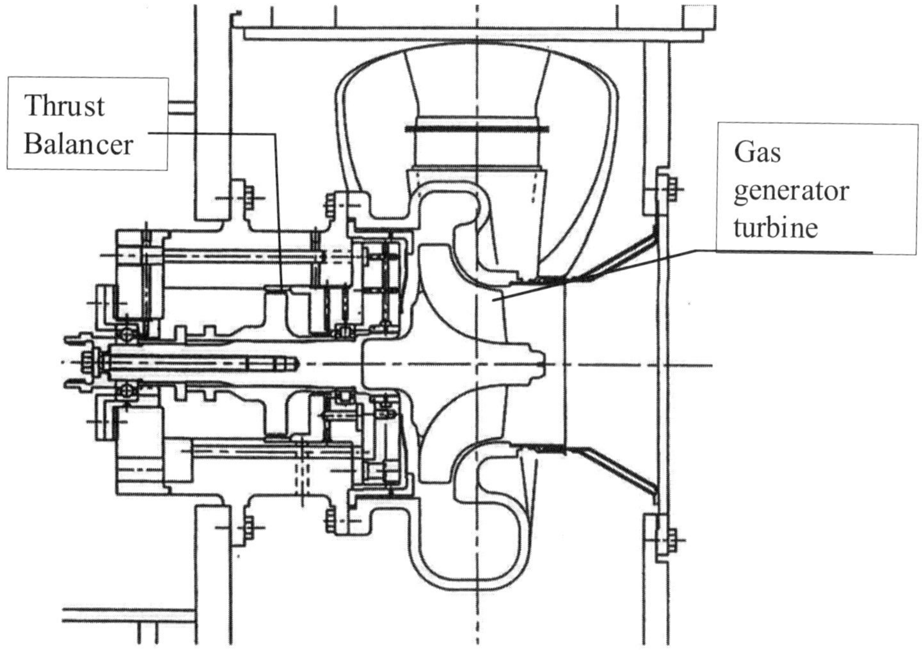

Figure 16.6 Hot Spin Test Rig (Reprinted from Ref. [1])

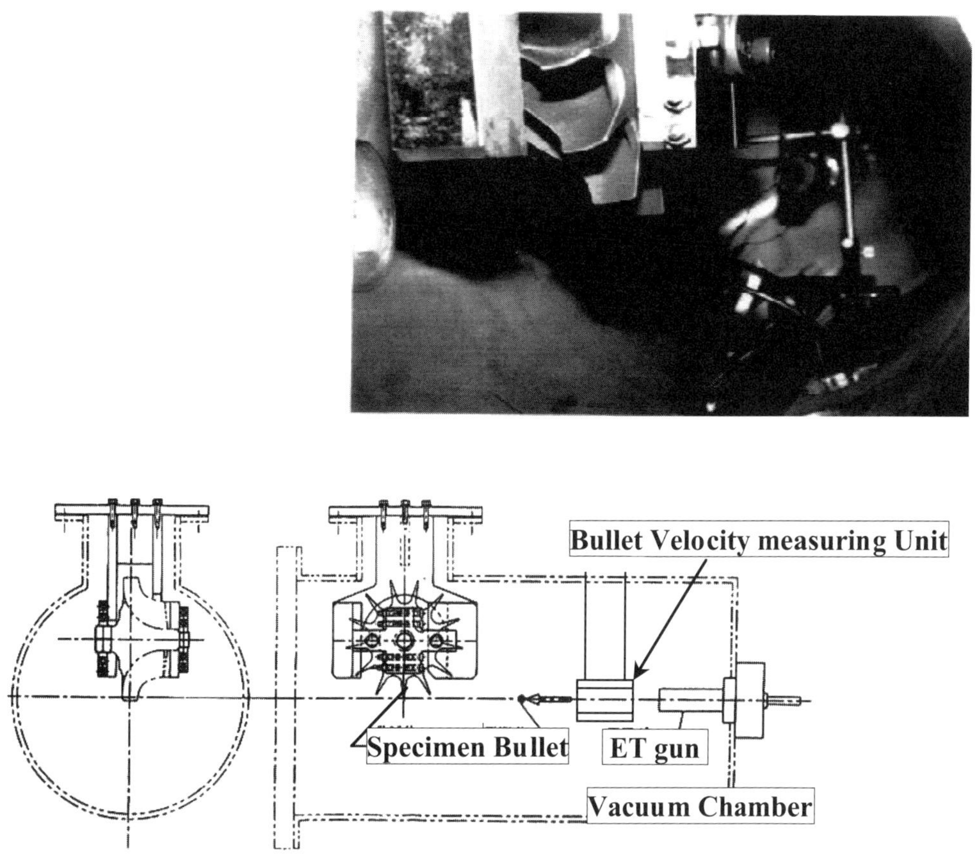

Figure 16.7 Foreign Object Impact Test Rig (Reprinted with permission from Yanmar Diesel Engine Company Limited)

Figure 16.8 Two Configurations of the GGT Rotor (Adapted from Ref. [8])

that the new blade had 1.3J strength of impact energy (Figure 16.9). Details of the impact study methodology and test results have been reported elsewhere [5–7, 11].

DEVELOPMENT OF STATIONARY CERAMIC COMPONENTS

Thermal cycling tests were introduced to establish the reliability of the stationary ceramic components prior to engine operation. In view of potential problems that could not be foreseen in a theoretical study, the test conditions were made as realistic as possible for simulation of actual engine operation: the same ceramic stationary components were used as in the engine itself, as was the engine case (see Figure 16.10). The same system adopted for the actual engine was used for supporting those stationary ceramic components.

In the thermal cycling test system, the combustion gas temperature was set in the range from 600°C to 1200°C (1473K). This controlled the temperature field so that nearly equivalent heat flux as in actual operation was obtained, and the gas velocity was set as in the actual engine as well.

The temperature schedule was established based on the operating data obtained on the metal gas turbine tested previously. Figure 16.11 shows the temperature schedule at the gas temperature of 1200°C (1473K). The temperature was held at the prescribed level for 15 minutes. It was revealed by other tests that the temperature of the components was stabilized during this time. During about 600 rounds of the thermal cycling tests, some ceramic components were broken, which necessitated a design or material change. On the other hand, no breakage occurred for such large-size components as the combustor or the scroll, the reliability of which had been our concern. Results of the study for the main stationary ceramic components, prior to use in the actual engine, are described hereunder.

Combustor

The combustor liner was subject to severe conditions due to the high temperature flame inside as well as the mechanical stress from the attachment design. As shown in Figure 16.12, the combustor liner was

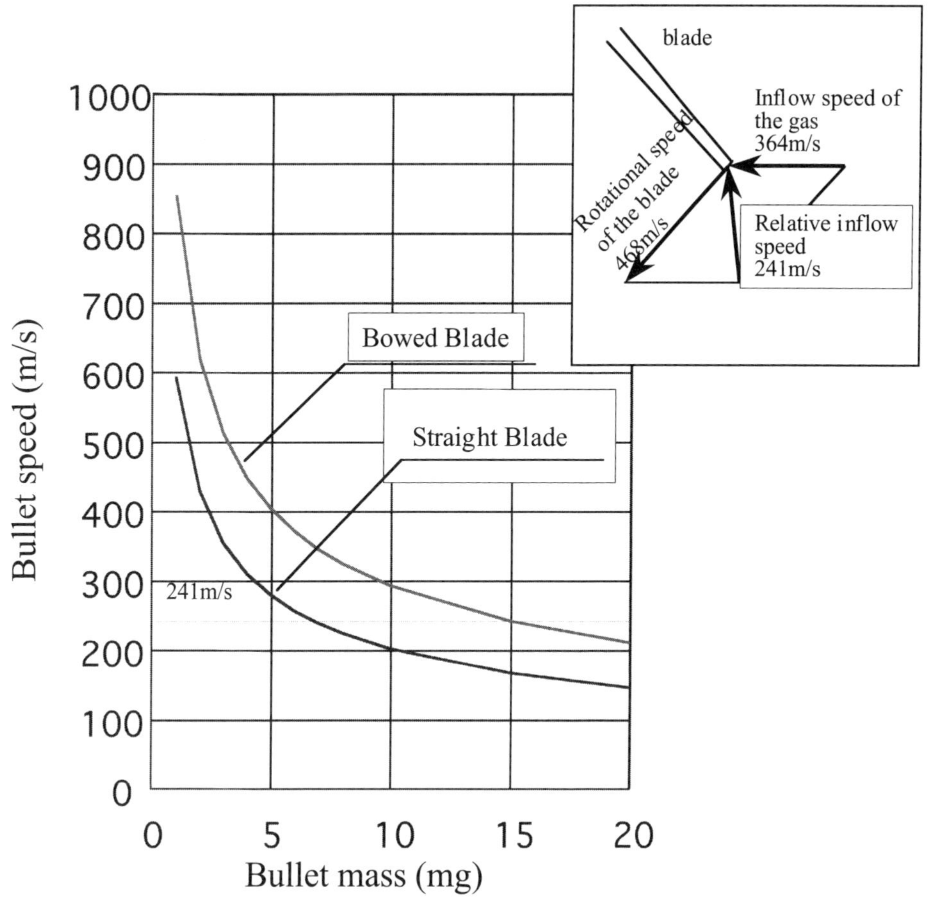

Figure 16.9 FO Impact Test Results (Adapted with permission from the Gas Turbine Society of Japan, Ref. [11])

divided into three segments, and each mating face had a spherical surface to allow the accommodation of any deviation in the center. The combustor was loaded downward by means of a spring positioned at the upper part thereof, and was supported at the four points on the tapered portion at the bottom of the combustor. Ceramic fiber yarn was placed between the combustor and the supports to absorb unexpected rattling. In case the engine was suddenly stopped, the highest tensile stress would occur around the diffusion ports. Transient analysis was performed and a reliability evaluation was carried out by means of NASA/CARES. As a result of the evaluation, a total fracture probability of 2.46×10^{-9} for the first fracture was obtained, which was low enough to expect reliable operation of the components. In support of this prediction, no case of breakage had been observed throughout the thermal cycling test conducted over about 600 rounds, and operation of the engine up to a TIT of 1350°C (1623K).

Scroll

The scroll was the largest component, over 400mm in diameter, yet with a thin (5mm) wall. It was necessary to split the scroll into segments to reduce the thermal stress. Studies to locate the best split lines, resulted in a decision to divide the structure into three sections (scroll A, B, and C) as shown in Figure 16.13.

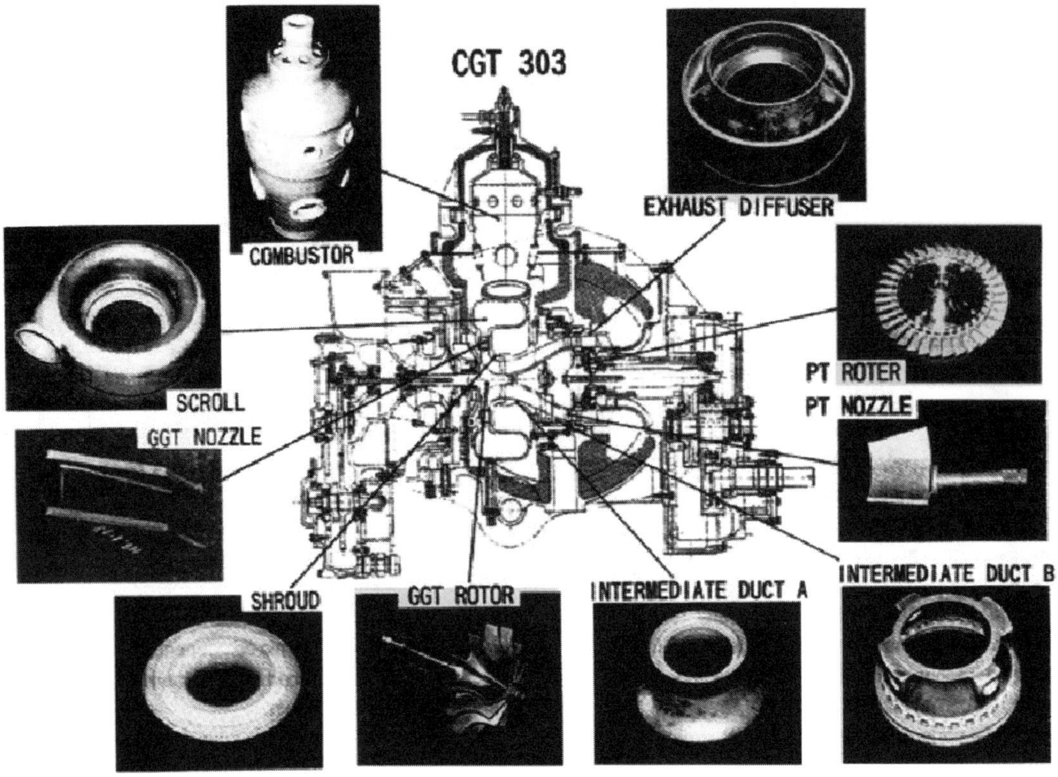

Figure 16.10 Ceramic Components in CGT303 (Reprinted with permission from the Gas Turbine Society of Japan, Ref. [4])

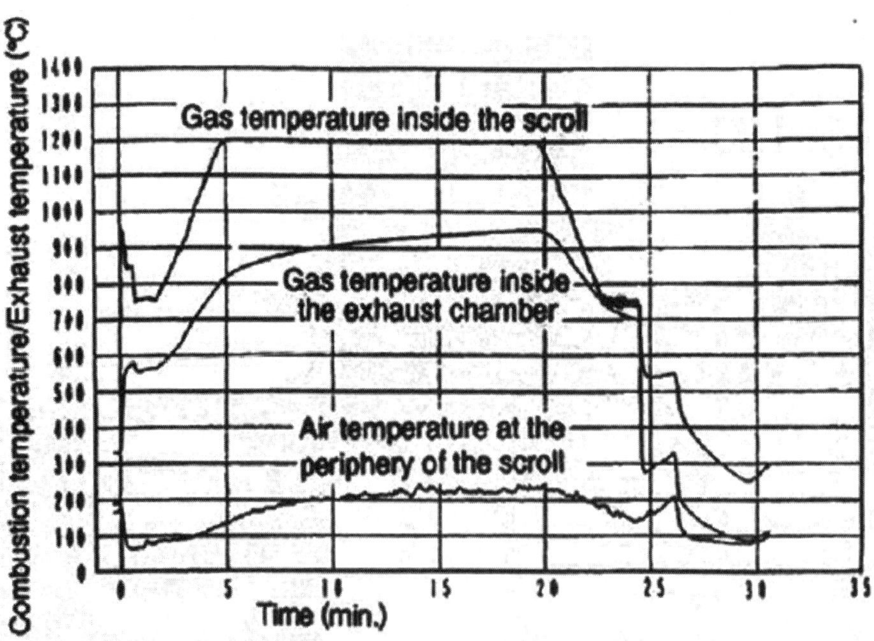

Figure 16.11 Heat Cycle Test Temperature Schedule (Reprinted with permission from the Gas Turbine Society of Japan, Refs. [3, 4])

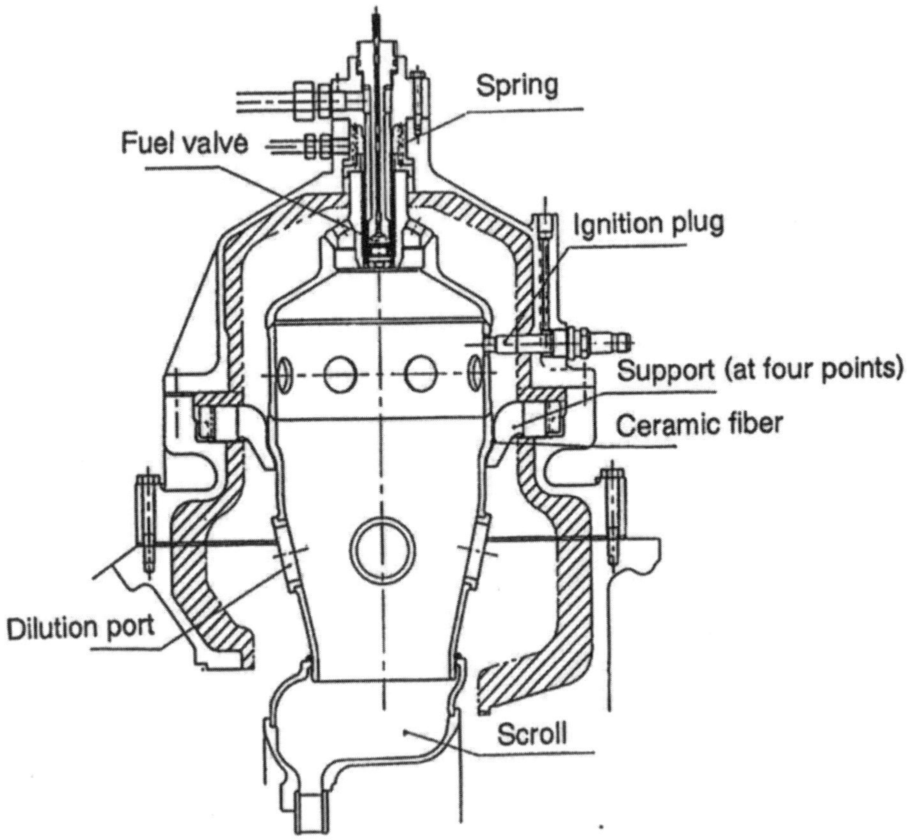

Figure 16.12 Combustor Liner Configuration (Reprinted with permission from the Gas Turbine Society of Japan, Refs. [3, 4])

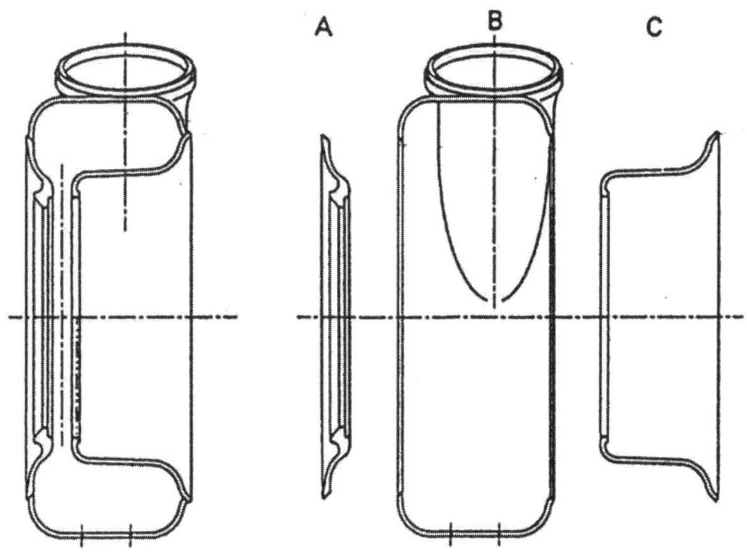

Figure 16.13 Scroll Configuration (Reprinted with permission from the Gas Turbine Society of Japan, Refs. [3, 4])

Figure 16.14 Intermediate Duct was Broken From a Locating Pin Hole (Reprinted with permission from Yanmar Diesel Engine Company, Limited)

Although heavy deformation was noted with the sintered product in the initial stage, it was improved through a substantial effort made by the ceramic manufacturer to fabricate a scroll of high dimensional accuracy. In the original design, the tapered fit system had been adopted for the scroll mating face, with the intent to have automatic concentric alignment of the scroll, the GGT nozzle, the shroud and the attached components by means of a spring load extended in the axial direction. Upon actual measurement of the movement of the scroll, however, it was found that scroll B could be tilted during the cooling process. It was established upon disassembly that the high rigidity of scroll B at the location of the fit with the combustor caused non-uniform deformation, which caused scroll A to turn around this fitting location. As it became clear that this would not only fail to hold them concentrically as originally envisioned, but that it would also be detrimental to the GGT nozzle, the taper butt-joint parts were modified to a spigot joint. Regarding scroll C, which was exposed high temperature gas from the combustor, thermal stress was of great concern. But no breakage was found as in the case of the combustor.

Intermediate Duct A

In the thermal cycling tests conducted at a gas temperature of 1000°C (1273K), the intermediate duct A had fractured. The fracture origin was at the hole for a locating pin as shown in Figure 16.14. It was assumed that the failure originated in a high thermal stress around the pin holes, which was aggravated by the loading of locating pin. This problem was not anticipated based on preliminary analysis. Subsequently,

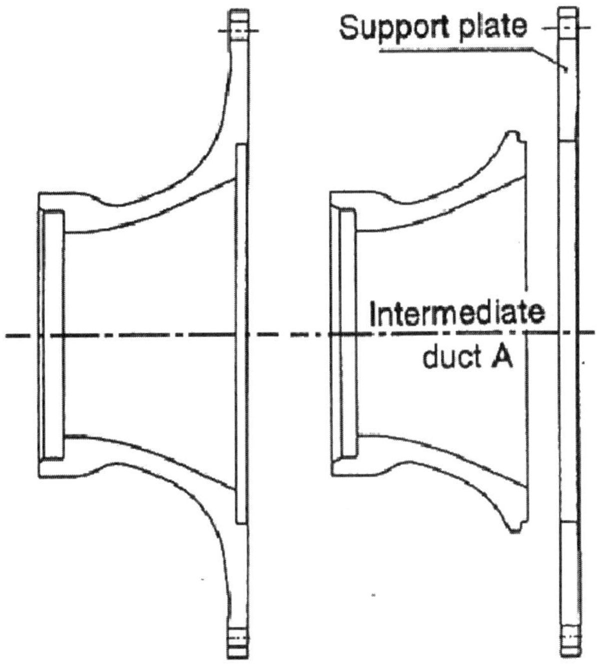

ORIGINAL DESIGN IMPROVED DESIGN

Figure 16.15 Intermediate Duct Design Improvement (Adapted from Refs. [3, 11], with permission from the Gas Turbine Society of Japan

this became an opportunity to measure the temperature and heat flux by using thermal paint and thermal couples which were buried into small drilled holes. The data obtained from actual components in the thermal cycling test rig showed that the heat flux from the high temperature gas to the compressor discharge air was a couple of times greater than predicted. To reduce the thermal stress the duct part was separated from the flange part, as shown in Figure 16.15.

Intermediate Duct B

As shown in Figure 16.16, the intermediate duct B was supported with the main support (metal) and positioning in the axial direction was done with two sets of ceramic locating pins. The main support was made to slide against the engine case in the axial direction, so that it would load the ceramic components with springs provided at the flange of the main support towards the front side of the engine with a compressive force of about 8000 N. The difference in thermal expansion between the ceramic components and the metal cases was absorbed by the sliding action of the main support.

In the thermal cycling test conducted at the gas temperature of 1200°C (1473K), the locating pin for the intermediate duct B had fractured as shown in Figure 16.17 and the portion around the pin hole was damaged. It was surmised that the cause of the damage was that the main support was stuck against the engine case, causing the locating pin to be subjected directly to the thermal expansion of the ceramic components. As a countermeasure, an ceramic fibrous insulator was inserted between the intermediate duct B and the

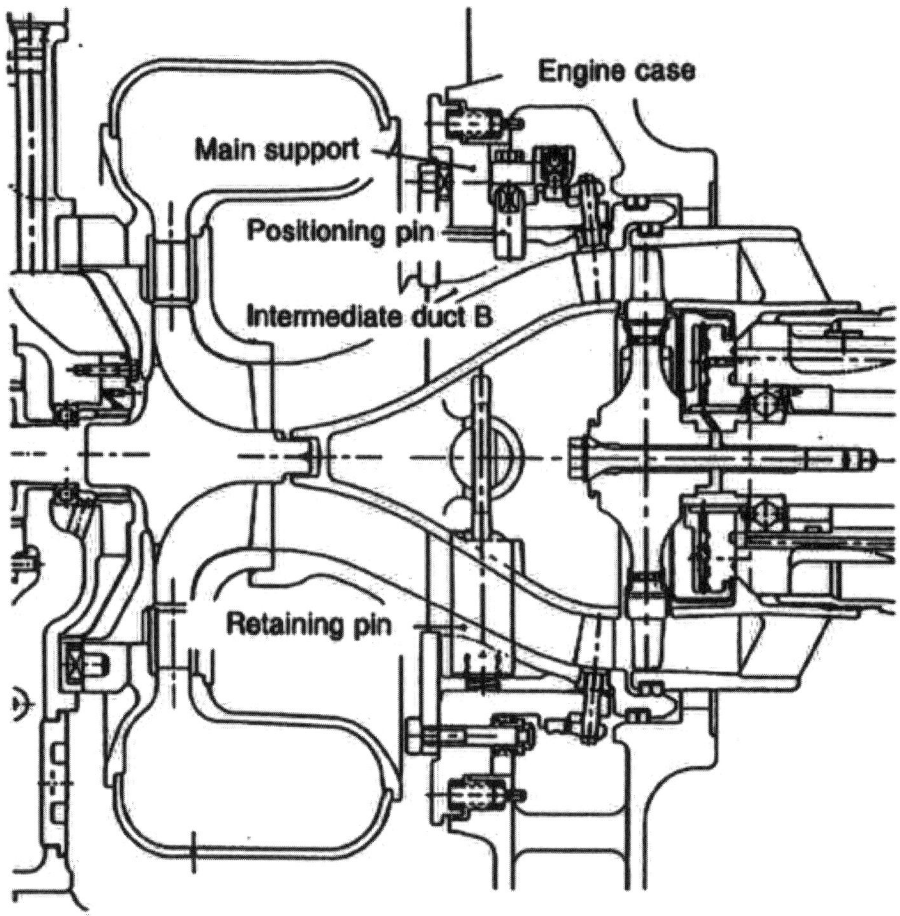

Figure 16.16 Ceramic Components Supporting System (Original Design) (Reprinted from Ref. [1])

main support to shield against the radiant heat, while fitting the solid lubricant to the peripheral portion of the main support to help the smooth sliding motion between the duct and the engine case.

Based on this experience, construction of the main support was modified as shown in Figure 16.18. In this modification, the function of pushing the upstream ceramic components including the intermediate duct A toward the front side of the engine was separated from the intermediate duct B and the variable mechanism for the power turbine nozzle, which were secured to the engine case by way of the main support A. The potential for deformation caused by sliding between the main supports A and B was reduced, and the problem attributable to sticking of these components was eliminated.

POWER TURBINE

The power turbine with a 230mm outer diameter consisted of 35 ceramic blades which were set in a nickel-base superalloy disk. A 0.15 mm thin nickel plate was inserted between the blade and the disk as a compliant layer. Initially in the hot spin testing, some of blades had fractured at the dovetail necks. The

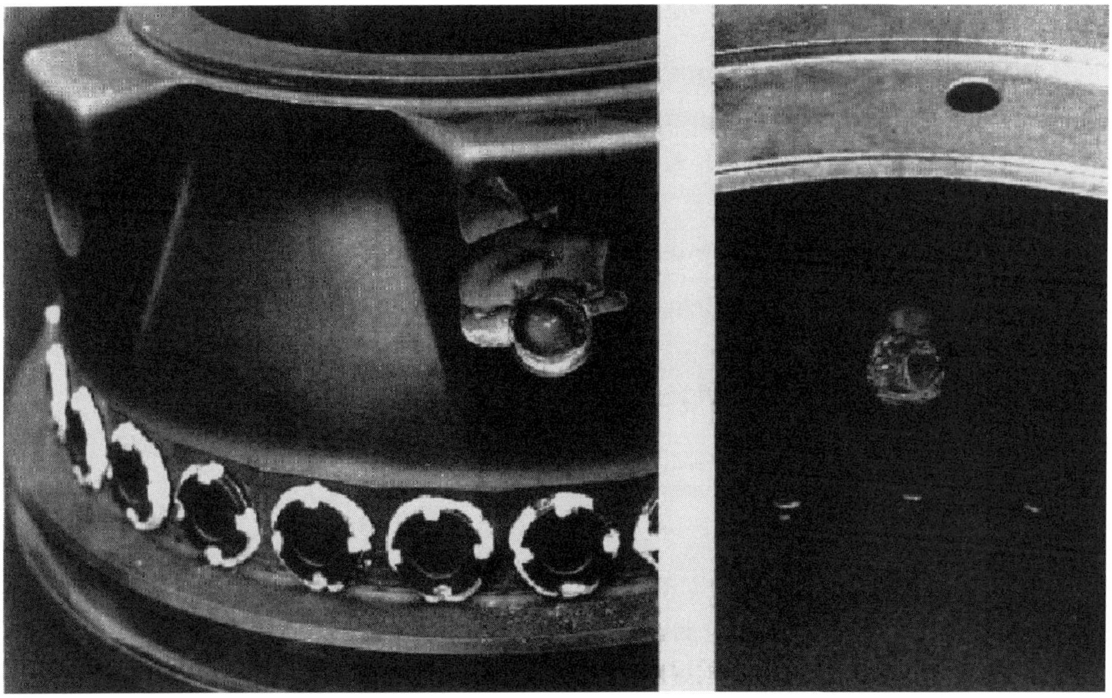

Figure 16.17 Intermediate Duct was Broken From Pin Hole (Reprinted with permission from the Gas Turbine Society of Japan, Refs. [3, 4])

observation revealed the remarkable sticking between blades, compliant layers, and the disk at the dovetail tapered faces. To reduce the deflected motion, boron nitride was coated on the sliding face of the dovetail.

For this hybrid turbine concept, the tip clearance was influenced by the taper angle, friction on a sliding surface, thermal expansion, and so on. Tip clearance measurements were carried out in the hot spin test rig. As shown in Figure 16.19, the tip clearance varied by 0.4 mm (from 0.6mm to 0.2mm) depending on the rotating speed.

ENGINE OPERATION

Based on improvements made to the ceramic components themselves and to adjoining metal components, we became confident that ceramic components would not fail even when exposed to operation at a TIT of 1200°C (1473K). In the next step, these components were incorporated into the engine and trial testing was performed. In the beginning, only no-load testing was conducted because the clearance of the rotors was too large (1.5mm). This clearance was reduced in a stepwise manner to 0.6mm, alongside with the elimination of the vibration of the GGT rotor mentioned previously. The total operating time at the TIT of 1200°C (1473K) was 21 hours. The maximum output attained was 165 kW which was only 72% of the target. The thermal efficiency was as low as 12% because no heat exchanger had been incorporated in the engine test.

Figure 16.20 shows the ceramic components used for operation at the TIT of 1200°C (1473K). Neither cracking nor chipping of the components was observed. Because of the delays with the development of the heat exchanger, the target of the CGT303 development program was changed into a GGT development program instead of whole engine development.

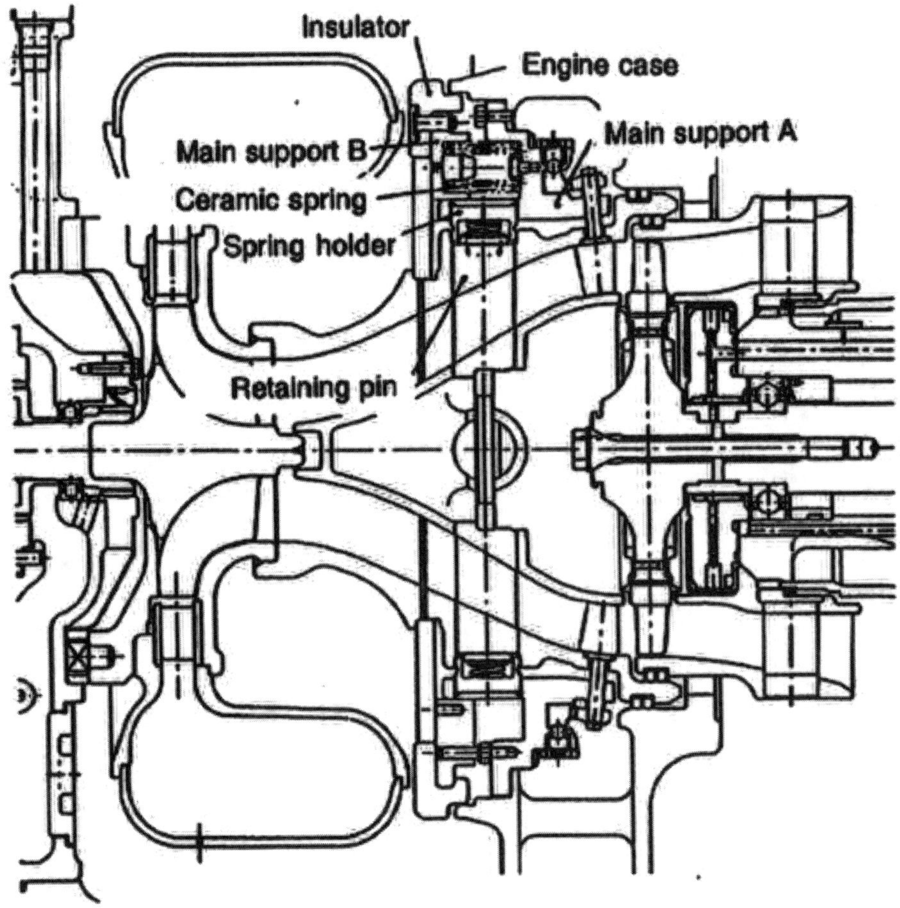

Figure 16.18 Ceramic Components Supporting System (Improved Design) (Reprinted from Ref. [1])

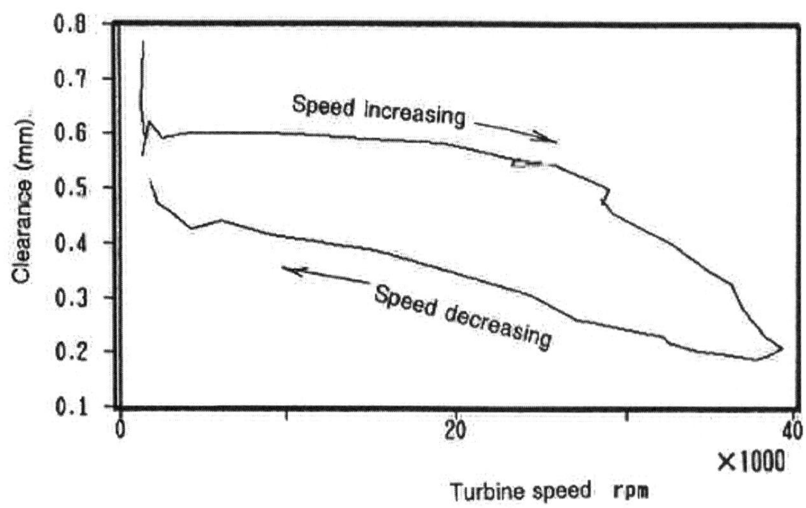

Figure 16.19 Hybrid Turbine Tip Clearance (Measured) (Reprinted from Ref. [1])

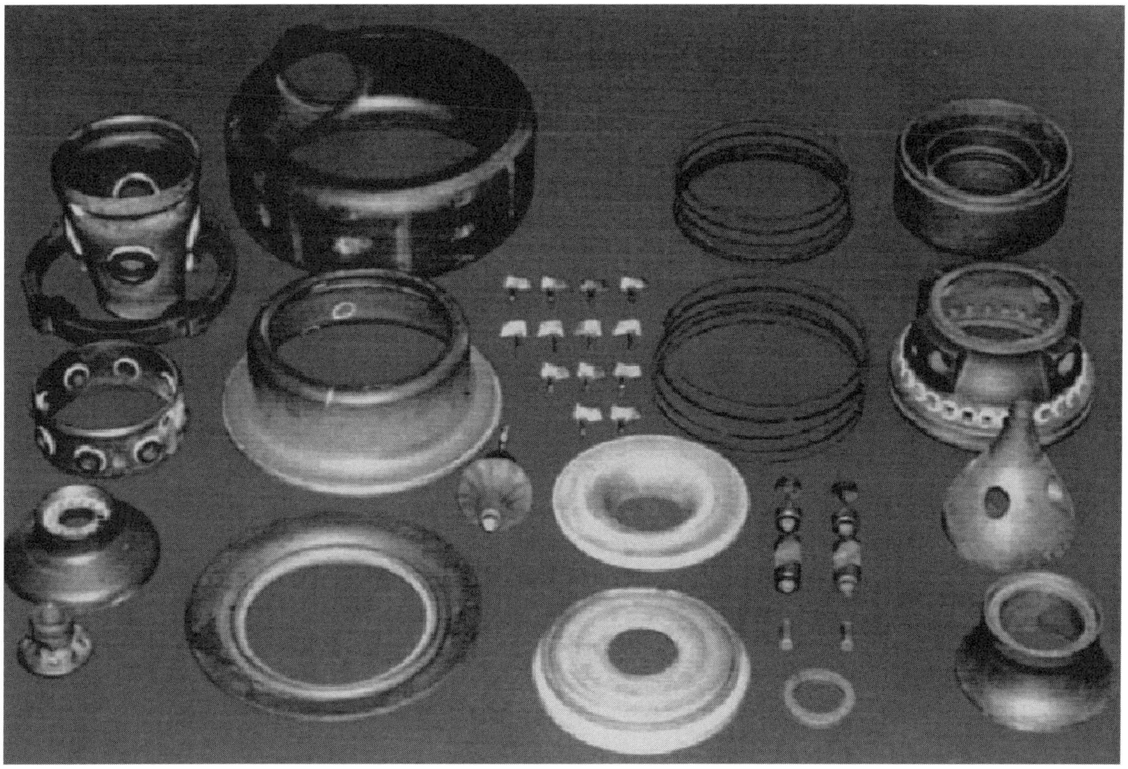

Figure 16.20 Ceramic Components After Operation at 1150°C (1423K) (Reprinted with permission from Yanmar Diesel Engine Company Limited)

To attain GGT operation at the ultimate program TIT of 1350°C (1623K), the engine case was redesigned and the power turbine module was removed as shown in Figure 16.21. The GGT rotor successfully operated at the TIT of 1350°C (1623K) and at 100% speed in this test rig for 3 hours in total (Figure 16.22). The GGT generated power was only used for the compressor drive, so that the turbine inlet gas temperature was changed by controlling the turbine discharge pressure.

CONCLUSIONS

Some lessons learned in the CGT303 development program are summarized as follows:

- Although ceramics do not favor a high gradient stress field, a strong contact force is required to fit the ceramic rotor to the metal shaft to ensure occurrence of an abnormal vibration. The CGT303 adopted a 0.75 diameter ratio with 3g/s cooling air and successfully completed the 1350°C (1623K) operation.
- Bowed shaped turbine blades showed greater resistance to foreign object damage than straight shaped blades.
- The vibrational mode and frequency of the turbine blade was measured by means of a laser beam instrument method. Good agreement was confirmed between the experimental and calculated results.
- Tip clearance of the hybrid turbine blade was successfully measured by means of an instrument for measuring electric static capacitance. The measured tip clearance showed some correlation with the rotational speed and gas temperature.

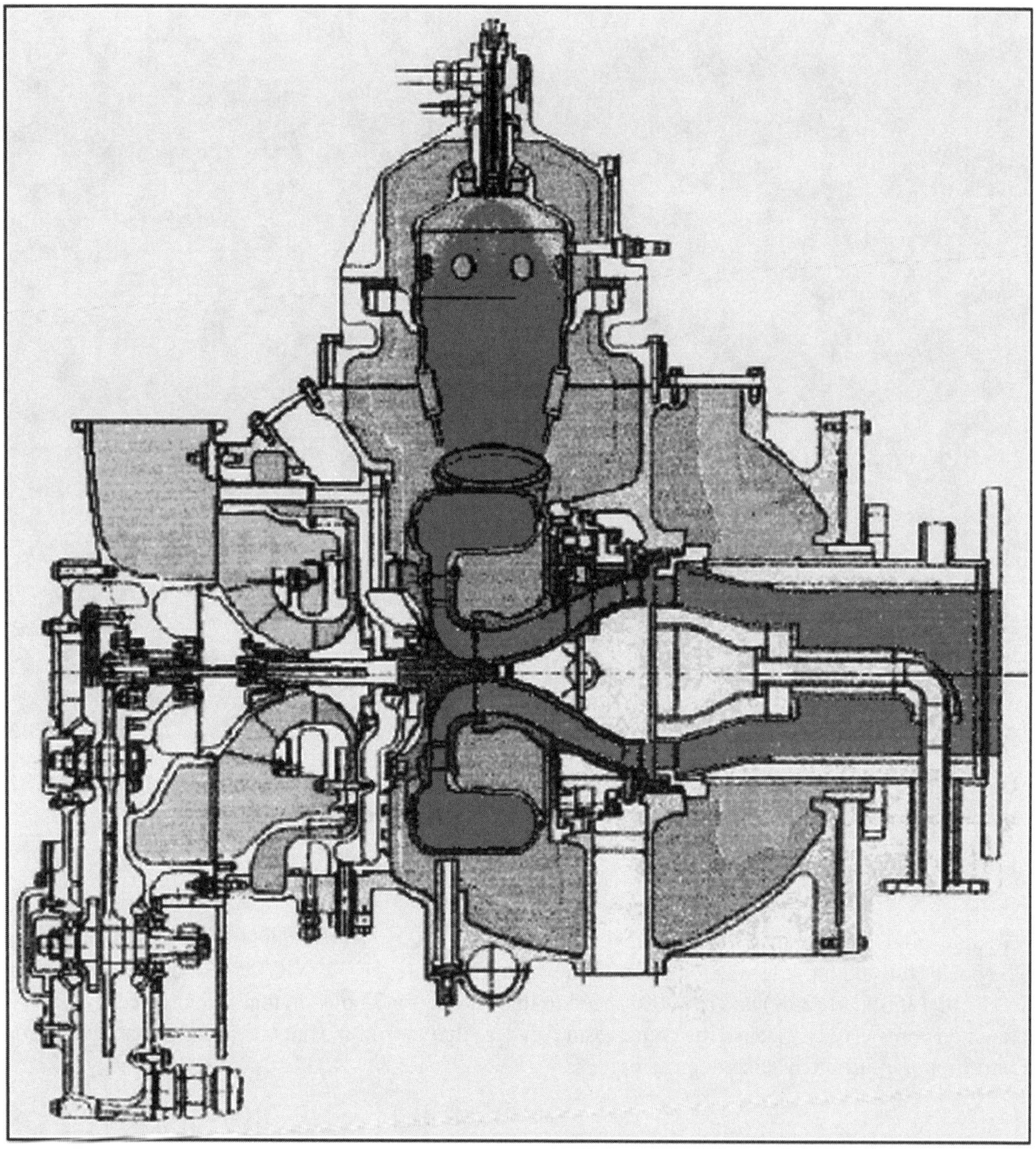

Figure 16.21 Engine Scheme of TIT 1623K Operation (Reprinted with permission from Yanmar Diesel Engine Company Limited)

- Thermal cycling testing was effective for the removal of unpredictable problems with the stationary ceramic components before installing them into an engine. The components of the CGT303 required some configurational change to ensure 1350°C (1623K) operation.
- The CGT303 successfully operated at a TIT of 1200°C (1473K) for 21 hours in total, and for 3 hours at a TIT of 1350°C(1623K) at 100% speed.
- The rotary heat exchanger was a great subject in the CGT303 project. Some problems due to thermal deformation of the heat exchanger and of neighboring components, prevented the evaluation of its performance when installed in the engine.

Figure 16.22 Gas Generator Turbine After the Operation at 1350°C (1623K) TIT (Reprinted with permission from the Gas Turbine Society of Japan, Ref. [11])

Ceramics are now reaching the stage where they could be used as gas turbine components. Further improvement are needed for better toughness and low cost

ACKNOWLEDGEMENTS

The authors wish to acknowledge the Agency of Industrial Science and Technology of the Ministry of International Trade and Industry, the New Energy and Industrial Technology Development Organization, and three national laboratories that participated in the CGT303 project. The authors also appreciate the efforts of the project members: Niigata Engineering Co., Ltd, Kyocera Corporation, and NGK Spark Plug Co., Ltd.

REFERENCES

[1] "Research and Development on Ceramic Gas Turbine (300 kW class)," FY 1994 Annual Report, New Energy and Industrial Technology Development Organization, 1994.

[2] Ohhashi, I., and Arakawa, S., "Development of 300 kW Class Ceramic Gas Turbine (CGT303)," ASME paper 94-GT-82, presented at the International Gas Turbine and Aeroengine Congress and Exposition, The Hague, The Netherlands, June 13–16, 1994.

[3] Murota, M., Ohhashi, I., Ito, Y., and Arakawa, S., "Research and Development of 300kW Class Ceramic Gas Turbine (Development of the Static Components for CGT303)," ASME paper 95-GT-200, presented at the International Gas Turbine and Aeroengine Congress and Exposition, Houston, Texas, USA, June 5–8, 1995.

[4] Ohhashi, I., Ito, Y., and Arakawa, S., "Research and Development of 300kW Class Ceramic Gas Turbine," 95-Yokohama IGTC-126, *Proceedings of the 1995 International Gas Turbine Congress,* Yokohama, Japan, October 23–27, 1995. Vol. I, I-157/I-163, 1995.

[5] Yoshida, H., Uematsu, K, and Burie, A., "Launching of Single, Small Particles for Impact FOD Testing of Ceramic Gas Turbines, ASME paper 95-GT-93, presented at the International Gas Turbine and Aeroengine Congress and Exposition, Houston, Texas, USA, June 5–8, 1995.

[6] Yoshida, H., Hoshly, Y., Uematsu, K., Kitazawa, Y., "A Single, Small Particle Launch System by Electrothermal Gun and Micro-sabot," Review of Scientific Instruments, Vol. 68, No. 1, pp. 178–183 (January, 1997).

[7] Yoshida, H., Ohhashi, I., Shimamori, T., Hara, K., Yoshida, M., "FOD Test Result of Ceramic Gas Turbine Component," *Proceedings of the 24th Congress of GTSJ*, (May 1996).

[8] "Research and Development on Ceramic Gas Turbine (300 kW class)," FY 1995 Annual Report, New Energy and Industrial Technology Development Organization, 1995.

[9] Hamada, Y., "A Two-shaft Regenerative Ceramic Gas Turbine for a Mobile Generator Set," Journal of the GTSJ Vol. 17 No. 67, (1989).

[10] Itoh, Y., Arakawa, S., Ohhashi, I., "Technology for Ceramic Static Component to be Applied to Gas Turbine," *Proceedings of the 9th Autumn Congress of GTSJ*, (November 1994).

[11] Ohhashi, I. and Itoh, Y., "Developing Design for CGT Ceramic Components," *Proceedings of the 25th Congress of GTSJ*, (May 1997).

CHAPTER

17

Application of Ceramics to a Power Generating Gas Turbine

Yutaka Furuse, Tetsuo Teramae, Toshiaki Tsuchiya

Energy and Environment R&D Center, Tokyo Electric Power Company, Yokohama, Kanagawa, Japan

Fukuo Maeda

Power & Industrial Systems Research & Development Center, Toshiba Corporation, Yokohama, Kanagawa, Japan

Yoshiaki Tsukuda

Turbine Engineering Department, Takasago Machinery Works, Mitsubishi Heavy Industries, Ltd., Takasago, Hyogo, Japan

Katsuo Wada

Hitachi Engineering and Product Division, Hitachi Ltd., Hitachi, Ibaraki, Japan

INTRODUCTION

In the middle of the 1980s, gas-steam combined cycle power plants were introduced for power generation in Japan. Since this system demonstrated high thermal efficiency and excellent load adjustment capability, it was expected to become a major means of thermal power generation. Also, it was believed that the coal gasification combined cycle power plant would be an effective method for the utilization of a large amount of coal. In order to achieve acceptable economics, it was necessary to develop a highly efficient combined cycle power plant.

The thermal efficiency of a combined cycle power plant is raised with an increase in the turbine inlet temperature (TIT: Temperature of combustion gas at combustor exit) of gas turbines. Increases in the TIT had been mainly achieved by the introduction of nickel- and cobalt-base superalloys along with advanced turbine cooling technology. The increase in TIT, however, might have undesirable side effects associated

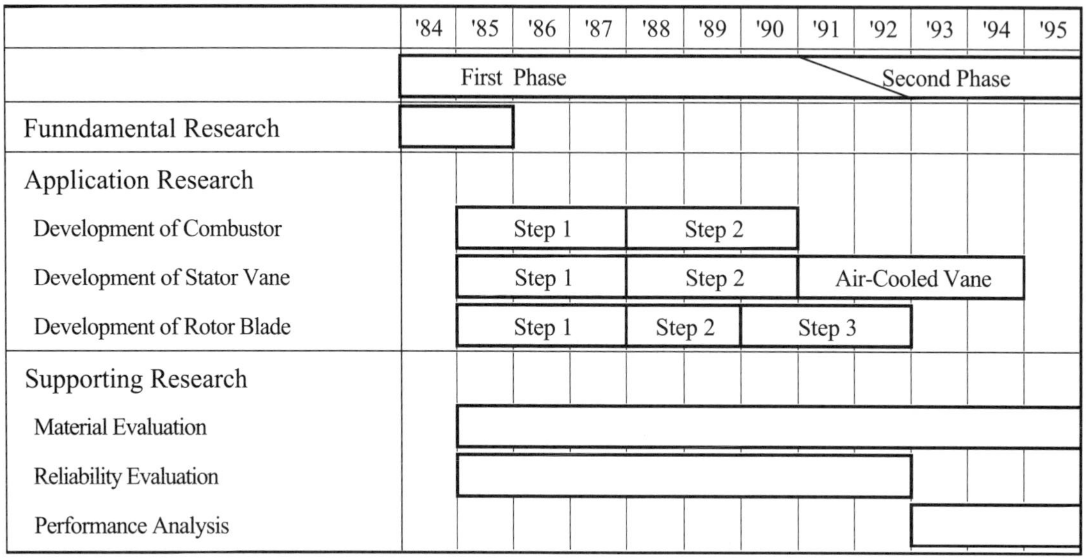

Figure 17.1 Master Schedule of the Program (Reprinted with permission from TEPCO)

with a resulting increase in NOx emissions. The lean-premixed combustion method was one of the most promising NOx abatement systems. In this method, a large portion of airflow was utilized for premixing with the fuel, and this led to a lack of cooling air. It seemed, therefore, that engine firing temperature increases would reach their limits as long as air-cooled metal components were used.

Ceramics, especially the silicon based non-oxide ceramics such as silicon carbide (SiC) and silicon nitride (Si_3N_4), offer a very attractive combination of high temperature capability to about 1300°C to 1400°C, high stiffness and high strength. Ceramics also offer advantages in terms of wear, erosion, and corrosion resistance. These excellent characteristics of ceramics enable an increase in TIT compared to uncooled or moderately cooled metal components, and reduce the amount of cooling air, resulting in significant fuel savings, increased output power, and reduced emissions. There was, therefore, a strong drive towards the introduction of ceramic components into power generating gas turbines.

During the period between 1984 and 1995, Tokyo Electric Power Co., Inc. (TEPCO) conducted cooperative research programs for the application of ceramics to a power generating gas turbine with Toshiba Corporation, Mitsubishi Heavy Industries, Ltd., and Hitachi, Ltd. [1, 2]. Furthermore, TEPCO conducted many cooperative research projects in support of this program.

This chapter will review the design of ceramic components for a power generating gas turbine and will present the results obtained from component tests and supporting research.

OBJECTIVES AND SCHEDULES

Figure 17.1 shows the master schedule for the program which consisted of the following two phases:

First Phase: Research for a 1300°C TIT Class Gas Turbine (1984–1992)
Second Phase: Basic Research for Gas Turbine with TIT higher than 1300°C (1991–1995)

Various research activities in support of the ceramic component development were also conducted. The objective of the first phase of this program was to verify the adaptability of silicon based monolithic ceramics for a gas turbine with an output power of 20MW and a TIT of 1300°C. The intent was to apply ceramics to hot section components such as the liners and transition pieces of the combustor, the first and second stage nozzle vanes, and the first stage rotor blade in a three-stage turbine. As for the rotor blade, parallel efforts were conducted for the application of thermal barrier coatings (TBC) on air-cooled metal blades [3].

In the first stage of the program, the preliminary design of a ceramic gas turbine was conducted and basic design specifications, such as the main dimensions and required material properties, were established for the ceramic components. On the basis of the results obtained in the preliminary design, efforts were initiated for the development of ceramic components such as the combustor, nozzle vanes and rotor blades. The development of these components consisted of the following three steps:

Step 1: Design, stress analysis, and tests under atmospheric conditions.

Step 2: Modification of the designs based on the results and findings obtained in step 1. Tests under full-pressure conditions for the combustor and nozzle vanes. Rotor blades were evaluated by room-temperature spin tests.

Step 3: Further development of the ceramic rotor blades. Hot spin tests were performed to validate the design.

Following the preceding development of the combustor and nozzle vanes, the development of the first stage ceramic rotor blade was successfully completed, and the initial target of the program was accomplished.

According to the initial master schedule of this project, an assembly test of a 20MW class gas turbine with ceramic components was planned after confirming the ceramic component designs and performance had been validated. However, during this research and development program, gas turbine technologies had progressed substantially, and 1300°C TIT class gas turbines for combined cycle power plants with metal components were being manufactured. Therefore, whereas it was still a valid assumption that ceramic components were superior to metal components in their effect on the performance and thermal efficiency of gas turbines, it seemed appropriate to change our target for the application of ceramic components to gas turbines with a TIT higher than 1300°C.

Further efforts were conducted for the application of ceramics to a gas turbine with a higher TIT, in the second phase of the program. The key technology was the development of high temperature resistant materials. In this phase, TEPCO conducted cooperative research to develop high temperature resistant structural ceramics. A parallel approach involved design modifications. To accommodate the limitation in the temperature resistance of existing ceramics, the utilization of a small amount of cooling air was studied for the first stage nozzle vane of a 1500°C class gas turbine. Studies were also carried out to assess the performance improvement of a combined cycle power generating system when ceramic components were applied to hot section parts of 1300°C and 1500°C TIT class gas turbines.

DEVELOPMENT OF CERAMIC COMPONENTS FOR 1300°C CLASS GAS TURBINE

Preliminary Design

Preliminary design of the gas turbine was conducted during the 1984–1985 period, and basic design specifications were established. On the basis of the results of this preliminary design, it was decided to apply ceramics to the liners and transition pieces of the combustor, the first and second stage nozzles and the first stage rotor in the three stage turbine shown in Figure 17.2. A 20 MW power output was selected as the program target because this size seemed to be suitable for the scale up of the developed techniques for practical power generation systems (100-200MW class) in the future. Furthermore, this size seemed to be suitable if we proceeded to the assembly test phase, because the three collaborating Japanese gas turbine manufacturers had already developed metallic gas turbines with an almost equivalent size. With regard to the fuel, a low calorific gas fuel was selected considering the future application to coal gasification combined cycles. Table 17.1 presents the specifications for the gas turbine design and Figure 17.3 shows the heat balance at the design point.

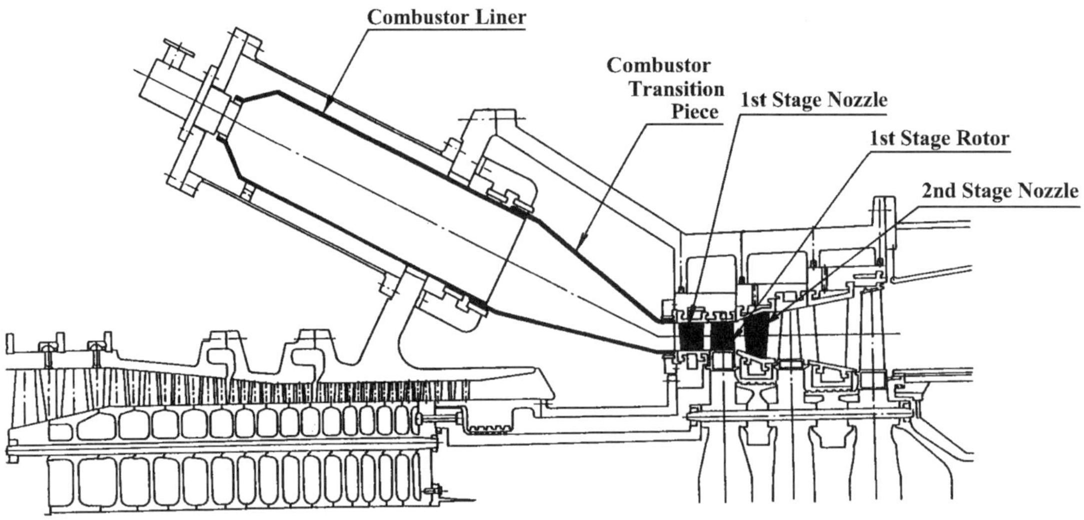

Figure 17.2 Application of Ceramics in a Gas Turbine (Adapted from Ref. [2])

TABLE 17.1 SPECIFICATIONS FOR THE CERAMIC GAS TURBINE	
Output	**20 MW**
Turbine Inlet Temperature	1300 °C
Exhaust Gas Temperature	666 °C
Pressure Ratio	15
Shaft Speed	10800 rpm

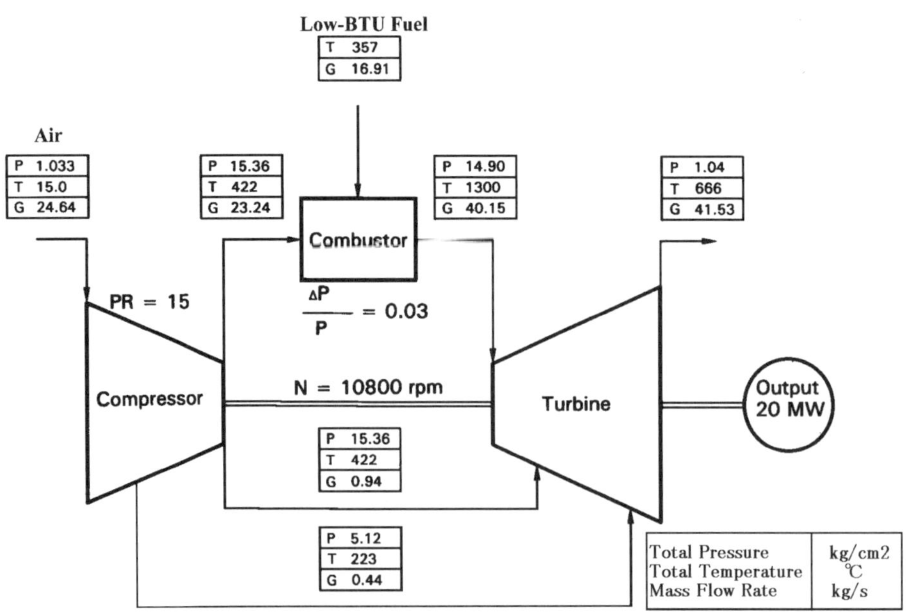

Figure 17.3 Heat Balance at the Design Point (Adapted from Ref. [2])

TABLE 17.2 OPERATING CONDITIONS OF THE CERAMIC COMPONENTS (APPROXIMATE VALUES)

		Nozzle		Rotor
	Combustor	First Stage	Second Stage	First Stage
Gas Temperature (°C)				
Max	1500	1400	1100	1250
Mean		1300	1000	1150
Gas Pressure (atm)	15	15	6	10
Gas Velocity (m/s)	>50	>700	>600	>600
Candidate Material	SiC	SiC	Si_3N_4	SiC
Surface Temperature (°C)	1400	1400	1100	1200
Thermal Stress (MPa)	60	400	200	400
Centrifugal Stress (MPa)				300

Basic Design Concept

Whereas ceramics have shown superior high temperature resistance, oxidation resistance, wear resistance, fracture strength, and creep strength at elevated temperatures, they have disadvantages such as lower fracture toughness, lower thermal shock resistance and a wider scatter of fracture strength compared to metallic materials. Therefore, the design of ceramic components was conducted carefully taking account of these disadvantages.

A hybrid construction was selected with metallic materials acting as the structural strength component and ceramic materials as the heat resistant component. Predicted operating conditions for each component are summarized in Table 17.2. In case of the first stage nozzle and the first stage rotor, estimated stresses seemed to be high compared with the strength of the candidate materials. On the other hand, it seemed to be much easier to apply ceramics to the combustor and the second stage nozzle.

Ceramic components should be designed to withstand the most severe operating conditions, but little information was available at this point for actual gas turbines. Therefore, an iterative design process was conducted, by comparing the experimental results obtained by material evaluations and component tests with the calculated results obtained by Finite Element Method (FEM) analyses.

Selection of Ceramic Materials

Monolithic ceramics, such as sintered silicon nitride (Si_3N_4) and sintered silicon carbide (SiC), were evaluated for their high temperature resistance, thermal shock resistance during emergency fuel shut down, strength, creep resistance, fracture toughness, as well as for their ability for fabrication of complex-shaped parts. At the beginning of this program, SiC seemed to be the most promising candidate for gas turbine applications because of its high strength and good oxidation resistance at elevated temperatures. Si_3N_4 had excellent strength, toughness and thermal shock resistance at temperatures below about 1200°C, whereas they had the tendency to degrade at temperatures above 1200°C.

On the basis of the results of temperature distribution analyses for each component, several sources of sintered SiC were selected as candidate materials for the combustor, the first stage nozzle vane and the first stage rotor blade, subjected to gas temperatures of 1200°C or higher. As for the second stage nozzle vanes which were exposed to a maximum hot combustion gas temperature of 1100°C, only one sintered Si_3N_4 was selected.

Meanwhile, there had been considerable progress in ceramic materials properties and fabrication technologies during this research period, and high performance Si_3N_4 ceramics were being developed. They demonstrated little degradation of strength and excellent oxidation resistance up to around 1400°C. Since

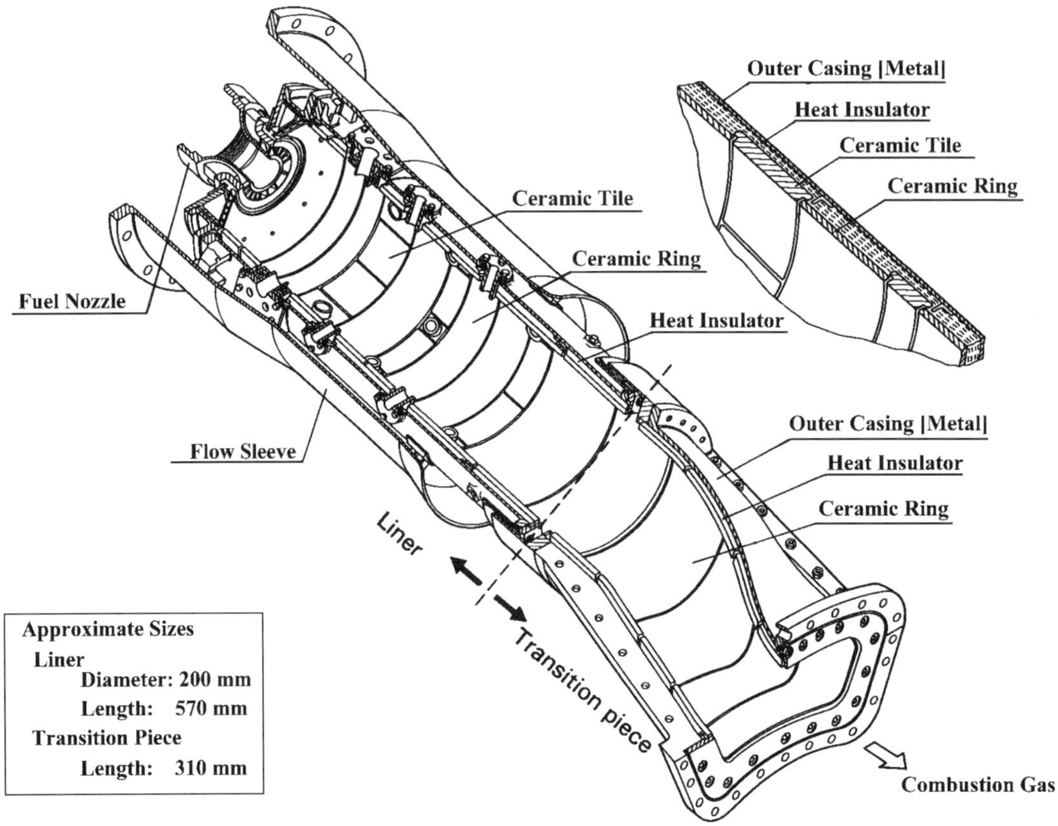

Figure 17.4 Combustor Assembly Drawing (Adapted from Ref. [2])

they had similarly high thermal conductivity as SiC, it was expected that the thermal stresses could be reduced by 30 to 50% when they were selected for the components instead of SiC. Therefore, two kinds of sintered Si_3N_4 ceramics were also selected as candidate materials for the first stage nozzle vane and the first stage rotor blade.

Development of Ceramic Combustor

Development of combustor was conducted in a cooperative research program between TEPCO and Toshiba Corporation [4].

In the first step of the ceramic combustor development, two types of combustors were designed and tested. One was a ring type wall and the other was a tile type wall. Atmospheric pressure combustion tests with low calorific heating gas fuel simulating coal derived gas were performed for both types of combustors. These tests had resulted in the final structure of the combustor as shown in Figure 17.4.

The combustor liner was composed of seven rows of ceramic rings with 5mm thickness. These ceramic rings were supported by heat insulating ceramic fibers inserted into the space between ceramic rings and outer metal casing. Ceramic tiles were adopted to reduce thermal stresses for the portion around air intake holes where the temperature gradient was supposed to be the steepest. As for the transition piece, ceramics were divided into four pieces to avoid generation of high thermal stress. SiC (Toshiba, SiC-1) was selected

Figure 17.5 Combustor Liner After a Trip Test (Reprinted with permission from TEPCO)

for combustor wall materials, because this material was fabricated by slip casting, which seemed to be the most suitable forming method for the complex shaped ceramic ring of the transition piece.

Full temperature and full pressure combustion tests were carried out at 1300°C and 15atm with a simulated coal gas having a composition and heating value of 930kcal/Nm^3, nearly equal to that of coal gasification fuel.[1] Failures of the ceramics were observed initially during the full pressure testing. After examination of the tested combustor it was deduced that these failures were caused by unexpected temperature distributions, such as local over-heating, probably due to combustion fluctuations and/or combustion gas passing between ceramic parts. It was necessary, therefore, to modify the detail design to improve the structural integrity against these temperature fluctuations.

The final evaluation was conducted in a series of one-day tests. First, a steady state test followed by normal shutdown and secondly a trip test. Figure 17.5 shows the combustor liner after the trip test. It was confirmed that the ceramic parts had remained intact after the trip test, in which the fuel was cut off instantaneously at the full load condition. Whereas the long term reliability of the ceramic materials and the stability of heat insulating materials still must be confirmed, there was no evidence of severe reaction or oxidation on the surface of the ceramic components.

In addition to confirming that the ceramic components had not suffered any noticeable degradation, it was also found that a high combustion efficiency, based on analysis of the constituents of the combustion

[1] These full pressure combustion tests with low calorific gas fuel were performed by courtesy of Yokosuka Laboratory of Central Research Institute of Electric Power Industry.

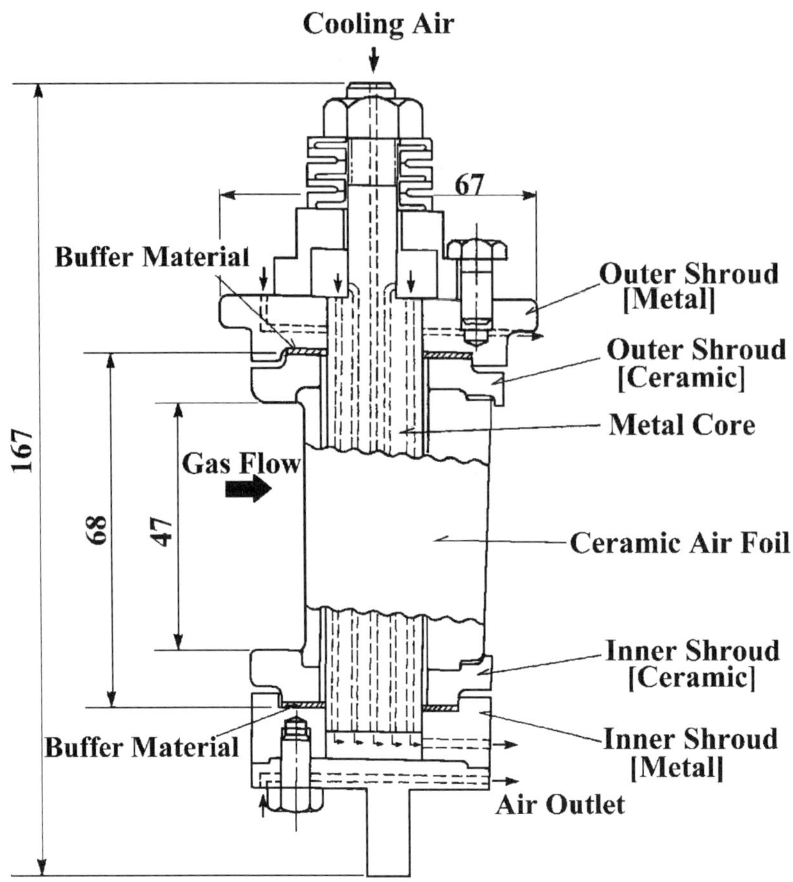

Figure 17.6 Assembly Drawing of the First Stage Nozzle (Adapted from Ref. [2])

gas, of more than 99.5% was obtained over a broad range of load conditions with a pressure loss of 3%, and an exit temperature non-uniformity of about 5%. Satisfactory low NOx emission characteristics including thermal NOx and fuel NOx were observed at the rated load conditions.

Development of Ceramic Nozzles

Development of the nozzle was conducted in a cooperative research program between TEPCO and Mitsubishi Heavy Industries, Ltd. [5, 6].

In the preliminary design and stress analysis, it was predicted that high thermal stress would be generated if a one-piece solid ceramic construction was adopted for the nozzle and shroud configuration. Therefore, the ceramic vane assembly was divided into three parts, i.e., airfoil, inner shroud and outer shroud, to reduce excessive thermal stresses generated by uneven temperature distribution. The calculated thermal stress of the three-piece construction was reduced by as much as 30% of the stress of a one-piece construction. A hybrid construction was adopted with metal shrouds and a metal core along with these ceramic parts. Figure 17.6 shows the assembly construction of the ceramic first stage nozzle. The ceramic airfoil and the inner and outer ceramic shrouds were assembled tightly with the compressive force by the metal core through metal shrouds of both sides. Buffer materials made of ceramic fiber cloth were inserted

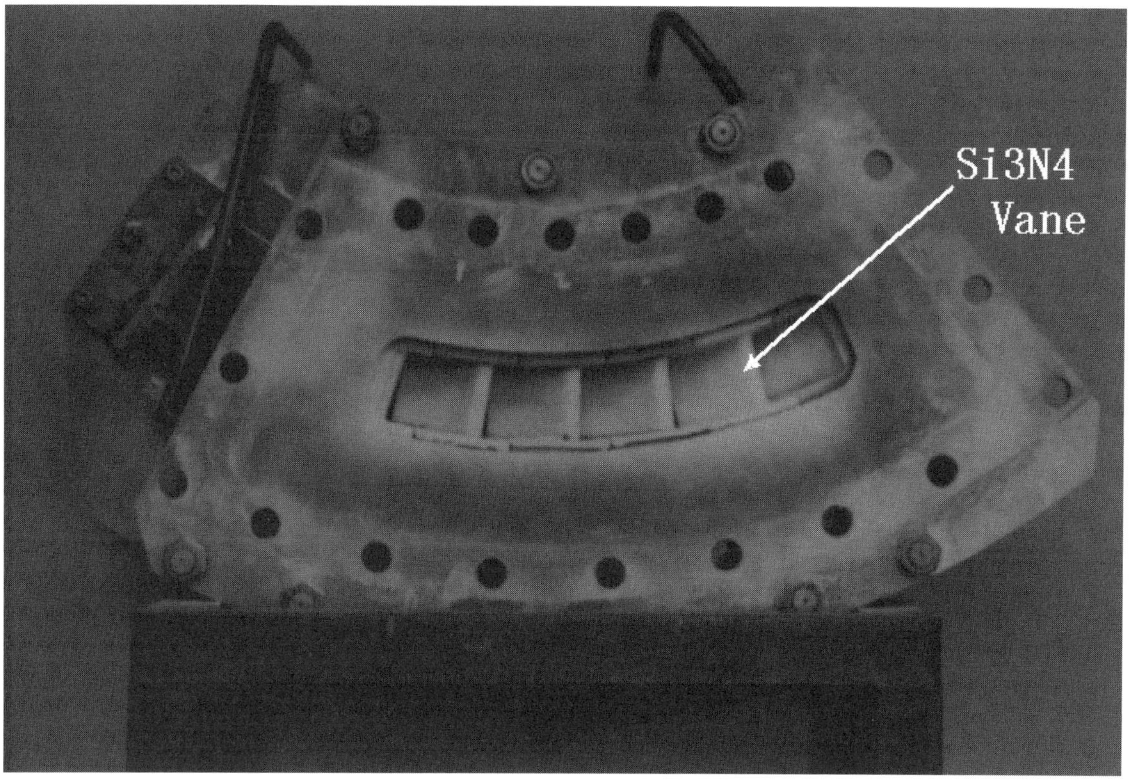

Figure 17.7 First Stage Nozzle After Trip Test (Reprinted with permission from TEPCO)

into the gap between these metal and ceramic shrouds to make the tightening load uniform and reduce heat transfer between them. Heat insulating ceramic fiber was also inserted into the space between the ceramic airfoil and the metal core.

To validate the nozzle vane design under actual gas turbine operating conditions, cascade tests were carried out by utilizing hot combustion gas flow. For the first stage nozzle vane, a series of intermediate pressure (6atm) cascade tests were carried out before conducting the full pressure (15atm) and full temperature (1300°C) cascade tests. The full pressure tests were conducted in two steps. First, a steady state test followed by a normal shutdown and secondly a trip test. Initially, SiC (Asahi Glass, C600) was selected as the candidate material for the first stage nozzle vane. Although no cracks were found and the ceramic nozzle vanes were intact following the steady state test, some ceramic vanes had experienced failures during the trip test, in which the ceramic vanes had experienced excessive thermal stress. Although the thermal stress depends on the boundary conditions, in this analysis, a maximum tensile stress of around 370 MPa was calculated 1 second after tripping. Since high temperature resistant Si_3N_4 ceramics had been developed, we studied the application of these materials to the first stage nozzle vane. When Si_3N_4 ceramics were used for the first stage nozzle vane instead of SiC, the thermal stresses could be reduced by 30 to 50%. After several modifications, first stage ceramic vanes made of SiC (Asahi Glass, C600B) and Si_3N_4 (NGK, SN-88, and Kyocera SN-252), demonstrated reliability against high thermal stress experienced in the trip testing. Figure 17.7 shows the first stage nozzle section after the trip test. Since the test duration was limited to a series of one-day tests, no signs of degradation, such as severe oxidation or corrosion, were observed by visual inspection. It was also confirmed that the amount of cooling air, which was used to cool the metallic parts, was about one tenth of that required for conventional metal vanes.

Similarly, cascade tests for the second stage nozzle vanes, made of Si_3N_4 (NGK, SN-84), were carried out under actual operating conditions (1000°C, 6atm) as well as over-load conditions (1200°C, 6atm). Thirty hours of endurance testing were also performed to validate the design and confirm the integrity of the ceramic parts.

Whereas the three-piece construction seemed to be effective in reducing thermal stresses, unknown factors at the contact surface of ceramics and problems associated with combustion gas and/or cooling air leakage between ceramic parts might result in unexpected failure of ceramics. In addition, delicate handling was required in the assembling process for the segmented construction. We therefore studied a one-piece first stage ceramic nozzle with improved Si_3N_4, because this Si_3N_4 had about twice the thermal shock resistance of SiC. Cascade tests were also carried out using the one-piece ceramic nozzle vanes and the design of these Si_3N_4 vanes was validated under actual gas turbine conditions.

Assembly Test of the Ceramic Combustor and Ceramic Nozzle Vanes

After having demonstrated the effectiveness of the ceramic combustor and ceramic nozzle vane designs separately, the ceramic combustor (SiC) and the first stage nozzle vanes (Si_3N_4) were assembled together and combustion tests were conducted under the full pressure (15atm) and full temperature (1300°C) conditions.

In this test series, natural gas fuel with steam injection was used instead of the simulated coal gasification fuel with low heating value, due to restrictions of the test facility, and this led to more severe trip conditions than would be expected for the actual gas turbine. Initially the ceramic combustor experienced failure probably due to fluctuations of the combustion gas and/or unexpected inflow of steam during trip operation. After several modifications of the combustor design and test facility, the reliability of the combustor (Asahi Glass, C600) and the ceramic nozzle vanes (NGK, SN-88) was verified from these assembly tests. No damage was observed for either component. However, the slight rainbow pattern on the surface of the SiC indicated the formation of a thin oxide layer.

Development of a Ceramic Rotor

Development of the rotor was conducted in a cooperative research program between TEPCO and Hitachi Ltd. [7, 8]. At the beginning of the development, several kinds of SiC were considered as the candidate materials for the first stage rotor blade. In the preliminary design, the condition at the inlet of the first stage rotor blade was assumed to be the average gas temperature of 1150°C and a gas pressure of 9.5 atm. The pitch circle diameter and the blade height were specified as 790 mm and 58 mm, respectively. For the first stage rotor blade, it was decided not to adopt an all-ceramic rotor but a hybrid type rotor which consisted of ceramic blades and a metal disk. The fundamental challenge for this hybrid type structure was how to mount ceramic blades to the metal disk. This structure should satisfy the following requirements:

(1) Avoiding a high local stress concentration at the root of the ceramic blade.
(2) Heat insulation between the ceramic blades and the metal disk.
(3) Adequate damping against blade vibration.
(4) A sealing mechanism to prevent the inverse flow of hot combustion gas.

Among these requirements, the most important problem in developing a ceramic rotor blade was how to minimize the stress on the blade induced by the centrifugal forces of rotation. Moreover, if the ceramic blade was designed to be in contact with the metal, the temperature of the ceramic should be limited to below around 600°C at the contact area to avoid a chemical reaction between them. Taking these points into consideration, three types of hybrid structure, as shown in Figure 17.8, were investigated. Ultimately, the root attachment Type-C was selected. Two metal pads were inserted symmetrically in the fitting layer to transfer the centrifugal force of the rotating blades directly to the metal disk.

The configuration of the root was investigated by FEM stress analyses and experiments. These studies showed that a decrease in the angle between the contact surface and radial direction resulted in a decrease in

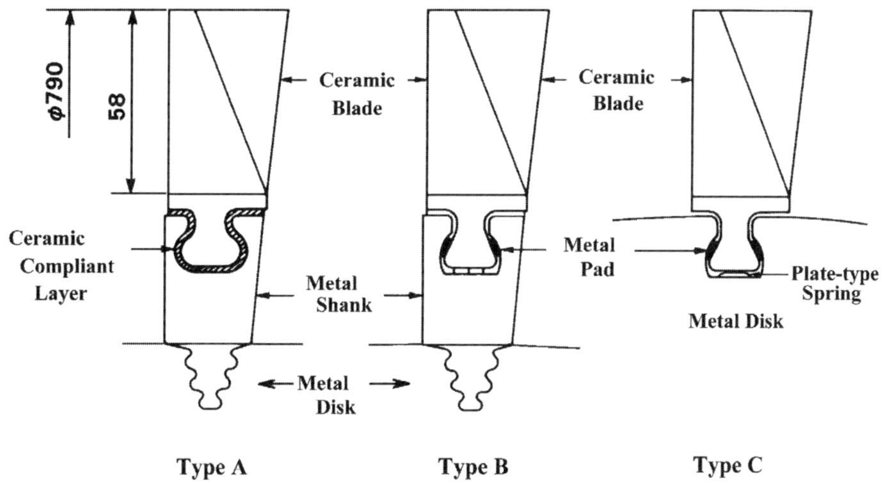

Figure 17.8 Hybrid Structures of Ceramic Rotor Blade (Adapted from Ref. [8])

peak stress at the root of the dovetail and an increase in contact stress. The angle of the contact surface was finally selected as 45 degrees by considering the above analytical results, and the balance between the normal force and shearing force at the contact surface.

Figure 17.9 shows the detail of the type C structure. A small amount of cooling air was introduced between the ceramic blade and the metal disk in order to provide heat insulation and to prevent the inverse flow of hot gas. Moreover, gaps along the edges of the blade platform were automatically sealed by cylindrical ceramic pins, which were pushed and held automatically at the proper positions by their own centrifugal forces. In order to locate blade and pads at their correct positions under static conditions, a plate-type spring was inserted between the bottom of the blade and the metal disk.

Two kinds of Si_3N_4 (NGK, SN-88 and Kyocera SN-252) were finally selected as the candidate materials for the ceramic blade, since much progress had been made with the strength and heat resistance of Si_3N_4 at elevated temperatures. Tensile tests were performed both at room temperature and at 600°C to examine the strength of the dovetail. All specimens exceeded the strength corresponding to 140% of the rated speed. Room temperature spin tests were performed to confirm the reliability of ceramic blades subject to centrifugal force. The test results showed that most of the type C specimens were able to survive the centrifugal force up to 140% of rated speed.

Thermal loading cascade tests were carried out under the design temperature and pressure for the gas in a static test rig. A cascade of five ceramic blade specimens with two side ceramic blades was set up into a metal base which had the same root configurations as the rotor disk. Two types of tests were conducted. First, a steady state test followed by normal shutdown and secondly a trip test. After each test, the cascade was disassembled and the ceramic blades were examined carefully by the visual appearance check and by fluorescent dye penetrant inspection. No cracks were found and it was confirmed that the ceramic blades were able to survive the severe thermal transient load caused by the trip.

Since the strength reliability of the type C structure was examined under the conditions of centrifugal force and thermal transient stress, it was decided to conduct high temperature spin tests using the high temperature gas turbine development unit (HTDU). Table 17.3 shows the fundamental specifications of the HTDU test. Figure 17.10 shows the setup of the ceramic blades into the metal disk. A total of six ceramic blades were mounted in the metal disk. To maintain the rotating balance, two sets of three blades were placed axi-symmetrically. A total of four tests were performed. At first, a spin test with compressed air was performed up to the rated speed, and the vibrations of the rotor shaft and ceramic blades were found to be very

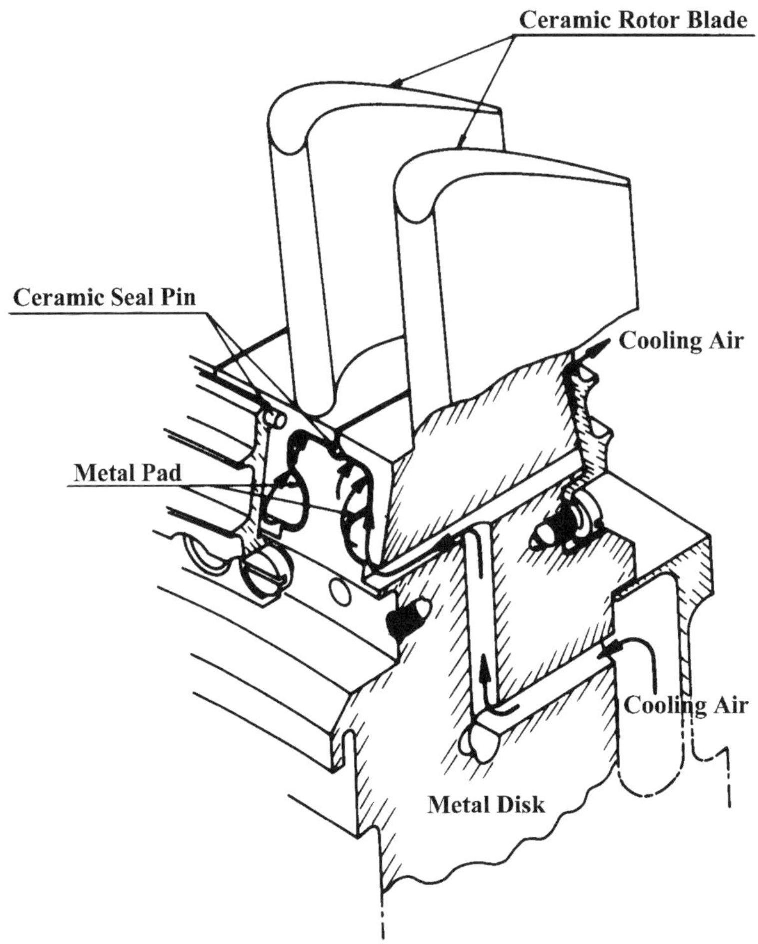

Figure 17.9 Design of C Structures Associated With Ceramic Rotor Blade. (Adapted from Ref. [8])

small and well below the allowable limits. The other three tests were conducted using combustion gas up to the design rated speed. After these tests, the rotor was disassembled and inspected. There were no detectable cracks or damage to the ceramic blades. The plate-type spring showed, however, some permanent deformation due to its own centrifugal force. So, it was necessary to make some design modifications for this spring.

SUPPORTING RESEARCH

Many research projects in support of the ceramic component development were conducted as cooperative research between TEPCO and Japanese and US companies and as in-house research of TEPCO. Typical research projects are listed in Table 17.4.

Material Evaluation

More than 30 kinds of Si-based ceramic materials, which were commercial grade and/or development grade, were evaluated at TEPCO to determine their strength, fracture toughness, thermal shock resistance,

TABLE 17.3 COMPARISON OF THE DESIGN CONDITION AND THE HTDU CONDITION		
	Design	**HTDU**
Average gas temperature at combustor exit	1300°C	1300°C
Gas pressure at combustor exit	15 atm	3 atm
Average gas temperature at rotor inlet	1150°C	1080°C
Rated rotational speed	10800 rpm	11264 rpm
Circumferential speed at PCD	413 m/s	413 m/s

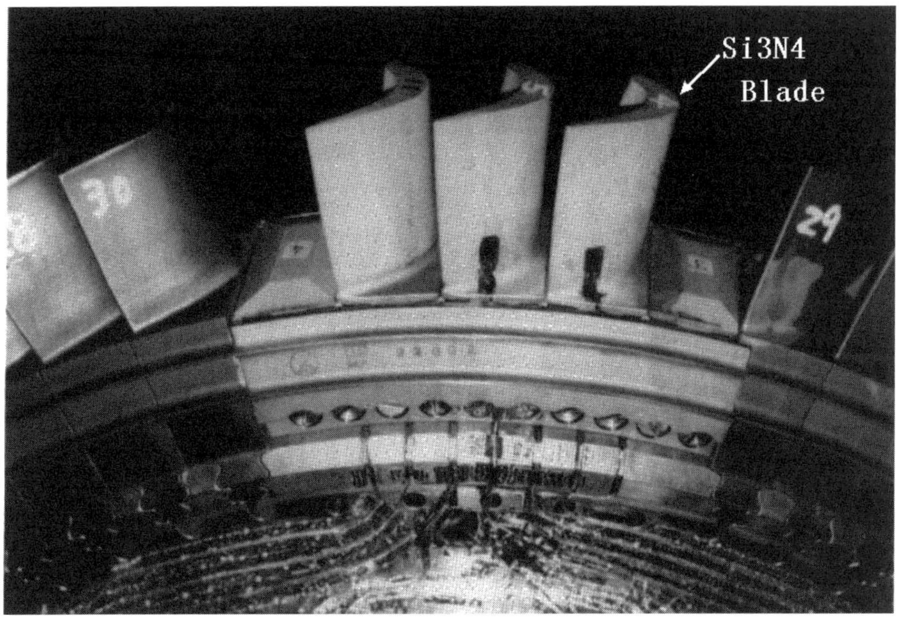

Figure 17.10 Set up of Ceramic Blades in Metal Disk Prior to HTDU Test (Adapted from Ref. [8])

TABLE 17.4 TYPICAL RESEARCH PROJECTS IN SUPPORT OF THE PROGRAM		
Research Items	**Collaborators**	**References**
Mechanical properties of ceramics	in house	[9–11]
FOD/Ball impact damage	Asahi Glass	[12]
	IHI	[13]
	Hitachi	[14]
	in house	[15–17]
Corrosion behavior under combustion gas flow	MES	[18–19]
Blade attachment	Asahi Glass	[20]
Ceramic to metal joining	Kobe Steel	
Non destructive evaluation	NSC	
Thermal stress testing	Toshiba	[21]
Ceramic protective coating	NGK	[19]
Reliability analysis code "G-FICES"	IHI	[22]
Cycle performance of ceramic gas turbines	MHI	[23]
	GE	[24]
SiC particle and platelet reinforced ceramics	NGK	[25]
SiO_2 doped Si_3N_4	NGK	[26]

IHI: Ishikawajima-Harima Heavy Industries Co., Ltd., MES: Mitsui Engineering & Shipbuilding Co., Ltd., NSC: Nippon Steel Corporation, GE: General Electric Co.

TABLE 17.5 APPROXIMATE PROPERTIES OF CANDIDATE MATERIALS (MEASURED AT TEPCO)

Supplier/Product No. or Designation	Properties				
	Flexural Strength [MPa] (4 Point Bending) R.T/ 1200/1400°C	R.T. Fracture Toughness [MPa.m$^{1/2}$] (SEPB)	Thermal Shock Resistance [°C] (Water Quench)	Oxidation Resistance [mg/cm^2] (Weight Gain) 1400°C, 100 hours	Fatigue (Rupture Strength) [MPa] 1400°C, 1000 hours
SiC					
Toshiba/SiC-1	380/390/400	2	350	0.2	Tensile:220
Asahi Glass/C600	480/450/440	5	400	0.3	Tensile:200
Asahi Glass/C600B	770/600/580	5	—	0.3	Bending:500
NGK Insulators/SC-21	630/620/620	2–3	400	0.2	Bending:300
Si_3N_4					
NGK Insulator/SN-88	800/750/720	7	1050	0.2	Bending:300
Kyocera/SN-252	670/550/500	7	800–900	—	—
NGK Insulators/SN-84	940/850/—	6	1050	—	—
Kyocera/SN-281	830/700/600	6	—	—	—

oxidation resistance, fatigue behavior, as well as their potential for fabrication of complex-shaped parts. Table 17.5 summarizes the approximate properties of monolithic ceramics which were selected as candidate materials for the component tests. During this research period, high performance Si_3N_4 ceramics had been developed, which demonstrated little degradation in strength and excellent oxidation resistance up to around 1400°C, along with high thermal shock resistance and high fracture toughness. Therefore, the advantage of SiC seemed to be of less interest, as long as the TIT of the gas turbine was below 1300°C.

The component test results indicated that high value of fracture toughness and thermal shock resistance was desirable to overcome unexpected failure. Foreign object damage (FOD) was considered to be one of the great problems for ceramic gas turbine components. While we had not experienced failures caused by FOD in component tests, we observed the trace of particle impact on the surface of ceramics. Much research has been conducted to study the damage of ceramics by solid particle impact [12–17].

No serious damage from oxidation or corrosion was observed on the surface of the ceramics after component tests because the duration was limited to only a series of one-day tests, except for the second stage nozzle vane which was tested for 30 hours. To clarify the long term behavior of ceramics, a series of endurance tests were conducted simulating the actual gas turbine environment, in which ceramics were exposed to high temperature, high pressure and high velocity gas flow[2]. The changes in flexural strength at room temperature were negligibly small after the tests. However, unacceptable amounts of weight-losses were observed on specimens made of SiC and Si_3N_4, when tested at temperatures above 1300°C and at pressure above 6 atm [18]. The weight loss was also observed on specimens coated with CVD-SiC (Chemical Vapor Deposited SiC), which was regarded as the highest oxidation resistant Si-based ceramic. While sufficient studies were not conducted to conclusively clarify the mechanism of this recession, the results of the tests suggested the following two mechanisms: the formation of volatile SiO in locally reducing atmospheres and/or the evaporation of SiO_2 resulted in the absence of a stable surface layer.

To prevent the weight loss and to improve the corrosion resistance of non-oxide ceramics, various kinds of specimens with oxide ceramic coatings, such as Al_2O_3, ZrO_2 and $ZrSiO_4$, were fabricated by plasma

[2] These high pressure endurance tests were performed by courtesy of Mitsui Engineering & Shipbuilding Co., Ltd.

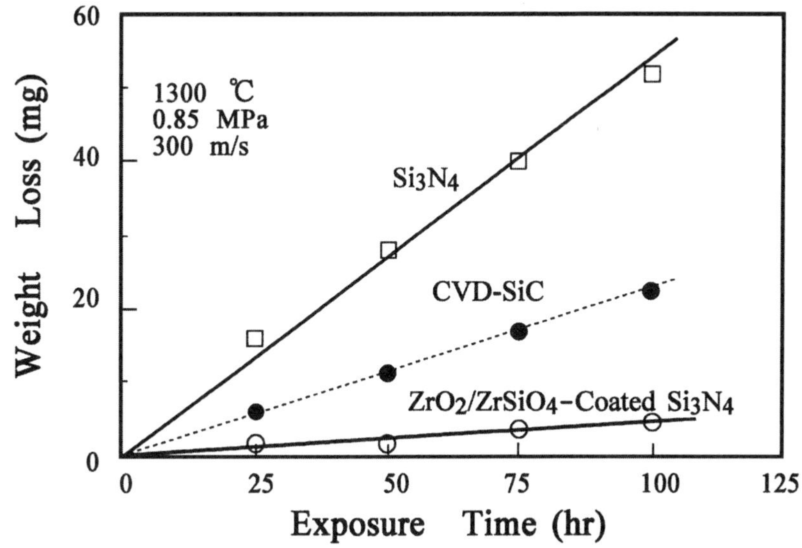

Figure 17.11 Relationship Between Exposure Time and Weight Loss in Combustion Gas Flow (Reprinted with permission from TEPCO)

spraying and tested [19]. A graded coating consisting of a surface layer of ZrO_2 and an under-layer of $ZrSiO_4$ on a Si_3N_4 substrate had demonstrated a significant effect on reducing the weight-loss of the substrate, as shown in Figure 17.11. This coating system showed the best stability in a combustion gas flow compared with other systems. However, we observed peeling of the coating layer when tested for 100 hours at temperatures in excess of 1400°C. The long term stability and the reliability of this protective coating, therfore, should be further improved.

Reliability Analysis

As a part of this project, TEPCO developed a program GFICES (Gas turbine—Fine Ceramics Evaluation System) with Ishikawajima-Harima Heavy Industries Co., Ltd. which evaluated the strength of structural ceramic components on the basis of statistical strength theory using a two-parameter Weibull probability distribution [22].

GFICES was able to analyze the failure probability for fast fracture and time-dependent fracture due to slow crack growth. The program had additional features, e.g. the effect of proof testing could be taken into account for fast fracture analysis. Moreover, the effect of slow crack growth could be incorporated into the probability estimation of fast fracture.

Stress analysis was conducted for the silicon nitride rotor blade. Centrifugal stress at the normal (steady state) design condition was 220 MPa at the root of the dovetail. Thermal loading yields a peak thermal stress of 322 MPa near the center of the leading edge during the design trip condition. The reliability analyses were conducted using GFICES. Table 17.6 summarizes the safety factor which was defined as the ratio of the load corresponding to a failure probability of 50% to the operating load. As shown in this table, the safety factor for the normal design condition was raised almost twice in the 120% spin test at room temperature. The safety factor for the design trip condition was 1.57 which was the most severe value for the ceramic rotor blade. It was believed to be quite necessary to establish a proof testing method for the thermal condition of a gas turbine in order to raise the reliability of ceramic gas turbine components.

TABLE 17.6 NUMERICAL RESULTS ON THE SAFETY FACTORS OF THE CERAMIC BLADE

Condition	Proof Test Condition	Safety Factor
Room Temperature Spin Test		
120% of normal design speed	————	1.54
Thermal Loading Test		
Steady state	————	4.90
Trip condition	————	1.87
High Temperature Spin Test		
Steady state	————	2.30
	120% room temperature spin test	4.31
Trip condition	————	2.42
Design Condition		
Normal condition	————	2.21
	Steady state of thermal loading test	2.21
	120% room temperature spin test and steady state of thermal loading test	4.27
Trip condition	————	1.57
	Trip condition of thermal loading test	2.11
	120% room temperature spin test and trip condition of thermal loading test	2.16

Performance Analysis for the Ceramic Gas Turbine

A study was carried out in a cooperative program between TEPCO and Mitsubishi Heavy Industries Ltd. to assess the performance improvement of a 1500°C class combined cycle power plant when ceramic turbine components were applied [23]. The performance improvement was obtained as a result of reduced blade cooling air. In this study four different kinds of combined cycles were investigated.

Results of this study indicated that the combined cycle with a simple gas turbine was the most practical of the four cycles studied with an efficiency of higher than 60%. The combined cycle with a reheat gas turbine had the highest efficiency if a higher compressor exit air temperature and a high gas temperature (over 1000°C) to reheat the combustion system was used. A higher pressure ratio was required to optimize the cycle performance of the combined cycle with the ceramic turbine components than that with the metal turbine components because of reduced blade cooling air. To minimize leakage air for these higher pressure ratios, advanced seal technology must be applied to the gas turbines. A comparison of steam cooling and a simple ceramic gas turbine was also conducted and it revealed that the ceramic gas turbine had benefits in cycle efficiency, cost, and was simple to control when compared with the steam cooled gas turbine cycle.

A study was also conducted in a cooperative program between TEPCO and General Electric Co. to assess the benefit of ceramic components when adopted in an existing 1300°C class gas turbine (MS9001FA) [24]. In this study two levels of usable temperature limits were set for SN-88 ceramics. One was an 'Optimistic' case of 1315°C and the other was a 'Less Optimistic' case of 1204°C. A small amount of cooling air was utilized, if necessary, to maintain the maximum ceramic temperature below its usable temperature limit. Results of this study indicated that combined cycle efficiency was improved up to 3% in the 'Optimistic' case and up to 2% in the 'Less Optimistic' case relative to the baseline.

BASIC RESEARCH FOR HIGHER TIT

When the program started in 1984, cooled metal components limited the TIT to about 1100°C, while it was believed that uncooled ceramics could be used at TIT's up to about 1300°C. In the early 1990's technology had improved to the point that TIT's of 1300°C were achievable with cooled metal components and

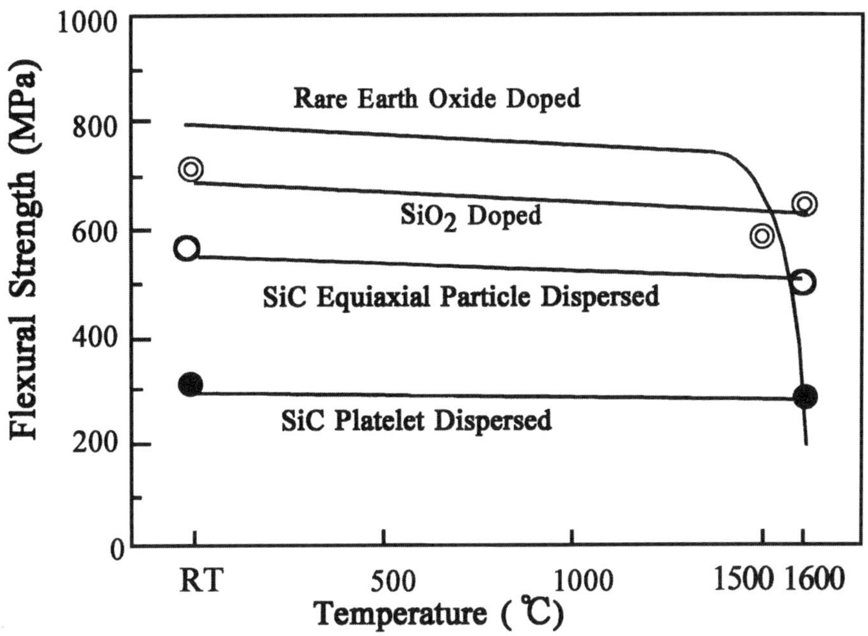

Figure 17.12 Temperature Dependence of the Strength of Silicon Nitride Ceramics (Reprinted with permission from TEPCO)

the incremental gains made with ceramic components were less significant. So the program goals were modified and a TIT target of 1500°C was adopted.

Improvement of Temperature Resistance of Ceramics

Since the efficiency of the gas turbine increases with TIT, there was strong requirement for a high temperature resistant structural material. Some B-C doped SiC demonstrated excellent high temperature strength, high creep resistance and high oxidation resistance in a static environment. However, it seemed difficult to apply these materials to gas turbine components due to their low fracture toughness. The fracture toughness of Al_2O_3 doped SiC was high compared with B-C doped SiC, and this material showed relatively good results in component tests. This material, however, showed strength degradation and less resistance to oxidation and creep deformation above 1400°C. We tried to improve these properties by optimization of the amount of additives. However, we were unable to prepare high temperature resistant SiC compatible with high fracture toughness. TEPCO conducted a cooperative research program with NGK Insulators, Ltd., to develop a high temperature resistant Si_3N_4 [26]. SiO_2 doped Si_3N_4 also showed excellent properties up to 1600°C, as shown in Figure 17.12. However, its low fracture toughness did not improve as was expected when SiC platelets or SiC particles were dispersed as reinforcements. As a result we concluded that existing ceramic materials were not applicable for the hot section parts of a 1500°C class gas turbine without cooling.

Research on Air-Cooled Ceramic Nozzle

The application of ceramics to the first stage nozzle vanes of a 1500°C TIT class gas turbine was carried out in a cooperative research program between TEPCO and Mitsubishi Heavy Industries, Ltd. [27, 28].

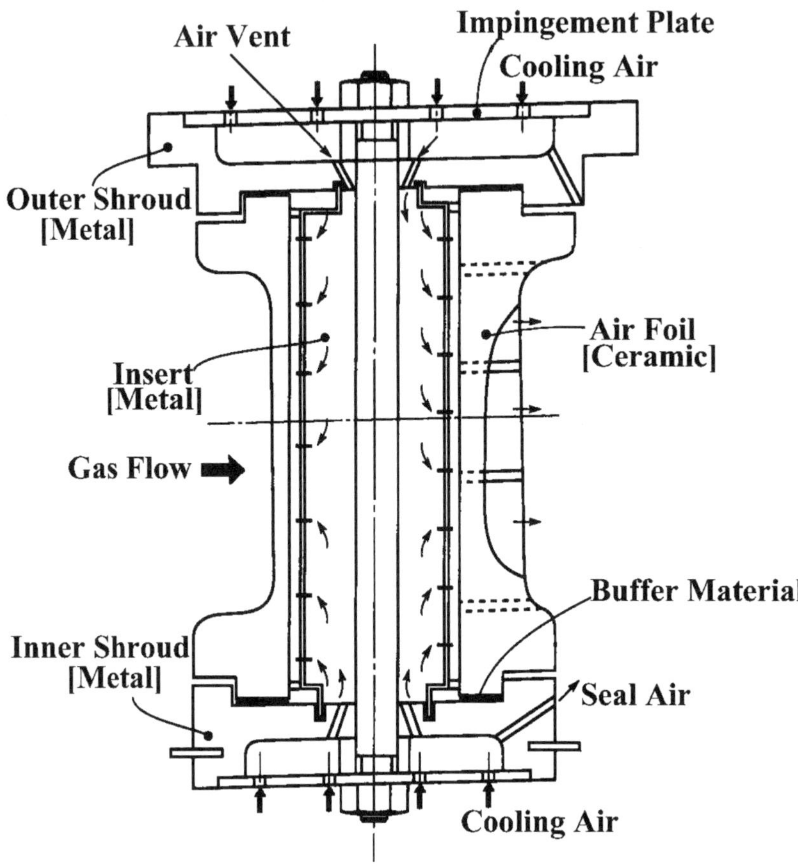

Figure 17.13 Assembly Drawing of the Cooled Ceramic Nozzle (Adapted from Ref. [27])

A 20 MW class gas turbine with TIT of 1500°C was selected as the first target. For a 1500°C TIT class gas turbine, the material should be able to withstand high temperature gas flow above 1600°C, considering the temperature distribution of the combustion gas.

A preliminary cascade test was conducted for uncooled first stage nozzle vanes made of SiC and Si_3N_4 under 1500°C and intermediate pressure (6atm) conditions to measure the temperature distributions. Some cracks were observed on SiC vanes and unacceptable erosion was observed on the trailing edge section of Si_3N_4 vanes.

To compensate for the limited temperature resistance of existing materials, a one-piece vane construction with a small amount of cooling air, had been studied. Two kinds of Si_3N_4 (NGK, SN-88 and Kyocera SN-281) were selected for this study. The assembly construction of the designed air-cooled ceramic nozzle vane is shown in Figure 17.13. Whereas the three-piece construction for an airfoil and outer/inner shrouds seemed to be effective to reduce thermal stress, a one-piece construction was chosen in the present case. This was to avoid the unknown factors at contact surfaces of ceramics and problems associated with gas leakage between ceramic parts which might cause unexpected failure of ceramics. The inner face of the vane was cooled by air impingement and this air was mixed into the main gas flow through a cooling slit exit at the trailing edge of the vane. In determining the amount of cooling air, two levels were considered. One was the standard cooling in which the amount of cooling air was determined to maintain the maximum ceramic temperature at 1300°C. The other was mild cooling in which the amount of cooling air was reduced to maintain a 1400°C maximum ceramic temperature. The amount of cooling air was 1.4% of the main gas flow for the standard cooling and 1.2% for the mild cooling.

Cascade tests were carried out for the air-cooled Si_3N_4 vanes under intermediate pressure (6atm) and 1500°C conditions. Furthermore, a series of cascade tests were carried out under the full pressure (13.5atm) condition. It was confirmed that the maximum ceramic temperature at the airfoil section could be maintained below 1300°C by a small amount of cooling air. It was about one fifth of that of a conventional air-cooled metal vane.

In spite of the increased thermal stresses by local cooling, all vanes survived in the intermediate pressure cascade tests including an emergency shut down test. However, some cracks were found after both steady state tests and emergency shut down tests under full pressure conditions. We made substantial efforts to identify the cause of this failure. However, the cause could not be conclusively identified. The calculated thermal stresses were 210 MPa for steady state conditions and 370 MPa for trip conditions at most. However, we considered there was a possibility that the test conditions were much more severe than intended. Whereas we were not able to obtain full clarification and to confirm the validity of the air-cooled ceramic vane at full pressure conditions, we believed that the utilization of a small amount of cooling air was an effective remedy for the application of ceramics to a gas turbine with higher TIT.

SUMMARY

In this chapter, the authors reported on a research program for the application of ceramics to a power generating gas turbine. During the first phase of this program, Si-based ceramics, such as SiC and Si_3N_4, were evaluated to confirm that uncooled ceramic components could be incorporated into a concept gas turbine with an output power of 20MW and a TIT of 1300°C.

(1) The combustor liner design with a hybrid construction of ring and tile type liner was confirmed. Ceramic parts were segmented into several pieces to avoid excess thermal stress due to the gas temperature distribution.
(2) The validity of nozzle vane designs with a metal load-bearing core and a heat-resistant ceramic outer airfoil were demonstrated. The three-piece nozzle vanes made of SiC and Si_3N_4, and the one-piece nozzle vanes made of Si_3N_4 survived in cascade tests.
(3) In addition to the preceding development of combustor and nozzle vanes, the development of a first stage ceramic rotor blade was successfully completed. To keep the contact stresses to an acceptable level for the blade inserted into the metal disk, a 45 degree dovetail angle was selected rather than a higher angle that would increase interface shear.
(4) The initial target of the project was accomplished and the progress of the research program indicated a positive prospect for an application of existing ceramic materials to uncooled components of a 1300°C TIT class gas turbine.
(5) The progress with material properties was significant. Improved Si_3N_4 performed well in uncooled component tests up to 1300°C in air, but exhibited unacceptable weight loss when tested in an actual high pressure turbine environment above 1300°C. Oxide surface coatings, especially a graded coating consisting of a surface layer of ZrO_2 and an under-layer of $ZrSiO_4$, substantially reduced the weight loss, but much additional work is needed to achieve a material system that will provide the life required for an industrial gas turbine.
(6) It was also confirmed that ceramics had potential for nozzles for a gas turbine operating at 1500°C TIT with a small amount of cooling air: that is about one-fifth the cooling required for a conventional metal nozzle.
(7) Performance analyses conducted for 1300°C and 1500°C TIT class gas turbines indicated that ceramics could significantly improve the efficiencies of existing and future combined cycle power plants.

It seems that further efforts will be needed before ceramic components can be put into practical use in a power generating gas turbine. There remain concerns about unexpected failure of ceramics due to their brittleness. The reliability of ceramic components must be improved by means of further development of the materials, designing methodologies, quality assurance technology, and so on. Still, these ceramic materials have potential for future applications once these concerns have been successfully addressed.

ACKNOWLEDGEMENT

The authors wish to thank heartily the persons in NGK Insulators, Ltd., Asahi Glass Co., Kyocera Co., Mitsui Engineering & Shipbuilding Co., Ltd., Ishikawajima-Harima Heavy Industries Co., Ltd., Kobe Steel, Ltd., Nippon Steel Corporation, Central Research Institute for Electric Power Industries, and General Electric Co., for their kind assistance and cooperation with this study.

REFERENCES

[1] Hara, Y., Hisa, S., Wada, K., and Tsuji, I., "Research and Development of Ceramic Components for Power Generating Gas Turbine," JSME Reports No. 870 [10], 111–116, 1987 (in Japanese).

[2] Hara, Y., Tsuchiya, T., Maeda, F., Tsuji, I., and Wada, K., "Development of Ceramic Components for a Power Generating Gas Turbine," ASME paper 91-GT-319, presented at the International Gas Turbine and Aeroengine Congress and Exposition, Orlando, FL, USA, June 3–6, 1991.

[3] Kojima, Y., Wada, K., Furuse, Y., and Teramae, T., "Highly Durable Thermal Barrier Coating for Heavy Duty Gas Turbine," *Proceedings of the International Thermal Spray Conference '95*, Kobe, Japan, 22–26 May, pp. 95–99, 1995.

[4] Maeda, F., Okabe, N., Tsuchiya, T., and Watanabe, N., "Development of Ceramic Combustor for Industrial Gas Turbines," *Proceedings of the International Gas Turbine Congress,* Yokohama, Japan, October 27-November 1, 1991, 91-YOKOHAMA-IGTC-44, pp. III-1—III—8.

[5] Miyazaki, S., Aoki, S., Tsuji, I., Hara Y., Furuse, Y. and Hamada, S., "Cascade Testing of Ceramic Vanes for Industrial Gas Turbines," ASME paper 91-GT-311, presented at the International Gas Turbine and Aeroengine Congress and Exposition, Orlando, FL, USA, June 3–6, 1991.

[6] Miyazaki, S., Tsuji, I., Mori, M., Hara, Y., Furuse, Y. and Hamada, S., "Development of Ceramic Vanes for an Industrial Gas Turbine," *Proceedings of the International Gas Turbine Congress*, Yokohama, Japan, October 27-November 1, 1991, 91-YOKOHAMA-IGTC-101, pp. I-149—I-155.

[7] Machida, T., Miyata, N., Wada, K., Furuse, Y. and Teramae, T., "Development of Ceramic Rotor for Industrial Gas Turbine," *Proceedings of the 1991 International Gas Turbine Congress*, Yokohama, Japan, October 27-November 1, 1991, 91-YOKOHAMA-IGTC-103, pp. I-165—I-170.

[8] Teramae, T., Furuse, Y., Wada, K., and Machida, T., "Development of Ceramic Rotor Blade for Power Generating Gas Turbines," ASME paper 94-GT-309, presented at the International Gas Turbine and Aeroengine Congress and Exposition, The Hague, The Netherlands, June 13–15, 1994.

[9] Tsuruta, H. and Furuse, Y., "Fracture Behavior of Advanced Ceramics for Gas Turbine," *Proceedings of the 1st International Conference on the Science of the Engineering Ceramics*, Aichi, Japan, pp. 577–582, 1991.

[10] Tsuruta, H. and Furuse, Y., "Investigation of R-Curve Behavior and its Effect on Strength for Advanced Ceramics," International Symposium on Fracture Mechanics of Ceramics, Nagoya, 1991, *Fracture Mechanics of Ceramics*, Vol. 9, pp. 251–263, 1992.

[11] Iyogi, G., Furuse, Y. and, Teramae, T., "High Temperature Strength Properties of Structural Ceramics," Proceedings of M & P'93, JSME, pp. 113–114, 1993 (in Japanese).

[12] Hara, Y., Hamada, S., Minematsu, T., Takahashi, H. and Abe, H., "Impact Damage of Silicon Carbide and Silicon Nitride," *Proceedings of the 3rd. International Symposium on Ceramic Materials & Components for Engines*, Las Vegas, NV, USA, November 27–30, 1988. V.J. Tennery, ed., The American Ceramic Society, Inc., 1989. pp. 1179–1188.

[13] Suzuki, A., Natsumura, T., Sugiyama, S., Teramae, T. and Hamada, S., "Particle Impact Damage in Silicon Carbide," Key Engineering Materials, Vol. 51/52, pp. 167–172, 1991.

[14] Nakayama, M., Machida, T., Teramae, T. and Hamada, S., "A study on Fracture Behavior in SiC by Projectile Impact," *Proceedings of the Asian Pacific Conference on Fracture and Strength '93*, Tsuchiura City, Japan, July 26–28, pp. 723–727, 1993.

[15] Hamada, S. and Teramae, T., "Fracture Analyses of SiC and Si_3N_4 Ceramics Impacted by Solid Particles," *Proceedings of the International Gas Turbine Congress*, Yokohama, Japan, October 27-November 1, 1991, 91-YOKOHAMA-IGTC-87, pp. III-327—III-331.

[16] Iyogi, G., Hamada, H. and Teramae, T., "Strength Recovery of Ball Impacted Ceramics by Heat Treatment," *Proceedings of the 5th. International Symposium on Ceramic Materials & Components for Engines*, Shanghai, People's Republic of China, 1994. pp. 498–501.

[17] Hamada, S., Iyogi, G., and Teramae, T., "Effects of Static Initial Stress on Foreign Object Damage of Ceramics," J. Soc. Mat. Sci., Japan, Vol. 43, pp. 1195–1199, 1994 (in Japanese).

[18] Furuse, Y., Tsuchiya, T., Handa, S., and Takahashi, N., "Oxidation Behavior of Structural Ceramics in Combustion Gas Flow," *Proceedings of the 2nd Fall Symposium of Ceramic Society of Japan*, pp. 178–179, 1989 (in Japanese).

[19] Endoh, Y., Tsuchiya, T., and Furuse, Y., "Corrosion Behavior of Ceramics for Gas Turbine Application," *Proceedings of the 1993 PAC RIM Meeting*, Honolulu, HI, USA, November 7–10, pp. 319–326, 1993.

[20] Kitajima, H., Furuse, Y., Toriyama, A and Ono, T., "Development of a Mechanical Attachment for Ceramic Blades and Metallic Wheel (Part 1)," Bulletin of the GTSJ, pp. 39–39, 1990.

[21] Matsuhita Y., Furuse, Y., Yasunaga, T. and Kawakami, T., "A New Method for Thermal Stress Testing of Structural Ceramics," *Proceedings of the 2nd Meeting of PAC RIM Ceramic Societies*, Cairns, Australia, 15–17 July, 1996, (CD-ROM).

[22] Teramae, T., Hamada, S., and Hamanaka, J., "Probabilistic Structural Analyses of Ceramic Gas Turbine Components," *IUTAM Symposium on Probabilistic Structural Mechanics—Advances in Structural Reliability Methods*, pp. 518–533, 1994.

[23] Sugishita, H., Mori, H., Chikami, R., Tsukuda, Y., Yoshino, S., Tsuchiya, T., and Furuse, Y., "A Study of Cycle Performance of Ceramic Gas Turbines," ASME paper 96-GT-492, presented at the International Gas Turbine and Aeroengine Congress and Exposition, Birmingham, U.K., June 10–13, 1996.

[24] Grondahl, C.M. and Tsuchiya, T., "Performance Benefit Assessment of Ceramic Components in an MS9001FA Gas Turbine," ASME paper 98-GT-186, presented at the International Gas Turbine and Aeroengine Congress and Exhibition, Stockholm Sweden, June 2–5, 1998.

[25] Sakai, H, Matsuhiro, K. and Furuse, Y., "Mechanical Properties of SiC Platelet Reinforced Ceramic Composites, Ceramic Transaction, Vol. 19, Advanced Composite Materials, pp. 765–771, 1991.

[26] Takahashi, T., Isomura, M., Endoh, Y. and Furuse, Y., "SiO_2 Doped Si_3N_4 Ceramics," presented at Silicon Nitride '93, Stuttgart, 1993. Key Engineering Materials Vol. 89–91, pp. 225–228, 1994.

[27] Tsuchiya, T., Furuse, Y., Yoshino, S., Chikami, R., Tsukagoshi, K. and Mori, M., "Development of Air-cooled Ceramic Nozzles for a Power Generating Gas Turbine," ASME paper 95-GT-105, presented a the International Gas Turbine and Aeroengine Congress and Exposition, Houston, TX, USA, June 5–8, 1995.

[28] Yoshino, S., Tsuchiya, T., Furuse, Y., Chikami, R., Kasai, Y. and Tsukuda, Y., "Design and Test of an Air-Cooled Ceramic Nozzle for a Power Generating Gas Turbine," *Proceedings of the International Gas Turbine Congress*, Yokohama, Japan, October 22–27, 1995, 95-YOKOHAMA-IGTC-80, pp. III-169—III-176.

CHAPTER

18

DEVELOPMENT OF CERAMIC GAS TURBINE FOR POWER PLANTS

Tohru Hisamatsu and Isao Yuri
Central Research Institute of Electric Power Industry, Yokosuka, Kanagawa, Japan

Takashi Machida
Mechanical Engineering Research Laboratory, Hitachi Ltd., Tsuchiura, Ibaraki, Japan,

Katsuo Wada
Hitachi Works, Hitachi Ltd., Hitachi, Ibaraki, Japan

INTRODUCTION

Recently the remarkable advancements in gas turbine technology have resulted in about 49% of overall thermal efficiency in Liquefied Natural Gas (LNG) combined cycle power plants using 1300°C class gas turbines. And about 53% of overall thermal efficiency is expected in LNG combined cycle power plants using 1500°C class gas turbines. Further, in the integrated coal gasification combined cycle (IGCC) power plants, about 47% of overall thermal efficiency can be expected with the 1300°C class gas turbines. Taking account of these high efficiencies, combined cycle power plants using LNG or coal gasified fuel are expected to play a leading role in future thermal power generation scenarios.

In order to improve the thermal efficiency of combined cycle power plants, higher turbine inlet temperature (TIT) and higher performance are indispensable in gas turbines. So far, the increase of TIT has been achieved by the development of heat resistant alloys and the progress in cooling technology. In conventional air-cooled metallic gas turbines, however, higher efficiencies will be constrained because a higher TIT leads to an increase in the amount of cooling air.

On the other hand, because of the excellent heat resistance of ceramic gas turbine hot section parts, a drastic increase in thermal efficiency can be expected in ceramic gas turbines because a higher TIT and a reduction in the amount of cooling air can be achieved simultaneously.

However, ceramics are brittle materials and their strength is not well defined. In other words, they lack reliability as structural materials. When applying ceramics to the combustors and turbine blades in

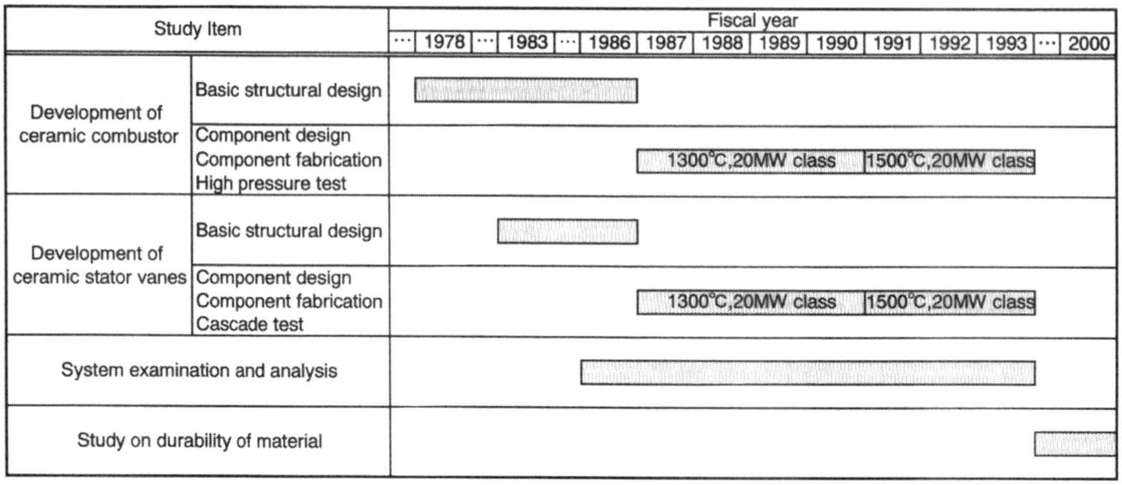

Figure 18.1 Development Schedule for the Ceramic Gas Turbine for Power Plants (Reprinted with permission from CRIEPI)

gas turbines for power plants, it is generally believed that simple substitution of metallic parts with ceramic components will be difficult, because the probability of failure (POF) of ceramic parts may become unacceptably high, and their average strength deteriorates due to an excessive effective volume. At the same time, when considering the extremely severe gas turbine environments characterized by high temperature, high pressure, and high velocity combustion gas flow conditions, in order to ensure the reliability of ceramic parts, it is necessary to develop a structural design technology suitable for ceramics.

The Central Research Institute of Electric Power Industry (CRIEPI) and Hitachi Ltd. have been seeking such design technologies through the development of ceramic gas turbines for power plants since 1983.

Ceramic gas turbines for power plants are suitable for topping of combined cycles because of the potential improvement in the thermal efficiency. As the topping cycle, the development target of the ceramic gas turbine must be a 1500°C class large-scale gas turbine with more than 100MW output power. Considering the risk and cost for development, however, it is an appropriate near term target to develop a smaller gas turbine, of about 20MW output power, which has the same structure as a large scale one. And considering the benefits of the application and the technological difficulty, it is believed that stationary components such as the combustor and stator vane in gas turbines are suitable components for the application of ceramics.

In the study reported here, we have examined the effect of applying ceramics to gas turbines for power plants and we have evaluated structural design concepts for ceramics for improved strength reliability. Based on the structural design concepts, a ceramic combustor and stator vanes for a 20MW, 1300°C class gas turbine were developed. The application of ceramics and the strength reliability of a ceramic combustor and stator vane were demonstrated in high-pressure tests. In order to examine the potential application of ceramics for higher temperature gas turbines, a ceramic combustor and stator vane for a 20MW, 1500°C class gas turbine were designed, developed and tested under high-pressure conditions. Furthermore, we have studied the durability of ceramic materials for the gas turbines through exposure tests in combustion gas flow.

CRIEPI was founded as a central research institute for the electric power industry of Japan. Considering the importance of coal as a future energy resource and the potential of the IGCC, CRIEPI has been performing a number of R&D programs for IGCC power generation technology. The focus of the study reported here was the combustor design and high pressure testing performed by using coal gasified fuel, and the development of a ceramic gas turbine, starting with a 1300°C engine. The time schedule for this study is shown in Figure 18.1. This study was done as a joint study between CRIEPI and Hitachi, Ltd. with the funding and facilities of both organizations.

EFFECT OF APPLYING CERAMICS

Combustor Application

The air supplied to the gas turbine is used as combustion air playing a main role in the combustion reaction, as dilution air for the adjustment of the combustor outlet temperature, and as cooling air for combustor wall temperature control. Although the relative amount of each air stream is generally determined by the combustor outlet temperature and the cooling effectiveness of the combustor wall, it is desirable to reduce the amount of cooling air and to increase the amount of combustion air to achieve optimal combustion performance.

In combustors with ceramic liners, combustion performance is improved because cooling air requirements are eliminated or significantly reduced, and the amount of air available for combustion is drastically increased. This feature is especially beneficial for lean premixed combustors which require large amounts of combustion air. In conventional metallic combustors, the combustion reaction is constrained in the vicinity of the combustor liner because the wall temperature must be below the maximum use temperature of the metallic material. On the other hand, in a ceramic combustor, because of a higher allowable combustor wall temperature, combustion stability can be expected to be increased by the acceleration of the combustion reaction.

Moreover, the thermal stress on turbine stators and blades can be reduced because the combustor outlet gas temperature distribution can be more uniform.

Application to Turbine Blades and Vanes

While conventional air cooled metallic blades and vanes require substantial amounts of cooling air, ceramic blades and vanes either require no cooling air, or the requirement for cooling air is drastically reduced. As a result, the application of ceramics to turbine blades and/or vanes leads to an improvement in the thermal efficiency.

Figure 18.2 shows the improvement in the thermal efficiency when a ceramic gas turbine is incorporated into combined cycle power plants using LNG and coal gasified fuel. Ceramics are applied to the first and second stage stator vanes or the first and second stage stator vanes and first stage rotor blades. Also, the application of a supercritical pressure boiler as heat-recovery steam generator (HRSG), and the reduction of sealing air due to an improvement in technology are considered. For comparison, the thermal efficiency is also estimated for the case where the turbine inlet temperature (stator vane inlet temperature) is varied in a conventional 1300°C class air-cooled metallic gas turbine.

In ceramic gas turbines, the thermal efficency is not only increased due to the reduction of cooling air, but also, because the bottoming cycle benefits from a higher exhaust gas temperature from the gas turbine and from the application of a supercritical pressure boiler (HRSG). As a result, even in the case of a ceramic gas turbine with 1300°C TIT, the thermal efficiency can be markedly higher than in the case of an air-cooled metallic gas turbine. The thermal efficiency of the entire plant is increased because of the higher TIT.

Figure 18.2 shows that it is not necessary to apply ceramics to all of the gas turbine hot section components, and that a noticeable benefit can be expected by applying them only to some components. Ceramics can be introduced step by step according to the progress in ceramic technology.

STRUCTURE OF COMBUSTOR AND STATOR VANE

Concepts of Structural Design for Ceramic Components

In order to overcome the brittleness of ceramics and to improve the strength reliability of ceramic components, the stress on ceramic parts must be reduced as much as possible in the following ways [1]:

(1) ceramic/metal hybrid structure
In the ceramic/metal hybrid structure, ceramics with excellent heat resistance are utilized for those constituents of the component which are exposed to high temperature gas, and metal with superior strength

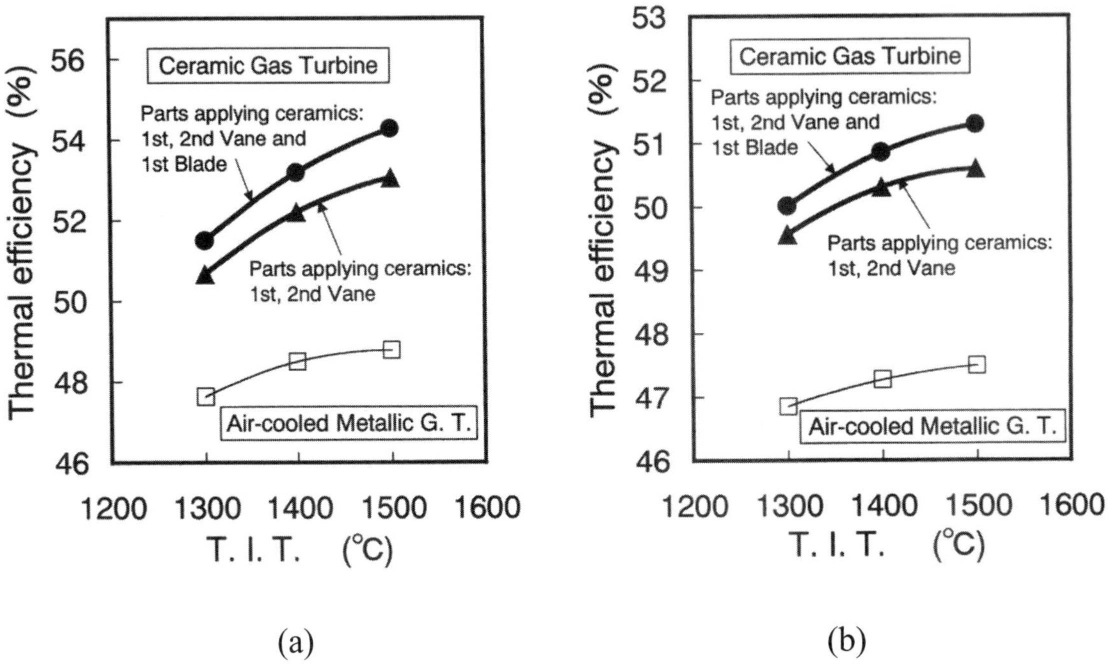

Figure 18.2 Effect of Applying Ceramics on the Overall Thermal Efficiency of Combined Cycle Power Plants (Reprinted with permission from CRIEPI and Hitachi, Ltd.) (a) LNG Combined Cycle Power plant (b) IGCC Power Plant

reliability is utilized for those constituents of the component which bear the mechanical load. In this structure, the mechanical stress on the ceramic parts can be substantially reduced.

(2) Segmentation of a ceramic part
By adopting the ceramic/metal hybrid structure, the stress on the ceramic parts is mainly thermal stress. When a ceramic part has a complex shape or when its size is large, the thermal stress on the ceramic part can be life limiting. By segmenting a ceramic part into smaller simply shaped subcomponents, the thermal stress can be reduced significantly and the stress concentration can be prevented.

(3) Heat insulation and buffering for fasteners for the ceramic part
The direct contact of a ceramic part with another one, or with a metallic part of a different temperature should be avoided as much as possible. Otherwise, interference between the parts may occur, e.g. as a result of thermal stresses due to localized cooling of ceramic parts, or stress caused by differential thermal expansion. In order to prevent or minimize these stresses, it is required that a heat insulating or buffering layer be inserted between the parts fastening the ceramic part.

Basic Structure

Based on the concepts of structural design, the basic structure of a ceramic combustor and stator vane for a 20MW, 1300°C class gas turbine was examined.

Ceramic Combustor. The basic structure of a ceramic combustor for a 20MW, 1300°C class gas turbine is shown in Figure 18.3 [2–4]. Although the figure shows a cross section of the combustor liner, the same

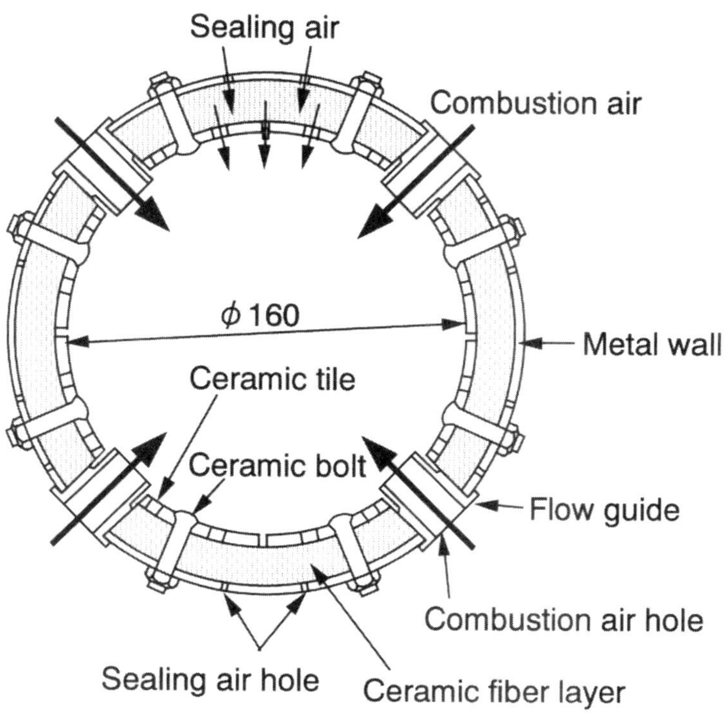

Figure 18.3 Basic Structure of the Ceramic Combustor for a 1300°C, 20MW Class Gas Turbine (Adapted from Ref. [4])

structure is adopted in the transition piece. The outer wall is made of metal and the inside of the metal wall is covered with a ceramic fiber layer. The inner wall of the combustor is lined with ceramic tiles. The ceramic fiber layer is protected from high temperature gas flow by the ceramic tiles. In order to reduce the thermal stress on the ceramic tiles of the combustor liner, which originates from the positions of combustion air holes, the ceramic tiles are divided into eight parts in the circumferential direction and eight parts in the axial direction. By feeding a small amount of sealing air which prevents the penetration of high temperature gas into the fiber layer and shields the heat from the ceramic tiles, the temperature of the metal wall is kept at the same level as the combustion air temperature. About 3% of the total amount of air supplied to the combustor (about 1/10 of the air-cooled metallic combustor) is supplied as sealing air to the ceramic fiber layer. Simultaneously, the fiber layer functions as a buffer layer to prevent the distortion between the ceramic tiles and the metal wall. Ceramic nuts and bolts are used to fasten the ceramic tiles and the ceramic fiber layer to the metal wall.

Ceramic Stator Vane. Following the concepts for structural design for ceramic components a ceramic/metal hybrid structure has been adopted for the stator vane. Ceramics are utilized for the parts which are exposed to high temperature gas, and metals are utilized for the parts which bear the mechanical load.

Figure 18.4 shows the basic structure of the first stage ceramic stator vane for a 20MW, 1300°C class gas turbine [5–7]. A metallic core penetrates the inside of the ceramic shell. In order to reduce the thermal stress and for ease of fabrication, the ceramic parts are segmented into shell parts and side-wall parts, and moreover, the shell parts are segmented into two parts in order to improve the thermal shock resistance. These ceramic parts are fitted by an interlocking structure.

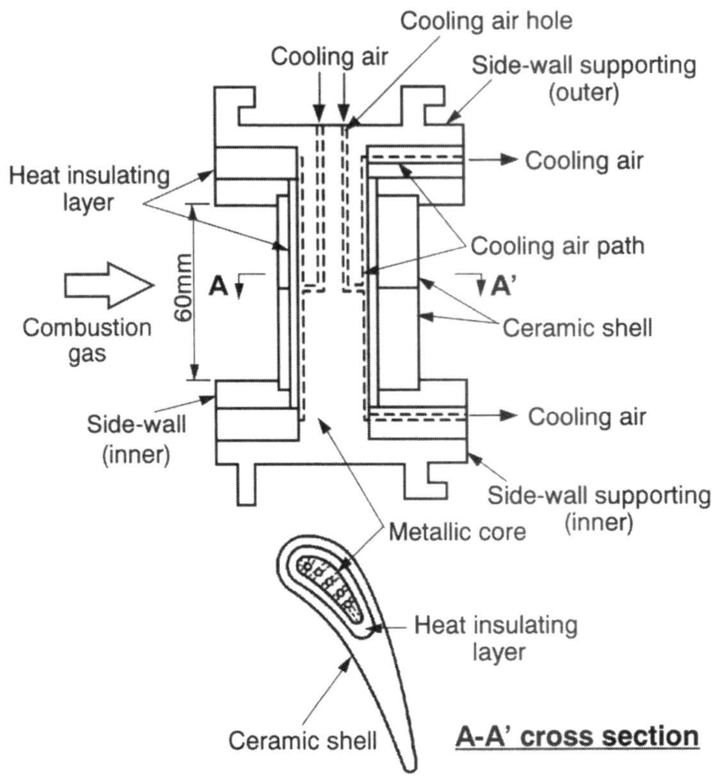

Figure 18.4 Basic Structure of the Ceramic Stator Vane for a 1300°C, 20MW Class Gas Turbine (Adapted from Ref. [7])

A heat insulating layer, which also functions as a buffer layer to prevent deformation, is placed between these ceramic and metallic parts. Particularly, since the heat insulating layer inserted between the ceramic shell and the metallic core is only 3mm thick, we adopted a cooling method in which the cooling air absorbs the heat conducted from the heat insulating layer in order to provide effective thermal shielding of the metallic core. The inner circumference (metallic side) of the heat insulating layer was composed of porous materials (i.e. ceramic fiber), and a small amount of cooling air is supplied to the porous layer through the cooling air hole incorporated into the metallic core.

DEVELOPMENT OF 1300°C CLASS COMBUSTOR AND STATOR VANE AND HIGH PRESSURE TEST RESULTS

In order to examine the strength reliability of the basic structure, we developed a prototype combustor (liner and transition piece) and stator vanes for a 20MW, 1300°C class gas turbine and conducted high pressure testing simulating the operating conditions in the gas turbine.

Silicon nitride (Si_3N_4, SN-252, Kyocera) was chosen as the ceramic material for the 1300°C class combustor, transition piece, and stator vane, and alumina (Al_2O_3, SILTECS-1400, Nichias) fiber was used as heat insulator. The outlet shape of the transition piece is a circular arc similar to that of the conventional metal gas turbine. The size of the combustor liner is 160 mm in inner diameter and 500 mm in length, the transition piece is 360 mm in length, and the stator vane is 60 mm in height.

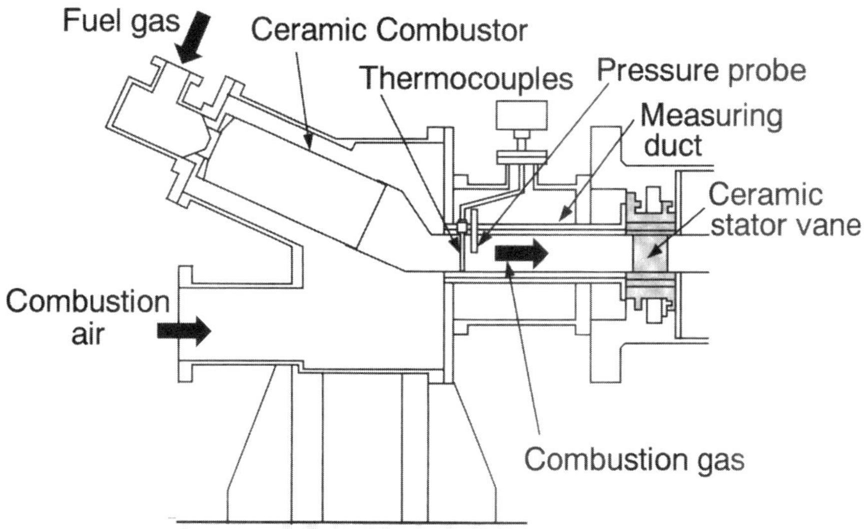

Figure 18.5 Sectional View of the Ceramic Combustor and Ceramic Stator Vane Test Equipment (Adapted from Ref. [7])

Figure 18.5 shows a sectional view of the test equipment for the combustor and stator vanes. After measuring the combustion gas temperature distribution, pressure and gas composition, the combustion gas flow exhausted from the combustor was introduced to five stator vanes arranged in a circular arc. In order to measure the temperatures at the various positions of the combustor and the stator vanes, thermocouples were embedded in the ceramic and metallic parts. A low BTU gaseous fuel was utilized (lower heating value: about 3.9 MJ/Nm^3), with almost the same composition as that of coal gasified fuel.

Figure 18.6 shows the temperature distribution of each component under rated load conditions (outlet gas temperature and pressure: 1300°C and about 1.42 MPa) [4]. The horizontal axis shows the axial distance through the combustor (liner and transition piece) from the swirler. All ceramic tile temperatures are over 1000°C and the maximum temperature, which exists at the center of the combustor liner, reaches approximately 1400°C. On the other hand, the all-metal wall temperatures in the combustor are kept the same as those of the combustion air. This shows that the performance of both the insulation and the heat shield are excellent.

The pattern factor (P.F.), which is an index of uniformity in the combustion gas temperature distribution, is about 6.5% under the rated load condition. This value is about half of that of a conventional metallic combustor. It is clear that, by applying ceramics to the combustor, it will be possible to make the gas temperature distribution more uniform [4].

Figure 18.7 shows the relationship between the amount of cooling air and the temperatures at various positions of the stator vane under the rated load condition. In the figure, the amount of cooling air is expressed as the cooling air flow ratio, which is the ratio of the amount of cooling air required for all of the first stage stator vanes to the total amount of compressed air in a gas turbine (air flow: 54kg/s at the rated load condition). Although the temperature of the metallic core decreases according to the increasing amount of cooling air, the temperatures of the ceramic shell are almost constant and close to the gas temperature regardless of the cooling air flow ratio because the heat insulating layers function effectively. The cooling air temperature in this experiment is lower than that of conventional gas turbines because of the test conditions and equipment parameters. When the cooling air temperature for conventional gas turbines is about 400°C and the allowable temperature of the metallic core is 800°C, a sufficient amount of cooling air is necessary

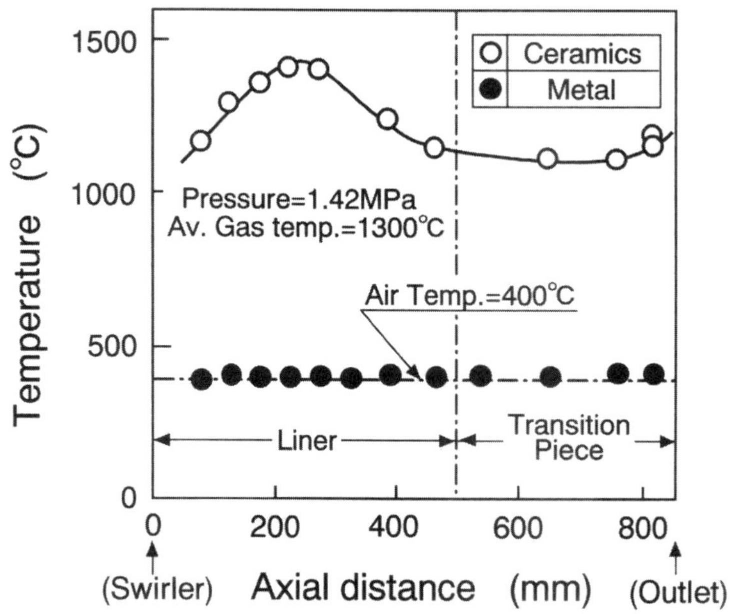

Figure 18.6 Temperature Distribution of Each Component Under Rated Load Condition (Adapted from Ref. [4])

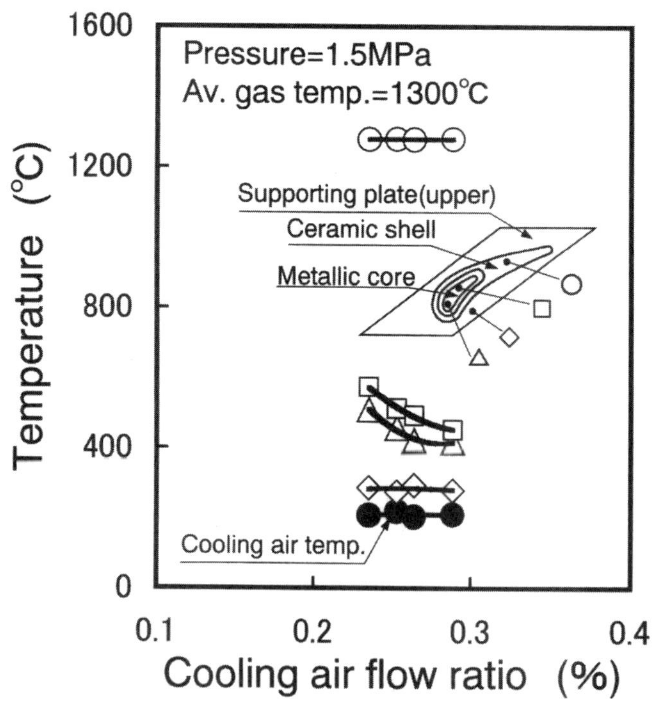

Figure 18.7 Relation Between Cooling Air Flow Ratio and Temperature of Each Stator Vane Component Under Rated Load Condition (Reprinted with permission from CRIEPI and Hitachi, Ltd.)

TABLE 18.1 MAXIMUM THERMAL STRESS ON CERAMIC SHELL

TIT (°C)	1300		1500	
Ceramic material	Si_3N_4	SiC	Si_3N_4	SiC
Maximum thermal stress (MPa)	170	235	198	276

so that the temperature difference between the metallic core and cooling air can be less than about 400°C. In the figure, the necessary cooling air flow ratio is 0.23~0.25% for stator vanes. This amount is extremely small and equivalent to about $1/20^{th}$ of state-of-the-art air-cooled metallic vanes [5].

In order to evaluate the strength reliability of ceramic parts, a rapid fuel cutoff test was performed from the rated load condition, simulating an emergency stop of gas turbines. After the test, no damage such as cracks were detected in the ceramic parts by visual and dye penetrant inspection. This ensured that the structural design was effective even for the emergency stop of gas turbines, which is the severest condition for ceramic parts [4–7].

STRUCTURAL IMPROVEMENTS FOR 1500°C CLASS COMBUSTOR AND STATOR VANE

Ceramic Material

Table 18.1 shows the effects of gas temperatures and candidate materials on the maximum thermal stress in the ceramic shell for the basic structure of the ceramic stator vane. When the gas temperature increases from 1300°C to 1500°C, in the case where the same material is applied, the maximum thermal stress increases about 20%. Because of the difference in material properties, the maximum stress for SiC (Hexoloy SA, Hitachi Chemical) stator vanes is about 1.5 times greater than that of Si_3N_4 (SN-252, Kyocera) vanes.

On the other hand, in terms of the strength at high temperature and the oxidation resistance at 1500°C, SiC was more promising than Si_3N_4 as a material for combustors and stator vanes for the 1500°C class gas turbines. However, since the fracture toughness of SiC is lower than that of Si_3N_4, and SiC is extremely brittle, a structural design that allows a low fracture toughness and improves the strength reliability is necessary when applying SiC for a combustor and for stator vanes for the 1500°C class gas turbines.

Ceramic Combustor for 1500°C Class Gas Turbine

In a ceramic/metal hybrid structure, the device used to fasten the ceramics to the metal is most important. This was achieved by using ceramic nuts and bolts. When applying SiC to the basic structure of the ceramic combustor, a fastening structure which allows a low fracture toughness of SiC and improves the strength reliability is necessary. Structural improvements were made as detailed below. The improved structure is shown in Figure 18.8 [8, 9].

(1) Heat insulation at the fastening part
The ceramic bolt passes from the inside of the combustor, which has a temperature of over 1000°C, through the metal wall which is cooled by combustion air. Thus, to reduce the thermal stress of the ceramic bolt, the fastening part of the nut and bolt are insulated.

(2) Insertion of a deformation buffer layer
To reduce the stress on the nut and bolt caused by the difference in thermal expansion between the ceramic fiber layer and the metal wall, a buffer layer made of a ceramic fiber layer, and a plate spring made of metal were set between the nut and the metal wall.

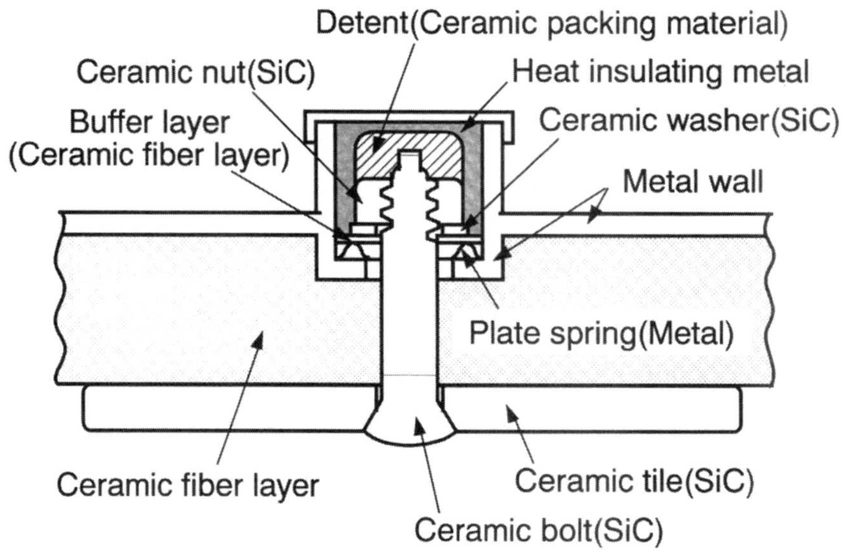

Figure 18.8 Structure of Ceramics Fastening Part of Combustor (Adapted from Ref. [8])

(3) Screw shape of nut and bolt
To improve the strength of the screw thread and prevent stress concentrations at the contact between bolt and nut, the shape of the screw was changed from a coarse screw thread, which is used for a 1300°C class ceramic combustor with Si_3N_4 liners, to a trapezoidal screw thread with a long pitch.

(4) Detent of ceramic nut and bolt
To apply the detent on the ceramic nut and bolt, the tip of the ceramic bolt is flattened and covered with ceramic packing materials.

A prototype 1500°C class SiC ceramic combustor for coal gasified fuel was developed and high pressure tests (rated load condition: 1500°C, about 1.52 MPa) were performed. The total time of the tests was about 8 hours, of which 2 hours were under the rated load conditions. Figure 18.9 shows the combustion characteristics of the combustor. Combustion efficiency increases in relation to the gas turbine load (G.T.L.), and reaches almost 100% when the G.T.L. is over 25%. It indicates the effect of more complete combustion due to the higher temperature of the ceramic wall. The NO_x conversion ratio (conversion ratio from NH_3 in the fuel to NO_x) is as low as about 40% at the rated load condition (1000 ppm of NH_3 and 2.5% of CH_4 in the fuel). It indicates that the combustor has excellent combustion performance [8, 9].

A fuel cutoff test, which simulated the conditions in a gas turbine trip from rated load, was conducted to examine the strength reliability. Figure 18.10 shows the changes of the conditions after fuel cutoff. In the actual gas turbine trip, the pressure and the combustion air flow drop rapidly. However, in these tests, the pressure (P) drop was small and the combustion air flow (Ga) increased after the fuel cutoff, because the opening of the pressure control value and the combustion air flow control value were constant. Consequently, the heat transfer coefficient at the surface of the ceramic parts after fuel cutoff would be higher than that in an actual gas turbine trip, and it is considered that the conditions for the ceramic parts were more severe. Although the measured maximum drop rate of the gas temperature (Tg) was about 400°C/s, the actual temperature drop rate was larger because the response time lag of the thermocouples was not compensated. The maximum rate of the change of the ceramic tile temperature (Tc) was about 50°C/s.

Figure 18.11 shows the appearance of the SiC ceramic combustor after the rapid fuel cutoff test from the rated load conditions. After inspection, no damage was observed on the ceramic component. Thus, it was

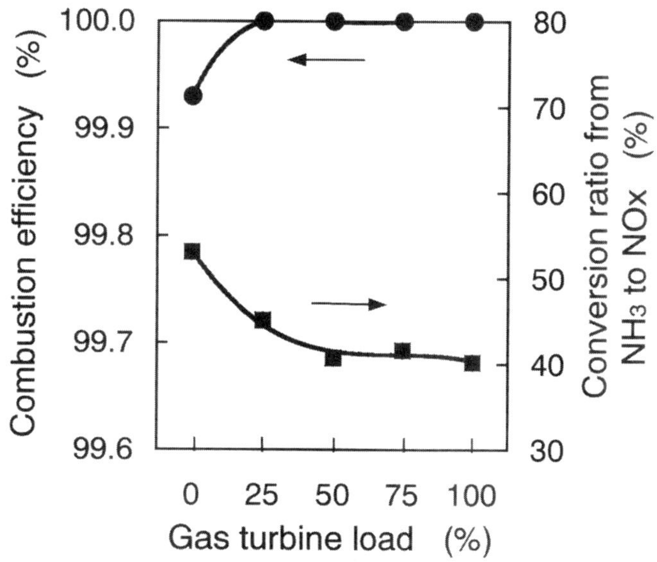

Figure 18.9 Combustion Characteristics for the 1500°C Class Ceramic Combustor (Adapted from Ref. [8])

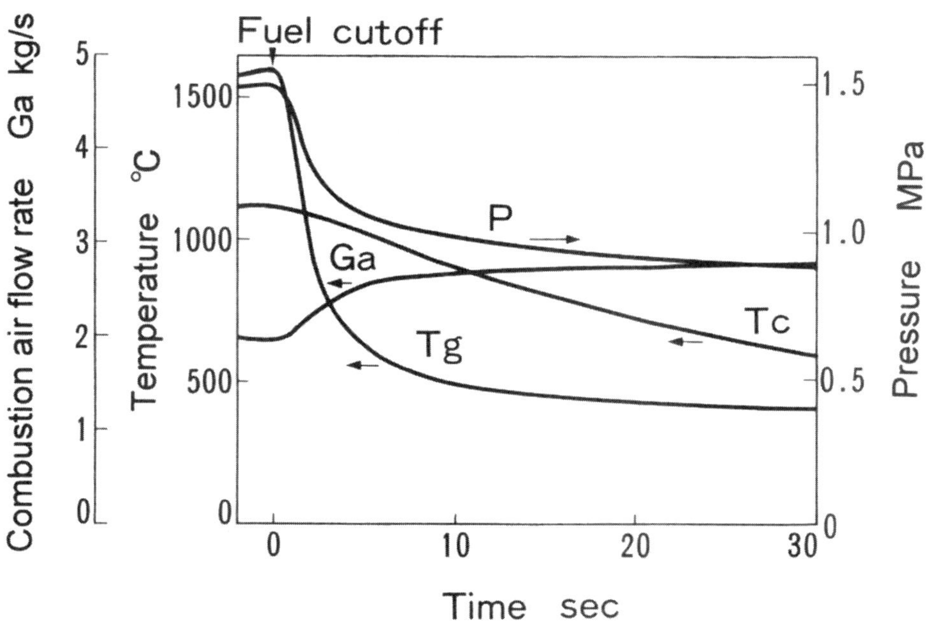

Figure 18.10 Changes of Conditions After Fuel Cutoff (Adapted from Ref. [8])

Figure 18.11 SiC Ceramic Combustor After Rapid Fuel Cutoff Test (Reprinted with permission from CRIEPI and Hitachi, Ltd.)

confirmed that the brittleness of SiC can be overcome with this structural design, and that SiC can be applied as a structural material in a 1500°C class gas turbine combustor [8, 9].

Ceramic Stator Vanes for 1500°C Class Gas Turbine

In the basic structure of the stator vanes described above, thermal stress increases at the inner hole of the trailing edge during the emergency gas turbine stop. In 1500°C class SiC stator vanes, the increase in the thermal stress at the inner hole of the trailing edge is minimized by expanding the hole to the trailing edge side and leveling the thickness of the ceramic shell.

Figure 18.12 shows the change of the maximum thermal stress at the inner hole of the trailing edge in the rapid fuel cutoff from the rated load condition. In the shape of the inner hole of the trailing edge adopted for the 1300°C class stator vane (R = 3.85mm), the thermal stress increases rapidly after fuel cutoff. On the other hand, by changing the shape of the inner hole of the trailing edge to R = 2.75mm and by leveling the thickness, the thermal stress hardly increases even after fuel cutoff. Considering the current state of ceramic processing technology, the inner hole of the trailing edge is expanded to R = 2mm [10].

Moreover, by increasing the effective thickness of the heat insulating layer inserted between the ceramic shell and the metallic core, the thermal stress on the ceramic shell is reduced. At the same time, in order to maintain a reliable cooling air passage, counter-flow cooling in the outer circumference of the metallic core is adopted as the method for cooling the inner circumference of the heat insulating layer, as shown in Figure 18.13 [10].

Figure 18.14 shows the relationship between the metallic core temperature and the maximum thermal stress on the ceramic shell. As the core temperature increases, the thermal stress is reduced. Decreasing the amount of cooling air not only improves cooling effectiveness but also the strength reliability of the ceramic shell. When the allowable temperature of the metallic core for heat resistance is about 800°C, the thermal

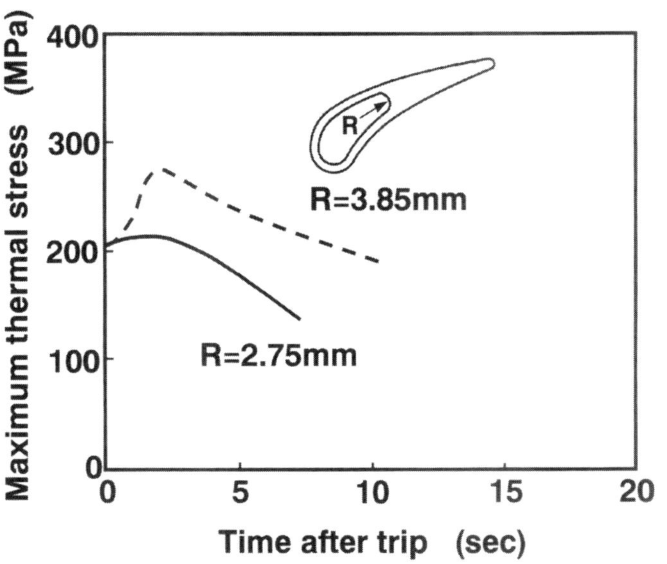

Figure 18.12 Changes in Maximum Thermal Stress on the Ceramic Shell with Time After Fuel Cutoff (Reprinted with permission from CRIEPI and Hitachi, Ltd.)

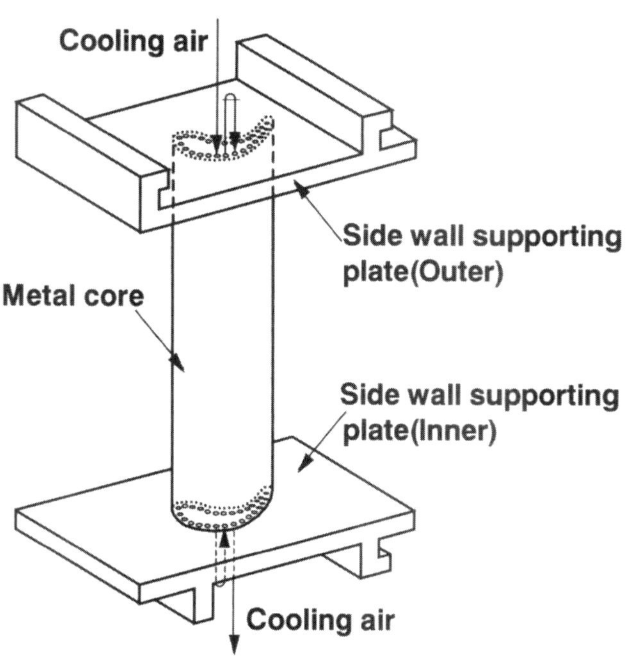

Figure 18.13 Cooling Structure of the Metallic Core (Adapted from Ref. [10])

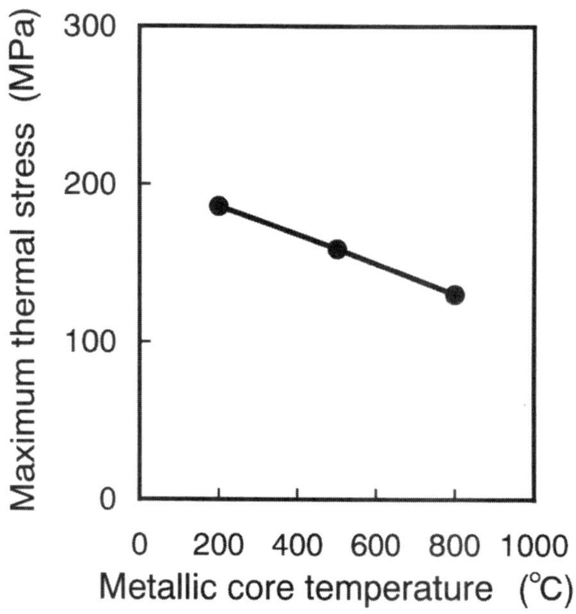

Figure 18.14 Relation Between Metallic Core Temperature and Maximum Thermal Stress on Ceramic Shell (Adapted from Ref. [10])

stress of the ceramic shell with the new core structure is about 130 MPa, a 40% decrease compared to the thermal stress of the ceramic shell with the conventional structure [10].

Five prototype 1500°C class SiC stator vanes were arranged in an arc, and a high pressure test (rated load condition: 1500°C, about 1.52 MPa) was performed. The total time of the test was about 4 hours, of which 1 hour was under the rated load conditions.

Figure 18.15 shows the relationship between the gas turbine load and the amount of cooling air required for the vanes. The necessary amount of cooling air (for all the first stage stator vanes) is about 0.4% of the total amount of compressed air under the rated load condition, and it is less than 1/10 required for state-of-the-art air-cooled metallic vanes [10].

Figure 18.16 shows the appearance of the ceramic stator vanes after the thermal loading cascade test under rated load conditions. No cracks attributable to thermal stress were observed in the ceramic shell when the temperature of the vane core was kept at about 800°C. Thus, by controlling the metallic core temperature at the appropriate level, it is expected that even extermely brittle SiC can retain strength reliability for short periods.

STUDY ON DURABILITY OF CERAMIC MATERIALS

From the viewpoint of the long term reliability of the ceramic gas turbine, it is necessary to evaluate the durability of ceramic materials in the combustion gas flow at high-temperature and speed. Since 1994, CRIEPI has studied the durability of ceramic materials for gas turbines through exposure tests in combustion gas flow [11, 12].

Figure 18.17 shows a cross sectional view of the exposure test equipment. As a source of combustion gas flow at high temperature and speed, a ceramic combustor for the gas turbine is applied. About 200 ceramic specimens installed in ceramic specimens holders are exposed to the combustion gas flow in the test section. The exposure test conditions at the entrance of the test section are: 1500°C, 150 m/s and 0.18 MPa.

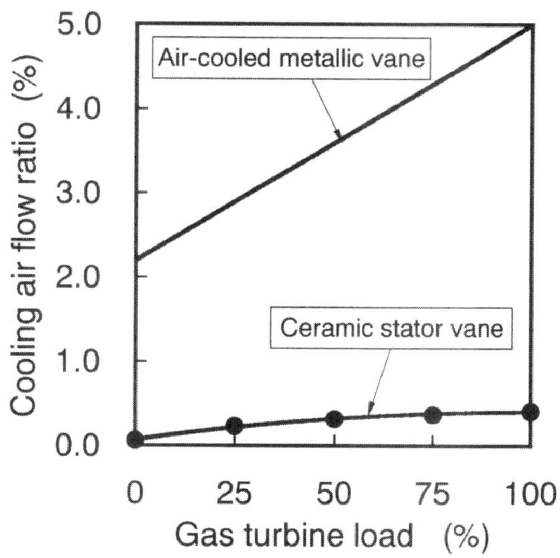

Figure 18.15 Relation Between Gas Turbine Load and Cooling Air Flow Ratio (Adapted from Ref. [10])

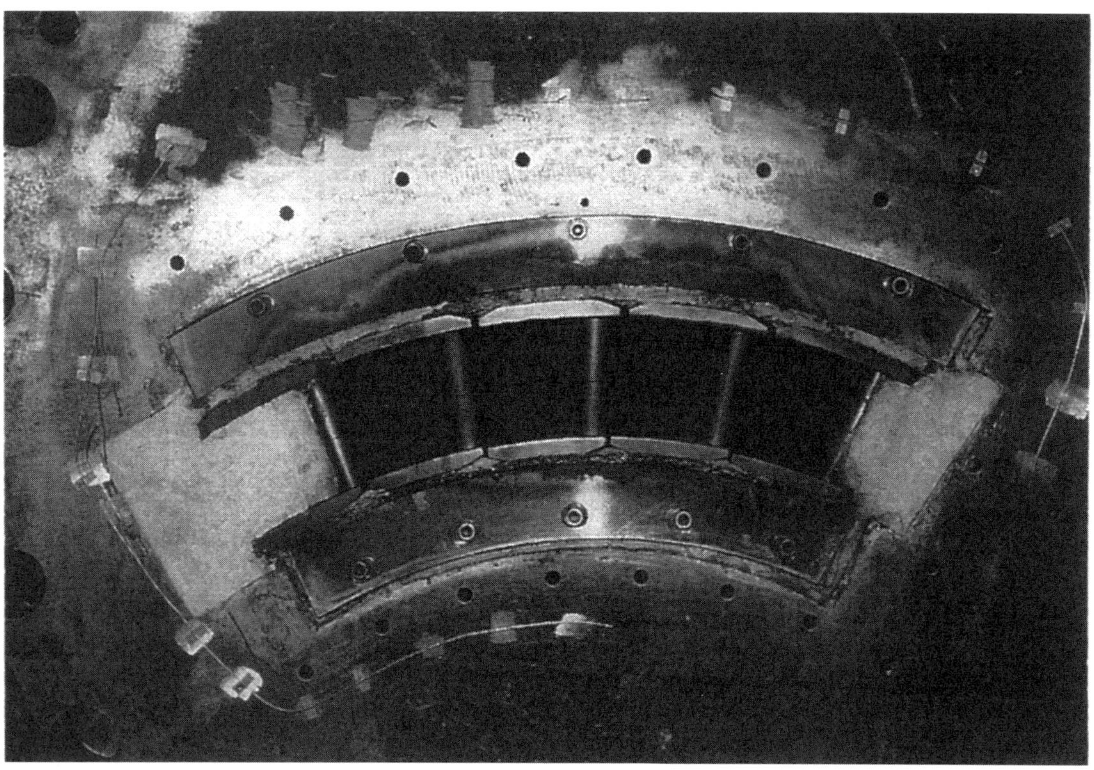

Figure 18.16 SiC Ceramic Stator Vane After Cascade Test (Adapted from Ref. [10])

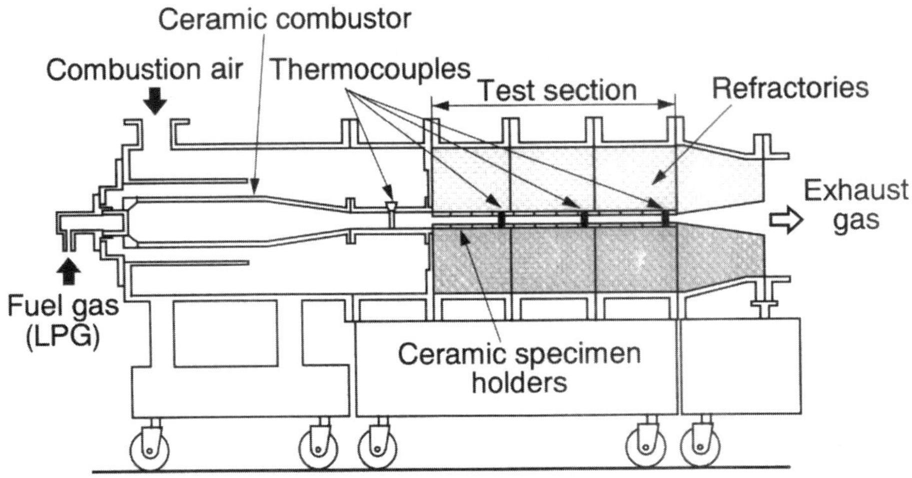

Figure 18.17 Sectional View of Exposure Test Equipment (Reprinted from Ref. [11])

The concentrations of oxygen and water vapor in the combustion gas are about 8 vol% and 9.5 vol%, respectively. The specimen geometry is 4mm wide × 3mm high × 60mm long (the length of the hot area is 40mm).

The weight before and after the exposure was measured. The weight of the substrate after having removed the oxide scale was used in order to determine the weight change of the substrate during the exposure. The morphology and the properties of the oxide scale were analyzed at the specimen surface and for a section by X-ray diffraction, SEM and EPMA. Furthermore, the flexural strength at room temperature (four point bending) was measured to determine the strength deterioration.

Strength Deterioration and Weight Change Characteristics

A 1000 hour exposure test was conducted to determine strength deterioration and weight change. Figure 18.18 shws the relation between the exposure time and the flexural strength after the exposure when SiC (Hexoloy SA, Hitachi Chemical) specimens were exposed for 1000 hours. Although the measured strength shows substantial dispersion, the strength does not depend on the exposure time and is almost fixed. From the observation of a section of the specimen by SEM, it was found that the fracture origin is an internal defect of the materials. It became clear that the exposure in the combustion gas flow conditions did not have such a significant effect on the strength deterioration of the ceramic materials.

Figure 18.19 shows the relation between the exposure time and the specific weight change when SiC specimens were exposed for 400 hours. The substrate weight change shows linear weight loss. And it was observed that the oxide thickness after the exposure is thin (1–2 μ m) and the thickness is independent of the exposure time. The measurement of the geometry (wide, height) of the substrate after having removed the oxide scale decreased also linearly with the exposure time, and recession occurred. From these results, we found that the surface recession of ceramics occurs in flowing combustion gas. Since the oxide scale disappeared it could not act as a protective layer in the combustion gas flow. Among a list of potential mechanisms for the disappearance of the oxide scale, the formation of volatile reaction products, SiO and $Si(OH)_4$ was considered [11].

Factors Influencing Degradation

In order to investigate the mechanism of the surface recession and to estimate the durability of ceramic components for gas turbines, it is important to clarify the surface recession behavior of the ceramics and the

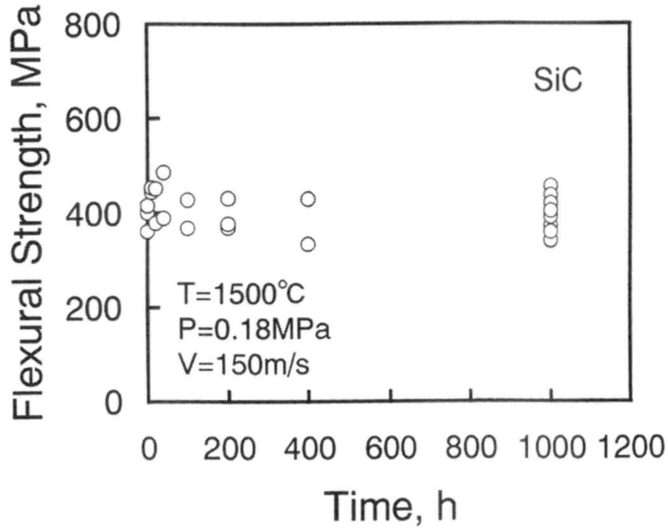

Figure 18.18 Relation Between Exposure Time and Flexural Strength (Reprinted with permission from CRIEPI and Hitachi, Ltd.)

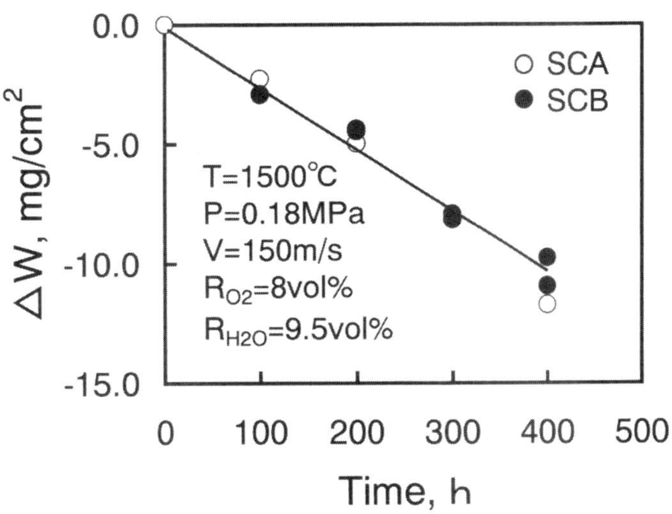

Figure 18.19 Relation Between Exposure Time and Specific Weight Change (Reprinted from Ref. [11])

effects of various basic factors on surface recession. The effects of various important factors such as sintering aids and combustion gas flow conditions (oxygen partial pressure, water vapor partial pressure, gas temperature, pressure, and gas flow velocity) on the weight loss of the substrate were experimentally investigated through exposure testing of SiC specimens.

Figure 18.19 also shows the effect of sintering aides on specific weight change. SCA is a sintered silicon carbide containing B and C as sintering aides. On the other hand, SCB is hot-pressed silicon carbide

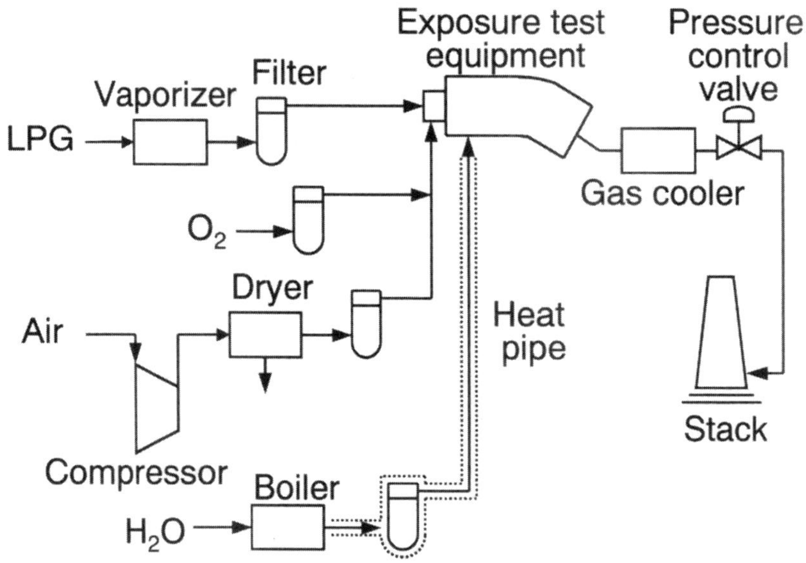

Figure 18.20 Schematic Diagram of Test Facility (Reprinted from Ref. [12])

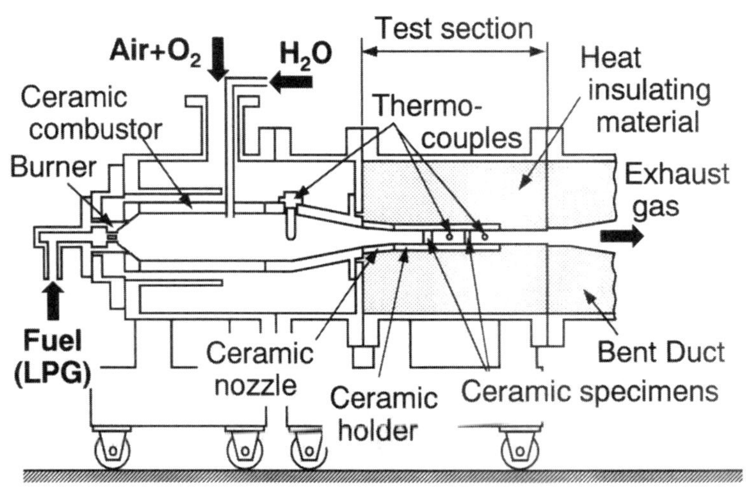

Figure 18.21 Sectional View of Exposure Test Equipment (Reprinted from Ref. [12])

without sintering aids. Comparing the weight loss rate of SCA with that of SCB, it is clear that the weight loss rates of both specimens are the same. From SEM observation of the surface of specimens after the exposure, both surface morphologies were the same. As for the oxide thickness after the exposure, the oxide layers on both specimens are thin (1–2 μ m) and the thickness are independent of the exposure time. Over the range of test conditions, the sintering aids did not have such a significant effect on the substrate weight loss rate.

Figures 18.20 and 18.21 show a schematic diagram of the test facility and a cross sectional view of the exposure test equipment, respectively. This facility was utilized to investigate the effects of various

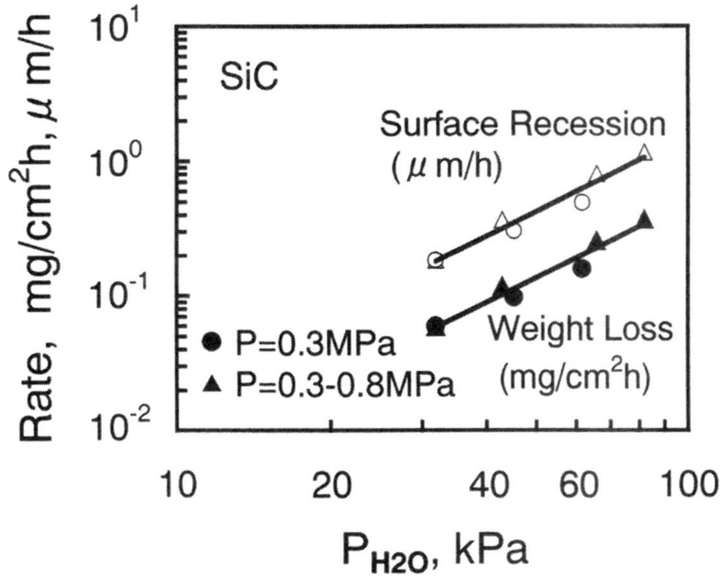

Figure 18.22 Relation Between Water Vapor Partial Pressure and Weight Loss Rate (Reprinted from Ref. [12])

basic factors (oxygen partial pressure, water vapor partial pressure, gas temperature, pressure and gas flow velocity) of combustion gas flow conditions on the degradation behavior of SiC. In order to derive the effects of oxygen partial pressure and water vapor partial pressure in combustion gas independently, the oxygen and water vapor supply systems were attached to the test equipment. The pressure was adjusted by a pressure control value located at the outlet of a gas cooler. The gas temperature was adjusted by the equivalent ratio, and the gas flow velocity was primarily adjusted by the combustion intensity and a pressure control valve.

The exposure tests were performed for widely varying experimental parameters of the gas temperature (T = 900–1500°C), pressure (P = 0.3–0.8 MPa), gas flow rate (V = 50–250m/s), oxygen partial pressure (P_{O2} = 24-44kPa) and water vapor partial pressure (P_{H2O} = 32-82kPa). In the exposure tests, the conditions of other parameters were fixed as much as possible except for the parameter being investigated. The exposure time was identical (10 hours) for all test conditions. The weight before and after the exposure was measured. The weight of the substrate after removal of the oxide scale by acid treatment was used as the weight after the exposure.

From the results of the exposure tests, the effects of water vapor partial pressure and gas flow velocity, shown in Figure 18.22 and Figure 18.23, were determined to be the main influencing factors. Figure 18.22 shows the relation between the water vapor partial pressure and the weight loss rate. As shown in Figure 18.22, the approximate surface recession rate converted by the density (3.21g/cm^3) of SiC is also shown for reference. The weight loss rate increases with the increase of the water vapor partial pressure, and the water vapor partial pressure dependence of the weight loss rate is remarkably strong as compared to dependence on other parameters. And Figure 18.23 shows the relation between the gas flow velocity and the specific weight change. The weight loss depends on the gas flow velocity, but the increase rate of the weight loss becomes less with increasing gas flow velocity. Consequently, each parameter dependence of the substrate weight loss rate almost agrees with the relation based on $Si(OH)_4$ volatile species, described

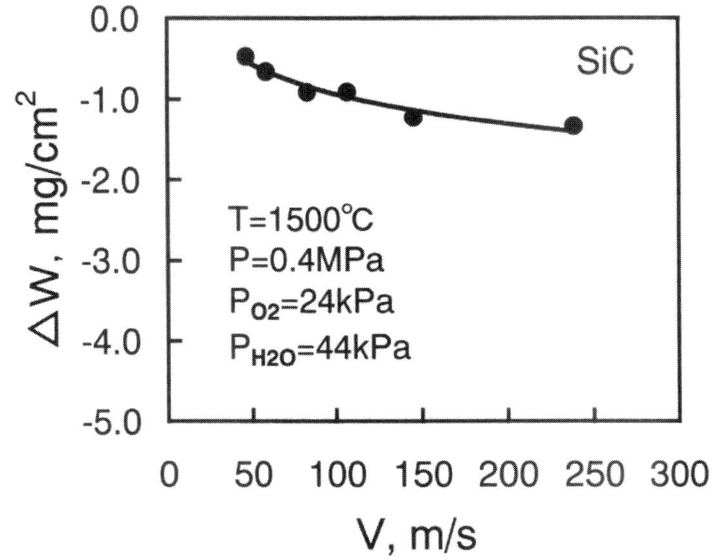

Figure 18.23 Relation Between Gas Flow Velocity and Specific Weight Change (Adapted from Ref. [12])

by Opila and Hann [13]. In future, it is necessary to clarify the detailed surface recession mechanism and develop the technology for reducing the recession rate of ceramic materials, e.g. by using a protective environmental barrier coating (EBC).

CONCLUSIONS

The application of ceramics for the design of combustors and stator vanes for gas turbines was investigated. Prototype combustors and stator vanes for 1300°C and 1500°C class gas turbines were developed, and the performance and the strength reliability were evaluated through high-pressure testing. The test results confirmed that the performance of each component can be improved by applying ceramics. It was also demonstrated that the strength reliability for short periods can be ensured by adopting the structure for stress reduction described in this chapter. On the other hand, it became clear that surface recession occurred in combustion gas flow, and it is important to clarify the surface recession mechanism and develop the technology to reduce the recession rate of ceramic materials. The primary objective of this study, was to examine the potential application of ceramics for the gas turbine. And economic assessment for commercialization of ceramic gas turbines needs to be addressed in the future.

ACKNOWLEDGEMENTS

The authors acknowledge the support from Dr. Toshio Abe, Mr. Noriyuki Mori and Mr. Kazunori Watanabe of the Central Research Institute of Electric Power Industry, and from Dr. Hiroshi Miyata, Mr. Kimiaki Nakakado and Mr. Masato Nakayama of Hitachi, Ltd. for the cooperation in the performance of this research project.

REFERENCES

[1] Miyata, H., Iijima, S., Ohshima, R., Abe, T., Hisamatsu, T., and Hamamatsu, T., "Application Technology on Ceramics for Structural Components of High-Temperature Machines," JSME Int. J. Series I, Vol. 32, No.4, pp. 596–604, 1989.

[2] Machida, T., Miyata, H., Iwai, K., Nakakado, K., Oosima, R., Abe, T., Hisamatsu, T., Mori, N., and Yuri, I., "Performance Test on Ceramic Combustor for High Temperature Industrial Gas Turbine," Proc. Fossil Fuel Combustion Symposium 1990, ASME PD-Vol. 30, pp. 183–187, 1990.

[3] Iwai, K., Machida, T., Urusidani, H., Wada, K., Abe, T., Hisamatsu, T., Mori, N., and Yuri, I., "Development of the Ceramic Combustor for Gas Turbine at High Specific Combustion Intensity," 19th Int. Congress on Combustion Engines, CIMAC, 1991.

[4] Mori, N., Hisamatsu, T., Yuri, I., Machida, T., Nakakado, K., Miyata, H., and Iwai, K., "High Pressure Test of a Ceramic Combustor for a High Temperature Industrial Gas Turbine," Proc. Emerging Energy Technology, ASME PD-Vol. 50, pp. 95–100, 1993.

[5] Mori, N., Hisamatsu, T., Yuri, I., Miyata, H., Nakakado, K., and Machida, T., "Development of Ceramic Stator Vanes for Power Plant Gas Turbines," *Proceedings of the 1991 Yokohama International Gas Turbine Congress*, Vol. I, pp. 143–147, 1991.

[6] Nakakado, K., Machida, T., Miyata, H., Hisamatsu, T., Mori, N., and Yuri, I., "Development and Reliability Evaluation of Ceramic Stator Vanes for Industrial Gas Turbine," *Proceedings of the 4th International Symposium on Ceramic Materials and Components for Engines*, Göteburg, Sweden, June 10–12, 1991, pp. 985–992. Elsevier Applied Science, R.Carlsson, T. Johansson, and L. Kahlman, eds.

[7] Nakakado, K., Machida, T., Miyata, H., Hisamatsu, T., Mori, N., and Yuri, I., "Strength Design and Reliability Evaluation of a Hybrid Ceramic Stator Vane for Industrial Gas Turbines," *Transactions of the ASME, Journal of Engineering for Gas Turbines and Power*, 117[2], pp. 245–250, 1995.

[8] Yuri, I., Hisamatsu, T., Watanabe, K., and Etori, Y., "Structural Design and High-Pressure Test of a Ceramic Combustor for 1500°C Class Industrial Gas Turbine," ASME-Paper 96-GT-346, presented at the International Gas Turbine and Aeroengine Congress and Exposition, Birmingham, U.K., June 10–13, 1996.

[9] Yuri, I., Hisamatsu, T., Watanabe, K., and Etori, Y., "Structural Design and High Pressure Test of a Ceramic Combustor for 1500°C Class Industrial Gas Turbine," *Transactions of The ASME, Journal of Engineering for Gas Turbines and Power*, 119[4], pp. 506–511, 1997.

[10] Machida, T., Nakayama, M., Wada, K., Hisamatsu, T., Yuri, I., and Watanabe, K., "Development of Ceramic Stator Vane for 1500°C Class Gas Turbine," ASME-Paper 96-GT-459, presented at the International Gas Turbine and Aeroengine Congress and Exposition, Birmingham, U.K., June 10–13, 1996.

[11] Etori, Y., Hisamatsu, T., Yuri, I., Yasutomi, T., Machida, T., and Wada, K., "Oxidation Behavior of Ceramics for Gas Turbines in Combustion Gas Flow at 1500°C," ASME-Paper 97-GT-355, presented at the International Gas Turbine and Aeroengine Congress and Exposition, Orlando, Florida, USA, June 2–5, 1997.

[12] Yuri, I., Hisamatsu, T., Etori, Y., and Yamamoto, Y., "Degradation of Silicon Carbide in Combustion Gas Flow at High-Temperature and Speed," ASME-Paper 2000-GT-664, presented at the International Gas Turbine and Aeroengine Congress and Exhibition, Munich, Germany, May 8–11, 2000.

[13] Opila, E.J. and Hann, R.E., "Paralinear Oxidation of CVD SiC in Water Vapor," J. Am. Ceram. Soc., 80[1], pp. 197–205, 1997.

CHAPTER

19

A Design and Experimental Evaluation of Ceramic Stator and Rotor Blades for 1300°C Class Medium Gas Turbine

Takanari Okamura, Tsuneo Hijikata, and Akinori Koga

Power & Industrial Systems Research & Development Center, Toshiba Corporation, Yokohama 230-0045, Japan

INTRODUCTION

From the viewpoint of effective energy utilization advanced gas turbine development has made great progress. Increasing the turbine inlet temperature (TIT) is being promoted to increase thermal efficiency. On the other hand, the suppression of NOx emission in the exhaust gas from the engine is an important subject demanded for conservation of the environment. The high temperature gas turbine has been developed through progress in cooling technologies and materials for combustor and turbine hot section components. The supply of cooling air to the hot section components is indispensable for ensuring high temperature fracture strength and oxidation resistance of the structural materials. However, the consumption of cooling air is the major factor that limits the efficiency improvement. The application of ceramics to the hot section components is attractive for two reasons. One is the efficiency due to the reduction of cooling air, because ceramics have high strength and oxidation resistance even at high temperature. The other is the potential reduction of NOx emissions by minimizing the combustion temperature in the gas turbine. However, the simple replacement of the turbine components with ceramic parts in the same structure does not assure the reliability of the gas turbine. The monolithic ceramics are inherently brittle and susceptive to damage unlike metals and the thermal expansion coefficients of the candidate ceramic materials are far smaller than those of the metals. Therefore, ceramic hot section components need to be designed with consideration for the above characteristics of ceramic materials. In addition, the evaluation and improvement through component proof tests, simulating the operating conditions of the actual engine, are indispensable for ensuring high reliability of the ceramic components.

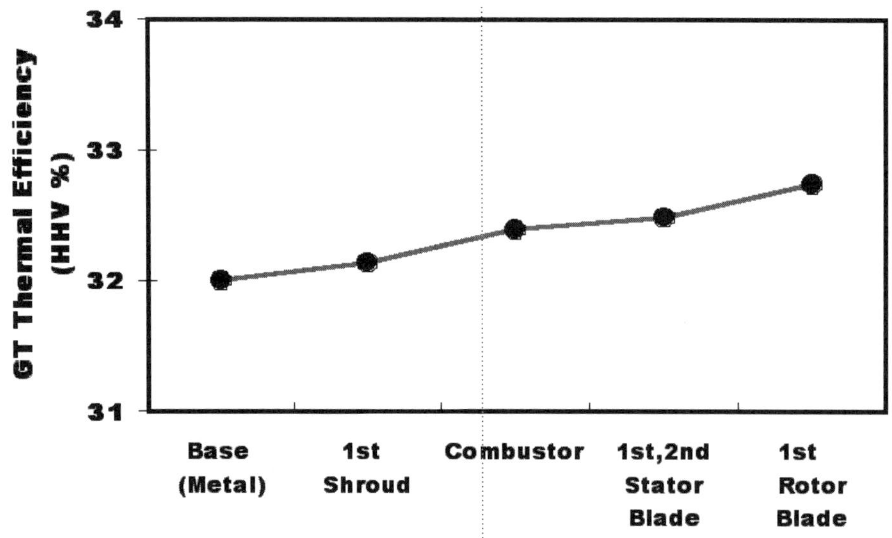

Component	*Ceramics Application*				
1st Shroud	*NO*	*YES*	*YES*	*YES*	*YES*
Combustor	*NO*	*NO*	*YES*	*YES*	*YES*
1st,2nd Stator Blade	*NO*	*NO*	*NO*	*YES*	*YES*
1st Rotor Blade	*NO*	*NO*	*NO*	*NO*	*YES*

Figure 19.1 Gas Turbine and Combined Cycle Thermal Efficiency by Ceramic Use (Reprinted with permission Toshiba Corporation)

The author's company has conducted investigations into the manufacturing processes, material properties, strength evaluation methods, and application techniques of ceramic materials for gas turbine hot section components. The R&D of the ceramic combustor was started in the mid-1980s under a collaborative project with Tokyo Electric Power Company (TEPCO). R&D for ceramic stator blades was started with internal funding following the above collaborative project. CMC (ceramic matrix composite) technology was developed in the 1990s and a CMC shield plate was first applied in a hot wind tunnel test of ceramic stator blades. This chapter presents the design concept for the first stage ceramic turbine blades and the experimental results of the stator blade hot wind tunnel cascade test, and the rotor blade hot spin testing [1–3]. A high probability of success (POS) of the ceramic turbine stator and rotor blades was confirmed by component proof testing.

EXPECTATION OF CERAMIC USE

Figure 19.1 shows the thermal efficiency increase of the gas turbine by the ceramic components application for shroud, combustor, stator blade, and rotor blade. In case of the ceramic shroud, combustor, stator blade, and rotor blade, thermal efficiency increases with the reduction of cooling air to these components. As far as the ceramic combustor is concerned, the thermal efficiency also rises with the reduction of the pressure loss around the combustor liner and transition piece. The total thermal efficiency gain in the gas turbine is roughly 1%.

Our company has developed gas turbine hot section components such as combustors and turbine stator and rotor blades as shown in Figure 19.2. The development of the ceramic combustor was a collaborative

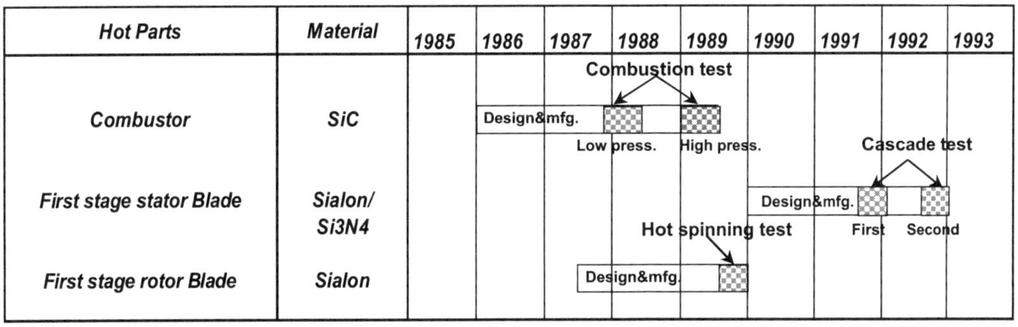

Figure 19.2 Timeline of Ceramic Parts Development for Gas Turbine Applications (Reprinted with permission from Toshiba Corporation)

effort under the TEPCO 20 MW ceramic gas turbine project (see chapter 17) and the ceramic turbine stator and rotor blades were developed with internal funding.

CONCEPTUAL DESIGN AND STRUCTURE

A 1300°C class air-cooled gas turbine T-GT15 with a thermal efficiency of 32% and output power of 15 MW has been installed in our factory for the supply of electrical power. The ceramic hot section component applications were planned to improve the engine's performance. Figure 19.3 shows the cross section of the gas turbine assembly for the first stage ceramic stator and rotor blades.

Stator Blades

The ceramic stator blades receive not only a high mechanical stress due to the gas reaction force but also a high thermal stress due to the base load, trips, etc. Besides, it is necessary to maintain a stable contact between the ceramic vanes and the metal components during the above operations. Therefore, an evaluation of the above stresses and the structural design for absorbing the difference in thermal expansion were performed for the design of the ceramic stator blade. An assembled stator blade and its constituents are shown in Figure 19.4.

The structural design concepts are the following:

(1) A continuous structure of the vane and the end walls is superior to a segmented one in order to smoothly transmit the gas reaction force from the vane to the end walls.
(2) A curved plate or shell structure is generally desirable for the ceramic parts from the viewpoints of thermal stress, fabrication, and quality control.
(3) From an aerodynamic point of view, a compound lean vane was adopted in order to suppress the secondary flow loss and also to optimize the radial distribution of the reaction percent in the stage.
(4) Each holder has plate springs. The thermal expansions of all the components are continuously absorbed to maintain a stable engaged condition through the elasticity of the springs.
(5) The cooling air impinging on the metal holder is blown downstream of the ceramic vanes, in order not to cool the ceramics.
(6) A heat shield plate with good insulating performance is inserted between the ceramic end walls and the metal holders so as to make the temperature gradient small and to release the thermal stress on the surface of the ceramics.

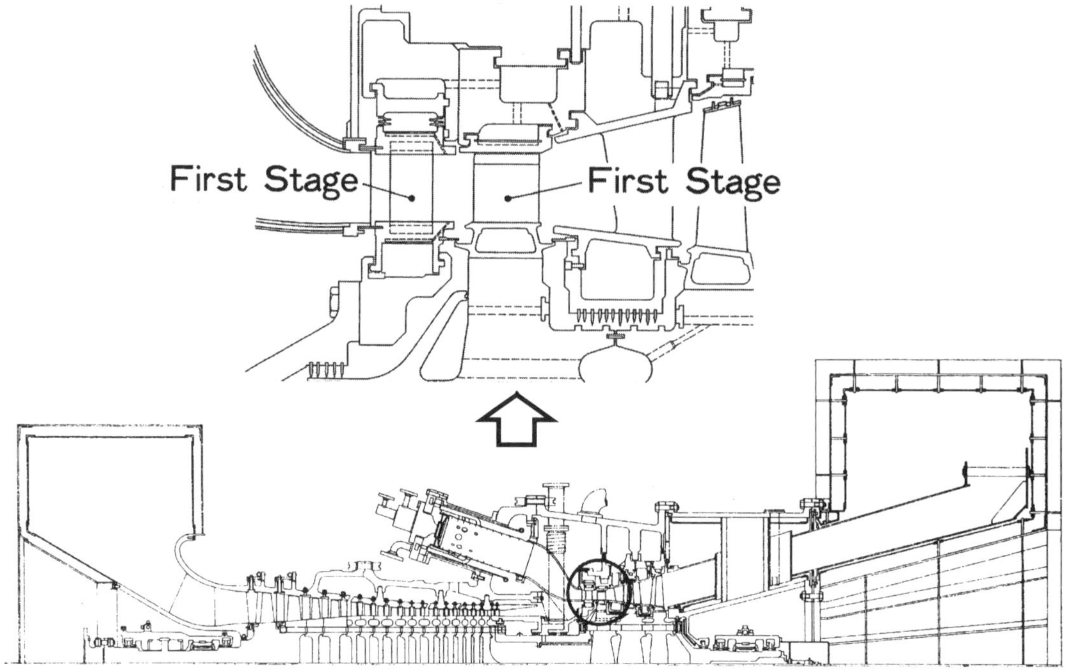

Figure 19.3 Cross Section of Ceramic Gas turbine (Adapted with permission from Ref. [1]

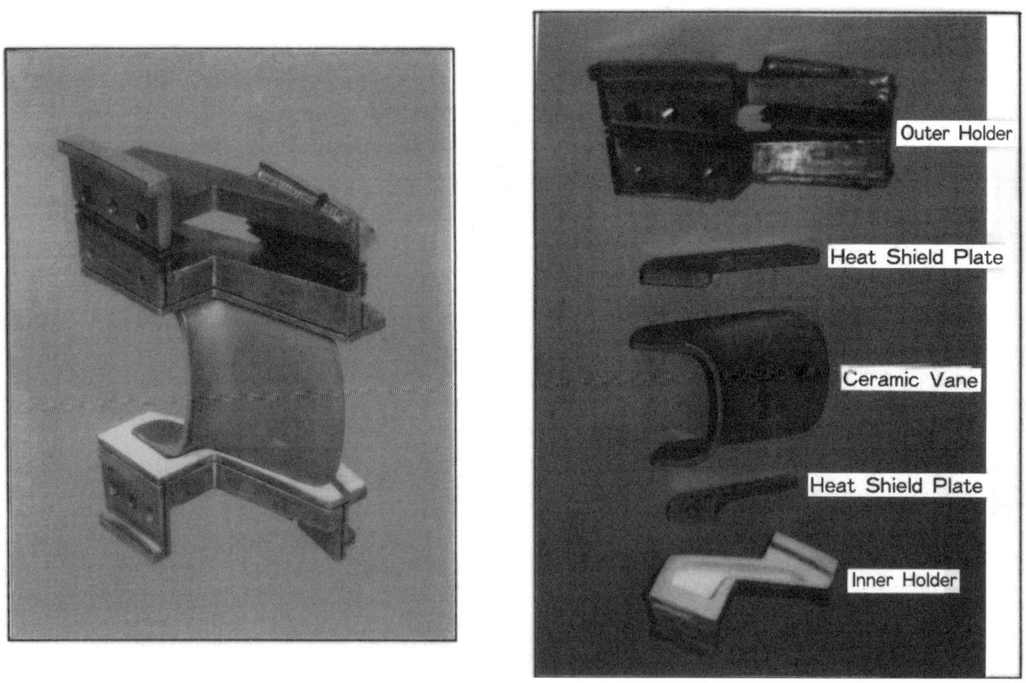

Figure 19.4 Assembled and Disassembled Components of Stator Blade (Reprinted with permission from Ref. [1])

Rotor Blades

The ceramic application for rotor blades is much more effective for the improvement of gas turbine efficiency. However, this application has its own challenges compared to the stationary hot section parts because of the following factors:

- The rotor blade has high tensile stresses because of the centrifugal forces
- Any extraneous foreign objects will collide with the blade surface at high velocity which is nearly equal to the peripheral speed of the rotor blade
- Inevitably, stress concentrations and high contact stresses occur at the dovetail-disk interface.

Therefore, it cannot be expected that just replacing the original metal blade with a ceramic part would retain the strength reliability. Moreover, as a large change in thickness occurs between the airfoil and the dovetail, a high thermal stress is generated under transient operating conditions.

Accordingly, a new design concept for the rotor blade is required for a successful ceramic application. A hybrid type rotor blade combining a ceramic sleeve with the metal core was proposed. The conceptual structure of the ceramic rotor blade is shown in Figure 19.5. This design concept is based on the use of the outstanding compressive strength of ceramics. The ceramic sleeve which acts as a heat shield is supported at the top of the metal core which places it under compressive stress. The centrifugal forces act on the metal core which has simple cooling holes from the attachment to the head. Because of the reduced heat flux it can be cooled by a minimal amount of cooling air. The cooling air is ejected from the head side through many cooling passages to the hot gas path. It is expected that the cooling air flow for this rotor blade decreases from 4.7% for an all-metal blade to 1.8%. A flexible rib structure is adopted at the interface where the ceramic sleeve contacts the head of the metal core. This structure can absorb the extra tensile stress of the ceramic sleeve caused by the difference in thermal expansion between the ceramic sleeve and the core metal.

Two specific control systems are incorporated in the ceramic gas turbine. One is a system to reduce the transient thermal stresses of the ceramic parts under gas turbine trip conditions by adding a compressed air bypass line and by using dual burners for firing temperature control. The other system is a ceramic-metal matching system for reducing the frictional stress near the contact surfaces of the ceramic sleeve in the rotor blade due to the difference in thermal expansion. The ceramic sleeve type has the following features:

- Fairly low thermal stress because of the low thickness of the sleeve
- It is easily manufactured and the quality control process is straightforward because of its simple structure
- Impingement forces from foreign objects are mitigated because the part is in compression.

Specifications for the ceramic stator and rotor blades are shown in Table 19.1

SELECTION OF MATERIAL

The requirement for the ceramic materials for gas turbine hot parts is as follows. The strength must be sufficient to withstand the mechanical stress caused by the centrifugal and gas forces, the stress caused by the difference in thermal expansion at the contact areas with metals, the thermal stress under gas turbine steady state or transient operations, and the stress induced at the ceramic surface due to the temperature gradient at the metal contact surface. In addition, the ceramic materials are required to have sufficient fracture toughness against impinging foreign objects and high oxidation and corrosion resistance. In view of the above, the properties required for the ceramic materials are summarized on a radar chart in Figure 19.6. The authors' company has developed the Sialon material SIN-E that nearly satisfies the targeted material properties. SIN-E is a HIPed sintered ceramic material that has improved strength with the optimum "z" quantity selection in the conventional $Si_{(6-z)}Al_zO_zN_{(8-z)};0<z<4.2$ and the sintering ability by the addition of HfO_2 to it [4].

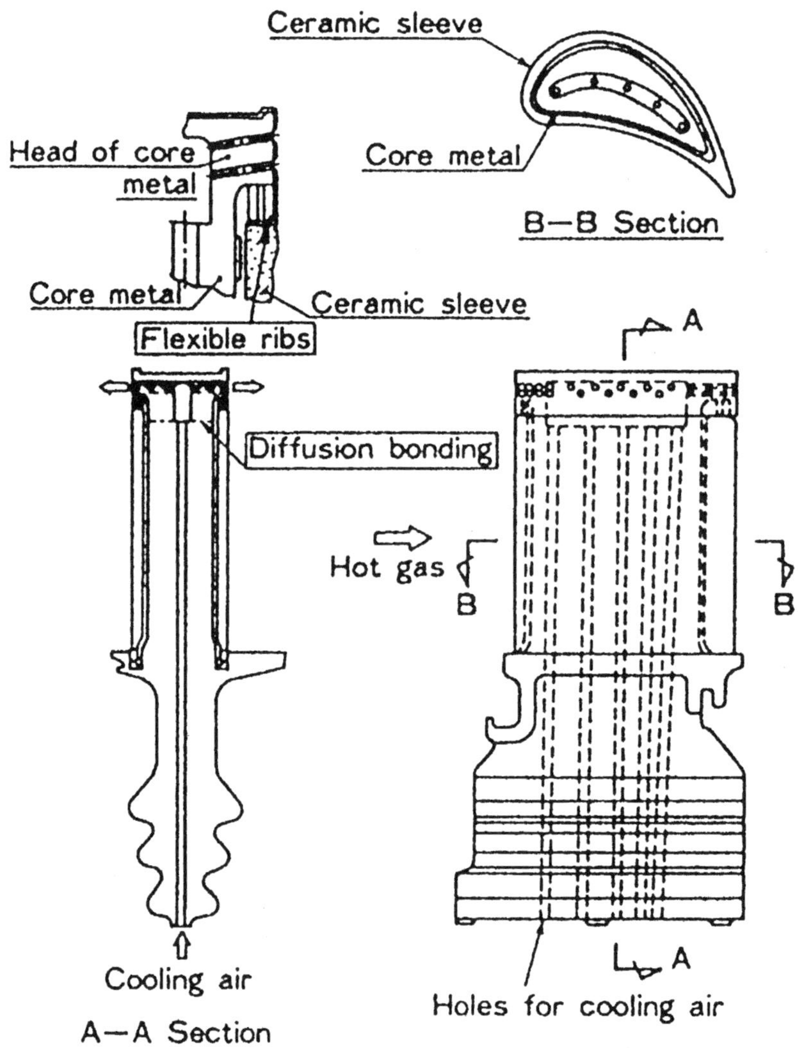

Figure 19.5 Ceramic Sleeve Type First Stage Turbine Rotor Blade (Part of Figure Reprinted from Ref. [6])

TABLE 19.1 SPECIFICATIONS FOR THE FIRST STAGE TURBINE STATOR AND ROTOR BLADES

	First Stage Stator Blade	First Stage Rotor Blade
Type	Compound lean plate type	Hybrid type
Blade height	63.2 mm	63.2 mm
Chord length	59.7 mm	54.4 mm
Thickness	3.9 mm	2.0 mm
Material Ceramic	Airfoil: Sialon (SIN-E:Toshiba) Si_3N_4: Kyocera	Sleeve: Sialon (SIN-E:Toshiba)
Metal	End wall: FS-414 (SUS)	Core metal: MAR-M247

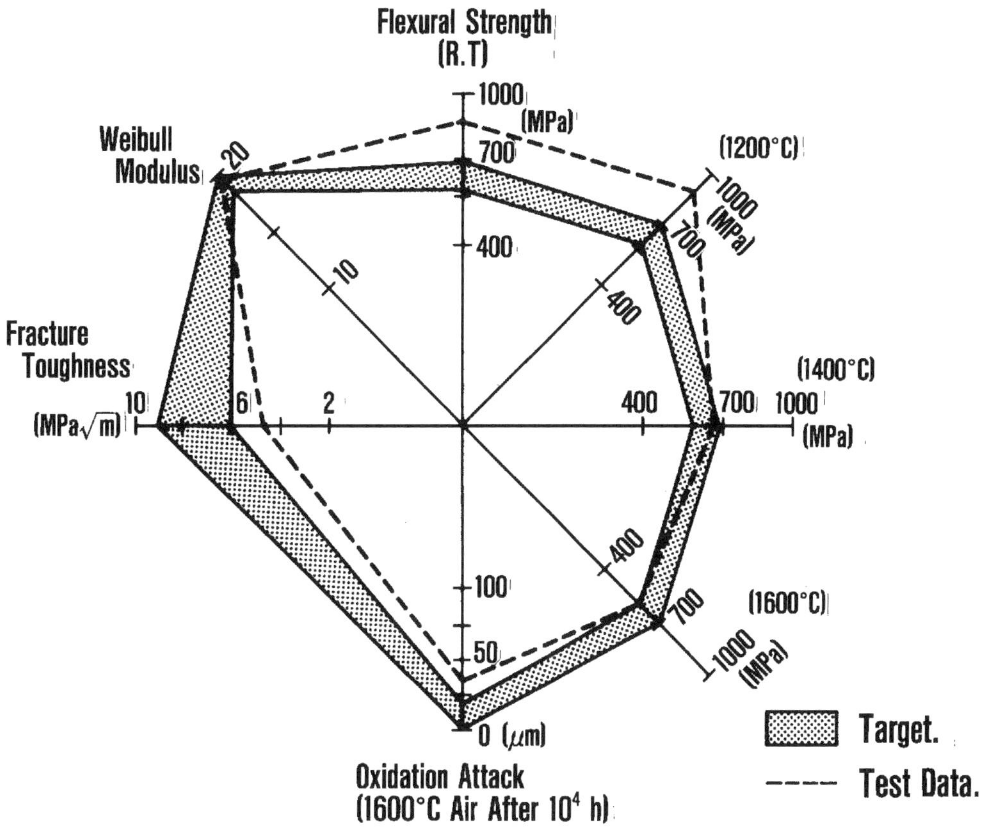

Figure 19.6 Material properties of target and test data for Sialon ceramics (Adapted from Ref. [6], Reprinted with permission from Ref. [1])

In the fabrication of a ceramic turbine blade, the following factors are necessary: a defect free part, good dimensional accuracy, reproducibility, and low manufacturing cost. Moreover, adequate technical potential for the fabrication of large size, complex shaped parts such as a near net shape forming process are required.

RELIABILITY ANALYSIS

Strength Estimation Method

For ceramic applications in the gas turbine, it is important that the candidate ceramics are analyzed and evaluated with respect to low cycle fatigue activated by thermal stresses generated during start/stop transients. High cycle fatigue and creep can take place at appropriately high temperature due to mechanical stresses resulting from centrifugal forces, gas pressure, thermal stress due to trip operation, and so on.

Ceramics usually have the following generic strength properties:

(1) Typical ceramics are brittle with much lower fracture toughness and much higher scatter in strength than metals

(2) Fast fracture, creep rupture, and fatigue failure dependent fracture is governed mainly by slow crack growth from one of innumerable initial flaws, and so the ceramic's fracture strength depends on the stress rate and stress hold time

(3) The fracture strength, depending on the size of the ceramic components and stress distributions, is affected by the volume dependence governed by the probability of existing initial flaws
(4) In ceramics, compressive strengths are typically about 5 times higher than tensile strengths.

Considering the matters as described above, the authors referred the unified estimation method as the means evaluating ceramic strength [5–7]. This method can evaluate in a unified way all the data obtained by fast fracture, cyclic fatigue, and creep repture tests based on tensile or bending stress. It is devised theoretically on the basis of crack growth from one of innumerable initial flaws existing in ceramics and derives the following probabilistic expression:

$$P_f = 1 - \exp(-((\sigma_{max}\ T_{eff}{}^{1/n} V_{eff}{}^{1/m})\sigma_o)^m) \tag{1}$$

Where n, m are the parameters of material properties and indicate crack growth rate index and Weibull modulus, respectively.

Specifically, this Eq.(1) indicates that the fracture probability of ceramics subjected to a maximum stress, σ_{max}, can be estimated to be P_f. Here, t_{eff} is the effective stress hold time expressed as follows:

$$t_{eff} = \int_0^{t_o} (\sigma\ (t)/\ \sigma_{max})^n dt \tag{2}$$

Where t_0 is the time until fracture, t_f, in the case of fast-fracture, and is the period time, t_0, in case of cyclic fatigue. On the other hand, V_{eff} is the effective volume expressed as follows.

$$V_{eff} = \int_V (\sigma\ /\ \sigma_{max})^m dv \tag{3}$$

Where σ is the stress arising at an arbitrary point in a ceramic of volume V.

Stress Analysis of Rotor Blade

Reliability analysis requires not only the maximum value of the principal stress, σ_{max}, but also the effective hold time, t_{eff}, and the effective volume, V_{eff}, when considering stress distribution. Thus, the effective hold time, t_{eff}, in a specific case such as a gas turbine trip is obtained by changing the maximum principal stress changing with time. Here, the following assumptions are made regarding the maximum principal stress criteria, where

- Each principal stress of an element contributes to the fracture based on crack growth from an individual initial flaw
- Compressive stress plays a negligible role in the above fracture.

The steady state and transient stress distributions including the centrifugal stress are calculated. The maximum stress is located at a concave corner near the trailing edge of the contact side. The maximum transient stress arises 1.3 s after a gas turbine trip. It is noticeable that the high stress is localized in a limited area.

The reliability of the ceramic component of the rotor blade was evaluated as described previously. The results of both steady state and transient thermal stress analyses are shown in Figure 19.7. Both transient (50 trips) and steady state (100,000 hours) of operation were considered in this reliability analysis. The fracture probability, P_f, is calculated using Eq. (1). The parameters assumed for transient thermal stress were: N_f = 50, t_{eff} = 0.87 s, and for steady state thermal stress: N_f = 1, $t_{eff} = 3.6 \times 10^8$ s. For both operating conditions the calculated fracture probability was very small when the SIN-E material was used.

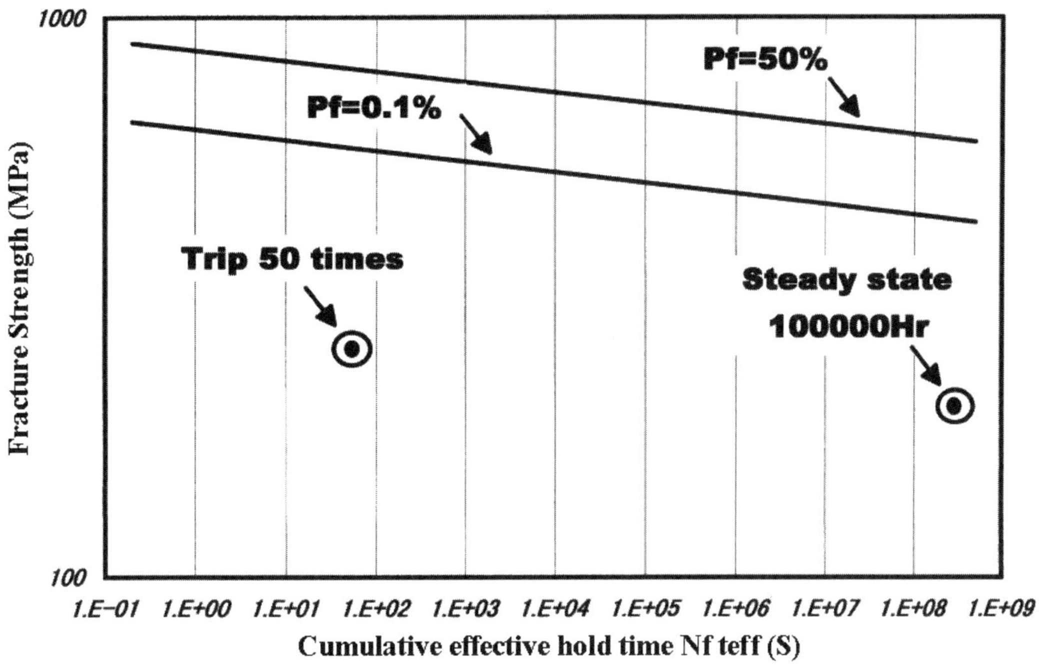

Figure 19.7 Evaluation of Fracture Probability Based on Unified Estimation Method for SIN-E Strength (Adapted from Ref. [6])

HOT WIND TUNNEL TEST OF STATOR BLADES

Experimental Apparatus

The test was carried out with the gas turbine component test apparatus. The facility consists of a high temperature wind tunnel cascade test rig, an exhaust gas cooler, and the system lines for fuel, combustion air, cooling air, and cooling water. The cross section of the test rig is shown in Figure 19.8. A dry low NOx combustor for operation on natural gas was installed. The combustion gas at 1.18 MPa and 1300°C was introduced into the ceramic stator blade cascade at the rated condition. Inlet gas conditions were measured in the transition duct formed of adiabatic material. The stator blades were observed though a window on the downstream wall of the vessel. Figure 19.9 shows the cascade section of the apparatus. Four ceramic vanes of the two materials, Sialon (SIN-E, Toshiba Corporation) and Si_3N_4 (SN252, Kyocera Corporation) were assembled in a rainbow arrangement in the same cascade. Both side wall sections of the cascade were composed of air-cooled dummy metal stator blades. The inner and outer rings supporting the holders were cooled to the same temperatures as they would be in an actual engine.

The test items and conditions of the wind tunnel were as follows. Steady state tests were carried out at 1300°C with a normal shutdown, and a trip test was performed by fuel cut off. In the wind tunnel steady state test at 1300°C the gas velocity distribution in proximity to the ceramic vane was adjusted to that in the actual engine. The flow ratio for the cooling air for the stator blade holders against the mainstream gas flow was also almost the same as that for the actual engine. The cooling effectiveness of the holders was measured in the steady state test at 1300°C. Although the stator blade inlet pressure of 1.18 MPa was slightly below that in the actual engine due to a limitation in the air supply equipment, the air inlet flow value increased about 24% after the fuel cutoff trip. Therefore, the heat transfer around the ceramic vane was maintained at a high level. The gas temperature dropped immediately from 1300°C to reach 400°C within a few seconds. It was

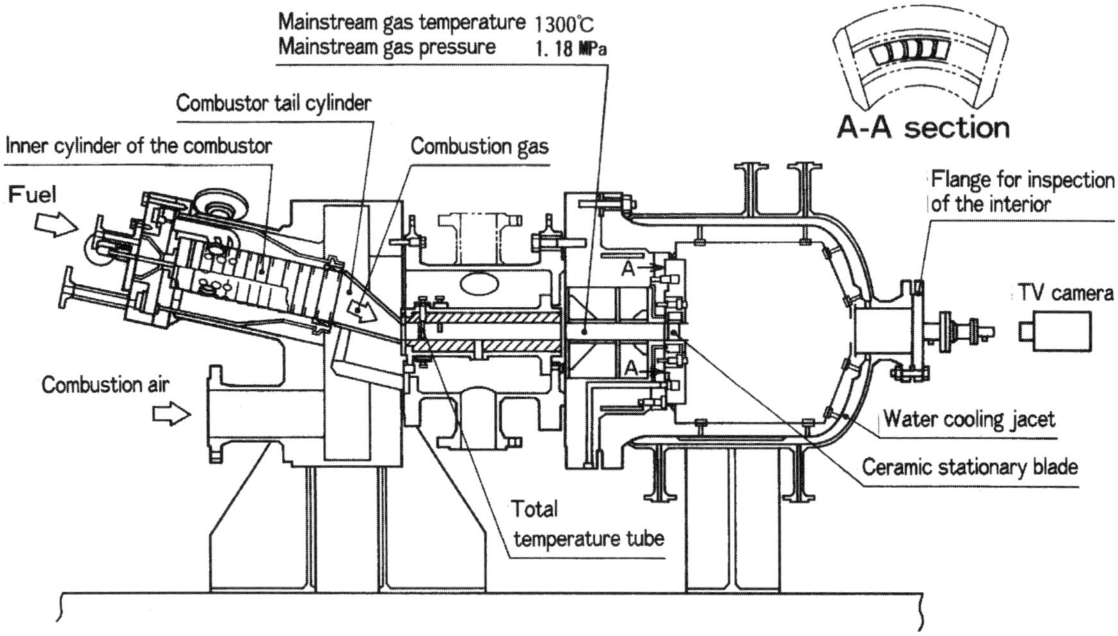

Figure 19.8 Cross Section of High Temperature Wind Tunnel Cascade Test Rig (Adapted with permission from Ref. [1])

confirmed that the thermal stress in the ceramic vane was almost equal to the predicted value during a thermal shock test in the actual engine.

Experimental Results

During the wind tunnel cascade testing the total operating time under firing conditions was 38 hours. The total time at steady state condition (1300°C) was 9.5 hours. The fuel cutoff trip test was performed three times.

Figure 19.10 shows the views at the outlet of the cascade during transient testing by fuel cutoff following operation at an average gas temperature of 1330°C. A maximum gas temperature of 1440°C was measured at the center of the cascade. Within 5 seconds after the fuel cutoff the temperature had dropped to approximately 500°C. Figure 19.11 shows a comparison of the transient gas temperature measured during the wind tunnel trip test and the predicted temperature in the actual engine. According to the transient thermal stress analysis for the ceramic vane under the design conditions for the actual engine, the thermal stress reaches a maximum about 3 seconds after the trip. When the measured temperature drop is compared with the predicted one for the first 3 seconds from the start of the trip, the gas temperature measured in the wind tunnel test decreases at the same gradient as the predicted gas temperature. The heat transfer coefficient around the ceramic vane is maintained at a high level, which is different from the condition for the actual engine. This is because the inlet flow increases by 24% compared with the steady state inlet flow with little pressure drop.

Transient thermal stresses were calculated for the ceramic vane for the above mentioned boundary conditions. The thermal stress in a Sialon vane is higher than in a Si_3N_4 vane. Figure 19.12 shows the maximum thermal stresses for the Sialon ceramic vane for the wind tunnel cascade trip test and for the actual engine. The maximum thermal stress in the wind tunnel trip test was approximately 240 MPa at 4 seconds after the trip started. The stress reached 96% of the stress level in the actual engine, i.e. almost the same.

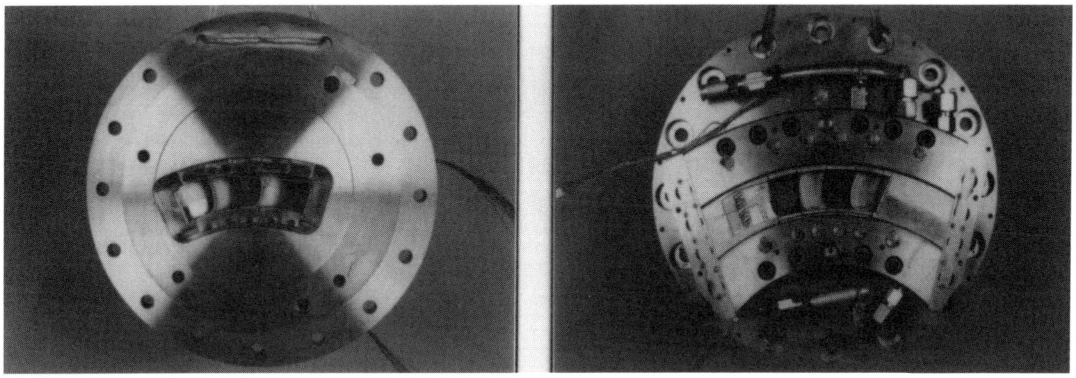

Figure 19.9 Cascade Section of Ceramic Stator Blades (Adapted with permission from Ref. [1])

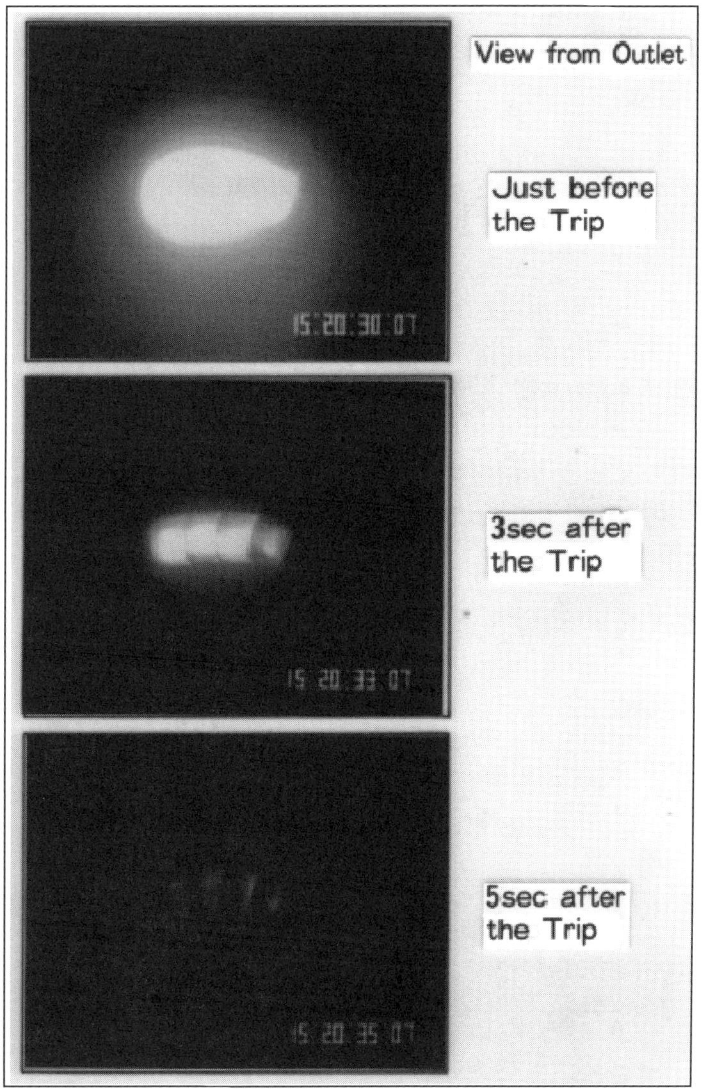

Figure 19.10 Views of Cascade at Trip (Reprinted with permission from Ref. [1])

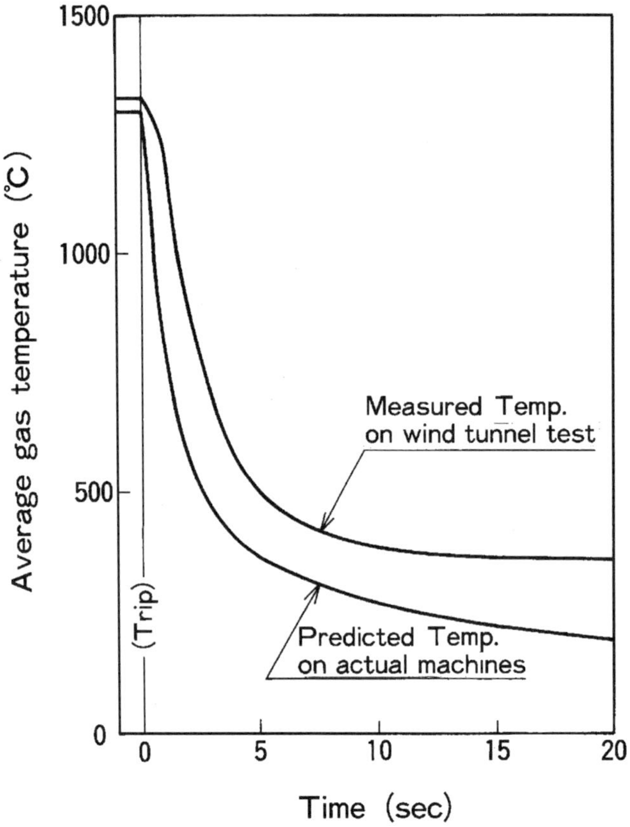

Figure 19.11 Temperature Transitions After Trip (Reprinted with permission from Ref. [1])

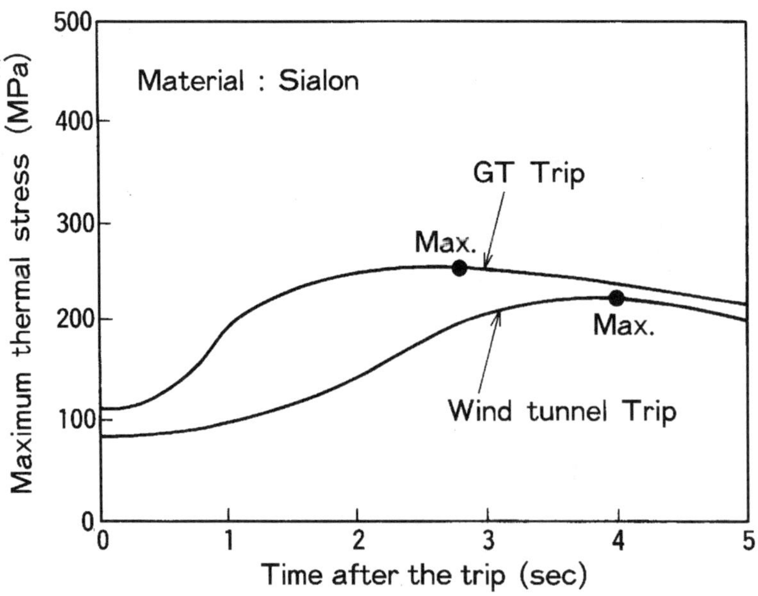

Figure 19.12 Maximum Thermal Stresses Under Trip Conditions (Reprinted with permission from Ref. [1])

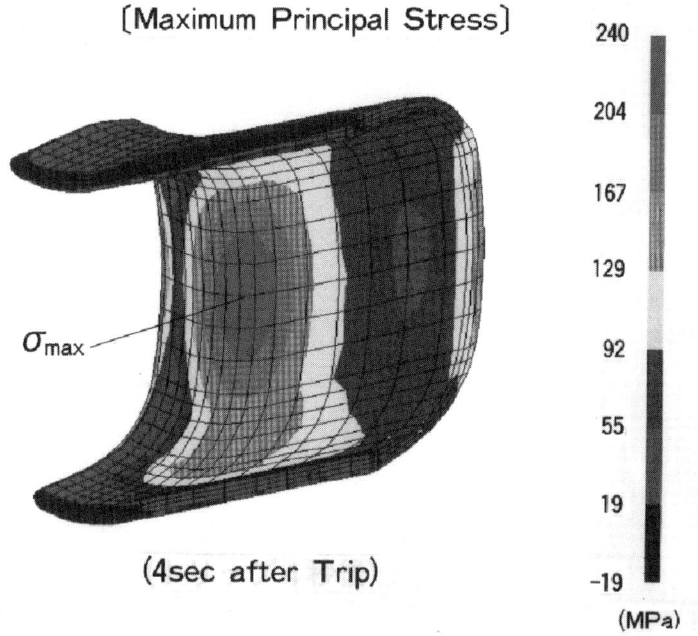

Figure 19.13 Maximum Thermal Stress of Ceramic Vane Under Trip Conditions (Reprinted with permission from Ref. [1])

The stress distribution for the ceramic vane at 4 seconds after the start of the trip is shown in Figure 19.13. Figure 19.14 shows the normalized stress of the transient thermal stress in the ceramic vane compared to the fracture strength of Sialon ceramics based on the unified strength estimation method [5, 7]. The design criterion for the actual engine is 50 trips. Compared to the material strength estimated at a fracture probability of 0.01%, the stress levels achieved as a result of the cumulative thermal shock impact after 3 trips in the wind tunnel test and after 50 trips in the actual engine were 49% and 59%, respectively. Within a reasonable estimate of error, they practically represent the same order of magnitude. Fluorescent dye penetrant inspection showed that the Sialon and Si_3N_4 vanes were intact without any evidence of cracking.

As far as the heat shield plate was concerned, the original 8 wt.% yttria-stabilized zirconia (ZrO_2-8%Y_2O_3) was replaced by a ceramic matrix composite (CMC: Nicaloceram of Nippon Carbon Co., Ltd.) as is shown in Figure 19.15. The yttria-stabilized zirconia heat shield was damaged during the normal shutdown. A detailed thermal stress analysis calculated a 292 MPa surface stress, which was in excess of the fracture strength of the material at the rated conditions. A similar thermal stress calculation for the CMC material resulted in a stress of 40 MPa which is less than the fracture strength for this material. The main contributors to the reduction in thermal stress for the CMC material are the coefficient of thermal expansion and Young's modulus which are much lower than those of yttria-stabilized zirconia. The successful application of the CMC as a structural material for a gas turbine hot section component proved its potential as a candidate material.

HOT SPIN TESTING OF ROTOR BLADES

Experimental Apparatus

The cross section and the specifications for the hot spin test equipment are shown in Figure 19.16 and Table 19.2, respectively.

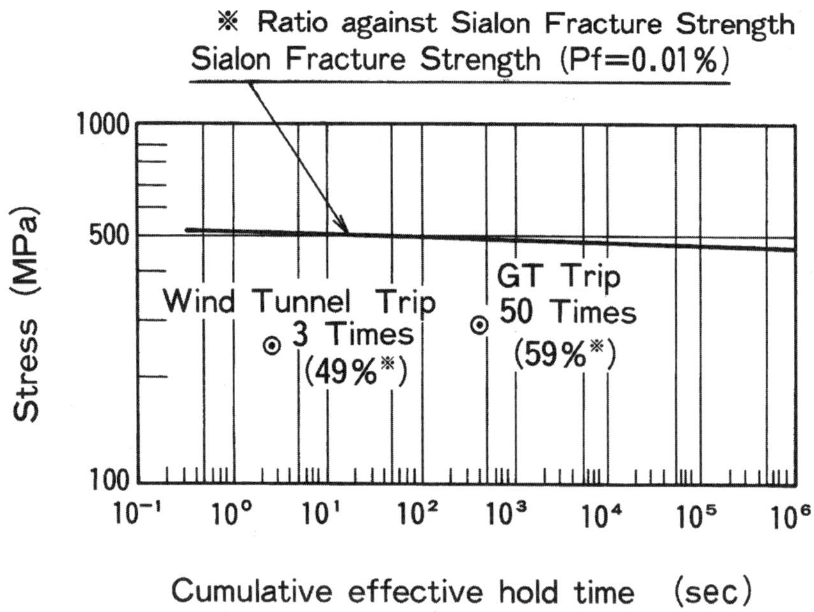

Figure 19.14 Evaluation of Fracture Probability Based on Unified Estimation Method (Reprinted with permission from Ref. [1])

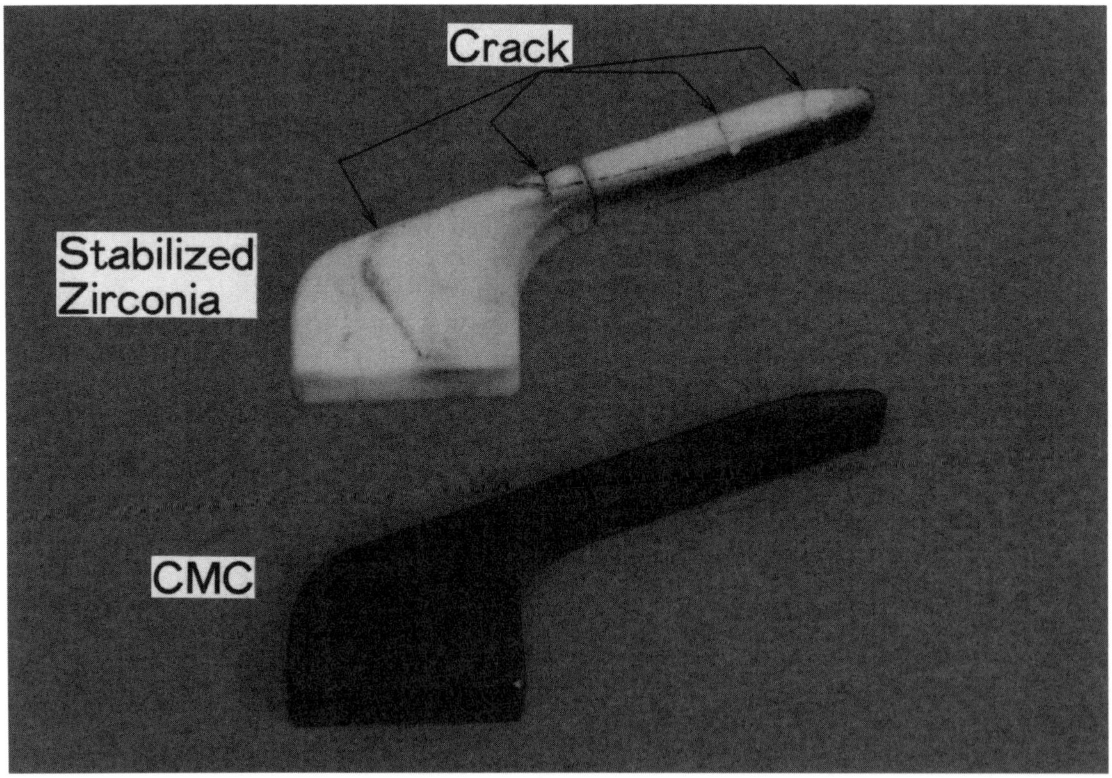

Figure 19.15 Heat Shield Plates of Stabilized Zirconia and CMC (Reprinted with permission from Ref. [1])

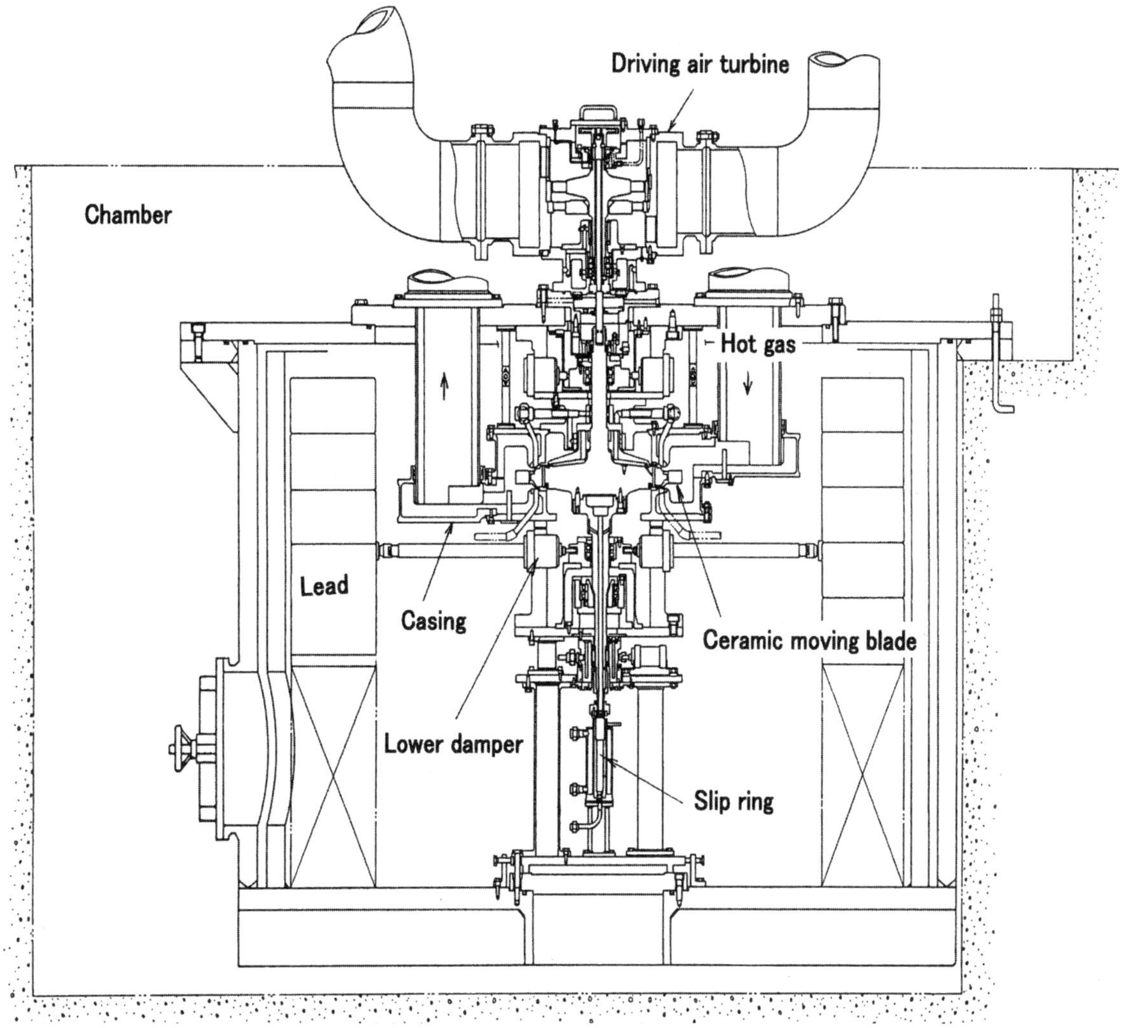

Figure 19.16 Cross Section of Hot Spin Test Equipment (Adapted from Ref. [6])

The air delivered from the air compressor was supplied to the driving air turbine. Air and municipal gas were supplied to the combustor, and hot gas was led to the test blade hot section and then exhausted to the atmosphere. Air was also distributed to blade cooling, packing sealing, and pit cooling. The blade cooling air temperature was controlled to maintain the same condition as in the actual engine. The two ceramic test blades were assembled at a 180° symmetrical circumferential position in the test rotor which was driven by the air turbine. The slip ring was attached at the end of the test rotor to measure the temperature of the core metal in the ceramic blade.

Experimental Results

First, the gas turbine start and stop operating simulation tests were carried out. The flexible rib structure was adopted to absorb the extra tensile stress at the head of the core metal contacted by the ceramic

TABLE 19.2 SPECIFICATIONS FOR THE HOT SPIN TESTING EQUIPMENT

Driving air turbine	Type	Vertical Single Stage Air Turbine
	Output	400 kW
	Revolution speed	12,240 rpm
Combustor	Gas temperature	1,400°C
	Gas pressure	1.0432 atm
	Fuel	Municipal gas
Air compressor	Driving power	2,750 kW
	Mass flow	15 kg/s
	Pressure	3.10 atm
Heater	Outlet temperature	Room temperature – ~360°C
	Mass flow	0.1 kg/s

sleeve. The operating schedule for the hot spin test, which simulated the operations of the ceramic gas turbine, is shown in Figure 19.17. The temperature of the metal core is controlled to 500°C at the rated rpm, and at 750°C at the rated hot gas temperature, thereby maintaining a 250°C rise in core metal temperature. In the beginning of the spin testing foreign object damage was found at the suction side of the blade airfoil. This damage might have occurred due to the collision with a piece of heat insulating material which was installed in the test equipment. It was estimated that the crack could not propagate in the entire airfoil area, because the ceramic sleeve in the rotor blade was under compressive stress. The hybrid sleeve type rotor blade demonstrated to be resistant to foreign object damage. To avoid the interference from foreign objects, protective metal blades were installed in front of the newly inserted ceramic rotor blades in the direction of rotation. These operating simulation tests were repeated ten times. Next, tests were carried out to establish limiting conditions such as a spin test at 50°C excess gas temperature and spin tests at 110% overspeed. There were 42 starts and the total operating time was about 72 hours.

The ceramic test blades were inspected after the operation simulation tests and after the limitation tests, respectively. The appearance of a test blade after the limitation test is shown in Figure 19.18. No abnormal indications were detected on the blade surface by fluorescent dye penetrant inspection. The reliability of the ceramic sleeve type rotor blades was fully demonstrated in these tests, including the strength of the flexible rib construction and the diffusion bonding between the head and the support of the metal core.

CONCLUSIONS

Experimental work was conducted on first stage ceramic turbine stator and rotor blades for a 1300°C class medium size gas turbine. The following conclusions were reached.

Stator Blade

(1) The transient thermal stress in the ceramic stator blade was calculated to be 240 MPa, corresponding to the same stress this blade would have in an actual engine. The normalized stress was about half of the level of the fracture strength estimated at a fracture probability of 0.01%. Since the inspection showed the ceramic vanes to be intact, the reliability of these parts was assumed to be assured against the thermal shocks of an actual gas turbine.

(2) The CMC application demonstrated that this material had excellent performance as a heat shield plate. This material proved to be a candidate material for the gas turbine.

(3) The operation test at the 1300°C design point confirmed that the stator blade holders and the heat shield plates were sufficiently cooled for the predicted cooling effectiveness.

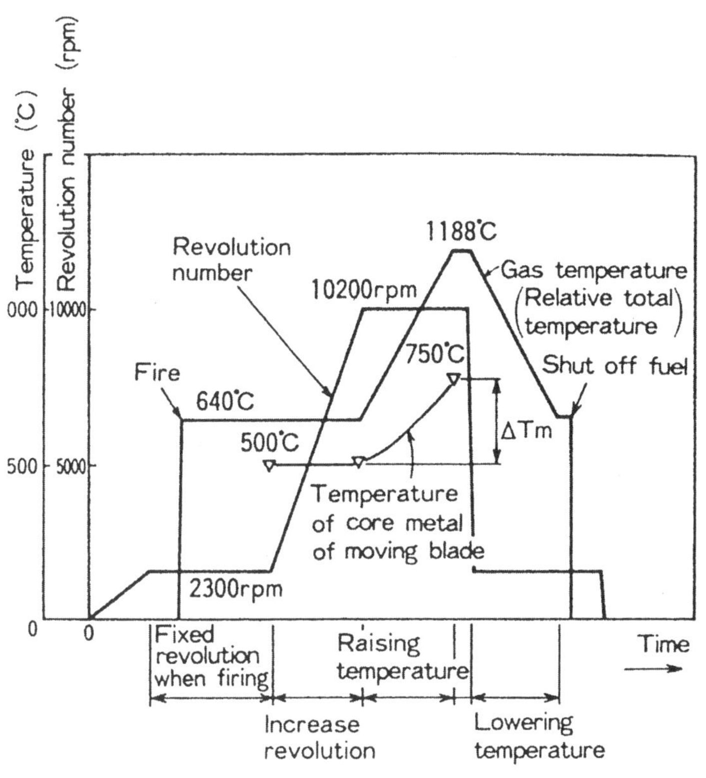

Figure 19.17 Operating Schedule of Hot Spin Test (Adapted from Ref. [6])

Figure 19.18 Appearance of Ceramic Turbine Rotor Blade in Hot Spinning Test (Adapted from Ref. [6])

Rotor Blade

(1) It was verified that the hybrid sleeve type rotor blade could avoid serious damage due to foreign objects and the soundness of the ceramic blade was confirmed in hot spin tests simulating actual gas turbine operation.
(2) The cooling air flow for the ceramic rotor blades could be reduced to about one third of that of the original air-cooled metal blade and the core metal temperature was kept at the design point of 750°C.
(3) The design concept of the hybrid turbine rotor blade which combines the ceramic sleeve with the metal core was verified.

The R&D efforts of the authors throughout the 1980s and 1990s verified that monolithic ceramic components were advantageous for hot section applications. The basic design methodology for the ceramic components was almost established. But when it comes to applying the ceramic hot section technology to the actual gas turbine, it is evident that it is necessary to improve the material properties, especially the fracture toughness to ensure adequate reliability. The authors expect that further improvements in the properties of monolithic ceramics will take place. The second wind tunnel test in 1992 demonstrated the potential of CMC for gas turbine hot section applications, and R&D of CMC materials has been conducted at Toshiba Corporation since the early 1990s. The results of these studies are described in chapter 18 of Volume II of this text, entitled "Development of Continuous Fiber Reinforced Reaction Sintered Silicon Carbide Matrix Composite for Gas Turbine Hot Parts Application."

NOMENCLATURE

σ_o = Normalized strength
σ = Stress
σ_{max} = Maximum stress or maximum principal stress
m = Crack growth rate index
n = Weibull modulus
t = Time
t_{eff} = Effective stress hold time
t_o = Time until fracture
V_{eff} = Effective volume
V = Whole volume
P_f = Fracture probability
N_f = Number of cycles to fracture

REFERENCES

[1] Hjikata T., Koga A., Asakuma K., "Reliability of Ceramic Stator Blade Tested in Hot Wind Tunnel", (IGTC-79), *Proceedings of the 1995 International Gas Turbine Congress*, Yokohama, Japan, October 23–27, 1995. Vol. III, 161–167, 1995.

[2] Nakano H., Okuno K., Niizeki Y., Kobayasi T., and Koga A., "Hot Spin Test for Ceramic Moving Blades," Transactions of the JSME, 50B[557], 310–316, 1993 (in Japanese).

[3] Hjikata T., Koga A., and Asakuma K., "Reliability of Ceramic Stator Blade Tested in Hot Wind Tunnel" *Proceedings of 8th Regular Meeting of the GTSJ,* p. 299–305, 1993 (in Japanese).

[4] Kameda T., and Asayama Y., "Sialon Ceramic Matrix materials for Heavy Duty Gas Turbine," Fine Ceramics Report No. 10[8], 267–269, 1992 (in Japanese).

[5] Okabe N., Ikeda T., Okamura H., Koga A., Izumi A., and Hirata H., "An Experimental Study on Ceramic Moving Blades for Heavy Duty Gas Turbine," IGTC-102, *Proceedings of the 1991*

International Gas Turbine Congress, Yokohama, Japan, October 27–November 1, 1991. 91-Yokohama-IGTC-102, pp. 157–164, 1991.

[6] Ikeda T., Okuma H., Okabe N., "Study of Ceramic Application in Heavy Duty Gas-Turbine Blades," ASME paper 91-GT-372, presented at the International Gas Turbine and Aeroengine Congress and Exposition, Orlando, FL, USA, June 3–6, 1992.

[7] Okabe N., "Unified Estimation Method as the Means for Evaluating Ceramic Strength," Research of Machines, 42[1], Japan (1990) (in Japanese).

CHAPTER

20

EVOLUTION OF CERAMIC GAS TURBINE DEVELOPMENT PROGRAMS AT ENGINE MANUFACTURERS IN WESTERN EUROPE

Mark van Roode

Solar Turbines Incorporated, San Diego, California, USA

INTRODUCTION

The countries with significant ceramic gas turbine development activities in Western Europe had established automotive and gas turbine equipment industries as well as a mature traditional ceramics industry. These countries were the United Kingdom, Germany and Sweden. There have also been ceramic gas turbine related activities in France and more recently in Italy.

A new element has been the role of the European Union (EU), a supra-national body which controls a substantial budget and which is actively promoting international scientific and technical collaboration between its member states. Over the past ten years several ceramic gas turbine related development programs have been directly funded by the EU.

This chapter introduces the ceramic development programs of the gas turbine and automotive manufacturers and other institutions with past and/or current development programs involving the design and testing of ceramic components in rigs and engines in Western Europe. The work of these companies is being discussed in detail in Chapters 21–29 of this text. Their activities span a broad range of interests from automotive to industrial to utility size gas turbines. The work at the Western European companies that are involved with the fabrication of ceramic components has been summarized in the introductory chapter to Volume II of this text and is being detailed in Chapters 9–11 of that volume. Work of a scientific nature is included in several of the chapters in the second half of Volume II (chapters 19–29).

UNITED KINGDOM

Rolls Royce (Chapter 26 of this volume) had programs for ceramic gas turbine development since the 1970s. The early work involved the development of blades, nozzles and combustor liners for the company's aeroegines. Lumby et al. mentioned testing of hot pressed silicon nitride (HPSN) turbine blades for 2 hours at 70,000 rpm, and the fabrication of Syalon blades [1]. Since the mid 1980s Rolls Royce has pursued the development of ceramic matrix composites (CMCs) for use in aircraft gas turbines. A range of materials was explored for vairous component applications. Carbon/carbon materials and subsequently SiC/Al_2O_3 were evaluated for use in hot structural components, in particular reheat (afterburner) flameholders. SiC/Al_2O_3 combustor liners with a thermal insulation system and turbine shroud seals with an abradable coating for cooling air reduction and clearance control were also tested. Other applications studied at Rolls Royce were glass and glass ceramic composites for compressor airfoil and casing applications, SiC/SiC exhaust guide vanes, SiC/SiC nozzles and oxide/oxide combustor tiles.

Alstom (formerly European Gas Turbines—EGT) (Chapter 27) has been the prime contractor since 1997 on a three year 2.6M Euros program, 50% funded under the EU's BRITE/EURAM umbrella. The project aimed at the development of a new design methodology for CMCs for gas turbine hot section components. The major program elements were to establish CMC gas turbine hot section component design data and methodologies. A prototype SiC/SiC CMC combustor with a protective oxide-based thermal protection system was rig tested.

GERMANY

BMFT program

The primary ceramic gas turbine development efforts in Germany focused on developing the technologies for vehicular gas turbines. Key drivers for this program were the desire to lower the burden on the environment (increase fuel efficiency and lower emissions of NOx) and to improve engine performance and. The first major development effort in the Federal Republic of Germany was the "Ceramic Components for Vehicular Gas Turbines" ("Keramische Komponenten für Fahrzeug-Gasturbinen") program (1974–1983), sponsored by the German Federal Ministery for Research and Technology (Bundesministerium für Forschung und Technologie—BMFT) with significant cost sharing by German industry. It has been estimated that about 142.8 million DM was spent, about half of it funded, between 1974 and 1986 on the BMFT and related programs [2].

The program participants included automotive (Daimler-Benz, Volkswagen AG) and gas turbine manufacturers (MTU), six ceramic component suppliers and research institutes providing supporting technologies. Progress summaries have been published in German [3–5] and English [6, 7]. The BMFT program was a major factor in the improvement of the properties of monolithic silicon based ceramics, witnessed by the flexure strengths of hot pressed silicon nitride (HPSN) and reaction bonded silicon nitride (RBSN) which almost doubled between 1974 and 1983. The components included single rotor blades, rotor rings, and integral rotors, single nozzles, nozzle segments and nozzle rings, nose cones, combustor liners, scrolls and recuperator elements. Details of the program are reviewed in Chapter 21 of Volume I and in Chapter 9 of Volume II of this text.

Components based on these concepts were tested in cold and hot spin rigs for rotating components, in gasifier rigs, in cascade test rigs and in engines. A Daimler-Benz car with a ceramic gas turbine developed under the BMFT sponsored "Car 2000" program (1978–1982) was shown to the public in 1984. And in 1990, a Mercedes-Benz car with a ceramic gas turbine successfully covered the distance from Stuttgart, Germany, to Brussels, Belgium, about 600 km.

The BMFT sponsored several other programs after the "Ceramic Components for Vehicular Gas Turbines" initiative. The BMFT "Structural Ceramics" program from 1983 to 1986 broadened the application range to include engines other than gas turbines. Daimler-Benz under this program worked on the development of commercially viable processes for the fabrication of sintered silicon nitride (SSN) rotors. Subcontractors were GTE in the USA, and Kyocera Corporation in Japan. In the 1990s Daimler-Benz started to focus its interests more on conventional automotive engine applications. A silicon nitride valve has been evaluated for Mercedes-Benz cars.

MTU participated in other programs aimed at the development of SiC materials and components. Two of these programs, focused on developing a pressureless sintered SiC with high strength and reliability were: "Basic Program for the Development of a High Temperature Structural SiC Ceramic" ("Grundlagenprogramm zur Entwicklung einer Hochtemperatur-Konstruktionskeramik aus SiC,") (1983–1987) and "Development of a Ceramic Material for Highly Loaded Combustor Liners and Hot Gas Tubing," ("Entwicklung eines keramischen Werkstoffes zur Auskleidung thermisch hochbelasteter Brennräume und HeiBgasführungen") (1987–1990). For this purpose MTU built a ceramic laboratory for SiC fabrication by pressureless sintering. MTU applied the process to fabricate a cooled hybrid metal-ceramic nozzle for a jet aircraft engine application under two programs sponsored by the German Ministery of Defense entitled: "Metal-Ceramic Hybrid Nozzle (Feasibility Study)" ("Metall/Keramik-Hybridschaufel [Machbarkeitsstudie],") (1980–1982") and "High Temperature Turbine Nozzle," ("Hochtemperatur-Turbinenleitkranz") (1985–1989). A cooled hybrid SiC blade was also developed [8]. The nozzle and blade development at MTU has been described in chapter 24. Between 1989 and 1999 MTU considered the emerging class of CMC materials (C/C, C/SiC, SiC/SiC). Still, the CMC's did not meet the design requirements and MTU suspended all ceramic activities in 1999 for the time being, to be resumed if and when newer and improved ceramic materials were to become available.

Asea Brown Boveri

Asea Brown Boveri (ABB) (currently Alstom Power Support GmbH), conducted an in-house evaluation of ceramic combustor materials and components between 1992 and 1997. The goal of the study was to develop a material and design that would enable operation at a surface temperature of 1400°C and higher for heavy duty gas turbine (160–240 MW output power) development, in order to be an improvement over thermal barrier coating (TBC) technology. The outer diameter of the annular combustor is on the order of 4 m, which necessitates the use of a liner consisting of tile segments. ABB selected a bolt/tile concept for evaluation. A number of commercially available materials were screened. While oxides (zirconia, mullite, alumina) were attractive in terms of chemical stability, silicon carbide was superior to these materials in terms of strength. Strength became the dominant factor and Hexoloy SA SiC was selected based on oxidation resistance and mechanical reliability. This work is detailed in Chapter 29 of Volume I.

In the AG Turbo program, supported by the German Federal Department of Commerce, ABB rig tested a 20 MW combustor in which SiC tiles with a thermal insulation package were mounted. The 50 hour test included 35 hours at full load and 5 trips. ABB extrapolated that as much as 0.8 mm of wall recession could be incurred in 8,000 hours of operation based on erosion/volatilization. ABB acknowledges the deleterious effect of water vapor in the combustion atmosphere, a phenomenon that has been reported on extensively in the US ceramic gas turbine development work since 1997 (see Chapter 26 in Volume II).

Siemens KWU

One rather successful application of monolithic ceramics for gas turbines are the combustor tiles used in Siemens VX4.2 gas turbines. The tiles consist of high-alumina refractories based on a corundum-silimanite mixture with a relatively high silica content. The combustor tiles have been used for many years and over 8 million operating hours have been accumulated. More recently Siemens has been investigating alternative tile materials to meet the requirements for hotter combustor liner conditions. Chapter 28 in Volume I describes the Siemens experience in detail.

Universities

A number of German universities and scientific institutes have been involved with ceramic gas turbine development work. In particular, they provided support work under the BMFT "Ceramic Components for Vehicular Gas Turbines" program between 1974 to 1983. The Max Planck Institut (Powder Metallurgy Laboratory) in Stuttgart supported the program with basic research, the Technical University in Berlin conducted studies in powder technology, the Technical University in Clausthal-Zellerfeld did work in joining and the Fraunhofer IzfP Saarbrücken contributed in the area of nondestructive testing. The Technical University of Karlsruhe performed materials property studies as did the German Aerospace Institute in Cologne.

The Institute for Thermal Turbomachinery (ITT) of the University of Karlsruhe has conducted a number of development studies for stationary gas turbine components in the 1990s. The studies involved a segmented siliconized SiC (SiSiC) flame tube as well as several integral CMC liners for a small gas turbine, a cooled nozzle, and a scroll for an industrial gas turbine. Fracture mechanics and fracture statistical methods and life prediction figured prominently in the work. Details of the studies at the University of Karlsruhe are described in Chapter 25 of Volume I.

Sweden

The ceramic development programs in Sweden have been summarized by Kronogård and Malmrup [9] in 1983 and by Pejryd in 1988 [10]. Based on Sweden's background in engines, particularly in gas turbines and Sterling engines and with an indigenous ceramics industry, the Swedish Board for Technical Development (STU) in 1976 decided to financially support the work in ceramic materials for heat engines. The program encompassed two parallel activities: one aimed at material development, the other at engine development. The long term goal of the material development was the domestic production of ceramic heat engine parts. The materials and process development work was started at R&D institutes with the ceramic suppliers becoming progressively more involved. The engine development was performed by the engine companies.

The early ceramic gas turbine work was performed by United Turbine a subsidiary of Volvo. The development engine was the 100kW automotive MK1. At the final design TIT of 1370°C this engine would have multiple ceramic parts, including the combustor, scroll, gasifier and power turbine nozzles, turbine rotor, and regenerator disks. In fact, only the ceramic wheel and regenerators were tested in the engine. The wheel was fabricated by ASEA Cerama in Robersfors from HIPed (hot isostatically pressed) sintered silicon nitride. A milestone was achieved when in March 1982 professor Kronogård drove a modified Volvo 760 powered by a MK1 turbine with the silicon nitride rotor on public roads. ASEA licensed its HIP encapsulation technology to several other companies including Norton Company and AlliedSignal Ceramic Components in the USA during the 1980's for their HIPSN materials.

Sweden also participated in several cooperative programs under the auspices of the EU. One collaborative effort (EUREKA R&D program) between Volvo Flygmotor AB (Sweden), MAN Technologie GmbH (Germany), Alfa Romeo Avlo (Italy), and SNECMA and SEP (France) was a feasibility study focused on the application of CMCs in gas turbines of 5–15 MW power rating. The program addressed the development and testing of gas turbine hot section components such as stator vanes, turbine blades, and combustors. The material to be used was CERASEP, a SiC/SiC CMC made by chemical vapor infiltration by SEP [10].

A second EU EUREKA project was the Advanced Gas Turbines for Automobiles (AGATA) program. Volvo/United Turbine collaborated with automotive companies in Germany (Daimler Benz, BMW, Volkswagen AG) and France (Peugeot), a gas turbine manufacturer (Garrett), a French National R&D organization (ONERA), and a number of ceramic suppliers in Sweden, Germany and France. The aim of the project was to develop ceramic components and systems for automobiles and other applications. The ceramic components of interest were an axial turbine wheel and variable stator [10].

This program evolved into a second Swedish-French European AGATA EU funded program which was conducted from 1992–1998. This program which was a collaborative effort of a number of Swedish and French companies in the automotive, catalysts, and ceramic fabricating fields with the support of research institutes, had the goal to develop three components for a hybrid electric vehicle gas turbine with a 60 kW output. Volvo Aero Corporation was the leading Swedish gas turbine manufacturer. The components were a ceramic radial turbine wheel, a catalytic combustor and a ceramic heat exchanger. These components were evaluated in rigs. This was primarily a component development program. This AGATA program has been described in Chapters 22 and 23 of this volume.

FRANCE

The French National Aerospace Research Institute (Office National d'Etudes et Recherches en Aérospatiale—ONERA) played a leading role in the French ceramic gas turbine development related work. Starting in 1983 ONERA developed expertise in ceramic recuperators for fuel efficient gas turbines for automotive and aircraft turbine applications. ONERA's expertise was in the design and testing of the recuperators. The fabrication was mostly carried out by the fabricator Céramiques et Composites (C&C) and its spin-off, Boostec Industries. In 1983 ONERA and C&C embarked on a program to develop a tubular ceramic recuperator for a turbine cycle for aircraft engines. The heat exchanger was initially fabricated from SSiC, but later Sialons were used because of their superior thermomechanical resistance. For the AGATA program ONERA participated in the design and testing of the heat exchanger, catalytic combustor and turbine wheel.

ONERA also participated in a collaborative development program with the Research and Technological Institute for Power Engineering (NITI EM) in St. Petersburg, Russia. The collaboration involved theoretical studies of ceramic heat exchangers for automotive and industrial gas turbines. ONERA also contributed to the development of the SOUGRAV high temperature ceramic materials and components. The ONERA development work has been described in Chapter 23 of this volume.

ITALY

Italy is a newcomer to the ceramic gas turbine development arena. Its activities are primarily at Nuovo Pignone, an industrial gas turbine manufacturer which is a subsidiary of General Electric of the USA. It is currently leading the development of stationary ceramic components (combustor, shroud) under a four year program funded by the European Union that started in the fall of 1998. It is envisioned that oxide-based CMC components will be tested in Nuovo Pignone's 2 MW GE-2 and 5 MW GE-5 (formerly PGT-2 and PGT-5) engines. Other participants on the program are a ceramic materials supplier (Dornier, Germany), a research laboratory (Joint Research Center of the EU, The Netherlands), and the University of Patras (Greece). Nuovo Pignone has also tested SiC/SiC CMC shrouds in the GE-2 engine. A short summary of some of the Nuovo Pignone activities is included in the General Electric chapter in this volume (chapter 9).

SUMMARY

Table 20.1 summarizes the major ceramic gas turbine programs in Western Europe, together with their objectives, targets, plateaus reached and challenges remaining. Details for most of these major programs can be found in the chapters mentioned previously.

TABLE 20.1 SUMMARY OF MAJOR CERAMIC GAS TURBINE PROGRAMS IN WESTERN EUROPE

Programs Time frame	Major Participants	Major Objectives	Targets	Plateaus Reached	Key Challenges Remaining
United Kingdom					
CMCs for aircraft GT's 1985–2000	Rolls-Royce	Apply CMCs to flameholders, combustors, nozzles, compressor casings	Improve component durability	* SiC/Al_2O_3 CMC flameholder and combustor tested in rigs and engine * Oxide/oxide combustor tiles survived * SiC/SiC nozzles tested for several hundreds of cycles	* Thermal insulation spalled from combustor * Cost oxide fibers
CMCs for combustors 1997–2000	Alstom, JRC, Ansaldo, MYDSYSYS	CMCs for combustors, develop CMC design, analysis, lifing methods	Improve componeng durability	* SiC/SiC CMC + TPS tested in rig	* Improve durability of TPS which cracked
Germany					
BMFT Ceramic Components for Vehicular Gas Turbines 1974–1983	Daimler-Benz, MTU, Volkswagen, Suppliers, Universities	Develop high performance for automotive gas turbines	1350°C TIT ceramic rotor, nozzle, combustor, scroll, recuperator	* Components were tested for hundreds of hours in rigs and engines * HPSN and RBSN strength doubled in 9 yrs * Mercedes-Benz car with ceramic gas turbine covered 500 hours of road testing at 1250°C TIT and 61,500 rpm	* Improve component reliability (strength, Weibull modulus, fracture toughness) * Long term durability and reliability at 1350°C TIT, cost effective fabrication methods
BMFT Car 2000 1978–1982	Daimler-Benz	Demonstrate ceramic gas turbine in road test	1350°C TIT ceramic gas turbine road test	* 58 hours at 1350°C TIT and 61,500 rpm in gasifier rig * Stuttgart-Brussels (600 km) test drive	* Demonstrate long term durability and reliability at 1350°C TIT, cost effective fabrication methods
BMFT Structural Ceramics 1983–1986	Daimler-Benz et al.	Develop mass production process for rotor	1350°C TIT 70,000 rpm	* GTE injection molded SSN rotor spin tested at 114% of design speed * Kyocera slip cast SSN rotor: 50% survived at 90,350 rpm * 65 hours in engine test – 23.5 hours at 90%+ load	* Long term durability and reliability, meet cost targets
Ministry of Defense "Metal-Ceramic Hybrid Nozzle (Feasibility Study)" 1980–1982	MTU	Develop and test cooled hybrid metal-ceramic (SSiC) nozzle for jet engine apps.	1500°C+ TIT	* Nozzle cascade testing for 300 hours up to 1627°C * Cyclic testing for hundreds of cycles	* Meet long term durability and reliability goals
"High Temperature Turbine Nozzle" 1985–1989	MTU	Develop cooled hybrid metal-ceramic (SSiC) blade			

Combustor development 1992–1997	ABB	Develop ceramic combustor for 160–240 MW engines	1500°C TIT	* 50 hour test – 35 hours at full load of Hexoloy SA tiles	* Reduce surface recession
Combustor development	Siemens	Improve ceramic combustor durability of Siemens large engines	1500°C TIT	* Established high-alumina technology with >8 million hours experience	* Develop improved combustor liner technology for 1500°C TIT
KHD T216 (~70 kW) Combustor and cooled nozzle development – 1990's	University of Karlsruhe	Develop durable ceramic combustor		* SiSiC segmented combustor survived 150 hours/1027°C max. liner temp. * SiC/SiC CMC combustor survived 90 hours/1050°C max. liner temp * SSiC nozzle evaluated in cyclic rig testing	* Demonstrate long term durability and reliability
Sweden					
STU Swedish national program 1978–1980's	United Turbine ASEA Cerama	Develop 100 kW automotive GT	1370°C TIT	* Road test of Volvo car with MK1 engine with ceramic rotor and recuperator in 1982	* Long term durability and reliability and component cost
AGATA program 1992–1998	Volvo Aero Corp., AC Cerama French companies	Develop rotor, catalytic combustor, recuperator for 60 kW hybrid vehicle GT	1350°C TIT	* Catalytic combustor was tested for 100 hours at TIT in excess of design TIT – NOx emissions <4 ppm * Max. burst speed of rotor in cold spin test: 91,000 rpm	* Demonstrate long term durability and reliability, improve thermal shock resistance of mullite/zirconia substrate * Meet design target of 125,000 rpm
France					
DRET Ceramic recuperator development 1983–2000	ONERA C&C	Develop ceramic recuperators for automotive and aircraft GT's		* Various recuperators were evaluated and theoretical studies were conducted	
AGATA program 1992–1998	ONERA C&C Other French and Swedish companies	Develop ceramic recuperator for 60 kW hybrid vehicle GT	1011°C gas inlet temp., 0.26 kg/s mass flow	* Full scale and quart scale recuperators were tested at design conditions	* Leakage between matrix and casing needs to be improved
Italy					
Stationary gas turbine development 1990's	Nuovo Pignone Dornier, JRC, U. of Patras	Develop combustor and shroud for stationary gas turbine	870°C shroud temperature	* SiC/SiC MI CMC shroud was tested in PGT-2 engine for over 100 hours	* Demonstrate long term durability and reliability

ACKNOWLEDGEMENTS

The author wishes to acknowledge the support of Mr. Gerhard Andrees who provided information on the development programs at MTU, and the support of Dr. Eberhard Seitz of the Research Center Jülich Ltd. who reviewed the information on funded German ceramic gas turbine related research and development.

REFERENCES

[1] Lumby, R.J., Hodgson, P., Cother, N.E., and Szweda, A., "Syalon Ceramics for Advanced Engine Components," SAE paper 850521, presented at the International Congress and Exposition, Detroit, Michigan, USA, February 25–March 1,1985.

[2] Seitz, E., "Advanced Ceramics for Structural Applications—R&D in the Federal Republic of Germany," in The Institute of Ceramics, *Proceedings 46, "Advanced Engineering with Ceramics,"* ISSN 0268-4373, 1990, pp. 71–83.

[3] "Keramische Komponenten für Fahrzeug-Gasturbinen" ("Ceramic Components for Vehicular Gas Turbines,") W. Bunk and M. Böhmer, eds., Project Status Workshop, 18 and 19 April, 1978, Bad Neuenahr, Germany, Springer Verlag, Berlin, 1978 (in German).

[4] "Keramische Komponenten für Fahrzeug-Gasturbinen II" ("Ceramic Components for Vehicular Gas Turbines II,") W. Bunk and M. Böhmer, eds., Project Status Workshop, 24–26 November 1980, Bad Neuenahr, Germany, Springer Verlag, Berlin, 1981 (in German).

[5] "Keramische Komponenten für Fahrzeug-Gasturbinen III" ("Ceramic Components for Vehicular Gas Turbines III,") W. Bunk, M. Böhmer, and H. Kiβler, eds., Project Status Workshop, 13–15 February 1984, Germany, Bad Neuenahr, Springer Verlag, Berlin, 1984 (in German).

[6] Bunk, W., "Overview on the German Program on Ceramic Components for Vehicular Gas Turbines," in *Proceedings of the First International Symposium on Ceramic Components for Engines*, Shigeyuki Sōmiya, Eizo Kanai, and Ken-ichiro Ando, eds., Hakone, Japan, October 17–19, 1983, pp. 9–20.

[7] Bunk, W., Gugel, E., and Walzer, P., "Overview of the German Ceramic Turbine Program," in "Ceramics for High-Performance Applications—III Reliability", Proceedings of the Sixth Army Materials Conference," Edward M. Lenoe, R. Nathan Katz, and John J. Burke, eds., Plenum Press, New York, 1983, pp. 29–50.

[8] Andrees, G., MTU, personal communication.

[9] Kronogård, S.O. and L. Malmrup, "Ceramic Engine Research and Development in Sweden," in *Ceramics for High-Performance Applications—III Reliability*, Proceedings of the Sixth Army Materials Conference," Edward M. Lenoe, R. Nathan Katz, and John J. Burke, eds., Plenum Press, New York, 1983, pp. 51–80.

[10] Pejryd, "National Program in Sweden for Engine Ceramics," *Proceedings of the Third International Symposium on Ceramics Materials and Components for Engines*, V.J. Tennery, editor, Las Vegas, Nevada, USA, November 27–30, 1988, pp. 1508–1514.

CHAPTER

21

German Automotive Ceramic Gas Turbine Development

Mark van Roode

Solar Turbines Incorporated, San Diego, California, USA

INTRODUCTION

The drive towards developing gas turbines for vehicular applications originated in the 1960s and 1970s in the USA, Europe and Japan, primarily out of a desire to lower the emissions of NOx, CO, and unburned hydrocarbons (UHCs) from automotive engines. The gas turbine has significantly lower emission levels than the internal combustion engine, because it is based on a continuous combustion process. Improving gas turbine performance became a secondary goal.

Figure 21.1 is a flowchart illustrating the operation of an automotive gas turbine [1]. Air is compressed in a radial compressor from atmospheric to about 5 bar, after which it is pre-heated in a heat exchanger utilizing the exhaust waste heat and subsequently introduced into the combustor. A part of the compressed air is used to burn the fuel in the combustor. Another part of the air cools the combustor gases down to about 1127°C (1300 K) before these gases enter the turbine section. About two thirds of the useful energy from the combustion gases is used to drive the gasifier turbine and the compressor. The remaining one third drives the power turbine which in turn powers the vehicle. The residual heat leaving the exhaust enters the heat exchanger, typically a rotating regenerator. The continuous combustion process takes place with an excess of air. The load is varied by controlling the temperature of the combustion gases and by adjusting the positions of the nozzles in the compressor and the turbine. The term "gas" turbine is generic since for automotive applications the fuel is usually liquid.

Industry had a strong interest in the automotive gas turbine and the car manufacturers funded programs for its development. As a result a number of gas turbine powered vehicles were operating in various parts of the world on a pilot basis. In the Federal Republic of Germany (FRG) Volkswagen AG (VW) developed two all-metal research gas turbines, the GT 70 and GT 150, which were demonstrated in a car. Motoren Turbinen Union (MTU) developed an all-metal gas turbine (MTU-7042) for trucks which was also operated in road tests. Daimler-Benz AG (DB) pursued automotive gas turbine R&D as well.

While these early metallic gas turbines had a number of advantages compared to gasoline and diesel powered engines their fuel efficiency was inferior because the metal component temperatures could not exceed

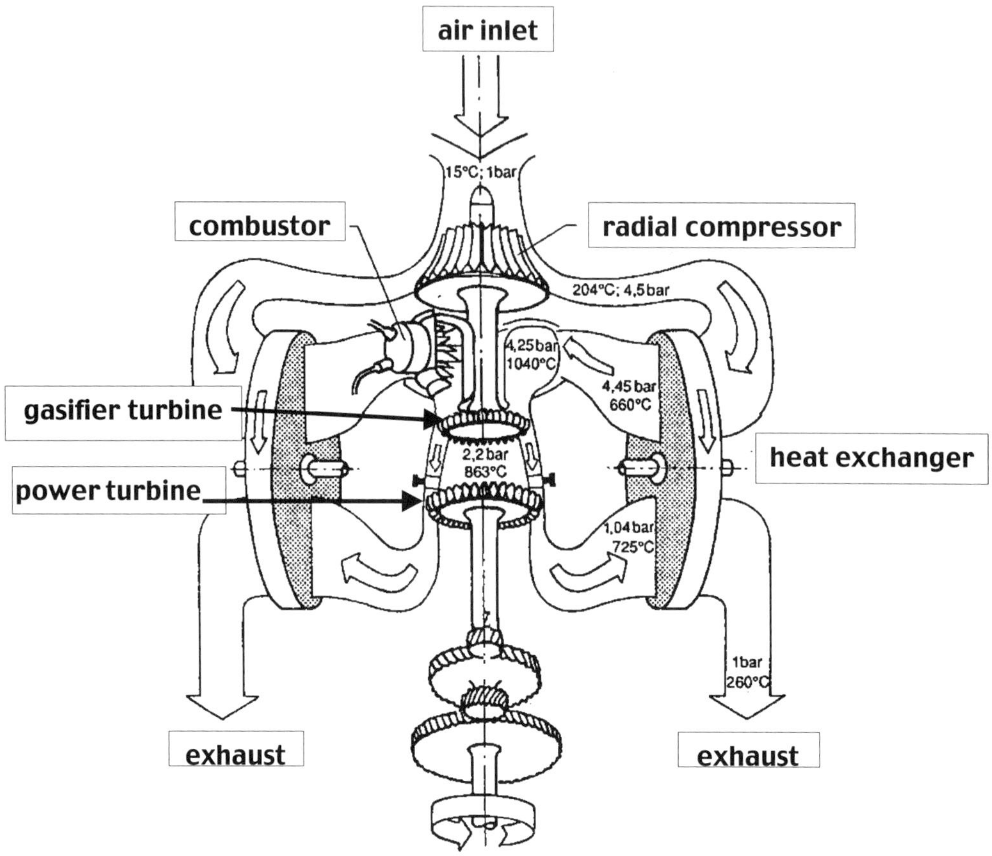

Figure 21.1 Flowchart for an Automotive Gas Turbine (Adapted with permission from Ref. [1], Figure 1–9)

~1000°C. Enabling operation at higher turbine inlet temperatures (TIT) by cooling of the components was not an option because of their small size. However, by replacing the metal components with ceramic parts, gas temperatures could be increased by about 300°C with significant savings in fuel consumption. It was estimated that by increasing the turbine inlet temperature (TIT) from 1027°C to 1327°C (1300 to 1600 K) the theoretically achievable fuel efficiency could be increased from about 31 to 40% [1] which makes the gas turbine competitive with conventional automotive engines. The gas turbine therefore became the focal point for the advanced material and power generating equipment development policies in the FRG and other developed nations. The goal became to improve its performance by operating it at high firing temperatures through the incorporation of high temperature ceramics as structural materials for uncooled hot section components such as combustor liners, nose cones, rotating blades and disks, stationary vanes, and as heat exchanger elements.

The ceramic gas turbine development efforts in the FRG culminated in a large national program which was conducted between 1974 and 1983. This program, entitled "Keramische Bauteile für Fahrzeug-Gasturbinen" ("Ceramic Components for Gas Turbines), was initiated by the Federal Ministery for Research and Technology (Bundesministerium für Forschung and Technologie—BMFT) and managed by the German Aerospace Research Establishment (Deutche Forschungs-und Versuchsanstalt für Luft- und Raumfahrt—DFVLR). The development efforts were structured in three phases (Phase I: 1974–1977; Phase II: 1977–1980; Phase III: 1980–1983). The participants included car manufacturers (VW and DB), an engine

TABLE 21.1 ACTIVITIES OF THE PARTICIPANTS FOR THE BMFT PROGRAM

	Combustor	Stator	Rotor	Heat Exchanger
Materials suppliers	Annawerk Rosenthal SIGRI	Annawerk Rosenthal SIGRI	Annawerk Degussa ESK Feldmühle	Rosenthal
Gas turbine and engine manufacturers	MTU Volkswagen	MTU Volkswagen	Daimler-Benz MTU Volkswagen	Daimler-Benz

TABLE 21.2 ACTIVITIES OF RESEARCH INSTITUTES

Institute	Activities
Max-Planck-Institut – Stuttgart	Basic research
Technical University – Berlin	Powder technology
Technical University – Clausthal-Zellerfeld	Joining
Fraunhofer – Izfp Saarbrücken	Nondestructive testing
Technical University – Karlsruhe	Creep and oxidation properties
DFVLR Cologne	Material characterization
	Project management

manufacturer (MTU), German ceramic materials and component suppliers, and research institutes. Details on the program's progress can be found in the proceedings from program reviews [2–4], in contractor progress reports [5–9], and in summaries presented at conferences [10–12]. The companies' work was 50% funded by the FRG government and 50% by industry, while the research institutes were fully funded. Cumulative program spending was about 100 million DM [10].

The car and engine manufactureres had their own objectives. DB had targeted the development of an all-ceramic 110 kW passenger car gas turbine. VW was developing a 50–100 kW gas turbine for a passenger car with ceramic components. And MTU was interested in a truck gas turbine with ceramic parts in the 250–300 kW range. The DB project probably received the highest visibility since it actually demonstrated several automotive ceramic gas turbines in road tests. Its activities continued beyond the BMFT program up till about 1990. VW and MTU demonstrated the ceramic components in rigs but not in actual engines in road tests. Their automotive ceramic gas turbine activities were concluded when the BMFT program ended in 1983.

Each of the BMFT program participants focused on a limited number of components. The gas turbine and engine manufacturers performed the component design and testing (some of them also became involved in ceramic fabrication), the ceramic suppliers focused on processing and component fabrication, and the laboratories and universities provided support in the areas of materials property evaluation, development of testing methodologies, and life prediction. Table 21.1 lists the program participants and their components, and Table 21.2 lists the research institutes providing technology support [13].

At the early stage of the development there were many challenges. First, there was the materials aspect. Materials processing and fabrication technologies had to be developed for components for engines with up to 1350°C operating capability, using multiple fuels. Novel conceptual designs for demanding dynamic and temperature conditions had to be developed using the latest analytical design methodologies. Furthermore, these designs had to reconcile the properties of the interfacing metallic and ceramic components in the gas turbine. In this chapter the development efforts of VW, DB, and MTU will be described by focusing primarily on the combustor, nose cone, nozzle ring, rotor and heat exchanger. The work of the German ceramic component suppliers is described in Chapter 9 of Volume II of this text. The program will be referred to as the "BMFT" program for the purpose of the discussion.

CERAMIC COMBUSTOR DEVELOPMENT

The ceramic combustor was the key component for lowering exhaust emissions in the gas turbine. The combustor walls cannot be cooled because of the high temperature of the pre-heated air entering the combustor from the heat exchanger. A temperature gradient is avoided by cooling the combustion gases after they pass from the hot primary zone to the dilution zone. Incorporating an uncooled ceramic combustor saves cooling air which can be used to lean out the primary zone. This, in turn reduces NOx because of lower flame temperatures, and lowers CO because of the "hot" uncooled combustor wall.

MTU and VW were responsible for ceramic combustor development and they jointly presented their work at the first program review in Bad Neuenahr, Germany [14], but the combustors for passenger cars (VW) and trucks (MTU) were different since the load profiles varied. Low fabrication cost was the primary driver for cars, especially for complex low emission combustors that are subject to extreme cyclic loading. For trucks, durability, crack resistance and simple cost effective design are important. MTU's design guidelines for the ceramic combustor specified a small volume in the primary zone, no sudden changes in surface profile, an even and thin liner, and adequate distance between air holes. The MTU combustor was designed for a gas turbine with a heat exchanger, with a TIT between 1277°C and 1327°C, and a pressure and temperature at the compressor exit of 5 bar and 827°C to 836°C, respectively [15]. The VW engine had a slightly higher TIT of 1350°C. MTU continued the work into Phases II and III of the German national program [15, 16]. A comprehensive overview of the MTU combustor work in the English language is available [17]. The combustor rig testing for all three program phases has been summarized in Table 21.3.

Phase I development (1974–1977)

VW and MTU evaluated cylindrical and conical liner designs (see Figures 21.2 and 21.3). The conical design had the advantage that it minimized the distance between the combustor wall and the flame zone and thereby prevented the emergence of higher temperature gradients along the walls. The design envisioned the liner wall to be as thin as possible [14].

Rig Testing. The combustor performance and durability at the conditions that are typically encountered in vehicular operation (cold starts, stationary operation, acceleration/deceleration, stop after hot running) were evaluated in hot rig tests. Figure 21.4 shows the schematic for a hot gas rig that MTU used to test stationary parts including combustors. Figure 21.5 shows MTU's hot gas rig in operation. A *stationary* test without thermal shock evaluated the capability of the ceramic combustors to survive the temperature gradient from the primary zone to the combustor exit. It was observed that combustors made of sintered silicon carbide (SiSiC) withstood the rig test without crack formation while combustors made of reaction bonded silicon nitride (RBSN) were prone to cracking at elevated temperatures [14, 17].

Subsequently, the ceramic combustors were subjected to a *cold start,* a condition that is accompanied by an extremely severe thermal shock since the temperature rise from ambient to the flame temperature takes only fractions of a second. A number of slipcast SiSiC combustors exhibited thermal cracking, but isostatically pressed SiSiC combustors withstood the thermal shock treatment [14, 17].

The next step was to experiment with various *stationary* engine operating conditions, such as no load idling and operation under full load. Figure 21.6 shows the temperatures of the cooling air, the flame and the outer wall of a VW combustor during test. A temperature peak occurred in the zone where the combustor diameter had a sudden expansion. A little further downstream the temperature dropped, because of the stabilizing effect of the wider combustor chamber. The combustor withstood a temperature gradient of up to 300°C without damage. The temperature change in the MTU combustor was much more gradual. The combustor was designed to prevent high temperature gradients. The combustion in the vicinity of the second row of liner holes was mostly completed and the heat transfer to the combustor wall increased as a function of the inlet cone design [14, 17, 18].

TABLE 21.3 SUMMARY OF COMBUSTOR RIG AND ENGINE TESTS AT MTU AND VW [14–16]

Test Type	Comp.	T_{inlet}(°C/K)	T_{exit}(°C/K)	Pressure (bar)	Comments
PHASE I (1974–1977)					
Stationary tests without thermal shock	VW(I)	10 min to 527 (800) (80°C/min)	3 min to 727 (1000) 3 min to 957 (1230) 45 min at 957 (1230) slow shutdown	2.8	SiC combustors: no cracks
	MTU	777(1050) (procedure similar to VW)	1327 (1600) (procedure similar to VW)	1.0–4.9	SiSiC combustors: no cracks RBSN combustors: cracks at elevated temperatures
Cold start tests	WV(II)	As stationary test—temperature increase from room temperature to operating in fraction of seconds			Slipcast SiSiC combustors: cracks Isopressed SiSiC combustors: no cracks
Simulated Engine operation Cyclic tests	VW MTU	632 (905) 607 (880)	927 (1200) 1367 (1640)	2.8 1.0	No results reported – proof test to screen combustors for cyclic tests
	VW(III)	As in Series I/II	10 cycles: 627 (120s)->1227 (120s) (900–1500)	1.6	VW(III)-combined with VW(I,II) Isopressed SiSiC (Refel): cracks
	VW(IV)	4hour cycle: 27(300)(30min)->527(800)(30min) ->927(1200)(1hour)->527(800)(30min) ->927(1200)(1hour)		4.5	Isopressed SiSiC (Refel): cracks Slipcast SiSiC: cracks
	MTU	607 (880)	50 cycles: 827(60s)->1327(60s) (1100->1600)	1.02	Condition of parts not reported
		607 (880)	1082 (1355)	1.0->4.9	Norton SiSiC: cracks SiSiC (Refel): no cracks
		607 (880)	200 cycles 792(60s)->1082(60s) (1065->1355)	4.9	SiSiC (Refel): fine crack between holes in 2nd row – fine cracks at inner wall
Engine test		618 (891)	982 (1255)	3.67	SiSiC (Refel) combustor: cracked because of mechanical overload
PHASE II (1977–1980)					
50 hour proof test	MTU	607 (880)	90 min. ramp up to 5 bar, 30 min. hold at 5 bar and 897 (1170), Cycle I: 897°C->1157°C Cycle II:897°C->1277°C Fuel cutoff: 1307->827 (1580->1100)	1–5	SiSiC (Refel) combustor lasted 685 cycles (44 hours) of cycle I tests – and was damaged during ramp up to cycle II SiSiC combustor (Rosenthal) survived 110 cycles (5 hours) of cycle II tests – test was stopped because of damage to nozzle ring
Start-stop test		27 (300)	1000 start-stop cycles to 1127(1500)-500 hours	1.0	SiSiC combustor (SIGRI) survived 500 hour cyclic test
PHASE III (1980–1983)					
Start-stop test	MTU	27 (300)	100 start-stop cycles to 1352 (1625)-50 hours	1.0	SiSiC (Rosenthal, SIGRI, BNFL) and SSiC (Annawerk) combustors were tested
Simulated engine tests		837 (1110)	1287 (1560)	4.9	SiSiC combustors: overheating, cracks SSiC combustors: cracks, spallations

Figure 21.2 Volkswagen Combustor Liners: From Left to Right: SiSiC, Slipcast—SiC-mullite (saturated), isostatically pressed—SiSiC (Refel), isostatically pressed (Reprinted with permission from Volkswagen AG)

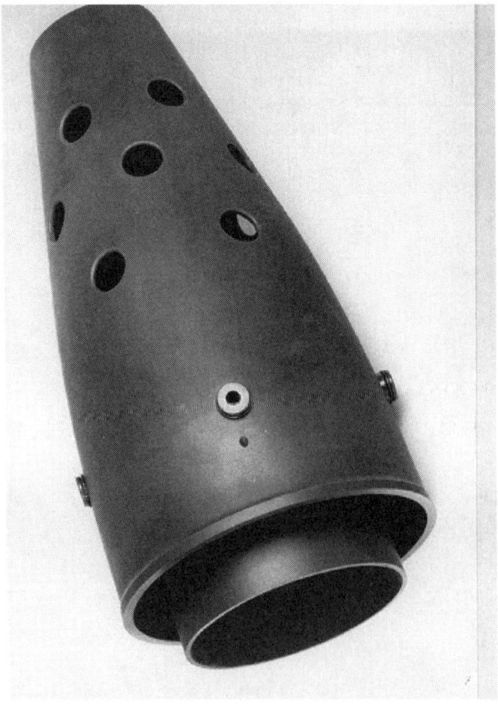

Figure 21.3 MTU combustor liner: SiSiC (Refel)—Isostatically Pressed, Configured with Shroud and Nose Cone (Reprinted with permission from MTU)

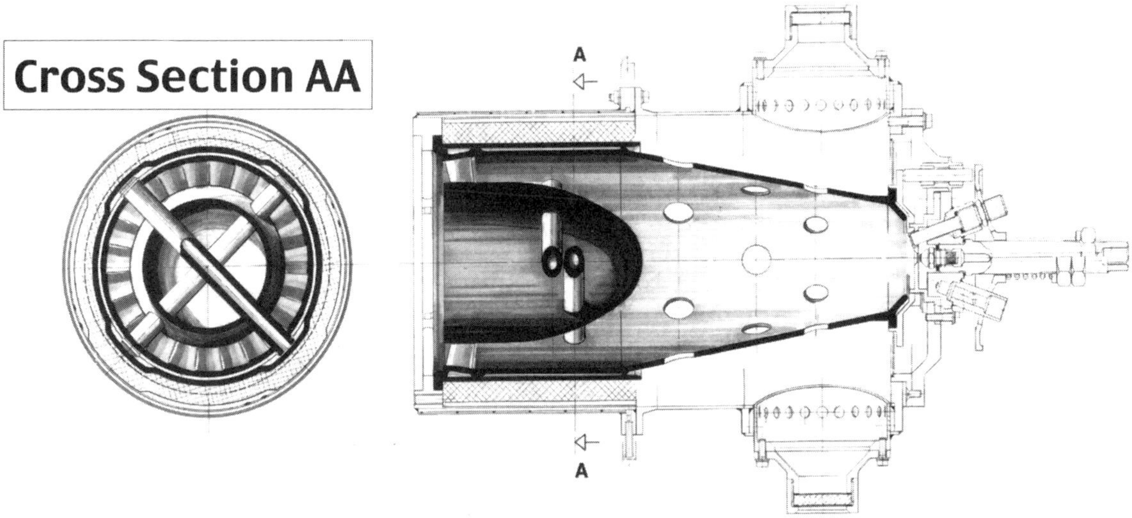

Figure 21.4 Schematic of MTU's Ceramic Stationary Component Rig (Reprinted with permission from MTU)

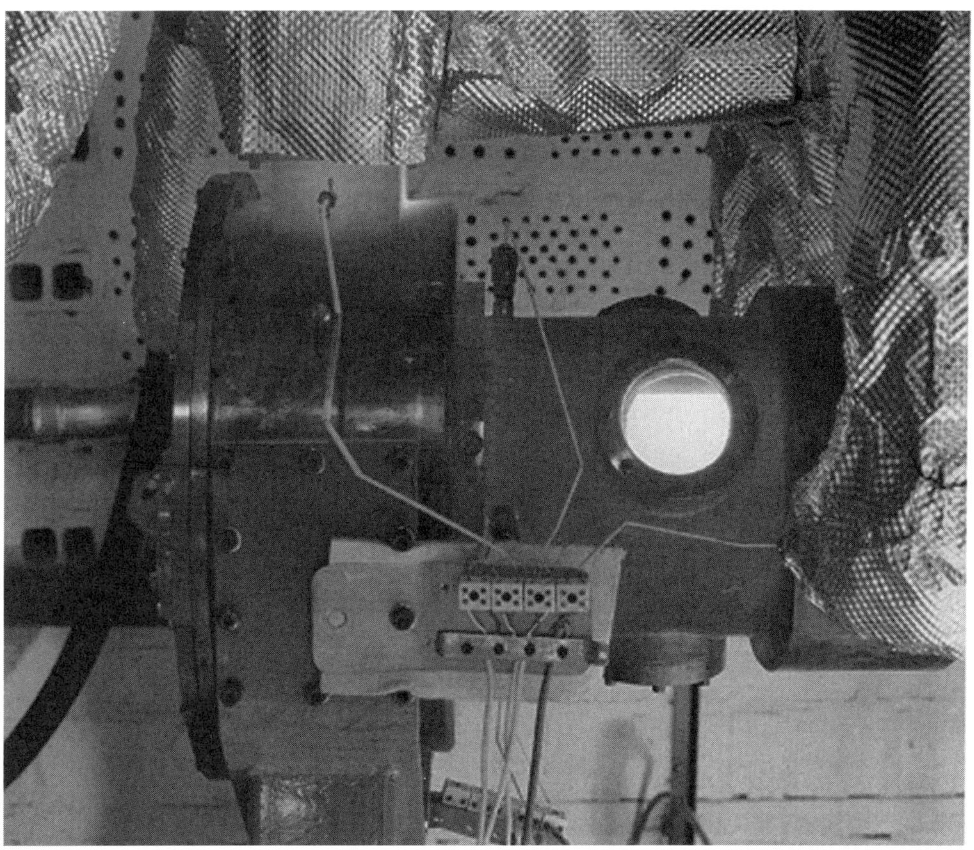

Figure 21.5 Hot Gas Test Rig for Stationary Ceramic Components (Reprinted with permission from MTU)

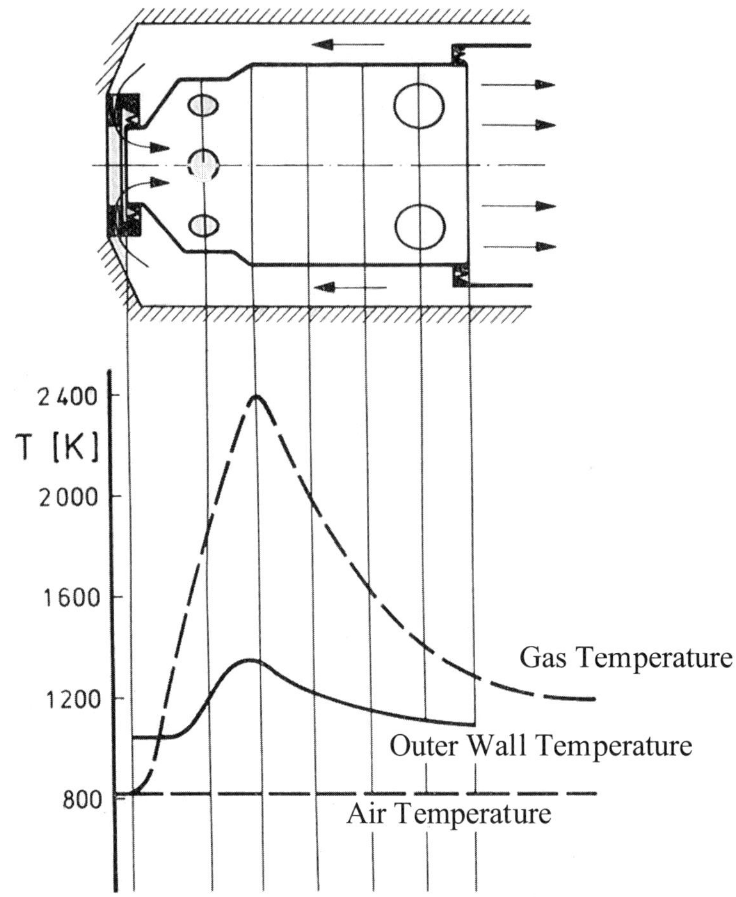

Figure 21.6 Temperature Distribution Over the Length of a VW Ceramic Combustor (Adapted with permission from Ref. [18])

At MTU the SiSiC combustors that survived the stationary tests and the cold start tests without cracking were then subjected to *cyclic* testing. Conditions included temperature cycling at atmospheric pressure and testing at constant inlet and exit temperatures but with stepwise increases in pressure from 1.0 to 4.9 bar. Norton combustors showed cracks originating from the first row of holes while BNFL (Refel) combustors did not show cracking. Silicon nodules were observed at the hot surface of both types of combustors. A Refel combustor was subjected to 200 test cycles between 792°C and 1082°C at the full pressure of 4.9 bar. The combustor was intact upon completion of the test, but there was a fine crack between two adjacent holes. Apparently, the crack did not interfere with the performance of the combustor. The inner wall was discolored and showed strong separations of Si next to a few fine cracks. The test showed that as long as temperature peaks were avoided along the surface of the combustor, cyclical temperature loads could be tolerated under full load conditions. The Refel SiSiC combustor following the test is shown in Figure 21.7 [14, 17].

At VW three conditions (cold start, stationary operation, cyclic operation) were combined in a single simulated engine test. A particular severe cyclic test at VW was conducted at a pressure of 4.5 bar, and incorporated a temperature jump from 27°C to 927°C. Isostatically pressed SiSiC (Refel) survived the cold start at 4.5 bar without damage. However, both isostatically pressed and slipcast SiSiC combustors were observed

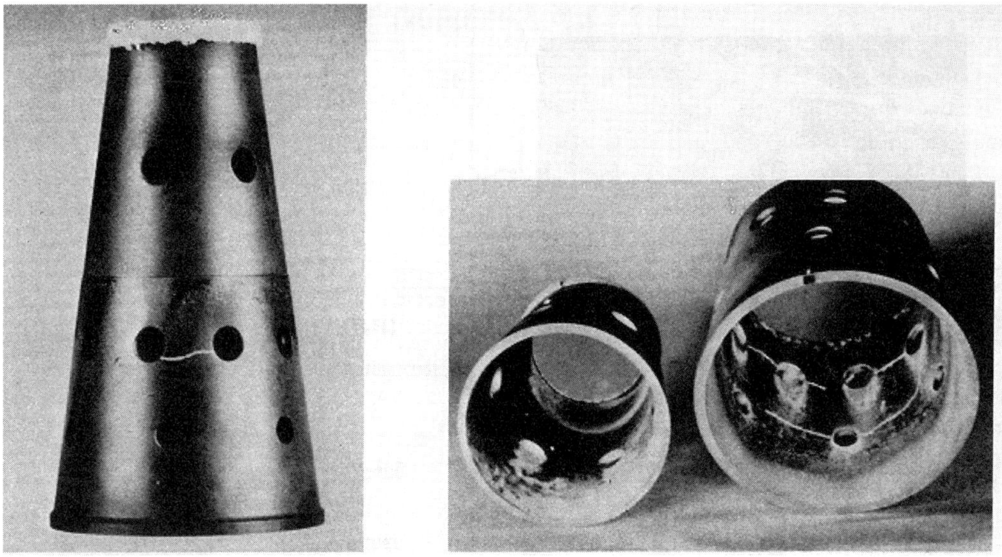

Figure 21.7 Refel SiSiC MTU Combustor Following 200 Cycle Test (Reprinted with permission from Refs. [14, 17])

to be cracked after the test. In both combustors the cracks originated at the holes, and in the case of the slip-cast combustor the cracks followed more or less the peak surface temperatures [14, 17].

Engine Testing. To investigate the performance of ceramic materials in an engine a partial ceramic combustor was incorporated into a MTU-7042 gas turbine. The combustor consisted of a metallic dome, a ceramic SiSiC (Refel) cylindrical part, and a metallic transition piece. Refel had performed best in the MTU rig tests. The ceramic part was sealed against the metal components which carried the mechanical loads via attachments that were placed in tension. The air inlet temperature was 618°C, and the combustor exit temperature was 982°C. The exit pressure was 3.67 bar. Cracks in the ceramic section observed during the engine assembly were attributed to the sealing at the transition between the metal and ceramic parts, presumably because of the difference in thermal expansion between these components. It was felt that this problem was solvable with an appropriate design. VW had similar experiences with a mullite soaked SiC combustor. In spite of the cracks the combustor remained functional throughout the test. A 13-point California test cycle was driven and the emissions were measured at the turbine exit. The emissions were significantly below the California guidelines [14, 17].

Phase II Development: 1977–1980

The development work reported at the second Bad Neuenahr seminar focused primarily on the MTU conical combustor design (wall thickness of 4 mm). Notwithstanding the initial unfavorable results with Si_3N_4 combustor materials, MTU introduced RBSN to be evaluated alongside with SiSiC into its development program.

The MTU hot gas rig incorporated not only the combustor but also a ceramic transition piece and nozzle ring, which were all evaluated together in the test. These components had all been screened in a 50 hour proof test prior to the rig test. To simulate the heat exchanger, air was preheated to 897°C and pressurized to 5 bar. After ignition and ramp-up to the design combustor inlet temperature of 607°C, the pressure was increased stepwise from 1 to 5 bar at constant air inlet temperature over a 90 min. period, and the load was then maintained for 30 minutes. The rig was subsequently operated in one of two modes, both cycling between no-load idle (1 min.) and full load (1 min.). The difference between the two cycles (ΔT) was 260°C in cycle I and 380°C in cycle II. The fuel was suddenly cut off at the end of the test to simulate a rapid drop in temperature for the nozzle ring from 1307°C to 827°C.

Initial testing resulted in severe damage to the combustor primary zone, which followed the contour of a soot deposit. Following improvements to the combustion system a BNFL SiSiC (Refel) combustor was tested for 44 hours, accumulating a total of 685 cycles from no-load idle to full load (cycle I) at 4.9 bar after the initial start-up cycle. During the ramp-up to cycle II cracking was observed and the combustor was replaced with a new SiSiC combustor from Rosenthal. An additional 110 cycles was run under cycle II conditions (the pressure had been lowered to 3.9 bar) over a 5 hour period with this combustor, but the test had to be interrupted because of damage to the nozzle ring. In parallel, a SiSiC combustor fabricated by SIGRI was tested for 1000 cycles and 500 hours in an atmospheric rig test, the longest component test at MTU [16].

Phase III development (1980–1983)

The limitations of the SiSiC material became fully evident in Phase III. SiSiC liners fabricated by Rosenthal, SIGRI and BNFL first underwent the standard atmospheric rig screening test of 100 start-stop cycles over 50 simulated engine hours at a combustor exit temperature of 1352°C. The combustors were subsequently tested in the stationary component test rig under simulated engine conditions at a pressure of 4.9 bar, an air inlet temperature of 837°C, and a combustor outlet temperature of 1287°C. All of the SiSiC combustors showed evidence of overheating in the primary zone area after one half to one hour of testing. Localized melting of free silicon at the inner wall and cracks between the holes in the first row were observed. These tests revealed that under conditions of full load a highly intense energy conversion took place in the primary zone whereby the temperature of the inner wall eventually exceeded the melting point of free silicon (~1417°C) [15].

The many failures during the combustor testing can be appreciated by realizing that the materials used at the time still had relatively low strength. For example, the flexure strengths of SiSiC and RBSN were often below 300 MPa, and the Weibull moduli were typically below 12. Because of these observations further development work was carried out with pressureless sintered silicon carbide (SSiC: Annawerk). The thermal limit of this material for service was between 1627°C and 1727°C. Cracks formed in the primary zone of the first SSiC combustor during ultrasonic drilling of the holes. The nature of the cracks indicated high characteristic stresses in the component. A second component was machined without any defects and tested in the component test rig. Evaluation under atmospheric pressure showed no evidence of damage. During testing under engine conditions (5 bar) cracks and small spallations were observed in the primary and intermediate zones. It was concluded that the fabrication process of SSiC combustors had to be further improved to mitigate excessive stress levels [15].

CERAMIC NOSE CONE DEVELOPMENT

The nose cone forms the transition between the combustor exit and the nozzle section. Its temperature is less than that of the combustor liner in the primary zone since the mixing and secondary air lowers the temperature to that for the inlet into the nozzle ring. Figure 21.8 shows a sequence of ceramic components as they were inserted in MTU's demonstrator engine which consisted of the stationary components as well as a ceramic rotor. The fully assembled demonstrator engine can be seen in Figure 21.9. The nose cone design in Figure 21.10 shows that the inner cone is held inside the outer ring with two perpendicular radial bolts. This design enabled stress-free expansion of the parts of the component under conditions of unequal

Figure 21.8 Ceramic Gas Turbine Components for the MTU Program—From Left to Right: Metal Disk with Inserted Ceramic Blades, Nozzle Ring, Nose Cone, and Shroud, Combustor Liner with End Plate (Reprinted with permission from MTU)

Figure 21.9 MTU's Ceramic Demonstrator Engine (Reprinted with permission from MTU)

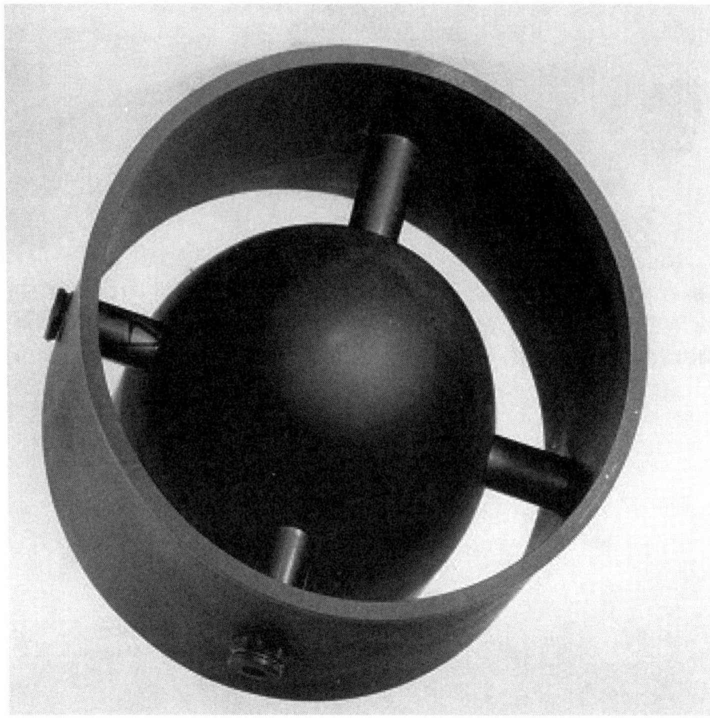

Figure 21.10 SiSiC Nose Cone in Bolted Version (Reprinted with permission from MTU)

temperature while assuring that the nose cone remains centered in the assembly. Figure 21.3 shows the nose cone configured with a MTU conical combustor. The materials performance showed similar trends as for the combustor. SiSiC performed significantly better than RBSN. Cones out of SiSiC (Refel) never failed during testing in combination with the combustors, accumulating several hundred hours of testing and several thousand start stop cycles [5, 6, 15].

CERAMIC NOZZLE DEVELOPMENT

MTU and VW were responsible for the development of a nozzle section for the automotive ceramic gas turbine. The two companies jointly presented their data at the first and second Bad Neuenahr seminar [16, 19].

The nozzle ring, a stationary part, is subject to relatively low mechanical loads, but the thermal stresses are considerable. Through careful design efforts lower strength ceramics such as RBSN, SiSiC, SSN, or SSiC could be used. Injection molding or slipcasting were used for forming of complex shapes with high density and low porosity. These parts did not require finishing.

The primary considerations for nozzles are thermal shock resistance and oxidation/corrosion resistance. The thin vanes are subject to gas temperature changes of 500°C or more. The long term strength and durability are mainly determined by the oxidation resistance and structural stability of the material under thermal loads. For materials with low density such as RBSN the porosity plays a role as well. Testing at VW showed a marked drop in strength with time and an increasing disruption of the grain structure that result in degradation of some of the silicon nitrides in the absence of external influences [19].

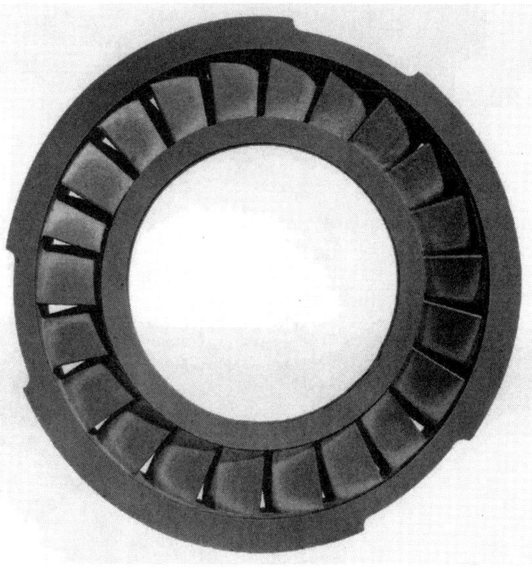

(a)

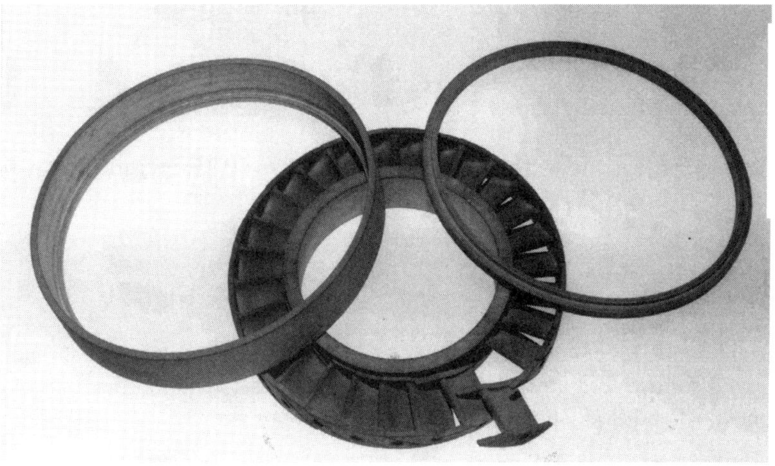

Figure 21.11 Three Different Nozzle Ring Concepts: (a) Single Shroud Integral Design (b) Dual Shroud Integral Design (c) Insertion Design (Reprinted with permission from MTU)

Phase I Development (1974–1977)

Design. Figure 21.11 shows parts for three nozzle assemblies based on different design concepts [5, 19, 20]. The integral single shroud design in Figure 21.11(a) was fabricated as a single piece by slipcasting of SiSiC. The inner shroud ring is an integral part of the inlet cone. In a variant of this design single shroud segments containing four vanes were also used. Figure 21.11(b) shows an integral design of inner and outer shrounds and vanes fabricated as a single piece by slipcasting RBSN or SiSiC. In the insertion design in Figure 21.11(c) separate hollow single vanes were loosely held between the outer and inner

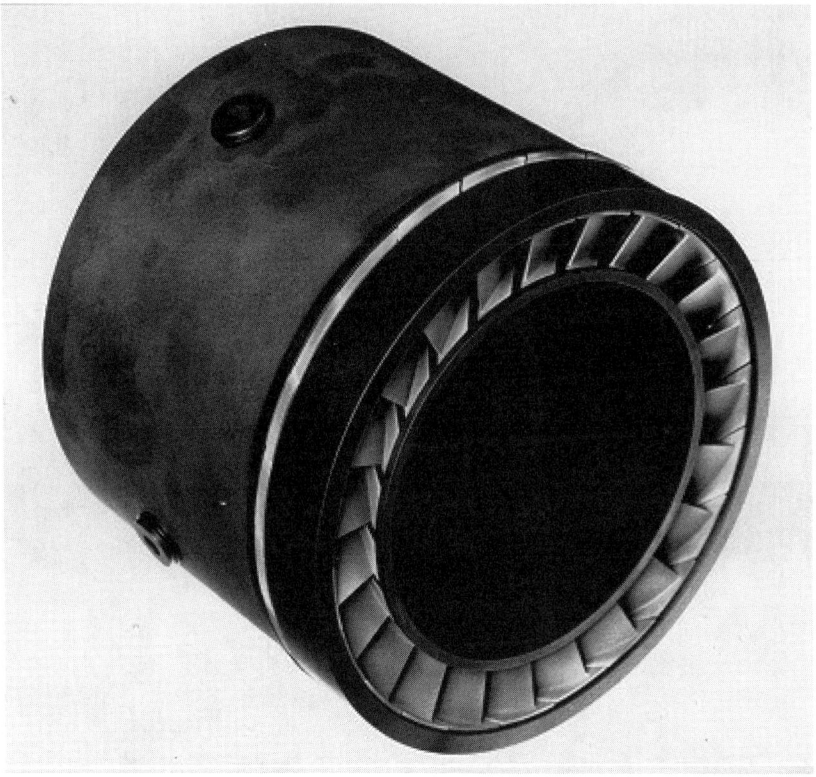

Figure 21.12 Nozzle Configured with the Nose Cone Shroud (Reprinted with permission from MTU, and from Ref. [15], Figure 8)

shroud rings. In a variant on the latter design the vanes were positioned in slots in the shroud rings. Figure 21.12 shows a nozzle ring configured with the nose cone shroud. Additional modifications of each of these designs were also conceived and evaluated.

Rig Testing [19, 20]. The gasifier turbine nozzle in a high temperature vehicular gas turbine will experience temperatures in excess of 1300°C during engine operation. Additionally, thermal stresses originating in uneven temperature distributions and mechanical loads, such as reaction forces from pressure drops and flow splits must be carried. Also, attachments can act as localized heat sinks where large temperature gradients build up. The highest thermal stress in the nozzle occurs during a cold start. The air in the engine which is still cold is rapidly mixed with the fuel and reaches a temperature of about 1230°C. The temperature profile in the combustor is very uneven since the recuperator delivers no heat to the air during a cold start. Later, temperature changes of 500°C/s take place during accelerations. When the engine is stopped sudden temperature drops occur which are not as high as during a cold start because the recuperator air is still hot.

Hot gas rig tests at VW and MTU were used to compare various nozzle rings for their thermal load carrying capability. However, realistic engine conditions were not simulated in the absence of pressure ratios and associated limited air flows. The nozzle ring was accommodated by springs, facilitating a suspension free from mechanical loads.

The test program for thermal shock and thermal cycling resistance covered a full load spectrum that included cold starts, accelerations, steady state operation, and shut off of the fuel supply. Figure 21.13 shows

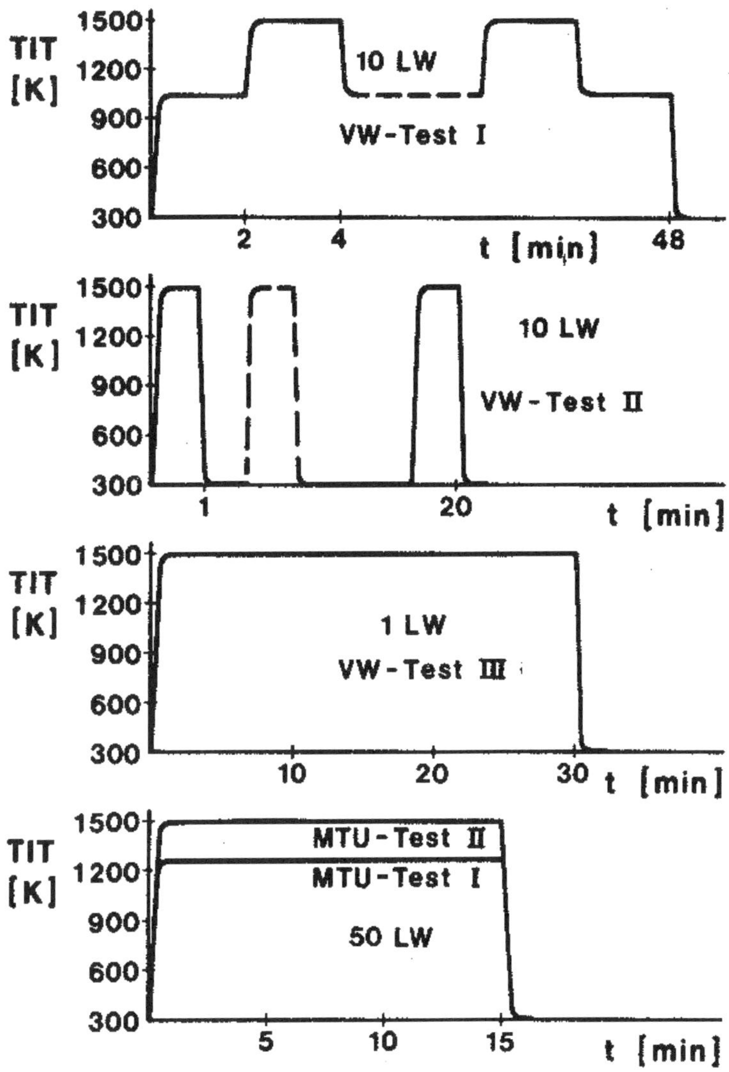

Figure 21.13 Test Procedures at VW and MTU (Reprinted with permission from Ref. [19], Figure 4)

the test procedures at VW and MTU. The two MTU tests totaled about 25 hours for each component with average combustor exit temperatures of 997°C (1270K) and 1247°C (1520K). In contrast, the tests at VW were purely short term tests that simulated cold starts, accelerations and stationary operation. The average turbine inlet temperatures (TITs) at VW were 777°C (1050K) and 1227°C (1500K). The non-uniform temperature distribution was not unlike what occurs during driving a car.

The independently performed tests at MTU and VW gave similar results. Typical failure patterns can be seen in Figures 21.14 and 21.15. A design with loosely inserted single nozzles, an integral inner shroud ring and a segmented outer shroud ring proved to be superior in testing at MTU. The 33 operating hours, included 73 starts up to 997°C, and 15 additional starts up to 1247°C. The vane cracks initiated at the area of high tensile stress in the trailing edge, predicted by Finite Element (FE) analysis. This design survived for only 7.5 hrs. in testing at VW (see Figure 21.14 (a)).

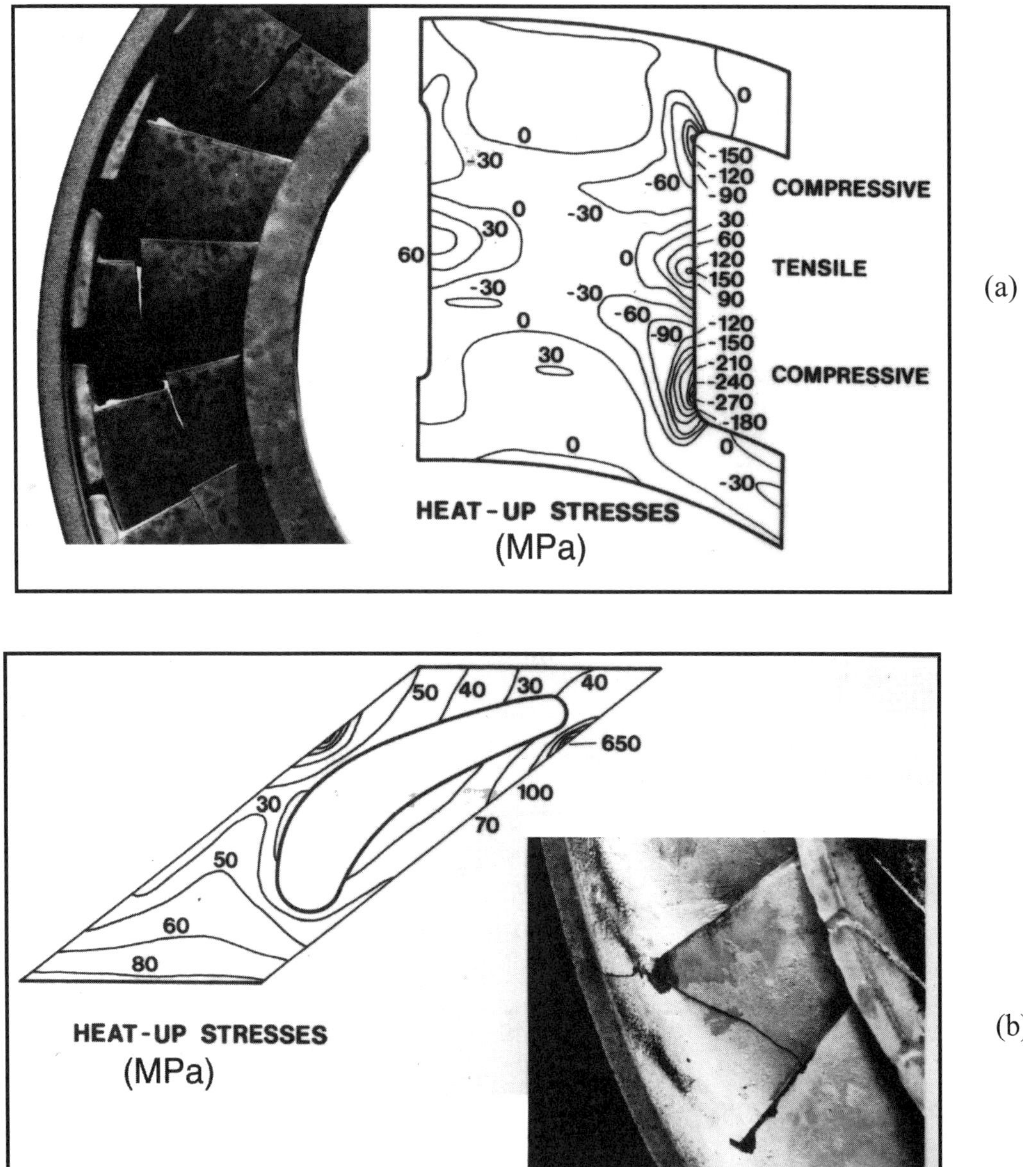

Figure 21.14 (a) Typical Airfoil Trailing Edge Failure (b) Typical Shroud Failure (Reprinted with permission from Volkswagen AG)

Figure 21.14(b) shows a thermal shock failure in a design which is somewhat similar to the single shroud integral design of Figure 21.11(a), except that each of the vanes also had a separate integral inner shroud segment. The failure originated at a highly stressed area on the backside of the outer shroud ring and subsequently progressed into the vane section. Figure 21.15 shows the characteristic failure of an integral dual shroud ring. Parts did not survive more than one cycle at full load in this design. Failures typically

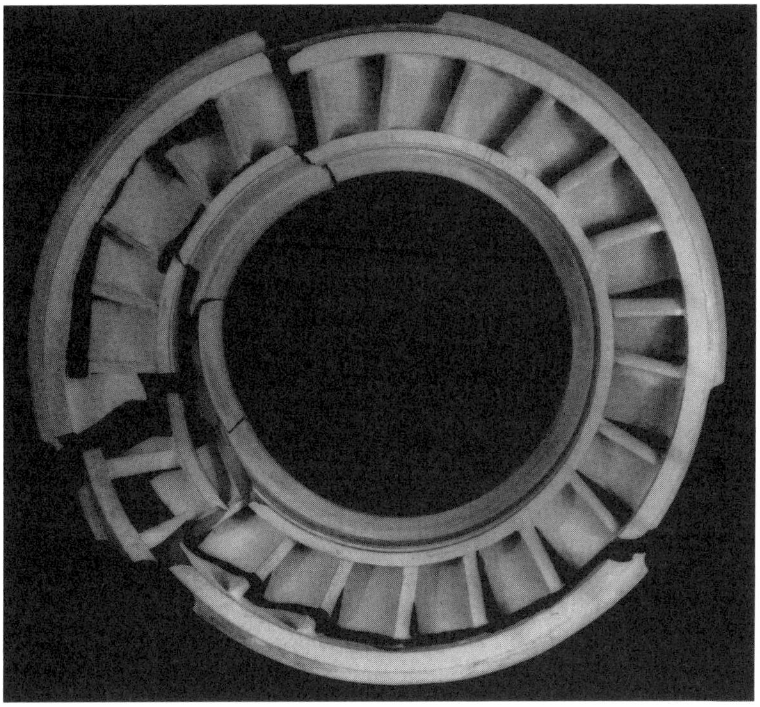

Figure 21.15 Integral SiSiC Nozzle Ring Failure After Hot Gas Rig Test (Reprinted with permission from MTU)

initiated at the sharp transition between the shroud and the vanes. RBSN components performed somewhat better than SiSiC parts.

Phase II Development (1977–1980)

Design. In Phase II FE analysis was increasingly used to predict stresses and temperatures for various nozzle ring designs. VW used a 3-D finite element (FE) model to study the nozzle-shroud ring transitions as well as the various nozzle joints. The case illustrated here is for a cold start followed by acceleration to full load operation. Six different designs were investigated. The approximate maximum stresses for SiC materials are listed in Table 21.4. It can be seen that segmentation of the shroud rings, either in full or in part, results in a dramatic reduction in the maximum stress state, up to a factor 10, compared to a full integral design. A 2-D FE model used by MTU calculated a stress of ~ 500 MPa for a solid SiC nozzle and ~ 130 MPa for a solid RBSN single nozzle (at the trailing edge). This stress dropped to ~ 100 MPa for a hollow RBSN nozzle.

MTU selected the integral nozzle/outer shroud ring design as well as a single hollow vane nozzle design for further evaluation. Solid single vane nozzle designs fabricated from SiSiC and RBSN were also procured to compare materials. In the VW design single nozzles were brazed onto outer and inner shroud rings that were partially cut to reduce thermal stresses through equalization of heat flows.

TABLE 21.4 APPROXIMATE MAXIMUM PRINCIPAL STRESS FOR 6 DIFFERENT NOZZLE RING DESIGNS PREDICTED BY VW [16]

		Maximun Stress (MPa)		
No.	**Design**	**Outer Ring**	**Inner Ring**	**Nozzle Center**
1	Integral nozzle/outer/inner shroud ring	~ 600 MPa	~ 430 MPa	~ 130 MPa
2	Fully segmented inner shroud ring	~ 330 MPa	~ 40 MPa	~ 170 MPa
3	Fully segmented outer shroud ring	~ 40 MPa	~ 160 MPa	~ 160 MPa
4	Fully segmented inner/outer shroud ring	~ 40 MPa	~ 40 MPa	~ 140 MPa
5	Cut in center of inner/outer shroud ring	~ 50 MPa	~ 40 MPa	~ 120 MPa
6	Cuts in edges of inner and outer shroud ring	~ 60 MPa	~ 50 MPa	~ 150 MPa

Rig Testing. Nozzle tests at VW were performed in an atmospheric rig test with automatic fuel control. New nozzle assemblies were first proof tested in a 25 hour stationary pre-test. A RBSN nozzle assembly (Degussa) survived the first 5 hours of the pre-test at 800°C (1973K) but separation at two joints was observed in the second 20 hours of the test at 1200°C (1473K) because of oxidation of the cement. Nozzle assemblies that passed the pre-test were subsequently evaluated in Tests I–III of the procedure illustrated in Figure 21.13. A typical test sequence involved 130 cycles with 13 cold starts and hot stops over a 6.4 hour period in Test I, followed by 100 start/stop cycles over 3.2 hours in Test II, and a 75 hour endurance test with 13 start/stops in Test III. Four nozzles were rejected over a cumulative test time of 109.6 hours and 139 start/stops. Three of these had incurred damage during installation or teardown for inspections, the fourth nozzle was incompletely nitrided in the trailing edge. Figure 21.16 shows a failed nozzle ring with integral outer shroud and separate inner shrouds. The failure initiated at the outer shroud ring and segmented the ring in three pieces [16].

Nozzle ring assemblies at MTU were tested alongside with combustor liners and nose cones according to the test protocol shown in Table 21.3. (Phase II—MTU—50 hour proof test). A RBSN (Degussa) nozzle ring failed during the initial stationary test (total 10 hours, 8 hours at 4.9 bar) through cracking which was initiated at the eyebrow slots of the outer shroud ring and continued through the nozzle vane. An integral nozzle ring of SiSiC (Annawerk) accumulated 325 cycles over 19 hours (17 hours at 4.9 bar). The outer shroud ring broke at four locations in this test. A second nozzle ring of the same material was deliberately sectioned in three locations, between the nozzle vane and the joint with the shroud ring. This nozzle survived 360 cycles of type I ($\Delta T = 260°C$), followed by 110 cycles of type II ($\Delta T = 380°C$), and a rapid stop which resulted in $\Delta T = 380°C$. Inspection showed the nozzles to be fully intact while the outer shroud ring had separated at the sectioning locations. An additional crack in the outer ring was also noticed [16].

Phase III Development (1980–1983)

The work reported for Phase III was all performed at MTU. Three different nozzle configurations were fabricated and evaluated in Phase III: an integral nozzle without inner shroud ring (slipcast SiSiC—Annawerk), a segmented design with 4 nozzle vanes per segment (injection molded RBSN—Rosenthal), and a hollow single nozzle vane (injection molded RBSN—Degussa). These nozzle configurations were evaluated in a hot gas rig under atmospheric pressure, a gas temperature of 1352°C (1625 K), in a cyclic test (30 min. start-stop cycles). After 200 hours of rig testing of 4-vane RBSN nozzle segments cracks were observed in the shroud ring and in several RBSN nozzles. The damage was believed to have originated in the low density and in large grains in the material [15].

With the integral SiSiC nozzle ring cracks formed in the outer shroud ring. They were the result of an uneven temperature distribution and appeared to be unavoidable. Still, the component remained functional since the surrounding parts kept the component flawlessly in shape and centered. The researchers

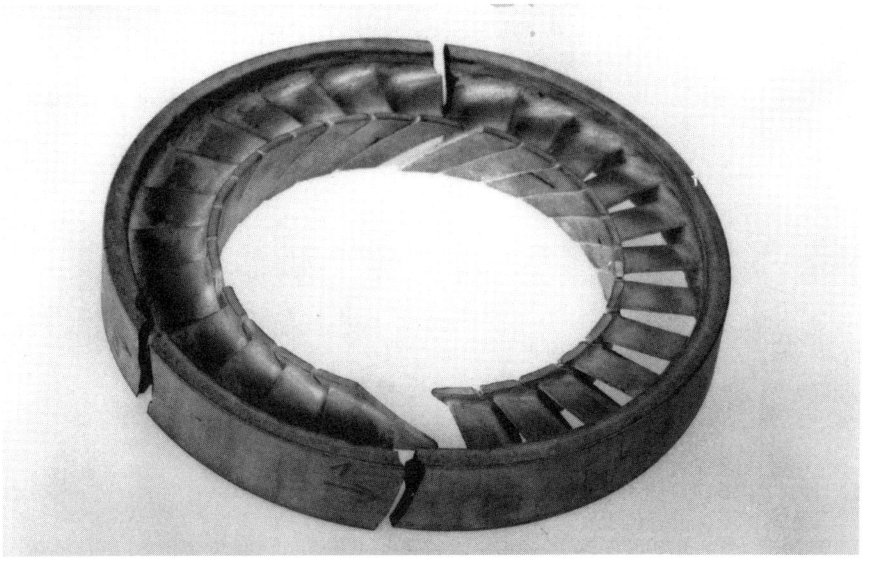

Figure 21.16 Typical Failure of RBSN Nozzle Ring (Reprinted with permission from Volkswagen AG)

Figure 21.17 RBSN Nozzle Ring Assembly Tested for 21 Hours in the MTU Hot Gas Test Rig (Reprinted with permission from Ref. [15], Figure 12)

turned the cracking problem into an advantage by inserting grooves in three appropriate locations which forced the formation of localized cracks. At the same time it prevented the formation of cracks in undesirable places. The benefits of this approach were demonstrated during a long term test (500 hours) in which the component retained its functionality. Figure 21.17 shows an assembly of individual hollow

TABLE 21.5 VARIOUS ROTOR DESIGNS INVESTIGATED BY THE GERMAN GAS TURBINE MANUFACTURERS

	Rotor Type	Daimler-Benz	Volkswagen	MTU
Duo-density ceramic rotor	Axial	✓	✓	
Hybrid metal-ceramic rotor	Axial		✓	✓
Integral ceramic rotor	Axial	✓	✓	✓
Radial ceramic rotor	Radial		✓	

nozzle segments that successfully survived 21 hours of cyclic testing in the hot gas rig under 5 bar pressure and temperatures between 927°C and 1252°C (1200 K and 1525 K) [15].

CERAMIC ROTOR DEVELOPMENT

The ceramic rotor was perhaps the most challenging component of the BMFT program. The main concerns for stationary components are thermal stresses and the stresses from interactions with interfacing metallic components. In the case of rotating components there is the additional stress contribution from the centrifugal forces of the rapidly rotating parts with the increased likelihood of life-limiting failures. Table 21.5 lists the various rotor design concepts that were explored by the German gas turbine manufacturers. All of these will be discussed here except for the radial rotor which played a minor role in the program

Duo-density Ceramic Rotor

Design. Cost effective component fabrication is a primary requirement for automotive applications, particularly for passenger cars. Assuming equivalent performance, ceramic part cost cannot exceed that of the equivalent metal component, and should preferably be less.

The most elegant solution to ceramic rotor design would be the use of a solid rotor entirely fabricated out of ceramics. At the time of the BMFT program the high stress levels in critical areas of an integral ceramic rotor required the use of hot pressed silicon nitride (HPSN) which had superior strength over a wide temperature range. However, this material doesn't lend itself easily to the forming of complex shapes such as that of a fully bladed rotor, and the only feasible way to achieve the desired geometry is by machining the part out of a relatively simple shape such as a disk, which is extremely expensive. Nevertheless, this approach was investigated by DB and VW as an initial step for demonstrating concept feasibility.

A potentially cost effective solution briefly pursued by DB, and more extensively by VW early on in the program, as well as by Ford in the U.S., was the duo-density concept. In this design the rotor hub is fabricated from a strong material (HPSN), and attached to an outer ring incorporating the rotor blades, fabricated from RBSN. The blade hub experiences the highest stresses for which HPSN is the ideal materials solution. The hub has a relatively simple shape and doesn't require elaborate machining. RBSN has a lower strength than HPSN, but the stresses in the blade region are relatively low and RBSN should have adequate margin to meet the design life criteria. Also, RBSN can be readily formed into complex shapes such as blades using injection molding [7, 8, 21].

The two parts of the rotor must be joined for the concept to be viable. The scenarios considered included: (1) hot pressing of the hub of Si_3N_4-powder into a pre-fabricated RBSN blade ring, (2) fabricating the entire rotor as a RBSN shape while hot pressing the hub portion later, or (3) joining a pre-fabricated HPSN hub and RBSN blade ring with a cement under relatively modest pressures. Each of these approaches

Figure 21.18 Development Stages for a Ceramic Bonded Rotor (Reprinted with permission from Volkswagen AG)

has its pros and cons. The various approaches were investigated by spin testing the shapes shown in Figure 21.18. These include a simple HPSN hub without attached ring, with an attached ring, with an attached ring with slits machined in, and finally a fully bladed rotor [11, 18, 21].

VW performed a FE calculation and estimated the probability of failure (POF) for a 131 mm diameter rotor in a cold start, a very severe transient condition for the engine. Assumed were a hot gas temperature of 1227°C (1500K) which corresponded to a TIT of 1350°C (1623K). The rotor speed of 55,000 rpm resulted in a blade tip speed of 377 m/s, conditions of interest for an automotive gas turbine. It was estimated that the highest stresses, about 400 MPa, which determine part life, occur in the neck and bore regions of the hub. Stresses in the blade root are much lower, on the order of 120 MPa maximum. For a good HPSN with average bending strength of 800 MPa and Weibull parameter m=10, POF is about 0.08 for these operating conditions [21].

Spin Testing. Table 21.6 shows the results of cold spin testing of the rotor configurations of Figure 21.18 [21]. Bonded rotors that had a simple attached ring showed satisfactory results, but the burst strength dropped markedly for fully bladed bonded rotors. Failure in the simple bonded ring occurred in the hub neck, so the joint was adequate. For the fully bonded rotor the failure was in the joint, indicating that the results of the simple configurations did not transfer to the fully bladed rotors. Two fully bladed rotors were also tested in hot spin testing for up to 30 min. The hot spin test was run up to 15,000 rpm, at a hot gas temperature of 1027°C (1300K), with 12 cooling cycles to 447°C (720K). The total test time was 30 min. for each rotor and no burst was observed [21]. Notwithstanding a partially successful study the duo-density route was abandoned by VW for the hybrid rotor concept that was investigated in parallel.

TABLE 21.6 RESULTS OF COLD SPIN TESTING ON VARIOUS STAGES OF BONDED ROTOR [21]

Test	Simple Disk	Hub + Ring	Hub + Split Ring	Full Bonded Rotor
Cold spin tests				
No. of tests	26	14	4	4
Av. Burst speed (rpm)	99,500	44,300	42,400	27,500
Std. Dev. (rpm)	8,800	21,500	8,700	12,000
Min. burst speed (rpm)	85,000	17,200	36,600	10,000
Max. burst speed (rpm)	114,700	76,200	55,300	35,500
Hot spin tests				
No. of tests				2
Av. Burst speed (rpm)				15,000
Time (min)				30 (no burst)

Figure 21.19 Volkswagen Turbine Rotor with Ceramic Blades in Metallic Disk (Reprinted with permission from Volkswagen AG)

Hybrid Metal-Ceramic Rotor

Design. The hybrid metal-ceramic rotor concept was developed by both MTU and VW [5, 6, 12, 15, 22–27]. The hybrid concept envisions ceramic blades with dovetail roots inserted into a metal disk as shown in Figure 21.19 for a VW rotor.

Because of the passenger car application, VW had a strong focus on low component fabrication cost. MTU's main concern was to minimize fuel consumption of its trucks which should not exceed that of a diesel

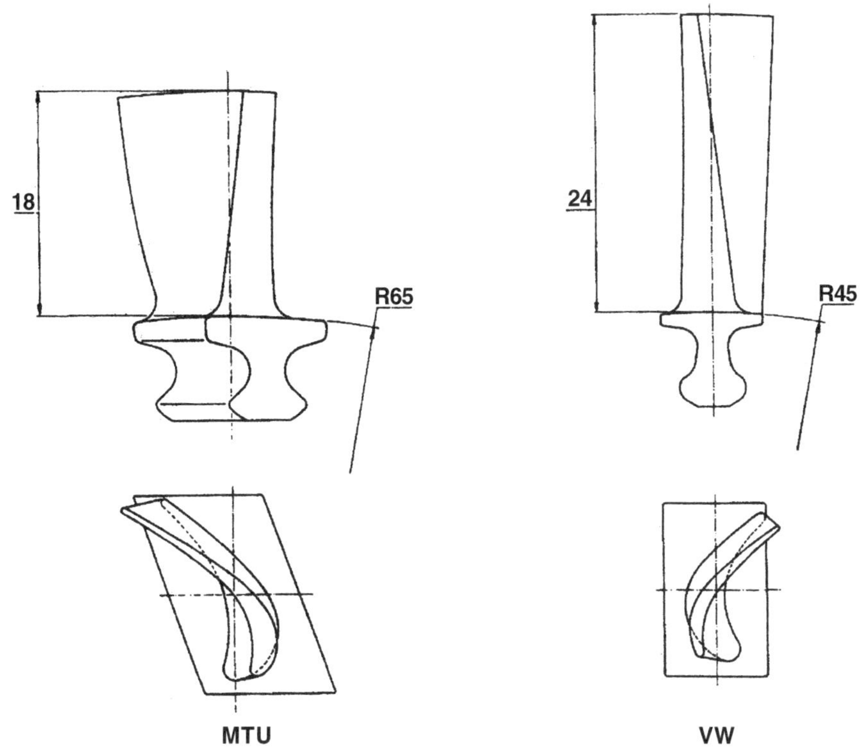

HPSN: Weight 4 g

RBSN: Weight 2.4 g

Figure 21.20 Ceramic Turbine Blades at MTU and VW (Reprinted with permission from Ref. [22], Figure 2)

motor. VW, therefore, focused on RBSN blades, since they can be formed inexpensively by injection molding. The low density (2.6 g/cm^3) and strength of RBSN makes it necessary to use a very lightweight design to minimize the load on the blade attachment. In the development project the VW blade design had a very slim airfoil shape, a low weight of 2.4 g, a small rotor radius of 48.6 mm, a rotor velocity at design point of 64,000 rpm, and a blade tip speed of 382 m/s [22].

MTU's truck engine concept places a high priority on performance. This translates into high blade tip speeds and a highly stressed blade attachment, which requires a strong hot pressed material such as HPSN. To save time consuming design efforts the profile of an existing metal turbine blade was selected. This blade is thicker and heavier than the VW blade. The mass of the MTU blade is 4 g when completely fabricated from HPSN with a density of 3.2 g/cm^3. The blade radius is 66 mm, the rotor speed is 50,000 rpm, and the blade tip speed is 387 m/s [22]. Figure 21.20 shows the MTU and VW blades with dimensions and weight.

The critical location in the hybrid design is the contact surface between the ceramic blade and the retaining slot in the metal disk. FE analysis was conducted for the MTU blade to optimize the root to disk interface. Assumed were a blade tip speed of 435 m/s which corresponded to a rotor speed of 50,000 rpm, HPSN blade material, and a Udimet 700 disk. Figure 21.21 shows the tensile surface stresses at various locations in the attachment for a contact angle of 60°C (i.e. the angle between the median plane of the dovetail and the contact surface). A small angle, say 30°, is very good for centering the wedge areas, but results in very high surface stresses in the contact area, especially when the dovetail becomes rigidly clamped inside the slot. For a large contact angle the contact surface stresses diminish, but the tensile stresses in the curved areas of the

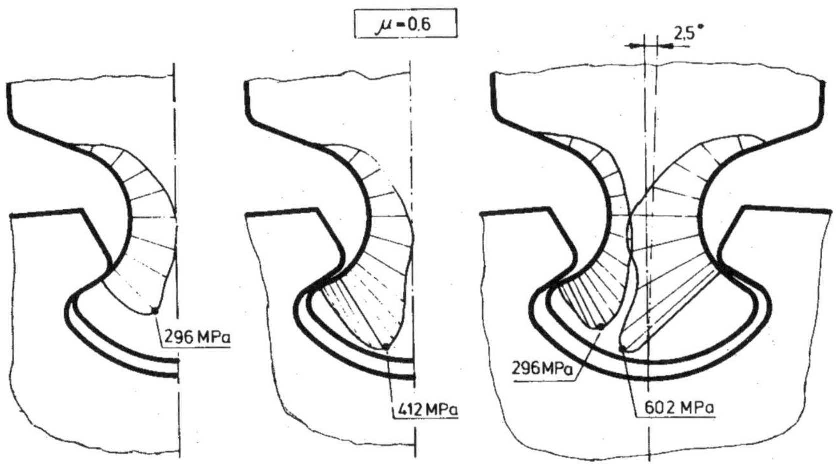

(a) (b) (c)

Figure 21.21 Ceramic Blade in metal disk slot* Stress Distribution at the Blade Surface at 50,000 rpm (a) Ideal surface contact, (b) Line contact, (c) Misaligned line Contact (Reprinted from Ref. [12], and with permission from Ref. [15], Figure 15)

root and the disk slot increase. Also, the ability to keep the blade centered is reduced for a large contact angle which increases the chance of skewed positioning. For these reasons, MTU selected a 45° contact angle.

A friction coefficient $\mu = 0.6$ was assumed for the MTU dovetail blade. The design in Figure 21.21 (a) is an example of an ideal accommodation between the surfaces of the blade attachment and the disk slot. The maximum tensile stress, 296 MPa, is located at the curved surface of the blade root near the contact point with the disk slot. A peak stress of 412 MPa was calculated for the case in Figure 21.21 (b) where the two opposing surfaces have a line contact for load transfer. The line contact is situated outside the ideal contact surface of the blade attachment. A very unfavorable situation occurs when the blade is in the misaligned position shown in Figure 21.21 (c). The unbalanced load distribution results in a peak stress of 602 MPa.

During spin testing rupture always occurs in the attachment at the location where the highest stress was calculated. Initially, skewed positioning of the blades in the slots was repeatedly observed. Figure 21.22 shows two cases of dovetail rupture. Figure 21.22 (a) shows the rupture in case of a normal load distribution. In figure 21.22 (b) the blade was misaligned in the slot (corresponding to 21.21 (c). To avoid blade misalignment, the root surface was flattened and designed with a corresponding flattening of the disk slot with a tolerance of 0.02 mm.

The condition of the attachment surface was also found to be important, particularly the effect of machining. The effect was demonstrated by machining flexure bars in different ways. Surfaces machined parallel to the load direction have greater flexure strength than those machined perpendicular to the direction of the applied load. Surface treatments such as radiation, heat treatment, or lapping restore the strength of the surface to that of the longitudinally machined condition.

To mitigate the friction forces which develop under load in the contact region, compliant layers such as a 0.1 mm thick platinum layer (MTU) [22] or a lenticular pad (VW [26]—see Figure 21.23) are inserted between the blade attachment and the disk slot at the location of the load transfer surfaces. In the latter case the blade contact surface moves under the centrifugal force which results in a rolling movement. An additional compliant layer in the contact region reduces stress concentrations resulting from tolerances and surface roughness.

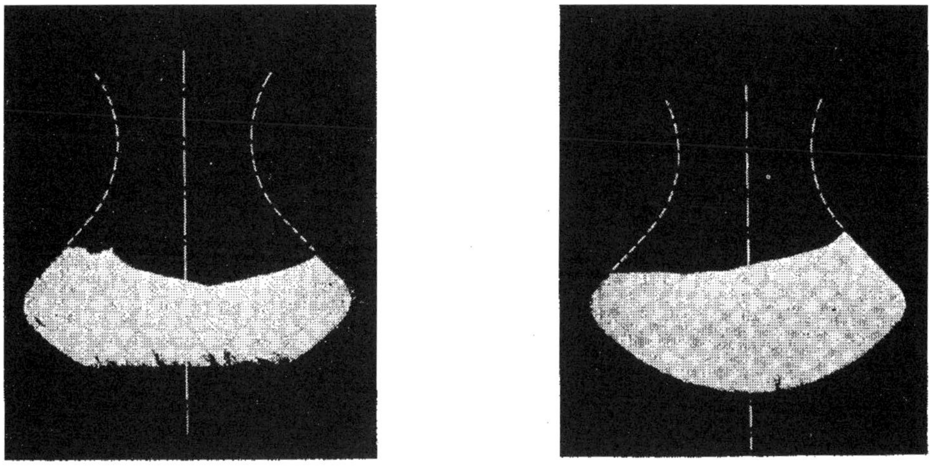

(a) (b)

Figure 21.22 Dovetail Rupture (a) Regular Load Transfer, (b) Load Transfer for Misaligned Positioning of the Blade (Reprinted with permission from Ref. [22], Figure 6)

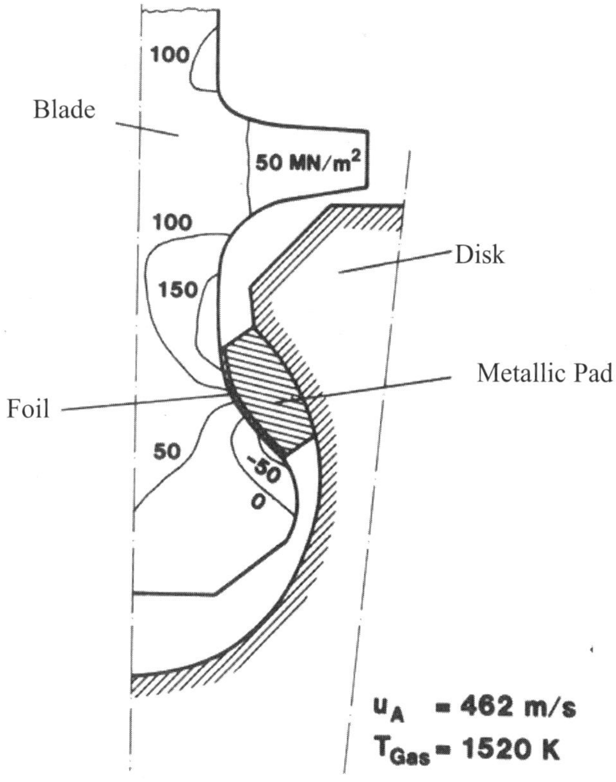

Figure 21.23 Stress Distribution in the Root of an Inserted Blade with Metallic Pad (Reprinted with permission from Ref. [26])

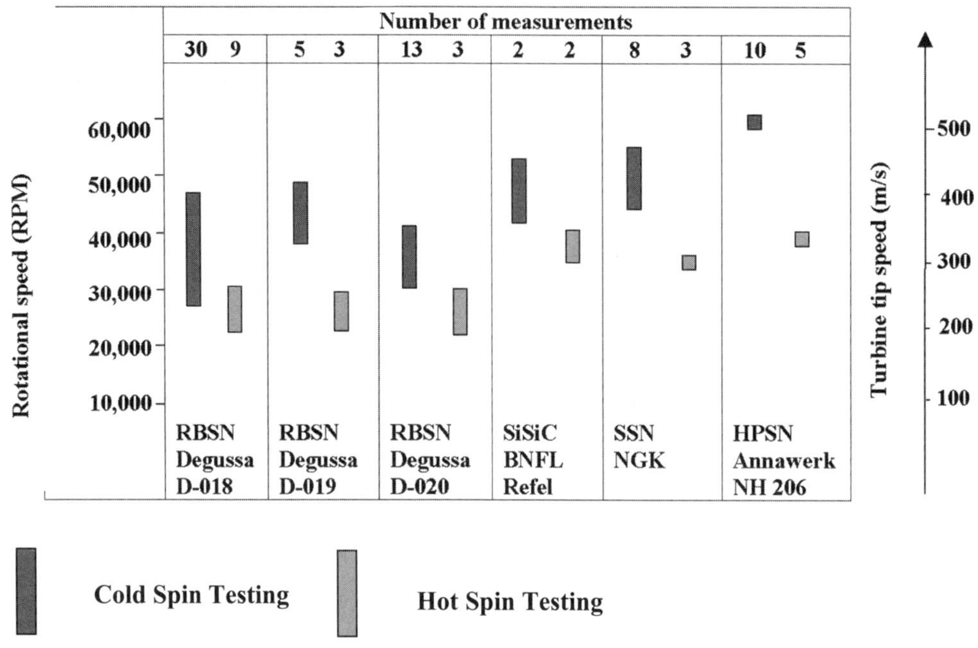

Figure 21.24 Results of Cold and Hot Spin Testing at MTU (Adapted with permission from Ref. [15], Figure 22)

Spin Testing. MTU evaluated silicon nitride blades fabricated by various suppliers, including RBSN injection molded blades of Degussa, SiSiC slipcast blades of BNFL, SSN slipcast blades made by NGK, and HPSN blades of Annawerk. Finishing of these blades by ultrasonic machining, polished and/or lapping was performed by MTU.

Figure 21.24 summarizes the results of cold and hot spin testing at MTU. Note that the HPSN blades did not rupture during the testing and easily met the target velocities of 60,000 rpm at room temperature, and 40,000 rpm in the hot gas. Blades of all other materials ruptured before the target velocities were reached. Subsequently, endurance testing of HPSN rotor blades was performed at the hot gas temperature of 1047°C (1320 K), a disk temperature of 697°C (970 K), and 40,000 rpm. No compliant layers were used for these tests. Part life in these tests was limited, and there was a wide spread in the data as rupture of the blades occurred between 55 minutes and 79 hours. The failure in all cases occurred in the area with the highest surface stresses in the blade. Possible causes were: reduction in material strength with increased temperature through subcritical crack growth, higher friction forces in the contact region between the blade attachment and the disk slot as a result of temperature related friction coefficient, occurrence of friction movements between the ceramic blade attachment and the metal disk slot as a result of relative thermal expansion differences, and disturbances in the contact interactions through thermal deformations and resulting incrased stresses in the blade attachment [15].

VW evaluated RBSN blades by cold spin testing at a rotor speed of 66,000 rpm (110% of design speed) and a blade tip speed of 435 m/s. Cyclic loading between 100% and 110% of design speed was performed after reaching the design speed. A lenticular pad was used at the blade-disk interface. Under these conditions about 26% of the RBSN blades failed the test. Using a sintering step to densify the RBSN parts to sintered RNSN (SRBSN) roughly doubled the strength from about 300–320 MPa to about 650–750 MPa. Incorporating an additional heat treatment at 1260°C for 100 hours enabled 95% of the SRBSN blades to pass the proof test. VW subsequently cold spin tested fully bladed SRBSN rotors (32 blades per rotor) with lenticular type inserts. An axial safety spring ensured that the blades did not change their position under the gas load. Blade tip speeds were 400 m/s. Rotors that passed this test were subjected to hot spin tests with stepwise increases in tip speed and temperature. Four rotors were tested. Two failed early on as the conditions of high temperature and tip speed were implemented, in about 1 and 2 hours, respectively at 1027°C (1300K)/360 m/s. A third rotor failed at 15 hours at 927°C (1200 K)/290 m/s, and the fourth rotor failed at 18 hours and 1027°C (1300 K/320 m/s). Detailed analysis showed the failure invariably to be the result of the debris from failure of the first blade impacting the other blades leading to a domino effect. None of the rotors met the 25 hour goal, and planned tests under engine conditions were therefore not implemented [25].

It appeared that the materials properties and/or designs in the VW and MTU programs were not adequate to meet the demanding requirements of high temperature hybrid metal-ceramic rotors within the schedule constraint of the program, and were not pursued further.

Integral Rotor

The closest design that technically met the requirements of a rugged component was the integral silicon nitride rotor developed by DB under the BMFT program. This rotor concept passed through the gates of materials property evaluation, finite element design, cold and hot spin testing, and finally engine testing including testing in the ultimate application, a passenger car that covered a significant distance under typical driving conditions. DB briefly considered a duo-density approach to rotor design because of it being economically attractive, and even spin tested some rotors. However, originally unforeseen problems let DB to develop a 100% HPSN rotor. DB developed a process for machining rotors to the final design out of cylindrical solid HPSN castings, procured from the ceramic suppliers. This fabrication process doesn't lend itself to mass production because of the expense of the machining operation, but since the project only required a feasibility demonstration, this was not considered a deterrent at the time.

DB developed such a high level in expertise with this approach that later on it was able to produce fully finished newly designed ceramic turbocharger wheels from castings within one week [28]. One of the major impediments to rapid progress in ceramic development is the time it takes to express a finished design into finished parts, typically 6 to 18 months, because of the bottlenecks of tooling, fabrication and design and process iterations. DB's approach was an early approach to rapid prototyping, which significantly shortened the iterative design-fabrication-test cycle.

While HPSN has excellent fast fracture strength (600–840 MPa), it does have some drawbacks, notably its creep behavior at elevated temperatures and stresses because of a small amount of MgO, typically present. Also, there are inherent stresses because of the manufacturing process of cylindrical and profile pressed solids. Additionally, subcritical cracks can grow under the influence of temperature and stress, which may lead to premature failure. The oxidation behavior in the hot turbine environment was not fully understood either, so the effect of surface damage on the degradation in strength was not completely known [29].

Design. Table 21.7 shows the cycle data for a simple cycle concept turbine for the developed rotor as well as for the two rotors of an experimental ceramic gas turbine for the "Car 2000" project to be discussed further on in this section [30].

Stresses and temperatures in the component were estimated using finite element analysis. Stresses were calculated for a turbine inlet temperature (TIT) of 1350°C, at a rotor speed of 66,000 rpm (110% of design speed). The maximum steady state stress in the critical neck region was estimated to be 352 MPa at a temperature of 1060°C. In the bore region the stress was 348 MPa at a temperature of 775°C. The highest stresses were observed under transient conditions but at lower part temperatures. The maximum stresses and temperatures in the neck region were 415 MPa and 760°C, and in the bore region 420 MPa and 290°C, respectively, under transient conditions [31].

TABLE 21.7 CYCLE DATA FOR MERCEDES-BENZ GAS TURBINE [29]

	Typical Simple Cycle Turbine Model	Daimler-Benz Gas Turbine	
		Gasifier Turbine	Power Turbine
TIT (°C)	1350		
Phase I		1250	1060
Phase II		1350	1155
Rotor speed (rpm)	60,000	61,500	55,000
Number of blades	37	57	53
Disk diameter (mm)	90	90	112
Stage pressure ratio	1.893		
Compressor pressure ratio		>4	
Nominal turbine performance (kW)	184		
Phase I		94	
Phase II		110	

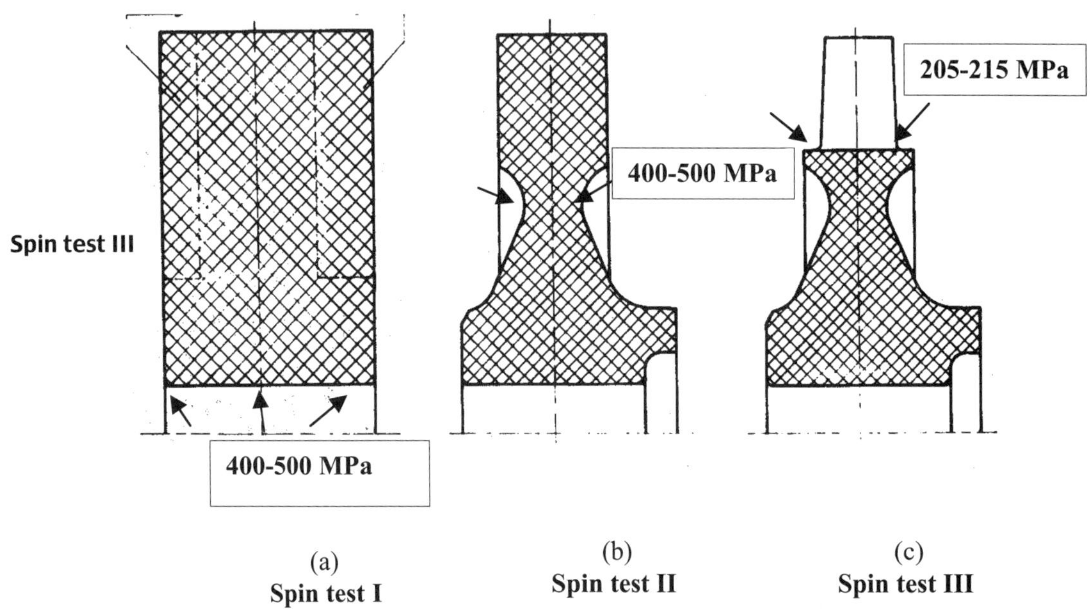

Figure 21.25 The Three Stages of Fabrication of HPSN Components Used for Testing (a) Cylindrical Disk, (b) Profile Disk, (c) Finished Rotor (Adapted from Ref. [9])

Testing. DB had a sequential approach to testing, incorporating progressively more challenging design aspects and conditions. Starting with thermal shock testing of a three blade segment (3700 cycles, 1350°C/60s → 70°C/60 s, hot gas velocity: 250 m/s, air/gas pressures of 1 bar), no damage to the test pieces was observed, but microscopic analyses indicated a migration of MgO towards the blade surface (Mg at surface was nearly 6x average Mg in the bulk). It was unknown whether this would lead to degradation over time.

Next, a fully bladed integral HPSN rotor was developed in a three-step process (see Figure 21.25 (a)–(c). Starting with a simple cylindrical casting (Figure 21.25 (a)), sections were cut off (these were turned into flexure bars) to give the profiled disk shape shown in Figure 21.25 (b), and the latter was machined into the finished rotor of Figure 21.25(c) by a combination of ultrasonic and diamond grinding. The test shapes

Figure 21.26 Three HPSN Turbine Rotors—Left: for "Ceramic Components for Vehicular Gas Turbines," Top: Gasifier Turbine for "Car 2000" Program, Right: Power Turbine for "Car 2000" Program (Reprinted with permission from DaimlerChrysler)

were bored so they could be attached to a shaft for spin testing. Further on in the program profiled disks were obtained from the supplier. This significantly reduced the machining time and only the profiled shape and the fully machined rotor had to be tested [9, 29, 32]. Figure 21.26 shows a finished rotor fabricated under the BMFT program and two rotors fabricated for the "Car 2000 program".

Each of the three shapes of Figure 21.25 was subjected to cold spin testing, prior to further machining (for shapes (a) and (b)) or component testing (for shape (c)). The cold spin proof tests were carried out at rotor velocities that resulted in maximum stresses of 500 MPa (bore region of (a)), 500 MPa (neck region of (b)), and 215 MPa (disk to blade transition for (c)) [29].

Conducting the proof test was believed to improve the reliability of the components. Table 21.8 shows the probability of survival (POS) for a HPSN (NH 206) gasifier rotor with and without proof test (cold spin testing). The data also shows that the effect of surface defects cannot be ignored. A complex road cycle was also analyzed. After 3500 hours (whcih equates to about 150,000 road km) the POS of the rotor was 0.9965 [29].

Of the profiled disks only one out of 13 parts failed during cold spin testing in the neck area. This represents a failure rate of 7.7%. Using Weibull statistics and assuming an average flexure strength of 700 MPa and Weibull factor of 17, a POS of 0.96 was determined [29]. The correspondence between the two results was believed to be surprisingly good by the DB investigators. Notwithstanding these optimistic predictions failures during subsequent hot gas rig testing indicated that probably not all relevant factors had been incalculated in the reliability predictions.

Having completed the cold spin proof test the finished rotor was subjected to a stationary thermal shock test and a hot spin test. The thermal shock test, which simulated the start up of the engine, generated

TABLE 21.8 PROBABILITY OF SURVIVAL (POS) FOR A HPSN GASIFIER ROTOR [9]

Operating Conditions	Designation	Full Load New Material		Road Cycle	
	Operating time	0 hours	1000 hours	1000 hours	3500 hours
No proof test	Volume	0.9992	0.9329		
	Surface	0.9993	0.9662		
	Volume + Surface	0.9985	0.9013		
With proof test	Volume	1.000	0.9868	0.9987	0.9975
	Surface	1.0000	0.9917	0.9995	0.9991
	Volume + surface	1.0000	0.9786	0.9982	0.9965

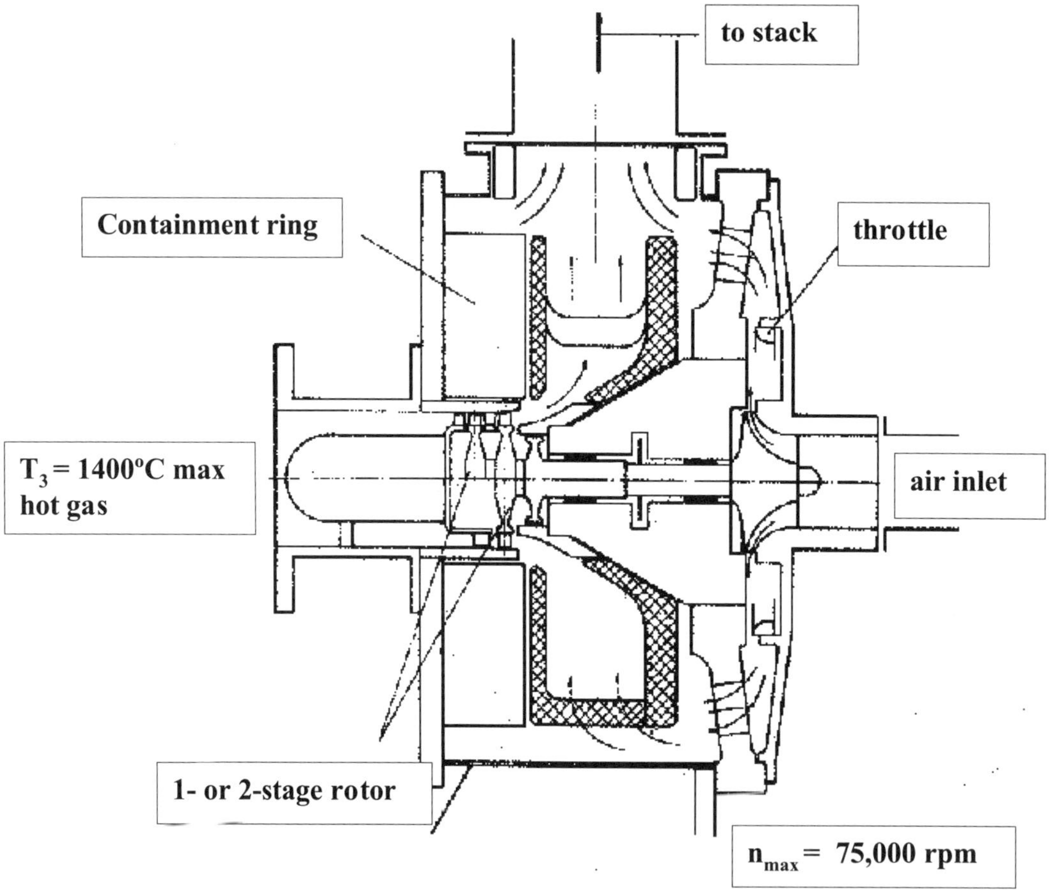

Figure 21.27 Hot Gas Spin Rig for Daimler-Benz Rotor Testing (Adapted from Ref. [7])

the highest stresses in the component. The rotor is introduced into a hot gas stream and subsequently moved into the ambient air. The maximum temperature ranges from just under 700°C in the bore to about 1300°C at the blade tip.

A schematic of the hot gas spin rig is shown in Figure 21.27. The rig has a mass flow up to 1 kg/s, a pressure ratio up to 3, it can be used at temperatures up to 1400°C, and reaches a rotor speed as high as 85,000 rpm [7]. Many HPSN rotors were evaluated, and a great number of these failed because of excessive

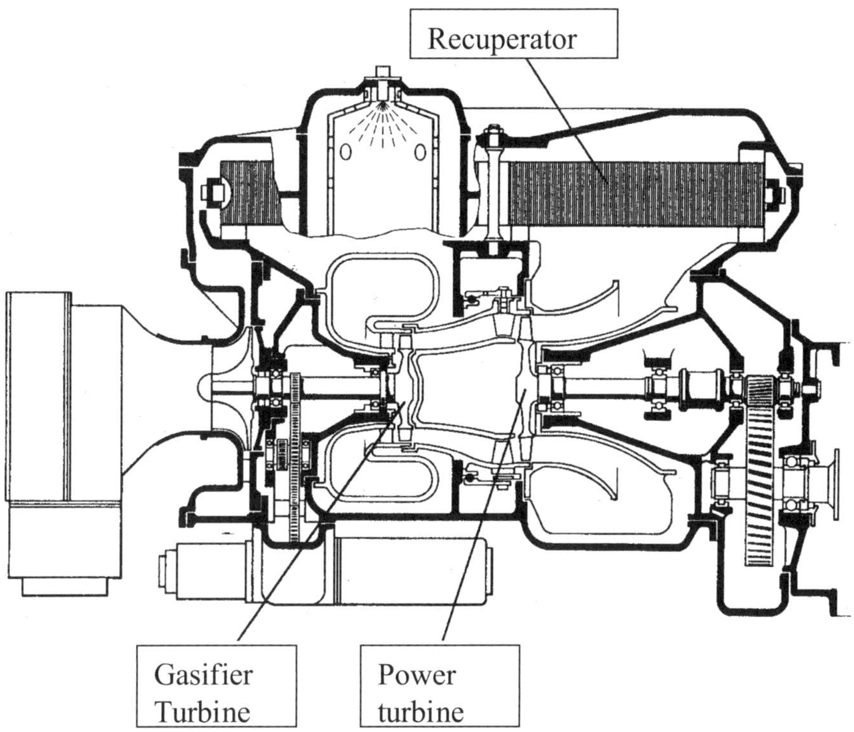

Figure 21.28 Schematic of the Daimler-Benz Research Gas Turbine PWT 110 (Adapted from Ref. [9])

thermal stresses operating on pre-existing flaws in the material as a result of casting processing or finish machining. In some cases the high thermal stresses resulted from operating problems in the rig. One of the rotors was tested for almost 70 hours, 10 hours of which at the design rotor speed of 60,000 rpm at a hot gas temperature of 1250°C. The cumulative total running time for all rotors was about 484 hours by the end of the BMFT program. However, none of the rotors met the design targets, a life of 50 hours at a hot gas temperature of 1350°C and a rotor speed of 60,000 rpm [29].

Car 2000 Program

In 1978 DB decided to develop a gas turbine as part of the "Car 2000" research program of the Ministry of Research and Development of the Federal Republic of Germany. The project engine was the Daimler-Benz Research Gas Turbine (PWT 110), which is schematically represented in Figure 21.28. This two-shaft engine had a target output of 94 kW at a TIT of 1250°C for Phase I of the program. The intent was to raise these performance targets to 110 kW and 1350°C in Phase II [33]. Both the gasifier turbine and the power turbine rotors, as well as the inlet nozzle and the regenerator, were to be fabricated out of ceramics. The scroll was also considered. The performance of the engine and data for the two turbine stages are supplied in Table 21.7 [29]. All critical components for the hot section were subjected to intensive pre-testing to reveal any weak points.

Rig Tests. Preliminary rig testing was conducted using the procedures developed under the German National Program. HPSN shapes similar to the configurations of Figure 21.25 were cold spin tested after

Figure 21.29 Impact Failure of Gasifier Rotor After 5 Hours and 56 Minutes at 1000°C and 42,000 rpm (Reprinted from Refs. [9,12])

each of the three fabrication steps, and the finished rotors were evaluated in a thermal shock rig. The turbine rotors were subsequently tested in a gasifier rig which included components from the PWT 110 engine (compressor, primary combustor, turbine). Its operating conditions closely matched those in the engine. The test rig was designed for the following operating conditions—speed: 65,000 rpm, TIT: 1350°C, PR: > 4 bar, delta T during cycling: up to 500°C/s [33]. Tests were initially performed with moderate temperature and speed gradients. The intent of the tests was to eventually achieve the normal start conditions of the Daimler-Benz Research Gas Turbine PWT 110 in the EPA-75-US Emissions Test.

One turbine wheel was tested for a total running time of 58 hours, at temperatures up to 1350°C and maximum speed up to 61,500 rpm. Another turbine wheel was tested for 9.4 hours at a TIT of 1350°C and a speed of 50,000 rpm [33].

Foreign object damage (FOD) posed a particular problem in gasifier rig testing. In a number of tests, particulates (from the test rig liner or from the combustor) caused failure of the turbine wheels. Figure 21.29 shows a damaged HPSN (NC 132) gasifier rotor which was tested in the gasifier rig. The component had initially survived for 5 hours and 56 min at a maximum speed of 53,000 rpm and gas temperatures up to 1060°C. The failure was attributed to impact from a piece of metal from the combustor section that had melted after localized overheating. The failure occurred at a speed of 42,000 rpm and a gas temperature of 1000°C [9, 12].

Engine Tests. The engine tests were conducted with the Daimler-Benz Research Gas Turbine PWT 110. A careful engine test strategy was adopted. Initially, the tests were relatively restrained as only manual electronic control was available, and the stresses on the ceramic components were relatively low. The stresses on the ceramic parts during emissions control tests (US-EPA-75) were significantly higher.

Figure 21.30 (a) and (b) shows sections of a gasifier turbine and a power turbine that were partially damaged after 11 hours of operation at a maximum speed of ~ 35,000 rpm and a maximum gas temperature of ~ 1200°C. The damage of the gasifier turbine is at the leading edge near the blade tips and of the power tur-

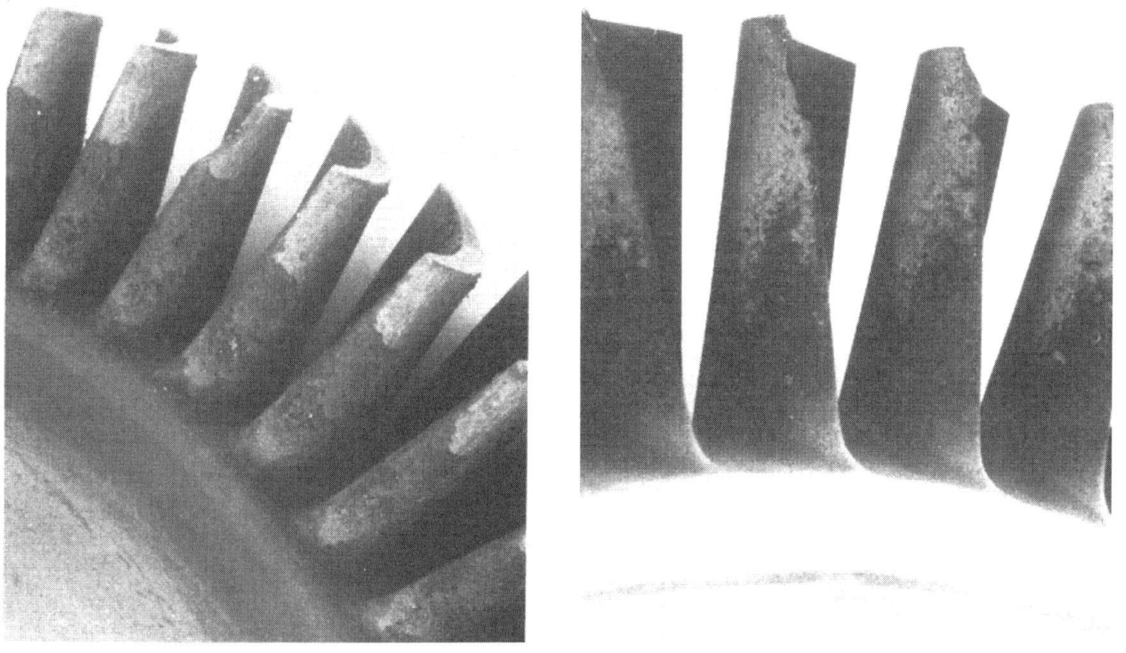

(a) (b)

Figure 21.30 Damaged Blades of Gasifier Turbine (a) and Power Turbine (b) Following 11 Hours of Engine Testing (Reprinted from Ref. [9])

bine at the trailing edge near the blade tips. While the cause that precipitated the failure could not be exactly determined it was surmised that the damage to the power turbine was from impact of debris that had separated from the gasifier turbine following impact [9].

DB analyzed the frequency of damage of engine builds against the causes of damage. Of the disassembly operations, 42% were not related to damage but were routine examinations, and inspections to study engine modifications. The inadequate strength (< 500 MPa) and the design concept (e.g. nozzle mounting) were factors believed to be responsible for damage to the nozzle. Damage to the turbine wheel was attributed to the quality of the material. There was no clearly identifiable cause for 15% of the engine damage. A slipcast SSN turbine wheel ran for about 66 hours at different loads before damage occurred. The maximum speed was 61,500 rpm [33].

When the "Car 2000" program was concluded in 1982, the gas turbine engine had been demonstrated at a TIT of 1250°C and a gasifier rotor design speed of 61,500 rpm. Figure 21.31 shows a ceramic gas turbine installed in the Mercedes-Benz research car 2000. The car was shown to the public in 1984 on the occasion of the third BMFT program status seminar in Bad Neuenahr. In 1990, a W124 Mercedes-Benz (200 to 300 model E range) car with a ceramic gas turbine successfully covered the distance from Stuttgart, Germany, to Brussels, Belgium, about 600 km. With these results the research work on ceramic gas turbine at Daimler-Benz Research was concluded. The total running time of the Daimler-Benz Research Gas Turbine equipped with two hot-pressed silicon nitride rotors was more than 500 hours on test rigs, in the car and on the road [28].

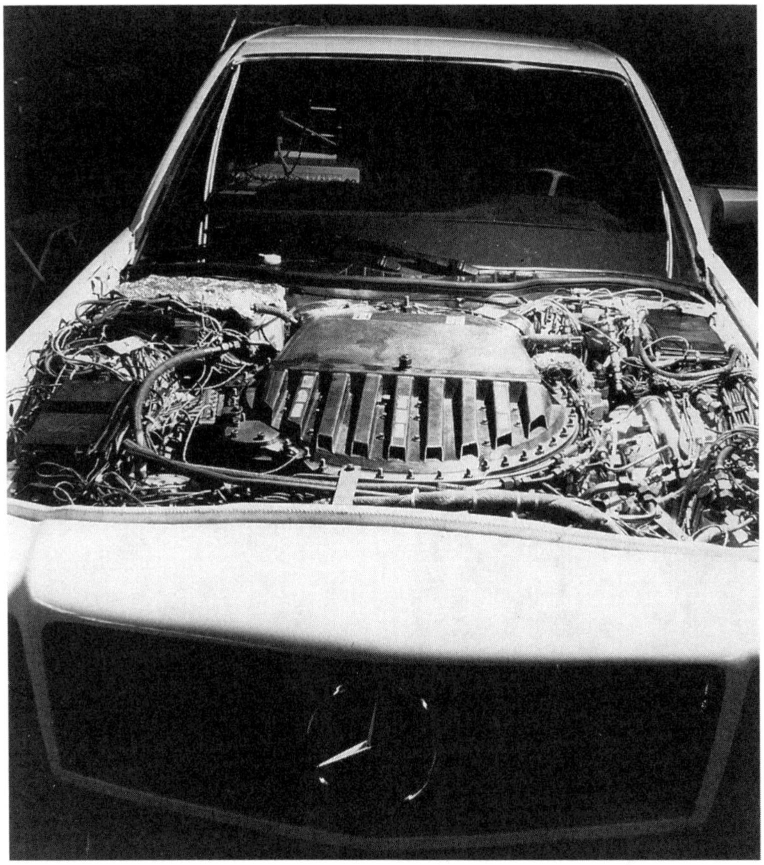

Figure 21.31 Ceramic Gas Turbine Installed in Daimler-Benz Research Car 2000 (Reprinted with permission from DaimlerChrysler, see also Ref. [9])

Structural Ceramics Program

Having developed the process for fabricating HPSN rotors, Daimler-Benz, in 1983, under another BMFT-sponsored and DFVLR-managed three year program "Structural Ceramics" started to develop a ceramic turbine rotor suitable for mass production technologies. The materials selected were sintered silicon nitride (SSN) and hot isostactic pressed silicon nitride. The main subcontractor was GTE Laboratories, of Waltham, Massachusetts, USA. After two years of development, injection molding was the method chosen for forming the complex-shaped ceramic rotor. The densification was performed by sintering or hot isostatic pressing [30]. The GTE fabrication efforts have been reported [34].

Rotor Design. The design objective was to develop a ceramic gasifier rotor suitable for an engine of the size of the PWT 110 but with an increased maximum rotor speed of 70,000 rpm. The maximum TIT was 1350°C. The hub diameter was 90 mm. The PWT 110 gasifier rotor was the starting point for the development. In the PWT 110 the blade count for this first stage rotor was 57. This number was reduced to 39 for two reasons. First, the more blades, the more difficult it is to remove them from the mold without breaking them in the green state. Secondly, to resist impact damage the blades had to be made thicker. Both injection molding and slipcasting seem to result in a similar strength of the sintered silicon nitride.

FE analysis showed that the maximum temperature in the turbine wheel was almost 100°C less than the TIT of 1350°C. The maximum stress, about 350 MPa, was in the bore. The largest stress in the neck was about 300 MPa, and about 160 MPa in the blade root. Transient conditions which cause higher stresses in the bore, neck and blade root, were also taken into account in the design of the turbine rotor. In 1983, the material behavior of SSN was still not as good as that of HPSN, the best available material at that time. But most of the materials properties assumed for the design were those of HPSN in the expectation that SSN would achieve those levels in the not too distant future.

In a parametric study the reliability was calculated for various strength regions of a turbine wheel. The data used were: Weibull modulus = 15; running at full load: 175 hours; TIT = 1250°C; rotor speed: 61,500 rpm. The overall reliability was 0.619 with the major contributor being the critical bore region with a reliability of 0.738. This was later confirmed by hot gas testing [30].

Injection Molded Turbine Rotors. The first injection molded GTE rotors were fabricated from the SSN material AY6. Eventually, GTE switched to hot isostatic pressing of AY6 for bladed rotors. The injection molding process was attractive because it was flexible enough to allow modifications to be made to the hub contour, trailing edge thickness, etc. The development process for the injection molding of these large, complex shaped silicon nitride rotors took place in various stages based on the results of cold spin tests. The first generation of turbine rotors typically exemplified the problems arising out of the injection molding process: folds, cracks, and fracture of the trailing edges during demolding. The second generation rotors were significantly improved over those of the first generation with the average burst speed being about 102% of the design speed. The third generation rotors showed further improvement. The average speed was now 114% of design speed and the Weibull modulus was vastly increased over the value from the second generation rotors. Failures always occurred at the highly stressed neck surface. Entire rotors were sectioned into MOR bars to obtain a strength distribution [30].

Slipcast Turbine Rotors. Based on the experience gained, turbine rotors were subsequently also designed for near net-shape processing by slipcasting. The subcontractor for the fabrication of the slipcast rotors was Kyocera. The rotors had slipcast blades and machined disk contours. The rotors were subjected to cold spin testing. The 50% survival probability speed was 90,350 rpm and the Weibull modulus was 24.3. Some of the flaws observed were hollow-outs similar to pock marks, large grains near the bore as fracture origin, and large subsurface grains.

After thermal shock and cold spin testing slipcast silicon nitride rotors were tested in the hot gas test rig and in the gas turbine engine, either on the test bench or installed in a production car of the Mercedes-Benz mid-size series for road testing. For turbine conditions of 1250°C and 61,500 rpm a life of up to 32 hours, and at 1350°C and the same speed, a life of 4.5 hours was achieved. In these cases damage could be attributed to material defect failure occurring as predicted in reliability analyses in the bore region, probably due to subcritical crack growth. In a gas turbine engine one of these rotors achieved a total of 65 hours of which 23.5 hours were near full load conditions (90% or higher) [30].

CERAMIC HEAT EXCHANGER DEVELOPMENT

It was the consensus at the time of the German National Program that the vehicular turbine could only become commercially successful if it was as least as fuel efficient as the internal combustion engine. Increased performance could be achieved through recuperation, and Daimler-Benz had the responsibility for the development of a ceramic heat exchanger that would meet the requirements for the automotive engine of the future.

Phase I (1974–1977)

Design. The design requirements included [35]:

- minimum combustor exit outlet temperature of 1350°C
- recuperator thermal effectiveness: ≥ 90%
- sum of all the pressure losses: no more than 7%
- leakage rate: ≤ 0.5%
- thermal conductivity of the matrix materials: ≤ 10 W/m.K
- compactness: heat exchanger together with the gas turbine must fit in a vehicle
- TIT 1350°C
- mass flow: 0.5 – 1.0 kg/s
- inlet air pressure: 0.5 MPa
- air inlet temperature: 200°C
- air exit temperature: 1000°C
- hot gas inlet temperature: ~ 1200°C (turbine at full load and maximum rpm)
- hot gas exit temperature: ~ 400°C
- hot gas exit pressure: ~ 0.11 MPa
- hot gas composition: about 15 vol.% O_2, 6.5 vol.% CO_2 and 4.5 vol.% H_2O.

The highest stationary load on the heat exchanger takes place at full load operation. The greatest transient load on the heat exchanger occurs during start-up to reach full load, when the hot gas temperature increases from 400°C to 1200°C in a mere 3 seconds. The air pressure in this interval increases from ~0.2 to ~0.5 MPa. The heat exchanger operates at full load for a total of no more than 20 hours, i.e. for about 1% of its expected life. The highest transient load will probably be achieved no more than 200 times. However, a ceramic heat exchanger must be designed for this short duration and relatively few severe cycles, and not for the relatively benign requirements of 10,000 normal starts and stops of the gas turbine.

While DB had worked on a rotary regenerator since 1968, the decision was made to develop a stationary ceramic recuperator under the German National Program because of the problems inherent in regenerators (e.g. leakage), and the suspicion that these problems would be aggravated with increasing pressure ratio and temperature. Figure 21.32 shows the ceramic recuperator design concept. The design is derived from that of a metallic recuperator. The heat exchanger has a modified counter-current of gases with integrated tubular air channels.

In terms of size this recuperator was at a disadvantage compared to a rotary regenerator of equal performance. The flow channels for air and gas are longer than in a disk and, because of the 7% pressure loss limitation, a larger cross section of the matrix channels was required. This in turn increased the matrix volume, to about twice the size of a regenerator. A typical recuperator would be about 3.0 to 3.5 m tall, which would obviously not easily fit into a standard passenger car. To keep the recuperator compact the design envisioned six rectangular blocks, each 500 mm x 150 mm × 100 mm in size. The matrix structure would have to carry the pneumatic loads by itself since an impermeable housing would add to size and weight. Integrated air channels would be needed to minimize leakage losses.

Further specifications were:

- Hydraulic cross section of the gas channels: about 0.8 mm
- Hydraulic cross section of the air channels: about 0.6 mm
- Foil thickness: about 0.2 mm
- Rib width: about 0.2 mm
- Specific matrix surface: 2500 m^2/m^3

The foil thickness is particularly critical for the recuperator effectiveness as Figure 21.33 shows.

DB was convinced that a high temperature recuperator could only be realized using ceramics. The material of choice became RBSN. It has acceptable strength over a wide temperature range, relatively low

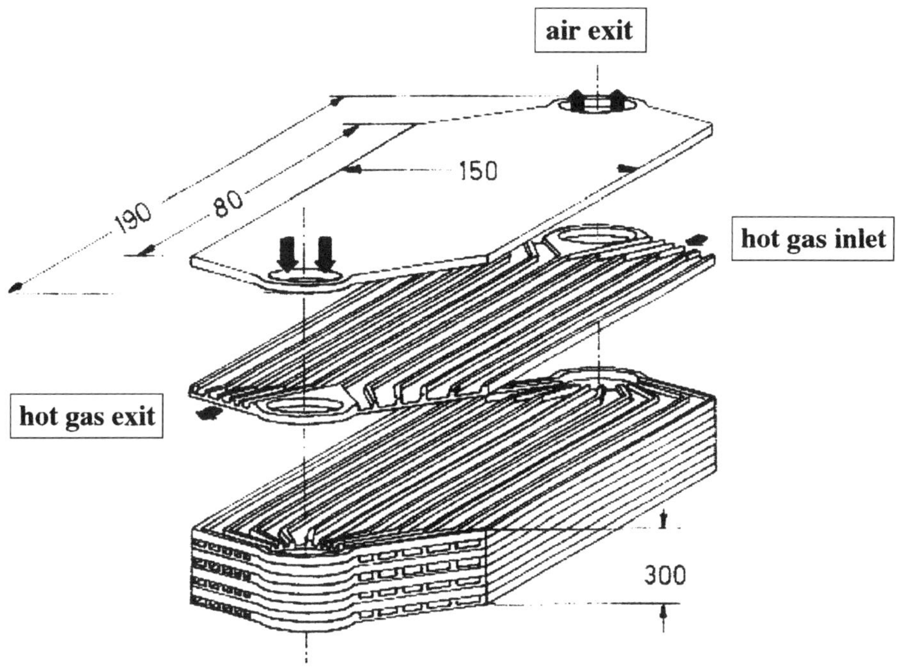

Figure 21.32 Ceramic Recuperator Design Concept (Adapted from Ref. [8]) (Dimensions in mm)

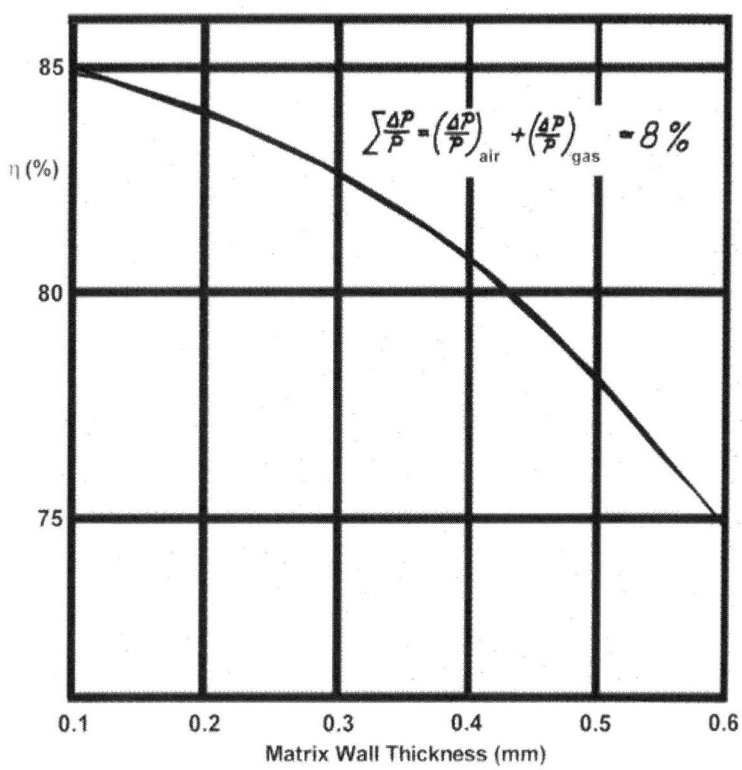

Figure 21.33 Relationship Between Foil Thickness and Thermal Effectiveness of the Recuperator (Adapted from Ref. [7])

elastic modulus, and suitable low thermal conductivity, about 10 W/m.K, four times lower than that of SiC. An additional advantage was its dimensional stability during fabrication.

All of the materials had some disadvantages. Because of its high thermal conductivity, SiC has a low recuperator effectiveness, particularly at part load. Cordierite and the glass ceramics (LAS/MAS) have outstanding performance, but LAS is chemically reactive, and MAS has a thermal expansion coefficient which is too high for its strength. Both of these glass ceramics shrink during fabrication, and both materials probably creep at temperatures around 1200°C (they need to be sintered between 1200 and 1500°C). RBSN's disadvantage is its porosity. Because of the leakage rate limitation of 0.5% in the matrix, the mass flow of the air cannot exceed 1.25 x 10^{-8} kg/s cm^2 [35].

Phase II (1977–1980)

The main emphasis in Phase II of the program was on the evaluation of critical properties of the recuperator matrix using specific test specimen configurations.

Matrix Test Specimen Fabrication. In 1974 DB, in collaboration with Rosenthal, started to investigate various techniques including lamination of wavy and flat foils and extrusion of foils. In both cases the joints between adjacent layers lacked strength and/or could not meet performance requirements. The eventual fabrication route became a modification of the laminating technique which was developed in the 1977/1978 timeframe. Sheets of Si and RBSN were fabricated which were used to fabricate cube type single- and multi-layer burst specimens. These specimens were used to investigate properties of interest such as: burst- and implosion resistance, air permeability, oxidation-, and thermal shock resistance (see Figure 21.34) [35].

Testing. *Exposure in air* at 1000°C resulted in damage to the RBSN specimens. Circular specimens deformed after 100 hours. Some specimens broke. Testing in air between 800 and 1050°C showed that after 1 hour at 950°C melting occurred in the channel openings of cube specimens. Various specimens were subjected to a *cyclic* test procedure which included: 3 × 2 hours between room temperature and 1000°C and a 16 hour hold at 800°C. Burst specimens showed no damage after 22 hours, but noticeable cracks were observed after 88 hours. Sealed collector tubes showed no degradation after 22 hours. A tube that had melted out after 200 hours of thermal exposure in air showed no further degradation after cyclic testing for 88 hours.

For *burst testing* of sealed and unsealed burst specimens the minimum requirement was the ability to withstand a pressure of 50 bar. Testing was conducted at pressures of several hundred bar. Unsealed specimens failed between 2 and 14 bar, which was unacceptable. The cause was insufficient bonding between the lamina. The performance was better for sealed burst specimens, but in only 2 out of 7 specimens did the burst strength exceed 50 bar. These were all counter flow tubes. All 5 tubes with parallel flow met the 50 bar requirement.

Air permeability was evaluated at $\Delta P = 1$ bar and 5 bar. For a wall thickness of 0.2 mm the specificied maximum allowable premeability was 1.25×10^{-1} g/m^{2}s. Only a single layer burst specimen, and a cross section of a parallel flow cube (in part) met this requirement, as well an air collector tube. However, the wall thickness for all specimens which ranged from 0.6 to 4.2 mm was greater than the specified value of 0.2 mm which implied that the data was actually less favorable than it appeared.

Thermal shock tests were performed on burst specimens, air collector tubes and matrix cubes. The tests results read like a litany of failures with a few promising results mixed in for good measure, and showed that, in general, the RBSN material could not consistently meet the requirements. Figure 21.35 shows burst specimens and air collector tubes following the tests. The test specimens were immersed in hot gas (to 1000°C or 1200°C) and cold air in an alternating fashion for 1 to 2 minutes. Two out of three sealed burst specimens withstood 300 cycles at a maximum temperature of 1000°C without damage. The third one broke after two cycles. A second lot of three other specimens broke after 15 s in the hot gas. The burst specimens ruptured

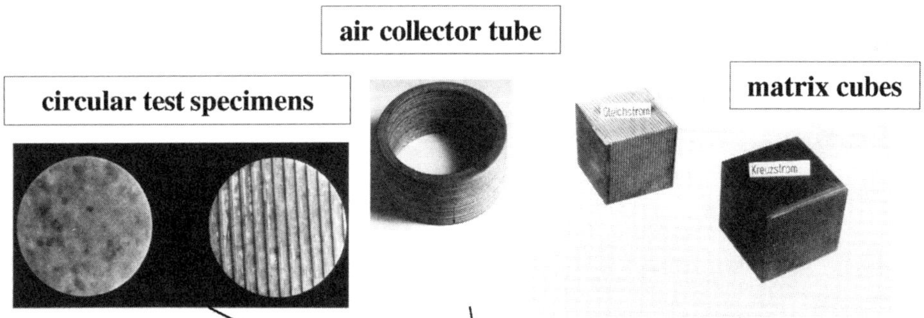

Density	X	X	X	X	X
Porosity	X	X	X	X	X
Oxidation resistance	X	X	X	X	X
Thermal shock resistance				X	X
Permeability		X	X	X	X
Burst strength					
Flexure strength	X				

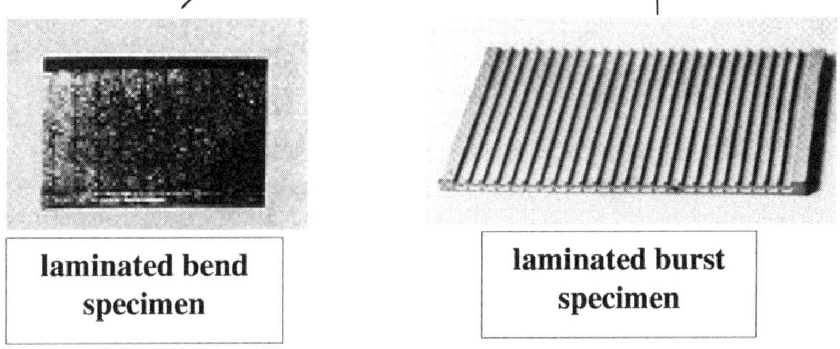

Figure 21.34 Test Specimens and the Tests They Were Used For (Adapted from Ref. [8], and with permission from Ref. [36], Figure 3)

at their centers. Three air collector tubes broke after a few cycles; two others broke after 193 and 195 cycles, respectively. The remaining four specimens withstood the required 200 temperature changes from 50°C to 1000°C/1200°C without damage.

Matrix cubes were tested in a dedicated rig. The thermal shock cycle had a hot gas temperature of 980°C and a cold air temperature of 70°C, the heating up and cooling down times were about 60s, the holding time at 980°C was 5 minutes, and the temperature after cooldown was 200°C. Two parallel flow and three cross flow cubes were investigated. The two parallel flow specimens broke after 2 and 3 temperature cycles, respectively. Defective joints were the cause of the failures. Two of the three crossflow cubes survived 9 and 10 temperature cycles, respectively, after which a significant crack formed. The third specimen showed small cracks after 10 cycles which did not intensify after 50 cycles. By the crossflow specimens the cracks traversed the foot print diagonally. The weak point was not the joint but the strength of the material.

(a) (b)

Figure 21.35 Thermal Shock Testing on Burst Specimens (a) and Air Collector Tubes (b) (Adapted from Ref. [8], and with permission from Ref. [36], Figure 3)

Phase III (1980–1983)

Table 21.9 summarizes the progress made throughout the recuperator development efforts. For eventual incorporation into a gas turbine all requirements had to be met without exception. The data in Table 21.9 indicate that close to the end of the program this was not the case [29]. Test specimens failed to meet the requirements with respect to oxidation resistance, burst performance and thermal shock resistance. Good progress had been achieved in meeting dimensional requirements and permeability. Still, the progress did not result in a commercially viable ceramic automotive recuperator at the time the program ended. Consequently, the ceramic recuperator was not used in DB's ceramic gas turbine road tests. Instead the ceramic engines were used in tandem with regenerators.

SUMMARY

Combustor

MTU and VW developed functional ceramic combustors for their specific applications, passenger cars and trucks, that met requirements of performance and emissions.

TABLE 21.9 PROGRESS MADE DURING CERAMIC RECUPERATOR DEVELOPMENT [29]

Parameter	DB Design Target	Status in 1980	Status in 1982
Dimensional (mm)	Matrix wall thickness: 0.2	~ 0.4	Thinnest: ~0.2
	Cell width: 0.2	0.5 –0.6	0.3, 0.45
Cell height:	0.7	0.68	0.6
Density (g/cm^3)	Maximum	2.3	2.38
Porosity (%)	Minimum	~28	~25
Oxidation resistance	No material degradation after exposure in air at: 1200°C/20 hours 1000°C/200 hours	Did not meet requirements	Did not meet requirements (crack formation upon furnace exposure)
Gas permeability	Maximum	$8.5 \times 10^{-1} - 1.0 \times 10^{0}$	$1.5 - 2.0 \times 10^{-3}$
at P=5 bar (g/m^{2}s)	1.25×10^{-1}	$7.0 - 8.0 \times 10^{-4}$	$2.5 - 7.5 \times 10^{-4}$
Burst performance	No damage at 50 bar	Approximately adequate	Inadequate (burst at 5 bar)
Thermal shock resistance	Cube test specimen: 70 × 70 × 70 mm No crack for ΔT=800 °C/3s	Inadequate- crack formation	Inadequate- crack formation

- Maintaining low temperature gradients along the combustor wall was found to be critical for avoiding cracking during start-up or cyclic operation.
- Ceramic combustors should be even and thin walled, without sudden changes in profile, and an adequate distance between the air holes.
- In general SiSiC performed better than RBSN as a combustor material.
- While initial results were favorable with SiSiC combustor liners, the residual Si easily separated out from the liner during rig operation and localized high silicon content was observed to promote crack initiation. The application of a silicon carbide material without residual silicon such as pressureless sintered silicon carbide (SiSiC) appeared to be preferable towards the end of the development program.

Nose Cone

The nose cone work was quite successful. A functional design was developed which was evaluated for several hundreds of hours and thousands of start/stop cycles in rig testing. SiSiC performed better than RBSN.

Nozzle

- The nozzle ring, a stationary part, is subject to relatively low mechanical loads. This allows the use of lower strength ceramics such as RBSN, SiSiC, SSN, or SSC. Stresses are noticeably lower in silicon nitride compared to silicon carbide materials.
- The primary considerations for nozzles are thermal shock resistance and oxidation/corrosion resistance. The thin vanes are subect to gas temperature changes of 500°C or more. The long term strength and durability are mainly determined by the oxidation resistance and structural stability of the material under thermal loads.
- The construction of the nozzle ring effects the life. Nozzle rings with inserted single nozzle segments and separate shroud rings, the outer one being segmented, survived the test without damage for a limited time. Damage typically occurs at the transition between nozzle and shroud ring and in the center of the nozzle airfoil.

Rotor

Volkswagen, MTU and Daimler-Benz developed and demonstrated several gas turbine rotor designs, including a duo-density rotor, a hybrid metal-ceramic design, and an integral rotor.

- The duo-density rotor consisting of a HPSN hub bonded to a RBSN blade ring, while in principal a cost-effective design, was abandoned because of insufficient strength of the bond between the hub and the blade ring.
- Significant progress was made with the concept of a hybrid metal rotor which is potentially also cost-effective. Both RBSN and HPSN blades were investigated. The position of the blade in the disk slot (not misaligned), the root angle (45° is preferred), the presence of a compliant layer or lenticular pad, and the machining direction of the root surface were all found to be important factors for blade life in service. HPSN blades performed best in spin testing. Sintered RBSN blades performed better than RBSN blades. None of the RBSN blades met the 25 hour life goal of hot spin testing. The materials and designs were felt to be inadequate to meet the conditions of vehicular operation.
- The use of an integral rotor was perhaps the most successful. Fully bladed HPSN rotors were tested in cold and hot spin rigs, a gasifier rig and an engine, and in road tests. Under the "Car 2000" program one HPSN rotor operated for 58 hours at temperatures up to 1350°C and a rotor speed of 61,500 rpm in a gasifier rig. An engine with HPSN rotors in a Mercedes-Benz car covered the distance from Stuttgart to Brussels, about 600 km, successfully. A total of 500 hours of rig and engine testing was accumulated. A commercialization development program was subsequently conducted with sintered silicon nitride (SSN), GTE's AY6 for injection molded, and a Kyocera's SSN for slipcast silicon nitride rotors. These rotors were also evaluated in rig and engine testing.

Recuperator

- Daimler-Benz selected a stationary ceramic recuperator for its automotive program. The matrix material was RBSN because of its conductivity characteristics and attractive cost.
- While the material met dimensional requirements specimen testing indicated severe deficiencies in oxidation resistance, burst strength and thermal shock resistance. These problems had not been solved by the end of the BMFT program and consequently, the stationary recuperator was not used in engine road tests.

Lessons Learned

Many valuable lessons were learned in the areas of component design and integration of these components into higher level assemblies and eventually an engine with a significant ceramic content in the hot section. Designs were tailored to the severe transient conditions of cold start, acceleration and deceleration, and sudden stop of the automotive cycle. The extensive rig testing for the combustor, nose cone, nozzle ring, rotor, and recuperator components showed which materials and designs performed best for the target applications. Silicon carbide became the material of choice for the combustor liner and nose cone while silicon nitride performed best for the nozzle ring, rotor and recuperator components. A lot of experience was gained with SiSiC for the combustor liners but the separation of silicon from the component resulted in crack formation and it was realized that a material free of residual silicon such as pressureless sintered SiC was needed for long term operation.

Relatively low strength materials such as RBSN and SiSiC could be used for the nozzle application. Various nozzle designs were evaluated. A nozzle design with inserted blades and a segmented outer shroud ring appeared to perform best. But none of the designs and materials evaluated had the required long term reliability needed.

Several promising rotor designs were thoroughly evaluated. Cost considerations led to explore designs that incorporated low cost fabrication techniques (duo-density and hybrid rotor designs). The important parameters for a functional hybrid rotor design were determined such as minimizing the friction coefficient between blade root and disk, the use of compliant layers or lenticular pads to mitigate contact stresses, the effect of machining of the root surface, the optimum root angle, etc. However, limitations in part processing and materials strength could not be overcome during the program to result in a successful low cost rotor design. Most successful was the work with an integral HPSN rotor and subsequent development of commercially attractive lower cost sintered silicon nitrides. While the test targets were not fully met various rotors were successfully tested in engine and road testing, the ultimate proof of concept demonstration. Given time this work could have been successful.

Daimler-Benz developed the technology for rapid prototyping of HPSN components such as turbochargers. The lessons learned were subsequently applied to the fabrication of ceramic components for other automotive applications such as values for diesel engines, etc.

In general, the German BMFT program was a concerted effort to solve some very difficult problems in gas turbine design, ceramic materials processing, and component fabrication. The magnitude of the effort was probably underestimated by most of the participants when the program started but was realized towards the end. A strategic decision was made to refocus efforts in areas that were more likely to have near term benefits such as components for conventional automotive engines.

ACKNOWLEDGEMENTS

The author is indebted to Mr. Manfred Langer of Volkswagen AG, Mr. Gerhard Andrees of MTU, and Dr. Klaus Mörgenthaler of DaimlerCrysler who kindly reviewed the chapter, supplied several of the figures from their companies' archives, and sent literature related to work conducted under the BMFT program. The author also wishes to thank Dr. Ernst Gugel, former head of R&D at CFI, who reviewed the chapter and supplied additional information on the BMFT program, and Dr. Peter Walzer of the University of Aachen, and Dr. Bjoern Schenk of Honeywell Engines, Systems & Services, both of whom provided background literature to the program.

REFERENCES

[1] Walzer, P., *Die Fahrzeug-Gasturbine* (The vehicular gas turbine,") © VDI-Verlag gmbH, Düsseldorf, Germany, 1991.

[2] *Keramische Komponenten für Fahrzeug-Gasturbinen,* ("Ceramic Components for Vehicular Gas Turbines,") W. Bunk and M. Böhmer, eds., Project Status Workshop, 18 and 19 April, 1978, Bad Neuenahr, Germany, Copyright © Springer-Verlag, Berlin, 1978 (in German).

[3] *Keramische Komponenten für Fahrzeug-Gasturbinen II,* ("Ceramic Components for Vehicular Gas Turbines II,") W. Bunk and M. Böhmer, eds., Project Status Workship, 24–26 November 1980, Bad Neuenahr, Germany, Copyright © Springer-Verlag, Berlin, 1981 (in German).

[4] *Keramische Komponenten für Fahrzeug-Gasturbinen III,* ("Ceramic Components for Vehicular Gas Turbines III,") W. Bunk, M. Böhmer, and H. Kißler, eds., Project Status Workshop, 13–15 February 1984, Germany, Bad Neuenahr, Copyright © Springer-Verlag, Berlin, 1984 (in German).

[5] Krüger, W., Kugenbuch, B., Peschel, W.-H., and Siebmanns, W., "Entwicklung von hochwarmfesten Keramikbauteilen für Gasturbinen," ("Development of High Temperature Ceramic Components for Gas Turbines,") Bundesministerium für Forschung und Technologie, Forschungsbericht T79-09, Technologische Forschung und Entwicklung, (Federal Department of Research and Technology, Research report T79-09, Technological research and development), MTU Motoren- und Turbinen-Union München GmbH, July 1979.

[6] Hüther, W., and Krüger, W., "Entwicklung von hochwarmfesten Keramikbauteilen für Gasturbinen," ("Development of high temperature ceramic components for gas turbines,") Bundesministerium für

Forschung und Technologie, Forschungsbericht T84-006, Technologische Forschung und Entwicklung, (Federal Department of Research and Technology, Research report T84-006, Technological research and development), MTU Motoren- und Turbinen-Union München, gmbH, January 1984.

[7] Tiefenbacher, E., and Räuchle, W., "Bauteile aus Keramik für Gasturbinen," ('Ceramic components for gas turbines,") Bundesministerium für Forschung und Technologie, Forschungsbericht T79-153, Technologische Forschung und Entwicklung, (Federal Department of Research and Technology, Research report T79-153, Technological research and development), Daimler-Benz Aktiengesellschaft, December 1979.

[8] Tiefenbacher, E., Girntke, H., Mörgenthaler, K., Heinze, R., Neubrand, F., Räuchle, W., Stute, M., and Sulzbacher, G., "Bauteile aus Keramik für Gasturbinen Phase II," ("Ceramic components for gas turbines Phase II,") Bundesministerium für Forschung und Technologie, Forschungsbericht T79-153, Technologische Forschung und Entwicklung, (Federal Department of Research and Technology, Research report T83-198, Technological research and development), Daimler-Benz Aktiengesellschaft, September 1983.

[9] Mörgenthaler, K., and Tiefenbacher, E., "Bauteile aus Keramik für Gasturbinen Phase III," ("Ceramic components for gas turbines Phase III,") Bundesministerium für Forschung und Technologie, Forschungsbericht T86–027, Technologische Forschung und Entwicklung, (Federal Department of Research and Technology, Research report T86-027, Technological research and development), Daimler-Benz Aktiengesellschaft, June 1986.

[10] Bunk, W., "Overview on the German Program on Ceramic Components for Vehicular Gas Turbines," in *Proceedings of the First International Symposium on Ceramic Components for Engines*, Hakone, Japan, October 17–19, 1983, Shigeyuki Sōmiya, Eizo Kanai, and Ken-ichiro Ando, eds., Elsevier Applied Science Publishers, Ltd., Barking, Essex, England U.K., 1986, pp. 9–20.

[11] Bunk, W., Gugel, E., and Walzer, P., "Overview of the German Ceramic Turbine Program," in *Ceramics for High-Performance Applications—III Reliability*, Proceedings of the Sixth Army Materials Conference," Edward M. Lenoe, R. Nathan Katz, and John J. Burke, eds., Copyright © Plenum Press, New York, 1983, pp. 29–50.

[12] Hagemeister, K., Tiefenbacher, E., and Walzer, P., "Ceramic Components for High-Temperature Vehicular Gas Turbines–State of the art of the German ceramic program," ASME paper 83-GT-205, presented at the 28th ASME International Gas Turbine & Aero Engine Congress and Exposition, Phoenix, Arizona, USA, March 27–31, 1983.

[13] Bunk, W., "Bericht Der Projektbegleitung Über Das Programm," ("Program Management Report,") in Ref. [2], pp. 3–10, 1978 (in German).

[14] Kappler, G. and Giesen, K., "Stand der Entwicklung von Brennkammern aus keramischen Werkstoffen," ("Status of the development of ceramic combustors,") in Ref. [2], pp. 299–319.

[15] Hagemeister, K., Hüther, W., and Langel, G., "Entwicklung von hochwarmfesten Keramikbauteilen für Gasturbinen," ("Development of high temperature resistant ceramic components for gas turbines,") in Ref. [4], pp. 329–359.

[16] Eggebrecht, R., and Langer, M., "Entwicklungsergebnisse mit keramischen Brennkammern und Turbinenenleitkranzen," ("Results of the development of ceramic combustors and turbine nozzle rings,") from Ref. [3], pp. 85–106.

[17] Kappler, G., "Status of combustor development from ceramic materials," in *Ceramics for High-Performance Applications–III Reliability*, Proceedings of the Sixth Army Materials Conference," Edward M. Lenoe, R. Nathan Katz, and John J. Burke, eds., Copyright © Plenum Press, New York, NY, USA, 1983, pp. 401–415.

[18] Walzer, P., Köhler, M., and Rottenkolber, P., "Hochtemperaturkeramik für Automobilgasturbinen," ("High temperature ceramics for automotive gas turbines"), MTZ Motortechnische Zeitschrift, Vol. 37 [11, 12], 1976. Reprint, pp. 1–12 (in German).

[19] Langer, M., and Krüger, W., "Stand der Entwicklung eines keramischen Turbinenleitkranzes," (Status of the development of a ceramic turbine nozzle ring") in Ref. [2], pp. 321–334.

[20] Burfeindt, H., and Langer, M., "Development of a Ceramic Turbine Nozzle Ring," *AGARD Conference Proceedings No. 276, "Ceramics for Turbine Engine Applications."*

[21] Walzer, P., and Siebels, J., "Stand der Entwicklung eines Turbinenlaufrades aus einem Siliziumnitrid—Verbund," ("Status of the Development of a Turbine Rotor Using a Silicon Nitride Joint,") from Ref. [2], pp. 335–351.

[22] Trappmann, K, and Rottenkolber, P., "Das Metall/Keramik-Turbinenrad" ("The Metal-Ceramic Turbine Wheel") in Ref. [2], pp. 353–371.

[23] Hüther, W., "Erprobung verschiedener keramischer Materialien für Turbinenlaufschaufeln im Kalt- und Heiβschleuderversuch," (Evaluation of various ceramic materials for turbine rotor blades in cold and hot spin testing,") in Ref. [3], pp. 243–254.

[24] Langer, M., Walzer, P., and Siebels, J.E., "Ceramic Component Development for Automotive Power Plants," in *Proceedings of the First International Symposium on Ceramic Components for Engines*, Hakone, Japan, October 17–19, 1983, Shigeyuki Sōmiya, Eizo Kanai, and Ken-ichiro Ando, eds., Elsevier Applied Science Publishers, Ltd., Barking, Essex, England U.K., 1986, pp. 100–109.

[25] Langer, M., Siebels, J., and Heinrich, H., "Turbinenrotoren aus nichtoxidischen keramischen Werkstoffen" ("Turbine Rotors from non-oxide ceramic materials") in Ref. [4], pp. 419–447.

[26] Walzer, P. and Langer, M., "Hochtemperaturkeramik für Verbrennungskraftmachinen" ("High Temperature Ceramics for Internal Combustion Engines"), MTZ Motortechnische Zeitschrift 44[6], pp. 225–228, 1983.

[27] Langer, M, Heinrich, H., and Siebels, J.E., "Development of ceramic components for car gas turbines," High Temperature Technology, Vol. 2 [4], pp. 225–228, 1984.

[28] Mörgenthaler, K.D., and Bühl, H., "Ceramics for Engines," in "*Tailoring of Mechanical Properties of Si_3N_4 Ceramics*", M.J. Hoffmann and G. Petzow, eds., Kluwer Academic Publishers, The Netherlands, 1994, pp. 429–441.

[29] Mörgenthaler, K.D, "Bauteile aus Keramik für Gasturbinen," ("Ceramic components for gas turbines,") in Ref. [4], pp. 389–417.

[30] Moergenthaler, K.D, Hempel, H., Jungbauer, L., Stute, M., Wiest, H., and Wunderlich, K., "Development and Testing of an Axial Silicon Nitride Gas Turbine Rotor," ASME paper 90-GT-96, presented at the Gas Turbine and Aeroengine Congress and Exposition, June 11–14, 1990, Brussels, Belgium.

[31] Tiefenbacher, E., "Turbinenrad aus heissgepresstem Siliziumnitrid," ("Turbine rotor out of hot pressed silicon nitride,") from ref. [2], pp. 373–386.

[32] Tiefenbacher, E., "Turbinenrad aus heissgepresstem Siliziumnitrid," (Turbine rotor out of hot pressed silicon nitride,") from ref. [3], pp. 217–226.

[33] Stute, M., Burger, H., Griguscheit, M., Holder, E., Mörgenthaler, K.D., Neubrand, E. and Radlof, M., "Testing of Ceramic Components in the Daimler-Benz Research Engine PWT 110," ASME paper 90-GT-97, presented at the Gas Turbine and Aeroengine Congress and Exposition, June 11–14, 1990, Brussels, Belgium.

[34] Bandyopadhyay, G., French, K.W., Sordelet, D.J., and Moergenthaler, K.D., "Fabrication and Development of Axial Silicon Nitride Gas Turbine Rotors," ASME paper 90-GT-47, presented at the Gas Turbine and Aeroengine Congress and Exposition, June 11–14, 1990, Brussels, Belgium.

[35] Tank, E., "Entwicklung eines Rekuperativ-Wärmetauschers aus Siliziumnitrid," ("Development of a silicon nitride-based recuperator,") in Ref. [2], pp. 283–298.

[36] Heinze, R. and Mörgenthaler, K.D., "Prüfung von Laminiertem RBSN-Matrixmaterial für recuperativ-Wärmetauscher," ("Demonstration of laminated RBSN matrix material for recuperators,") in Ref. [3], pp. 107–125.

CHAPTER

22

Ceramic Components for Automotive Gas Turbines – Experience at Volvo

Robert Lundberg, Rolf Gabrielsson

Volvo Aero Corporation, Trollhättan, Sweden

INTRODUCTION

Ceramic materials have been of interest for Volvo for the last 20 years. Especially for small gas turbines for automotive applications. These small gas turbines typically have a heat exchanger resulting in high temperatures for turbine and combustor parts. Target efficiency typically leads to a turbine inlet temperature of 1350°C. The small size of the components in an automotive gas turbine makes them more suitable for monolithic ceramics. This chapter outlines the development work done at Volvo Aero Corporation of ceramic components for automotive gas turbines. The focus is on design and evaluation. Component manufacturing for the Volvo gas turbines is described in more detail in another chapter of this publication [1].

EARLY CERAMIC GAS TURBINE WORK

A series of three shaft machines with an axial ceramic turbine were developed based on a concept of Prof. S-O. Kronogård [2–5] (see Figure 22.1). The first machine named MK1, a 100 kW gas turbine, was developed in the early 1970s. This gas turbine, which had an axial ceramic turbine wheel, was probably the first gas turbine with a ceramic turbine tested in a car (a modified Volvo 264 in the early 1980s [5]). The axial turbine wheel and heat exchanger core were the only ceramic parts tested in this engine. Injection molding of the wheel was developed in close collaboration with ASEA Cerama [1]. The material was HIPed silicon nitride, processed using the glass encapsulation method. The car installation was a direct gas turbine drive installation (see Figure 22.2).

The MK1 gas turbine was futher developed into the GT110 gas turbine. The GT 110 was also a 3-shaft machine, this time with an axial silicon nitride turbine wheel, ceramic Lean-Premixed Prevaporized (LPP) combustor and ceramic heat exchanger core. The GT110 gas turbine was built and bench tested in the mid-1980s. Again, the main supplier of turbine wheels was ASEA Cerama.

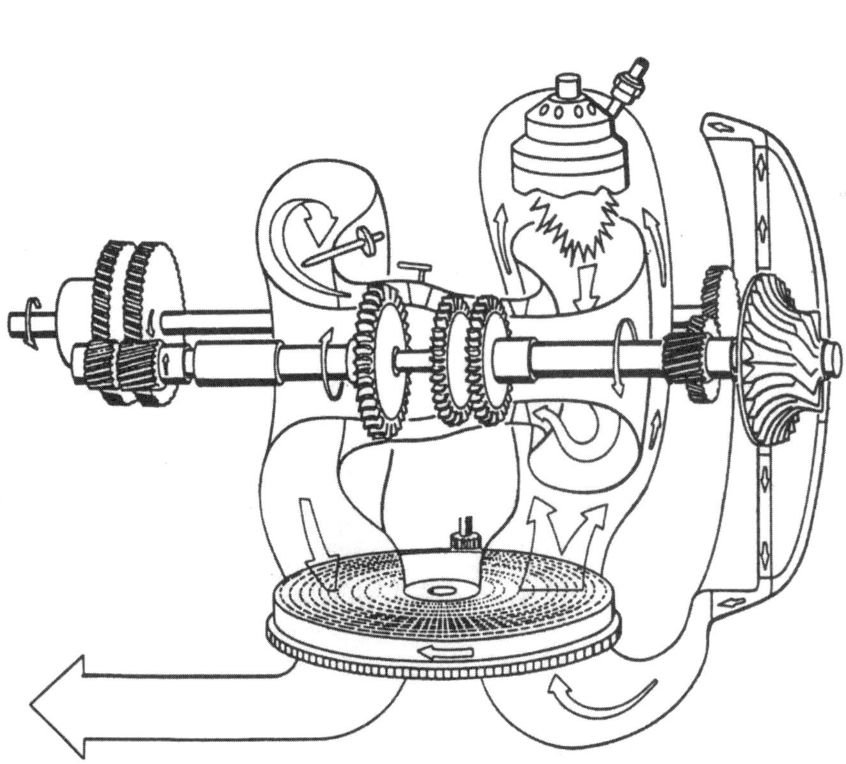

Figure 22.1 Schematic of Volvo 3-Shaft System (Reprinted with permission from Volvo Aero Corporation)

Figure 22.2 Installation of the MK1 Engine in a Volvo 264 Automobile (Reprinted with permission from Volvo Aero Corporation)

Figure 22.3 Silicon Nitride Combustor Tested in the GT110 Engine (Reprinted with permission from Volvo Aero Corporation)

Important lessons learned from these projects were for example: a procedure for developing rotors by injection molding and glass encapsulated HIP through a series of experimental geometries (spin disk→blisk with rough geometry→final geometry). A method of attaching and balancing for cold spin testing without balance machining was also developed. Shrink fitting was used as the joining method between ceramic wheel and metal shaft. A procedure for balancing and machining was established in collaboration with the ceramic manufacturer ASEA Cerama. Weibull statistics were included in the design calculations through a set of subroutines developed in-house in combination with commercial finite element (FE) software. In the combustor area a design for the assembly of a ceramic combustor through external spring loading was developed. Several different ceramic materials and suppliers were evaluated (for example silicon infiltrated SiC and slip cast and sintered Si_3N_4—see Figure 22.3) as combustor material.

A rotating ceramic heat exchanger with a ceramic core was developed. The same type of heat exchanger was later used for the Volvo Environmental Concept Car (ECC). This hybrid electric car had a 40 kW single shaft gas turbine (although without ceramics, except for the heat exchanger—see Figure 22.4).

AGATA

Volvo then continued in the European AGATA (*A*dvanced *GA*s *T*urbine for *A*utomobiles) project, running from 1992 to 1998 [6–17]. In this project three components for a 60 kW turbogenerator in a hybrid

Figure 22.4 Installation of 40 kW Turbogenerator in Volvo ECC Hybrid Electric Car (Reprinted with permission from Volvo Aero Corporation)

electric vehicle were developed: ceramic catalytic combustor, ceramic radial turbine wheel, and ceramic recuperator. These three components were tested in separate rigs. The AGATA partners were car manufacturers as well as companies and research institutes in the turbine, catalyst and ceramic material fields in both France and Sweden. The conceptual AGATA engine is shown in Figure 22.5. The AGATA specification is listed in Table 22.1

In AGATA it was possible to draw on the experience from the earlier three shaft machine, for example using the same balancing procedure, the same cold spin burst test rig and procedure, the Weibull approach adopted earlier as well as the spring mounting of the combustor.

AGATA—Catalytic Combustor

A catalytic combustor provides an attractive, alternative means of achieving lean air/fuel mixtures and low combustion temperatures, with the aim of reducing thermal NO_x emissions. The concept of the catalytic combustor was developed further as part of the AGATA project. The combustor design is shown in Figure 22.6.

The specification related to the catalytic combustor is summarized in Table 22.2.

The overall dimensions chosen for the catalytic section were:

- Catalytic section total length: 172 mm and diameter: 100 mm, built up in sections.
- The first section was an inactive substrate with a thickness of 12.5 mm as a monolithic guard for the residual droplets from the premixer during transient or low temperature operations.
- The active catalytic section was built up with 5 segments of 25 mm thickness and with a cell density of 350 cpsi. The pressure drop over the catalytic section was 3.3%.

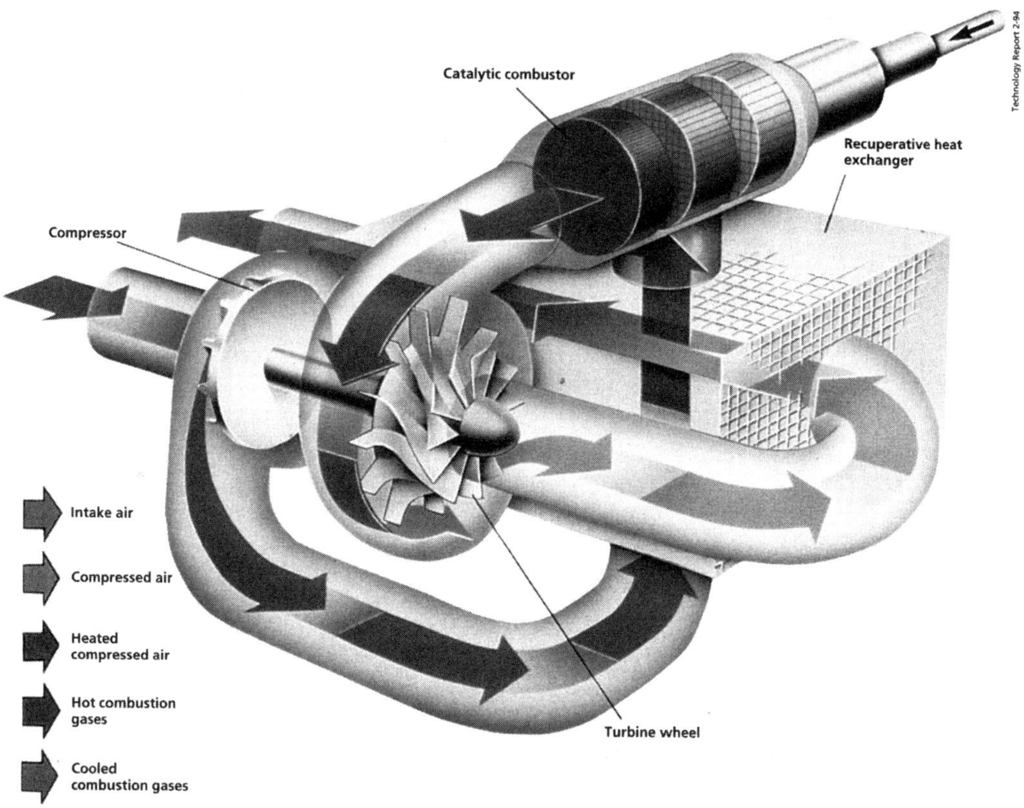

Figure 22.5 The Conceptual AGATA Gas Turbine (Adapted from Ref. [6])

TABLE 22.1 AGATA SPECIFICATION (ADAPTED FROM REF. [6])

Mechanical output	60 kW
Specific fuel consumption	200 g/k Wh
Turbine inlet temperature	1350°C
Emission standard	ULEV or equivalent European standard
Fuel	Diesel or alternative fuel
Application	Hybrid electric vehicle

The main emphasis of the technical development program was on ceramic structural components and catalytic combustion. All of the ceramic components—principally the combustor housing, honeycomb substrate and afterburner—were designed to operate without cooling at 1350°C. The housing and afterburner were manufactured respectively by Aerospatiale, France, and AC Cerama, Sweden, from high-strength, thermal shock resistant, high-performance Si_3N_4 ceramics with proven potential for operation at 1350°C. The housing of the catalytic sections was made as rings. These rings were made of SiC-whisker-reinforced Si_3N_4

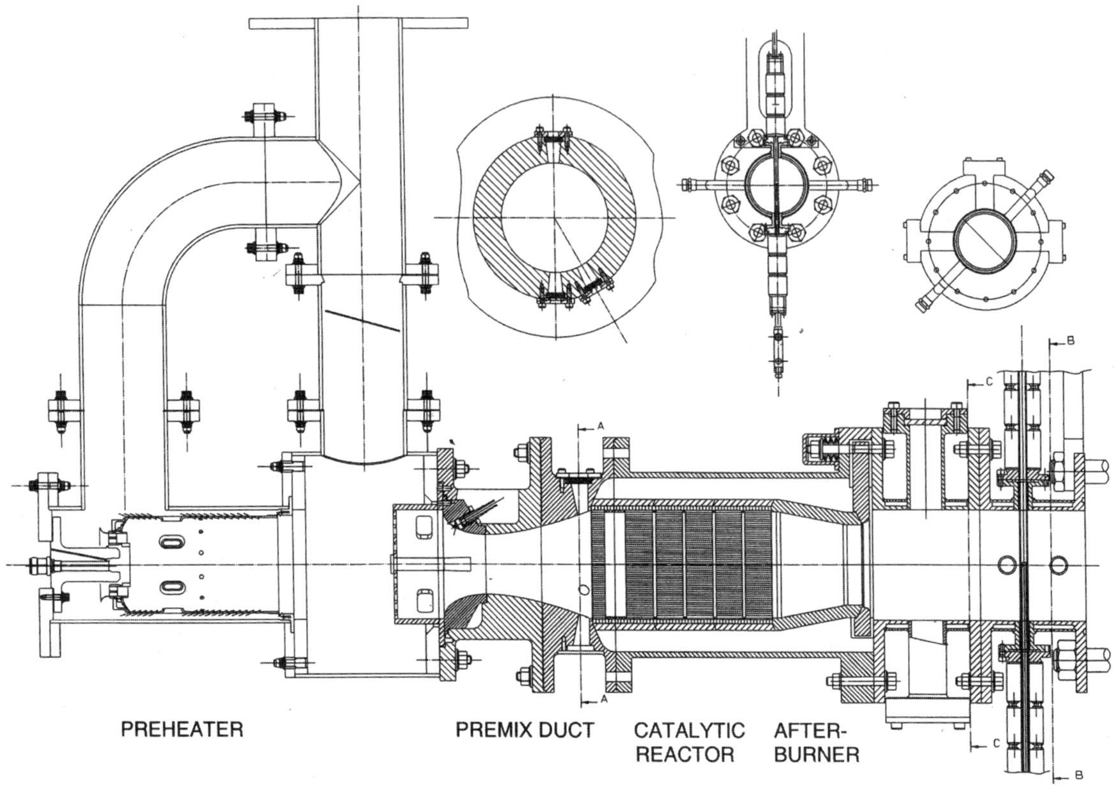

Figure 22.6 AGATA Catalytic Combustor Design (Adapted from Ref. [15])

TABLE 22.2 AGATA COMBUSTOR SPECIFICATION (ADAPTED FROM REF. [15])

Inlet temperature	935°C
Inlet pressure	3.9 bar
Fuel/air ratio	0.0128
Combustor pressure drop	6.5%
Outlet temperature	1350°C
Fuels	Diese/Ethanol

by Aerospatiale. Whisker reinforcement increases thermal shock resistance, toughness and creep resistance. The rings were manufactured by uniaxial hot pressing. The exit cone was manufactured by AC Cerama of HIPed CSN 101 grade Si_3N_4. The rings and cone were joined by polished conical contact surfaces and the combustor was held together by externally mounted springs.

The ceramic honeycomb catalyst substrates were developed by Céramiques & Composites, France. Extruded mullite-zirconia material, grade ZM3, was used. This type of material is widely used at an industrial scale for the manufacturing of honeycomb filters principally used in the refractory metallic alloy industry. The main geometrical objectives (wall thickness and cell density) met the AGATA specifications [16]. Palladium was chosen as the active catalytic material after extensive testing at the pilot scale. Ageing at high temperature (1270°C) was shown to have a strong effect on catalyst deactivation.

TABLE 22.3 AGATA CATALYTIC COMBUSTOR EMISSIONS RESULTS (ADAPTED FROM REF. [15])

	Measured emissions			
	100% load		50% load	
Emission	EI g/kg fuel	ppm @ 15% O_2	EI g/kg fuel	ppm @ 15% O_2
CO	0.99–1.26	22–28	1.69	38
HC	0.30–0.69	4–10	0.15	2
NOx	0.26–0.30	4	0.21	3

The full-scale testing of the combustor was successfully completed. The combustor testing was run in a number of steps in test rigs in France and Sweden, and included a comparison between diesel and ethanol fuels, as well as final testing at maximum design temperatures and pressure. The catalytic combustor was run on diesel fuel during the entire test period. A test campaign comparing exhaust emissions when running on ethanol fuel was performed at Volvo Aero. These results showed that the catalyst reaction rate and CO/HC/NOx emissions were similar. This means that the chosen catalytic combustor could be used as a dual fuel combustor for diesel/ethanol. The final test campaign at ONERA, France, was run up to temperatures slightly above the specified maximum design temperatures. The test rig has been described previously [11]. Inlet temperature was 962°C (design 935°C) and the exhaust temperature was 1362°C (design 1350°C). These tests showed that NOx emission levels below 4 ppm @15% O_2 were obtained when low CO and HC emissions levels were measured at full load conditions. Emissions data is given in Table 22.3.

During the full scale testing at ONERA and Volvo Aero Corporation the total running time of the catalytic combustor was approximately 100 hours. The individual catalytic sections have been run up to 20 hours. The catalytic segments and afterburner cone were facing the hottest environment. It should also be noted that there were no indications of catalyst reaction rate deterioration during the full scale test series.

The catalytic combustor material withstood transient tests and steady state tests at 1350°C. The catalyst substrates exhibited limited thermal shock cracking. The substrates were, however, held together by the slight compression of the mounting and the cracks were not detected until dismounting of the catalyst section. The silicon nitride afterburner showed no evidence of oxidation or mechanical damage. The catalytic section after test is seen in Figure 22.7. A more detailed description of the catalytic combustor and full scale testing has been presented elsewhere [9, 11, 14, 15].

AGATA Turbine Wheel

The aerodynamic and mechanical design of the wheel was done by Volvo Aero Corporation. The rotor was designed with a fairly high tip speed (125,000 rpm). The target efficiency was 86%. In a wheel efficiency test 84% was measured using a wheel of the final design. The overall geometry and blade thickness distribution was selected in cooperation with the ceramic manufacturers. Efforts were put into reduction of surface stresses at the blade roots. Thorough stress analyses, both transient and steady state, were carried out for several design iterations. Maximum stress, which was found to be around 290 MPa, occured during the cold start transient. Weibull statistics have also been included in the FEM program and failure probability has been calculated, using Volvo Aero Corporation developed subroutines. Life time calculations based on tensile creep experiments were also performed (tensile creep rupture results, see Table 22.4). The material selected was AC Cerama grade CSN 101 HIPed silicon nitride. The wheels were manufactured by injection molding. The wheel is shown in Figure 22.8.

Figure 22.7 AGATA Ceramic Catalytic Section After Test (Reprinted from Ref. [16])

TABLE 22.4 TENSILE CREEP RUPTURE (IN AIR) OF AC CERAMA CSN 101 (ADAPTED FROM REF. [13])

Bar#	Temperature	Stress	Test time	Observations
1	1350°C	50MPa	>1700 hours	No failure
2	1350°C	75 MPa	> 796 hours	No failure, load increased
3	1350°C	100 MPa	> 687 hours	No failure
4	1300°C	100 MPa	> 425 hours	No failure, load increased
5	1300°C	200 MPa	> 150 hours	No failure

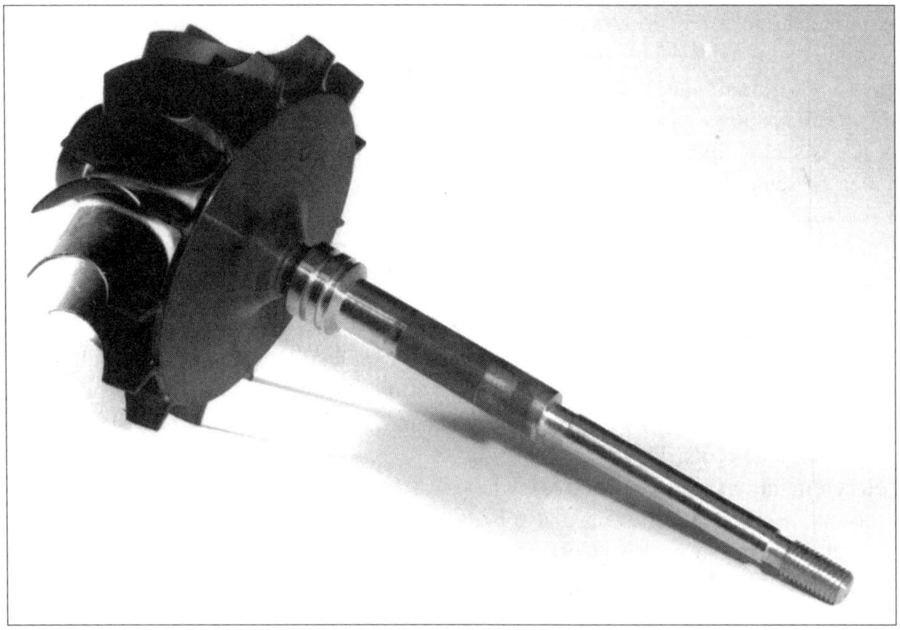

Figure 22.8 AGATA Turbine Wheel/Shaft Assembly (Reprinted from Ref. [16])

The ceramic wheel was to be joined to the metal shaft by brazing. A two-step brazing process was developed by CEA/CEREM, France, in which the ceramic surface is first metallized and then brazed to the metal shaft using an interlayer with intermediate thermal expansion.

Nondestructive testing is very important for ceramic component development. A three-dimensional X-Ray cone beam tomograph dedicated to the non destructive examination of ceramic turbine wheels was developed by CEA/LETI in partnership with Renault within AGATA. The principle of 3D X-Ray Cone Beam Tomography, the description and the working principles of the test set-up has been previously presented [13]. With this equipment it was shown that defects of about 150 μm could be detected and accurately positioned in ceramic turbine wheels. This tool was found to be useful in the development stage, since large processing flaws could be detected and their position determined. The resolution is not high enough though to make the method useful for the final production wheels. Here proof spin testing is probably the best method.

Both spin discs and wheels of the final design were cold-spin tested at room temperature in vacuum. During AGATA two bladeless dummy wheels and 13 wheels of the final design were tested to burst. The maximum burst speed reached was 91,000 rpm. An additional eight final design wheels were cold spin proof tested to 60,000, 80,000 and 90,000 rpm. This is far from the target operating speed of 125,000 rpm. Most wheels failed from fabrication defects. Typical defects were cracks created during injection molding/binder burn-out that became glass-filled rather than closed during HIP. Injection molding of the radial wheel with very thin blades and a thick hub was found to be more difficult than anticipated. Some design iterations were carried out such as increasing the blade thickness, as well as some processing iterations such as changing the mold filling location. A lesson learned is that a larger number of wheels need to be ready for spin testing earlier in the project so that more design iterations can be carried out. In AGATA we ran into unforseen ceramic manufacturing problems too late in the project to be able to optimize design and processing parameters, which is quite difficult for such a complex component as a radial turbine wheel.

Hot-spin tests in a gas turbine rig under actual pressure, temperature and aerodynamic loading conditions were planned within AGATA. These tests should take the form of start/stop, low-cycle fatigue tests. A hot spin rig was designed and built at ONERA. This rig was presented in a previous paper [11]. Hot spin testing may still be performed in a follow-on project to AGATA.

AGATA Silicon Nitride Development

To meet the demands of high strength and fracture toughness coupled with good creep resistance at 1350°C and good thermal schock resistance, AC Cerama AB developed the Si_3N_4 grade CSN 101. This material, containing high purity Si_3N_4 powder with 3 wt% Y_2O_3as sintering aid, can be green formed either by injection molding or by cold pressing. The densification is then performed using the proprietary glass encapsulation and hot isostatic pressing (HIP) technique as described in another chapter of this publication [1].

Mechanical properties of CSN 101 Si_3N_4 were evaluated within AGATA. Flexural strength from room temperature to 1350°C is shown in Figure 22.9. The Weibull modulus at room temperature was found to be as high as m=28. Tensile creep results are presented in Table 22.4.

Long term oxidation tests both in static air and in a high velocity gas burner were performed on bend specimens. After static oxidation exposure, a cracked glassy oxide layer covers the specimens. An average weight gain up to 0.26 mg/cm^2 was observed after the 1000 hour run. Burner rig hot corrosion environment tests were performed at Kyushu National Industrial Research Institute, Japan, with the following nominal conditions at one atmosphere:

- Fuel composition (vol%): C_4H_{10}: 21.9, O_2: 16.4, N_2: 61.7
- Gas/Air flow (Nm^3/h): 17/128
- Combustion gas velocity: 194–202 m/s

All burner rig exposed surfaces exhibited a powder like oxide surface layer easily removed by hand. With both types of oxidation we observed the growth of elongated crystals (assumed to be $Y_2Si_2O_7$) at the

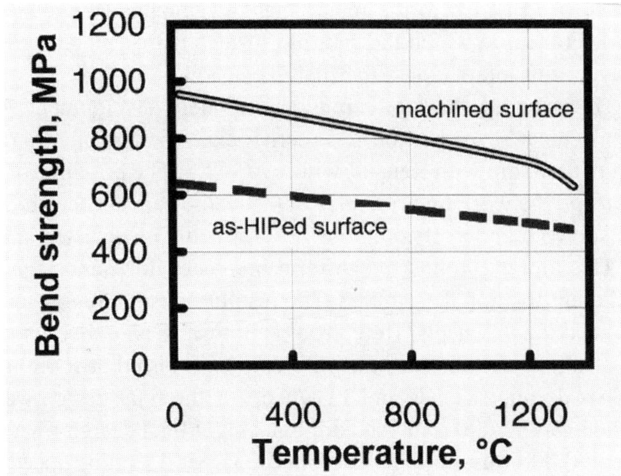

Figure 22.9 Bend Strength of AC Cerama Grade CSN101 HIPed Si_3N_4 (Adapted from Ref. [12])

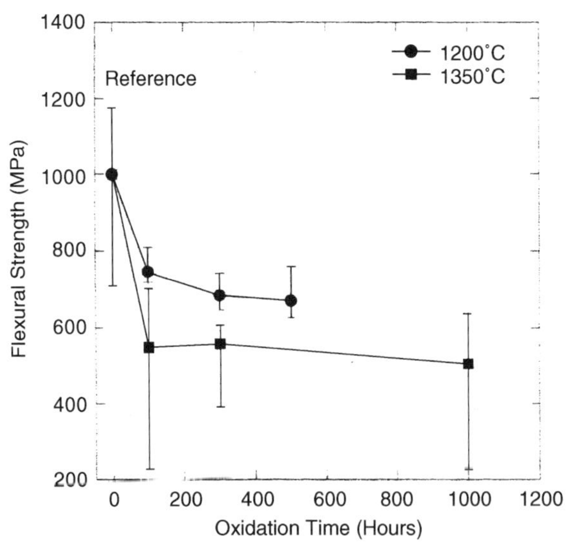

Figure 22.10 Flexural Strength After Static Oxidation (Bars Indicate the Maximum and the Minimum Values) (Reprinted from Ref. [13])

surface of the specimens. 4-point flexure tests were conducted after exposure. The residual 4-point bend strength, up to 1350°C in static air, is shown in Figure 22.10. After a fairly rapid drop the strength level was found to be rather constant up to 1000 hours. A similar rapid flexural strength drop is observed after the burner rig oxidation but the average strength is higher than with static oxidation (see Figure 22.11). These were considered encouraging results as the burner rig environment simulates more closely conditions within

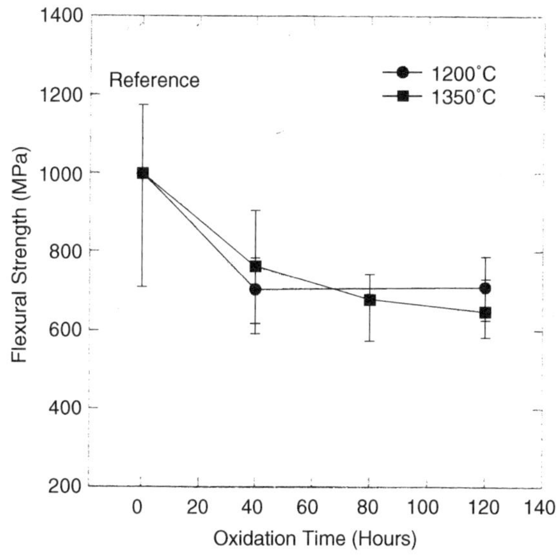

Figure 22.11 Flexural Strength After Burner Rig Exposure (Bars Indicate the Maximum and the Minimum Values) (Reprinted from Ref. [13])

TABLE 22.5 PROPERTIES OF AC CERAMA CSN 101 SI_3N_4 (ADAPTED FROM REF. [10])

Density	3.22 g/cm^3
Hardness	19 GPa
Flexural strength (RT 4-point)	955 MPa
Weibull Modulus (RT)	28
Flexural strength (1350°C 4-point)	630 MPa
Fracture toughness	6 MPa$\sqrt{m}$
Coefficient of thermal expansion	3.7.10^{-6}K^{-1} (20–800°C)
Thermal conductivity (RT)	40 W/m.K

gas turbines. Standard fractographic analysis has been conducted on the fracture surfaces of the specimens oxidized in the static and in the burner rig environment. For both the static and burner rig environment, the typical fracture cause was a kind of surface oxidation pit located at the tensile surface of the specimens. The shape was elliptical (from 30 to 180 μm long and as deep as 18 to 60 μm). The difference in strength was mainly caused by the position and size of the pit.

The burner rig results were very encouraging and showed that CSN 101 could be used for an automotive gas turbine application. It should be noted that the total operating life-time in an automotive application is significantly shorter than for an industrial or power generating gas turbine where oxidation/corrosion could be a more serious problem for nitride ceramics.

In summary it could be stated that the good room temperature strength, relatively high fracture toughness (6 MPa✓m, by the Chevron notch method) and excellent mechanical properties in the as-HIPed condition coupled with the ability to produce complex shapes at a low cost using injection molding, makes this Si_3N_4 grade a very interesting material also for other applications (a summary of material properties of CSN 101 is given in Table 22.5)

Figure 22.12 AGATA Ceramic Heat Exchanger (Reprinted from Ref. [16])

AGATA Heat Exchanger

A ceramic counterflow plate heat exchanger was developed within AGATA [16]. Alternating flat and sinusoidal cordierite sheets where bonded together during sintering. The heat exchanger is shown in Figure 22.12. The heat exchanger development is described in other chapters of this publication [18, 19].

CONCLUSIONS

From the MK1 and GT110, Volvo got a basic understanding for ceramic component design. This knowledge was subsequently used with in the AGATA project. The main results from the AGATA project were that a ceramic catalytic combustor was successfully tested both in transient conditions and at a steady state 1350°C turbine inlet temperature, demonstrating extremely low emissions. The mullite/zirconia substrates were compatible with the active catalyst and could withstand extended operation at 1350°C. For a future application, however, it is likely that a substrate material with better thermal shock resistance is needed. For the AGATA turbine wheel a new grade of HIPed silicon nitride was developed (CSN 101), showing very good mechanical properties. Encouraging oxidation/corrosion results were obtained showing that CSN 101 could be used for an automotive gas turbine application. It should be noted that the total operating life-time in an automotive application is significantly shorter than for an industrial or power generating gas turbine where oxidation/corrosion could be a more serious problem for nitride ceramics. Injection molding of a radial turbine wheel was developed, as well as a wheel/shaft braze joining technique that could withstand higher temperatures than shrink fitting. It was also shown within AGATA that a ceramic recuperator is feasible.

Concerning cost of ceramic components it is encouraging to see that ceramic catalyst substrates today are state-of-the-art in automotive exhaust catalysts. Silicon nitride parts are being introduced in wear parts in the valve mechanism of automotive piston engines, and silicon nitride balls find an increasing use in hybrid ball bearings. Estimates at Volvo Aero Corporation together with AC Cerama have shown that for example silicon nitride combustor parts manufactured by injection molding, with limited machining compares favorably with a metal design from a cost point of view.

In conclusion Volvo now has the level of competence necessary to introduce several ceramic components in series production gas turbines. The most likely path of introduction will be starting with small static silicon nitride parts, such as nozzle guide vanes. A ceramic catalytic combustor is also considered to be feasible at this point. When these components have been established, rotating components will follow.

ACKNOWLEDGEMENTS

The authors would like to thank the AGATA partners (AC Cerama AB and Volvo Aero Corporation in Sweden, Aerospatiale, AlliedSignal Catal. Environ. (ACE), AlliedSignal Turbo SA, CEA/CEREM, CEA/LETI, Céramiques & Composites, GRETH, IFP, ONERA, Peugeot SA and Renault, France, for their contribution to the technical reports; the French Ministries of Research and Industry as well as NUTEK (Sweden) for their sponsoring.

REFERENCES

[1] Larker, H.T., "Hot Isostatic Pressing of Ceramic Turbine Components to Shape—Experience at ASEA High Pressure Laboratory, ASEA Cerama AB, ABB Cerama AB and AC Cerama AB," Chapter 10 in *"Progress in Ceramic Gas Turbine Development," Volume II—"Ceramic Gas Turbine Component Development and Characterization,"* Mark van Roode, Mattison K.Ferber, David W. Richerson, eds, ASME PRESS, New York, USA, 2003.

[2] Kronogård, S.O. "Automotive Turbine—Advantages of Three Shaft Configurations," Gas Turbine International, Nov.-Dec. 1975.

[3] Kronogård, S.O., Malmrup, L., "Advanced Automotive Turbine Transmission System—Diesel Competitive Fuel Economy with Low Emissions," First Int. Automotive Fuel Economy Research Conference, Arlington, Virginia, USA, 1979.

[4] Holmqvist, G., Malmrup, L., "Component Development for the KTT MKI Automotive Gas Turbine" ASME paper 85-GT-204, presented at the Gas Turbine Conference and Exhibit, Houston, Texas, USA, March 18–21, 1985.

[5] Press release, March 27, 1982.

[6] Lundberg, R., "AGATA: A European Ceramic Gas Turbine for Hybrid Vehicles," ASME-paper 94-GT-8, presented at the International Gas Turbine and Aeroengine Congress and Exposition, The Hague, Netherlands, June 13–16, 1994.

[7] Lundberg, R., Gabrielsson, R., "Progress on the AGATA Project—A European Ceramic Gas Turbine for Hybrid Vehicles," ASME paper 95-GT-446, presented at the International Gas Turbine and Aeroengine Congress and Exposition, Houston, Texas, USA, June 5–8, 1995.

[8] Bauer, M., Lundberg, R., Gabrielsson, R., "Current Status of the European Ceramic Gas Turbine Program AGATA," *Proc. DOE Automotive Contractors Meeting*, Dearborn, Michigan, USA, 1995.

[9] Gabrielsson, R., Holmqvist, G., "Progress on the European Gas Turbine Program—AGATA," ASME paper 96-GT-362, presented at the International Gas Turbine and Aeroengine Congress and Exposition, Birmingham, U.K., June 10–13, 1996.

[10] Lundberg, R., Adlerborn, J., "Silicon Nitride Components for AGATA—A European Gas Turbine for Hybrid Vehicles," ASME paper 96-GT-287, presented at the International Gas Turbine and Aeroengine Congress and Exposition, Birmingham, UK, June 10–13, 1996.

[11] Avran, P., Leclair, A., Payen, G., "Facilities for Automotive Gas Turbine Ceramic Components," ASME paper 97-GT-464, Presented at the International Gas Turbine and Aeroengine Congress and Exposition, Orlando, Florida, USA, June 2–5, 1997.

[12] Lundberg, R., "HIPed Silicon Nitride Components for AGATA—A European Gas Turbine for Hybrid Vehicles," ASME paper 97-GT-322, presented at the International Gas Turbine and Aeroengine Congress and Exposition, Orlando, Florida, USA, June 2–5, 1997.

[13] Lundberg, R., Moret, M., Garguet-Duport, L., "HIPed Silicon Nitride Components for AGATA—Properties and Evaluation," ASME paper 98-GT-566, presented at the International Gas Turbine and Aeroengine Congress and Exposition, Stockholm, Sweden, June 2–5, 1998.

[14] Gabrielsson, R., Lundberg, R., Avran, P., "Status of the European Gas Turbine Program—AGATA," ASME paper 98-GT-392, presented at the International Gas Turbine and Aeroengine Congress and Exposition, Stockholm, Sweden, June 2–5, 1998.

[15] Gabrielsson, R., Payen, G., Avran, P., "Catalytic Combustor Development Within the AGATA Program," ASME paper 99-GT-442, presented at the International Gas Turbine and Aeroengine Congress and Exposition, Indianapolis, Indiana, June 7–10, 1999.

[16] Lundberg, R., Ferrato, M., "Ceramic Component Development for AGATA," ASME paper 99-GT-392, presented at the International Gas Turbine and Aeroengine Congress and Exposition, Indianapolis, Indiana, June 7–10, 1999.

[17] Gabrielsson, R., Avran, P., "The European Gas Turbine Program—AGATA," Paper OS-103, *Proceedings of the 7th International Gas Turbine Congress*, Kobe, Japan, November 14–19, 1999. Vol. I, pp 73–79.

[18] Ferrato, M., Martinez, G., "Ceramic Heat Exchanger Development Involving the French Ceramic Companies: Céraver, Céramiques&Composites and Boostec Industries," Chapter 11 in *"Progress in Ceramic Gas Turbine Development," Volume II—"Ceramic Gas Turbine Component Development and Characterization,"* Mark van Roode, Mattison K. Ferber, David W. Richerson, eds, ASME PRESS, New York, USA, 2003.

[19] Avran, P., Boudiques, S., "Ceramic Development at ONERA," Chapter 23 in *"Progress in Ceramic Gas Turbine Development," Volume I—"Ceramic Gas Turbine Design and Test Experience,"* Mark van Roode, Mattison K. Ferber, David W. Richerson, eds, ASME PRESS, New York, USA, 2002.

CHAPTER

23

CERAMIC DEVELOPMENT AT ONERA

Patrick Avran and Serge Boudigues

Office National d'Etudes et Recherches en Aérospatiales 29, avenue de la Division Leclerc, Chatillon-sous-Bagneux, France

INTRODUCTION

ONERA (Office National d'Etudes et Recherches en Aérospatiales), the French National Aerospace Research Institute, has been involved in a number of ceramic gas turbine and heat exchanger related programs since the early 1980s. The early R&D activities established expertise and capabilities in recuperation, combustion, and turbine development. Subsequently, because of its know-how and capabilities ONERA has participated in several collaborative programs with other institutes and companies in Europe and in Russia. ONERA has been active in both military and civilian programs. The military programs cannot be discussed in the context of this publication because of their classified nature, thus the activities in this chapter describe exclusively ONERA's civilian R&D efforts. The main focus of the chapter is on ONERA's heat exchanger work since it is in that area that the institute has unique expertise and capabilities. Other ceramic work related to catalytic combustion and gas turbine hot section development is cursorily discussed, primarily with reference to the work discussed in more detail in the chapters from Volvo Aero Corporation in Sweden (Volume I, Chapter 22), and the collaborative ceramic gas turbine and materials development work with NITI EM and related companies in Russia (Volume I, Chapter 32, and Volume II, Chapter 13). The principal programs discussed in this chapter are listed in Table 23.1.

HEAT EXCHANGERS FOR GAS TURBINES

The terms heat exchangers and recuperators are used interchangeably, as they connote different aspects of the same physical process in which heat from one stream of fluid (gas in the context of this chapter) is transferred to or recuperated by a second independent stream of fluid (gas). Typically, the first stream consists of hot exhaust gases from a combustion or incineration process, or a chemical process, while the second stream is clean air from a compressor. Figure 23.1 shows a typical arrangement of a recuperator for a gas turbine [1]. The exhaust gas exiting the turbine from stage T2 enters the hot gas path in a fixed recuperator while air from the compressor C enters the recuperator from the opposite site moving in a counterflow direction with respect to the hot gas stream. As the cold air moves through the recuperator the heat

TABLE 23.1 ONERA'S PARTICIPATION IN CERAMIC GAS TURBINE PROGRAMS

Timeframe	Program	Sponsor	Activities
1983–2000	Ceramic Recuperators	DRET	Recuperators for fuel efficient gas turbines
1993–1998	AGATA Program	European Union	Heat exchanger, catalytic combustion, ceramic gas turbine development
1993–2000	ONERA-NITI EM		Heat exchanger and ceramic gas turbine development

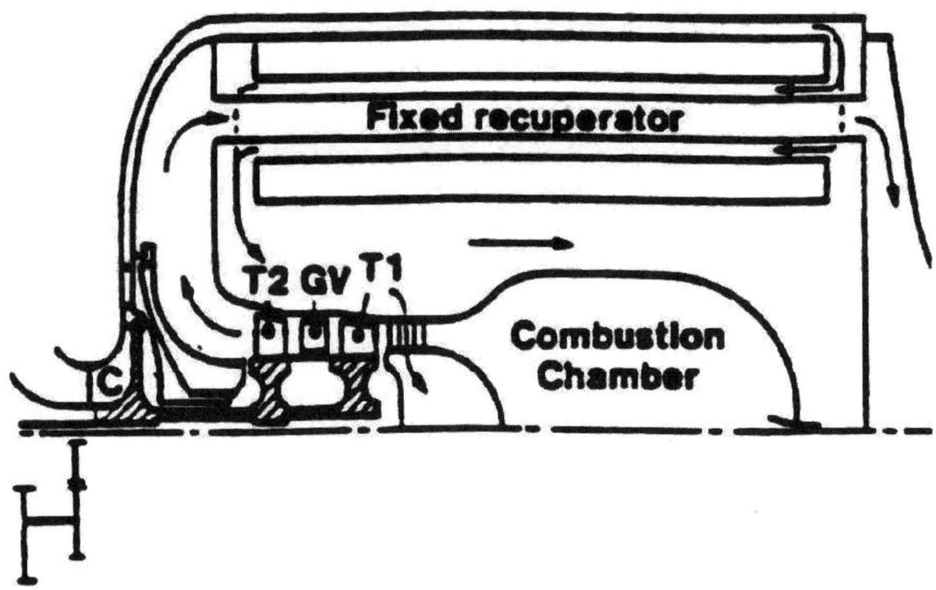

Figure 23.1 Recuperated Gas Turbine (Reprinted with permission from Ref. [1b])

transferred through the recuperator walls causes the temperature of the compressed air to rise. By the time the compressor air reaches the combustor it will have a much higher inlet temperature than compressor air without prior exposure to a recuperator. Therefore, less fuel will be required and the thermal efficiency of the engine is increased. Based on these principles various recuperator designs have been developed, including shell-and-tube types, plate-fin types, and regenerators. This chapter will primarily focus on the shell-and-tube type recuperator design.

Typically, in a simple cycle (non-recuperated) engine the compressor exit temperature increases with the number of stages. Since the recuperator injects additional heat to the air from the compressor exit fewer stages will be required to achieve the same effective compressor exit temperature in an engine with a recuperated cycle. The reduction in compressor stages is a cost-effective way to reduce engine cost, and may offset all or part of the increase in the cost incurred from the addition of the recuperator. The prerequisite for a recuperator material is that it provides an efficient conduction pathway for heat from the "hot" gas to the "cold" air stream. Both metals and ceramic materials with acceptable thermal conductivity have been used as recuperator materials.

Metallic Recuperators

ONERA has collaborated with MICROTURBO company of France to develop various fixed metallic recuperators, examples of which are given in Figure 23.2 to 23.4. The first one (Figure 23.2) uses a bank of simple

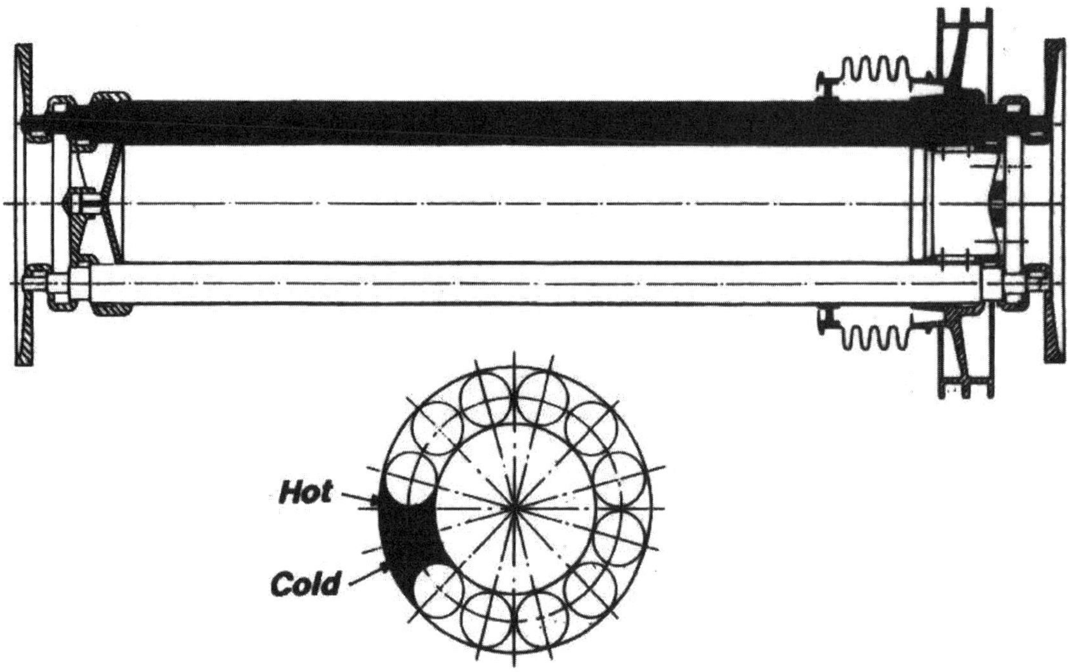

Figure 23.2 Type 1 Heat Exchanger (Reprinted with permission from Refs. [2a, b])

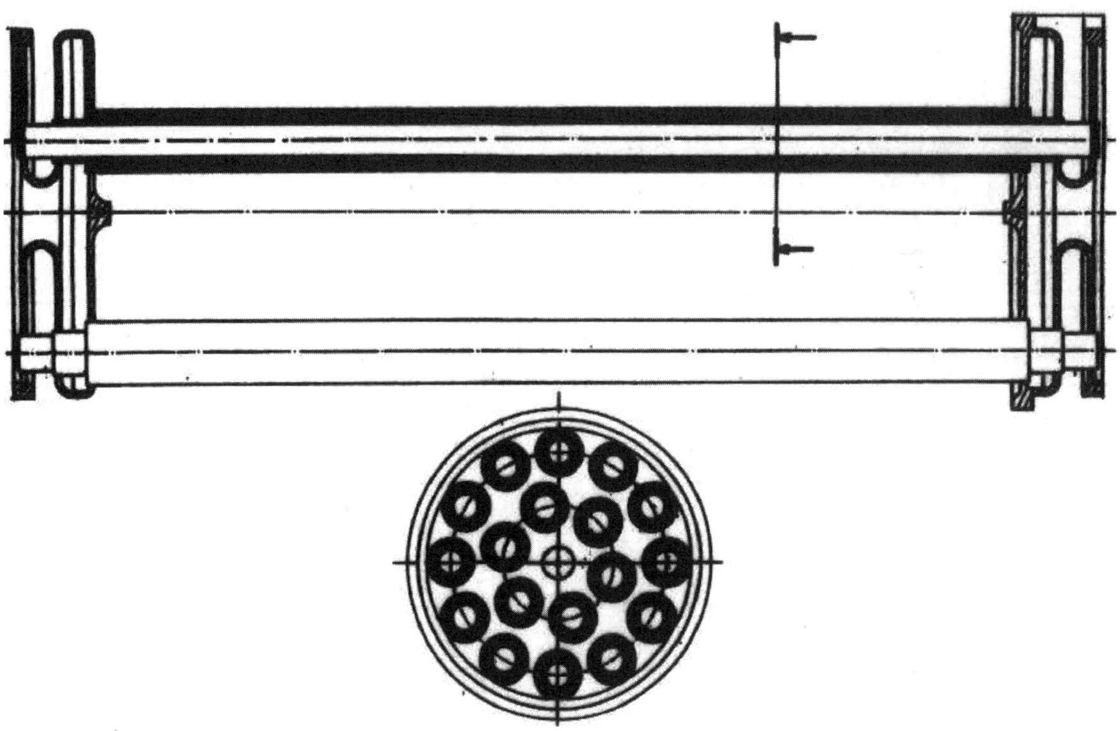

Figure 23.3 Type 2 Heat Exchanger (Reprinted with permission from Refs. [2a, b])

Figure 23.4 Type 3 Heat Exchanger (Adapted with permission from Refs. [2a, b])

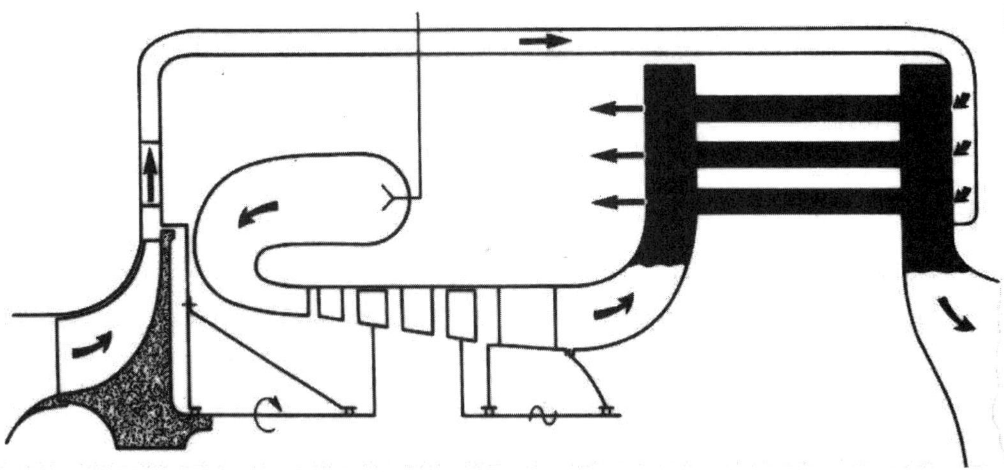

Figure 23.5 Traditional Recuperator Design Downstream from the Turbine (Adapted with permission from Ref. [3b])

concentric tubes with several circuits for the cold air and the hot gases. The second one (Figure 23.3) has two circuits, made of finned parallel tubes. The last one (Figure 23.4) consists of a bank of tubes placed on concentric cylinders with regards to the axis of rotation. As an example of the latter, a recuperator was manufactured with 240 primary tubes, each tube consisting of 33 smaller tubes placed in three concentric cylinders.

At an early stage excessive weight and voluminous spatial requirements were recognized as two major obstacles preventing the use of metal heat exchangers for road vehicles, and even more so, for aircraft applications. The traditional metal recuperator (Figure 23.5) is located downstream of the power turbine. The exhaust gases are at atmospheric pressure and the requirements for turbine cycle optimization necessitate a rather low compressor exit pressure. It is for these two reasons that the metal recuperator needs to be large, heavy, and therefore, expensive.

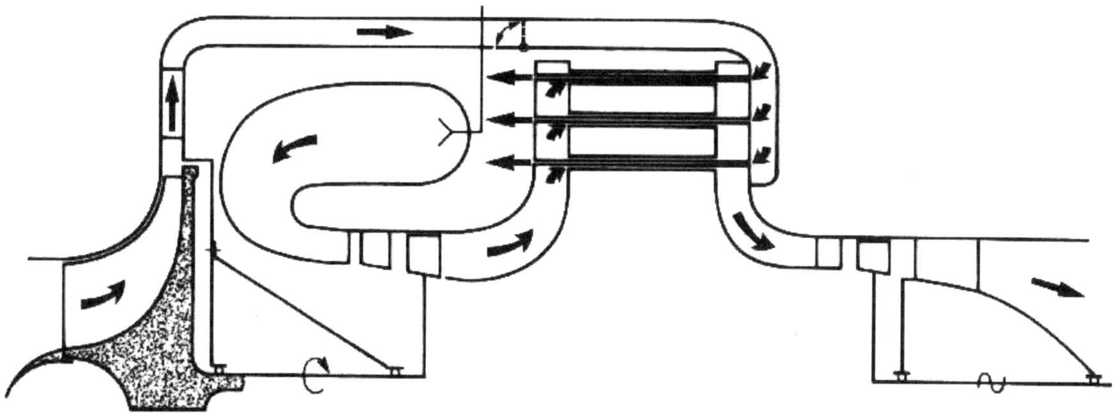

Figure 23.6 Intermediate Recuperator Design (Reprinted with permission from Ref. [3b])

One way to reduce the size, weight and cost of the recuperator is to locate it between the gas producer and power turbines (Figure 23.6). In this arrangement, the pressure of the gas entering the recuperator will be about three times atmospheric. Furthermore, optimizing the turbine cycle enables a much higher compressor exit pressure. With this dual strategy the recuperator size, weight and cost can all be reduced by a factor two compared to the traditional design, while obtaining the same overall system cycle performance. Of course, since the recuperator's performance is reduced, that of the gas turbine needs to be improved. This may actually be helpful, given that the turbine's performance is typically weak. Since the turbine is usually lighter and cheaper than the recuperator, the overall performance compares favorably for the intermediate design compared to that of the traditional concept.

The intermediate turbine design may increase performance without increasing the turbine inlet temperature. This could be particularly advantageous for helicopter turbines and for increased thrust for gas turbines in cruise missiles. Figure 23.7 shows the specific fuel consumption for a gas turbine in simple cycle (SC), for a gas turbine with intermediate recuperator (IE), and for a gas turbine with recuperator downstream from the turbine (DE). The data was generated for four recuperator efficiencies (0–0.6–0.7–0.8), and maximum cycle temperatures, T_{5i}, varying between 827°C to 1427°C (1100–1700K). It can be seen that even at low T_{5i} values, the recuperator has improved specific fuel consumption compared with the simple cycle with a very strong reduction in optimum compressor pressure ratio.

Overall one could say that for a gas turbine, recuperators (intermediate or downstream) give a slight improvement in specific fuel consumption, but, more so, they enable the same fuel consumption at much lower pressure ratios than for a simple cycle turbine. There is a trade-off between the added weight, space, and cost of the recuperator against the weight, space and cost savings from eliminating compressor stages. Note also, that for very low mass flows the recuperator route appears to be the only way to take advantage of high firing temperatures.

Ceramic Recuperators

ONERA's R&D under the sponsorship of DRET (Direction de Recherches, Etudes, et Techniques) resulted in significant contributions to the design, analysis and evaluation of ceramic heat exchangers for automotive and aircraft turbine applications. These development activities have been summarized in the literature [2, 3].

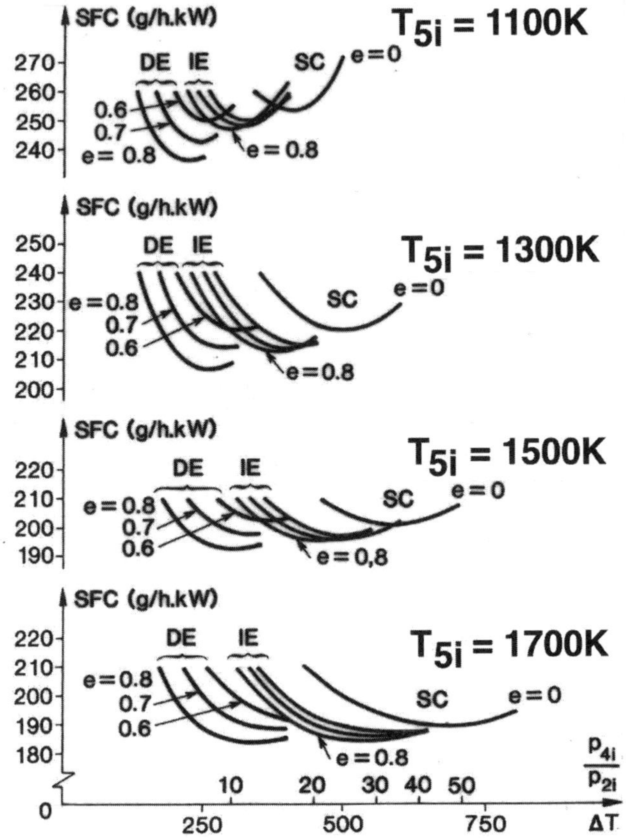

Figure 23.7 Comparison of Specific Fuel Consumption at Various T_{5i} and Recuperator Efficiencies (e)—Simple Cycle (SC)—Intermediate Recuperator (IE)—Downstream Recuperator (DE) (Adapted from Ref. [3b])

The incentive for ceramic heat exhanger R&D came from the realization that some silicon based advanced ceramics, with their low density, excellent thermal conductivity and superior high temperature durability, were prime candidate design materials for future automotive and aircraft applications requiring light weight and superior fuel economy. Figure 23.8 shows the range of properties for candidate materials for metallic and ceramic recuperators. From a durability perspective nickel- and cobalt-base alloys and superalloys are quite adequate for component temperatures below 1000°C, but for higher temperatures ceramic materials will be required. Properties of candidate ceramic materials have been listed in Table 23.2 [1]. Based on thermal conductivity which is a major property, silicon carbide ceramics are superior to silicon nitrides. As a result most of the ceramic heat exchanger development efforts have focused on silicon carbide materials.

Figure 23.9 shows a ceramic heat exchanger design developed by ONERA. Two design concepts, involving three or four concentric tubes, have been illustrated. Recuperator tubes of this concept were fabricated by Céramiques & Composites, a subsidiary of Rhônes-Poulenc, using an extrusion process. The concentric cylinders are in contact with each other through radial connections. The hot air and gas circulate in a counter-current flow pattern, demarcated by inter-tube connections and subsequent tube batteries. It is possible to improve the thermal utilization by further complicating the inlets and outlets

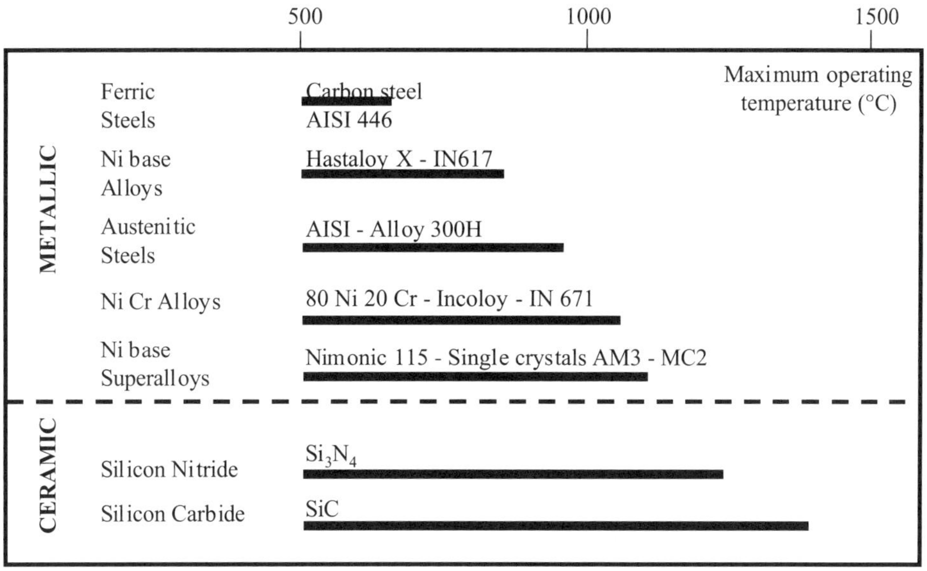

Figure 23.8 Operating Temperature Range for Candidate Recuperator Materials (Adapted with permission from Ref. [1b])

TABLE 23.2 PROPERTIES OF SILICON CARBIDE AND SILICON NITRIDE RECUPERATOR MATERIALS (ADAPTED WITH PERMISSION FROM REF. [1B])

Properties		Units	SiC	Si_3N_4
Physical	Open porosity	Vol. %	0	0
	Apparent Bulk Density	$g.cm^{-3}$	3.10	3.18
	Theoretical Density	$g.cm^{-3}$	3.21	3.28
Thermal	Coefficient of Linear Thermal Expansion			
	20–500°C	10^{-6}°C	4	2.4
	20–1000°C	10^{-6}°C	4.6	3
	20–1400°C	10^{-6}°C	4.9	3
	Thermal Conductivity at			
	20°C	$W.m.°C^{-1}$	170	20
	500°C	$W.m.°C^{-1}$	65	20
	1000°C	$W.m.°C^{-1}$	35	20
	Specific Heat at			
	100°C	$J.kg.°C^{-1}$	900	720
	500°C	$J.kg.°C^{-1}$	1230	
	1000°C	$J.kg.°C^{-1}$	1285	
Mechanical	Vickers Hardness	$kg.mm^2$	2400	1600
	Flexural Strength (3-pt)			
	20°C	MPa	450	650
	1000°C	MPa	430	620
	1200°C	MPa	430	600
	Compressive Strength at 20°C	MPa	2100	1900
	Young's Modulus at 20°C	GPa	420	290
	Elastic Modulus at 20°C	GPa	180	118
	Poisson's Ratio at 20°C		0.16	0.26
	K_{IC}	$MPa.m^{1/2}$	4.1	5.4
Electrical	Electrical Resistivity	'Ω-m	550	$<10^{13}$
	Dielectric Loss tgδ at 100kHz	10^{-4}	–	50
Service	Maximum Temperature in air	°C	1450	1200
	in inert atmosphere	°C	1800	1500

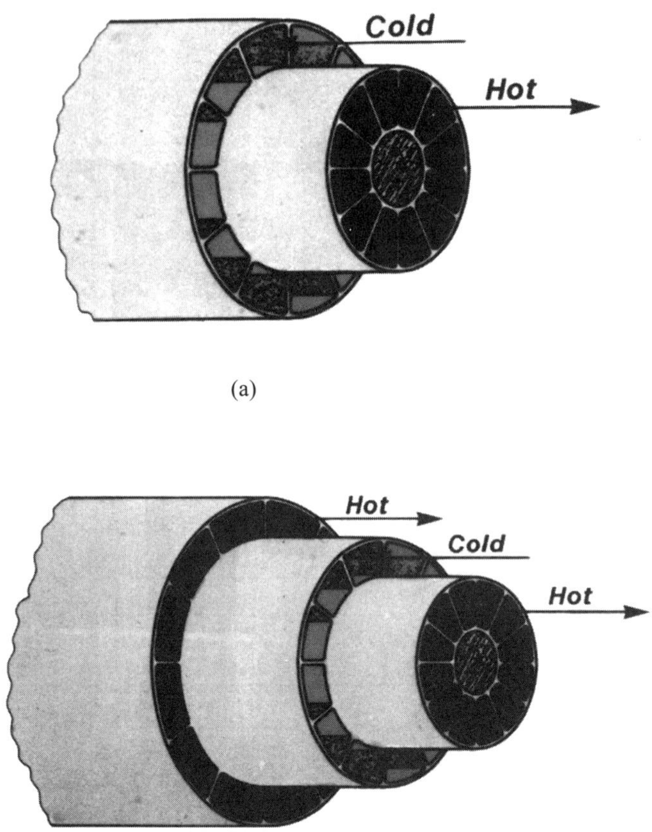

(a)

(b)

Figure 23.9 (a) Three and (b) Four Concentric Cylinder Recuperator Designs (Adapted with permission from Ref. [3b])

of the recuperator, and the number of concentric cylinders, but this will result in increased size, weight and cost. Three or four concentric tubes is the practical limit [3].

The bcst compromise between weight, length, size, and heat losses is obtained by finding an optimum arrangement of the number of tubes, the number of intra-tube connections, and their proximity. Figure 23.10 gives an example of a 450 kW recuperator design with an air flow of 1.6 kg/s, 8 bar air pressure, 3 bar gas pressure, a compressor exit temperature of 307°C (580K), and a recuperator inlet temperature of 877°C (1150K). The recuperator effectiveness is 0.7. It can be seen that the number of tubes (t), their interconnects (n), their length (L), their external radius (R), the weight (M), and the pressure drop ($\Sigma\varepsilon$) vary over a wide range. A small number of tubes and interconnects requires a small external radius (200 mm) and a heavy weight (75 kg), and a pressure drop of about 6%. On the other extreme, with a large number of tubes and interconnects, tube length can be low (250 mm), the weight low (45 kg), the pressure drop is about 4%, but the tube radius must be large (450 mm). For this example the weight of the gas turbine is about 90 kg.

The ceramic recuperator development at ONERA was conducted with subscale heat exchangers (330 mm tube length). Further design improvements envisioned involved modification of the tube external shape and internal configuration, and by changing the material from silicon carbide to SiAlON for greater thermo-

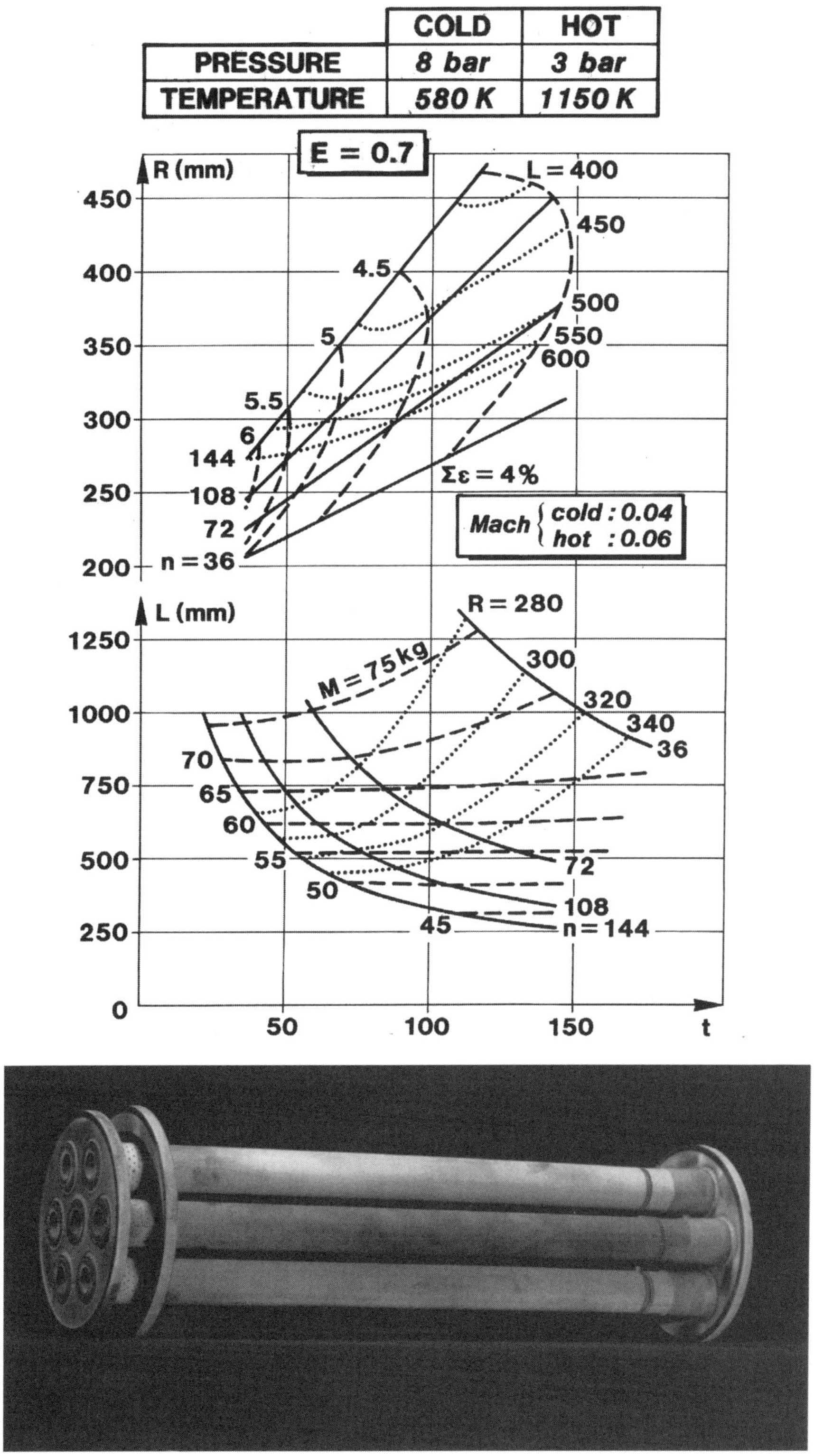

Figure 23.10 Range of Parameters for 450 kW Recuperator Design (Adapted with permission from Ref. [3b])

TABLE 23.3 CERAMIC HEAT EXCHANGER DATA (ADAPTED WITH PERMISSION FROM REF. [3B])

	Two-Circuit Recuperator		Three-Circuit Recuperator	
	Cold	**Hot**	**Cold**	**Hot**
Flow (g/s)	105	103	137	66 75
Inlet Pressure (bar)	9.61	2.85	8.55	3.05
Inlet Temperature (°C/K)	301/574	505/878	301/574	508/881
Pressure Drop (%) Measured	4.2	10	8.2	13.1
Pressure Drop (%) Calculated	3.5	12.1	4.0	6.9
Thermal Effectiveness Measured	0.701		0.720	
Thermal Effectiveness Calculated	0.695		0.707	

mechanical resistance. Improving the durability of the sealant and reducing its porosity would help to slow down the leakage of cold air to the hot gas stream. A major activity was the collaborative effort between Céramiques & Composites and ONERA to develop new sealants with improved mechanical adherence. The work involved modifying chemistries, fabrication processes and application methods.

Development work has moved on to longer prototype sub-assemblies (450 mm tube length) for full scale applications. Recuperator efficiencies are around 0.7 and further improvements are believed to be possible. Leakage rates remained at an unsatisfactory 2%. Table 23.3 shows experimental data for two-circuit and three-circuit recuperators. Measured thermal effectiveness data closely matches the calculated values (<2% difference), but the pressure loss discrepancies between measured and calculated values is not as good.

Critical Recuperator Parameters

ONERA conducted a theoretical comparative study for its fixed heat exchanger concept and the rotary heat exchanger widely investigated in ceramic gas turbine programs in the U.S. and Japan [2]. A number of conclusions were reached:

Mass Flow. Decreasing the mass flow improves recuperator effectiveness. By implication, recuperator effectiveness (e) increases for part load operation. For example, for a fixed recuperator e increases from 0.8 at full load to 0.930 at 30% load. For a regenerator e increases from 0.9 at full load to 0.968 at 30% load.

Leakage. Leakage will reduce recuperator effectiveness. For example, it is estimated that if e=0.9 for a regenerator at full load in the absence of leakage, then a 7% total leakage factor will reduce e to 0.853. Since recuperator effectiveness improves at part load, e at 30% part load increases to 0.918. Leakage is less detrimental for a fixed recuperator. Assuming that a fixed recuperator has no leakage the fixed recuperator's e is about 10% less than that of the regenerator at full load, but about 1% better than for the regenerator at 30% load. Total leakage is nearly proportional to pressure difference between air and gas. Attempts to counter leakages by increasing the pressure ratio are therefore counterproductive, since leakages will increase proportionally. To restore performance from leakages is expensive. For example, it is estimated that a regenerator with 7% leakage needs to be double in size, weight, and will be twice as expensive as a leak-free regenerator.

Another detrimental effect of leakage is that it tends to reduce the power of the turbine. It can be shown that the potential percentage loss in power always exceeds the leakage and can, in some cases, be almost double that of the percentage leakage. To compensate for the power loss, engine size and weight needs to be increased. It has been estimated that a 7% leakage factor requires increasing the engine weight by 21.7% or more to obtain the same output. Leakage also leads to an increase in fuel consumption.

TABLE 23.4 AGATA HEAT EXCHANGER SPECIFICATIONS (REPRINTED WITH PERMISSION FROM REF. [6])

Parameter	Value
Efficiency	90%
Mass Flow	0.26 kg/s
Air Inlet Temperature	208°C
Air Outlet Temperature	935°C
Gas Inlet Temperature	1011°C
Gas Outlet Temperature	324°C
Air Inlet Pressure	3.97 Bar
Air Pressure Drop	2%
Gas Inlet Pressure	1.07 Bar
Gas Pressure Drop	5%

It was shown that the cycle performance for an engine with a fixed metal recuperator with e=0.8 and zero leakage, operating at a turbine inlet temperature (TIT) of 1177°C (1450K), was similar to that of an engine with a ceramic regenerator with e=0.9 and 6.8% leakage, operating at a TIT of 1342°C (1615K). Thus, in spite of the TIT being 166°C higher, and a regenerator effectiveness of 0.9 vs. 0.8 for the fixed metal recuperator, the performance of both cycles was similar. Obviously, the leakage factor offsets the performance improvements resulting from the higher temperature capability of ceramics.

Ceramic Heat Exchangers for the AGATA Program

Because of the resident expertise in recuperator development ONERA became a partner in the European AGATA (*A*dvanced *GA*s *T*urbine for *A*utomobiles) program, dedicated to the development of three critical ceramic gas turbine components—a catalytic combustor, a radial turbine wheel and a ceramic heat exchanger—for a 60 kW turbogenerator in a hybrid electric vehicle. The AGATA program, initiated in 1993 was a feasibility study which addressed three components that are considered to be of critical importance to the achievement of low emissions and high efficiency in automotive applications. The ultimate objective of the AGATA program was to develop key technologies for hybrid vehicle applications, in which a catalytic combustor and gas turbine provides the power for long distance driving and an electric battery the power for urban driving. The gas turbine can charge the battery during highway driving. The AGATA program team consisted of automotive manufacturers as well as companies and research institutes in the turbine, catalyst and ceramic materials fields in both France and Sweden. The five year project commenced early in 1993 and finished in April 1998. For details of the AGATA program the reader is referred to the chapter by Volvo Aero Corporation (Volume I, Chapter 22) and by references to this chapter [4–7].

ONERA's role on the program was to conduct testing on the heat exchanger, turbine and catalytic combustor. ONERA brought to the program the resources of a well equipped facility and expert scientific and engineering staff that could conduct the sophisticated experiments required for the heat exchanger, gas turbine and catalytic combustor testing. The heat exchanger specifications are listed in Table 23.4.

The heat exchanger concept is based on stacking undulating layers separated by thin flat sheets (Figure 23.11). The thickness of the wavy layers is 0.2 mm, and of the flat sheets 0.3 mm, with the channel heights on the gas side 3 mm and on the air side 2 mm. The preferred ceramic material was cordierite which has good thermal conductivity and thermal shock resistance.

The bonding properties of the cordierite heat exchanger were evaluated during transient start-up and shut-down cycling at temperatures up to 1027°C (1300K) in the rig shown in Figure 23.12. These transient tests were run with 50 mm x 50 mm x 50 mm test samples. The bonding between the individual elements appeared to be very resistant to transient thermal stresses, although some microcracks were found in the air sealing bars at the sample edges.

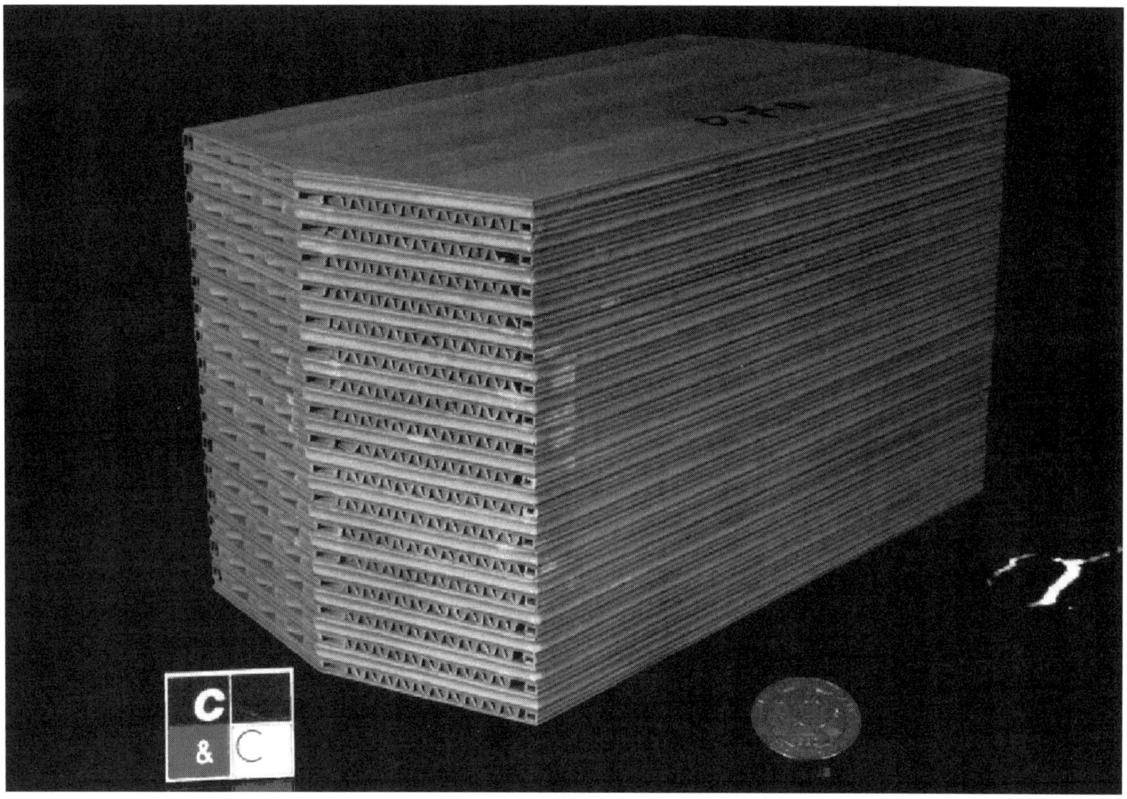

Figure 23.11 AGATA Cordierite Heat Exchanger on Scale 1:4 with Length Scale 1:1 (Reprinted from Ref. [4])

During the design phase a number of matrix configurations were evaluated. Volume and weight are essential parameters for automotive applications. Depending on the heat exchanger's hydraulic diameter and geometry, the matrix volume necessary to meet the AGATA specifications was varied from 10–67 liters when fabricated from cordierite. A configuration with 26 liters of matrix volume was chosen for testing. The recuperator configuration was also evaluated for SiC material which requires a volume of only 17 liters. Cordierite was ultimately selected because of the lower thermal stresses which were preferred notwithstanding the higher strength and lower volume of SiC. Total volume and weight including inlet/outlet zones and casing are about 35 liters and 50 kg, respectively.

The most critical aspect of recuperator design is the connection of the outlet and inlet ducts to the ceramic matrix to assure sealing of the air and gas circuits. The sealing was accomplished by springloading the air inlet and outlet sections (Figure 23.13).

ONERA's rig facilities were used to conduct one quarter scale and full scale testing of the AGATA heat exchanger. A one quarter scale heat exchanger with the inlet/outlet sections dismantled is shown in Figure 23.14(a), and installed in the test rig at ONERA in Figure 23.14(b). The heat exchanger test rig has two independent circuits, one for cold air, and the other for hot gases. Air is preheated by an independent recuperator that simulates the compressor outlet conditions. The hot gas is generated by a combustor. Its temperature is the same as that of the turbine exhaust which is led into the coupled heat exchanger. Pressure and mass flow are adjusted by gradual valves and sonic nozzles.

Transient and thermal shock conditions can also be simulated by this test rig. The various parameters are controlled during the test by thermocouples, pressure transducers, and data processing equipment. One

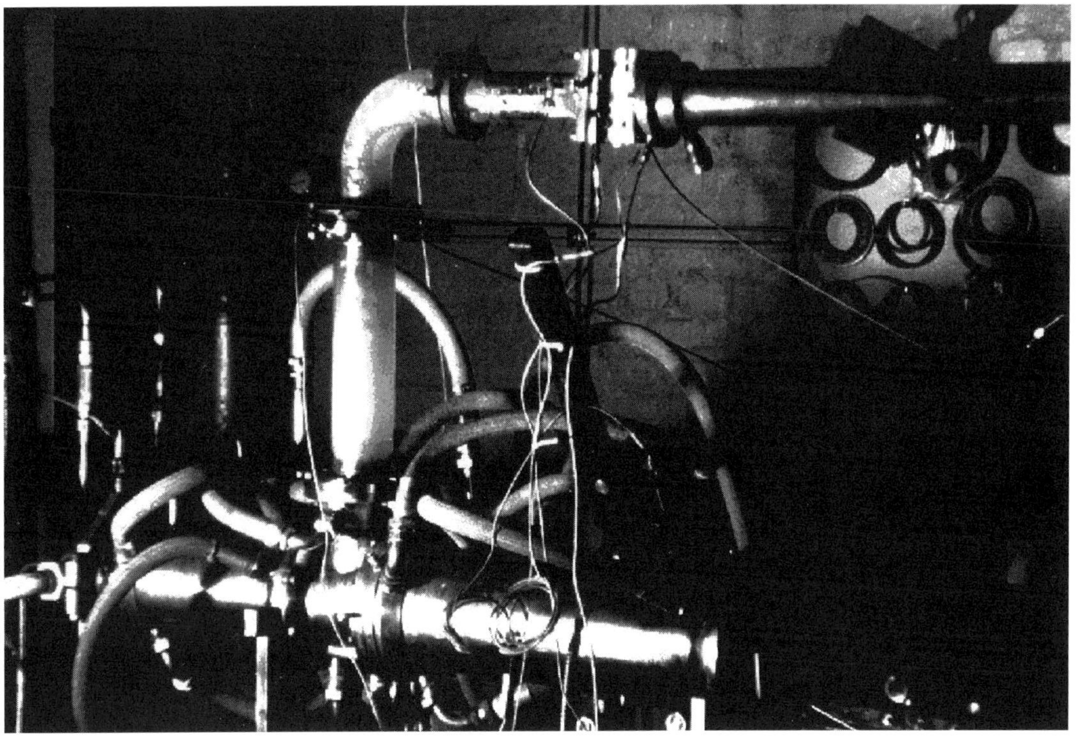

Figure 23.12 Apparatus for Stress-temperature Testing (Reprinted from Ref. [4])

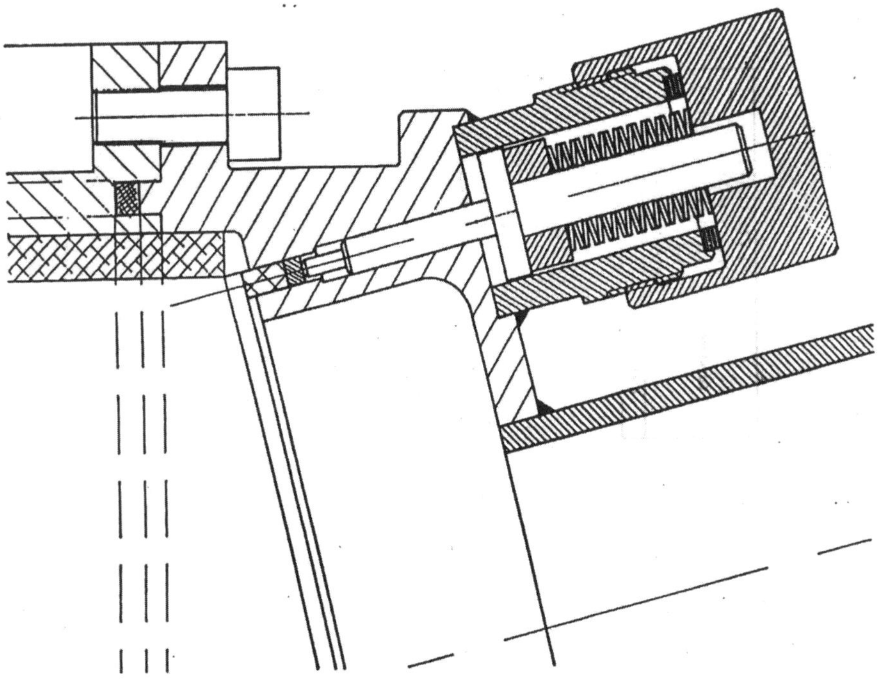

Figure 23.13 AGATA Heat Exchanger Air Inlet/Outlet Sealing Section (Reprinted from Ref. [4])

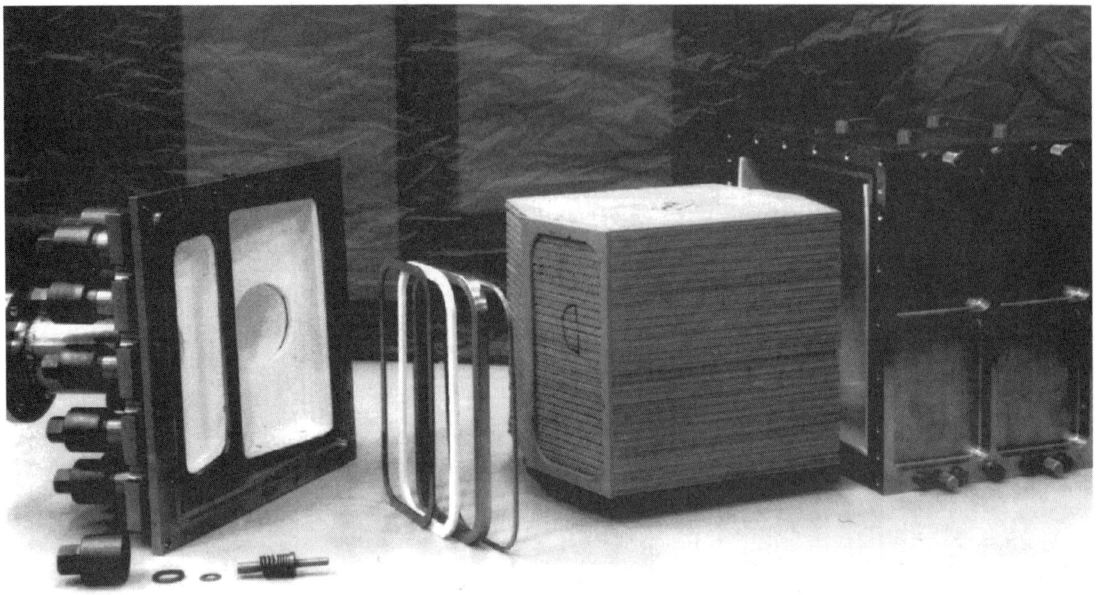

(a)

(b)

Figure 23.14 Heat Exchanger Hardware (a) Dismantled Inlet/Outlet Section, (b) Installed in the Test Rig (Adapted from Ref. [4])

of the critical aspects of recuperator design is the connection of the metal inlet and outlet ducts to the ceramic elements which ensures the airtightness of both the air and gas circuits. The incorporation of a metallic envelope helps to minimize potential leakage problems. The one quart scale heat exchanger was shown to be able to withstand the high temperature testing in the ONERA test rig. However, leakages were observed in the sealing section between the matrix and the casting [6].

ONERA/NITI EM Collaborative High Temperature Recuperator Study

Because of its recuperator expertise ONERA participated in a cooperative development program with researchers affiliated with the Research and Technological Institute for Power Engineering (NITI EM) in St. Petersburg, Russia. The collaboration involved theoretical studies about unsteady operating conditions of ceramic recuperative air heaters for automotive engines (mentioned previously), as well as design and testing of ceramic recuperators for a stationary gas turbine with a ceramic hot section. ONERA also contributed to the development of a new class of ceramic material for high temperature applications, SUGRAV, and the fabrication and testing of hot section components (rotor blades, nozzles, and combustor liners). The recuperator development efforts will be summarized in this section.

Automotive recuperators operate under unsteady conditions for considerable periods of time. The resulting variations in temperature lead to an increase in thermal stresses in the recuperator elements and assemblies, particularly in gas-gas ceramic heat exchangers. In a collaborative study, researchers of ONERA and of several institutes in St. Petersburg, Russia, analyzed the effects of duty and design factors on the temperature and stress distribution of ceramic recuperator matrix components [8]. A numerical analysis was performed using the thermal balance method applied to the compact plate and tube-fin heat exchanger. The time and zone with the maximum temperature variation were established, and the temperature and stress patterns in the components were determined. It was predicted that the stresses in the corner zones of the ceramic matrix were particularly high and a design change was recommended that would allow a decrease of critical temperature gradients and lowering of accompanying stress levels.

A modular ceramic heat exchanger was developed for automotive applications. Use was made of hybrid matrices, in which only the high-temperature section is fabricated from ceramics while the low-temperature section is fabricated from heat resistant alloys or steel. This mixed ceramic-metal hybrid design is cost-effective and minimizes recuperator size and weight. For the "cold" heat exchanger section a plate structure of profiled sheets with a surface structure of semi-spherical craters and bulges was developed. The "hot" section of each heat exchanger module is of the shell-and-tube counterflow type (Figure 23.15). The section matrix is made of ceramic tubes with longitudinal external ribbing. Tubes were fabricated of alumino-boron-nitride SUGRAV ceramics (see Table 32.2 in chapter 32 for materials properties) [9].

The SUGRAV ceramic has acceptable thermal conductivity and low shrinkage during heat treatment. The heat transfer surfaces of the tubes are relatively short ribs which extend the external heat transfer surface nearly three-fold. To enable heat transfer enhancement inside the tubes, ball turbolators are applied which allow a 3 to 5 times increase in the coefficient of convective heat transfer from gas to tube wall. These balls exert a hydrodynamic effect on the hot gas stream, but having been heated nearly to the temperature of the gas they radiate heat to the "cold" tube walls, thereby additionally enhancing heat exchange. To increase the energy effectiveness of the matrix inside path, the ball turbolators are arranged discreetly along the tubes. Such turbolator arrangement results in a sharp reduction in pressure losses. Further improvement in the heat transfer coefficient was achieved by using ribbed inserts separating the ball-turbolators (Figure 23.16). While these inserts increase pressure losses by 30–50% as a function of their structure and the surface extension factor, the net effect is still an increase in heat output by 15–25% without change in the external tube size [9]. In-series connection of the ceramic shell-and-tube section with the metal one enables to get a high gas turbine unit regeneration ratio (0.8–0.9). The recuperator compactness factor can be increased to 515–720 m^2/m^3, while the specific heat output can be increased up to 1.8–6.6 kW/kg.

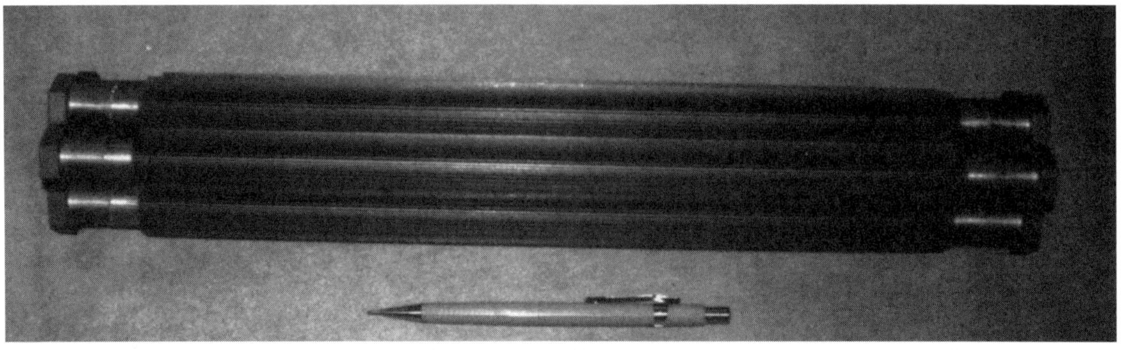

Figure 23.15 SUGRAV Ceramic Heat Exchange Element of the "Hot" Section of the Recuperator (Reprinted with permission from Ref. [9b])

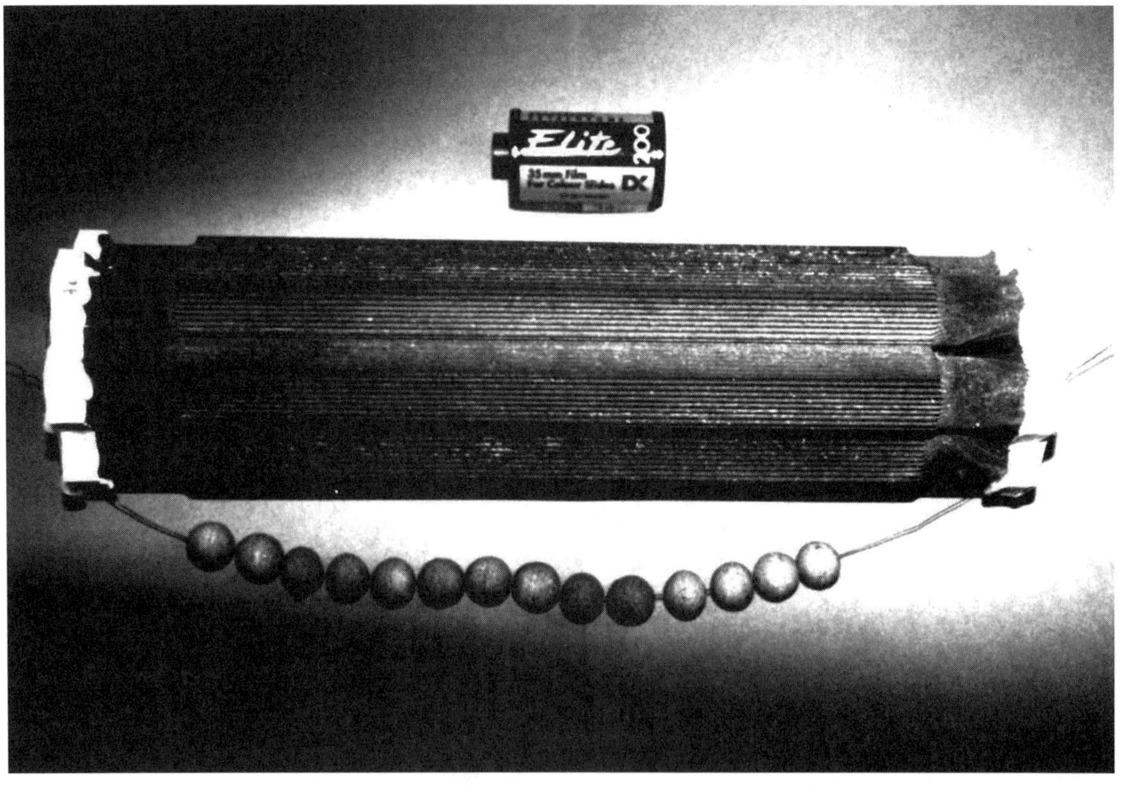

Figure 23.16 SUGRAV Heat Exchanger Module with Ball Turbolators (Reprinted with permission from Prof. Dr. A.S. Soudarev)

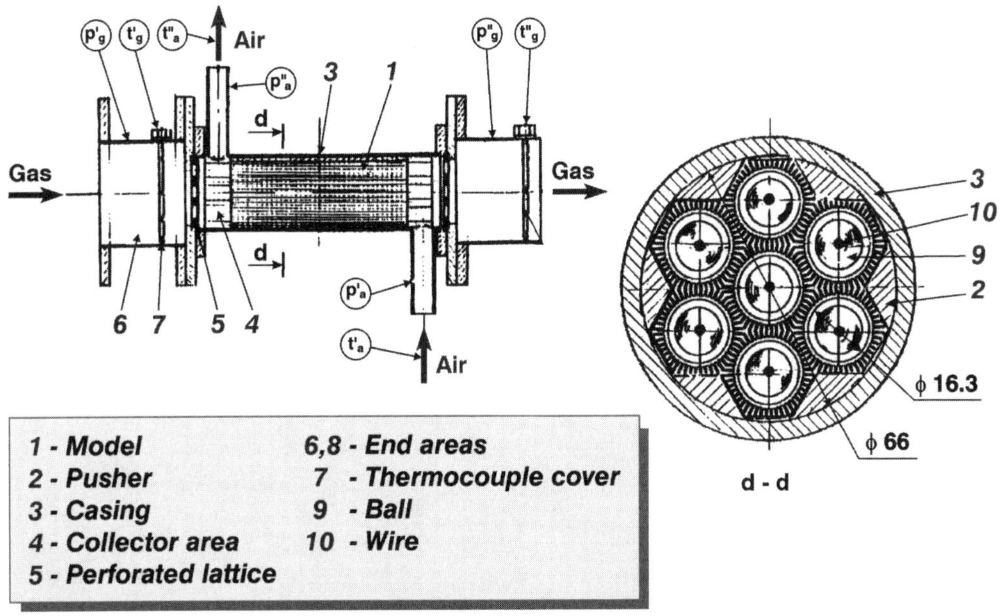

Figure 23.17 Model of Heat Exchanger with Seven-tube Ceramic Matrix (Adapted with permission from Ref. [9b])

Tests were conducted to evaluate pressure losses and actual heat transfer. In "cold" testing (tg < 120°C) the ceramic tube matrix was placed in an organic glass casing, while in "hot" testing it was inside a steel casing (Figure 23.17). Pushers placed over its perimeter ensured air delivery to the slot channels between the longitudinal tube ribs. Seven ceramic ball-turbolators were placed inside the tubes with the use of nichrome wire and at the same spacing over its length. The wire's ends were secured on the perforated grills limiting the ball package. Thermocouples, measuring orifices, and pressure gauges were used to monitor temperatures, gas flows, and pressure drops. The test rig schematic is shown in Figure 23.18. Test parameters are listed in Table 23.5. The "cold" and "hot" data agree well with each other and with models in the literature.

Ceramic materials are frequently sensitive to thermo-cyclic loads. Tubular specimens and tube fragments of SUGRAV ceramics were therefore exposed to "hot" gas temperature gradients. The maximum temperature was 1000°C, and temperature variations were between 100–250°C, and the rate of temperature change was from 2 to 7 °C/s. The material showed no signs of degradation under these conditions.

Recuperator Experience Summary

As a result of almost twenty years of recuperator development several important lessons were learned.

Recuperation is an effective method for improving the specific fuel efficiency of a simple cycle gas turbine, or attaining the same specific fuel efficiency at much reduced compressor pressure ratio. However, incorporating a recuperator comes at a price: the added cost of the heat transfer elements and the connecting piping, and the inlet and outlet sections to the gas turbine compressor and combustor. On the other hand, recuperation reduces the number of stages required for the compressor, thereby lowering the overall cost of the gas turbine.

The size, weight and cost of a recuperator configured with a gas turbine can be reduced by a factor of two by locating it between the gas producer and power turbines instead of aft of the power turbine as in the traditional design.

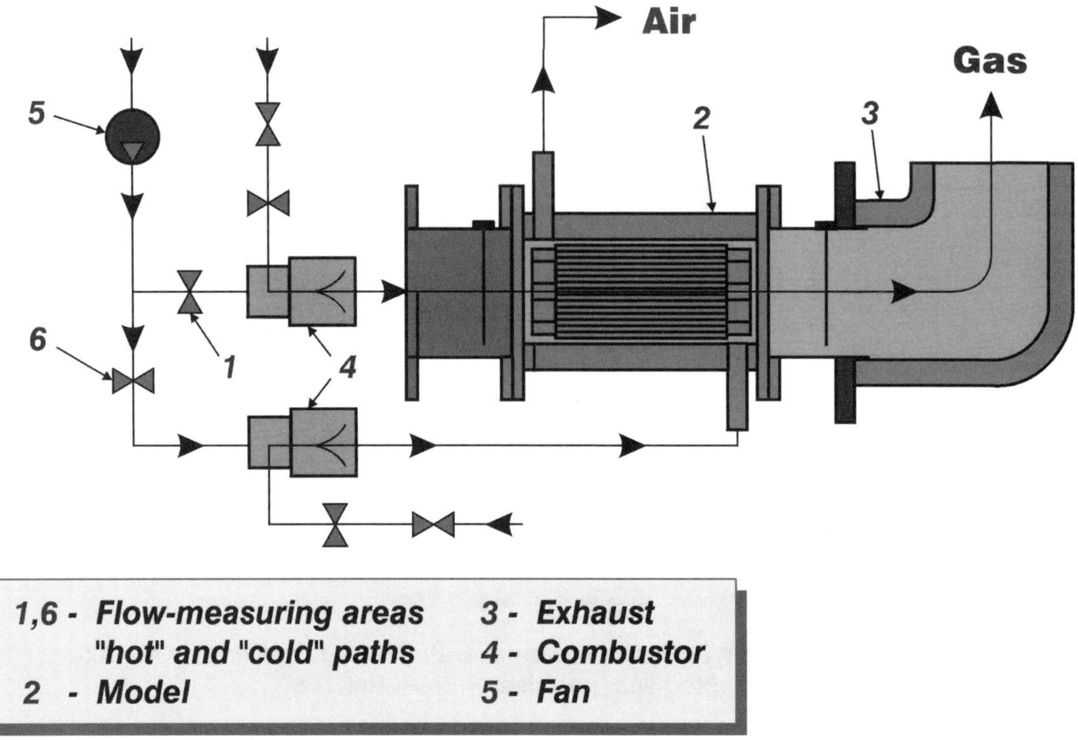

Figure 23.18 Test Rig Schematic (Adapted with permission from Ref. [9b])

TABLE 23.5 RECUPERATOR RIG TEST PARAMETERS (ADAPTED WITH PERMISSION FROM REF. [9B])

Parameter	Design	Units	Range
Gas flow	G_g	kg/s	6.4–12.0
Air flow	G_a	kg/s	6.0–10.6
Gas inlet temperature	T_g'	°C	1175–418
Air inlet temperature	T_a'	°C	27–16
Outlet gas temperature	T_g''	°C	608–99
Outlet air temperature	T_a''	°C	953–215
Inlet gas pressure	P_g'	MPa	0.102–0.100
Inlet air pressure	P_a'	MPa	0.100–0.99
Gas pressure drop	ΔP_g	kPa	3.2–0.15
Air pressure drop	ΔP_a	kPa	8.1–0.48
Wall temperature – Gas piping	$T_{f,g}$	°C	1070–810
– Air piping	$T_{f,a}$	°C	990–150
Barometric pressure	P_{bar}	kPa	98.5
Temperatures flow measuring areas – Gas	T_g'	°C	16–25
– Air	T_a'	°C	16–20
Excess pressure at flow measuring areas – Gas	P_g	kPa	0.102–0.100
– Air	P_a	kPa	0.102–0.100

The use of silicon-based ceramics for the heat transfer elements of the recuperator enables significantly increased operating temperatures of both the recuperator and the gas turbine.

Nickel- and cobalt-base alloys are limited to use below 1000°C. Above this operating temperature ceramics will be required. A hybrid recuperator design in which the lower temperature sections are fabricated from metallic components and the higher temperature sections from ceramics can be a cost-effective design by minimizing recuperator size and weight.

Optimizing the recuperator effectiveness and cost is the result of trade-offs between the various recuperator parameters. In a shell-and-tube type recuperator the number of tubes, their interconnects, their length, their external radius, their weight and pressure drop, and associated impacts on recuperator and overall system cost need to be considered in the trade-off study.

Recuperator effectiveness can be improved by modification of the external and internal shapes of the elements, adding fins to the external surfaces and ribs to the internal surfaces, as well as through the incorporation of ball-turbolators.

Decreasing mass flow through the recuperator improves its effectiveness. For this reason effectiveness improves at part load operation of the gas turbine.

Leakage greatly reduces recuperator effectiveness. The total leakage is proportional to the pressure difference between air and the hot gas. Increasing the pressure ratio is therefore counterproductive since leakages increase proportionally. Recuperator size needs to be increased to maintain its effectiveness in the presence of leaks. Leakages also reduce the output power of the gas turbine.

The ability to withstand thermal stresses is an important criterion for selecting recuperator materials. In the AGATA program cordierite was selected over SiC as the preferred recuperator material because of its superior thermal shock resistance.

Future of Recuperation for Gas Turbines

Recuperation is an established tool for improving the fuel efficiency of a gas turbine. Still, it is not widely used for this purpose, although some commercially available gas turbine manufacturers offer recuperation as an option for some of their models. Worldwide and regional fuel prices are major factors in the popularity of recuperation as an efficiency enhancement practice. There will be little incentive to consider recuperation as long as fuel prices are low. This is especially true for gas turbines used for oil & gas related applications because of the easy access to fuel. There will be more incentive to consider recuperation for industrial power generation or transportation applications in which fuel has to be transported to the gas turbine or carried with a vehicle. For the larger utility size gas turbines the widespread use of combined cycles offer a cost-effective alternative to recuperation which will be very expensive for large size engines.

Recuperation has the potential to become widely accepted for small gas turbines for high performance applications. Since cooling options are limited because of small hot section component size, a combination of increase in turbine inlet temperature (TIT) and recuperation will be required to achieve target thermal efficiencies of 40% or higher.

For automotive applications regenerators are the recuperators of choice. Regenerators have been extensively tested in the automotive ceramic gas turbine programs in the U.S.A., Japan and Western Europe. Leakages and cost factors have proven to be formidable hurdles that have yet to be overcome. For stationary applications fixed recuperators of the shell-and-tube type with substantial ceramic content are of interest. The current high cost of silicon-based ceramics and related materials and the limited experience with long term operation are still impediments for these applications where cost and durability are overriding factors. The direction of automotive development and the ever increasing targets for fuel efficiency, as well as the ability of designers and the suppliers of ceramics to simplify and ruggedize designs, to improve durability, and reduce costs, will ultimately determine the future of recuperation in gas turbine technology.

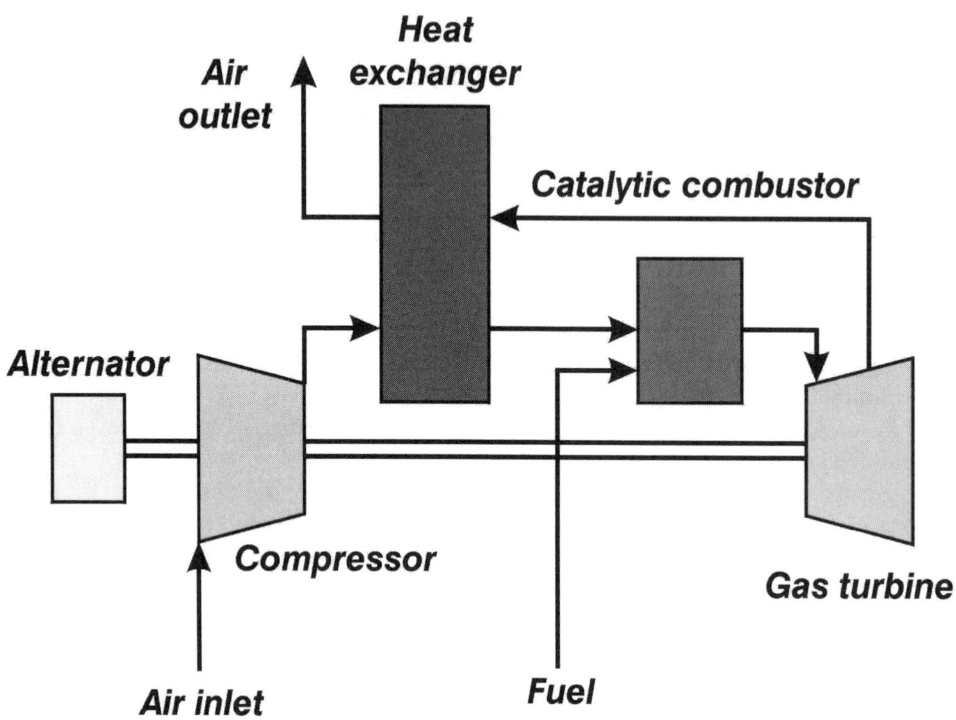

Figure 23.19 Diagram of the AGATA Concept (Adapted with permission from Ref. [7])

AGATA CATALYTIC COMBUSTOR DEVELOPMENT

ONERA had a major role in the testing of the catalytic combustor for the AGATA program. Figure 23.19 is a diagram that illustrates the AGATA concept. Most of the published work to-date has been on catalytic combustor development and the narrative in this chapter will focus on that aspect of ONERA's contributions to the AGATA program.

Figure 23.20 illustrates the AGATA catalytic combustor concept [7]. The preheater operates only during the first 60 seconds of the start-up and is subsequently by-passed for all other engine conditions. Its sole function is to increase the temperature of the air exiting the compressor to the light off temperature of the catalyst. The premix duct is based on the Lean Premixed Prevaporized (LPP) concept with a venturi configuration which has been optimized in order to obtain a good velocity profile at the inlet to the catalytic section. The catalytic reactor consists of catalyst substrates made of ceramic honeycomb of an extruded oxide ceramic material inside the catalytic section envelope made of silicon nitride material. The afterburner cone, for exhaust gases is designed for high strength and thermal shock resistance. Its temperature can be as high as 1400°C (1673K). It is made of silicon nitride.

ONERA has supported the AGATA catalytic combustor development work with rig testing and flow velocity modeling. Figure 23.21 shows the ONERA catalytic combustor test rig. In this rig the high pressure/temperature conditions during steady state and transient operation can be achieved. The operating temperature target is as high as 1400°C (1673K). During the first 9 seconds the preheater increases the turbine inlet temperature (TIT) to 1300°C (1573K). At the 9 second mark the preheater temperature is reduced to the catalytic combustor inlet temperature of 935°C (1208K) and the main fuel is being introduced. The tem-

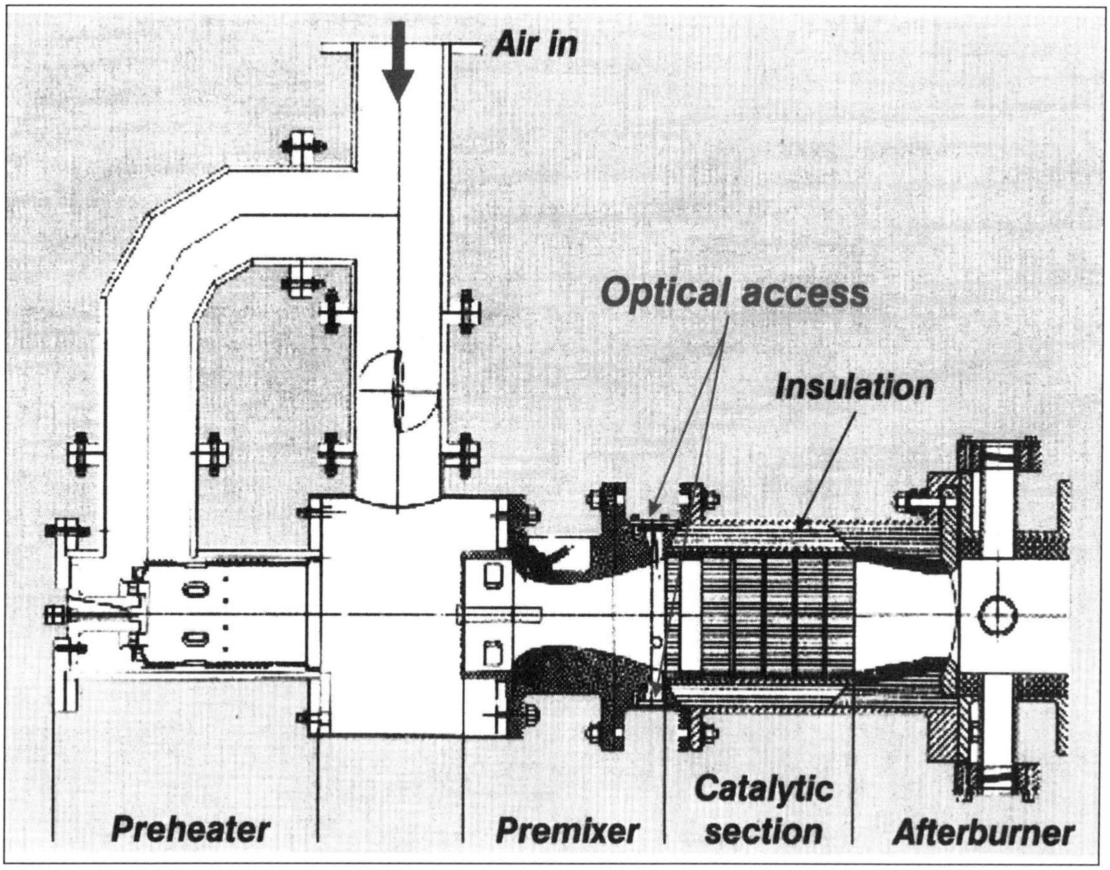

Figure 23.20 AGATA Catalytic Combustor with Exhaust Measurement Section (Adapted from Ref. [4])

perature profile at this stage of a cold start is relatively flat. Since the temperature of the recuperator is increasing, the temperature rise over the preheater is limited to 141°C [4].

In order to obtain high hetero/homogeneous combustion efficiency and low pollutants emission the gas flow needs to have an even velocity profile at the catalytic inlet section. An important aspect of the development work at ONERA focused therefore on catalytic combustion chamber inlet velocity field characterization. The trick was to simulate the engine conditions at the inlet of the catalytic combustor faithfully. At ONERA two parallel experiments were carried out, one using a cold set up in which the conditions were chosen with the same Reynolds number as for the hot steady state conditions. In the cold experiment the flow visualization is achieved by "seeding" the catalytic inlet section with olive oil particle droplets and inject them in the main flow before the premix duct. A Laser Doppler Velocimeter (LDV) is used to characterize the upstream flow at the inlet of the catalytic combustor In a parallel hot combustion test a hot wire probe in the test section was used for comparison with the LDV velocity measurements [7]. While the data plotted in Figure 23.22 suggest that the hot wire results are slightly lower than the results obtained with the LDV method, the two sets of data are in fact very close. The velocity profile with experimental mean velocity of 11.66 m/s corresponds to a mean velocity of 28 m/s at design point conditions. The measured RMS (root mean square values) indicated that a high turbulence level exists in the venturi of the catalytic inlet section, a highly desirable condition since it enhances the mixing and fuel vaporization process [4].

Figure 23.21 ONERA Catalytic Combustor Test Rig (Reprinted from Ref. [4])

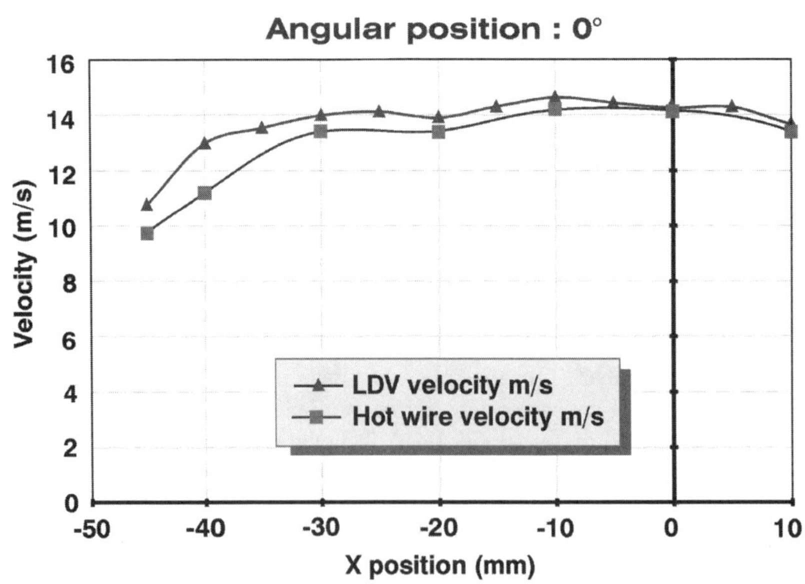

Figure 23.22 Comparison Between LDV and Hot Wire Results (Adapted with permission from Ref. [7])

TABLE 23.6 AGATA CATALYTIC COMBUSTOR EMISSIONS RESULTS (REPRINTED WITH PERMISSION FROM REF. [6])

	Measured Emissions			
	100% Load		50% Load	
Emittant	EI(g/kg fuel)	ppm(15% O_2)	EI(g/kg fuel)	ppm(15% O_2)
CO	0.99–1.26	22–28	1.69	38
HC	0.30–0.69	4–10	0.15	2
NOx	0.26–0.30	4	0.21	3

The final catalytic combustor test campaign for the AGATA program was conducted in the ONERA test rig. CO/HC/NOx emissions and O_2/CO_2 were measured with a traversible gas sampling probe at the combustor outlet. The test campaign was run with diesel fuel. Emissions and temperature profiles were evaluated from idle to full load conditions. The full load tests were run up to temperatures slightly above the specified maximum design temperatures (inlet temperature 962°C, compared with design 935°C, and exhaust temperature 1362°C, compared with design 1350°C). Table 23.6 shows the data for 100% and 50% load [6]. The NOx emissions are low while the CO emissions are still relatively elevated. Diagnostic temperature measurements indicated a cold streak at one side of the combustor. By optimizing flow dynamics and fuel/air mixture homogeneity further lowering of CO and HC appears to be possible.

Total operating time on the catalytic combustor at ONERA and Volvo Aero Corporation during full scale testing was approximately 100 hours. Individual catalytic sections were run up to 20 hours [5, 6].

CERAMIC GAS TURBINE DEVELOPMENT

Turbine Wheel Testing for the AGATA Program

As mentioned previously, ONERA became involved with the development work for the Swedish-French AGATA program because of its testing facilities. One capability that was planned to be used was the hot spin rig shown in Figure 23.23 [10, 11]. The test rig is able to duplicate engine conditions within the AGATA project, particularly start/stop. These conditions include a gas mass flow of 0.264 kg/s, an inlet temperature of 1350°C, an inlet pressure of 3.7 bar, and a rotor speed of 125,000 rpm. The base of the test rig is a modified turbocharger. The air supplied to the turbine side is independent of the air supplied to the compressor side. The compressor in the rig is used as a brake. The details of the AGATA turbine wheel design and cold spin proof testing have been described in the chapter of Volvo Aero Corporation (Volume I, Chapter 22), and in the literature [4]. As of the writing of this chapter testing of the AGATA turbine wheels had not been performed.

Cooperative Program with NITI EM

ONERA participated in a collaborative effort with the French company MICROTURBO and two Russian companies NITI EM (Research and Technological Institute of Power Engineering) and the research manufacturing enterprise TARC in a program funded by DRET (Direction des Recherches Etudes et Techniques). The chapters from NITI EM and associated companies (Volume I, Chapter 32; Volume II, Chapter 13) and a report in the literature detail the ceramic gas turbine hot section design and testing experience, and materials development work [12].

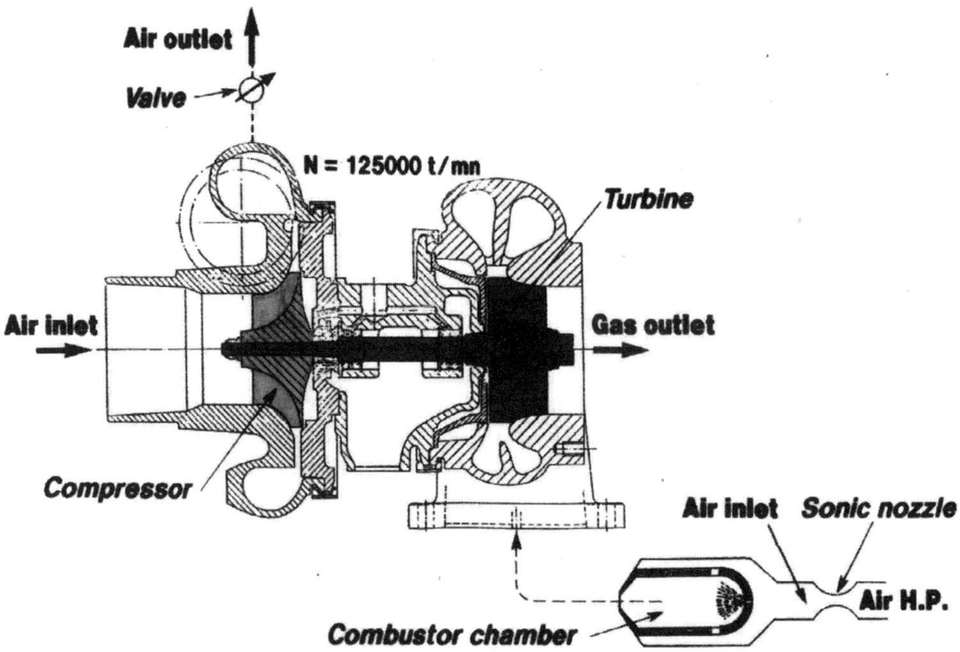

Figure 23.23 Ceramic Turbine Hot Spin Test Rig (Adapted from Ref. [10])

REFERENCES

[1] (a) Avran, P., and Bernard, P., "High Temperature Heat Exchangers for Gas Turbines and Future Hypersonic Air Breathing Propulsion," paper presented at the 1st International Conference on Aerospace Heat Exchanger Technology, February 15–17, 1993, Palo Alto, California, USA. (b) ONERA Report TP 1993–75, 1993.

[2] (a) Avran, P., and Boudiques, S., "Comparison of Regenerators and Fixed Heat Exchangers for Automotive Applications," *Proceedings of the International Gas Turbine Congress*, October 27–November 1, 1991. Yokohama, Japan, Volume II, pp. 283–290, 1991. (b) ONERA Report TP 1991–230, 1991.

[3] (a) Avran, P., and Boudiques, S., "Ceramics Heat Exchangers and Turbine Blades Theory and Experimental Results" (Echangeurs de Chaleur et Aubes de Turbine en Ceramique Théorie et Resultats Experimentaux". Paper presented at the 72nd PEP/AGARD *Instabilités de Combustion dans les Systèmes Propulsifs à Combustible Liquide*, Bath, U.K., October 3–7, 1988. (in French). (b) ONERA Report TP 1988–157.

[4] Gabrielsson, R., Lundberg, R., and Avran, P., "Status of the European Gas Turbine Program AGATA," ASME paper 98-GT-392, presented at the International Gas Turbine and Aeroengine Congress and Exposition, Stockholm, Sweden, June 2–5, 1998. ONERA Report TP 1998–67.

[5] Gabrielsson, R., Payen, G., and Avran, P., "Catalytic Combustor Development within the AGATA Program," ASME paper 99-GT-442, presented at the International Gas Turbine and Aeroengine Congress and Exposition, Indianapolis, Indiana, USA, June 7–10, 1999.

[6] Gabrielsson, R., and Avran, P., "The European Ceramic Gas Turbine Program—AGATA," *Proceedings of the International Gas Turbine Congress*, Volume I, pp. 73–79. Kobe Japan, November 14–19, 1999.

[7] Avran, P., Gicquel, P., Leclair, A., and Payen, G., 1999, "Upstream Flow Characterization of the AGATA Catalytic Combustor by 2D LDV Measurements," *Proceedings of the International Gas Turbine Congress*, Volume II, pp. 647–651. Kobe Japan, November 14–19, 1999.

[8] Avran, P., Sudarev, A.V., Sudarev, B.V., and Sudarev, V.B., "Unsteady Operating Conditions as Applied to Ceramic Recuperative Air Heater," *Proceedings of the International Gas Turbine Congress*, Volume II, p. 271–274. Yokohama, Japan, October 22–27, 1995. ONERA Report 1996–46.

[9] (a) Avran, P., Leclair, A., Soudarev, A., Soudarev, B., and Soudarev, V., "Shell-and-Tube Ceramic Heat Exchangers for High-Temperature GTEs Applications." *Proceedings of the 11th International Heat Transfer Conference*, Kyongju, Korea, August 25–28, 1998. (b) ONERA Report TP 1998–12.

[10] Avran, P., Leclair, A., and Payen, G., "Test Benches for Ceramic Components for Automobile Gas Turbines," ASME paper 97-GT-69, presented at the International Gas Turbine and Aeroengine Congress, Orlando, FL, USA, June 2–5, 1997.

[11] Avran, P., Leclair, A., and Payen, G., "Facilities for automotive gas turbine ceramic components," ONERA Report 1997–69, 1997.

[12] Avran P., Sudarev, A.V., Grishaev, W., and Podgorest, V.Y., "Complex Tests of Thermal-Stressed Elements in High Temperature Gas Turbines," *Proceedings of the 3rd International Symposium on Experimental and Computational Aerothermodynamics of Internal Flows (ISAIF)*, Beijing, China, September 1–6, 1996. ONERA Report 1996–103.

CHAPTER

24

Ceramic Gas Turbine Component Development at Motoren Turbinen Union

Wofgang Krüger

ATENA Gesellschaft für Engineering Services mbH Munich, Germany

Mark van Roode

Solar Turbines Incorporated San Diego, California, USA

INTRODUCTION

Motoren Turbinen Union (MTU) in Munich, Germany, had a number of active ceramic gas turbine programs from the 1970s through the late 1990s. A significant activity was MTU's participation in the automotive vehicular gas turbine component development between 1974 and 1983 under a program sponsored by the German Federal Ministery for Research and Technology (Bundesministerium für Forschung und Technologie—BMFT). The activities under that program are described in some detail in Chapter 21 of Volume I of this text.

The topic of this chapter are programs conducted by MTU and sponsored by the German Federal Ministry of Defense to develop a ceramic nozzle concept for a high temperature jet engine. The objective of the programs, which were conducted between 1980 and 1990, was to achieve a significant reduction in cooling air requirements of the high pressure turbine (HPT) nozzles of jet engines by introducing into the turbine hot section a new hybrid metal-ceramic nozzle. Summaries of the program results have been published [1, 2]. The work was performed under two projects:

(1) Metal-ceramic hybrid nozzle (feasibility study) 1980–1982
(2) High temperature nozzle assembly 1985–1989

Nozzles in aerojet engines in the 1980's were typically investment castings fabricated from nickel- or cobalt-base superalloys. The nozzle designs relied on sophisticated internal and film cooling schemes to mitigate the thermal stresses imparted by the gas stream entering the turbine. In military aero-engines the gas temperature exceeded the allowable wall temperature of the nozzle vanes by as much as 500°C to 700°C. Between 7 to 10% of compressor air flow was required to cool the nozzles to meet component design life targets.

Figure 24.1 Four-vane Hybrid Metal-ceramic Cascade (Reprinted from Refs. [1, 2])

To achieve savings in cooling air, and thereby improve engine performance and fuel efficiency, materials with considerably higher temperature capabilities than superalloys were required. Candidate materials included intermetallics, carbon-carbon with protective SiC coating, continuous fiber-reinforced ceramic composites, and monolithic ceramics. Novel vane concepts were needed to overcome the poor ductility of these high-temperature advanced materials.

The development of the MTU nozzle was based on the concept of a hybrid structure in which the metal components were load bearing at a relatively low temperature while the ceramic part acted as a heat shield enabling the operation at high turbine inlet temperature (TIT). The ceramic material selected for the nozzle design was pressureless sintered α-silicon carbide (SSiC) which has a temperature capability of 1627°C–1727°C (1900K–2000K) for short term operation. Following the monolithic development work the nozzle design was modified and the SSiC ceramic heat shield was replaced with a silicon-based ceramic matrix composite (CMC) heat shield. A limited development effort was also conducted for a hybrid metal-ceramic blade. This chapter will primarily address the hybrid metal-ceramic (monolithic) nozzle development work.

NOZZLE DEVELOPMENT FOR CASCADE TESTING

Figure 24.1 shows the fabricated hybrid nozzle assembly developed by MTU, ready for cascade rig testing. The design concept separated the functions of hot gas ducting and force transfer. The SSiC vane shells with their excellent high-temperature capability are in direct contact with the hot gas stream and act as heat shields. The superalloy load bearing structure absorbs the external forces of the secondary components interfacing with the nozzle assembly and the aerodynamic forces transferred from the gas loads. Each nozzle segment consists of two vane airfoils with the associated metal support structure.

Design Guidelines

Guidelines, developed under previous programs [3, 4], were applied to the nozzle design. The major ones were [2]:

- Simple shapes of constant wall thickness should be used to the greatest extent possible. This will minimize thermal stresses, especially under transient operating conditions.
- Notches, slots, and sharp edges, which are a frequent origin of cracks should be avoided. All component edges should therefore be chamfered.
- 3-D finite element (FE) stress/temperature calculations need to be performed for steady state and transient operating conditions.
- The component loads should be as low as possible to minimize the risk of failure. Tensile stresses should be minimized or avoided if possible.

To integrate the ceramic vane shell into the metal support structure the following rules were followed [2]:

- Non-interference with the thermal expansion of the ceramic parts. If interaction cannot be avoided, ensure that the resulting stresses are compressive rather than tensile.
- The use of large surface areas for force transfer, avoidance of point loads and minimizing Herzian surface stresses.
- Avoidance of shallow angled retention to prevent self-locking (sticking).
- Use of a simple supported design without double centering positions.
- Incorporation of thermal insulation to prevent direct contact between metal and ceramic at high temperatures and to reduce the thermal gradient in the contact area.
- Design the vane for fail-safe performance so that a crack does not inevitably result in failure of the component.

Nozzle Cascade Design

Figure 24.2 shows an exploded view of the nozzle vane cascade design. The nozzle design is a modification of the design of the RB 199 HPT (high pressure turbine) nozzle. The RB199, a military engine with a trust range of 70 to 80 kN, is mainly used for the Panavia Tornado. The engine was jointly developed and produced by MTU, Rolls-Royce and FiatAvio [5]. The nozzle design incorporated the guidelines mentioned above. The cascade consisted of four adjacent ceramic shells configured with the mating metal support structure. The hollow vane shells, consisting of an airfoil and integral inner and outer shrouds were fabricated from SSiC. The IN100 metallic support structure consisted of inner and outer platforms, a metallic core, and a heat shield between the metal core and the cavity of the ceramic shell. Partially stabilized zirconia (PSZ) insulators were inserted between the ceramic vane shrouds and the metal platforms. The ceramic shells floated in recesses in the metallic outer and inner platforms by means of the PSZ insulators. Each set of inner and outer platforms spanned two ceramic shell assemblies. Two sets of platforms constituted the four-vane cascade.

In the assembly process the two metal vane cores were first brazed onto the outer platform. The heat shields which had a reflective coating on the outside were then slipped over the vane cores. Next, the ceramic vane shells and the insulators were placed in the platform recesses. The inner platform was added and brazed to the vane cores. The structure was tightened with a washer and nut at the outer platform side and closing plate at the inner platform side.

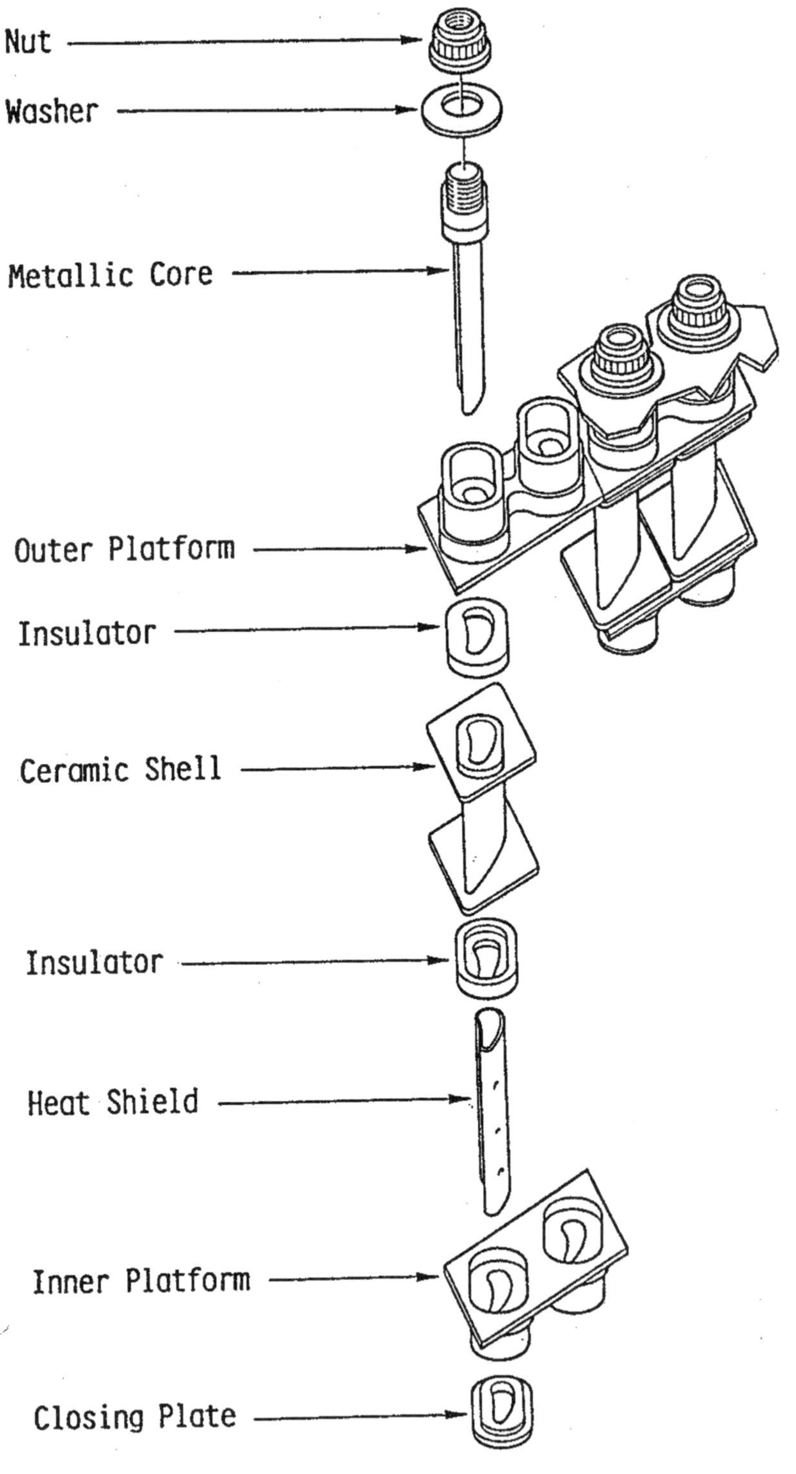

Figure 24.2 Schematic of the Metal-ceramic Vane Assembly (Reprinted from Ref. [2])

Figure 24.3 Mold for Cold Isostatic Pressing of the Ceramic Shell (Reprinted from Refs. [1, 2])

The ceramic shell is not cooled in this design, only the metal support structure. To cool the vane core cooling air is ducted through the clearance between the heat shield and the vane core. The metallic platforms are cooled externally by impingement cooling.

Fabrication of the Ceramic Vane Shell

The ceramic shells were fabricated from sintered α-silicon carbide by cold isostatic pressing and green machining of a preform which was then densified by pressureless sintering. This process gives a dense silicon carbide with high temperature durability. The nozzle was fabricated in a dedicated facility set up at MTU for this purpose. The fabrication process involved the following steps:

- *Slurry preparation.* A 60% SiC slurry was prepared by dispersing α-SiC powder (H.C. Stark A10) with 0.6% amorphous boron, 3% phenolic resin, and 3% organic binder in water. The slurry was spray-dried to pre-agglomerate the powder (mean agglomerate size: 40 μm).
- *Forming of green body.* The pre-agglomerated powder was poured into a mold (see Figure 24.3) with the shape of the inner profile of the vane and a press bag. The green body was formed by cold isosta-

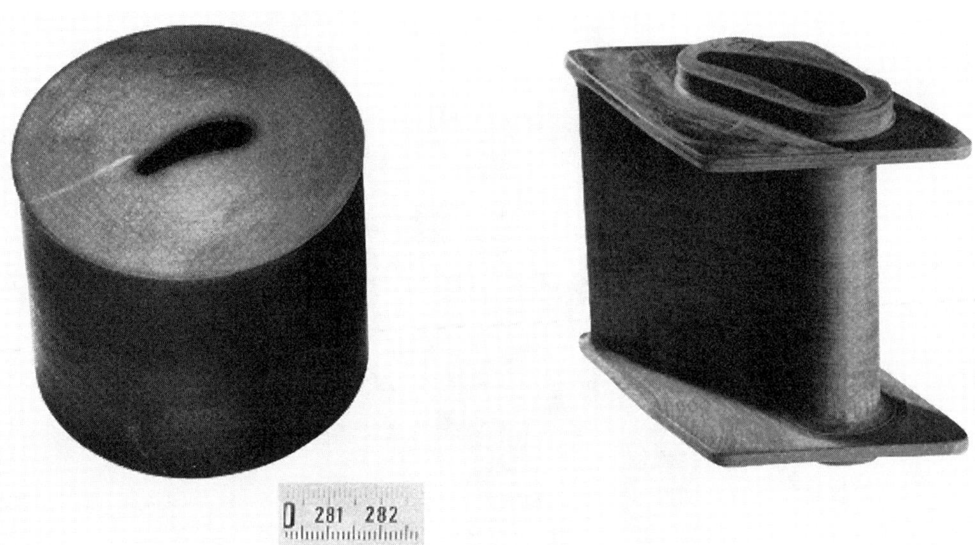

Figure 24.4 Vane shell preforms—(a) Green Body After Pressing, (b) Ceramic Shell Preform After Green Machining (Reprinted from Refs. [1, 2])

tic pressing at 2000 bar. The final contour of the ceramic shell was obtained by green machining (see Figure 24.4).
- *Binder burn-out.* The organic residues were incinerated by heating the preform in vacuum to 627°C over a 24 hour period.
- *Sintering.* The sintering process was performed in a graphite furnace at 2147°C in an argon atmosphere at 0.8 bar for 10 min. Greater than 98% theoretical density was achieved for the sintered parts. Linear shrinkage between the green machined preform and the final sintered part was about 15%.
- *Material characterization.* Flexure bars (3.5 × 4.5 × 5.0 mm) were fabricated by axial pressing and using the same steps as for the processing of the vanes. Room temperature 4-point flexure strength was 300 MPa with a Weibull modulus of m=10.

VANE CASCADE TESTING

Vane Cascade Test Facility

Figure 24.5 is a schematic showing the dedicated vane cascade test rig. The combustor, flow nozzle, nozzle vane cascade, and components of the heat exchanger (not shown) were all fabricated from SSiC. The test rig enabled evaluation of the ceramic nozzle at hot gas temperatures in excess of 1527°C (1800K). The combustor chamber was developed under the previously mentioned BMFT-sponsored automotive vehicular gas turbine component development program. The flow nozzle downstream from the combustor converted the hot gas flow exiting from the circular combustor outlet to a rectangular area at the cascade entry. The mach number at the nozzle entry was similar to that of the HPT nozzle of the RB 199 engine, but the Reynolds number was lower because the gas was expanded to atmospheric pressure. After the nozzle cascade the hot exhaust gas is passed through a tubular ceramic heat exchanger (not shown) which

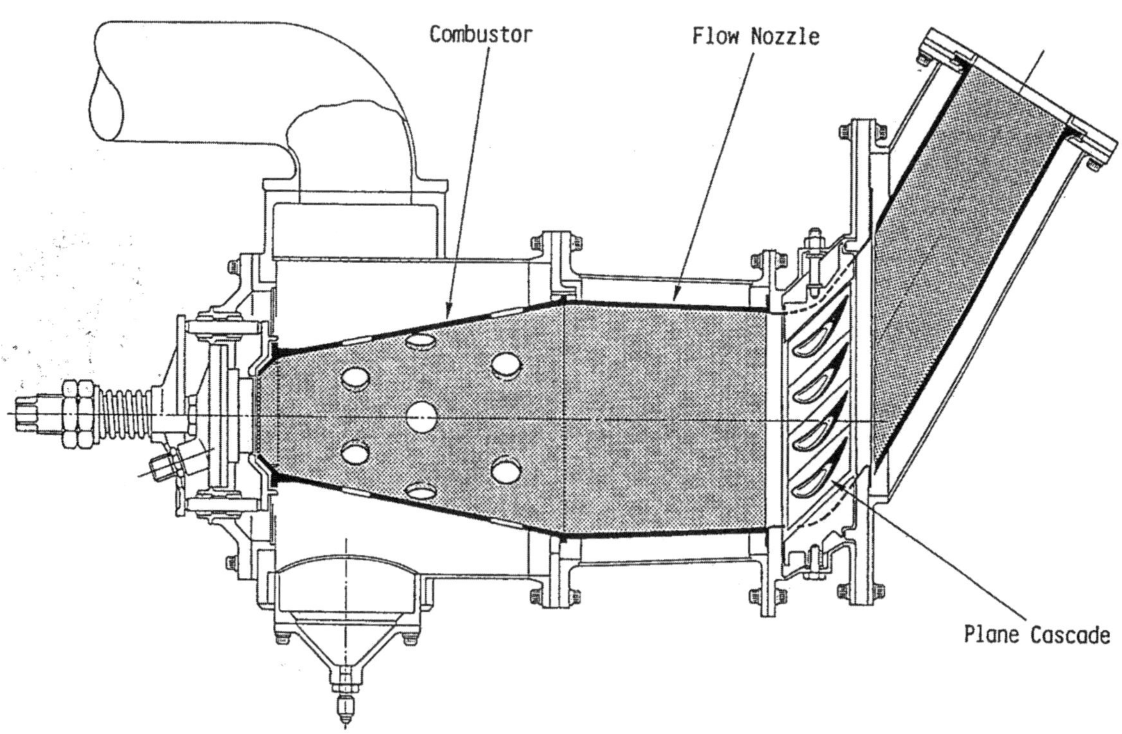

Figure 24.5 Vane Cascade Test Rig (Reprinted from Ref. [2])

heats up the compressed air prior to entering the combustor. The recuperation adds 700°C to 800°C to the compressor air exit temperature.

Test Results

The rig test data have been summarized in Table 24.1. The metal and ceramic components were intact with no evidence of degradation after initial stationary testing up to 1402°C and cyclic testing between 727°C-1402°C (Series #1). During testing at 1402°C the air cooling requirement was measured and scaled to the environment of the RB 199 with 34 vanes. A total of 2% of cooling air (metallic inner and outer platforms: 1.5%; vane core: 0.5%) would be required comparaed to about 7% for the RB 199 HPT nozzle. Subsequently, tests were conducted at higher stationary hot gas temperatures and a steeper cyclic temperature gradient. Yet, the metal and ceramic components displayed no evidence of degradation after stationary testing up to 1627°C and cyclic testing between 527°C and 1627°C (series #2). The PSZ insulators were cracked but they did not affect the functionality of the cascade during the test (see Figure 24.6). Testing at temperatures in excess of 1627°C was not successful. Testing at about 1700°C (~1970K) (series #3) resulted in fracture in the ceramic shells.

NOZZLE DEVELOPMENT FOR DEMONSTRATOR ENGINE

Following the successful nozzle cascade tests a hybrid metal-ceramic nozzle was designed for a demonstrator engine, similar to one of MTU's military propulsion engines (see Figure 24.7). The various parts of the nozzle assembly are shown in Figure 24.8. The design concept is similar to that of the cascade nozzle

TABLE 24.1 CASCADE RIG TEST DATA [1, 2]				
Test	**Hot Gas Temperature**	**Test Duration**	**Cycles**	**Comments**
Series #1	1327°C (1600K)	100 hours	–	No damage to ceramic or metal nozzle components
	1402°C (1675K)	200 hours	–	
	727°C–1402°C (1000K–1600K)	–	1500	
Series #2	1527°C (1800K)	200 hours	–	No damage to ceramic or metal nozzle components PSZ insulators cracked
	1627°C (1900K)	100 hours	–	
	527°C–1627°C (800K–1900K)	25 hours	100 av. heating rate: 90°C/min hold at 1627°C: 3 min av. cooling rate: 75°C/min	
Series #3	~1700°C(~1900K)	–	–	Fracture in ceramic shells

Figure 24.6 Metal-ceramic Vanes After Cascade Test at 1627°C (Reprinted from Refs. [1, 2])

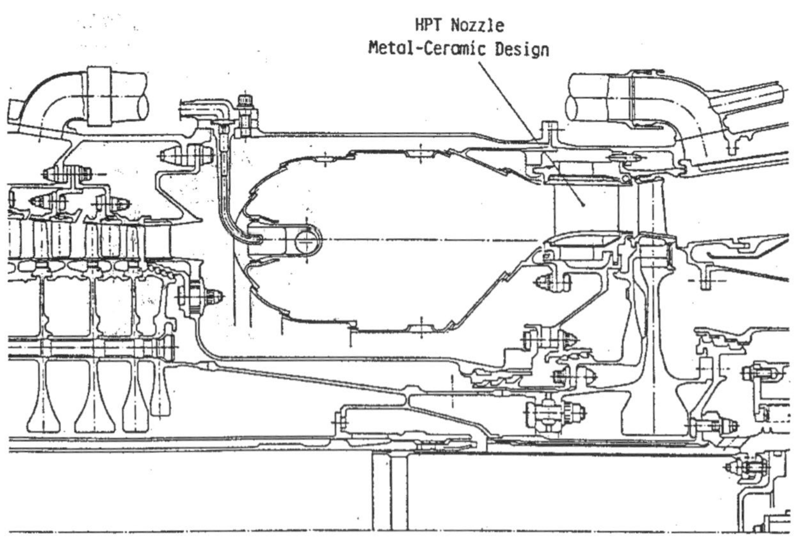

Figure 24.7 MTU Demonstrator Engine (Reprinted from Ref. [2])

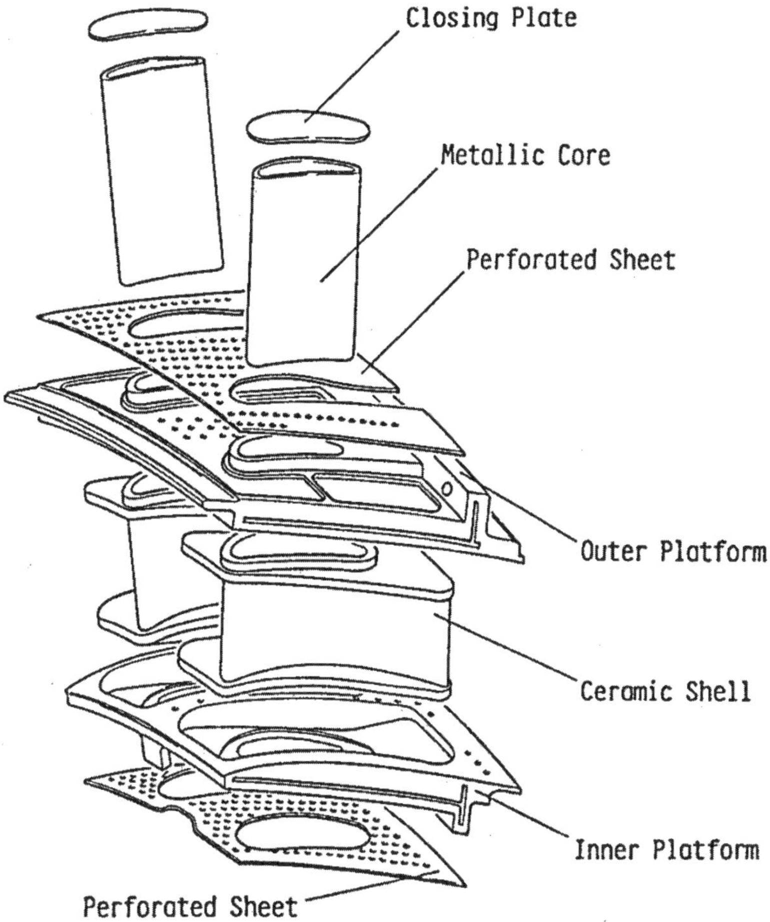

Figure 24.8 HPT Metal-ceramic Nozzle Design (Reprinted from Ref. [2])

design of Figure 24.2, with the uncooled ceramic shell acting as a heat resistant shield that protects the cooled metallic support structure. The aerodynamics and thermodynamics of the nozzle design were based on actual engine cycle data, except for an increase in the turbine inlet temperature (TIT) to 1577°C (1850K). Some improvements were made to the cascade design. The PSZ insulators were replaced with zirconia thermal barrier coatings (TBCs), plasma sprayed onto the inner and outer metal platforms. The gap between the metal and ceramic parts was minimized which reduced the leakage. The HPT nozzle arrangement had 17 twin-vane assemblies.

The cooling system was also redesigned (see Figure 24.9). As in the cascade design impingement cooling was used to cool the outer and inner platforms but now a double wall was used to improve the cooling effectiveness. The vane core which was heated by radiation from the ceramic vane shell was cooled with a number of longitudinal cooling holes and coated with a reflective coating adjacent to the ceramic vane. The additional cooling adequately protects the part from overheating.

Fabricated nozzles were tested in the modified hot gas cascade rig of Figure 24.7. After an initial 10 hour ramp up the hot gas test temperature was about 1552°C (1825K). A total of seven ceramic vane shells were tested, three for 100 hours, three for 200 hours and one for 300 hours.

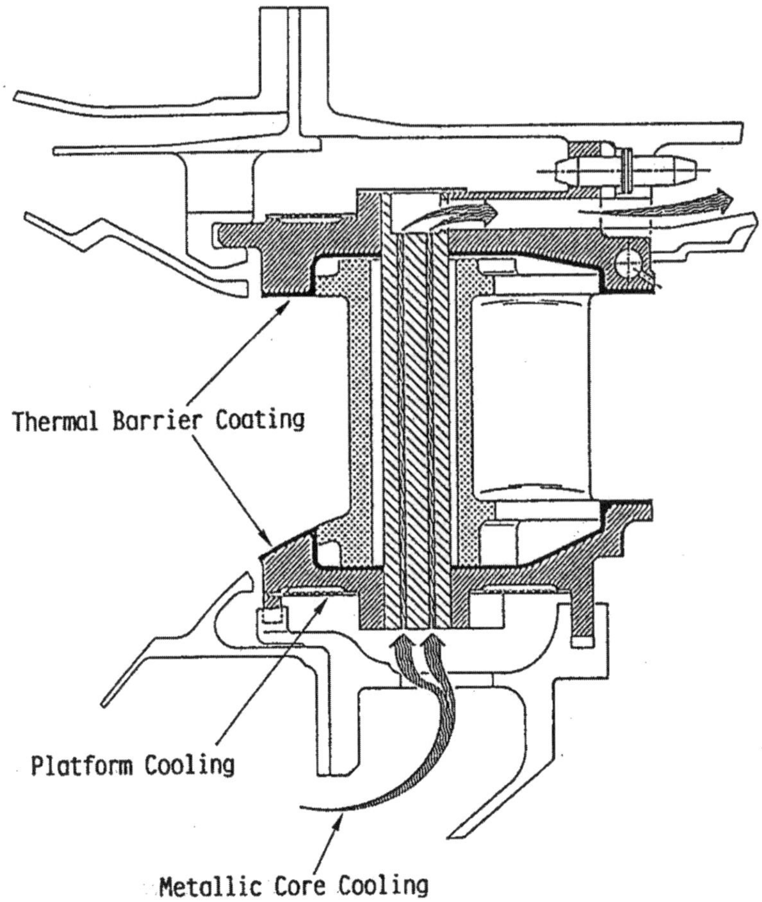

Figure 24.9 Cooling Systems of the HPT Nozzle (Reprinted from Ref. [2])

Cracking was observed in some of the vanes after the 200 hour test [6]. Figure 24.10 shows the four vanes with which the test regime below was initiated [5]:

(1) 1202°C (1475K) for 5 hours—no damage

(2) 1402°C (1675K) for 5 hours—one vane had a small crack because of a cold air spot on the vane, the result of a defect sealing strip. This strip was replaced for further testing. The other vanes were in satisfactory condition.

(3) 1552°C (1825K) for 200 hours—the initial crack had grown through the whole vane cascade, but the cascade remained functional. Figure 24.11 shows the cracked vane together with the three intact vanes. For the photograph the vanes had been assembled differently from their position in the cascade in which the cracked vane was in an outside position.

(4) 1552°C (1825K) for 100 hours—three vanes including the cracked vane had been replaced with new vanes. The vanes were found to be in satisfactory condition following the completion of the test.

It should be noted that all tests in the cascade rig were conducted at atmospheric pressure. Under actual turbine conditions the pressures are much higher, and the vanes will experience higher temperature gradients, especially under transient conditions. A finite element calculation showed a rather low probability of survival for the vanes for operation in an actual gas turbine environment. Maximum stresses were calculated

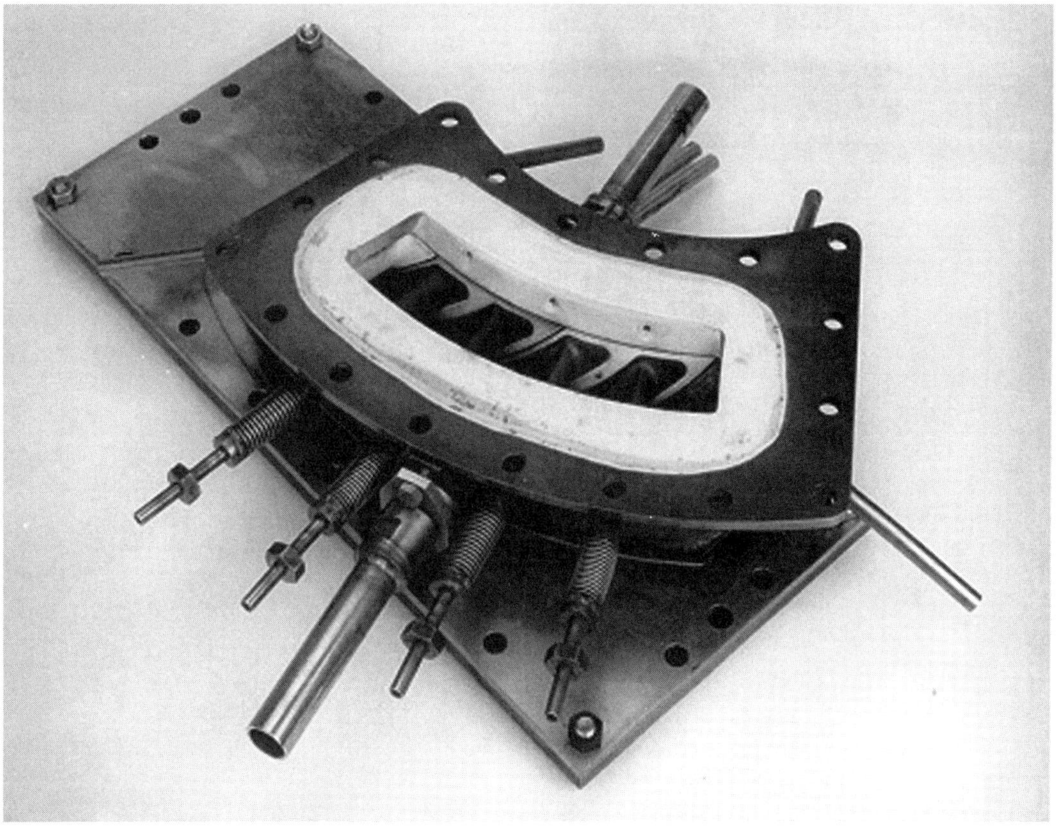

Figure 24.10 Four SSiC Vanes Mounted in the Cascade Rig Ready for Testing (Reprinted with permission from MTU)

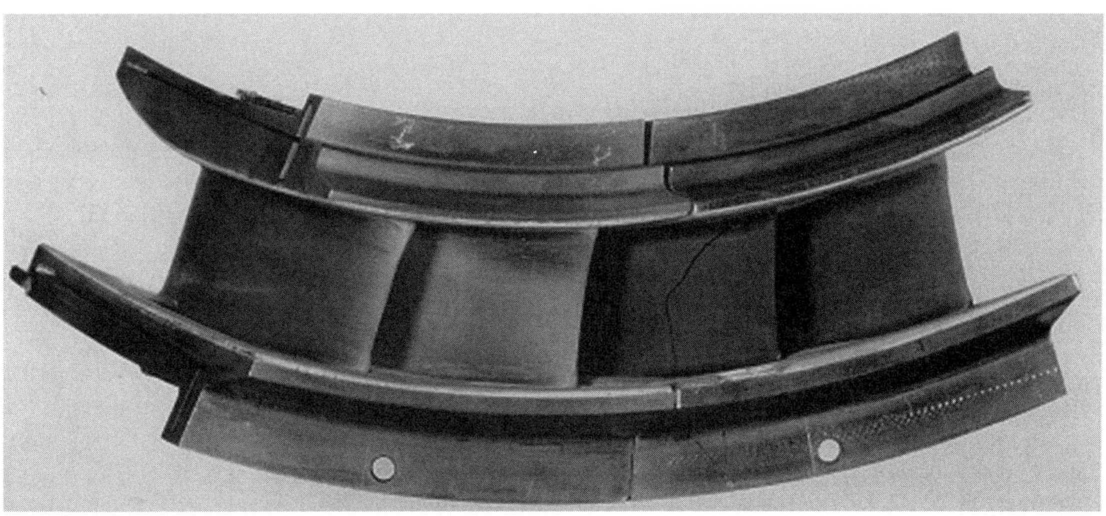

Figure 24.11 Four Vanes Tested for 200 hours at 1552°C (1825K) in the Cascade Rig (Reprinted with permission from MTU)

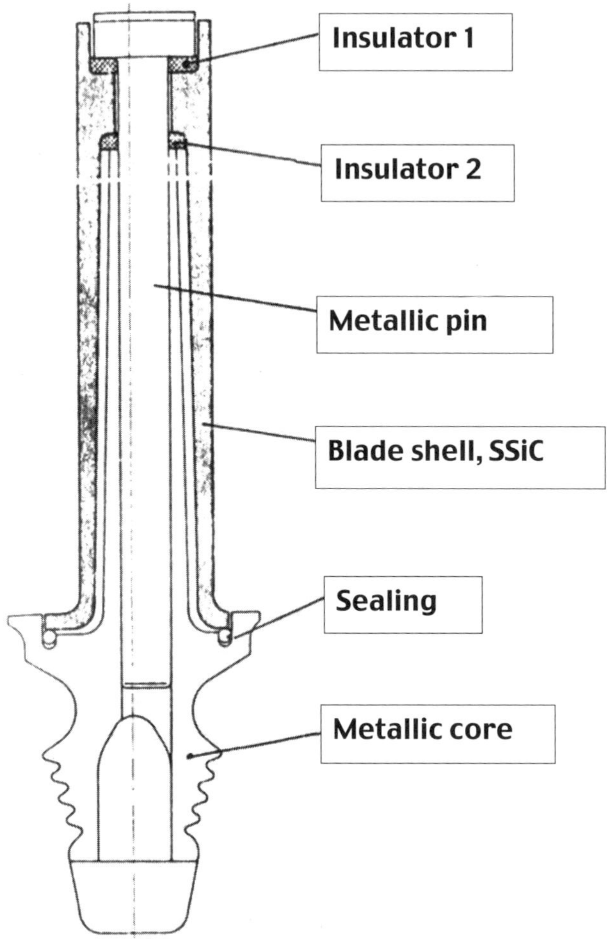

Figure 24.12 HPT Blade, Metal Ceramic Design (Reprinted with permission from MTU)

to be about 350 MPa. As a result MTU management considered the hybrid metal-ceramic (monolithic) nozzle design to have an unacceptably high risk and discontinued further development work. While the rig testing was successful the hybrid metal-ceramic nozzles were not tested in the demonstrator engine [5].

CERAMIC BLADE DEVELOPMENT

Conventional metal turbine blades for medium- and high performance gas turbines are precision castings with a sophisticated internal and film cooling system. As much as 5% of the compressor mass flow is required for cooling these blades. Significant cooling air savings can be realized by replacing these conventionally cooled metal blades with cooled hybrid metal-ceramic turbine blades.

Figure 24.12 shows the cooled metal-ceramic blade developed by MTU. In this design, the ceramic blade shell is supported over a two-shank pin in the upper part of the metallic blade core. The pin is fixed over the blade core by high-temperature brazing. Insulators were inserted between the ceramic shell and the load-bearing cooled metal support structure. Their function is to reduce the heat flow and the friction and

Figure 24.13 HPT Blade Assembly (Reprinted with permission from MTU)

compensate for peak stresses because of their low modulus of elasticity. The lower part of the airfoil fits into a recess in the blade platform. The gap between the ceramic shell and the platform is sealed. The seal acts centrifugally, i.e. it becomes effective upon rotation.

Figure 24.13 shows one of a number of hybrid SSiC blades that were manufactured, and tested in the cold spin test rig at circumferential speeds up to 350 m/s. Burst speeds scattered in a narrow range about the average value. By improving the design in the area of contact between the centrifugally loaded ceramic blade shell and the metal load-bearing structure the circumferential speeds can be further reduced.

The decision by MTU management to discontinue the monolithic nozzle development also affected the blade development. Blades were designed and manufactured, but not actually tested in an engine [5].

CERAMIC MATRIX COMPOSITE DEVELOPMENT

MTU shifted its focus from monolithic component development for engine hot section components to investigating the potential for fabricating the components from ceramic matrix composites (CMC's). Initial development work was reported at the 4th International Symposium on "Ceramic Materials and Components for Engines" in Göteborg, Sweden, in 1991. The problems inherent in monolithic ceramics were recognized, such as brittleness (low fracture toughness), and insufficient mechanical thermal shock resistance and damage tolerance characteristics. CMC materials were of interest because of their known superior fracture toughness with K_{IC} values as high as 30 MPa.m$^{1/2}$, versus 3-6 MPa.m$^{1/2}$ for monolithic silicon carbides and silicon nitrides of the early 1990s. MTU embarked on an evaluation of SiC/SiC CMC materials and reported some initial mechanical property data [6]. Subsequently, the hybrid metal-ceramic nozzle concept was adapted by replacing the SSiC heat shield with a CMC heat shield. Limited rig testing was conducted for the hybrid metal-CMC nozzle design, but no results have been published in the open literature. Ceramic development at MTU continued until the late 1990s, but was stopped in recognition of the need for radical breakthroughs in materials properties.

SUMMARY

MTU conducted innovative and pioneering work in ceramic nozzle and rotor blade design and testing for larger engines, starting in the early 1980s. Realizing the limitations and strength of metals and monolithic silicon-based ceramics, a design was developed that applied both classes of materials where they would be most effective. The ceramic was used for a high temperature non-load bearing heat shield that protected the minimally cooled metal support structure. The ceramic selected for the development was pressureless sintered silicon carbide (SSiC) and MTU set up a facility for fabricating prototype components of this material. Hybrid metal-ceramic nozzles were successfully tested for hundreds of hours and over as many as 1500 cycles in hot gas rigs at temperatures as high 1627°C (1900K). While some cracking was observed at the highest temperatures, the results were encouraging for short term operation. More detailed studies including finite element analyses revealed unacceptably high stresses and the program for monolithic ceramics was subsequently discontinued in the late 1980s. Limited development, fabrication and testing were also conducted for a hybrid metal-ceramic blade. Subsequently, MTU refocused its ceramic materials development work on CMC development. Materials studies were conducted and the monolithic nozzle design was adapted for a CMC heat shield. Some testing was performed but results have not been published in the open literature. MTU recognizes the need for substantial materials improvements before ceramics can be applied for commercial gas turbine applications.

ACKNOWLEDGEMENT

The authors gratefully acknowledge the support of MTU in reviewing the chapter, and for giving permission for the reprinting of the figures from previous MTU publications. The helpful comments of Mr. Gerhard Andrees of MTU, Munich, Germany, have been particularly helpful in clarifying the context in which the R&D was conducted.

REFERENCES

[1] Hüther, W., and Krüger, W., "Initial Results of Tests on Metal-Ceramic Guide Vanes," paper presented at the AGARD Conference on the Application of Advanced Material for Turbomachinery and Rocket Propulsion, Bath, U.K., North Atlantic Treaty Organization, Conference Proceedings, pp. 13-1/13-5. March 1989.

[2] Krüger, W, and Hüther, W., "Metal Ceramic Guide Vanes New Design Concept," ASME paper 89-GT-334, presented at the Gas Turbine and Aeroengine Congress and Exposition, Toronto, Ontario, Canada, June 4–8, 1989.

[3] Hüther, W., and Krüger, W., "Development of High-Temperature Ceramic Components for Gas Turbines," ("Entwicklung von hochwarmfesten Keramikbauteilen für Gasturbinen,") BMFT Forschungsbericht T84-006, 1984 (in German).

[4] Hüther, W., and Krüger, W., "Ceramics in Engine Design," ("Keramik in Triebwerkbau,") Mönch Verlag, Jahrbuch der Wehrtechnik, 1983 (in German).

[5] Andrees, G., MTU, private communication, 2000.

[6] Krüger, W., and Bunk, W., "Ceramic Materials and Component Design for Aerospace Applications," *4th International Symposium on Ceramic Materials and Components for Engines*, Göteborg, Sweden, June 10–12, 1991, Elsevier Science Publishers, Ltd., Barking, Essex, England. pp. 1001–1008.

CHAPTER

25

New Design Principles for Ceramic Hot Gas Components of Gas Turbines

Achmed Schulz

Lehrstuhl und Institut für Thermische Strömungsmaschinen
Universität Karlsruhe, 76128 Karlsruhe, Germany

INTRODUCTION

In continuation of the extensive research program on science and technology of ceramic components for automotive gas turbines, funded by the Federal Secretary of Science and Technology from 1974 through 1983, the Institute for Ceramics in Mechanical Engineering was founded at the University of Karlsruhe in 1985. The initiative for this institute came from Fritz Thümmler, then head of the Institute for Material Science and Solid Body Research at the Nuclear Research Center in Karlsruhe, who was actively involved in the previous program [1, 2]. The Ceramics Institute emerged from six institutes of the Department of Mechanical Engineering at the University of Karlsruhe and has been endowed with public funds from the state of Baden-Württemberg. The institutes involved comprise three material science institutes with differing main areas, an institute for engineering design and car manufacturing, an institute for reciprocating engines, and the Institute for Thermal Turbomachinery. The newly founded institute provided effective means to combine the existing experience of the material science institutes with the knowledge of the application and engine related institutes for integrated development of ceramic components. Moreover, the Ceramic Institute provided the educational tools to satisfy requests for broad education of ceramic engineers. In more recent years, the Ceramic Institute was the starting point for two other programs, the ceramic cooperation Karlsruhe-Stuttgart and the Graduate College within the Ceramic Institute. The Karlsruhe-Stuttgart cooperation in the ceramic area was proposed and funded by the State of Baden-Württemberg from 1989 to 1998 to bring together the wide spectrum of ceramic activities in Baden-Württemberg. Operating since 1990, the Graduate College is one of the funding modes of the Deutsche Forschungsgemeinschaft designed to promote high-quality graduate (doctoral) studies through participation of graduate students in interdisciplinary research programs. The program aims, among other things, at shortening the average time required to complete a doctorate degree to three years after the first degree. The Institute for Thermal Turbomachinery (ITS), as a member of the Ceramic Institute, was and still is involved in these programs as well as with a variety of other projects.

From the results and experiences of the ceramic automotive gas turbine program, it became very clear that successful application of ceramic components requires new design methodologies. Most importantly, initial design efforts should concentrate on non-rotating components of gas turbines. Therefore, at the beginning, work at the ITS concentrated on combustor flame tubes as thermally highly loaded components. In these studies, the suitability of a ceramic as wall material for advanced combustion technology (i.e. rich quench lean or lean premix prevaporize combustion) was demonstrated. In a first attempt, thermally induced stresses were reduced by segmentation of the flame tubes. To quantify the impact of segmentation on the failure probability and life times of ceramic components, a fracture statistical code (using stress distributions from finite element calculations as input) was developed in parallel to the experimental work. However, not only thermal and structural properties were found to be favorable but also the surface topology, primarily of porous ceramics, showed excellent performance when liquid fuel films were evaporated on the surface. From these early tests and calculations it became obvious that

(1) When designing a ceramic hot gas component, the entire operating cycle must be taken into account, since primarily transient temperature distributions cause critical stress states.
(2) New design principles, based on a local and time dependent adjustment of heat fluxes without violating the functionality of the component, have to be developed.

These two points will be treated in detail with special emphasis on the latter point. The design principles are based on careful adjustment of local thermal behavior of a ceramic structure through adjusting external and internal boundary conditions. As adjustments are made, the desire is a uniform temperature distribution within the ceramic component, which is realized by a hybrid design of the ceramic wall. The application of the design principles to various hot gas components, such as a flame tube for a small gas turbine, a nozzle guide vane for a ground based gas turbine, and a volute of a small radial gas turbine, will be demonstrated. The design principles and their necessary modifications will be discussed for all of these components, as they result in considerable reductions of thermally induced stresses.

CERAMIC RESEARCH COMBUSTOR

In developing new combustion chambers, the main effort is proper modeling of the combustion process to minimize pollutant emissions. In conventional direct spray combustors, pollutant reduction is limited because the fuel/air mixture generally reacts at locally stoichiometric conditions with high thermal NO_x formation. Therefore, new combustion concepts often consider reactions of homogeneous fuel/air mixtures under fuel-lean or fuel-rich conditions. Equivalence ratios of lean combustion systems are usually in the range of 0.25 to 0.50 depending upon combustor inlet temperature. NO_x formation rates are low due to the reduced flame temperature. Moreover, the excess air supports the oxidation of carbon monoxide and leads to low CO emissions. A primary requirement for lean combustion, however, is a homogeneous fuel/air mixture; otherwise hot spots inducing high NO_x formation will occur. To avoid hot spots, the fuel may be premixed and prevaporized in an external chamber with subsequent combustion under well-defined conditions. Potential difficulties, especially at high inlet temperatures as commonly occur in small regenerative gas turbines, are autoignition of the mixture within the premixing zone and flashback of the reaction zone.

To achieve low emissions with a rich-lean combustor, staged combustion with a rich initial reaction and lean after burning is employed. The production of thermal NO_x in the primary zone is low due to relatively low flame temperatures and oxygen deficiency, and carbon monoxide is oxidized in the secondary reaction zone at moderate temperatures. However, a very rapid and uniform transition from rich to lean conditions is required to prevent hot streaks with stoichiometric conditions and excessive NO_x formation. Successful applications of the rich/lear combustion concept are reported by Rizk and Mongia [3] and Novick et al. [4].

TABLE 25.1 NOMINAL DESIGN CONDITIONS

Parameter	Unit	Value
Combustor Inlet Temperature	°C	927
Turbine Inlet Temperature	°C	1377
Pressure Ratio		4.5
Specific Power	kJ/kg	261
Specific Fuel Consumption	kg/kWh	0.17
Reduced Mass Flow	kg.K/(s.bar)	2.21

As a consequence of the basic design concepts discussed, two-stage combustion with variable stoichiometric ratios within the primary zone, fuel-rich as well as fuel-lean, is utilized. The basic objective is evaluation of the emission characteristics of various combustion conditions. The nominal design parameters of the combustor are summarized in Table 25.1. The parameters represent a small regenerative gas turbine with a net power of about 70kW and a thermal efficiency of 50%.

In a fuel-rich primary zone, the flame tube cannot be film-cooled as the cooling air leads to locally stoichiometric conditions with corresponding high NO_x formation rates. On the other hand, adiabatic flame temperatures are far beyond 2000°C, inducing extremely high thermal loading of the wall. The presence of cooling air along the liner wall of the lean combustion zone leads to locally cold regions with a subsequent suppression of CO-oxidation. Therefore, the success of low-NO_x combustion concepts for gas turbine combustors with high inlet temperatures depends considerably on the applicability of temperature resistant ceramic materials. Figure 25.1 shows a schematic of the ceramic research combustor with a variable geometry flame tube.

A major intention of the ceramic research combustor, besides testing and developing new combustion concepts, is development of design strategies to apply ceramics in the hot gas path of gas turbines. Therefore, in designing the flame tube, the inherent material characteristics of ceramics must be considered. The failure behavior of brittle materials is characterized by fast fracture, i.e. local stress concentrations caused by volume or surface flaws cannot be reduced by plastic deformation as in the case of metals. Moreover, strength parameters cover a wide range of values and may only be described by statistical methods. To predict the failure behavior due to mechanical and thermal loads, fracture mechanics in combination with fracture statistical methods are necessary. Such fracture statistical analysis, based on finite element calculations of stress distributions, is performed with appropriate assumptions for flaw distributions and failure characteristics. With this approach, the failure probability and lifetime of the structure are determined and, with this information, critical regions may be identified. A summary of this procedure is given by Stürmer et al. [5], who describe the fracture statistical computer code CERITS-L for predicting life times for ceramic components.

In the design phase of the combustor, theoretical studies were performed using the computer code CERITS-L. From these analyses, it became apparent that the axial length of a cylindrical ceramic structure is the most significant geometric parameter for the determination of failure probability when considering typical combustor conditions. Decreasing length provides lower thermal stresses, and consequently, the risk of failure is reduced. In addition, circumferential segmentation of a ring further reduces failure probability. One-piece concepts for flame tubes (as utilized in the former ceramic automotive gas turbine program, Bunk [6]) may therefore not fulfill reliability requirements. As a consequence of these and other theoretical results, the flame tube of the combustor is subdivided in the axial and circumferential directions. The phase I flame tube with ribbed ceramic primary zone elements is shown in Figure 25.2. The transition piece and the secondary zone elements are made of metal.

Because siliconized silicon carbide (SiSiC) was first intended for use as a flame tube material, the primary zone elements are provided with cooling ribs to ensure that the inner surface temperatures do not exceed the melting temperature of silicon. The actual flame tubes, however, are made of sintered silicon

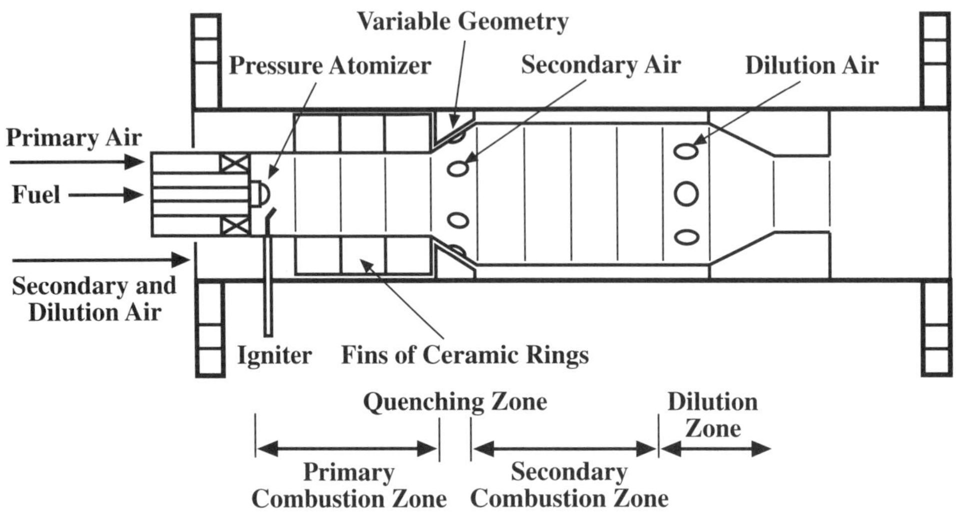

Figure 25.1 Schematic of the ITS Ceramic Research Combustor (Reprinted from Ref. [9])

Figure 25.2 Phase I Flame Tube with Ribbed Ceramic Primary Zone Elements (Reprinted from Ref. [9])

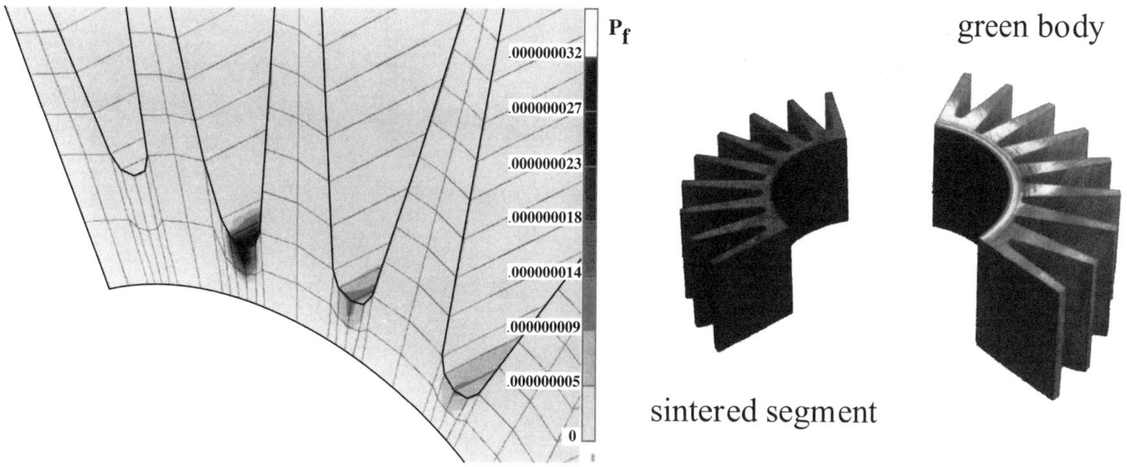

Figure 25.3 Failure Probability of the Ribbed Segments (left), Ribbed Combustor Liner Segments (Right) (Reprinted with permission from the University of Karlsruhe)

TABLE 25.2 MATERIAL PROPERTIES OF SINTERED SILICON CARBIDE.

Temperature T (°C)	Density (g/cm^3)	Thermal Conductivity (W/mK)	Specific Heat c_P (J/kgK)	Young's Modulus E(GPa)	Poison's Ratio	Coefficient of Thermal Expansion ($10^{-6}K^{-1}$)
20	3.13	72	700	433	0.16	3.24
200		64	840	430		
400		54	1030	426		
600		46	1125	423		
800		41	1200	419		
1000		39	1260	425		4.2
1200		36	1300	412		
1400		35	1350	408		
1600		34	1400	405		

Bending strength (20°C): σ_b=362 MPa Weibull Modulus (20°C): m=8.6

carbide (SSiC) and the rings of the phase II flame tube are fabricated without fins. This design change is suggested by results from failure probabilistic analyses, which reveal an increased failure probability in the areas between the fins. This is shown in Figure 25.3.

As indicated before, design efforts concentrate on non-rotating, thermally highly loaded gas turbine components. Therefore, sintered silicon carbide was chosen for various reasons, such as very low deterioration of strength up to temperatures of 1400°C, high thermal conductivity and thermal diffusivity resulting in a favorable thermal shock resistance, resistance against creep and oxidation, and availability. Availability and affordability were crucial at the beginning and the support of Motoren-und Turbinen Union, Munich (MTU), which provided ITS with ceramic green bodies for free, was highly appreciated. Machining of the green bodies was done at ITS and the sintering was performed by MTU. The properties of the sintered silicon carbide that is used for the combustors and turbine guide vanes to be discussed later are given in Table 25.2.

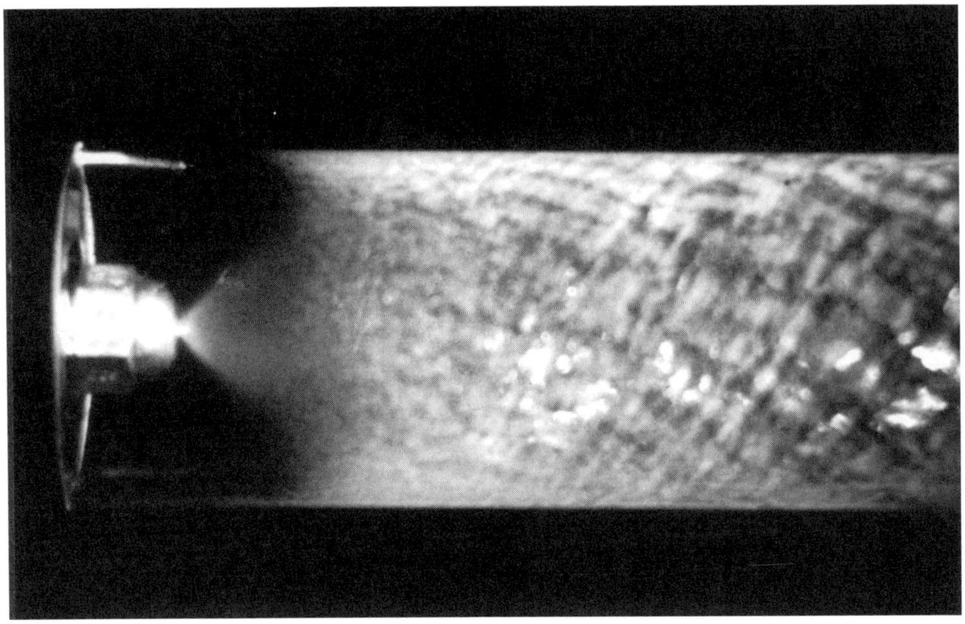

Figure 25.4 Liquid Film Established on the Inner Surface of a Tube by a Hollow Cone Pressure Atomizer (Reprinted from Ref. [9])

Besides the actual combustion process, an alternative method of fuel/air preparation, the evaporation of a shear driven liquid fuel film was investigated. Various successful applications of film evaporation in gas turbine combustors are described by Kuck [7], and Smith and Cowell [8]. In general, the formation of a homogeneous mixture, promoted by the intense heat and mass transport within the boundary layer and the relative slow heating of the liquid fuel, are considered to be responsible for reduced soot and pollutant emissions over a wide range of operating conditions. The basic objectives of film evaporation are high evaporation rates for large amounts of fuel on a small area, requiring intensive heat transfer both from the gas and the wall to the liquid film. Generally, the heat transfer, and thus, the vaporization intensity of a vaporizer tube can be increased by high temperature differences between gas flow and film, high free stream velocities, and additional heat transfer from the wall. The formation of a continous liquid film is a mayor challenge in utilizing film evaporation. Since wall coverage through discrete holes is difficult, spraying the fuel onto the inner surface of the vaporizer tube is used in the present combustor. The practicality of this technique was confirmed by visualization tests in a glass tube of the same size as the primary zone of the combustor. Figure 25.4 shows the shear driven wavy liquid fuel film in this glass tube.

The experimental investigations with the ceramic research combustor demonstrate that air-staging and film evaporation of the liquid fuel in the primary combustion zone lead to a considerable reduction in pollutant formation. Putting the results into their proper perspective can be achieved by considering the emission standards for passenger cars. Figure 25.5 shows the CO and NO_x emissions for a wide range of operating conditions. Two emission standards, the EC2000 (EURO III, light gray) and EC2005 (EURO IV, dark gray) are given for comparison. The respective emission regions are composed of the emission limits for gasoline and diesel engines. As can be seen from the figure, even the stringent limits of the future EC2005 standard can be met by a number of operating points without the need for further treatment of the exhaust gases. Further details on the ITS ceramicresearch combustor are given by Pfeiffer et al. [9, 10].

The experiences gained from operating the ceramic research combustor are summarized as follows:

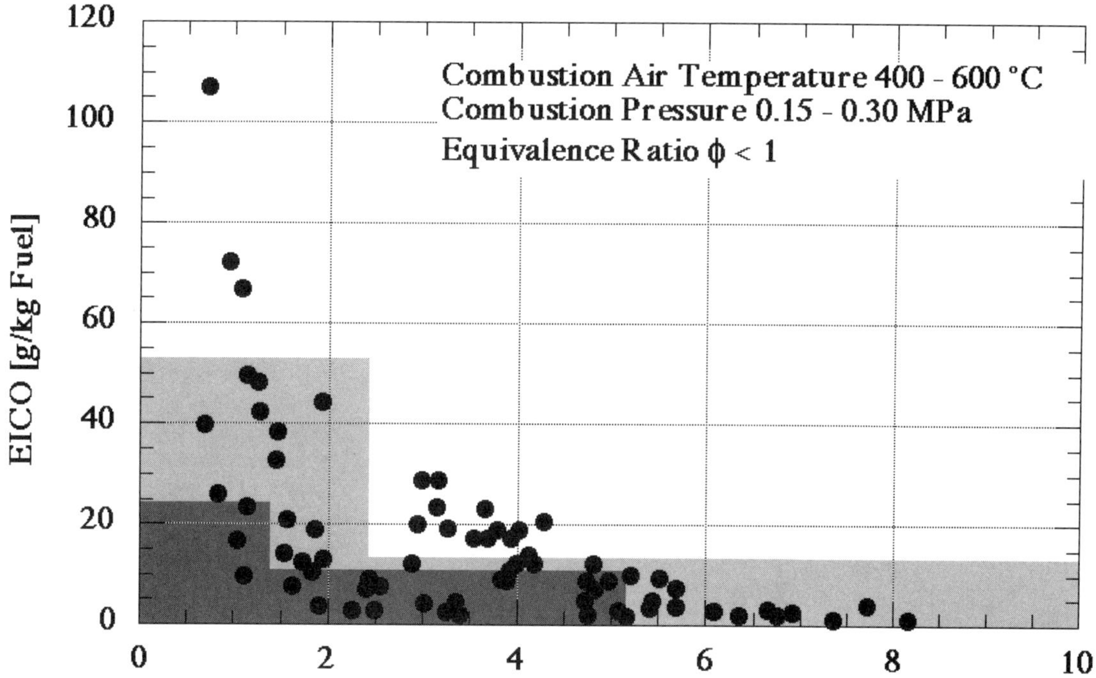

Figure 25.5 CO and NO_x Emissions of the ITS Ceramic Research Combustor (Reprinted with permission from University of Karlsruhe)

- The stress reducing effect of circumferential segmentation is weak compared to an axial segmentation of the flame tube as revealed by failure probability analyses.
- The fixation of the ceramic flame tube rings in the metal containment with spring-loaded plungers is very expensive and not applicable to a standard combustor.
- Film evaporation on a hot ceramic surface is feasible in principle. However, the formation of carbonaceous deposits must be avoided.

These points are also considered in two other research projects, one concentrating on the fixing of ceramic parts in a metal containment, and the other investigating liquid evaporation and deposit formation on heated ceramic surfaces.

The challenge of integrating ceramic parts into a metal casing is met by using a compliant layer between the ceramic flame tube rings and the metal casing, instead of the spring loaded plungers utilized in the research combustor. The method was tested in the combustor of a small 70kW commercial gas turbine (KHD T216). The three-layer design of the flame tube is shown in Figure 25.6.

The objective of the study is the demonstration of the applicability of the multi-layered segmented design under real engine conditions. Additionally considered is the impact on the combustion process of the non-film cooled wall with higher wall temperatures.

The flame tube is instrumented with a number of temperature sensors to measure local temperatures on the outer surface of the ceramic rings. These local temperatures are then used for adjusting the inner boundary conditions of a finite element calculation of the three-layered construction. The heat resistance of the joints of the ceramic rings caused by the topology of the respective surfaces is taken into account by thin insulating layers in the finite element calculation. The model is derived from the investigation by Yovanovich et al. [11] on contact heat transfer of rough surfaces. A typical outer surface temperature distribution of the

Figure 25.6 KHD T216 Flame Tube with a Flexible Alumina Silicate Fiber Layer Between SSiC Flame Tube Rings and Metal Containment (Reprinted from Ref. [13])

ceramic flame tube is given in Figure 25.7. The assumptions made for the calculation are confirmed by the good agreement between measured and calculated temperatures. The highest wall temperatures occur at the end of the primary zone and attain values of about 1000°C. Due to the insulating layer between ceramic flame tube and metal casing, the temperature differences in the radial direction of the SSiC rings are very small. The insulating effect of the joints can be clearly seen in areas with strong stream wise variations of the boundary conditions (flame temperature) as in the primary zone and in the dilution zone of the combustor. The stream wise steps in surface temperature lead to a considerable reduction of the temperature gradient of the respective ceramic rings and thus to reduced thermally induced stresses. Detailed stress calculations including transient operating conditions reveal maximum first principal stresses of 75MPa in ceramic rings without mixing air holes, and 90MPa at the edge of a mixing air hole (Münz et al. [12]).

The combustor has been successfully operated for more than 150 cumulative hours. Even though a few cracks occurred at the mixing air holes, the function of the combustor is not affected by these cracks. Since wall-cooling films can be omitted, combustion conditions to obtain complete reactions in the vicinity of the wall are improved. This is indicated by pronounced reductions of unburned hydrocarbons, carbon monoxide and soot as compared to the original combustor (Münz [13]).

The formation of carbonaceous deposits is of course an undesirable aspect of liquid fuel film evaporation on hot ceramiccombustor surfaces. It is also undesirable in conventional combustors due to its negative impact on combustor liner cooling. For this reason and because of the promising evaporation behavior of

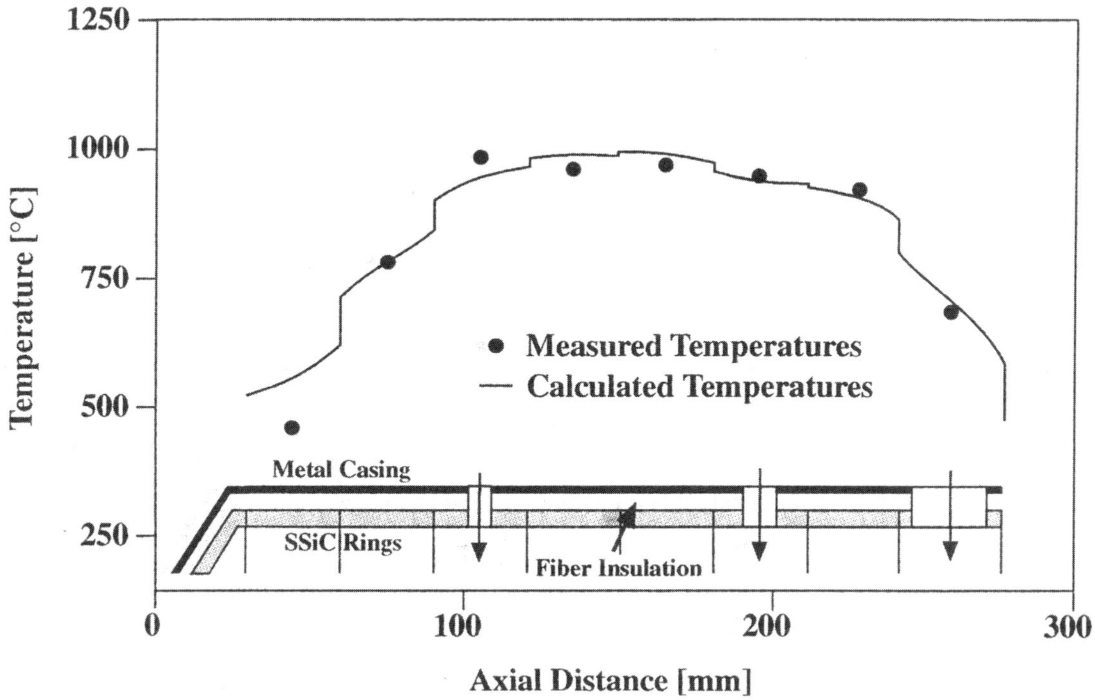

Figure 25.7 Temperature Distribution Along the Outer Surface of the Ceramic Rings (Adapted from Ref. [12])

porous ceramic surfaces, a comprehensive study on the evaporation of liquid fuel films and the formation of carbonaceous deposits on heated surfaces was undertaken.

For effective film vaporization, high wall temperatures leading to high heat fluxes in the liquid film and short vaporization times are imperative. Short evaporation times are necessary for suppressing the formation process of carbonaceous deposits due to fuel instability and can be achieved by very thin and short liquid films. The lower limit of a stable liquid film is determined by the surface material. Films with a high specific wetted area can be formed on porous ceramic surfaces. Moreover, vaporization intensity on such surfaces is considerably increased due to the higher Leidenfrost temperatures with accordingly higher wall heat fluxes. The Leidenfrost temperature represents the minimum temperature when an insulating vapor layer between liquid and wall surface is formed.

The formation of the carbonaceous deposits on combustor walls was investigated over a wide range of engine surface temperatures for a variety of materials and fuels (Brandauer et al. [14, 15]). Formation mechanisms were identified. Deposit formation takes place in a typical surface temperature range depending on the fuel's boiling curve and its composition of hydrocarbons. The temperature dependence of the deposit formation of fuel oil on an aluminum oxide surface of 25% porosity is given in Figure 25.8. No deposits are found for temperatures beyond 400°C. On plain metal surfaces this temperature is about 50–60°C higher. The vaporization characteristic of fuel oil is also shown in this figure. The vaporization time approaches its minimum at a temperature of 400°C, which coincides with the highest temperature for deposits. For higher temperatures, a drastic increase in vaporization times is observed indicating the formation of a vapor layer. With such results a surface temperature range for the operation of a combustor with liquid fuel film evaporation can be identified (Brandauer et al. [16, 17]). In a present study, these findings are applied to a small combustor with a new LPP combustion concept based on vaporizing a liquid fuel film on porous SiC surface.

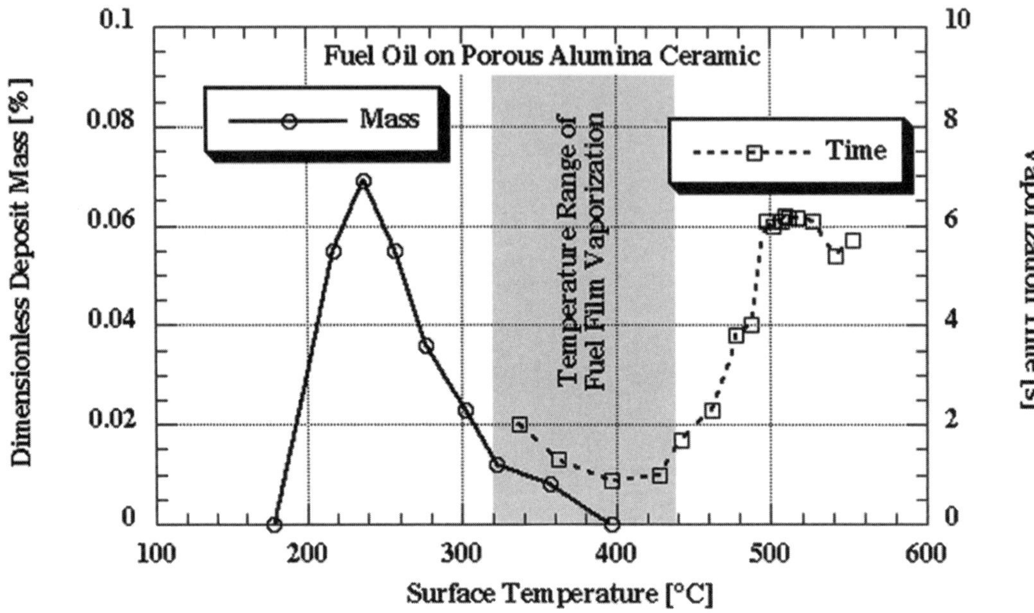

Figure 25.8 Vaporization Time and Deposit Formation of Fuel Oil on a Heated Alumina Oxide Surface with 25% Porosity (Adapted from Ref. [16])

Experimental Assessment of Fiber Reinforced Ceramic Combustor Liners

The advantage of monolithic ceramic materials, such as their high Young's modulus, high thermal conductivity, and excellent oxidation resistance, are partly offset by their relatively low fracture toughness. A wide range of additives and particulate reinforcements has been tried as toughening agents. Reinforcement with SiC or C continuous fibers yields the highest fracture toughness. These materials show an increase in fracture toughness by one order of magnitude (Helmer et al. [18]). Since the desired application involves extended periods of operation in oxidizing atmospheres, proven oxidation resistance of the composites is required. In the study four flame tubes, manufactured from three different composite materials were tested under real engine conditions in the combustion chamber of the small commercial gas turbine, described earlier (Filsinger et al. [19]). The specification of the flame tubes is summarized in Table 25.3. The carbon-reinforced composites are coated with SiC either by chemical vapor deposition (CVD) or a dross process. The SiC/SiC materials are protected by a SiC (CVD or dross) layer.

The gas turbine was run under various operating points, with highest outer wall temperatures of the flame tube of about 1050°C. This temperature was observed in the middle between the two rows of dilution holes. Initially, the accumulated testing time for the carbon-reinforced composites was limited to 10 hours.

All four flame tubes withstood the thermal load under an oxidizing atmosphere without severe damage. The Dornier tube, coated with the dross SiC layer, shows debonding on the inner surface in an area of highest thermal load (see Figure 25.9). The second Dornier flame tube coated with CVD-SiC, shows no damage on the inner surface. However, on the outer surface a small part of the coat chipped off. The thermally high loaded area shows an intensive discoloring, which is found on all flame tubes except the one coated with dross SiC. The discolorations are due to the formation of a SiO_2 layer, which results from the oxidation process and helps to suppress subsequent oxidation. The thin amorphous glass layer reflects the different

TABLE 25.3 FIBER REINFORCED CERAMIC FLAME TUBES

Composite	Interface Coating	Fabric Layup	Oxidation Protection	Manufacturer
C/C		Satin Weave (2D-0/9°)	CVD-SiC	Schunk
C/SiC	PyC	UD Laminate (2D-0/90°)	CVD-SiC	Dornier
C/SiC	PyC	UD Laminate (2D-0/90°)	Dross SiC	Dornier
SiC/SiC	PyC	Plain Weave (2D-0/90°)	CVD-SiC	MAN Tec

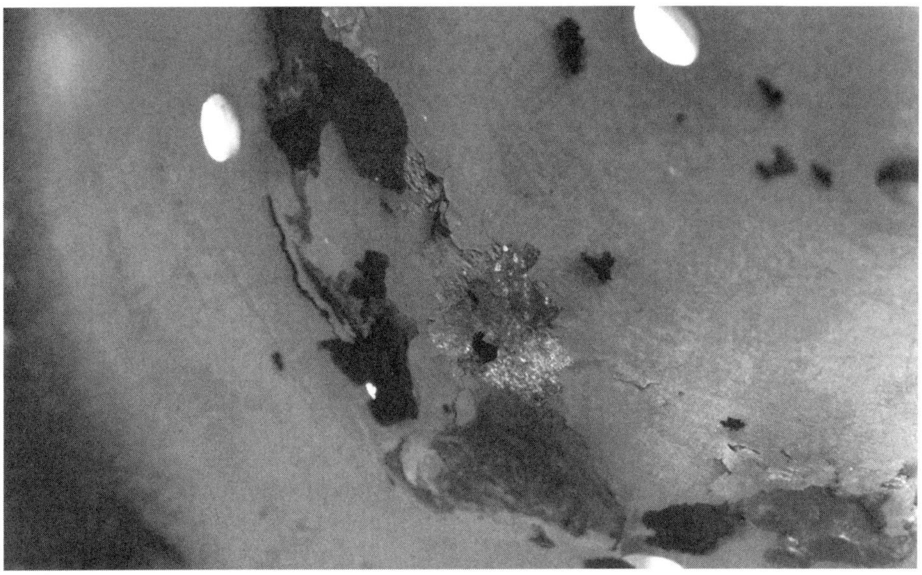

Figure 25.9 C/SiC Flame Tube with Dross SiC Coat After 10 Hours of Engine Operation (Reprinted from Ref. [19])

wavelengths of the incoming light depending on its thickness and as a result, the surface appears with rainbow like colors (Figure 25.10).

The C/C flame tube and the SiC/SiC flame tube withstood testing without structural defects. Therefore, the test program was extended. The C/C flame tube could be operated for 86 hours until the structure had disintegrated. The SiC/SiC combustion chamber, which had operated for more than 100 hours, exhibited no damage, although numerous start up and shut down processes occurred, causing critical loads due to high temperature gradients.

Nozzle Guide Vane For A Ground Based Gas Turbine

The major goals in using ceramic for a low-cooled nozzle vane for the first stage of a stationary gas turbine are the reduction of cooling air, and an increase of the turbine inlet temperature. Temperatures of 1400°C and higher are desired with an amount of cooling air of less than 1% of the total air flow for the first stage vane. For a metallic design operating under comparable conditions, at least 10% of the air is needed for cooling. Hüther and Krüger [20] describe initial attempts to develop a ceramic nozzle guide vane for a high-pressure turbine of a jet engine in the 1980s. In more recent years, activities concentrated on the replacement of metal first stage vanes of ground based gas turbines by various ceramic designs. In the present investigations, two load-oriented designs for a first stator vane are developed. The structure of the

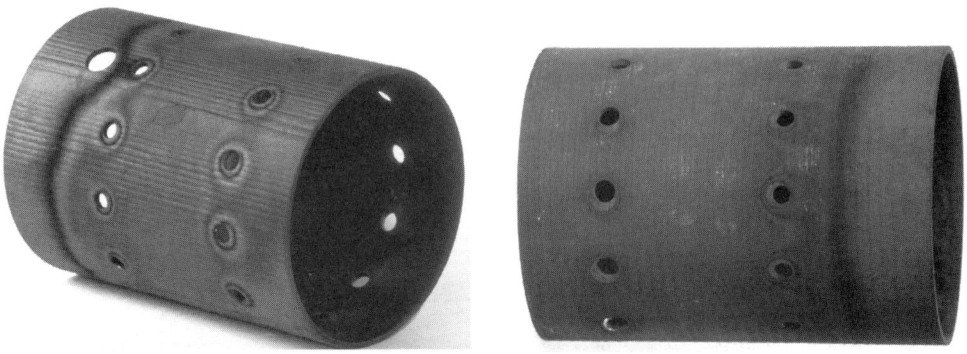

Figure 25.10 C/SiC Flame Tube with CVD-SiC Coat After 10 Hours (Left), SiC/SiC Flame Tube After 50 Hours of Engine Operation (Right) (Adapted from Ref. [19])

TABLE 25.4 SPECIFICATIONS OF GAS TURBINE AND FIRST STATOR VANE

Gas Turbine		Ceramic Nozzle Guide Vane	
Net Power	70MW	Untwisted, Integrated Sidewalls	
Pressure Ratio	17	Material	Sintered Silicon Carbide
Mass Flow	200kg/s	Coolant Mass	<2% of Total Mass Flow
Turbine Inlet Temp. (TIT)	1400°C	Vane Height	60mm
Coolant Temperature	400°C	Chord Length	58mm
Design Concepts:			
• Hybrid (three layer design)+Load Oriented+No Coolant Ejection			
• Directly Cooled + Load Oriented + Trailing Edge Ejection			

ceramic vanes is adjusted to local thermal loads at varying operating conditions. The specifications of the gas turbine and the stator vane are given in Table 25.4.

The first design is based on a three-layer airfoil wall (see Gutmann et al. [21]). An insulating fiber ceramic layer is embedded between an outer monolithic ceramic shell and an inner thin-walled metal structure. The local thickness of the ceramic shell and the insulating layer is adjusted to the local thermal boundary conditions to homogenize the mean temperatures of the ceramic shell. The second design employs direct convective cooling of the ceramic shell instead of a dividing fiber ceramic layer. Here too, the local thickness of the ceramic wall and the local intensity of the internal cooling are adjusted depending upon the thermal boundary conditions. Because these adjustments represent important departures from previous designs (see e.g. Nakakado et al. [22], Tsuchiya et al. [23], Wenglarz et al. [24], Day et al. [25]), additional details are therefore given in the following paragraphs.

Special emphasis is given to the treatment of the whole operating spectrum since critical loads occur during transient operating conditions. Consequently, a number of parameters must be specified as a function of time. These include temperature profiles of the hot gas flow and local heat transfer coefficients on the airfoil's surface and the platforms (Gutmann et al. [21]).

The principle of the hybrid design and the effect of varying wall thickness are now explained for a flat wall element in reference to Figure 25.11. The thickness of the outer ceramic layer may be varied within the constraints imposed by the overall geometric boundaries and manufacturing requirements. On the cooling air side, a high temperature alloy of constant thickness covers the fiber insulation. The thickness of the insulation can be adjusted to accommodate the thermal loading providing geometric limitations are kept in mind.

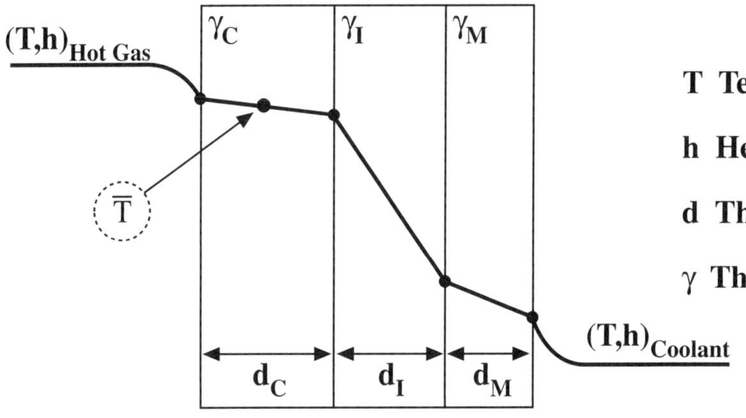

Figure 25.11 Model of the Three-layered Wall (Adapted from Ref. [21])

If all areas of the outer ceramic shell are maintained at the same mean temperature at every instant of the operating cycle, the only stresses, which will remain, will be the ones caused by temperature gradients normal to the wall. In the actual design, these gradients can be kept very small using the insulating layer for reducing the heat flux through the wall.

Under steady state operating conditions the local wall thickness can be derived from a simple heat balance through the wall given by

$$\dot{q} = \frac{1}{\frac{1}{h_{\text{Hot Gas}}} + \frac{d_C}{\gamma_C} + \frac{d_I}{\gamma_I} + \frac{d_M}{\gamma_M} + \frac{1}{h_{\text{Coolant}}}} \cdot \left(T_{\text{Hot Gas}} - T_{\text{Coolant}}\right)$$
$$\dot{q} = \frac{1}{\frac{1}{h_{\text{Hot Gas}}} + \frac{d_C}{2 \cdot \gamma_C}} \cdot \left(T_{\text{Hot Gas}} - \overline{T}\right) \tag{1}$$

At first, the hot gas temperature is regarded as constant. Accordingly, the ratio of the heat resistance between the middle of the ceramicshell and the coolant and the heat resistance between the hot gas flow and the middle of the ceramic shell must be constant when a constant mean temperature of the ceramic shell is the objective.

$$\frac{\frac{d_C}{2 \cdot \gamma_C} + \frac{d_I}{\gamma_I} + \frac{d_M}{\gamma_M} + \frac{1}{h_{\text{Coolant}}}}{\frac{1}{h_{\text{Hot Gas}}} + \frac{d_C}{2 \cdot \gamma_C}} = \text{const.} \tag{2}$$

From this relation, the temperature field under steady state conditions depends upon the thickness of the insulating layer since the heat resistance of the inner and outer boundary layer and of the ceramic shell and metal structure is much smaller than the heat resistance of the insulation. The thickness of the insulation is thus inversely proportional to the external heat transfer coefficient as expressed using

$$d_I \propto \frac{1}{h_{\text{Hot Gas}}} \tag{3}$$

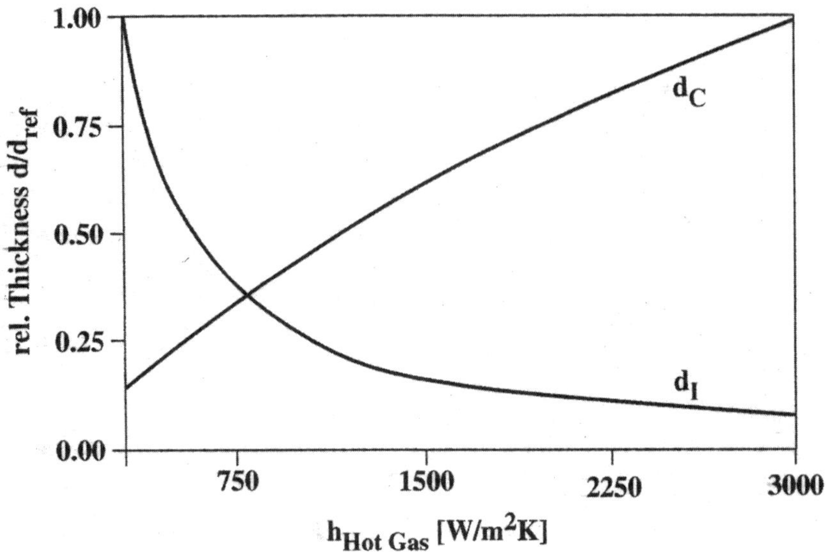

Figure 25.12 Optimal Thicknesses of the Ceramic Shell (d_C) and the Insulating Layer (d_I) (Adapted from Ref. [21])

Transient operating conditions are taken into account by a quasi-steady state approximation, which is of the form

$$\frac{\partial T}{\partial t} = a\frac{\partial^2 T}{\partial x^2} = \text{const.} \tag{4}$$

For low Biot numbers, this approximation describes the heating and cooling of the ceramic layer fairly well, providing the heat flux to the cooling air is neglected, which is permissible for short periods of time after a sudden temperature step. Integrating this approximation and introducing the heat resistance of the hot gas boundary layer, the mean temperature of the ceramic shell can be written as

$$T_{\text{Hot Gas}} - \overline{T} = \dot{q}\cdot\left(\frac{1}{h_{\text{Hot Gas}}} + \frac{d_C}{3\gamma_C}\right) \tag{5}$$

Again considering the necessity of uniform local mean temperature of the ceramic shell, the optimum thickness of the shell consequently depends on the outer heat transfer coefficient as indicated by the following relation

$$\frac{d_C}{d_{\text{Hot Gas}}} + \frac{d_C^2}{3\gamma_C} = \text{const.} \tag{6}$$

Figure 25.12 shows the normalized optimum thickness as dependent upon heat transfer coefficient for a plane wall.

According to Eqns. (2) and (6) the local thickness can be determined given a convective loading together with a wall thickness or a reference temperature. For variable main flow temperatures, e.g. over the blade height, the local temperature level at steady state operation must be changed to accomplish this. This can be achieved by optimizing the thickness of the insulating layer according to the following equation

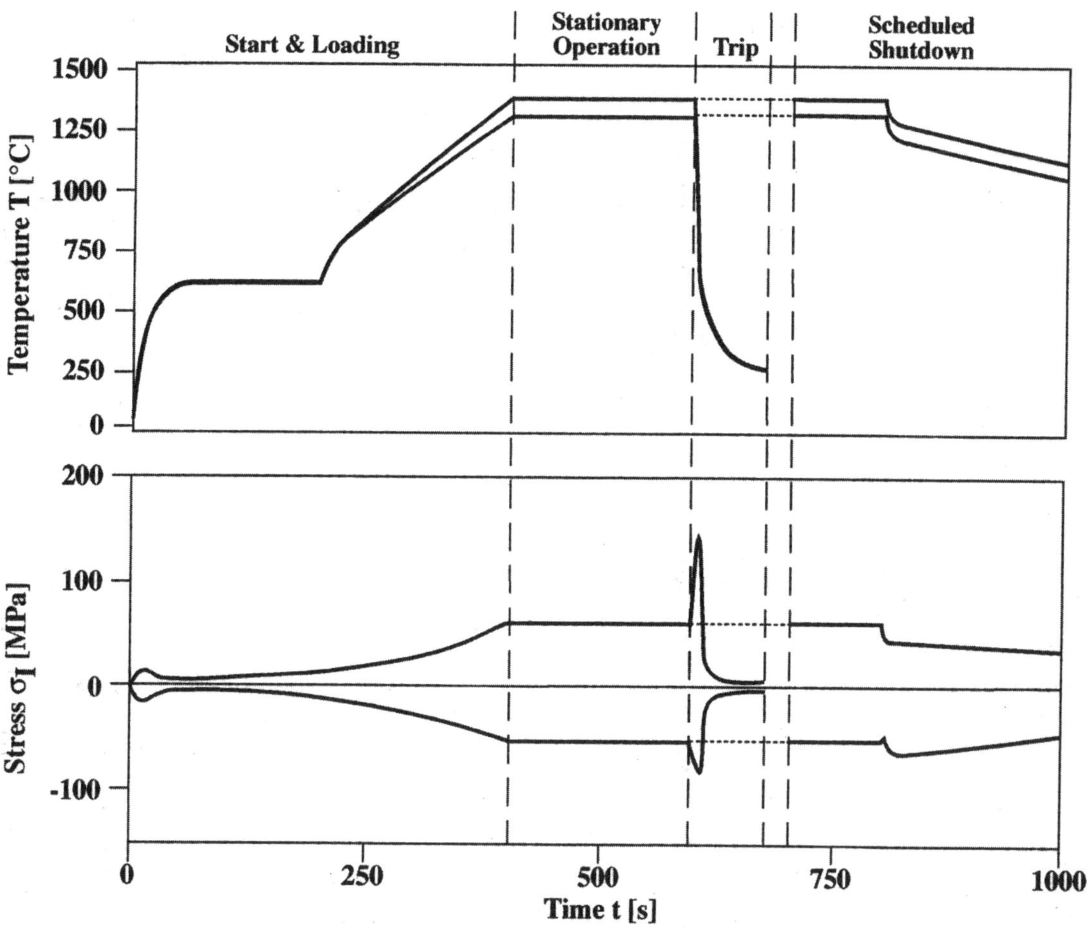

Figure 25.13 Maximum and Minimum Temperatures and Stresses in the Ceramic Shell of the Three-layered Hybrid Design (Adapted from Ref. [27])

$$T_{\text{Hot Gas}} - \frac{\dfrac{1}{h_{\text{Hot Gas}}} + \dfrac{d_C}{2\gamma_C}}{\dfrac{1}{h_{\text{Hot Gas}}} + \dfrac{d_C}{\gamma_C} + \dfrac{d_I}{\gamma_I} + \dfrac{d_M}{\gamma_M} + \dfrac{1}{h_{\text{Coolant}}}} \left(T_{\text{Hot Gas}} - T_{\text{Coolant}}\right) = \text{const.} \tag{7}$$

Also see Eqn. (1). With these equations, the optimization of the ceramic vane can be performed for all shell-shaped areas. Another optimization method, based on the compensation of thermal strains is employed for areas such as the trailing edge region of the airfoil, where cooling cannot be optimized according to the layer thickness method (see Gutmann [26]).

The optimization approach is introduced into a finite element numerical scheme. Based on local and time dependent boundary conditions, a three-layered finite element mesh with variable layer thicknesses is generated for calculating the temperature and stress distributions within the airfoil.

Figure 25.13 shows the calculated maximum and minimum temperatures of the ceramic shell and the corresponding first principal stresses. The maximum tensile stresses remain below 75MPa and 135MPa for continous operation and trip, respectively. Failure probability calculations demonstrate sufficient reliability for engine operation. The time dependent (slow crack growth) failure probability determination led to $P_f=0.84\times10^{-4}$ considering 10,000 hours of continuous operation.

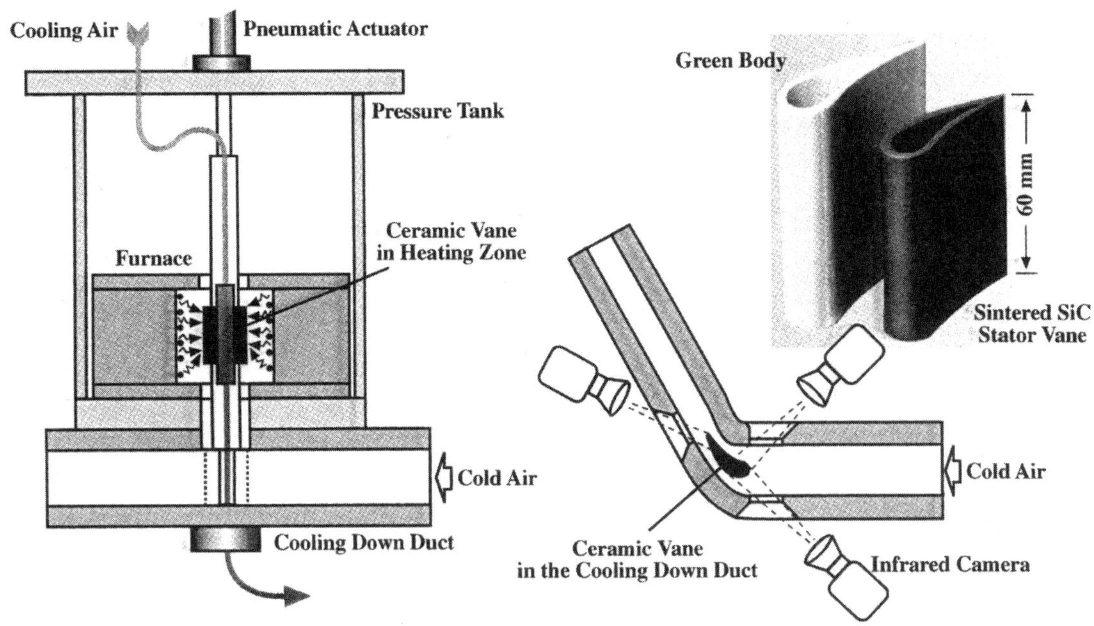

Figure 25.14 Schematic of Thermal Shock Test Section and Tested Stator Vane Profiles (Adapted from Ref. [27])

For an experimental verification of the design method, it is necessary to simulate the thermal loading that occurs during the trip of the engine. Therefore, a test rig, which causes thermal stresses in the ceramic-structure, equally to those induced by the temperature drop during the fuel cutoff in real engines, was constructed (Dilzer et al. [27]). A schematic of the test rig is shown in Figure 25.14. The ceramic nozzle guide vane is heated up to temperatures of 1500°C in an electric furnace installed in a pressure tank on top of the flow channel (cooling down duct). The inner cooling of the test specimen is achieved by a coolant flow through small tubes. A variation of the diameter and the distance of the cooling tubes makes it possible, to adjust the local temperature according to real engine continuous operation. After the heat up process, the tested vane profile is moved into the cooling down duct (T=20°C, p_{max}=1MPa) using a pneumatic transport mechanism. During the cooling down process the surface temperature mapping of the ceramic vane is recorded by an infrared camera.

Due to the fixed temperatures of the main flow and the coolant (20°C), the conditions for the test specimen are more severe than in real engines, provided the Reynolds number of the main flow and hence heat transfer is identical. If cooling down starts from the same temperature level the temperature drop in the test rig exceeds the one occurring in real engines after fuel cutoff. With regard to this, the temperature of the ceramic nozzle vane is increased step by step with increments of about 50°C. For each temperature level at least three thermal shock tests are performed, in order to observe the ceramic vane profile from each perspective (see Figure 25.14).

A finite element analysis of the time dependent temperature and stress distribution in the test specimen throughout the thermal shock test revealed, that the initial temperature of the shock test (furnace temperature) has to be set to 1050°C, for thermal loading in the test to be the same as in the real engine during a trip. For both cases the temperature of the ceramic vane profile changes by approximately 80°C/s within the first 10 seconds.

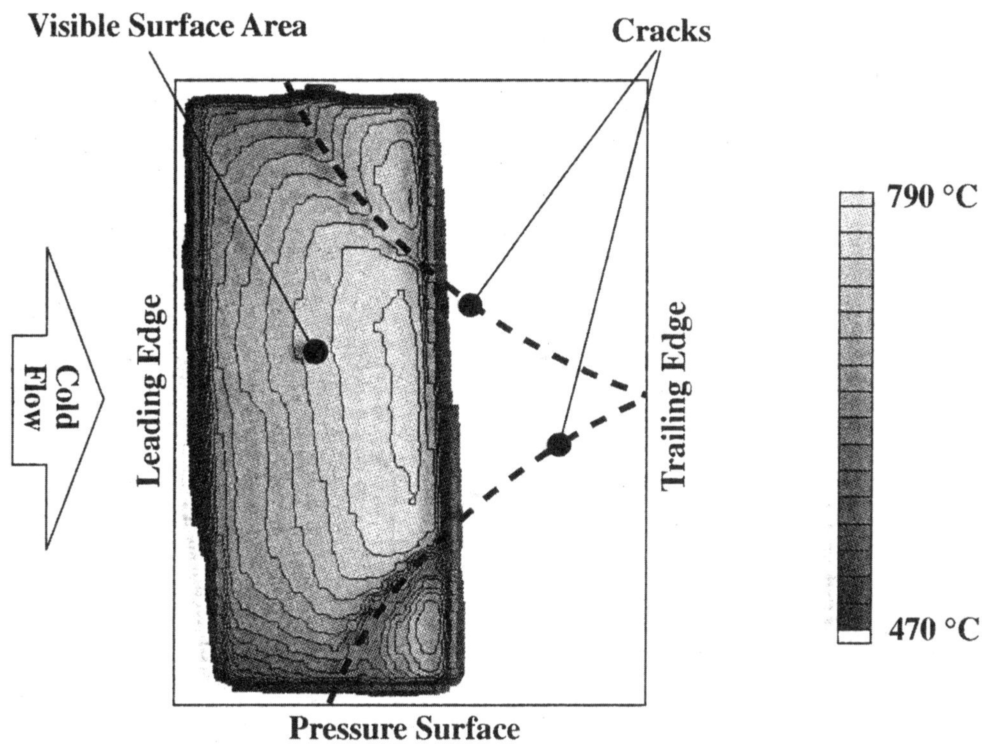

Figure 25.15 Temperature Mapping of Front Area of the Pressure Surface Just in the Moment of Fracture (Reprinted with permission from the University of Karlsruhe)

So far experiments have been carried out with three ceramic vane profiles only. One fractured as a result of misalignment of the pneumatic transport mechanism, after having endured some experiments with rather small stress loadings. The other withstood a large number of thermal shock tests with induced tensile stresses up to 150MPa, without any visible damage. This stress level is the maximum load that occurs during trip of the engine (see Figure 25.13). For one of the two test specimen the induced load was increased step by step until it fractured after the fourth experiment with induced stresses of approximately 250MPa. Figure 25.15 shows the surface temperature mapping just at the moment of fracture, 1.72 seconds after start of the shock test. The two cracks can be clearly seen by the distortions of the isotherms. They have their origin at the trailing edge mid-span, where highest tensile stresses are predicted by the finite element analysis (Dilzer et al. [27]).

When compared with the results of conventionally designed and directly cooled ceramic nozzle guide vanes with induced stresses of more than 400MPa, the superiority of the hybrid design becomes apparent. With this design, the maximum material temperature is very close to the hot gas temperature due to the low internal cooling, which is intended in order to reduce temperature gradients. This means that the present design can be applied only if the hot gas temperatures remain clearly below values representative for starting surface corrosion and degradation. Since today's gas turbines are run at turbine inlet temperatures far beyond temperatures at which surface corrosion begins, the ceramic used for the shell has to withstand corrosion or be coated. As an alternative to these measures, an engineeringapproach based on the previously discussed load oriented design with direct internal cooling and trailing edge ejection of the coolant has been developed.

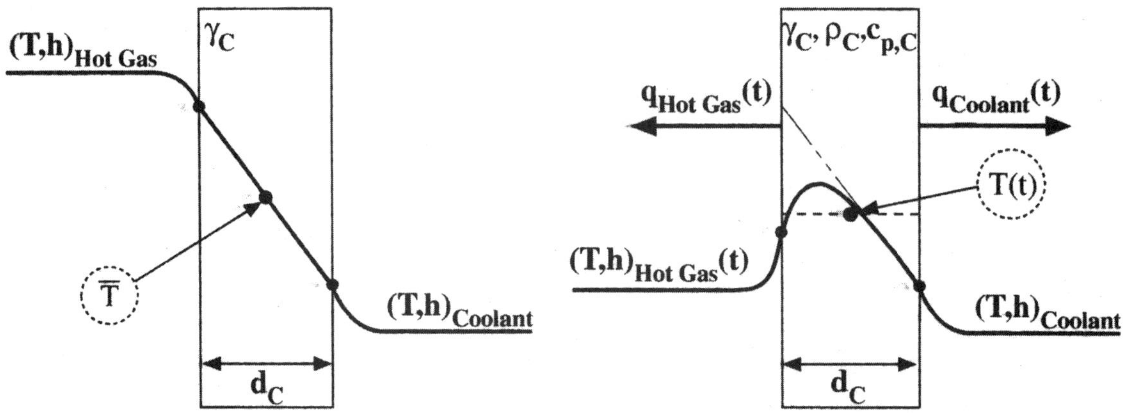

Figure 25.16 Flat Wall Model for Steady (Left) and Transient (Right) Operating Conditions (Adapted from Ref. [28])

The second design for a nozzle guide vane, which has the potential for very high turbine inlet temperatures, is derived from single layer flat wall considerations (Dilzer et al. [28]). Here, as well as in the previously discussed approach, a constant mean wall temperature is the objective. For constant operating conditions, this can be achieved by adjusting the inner heat transfer coefficient (coolant side) to the external boundary conditions. For transient operating conditions, the thickness of the ceramic shell has to be adjusted. Figure 25.16 shows the flat wall model for steady state and transient operating conditions.

For steady state conditions, the expression for the inner heat transfer coefficients reads as follows

$$h_{Coolant} = \frac{h_{Hot\,Gas}\left(\bar{T} - T_{Hot\,Gas}\right)}{\frac{d_C}{2 \cdot \gamma_C} \cdot h_{Hot\,Gas}\left(T_{Hot\,Gas} + T_{Coolant} - 2 \cdot \bar{T}\right) + T_{Coolant} - \bar{T}} \tag{8}$$

A graph of this relation is given in Figure 25.17. There is a strong dependence of the coolant side heat transfer coefficient on the hot gas heat transfer coefficient. The influence of the coolant temperature is comparatively weak. The optimization shown in Figure 25.17 is performed for a hot gas temperature of 1427°C, a mean temperature of the ceramic of 1227°C, a thermal conductivity of the ceramic of 38W/mK, and a constant thickness of the ceramic shell of 2.5mm.

As indicated earlier, it is imperative to maintain a homogeneous temperature distribution of the ceramic component for the whole operating cycle of the engine. For this reason, the flat wall considerations are extended to account for transient thermal boundary conditons. Some simplifications are introduced, such as neglecting of temperature gradients normal to the wall; assuming temperature and heat transfer coefficient on the hot gas side vary linearly with time for the first few seconds after fuel cutoff; and assuming thermal conditions on the coolant side do not vary within the first few seconds after fuel cutoff. Taking these assumptions into account, the wall heat fluxes can be described as

$$\begin{aligned} \dot{q}_{Hot\,Gas}(t) &= h_{Hot\,Gas}(t) \cdot \left(T_{Hot\,Gas}(t) - \bar{T}(t)\right) \\ \dot{q}_{Coolant}(t) &= h_{Coolant} \cdot \left(T_{Coolant} - \bar{T}(t)\right) \end{aligned} \tag{9}$$

and the local heat balance for the ceramic shell can be formed as follows

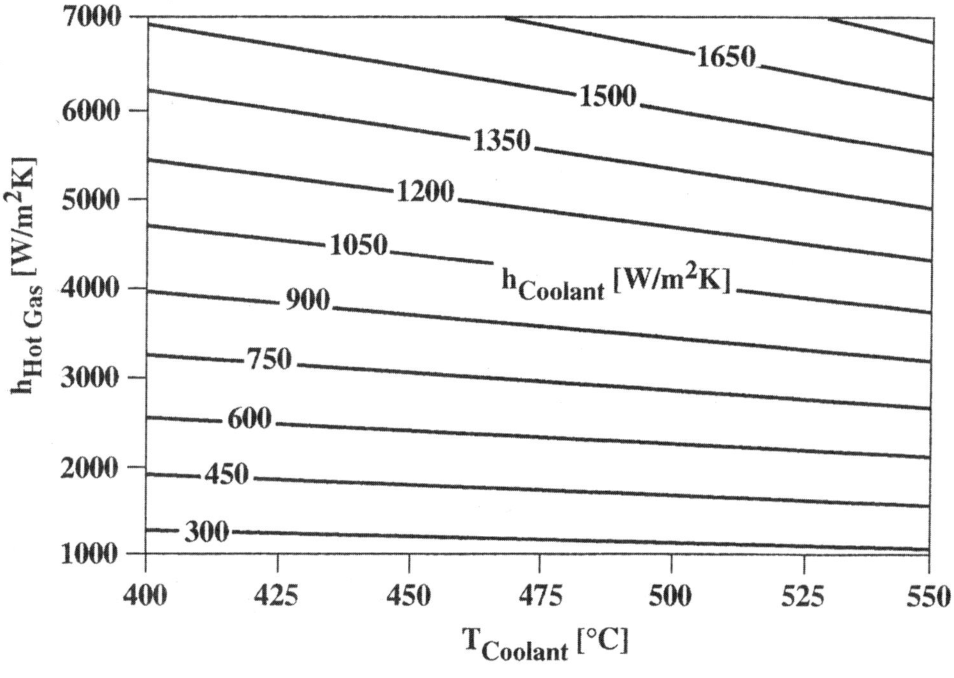

Figure 25.17 Dependence of the Inner Heat Transfer Coefficient on the Outer Heat Transfer Coefficient and Coolant Temperature (Adapted from Ref. [28])

$$\dot{q}_{Hot\ Gas}(t) + \dot{q}_{Coolant}(t) = d_C \cdot \rho_C \cdot c_{p,C} \cdot \frac{\partial \bar{T}(t)}{\partial t} \tag{10}$$

For a given temperature gradient of the mean wall temperature, the local wall thickness must be chosen according to

$$d_C = \frac{h_{Hot\ Gas}(t) \cdot \left(T_{Hot\ Gas}(t) - \bar{T}(t)\right) + h_{Coolant} - \bar{T}(t)}{\rho_C \cdot c_{p,C} \cdot \frac{\partial \bar{T}(t)}{\partial t}} \tag{11}$$

The thickness, of course, cannot vary with time. Therefore, the differential equation (11) is solved for a certain moment after fuel cutoff. Preliminary tests revealed that the optimum thickness of the ceramic shell should be adjusted for thermal boundary conditons occurring 2.5 seconds after fuel cutoff. The optimum thickness obtained for these conditions are shown in Figure 25.18. The optimum wall thickness d_c increases strongly with increasing hot gas heat transfer coefficient. The influence of the coolant side heat transfer coefficient is not as pronounced. The optimization again is performed for a hot gas temperature of 1427°C, the coolant temperature is fixed at 477°C, and the temporal gradient of the mean temperature is 50°C per second.

According to these design guidelines a finite element model of the vane was generated with wall thicknesses ranging from 2.17mm to 3.65mm at the stagnation point. After carefully adjusting the local inner heat transfer coefficient, analyses for a complete operating cycle were performed. The calculated maximum-stresses remain below 100MPa and 180MPa for steady operation and trip, respectively. The minimum temperature difference between hot gas and ceramic vane is more than 100°C. The stress level is somewhat

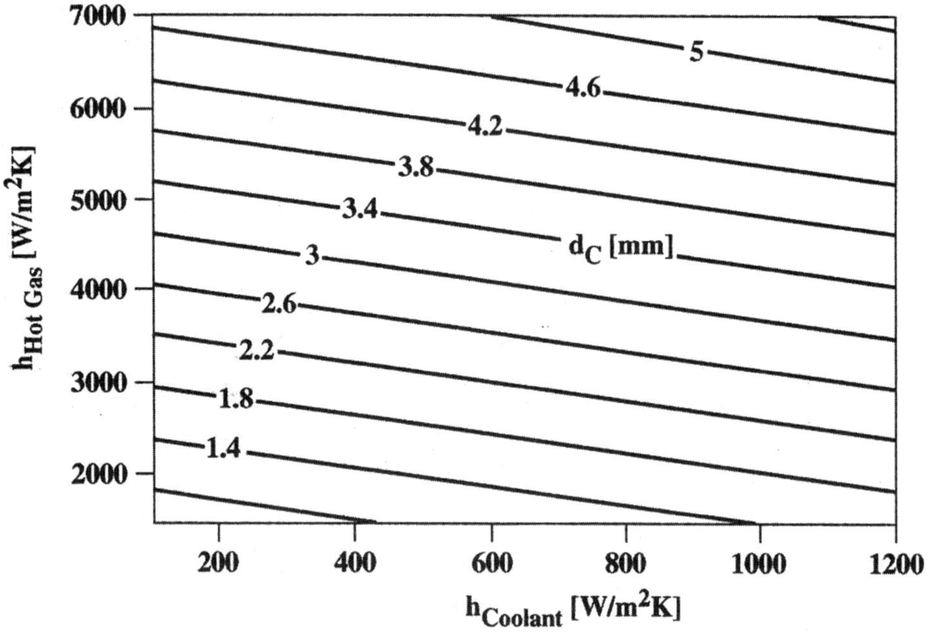

Figure 25.18 Optimum Wall Thicknesses of the Ceramic Shell (Adapted from Ref. [28])

higher than that for the three-layered hybrid design, but remains below 40% of the maximum stresses calculated by Tsuchiya et al. [23] and Day et al. [25].

Since inner heat transfer cannot be varied arbitrarily due to aerodynamic and geometrical limits, a transformation of the ideal internal cooling configuration to a technically feasible cooling setup has to be performed. To achieve an internal heat transfer distribution that is very close to the ideal heat transfer distribution, the inner surface is divided into three areas with different cooling methods. Figure 25.19 shows a cross sectional view of the guide vane with varying shell thickness and three different cooling schemes.

The calculation of the local internal heat transfer coefficients is performed by utilizing correlations for the different cooling methods. Heat transfer intensifying measures, such as ribs, pin fins, or dimples on the inner surface of the shell are not considered, since this would complicate manufacture of the ceramic shell. A comparison of the ideal internal heat transfer distribution and the heat transfer gained from correlations is given in Figure 25.20. There are two areas in the vane causing difficulties, the laminar-turbulent transition on the suction surface and the trailing edge of the vane. The non-ideal internal boundary conditions are reflected in the results of the finite element stress calculations. For steady operating condition the maximun stresses rise from 100MPa to 170MPa and for trip conditions the stresses rise by 10MPa to 190MPa.

The results of failure probability calculations for ideal and technically feasible internal cooling using the fracture statistical computer code CERITS-L are summarized in Table 25.5. As expected from the stress results, there is a considerable difference in failure probability for continuous operation. Under trip conditions, the failure probabilities are very similar. A detailed description of the design methodology and its application to a real nozzle vane is given by Dilzer et al. [28].

Ceramic Scroll for A Small Radial Turbine

In the low power range, radial inflow turbines have several advantages compared to axial turbines, such as simplicity, low cost, and performance. Effective cooling requires a high technical and developmental expen-

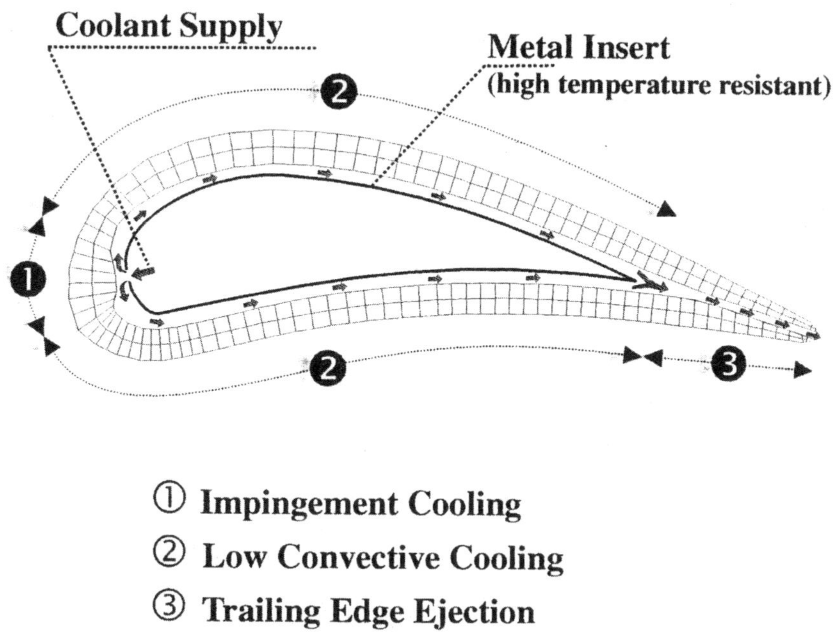

Figure 25.19 Cross-section of the Directly Cooled Guide Vane (Adapted from Ref. [28])

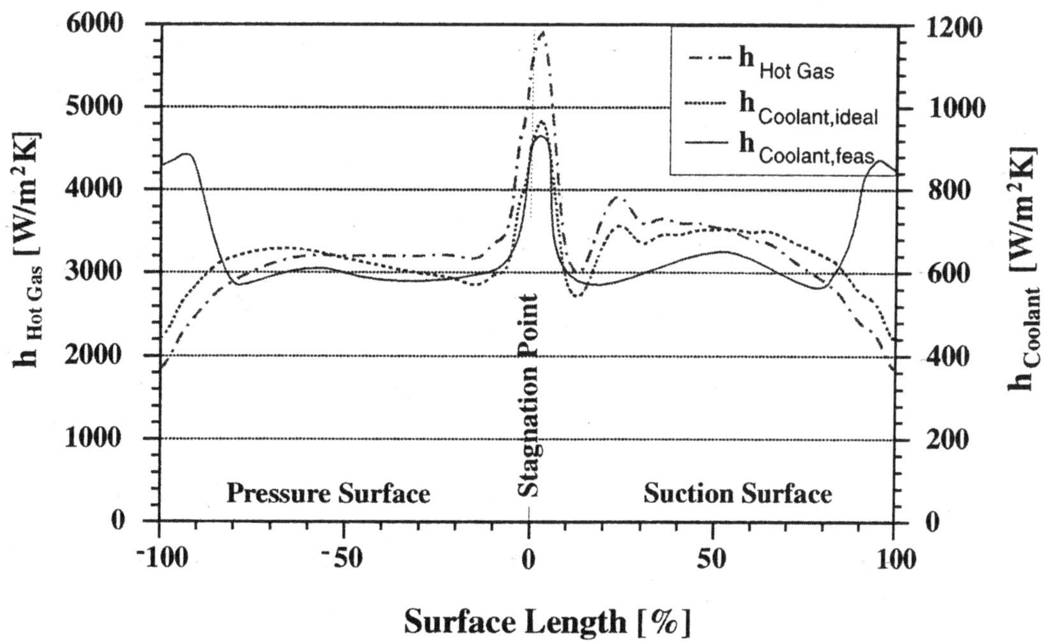

Figure 25.20 Inner and Outer Heat Transfer Distributions of the Directly Cooled Guide Vane (Adapted from Ref. [28])

TABLE 25.5 FAILURE PROBABILITIES FOR STEADY AND TRANSIENT OPERATING CONDITIONS

Internal Cooling Configuration	Operation (Fracture Mode)	Failure Probability
Optimum	10,000 hours Steady Operation (Slow Crack Growth)	1.5402×10^{-7}
	Trip (Fast Fracture)	6.2369×10^{-7}
Technically Feasible	10,000 hours Steady Operation (Slow Crack Growth)	6.9848×10^{-7}
	Trip (Fast Fracture)	9.6226×10^{-7}

TABLE 25.6 TYPICAL VALUES FOR A SMALL RADIAL INFLOW TURBINE

Small Radial Inflow Gas Turbine			
Mass Flow	200g/s	Inlet Diameter	56mm
Pressure Ratio	2.6bar	Outlet Diameter	60mm
Inlet Temperature	1100°C		

diture because of the small size of the components. Hence, ceramics allowing high material temperatures offer great potential for applications in high efficient small gas turbines (Lundberg and Gabrielsson [29]).

After the combustion chamber, the volute is the thermally most loaded part of the engine, and because of its complex geometry the scroll is a very demanding part for demonstrating the suitability of the design principles developed for static ceramic gas turbine components. In this theoretical study a vaneless volute is employed. It is designed on the assumption of free vortex flow and uniform mass flux in the radial direction. The design values of the radial inflow turbine are given in Table 25.6.

To quantify the mechanical behavior of the structure, time-resolved boundary conditions have to be specified. Preliminary calculations clearly showed that mechanical stresses caused by pressure differences could be neglected in comparison with thermally induced stresses. For the calculator of thermal stresses, the temperature distribution of the scroll, and hence the heat transfer to its walls, has to be known. Studies describing the flow in scrolls of radial turbines are used to determine the internal heat transfer coefficients (see e.g. Tabakoff et al. [30], Vu and Shyy [31]). On the outer surface of the scroll, free convection is assumed. The free stream temperatures are fixed to ambient temperature on the outside and to the turbine inlet temperature on the inside. Assuming an isentropic expansion of the hot gas, a linear reduction from turbine inlet temperature to exhaust temperature is presumed in the shroud region. Transient operating conditions are modeled by a variation of the hot gas temperature. In contrast to the ceramic guide vane, there is no distinction between trip and shutdown since there is no controlled shutdown for small engines. The boundary conditions are described in more detail in Filsinger et al. [32].

To begin with, a turbine scroll designed as a single-layered monolithic structure with a constant wall thickness of 4mm showed pronounced inhomogeneities in its temperature distribution, especially in the shroud region due to the drop of the hot gas temperature. This caused very high stresses under steady state operating conditions with an accompanying failure probability of 88%. Therefore, the design principles developed for ceramic nozzle guide vanes are worthwhile to consider. According to the calculated inner heat transfer, a three-layered scroll with variable local ceramic and insulation thicknesses was designed. In order to avoid impractical layer thicknesses, the outer heat transfer has to be increased beyond that of natural convection. Furthermore, the region near the tongue is excluded from the optimization because of spatial constraints. The thickness of the ceramic wall changes according to the increasing heat transfer coefficient from 2.7mm at the inlet to 7.0mm at the end of the spiral and at the shroud. The limits for the thickness of the ceramic are chosen by considering the current manufacturing skills for components made of sintered silicon

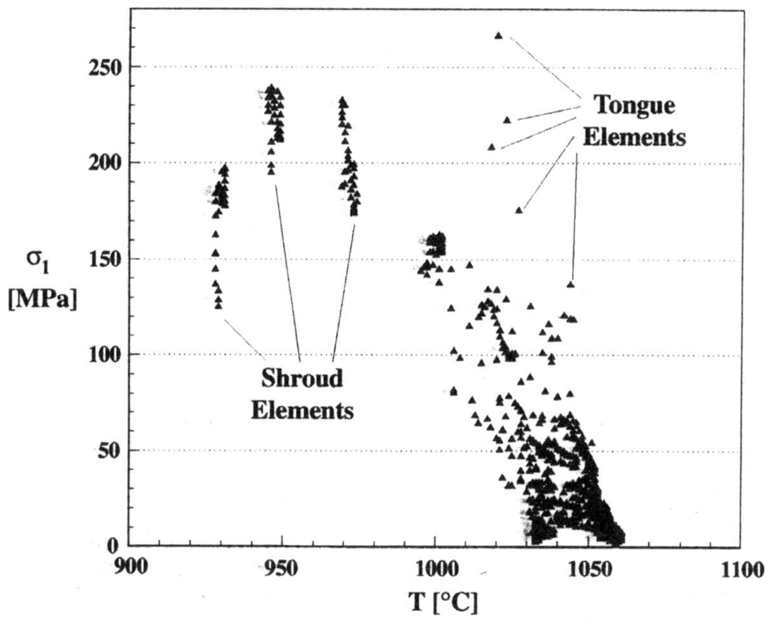

Figure 25.21 Calculated Stress and Temperature Loading in the Turbine Scroll for Each Finite Element Under Steady State Operating Conditions (Reprinted from Ref. [34])

carbide. The insulation layer correspondingly varies from 3.0mm at the inlet and 10.0mm at the end of the spiral, while the thickness of the metal casing is kept constant at 2.0mm.

Additionally, segmentation of the ceramic scroll has to be employed. The most important aspect for segmentation is to specify where free surfaces have to be created. To illustrate the procedure the calculated stresses under steady state operating conditions in the monolithic scroll are plotted in Figure 25.21.

It shows the maximum tensile stress for each finite element versus the maximum element temperature. The highly loaded regions in the shroud and the tongue can be easily identified. There are three rows of finite elements modeling the shroud at a homogeneous temperature level, which is considerably lower than the temperature level in the spiral casing. The thermo mechanical load in these elements, induced by the fixed-bonding to the hotter spiral, is quite high. Only a separation of the shroud from the spiral can effectively reduce the temperature differences within each part. Moreover, the stiffening effect of the shroud can be eliminated by the relative movement of the created free surfaces. Figure 25.22 shows the finite element grid of the segmented multi-layered turbine scroll.

The result of the finite element calculation for the optimized turbine scroll is given in Figure 25.23. The extremes of temperature and principal stresses occurring in the ceramic layer of the scroll and shroud for different operating conditions are shown. For all simulated operating conditions, a reduction of the tensile stress is achieved, especially in the highly loaded shroud, which has now only to resist an under-critical load. The highest principle stress, which occurs during acceleration, is now 60% of the highest value in the monolithic design. The stress value under steady operating conditions is reduced to 55%. This is achieved by reducing the local temperature gradients, which is confirmed by the drastically reduced temperature differences over the component. The effect of the separation joint can be seen in the difference between the temperature in the shroud and the temperature in the spiral (Figure 25.23, bottom). The reduction of the stresses is expressed in a considerably higher failure probability. Under fast fracture conditions the calculated failure probability for the optimized multi-component ceramic scroll is 0.8% (Filsingeret al. [33, 34]).

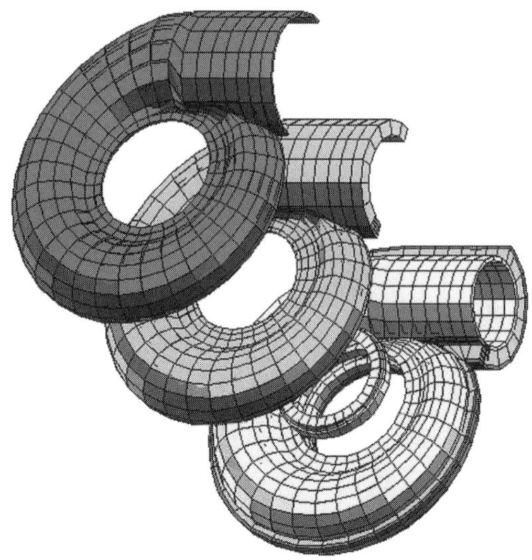

Figure 25.22 Three-layered Finite Element Grid of the Segmented Turbine Scroll. (Reprinted from Ref. [34])

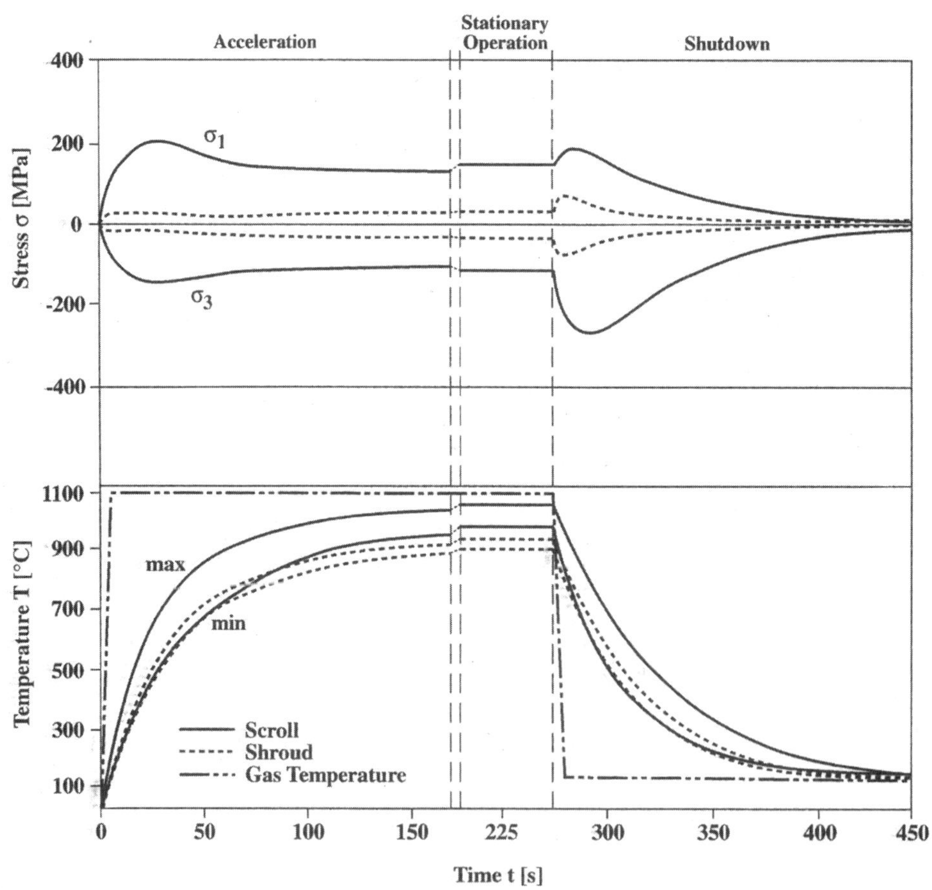

Figure 25.23 Calculated Maximum and Minimum Stress and Temperature in the Optimized Turbine Scroll for Different Operating Conditions (Reprinted from Ref. [34])

CONCLUSION

Various ceramic flame tubes were designed and tested. Multi-layered segmented designs were introduced to improve reliability of the flame tubes. The favorable impact of uncooled ceramic walls on the combustion process was demonstrated. Besides monolithic sintered silicon carbide three fiber reinforced ceramics were tested as combustor liner materials in a small gas turbine. The tests revealed that oxidation protection is imperative for carbon-reinforced materials.

Although numerical calculations cannot replace experimental investigations, they offer great potential for the preliminary design of ceramic components. The brittle behavior of ceramic materials and their failure caused by the most unfavorable combination of stress, size and orientation of flaws lead to the demand for the knowledge of the stress distribution in the component. To achieve sufficient reliability, the critical tensile stresses induced by thermal load have to be limited. Therefore, design principles have been developed, which aim at a homogeneous temperature distribution in the ceramic part. These principles, applied to various hot gas components of gas turbines, such as a first stator vane and a small turbine scroll, clearly demonstrate their general applicability and their excellent impact on survival probability of ceramic gas turbine components. Stress reductions up to 75%, compared to the initial design, can be achieved by utilizing the multi-layer hybrid approach with locally varying wall thicknesses. The multi-layer approach leads to a very uniform temperature in the ceramic structure, however at a high level due to the insulation by the fiber ceramic layer. Therefore, this approach is best suited for hot gas temperatures below values at which surface corrosion of monolithic silicon based materials becomes significant (1200°C to 1400°C), unless the ceramic component is provided with protection using environmental barrier coatings.

For very high hot gas temperatures, the fundamental relations of the multi-layer approach are transferred to a single layer design with varying internal cooling. With this design, the wall normal heat fluxes are intensified thus leading to reduced material temperatures. The stress reduction, however, is not quite as high as in case of the multi-layer approach.

An experimental verification of the hybrid stator vane design revealed good results. Three ceramic vane profiles were tested in a thermal shock facility, which induced stress distributions identical to those occurring in a real engine under trip conditions. The test specimen withstood a large number of thermal shock tests even at elevated loads. Although the limited number of test specimens did not allow for a statistical interpretation of the tests, the efficacy of the hybrid design was demonstrated.

Future efforts should be directed towards manufacture of ceramic nozzle guide vanes (hybrid and directly cooled) and testing in real engines.

REFERENCES

[1] Thümmler, F., and Grathwohl, G., "High Temperature Oxidation and Creep of Si_3N_4- and SiC-Based Ceramics and their Mutual Interaction," MRS Intl. Meeting on Adv. Materials, Tokyo, Japan, March 5,6, 1988, Materials Research Soc., pp. 237–253, 1989.

[2] Grathwohl, G., "Fatigue in Ceramics," in R.J. Brook (Ed.): *Concise Encyclopedia of Advanced Ceramic Materials*, Pergamon Press, Oxford, pp. 147–152, 1991.

[3] Rizk, N., and Mongia, H., "Ultra-Low-Nox Rich-Lean Combustion," ASME paper 90-GT-87, presented at the 35th ASME International Gas Turbine & Aeroengine Congress and Exposition, Brussels, Belgium, June 11–14, 1990.

[4] Novick, A., Troth, D., Notardonato, J., "Multifuel Evaporation of a Rich/Quench/Lean Combustor," ASME paper 83-GT-140, presented at the 28th ASME International Gas Turbine & Aero Engine Congress and Exposition, Phoenix, Arizona, USA, March 27–31, 1983.

[5] Stürmer, G., Schulz, A., Wittig, S., "Life Time Prediction for Ceramic Gas Turbine Components," ASME paper 91-GT-96, presented at the 36th ASME International Gas Turbine & Aeroengine Congress and Exposition, Orlando, Florida, USA, June 3–6, 1991.

[6] Bunk, W., "Overview on the German Program on Ceramic Components for Vehicular Gas Turbines," in *Proceedings of the First International Symposium on Ceramic Components for Engines*, Hakone, Japan, October 17–19, 1983, pp. 9–20, 1984.

[7] Kuck, H., "Experimentelle und theoretische Untersuchung der Vorgänge in einer zylindrischen Modellbrennkammer bei Wandauftragung des Kraftstoffs," (Experimental and theoretical investigation of the processes in a cylindrical model combustor under the influence of fuel on the liner) Dissertation, RWTH Aachen, Germany, 1972 (in German).

[8] Smith, K., and Cowell, L., "Experimental Evaluation of a Liquid-Fueled Lean-Premixed Gas Turbine Combustor," ASME paper 89-GT-264, presented at the 34th ASME International Gas Turbine & Aeroengine Congress and Exposition, Toronto, Ontario, Canada, June 4–8, 1989.

[9] Pfeiffer, A., Schulz, A., Wittig, S., "Principles of the ITS Ceramic Research Combustor Design: Segmented Flame Tube and Staged Combustion," ASME paper 91-GT-63, presented at the 36th ASME International Gas Turbine & Aeroengine Congress and Exposition, Orlando, FL, USA, June 3–6, 1991.

[10] Pfeiffer, A., Schulz, A., Wittig, S., "Design and Development of the ITS Ceramic Research Combustor," In: *Proc of the 1991 Yokohama International Gas Turbine Congress*, pp. III- 9-15, Gas Turbine Society of Japan.

[11] Yovanovich, M., Fisher, N., Saabas, H., "A Light Load Model Combining Surface Roughness and Waviness to Predict Thermal Contact Conductance," AIAA-83-1435, AIAA 18th Thermophysics Conference, 1983.

[12] Münz, S., Schulz, A., Wittig, S., "Evaluation of a New Design Concept for a Ceramic Flame Tube Under Engine Conditions," ASME paper 96-GT-222, presented at the 41st ASME International Gas Turbine & Aeroengine Congress, Birmingham, England, UK, June 10–13, 1996.

[13] Münz, S., "Entwicklung und maschinentechnische Bewertung neuer konstruktiver Konzepte für ein keramisches Flammrohr," (Development and mechanical engineering evaluation of new design concepts for a ceramic flame tube) Doctoral Thesis, Institute for Thermal Turbomachinery, University of Karlsruhe, Germany, 1995 (in German).

[14] Brandauer, M., Schulz, A., Pfeiffer, A., Wittig, S., "Experimental Study of Coke Deposition under Gas Turbine Combustor Conditions," Combust. Sci and Tech., Vol. 89, pp. 265–273, 1993.

[15] Brandauer, M., Schulz, A., Wittig, S., "Mechanisms of Coke Formation in Gas Turbine Combustion Chambers," ASME paper 95-GT-49, presented at the International Gas Turbine Aeroengine Congress and Exhibition, Houston, TX, USA, June 5–8, 1995. *J. of Eng. for Gas Turbines and Power,* Vol. 118[2], pp. 265–270, 1996.

[16] Brandauer, M., "Grundlegende Untersuchungen zur Bildung von Ablagerungenin Brennräumen unter gasturbinentypischen Bedingungen," (Fundamental investigations in the formation of deposits in combustors under typical gas turbine conditions) Doctoral Thesis Institute for Thermal Turbomachinery, University of Karlsruhe, Germany, 1993 (in German).

[17] Brandauer, M., Schulz, A., Wittig, S., "Optimization of Fuel Prevaporization on Porous Surfaces for Low-NOx-Combustion," *Proc. of the Third Int. Conf. on Combust. Technology for a Clean Environment*, Lisbon, Portugal, July 1995.

[18] Helmer, T., Peterlik, H., Kromp, K., "Coating of Carbon Fibers—The Strength of the Fibers," J. of the American Ceramic Society, Vol. 78, No. 11, pp. 133–136, 1995.

[19] Filsinger, D., Münz, S., Schulz, A., Wittig, S., Andrees, G., "Experimental Assessment of Fiber Reinforced Ceramics for Combustor Walls," ASME paper 97-GT-154, presented at the 42nd

ASME Internatinal Gas Turbine & Aeroengine Congress and Exposition, Orlando, FL, USA, 2–5 June 1997.

[20] Hüther, W., and Krüger, W., "Metal Ceramic Guide Vanes—New Design Concepts," ASME paper 89-GT-334, presented at the 34th ASME International Gas Turbine & Aeroengine Congress and Exposition, Toronto, Ontario, Canada, June 4–8, 1989.

[21] Gutmann, C., Schulz, A., Wittig, S., "A New Approach for a Low-Cooled Ceramic Nozzle Vane," ASME paper 96-GT-232, presented at 41st ASME International Gas Turbine & Aeroengine Congress and Exposition, Birmingham, UK, June 10–13, 1996.

[22] Nakakado, K., Machida, T., Miyata, H., Hisamatsu, T., Mori, N., Yuri, N., "Strength Design and Reability Evaluation of a Hybrid Ceramic Stator Vane for Industrial Gas Turbines," *Transactions of the ASME, J. of Eng. for Gas Turbines and Power*, Vol. 117[2], pp. 245–250, 1995.

[23] Tsuchiya, T., Furuse, Y., Yoshino, S., Chikami, R., Tsukagoshi, K., Mori, M., "Development of Air-Cooled Ceramic Nozzles for a Power Generating Gas Turbine," ASME paper 95-GT-105, presented at the 40th ASME International Gas Turbine & Aeroengine Congress, Houston, TX, USA, 5–8 June, 1995.

[24] Wenglarz, R., Calcuttawala, S., Pope, E., "Ceramic Vane Demonstration in an Industrial Gas Turbine," ASME paper 97-GT-15, presented at the 42nd ASME International Gas Turbine & Aeroengine Congress and Exposition, Orlando, FL, USA, June 2–5, 1997.

[25] Day, W., Kouris, K., Holowczak, J., Blair, M., Hale, S., Bornemisza, T., "Development of Cooled, Ceramic First Stage Vanes for the FT (Aeroderivative Gas Turbine," ASME paper 98-GT-133, presented at the 43rd ASME International Gas Turbine & Aeroengine Congress and Exposition, Stockholm, Sweden, June 2–5, 1998.

[26] Gutmann, C., "Die keramische Gasturbinenleitschaufel als hochbelastete Schalenstruktur: Neue Konzepte der konstruktiven Gestaltung," (The ceramic gas turbine nozzle as a highly loaded flat structure, new design concepts) Doctoral Thesis, Institute for Thermal Turbomachinery, University of Karlsruhe, Germany, 1994 (in German).

[27] Dilzer, M., Gutmann, C., Schulz, A., Wittig, S., "Testing of a Low Cooled Ceramic Nozzle Vane Under Transient Conditions," ASME paper 98-GT-116, presented at the 43rd ASME International Gas Turbine & Aeroengine Congress and Exhibition, Stockholm, Sweden, June 2–5, 1998.

[28] Dilzer, M., Schulz, A., Wittig, S., "Systematic Reduction of Stress Levels for a Thermally High Loaded Ceramic Nozzle Vane with Trailing Edge Ejection," ASME paper 99-GT-450, presented at the 44th International Gas Turbine & Aeroengine Congress and Exposition, Indianapolis, IN, USA, June 7–10, 1999.

[29] Lundberg, R., Gabrielsson, R., "Progress in the AGATA Project—a European Ceramic Gas Turbine for Hybrid Vehicles," ASME paper 95-GT-446, presented at the 40th ASME International Gas Turbine & AeroengineCongress, Houston, TX, USA, June 5–8, 1995.

[30] Tabakoff, W., Sheoran, Y., Kroll, K., "Flow Measurements in a Turbine Scroll," *Transactions of the ASME, J. of Fluids Engineering*, Vol. 102, pp. 290–296, 1980.

[31] Vu, T., Shyy, W., "Viscous Flow Analysis as a Design Tool for Hydraulic Turbine Components," *Transactions of the ASME, J. of Fluids Engineering*, Vol. 112, pp. 5–11, 1990.

[32] Filsinger, D., Schulz, A., Wittig, S., "Reliability Improvement of Static Ceramic Components for Small Gas Turbines," in Proceedings of the 29th International Symposium on Automotive Technology and Automation, Florence, Italy, pp. 669–676, Automotive Automation Ltd., 1996.

[33] Filsinger, D., Schulz, A., Wittig, S., "Reliable Design of Static Components for Small Gas Turbines," International Ceramics, Issue 1, 1996.

[34] Filsinger D., "Strukturmechanische Optimierung thermisch hochbelasteter keramischer Gasturbinenkomponenten," (Mechanical design optimization of thermally highly loaded ceramic gas turbine components) in Forschungsberichte aus dem Institut für Thermische Strömungsmaschinen, Band 2/1998, Cuvillier Verlag, Göttingen (in German).

CHAPTER

26

ROLLS-ROYCE EXPERIENCE IN CERAMIC MATRIX COMPOSITE DEMONSTRATION

Anthony G. Razzell

Rolls-Royce plc, Derby, United Kingdom

INTRODUCTION

Rolls-Royce has pursued ceramic matrix composites for use in aero gas turbines for the past 15 years, making use of new material developments in the process. Extensive work was carried out in the 1980s on monolithic ceramics (principally silicon nitride), with turbine blades, shroud rings and air bearing components demonstrated in a helicopter engine. Work was redirected towards ceramic composites due to the perceived risks associated with catastrophic failure of monolithics, and the greater likelihood of defects leading to failure being present in larger components. A number of fiber/matrix combinations were evaluated, a subset of which were manufactured in component form and described below.

CARBON/CARBON FLAMEHOLDERS

Carbon/carbon (C/C) materials were explored in the 1980s at Rolls-Royce for use in hot structural components, in particular reheat (afterburner) flameholders. The inherent poor oxidation characteristics of the material were addressed by the addition of coatings (based on SiC and boron containing compounds) and matrix additions. It was found that the protection was ineffective after representative thermal cycling, in some cases, the internal C/C composite was oxidized, leaving a shell of coating. Figure 26.1 shows an oxidation damaged flameholder where the leading edge has totally disintegrated. For short term, ultrahigh temperature use (e.g. spacecraft re-entry protection) C/C is an ideal material, however the long life/high cycle requirement of gas turbine engines rules C/C out for future applications.

SiC/SiC NOZZLE PETALS

The advent of SiC fiber (Nicalon) allowed the development of a SiC/SiC, a new class of composite, using generic chemical vapour infiltration technology developed for processing carbon/carbon. The material

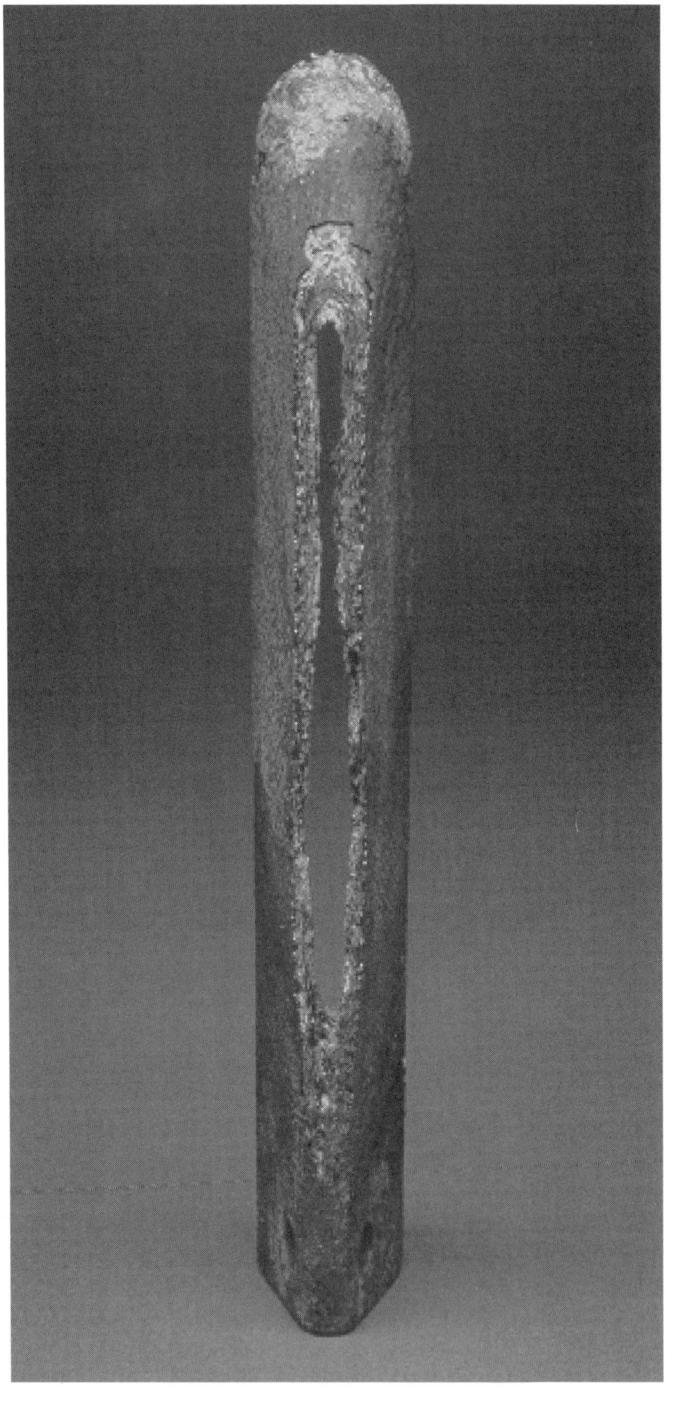

Figure 26.1 Oxidation Damaged Carbon-carbon Flameholder

Figure 26.2 SiC/SiC Nozzle Petals

has a slightly higher density compared to carbon/carbon, but still significantly below that of nickel based alloys. Rolls-Royce procured a set of nozzle petals in the first generation of SiC/SiC (Cerasep™) and carried out extensive material studies in support of the program. The nozzle petals have successfully run for several hundred cycles on bench engines. Figure 26.2 shows a set of petals installed in an engine.

GLASS CERAMIC COMPRESSOR AIRFOILS AND CASINGS

A significant effort was made in developing glass and glass ceramic composites for use in compressor airfoil and casing applications. The aim was to eliminate the fire risk associated with titanium airfoils, and reduce weight. Pyrex glass was investigated, and it was found that a carbon rich layer was produced via an in situ reaction between the Nicalon™ fibers and the matrix, giving the material composite behavior. It was found that the vibration characteristics of the airfoils were unacceptable due to the modulus and density of the composite, and the casing did not have acceptable containment characteristics. Glass and glass ceramic composites also have a degradation problem, especially at intermediate temperatures (800°C) due to oxidation of the carbon in the fiber/matrix interface, and formation of silica. Figure 26.3a shows a Nicalon/glass composite compressor blade, produced by hot pressing in a die. Figure 26.3b shows a model compressor casing, with integrally molded flanges.

SiC/Al_2O_3 FLAMEHOLDERS

SiC/Al_2O_3 composites were developed using the Lanxide™ process of directed metal oxidation (DIMOX™). Flameholders were procured and run in a reheat (afterburner) rig and a technology demonstra-

(a)

(b)

Figure 26.3 (a) Compressor Airfoil and (b) Compressor Case

tor engine. Due to a combination of poor material properties, the design of the mounting and vibration problems, it was found that the components failed when reheat was selected. The flameholders were redesigned (including the mounting system) for demonstration on an engine in a subsequent program, which is ongoing. Figure 26.4 shows a SiC/Al_2O_3 composite flameholder.

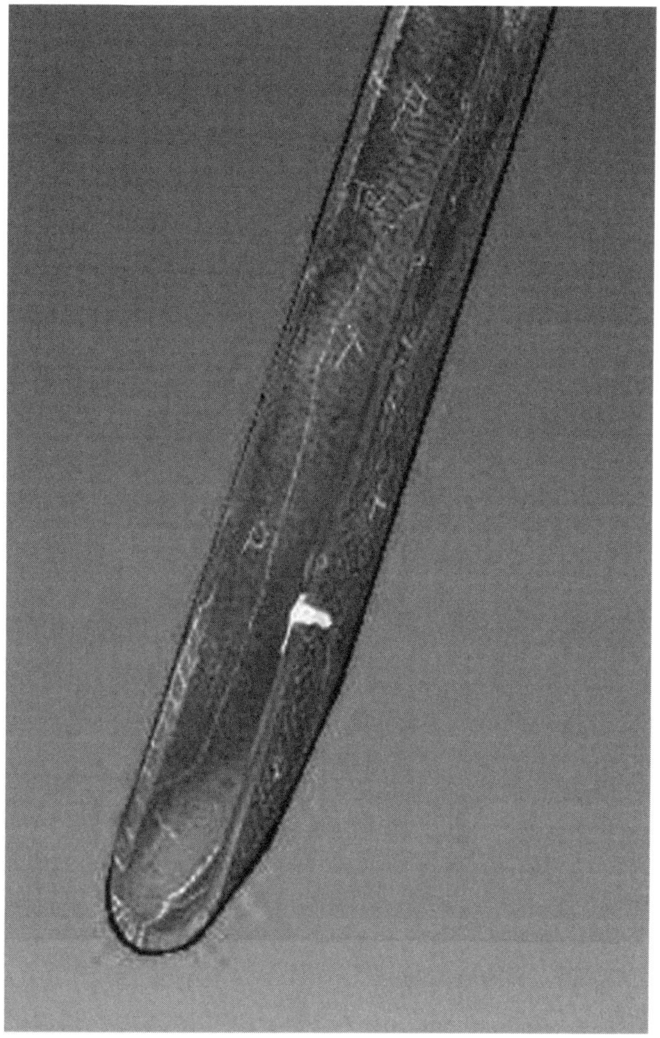

Figure 26.4 SiC/Al_2O_3 Composite Flameholder

SiC/Al_2O_3 COMBUSTION LINER

An outer combustion liner was manufactured and tested with a novel thermal insulation system. Laser drilling was used to produce angled effusion cooling holes, and material testing carried out to establish the debit in properties as a result of the holes. The insulation suffered from spallation during the test, and the CMC was found to suffer from fretting damage from a metallic antirotation pin at one end of the liner. Figures 26.5a and 26.5b shows the CMC liner after the test and a notch in the wall due to fretting, respectively. The latter was of particular interest, as wear of metallic parts by CMCs have been observed in other material configurations. In spite of these problems, combustion liners are attractive components for CMC application for reduced emissions and/or weight.

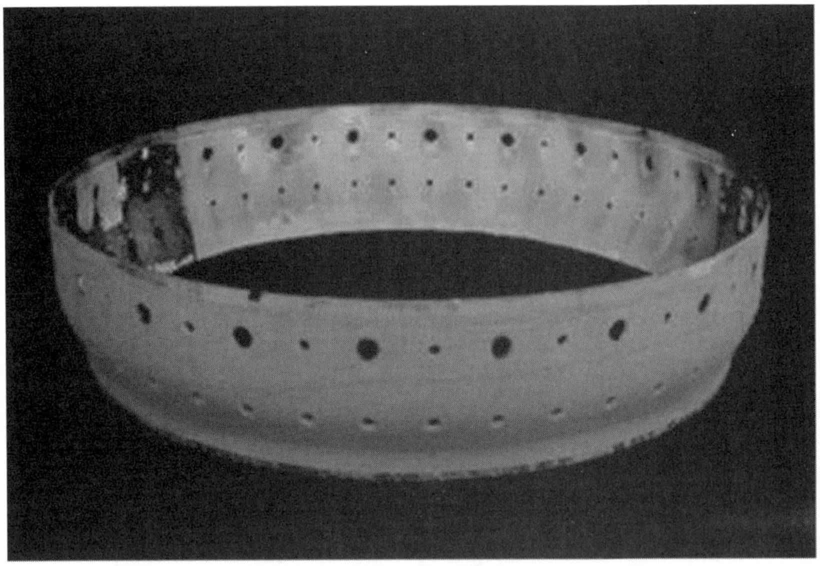

(a)

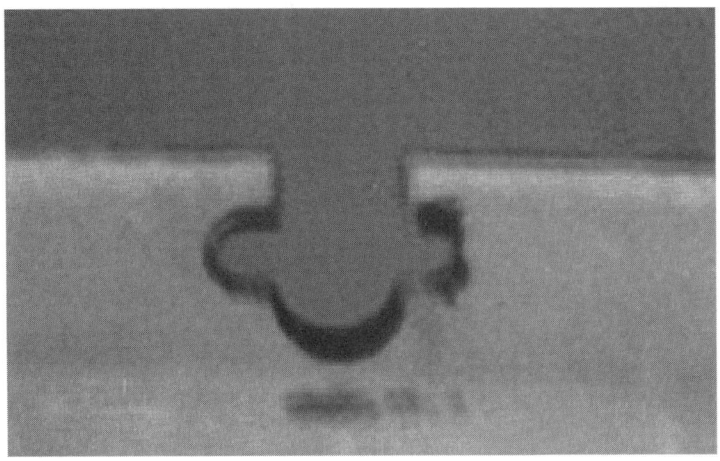

(b)

Figure 26.5 (a) CMC Combustor Barrel Post Testing and (b) Notch in CMC Wall Due to Fretting

SiC/Al_2O_3 SHROUD SEAL

Rolls-Royce has demonstrated a turbine shroud seal in an intermediate pressure turbine. The metallic shroud seal required cooling air from the compressor, which can be reduced by the use of a higher temperature material. The use of a CMC leads to an improvement in efficiency and a consequent reduction in fuel consumption. SiC/Al_2O_3 was chosen for use in the component after preliminary mechanical testing of a number of CMCs. A novel abradable material (HASSMAP) was developed, and was attached to the CMC backing plate by adhesive bonding. The HASSMAP abradable consists of hollow aluminosilicate

Figure 26.6 SiC/Al_2O_3 Turbine Shroud Seal with HASSMAP Abradable

(mullite) spheres in a matrix of mullite and aluminium phosphate (see Figure 26.6). This results in a material with a minimal thermal expansion mismatch with the SiC/alumina backing plate, with the required high temperature abradable properties. A set of shroud seals were successfully tested in a civil development engine [1, 2].

SiC/SiC EXHAUST GUIDE VANE

Exhaust guide vanes (EGVs) were manufactured by novel routes using knitted and braided fiber preforms (see Figures 26.7a and 26.7b). The preforms were infiltrated with SiC.The knitting process offers considerable shape flexibility, however it is difficult to obtain a high enough fiber volume fraction in the final composite to give adequate properties. In addition, the knitting process generates voids, causing localized areas of weakness. Braiding potentially offers a higher volume fraction, but it can be difficult to control fiber angle and volume fraction adequately. The vanes were produced with integral platforms (as in the metallic design), but high stress concentrations are generated in the airfoil roots in this configuration. The vanes were mechanically tested to simulate the loading experienced in the engine.

OXIDE/OXIDE COMBUSTOR TILE

The aim of this program (carried out in conjunction with Volvo Aero Corporation, SNECMA and ONERA) was to develop a model oxide/oxide composite material based on single crystal alumina monofilaments and demonstrate it in a representative gas turbine combustor environment. Reductions in

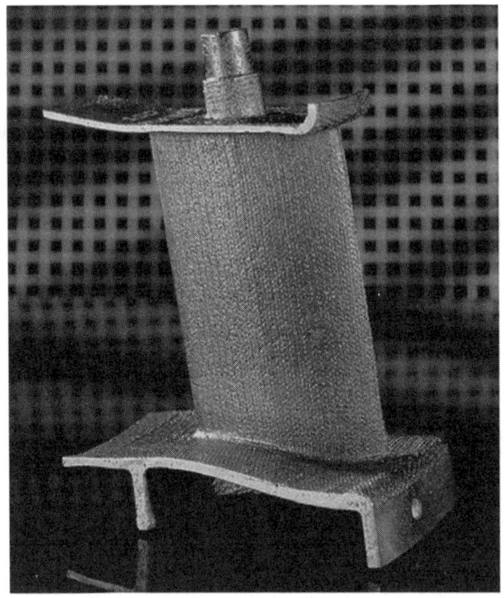

(a)

(b)

Figure 26.7 (a) Knitted EGV and (b) Set of Braided EGVs

emissions (e.g. NOx) due to elimination of combustor wall cooling are potentially achievable using ceramic matrix composites. A material processing method based on tape casting and hot pressing was developed, and basic characterization carried out. Some of the results of the program have been presented elsewhere [2, 3].

An interface coating of zirconia was applied onto the single crystal alumina fibers, being sufficiently adherent to survive the green forming and densification processes, but providing the necessary crack deflec-

Figure 26.8 Oxide/oxide Combustor Tile (with Dilution Holes) in Test Combustor (Reprinted from Ref. [4])

tion and fiber pullout behavior during fracture of the composite. A porous zirconia coating was chosen, which remains stable for long periods at high temperature (i.e. remains porous). The coating was applied by an in-line slurry deposition process during the manufacture of the sheets of pre-preg via tape casting. The slurry contained a fugitive binder that left the desired porous preform after removal by oxidation in air. Hot pressing or HIPping was used to densify 0°/90° balanced lay-ups of pre-preg, and conventional diamond sawing/grinding methods used to produce test pieces.

Aging at 1400°C for 1000 hours did not significantly affect the mechanical behavior of the material. The ultimate tensile strength (UTS) of the composite at 1400°C was 50% of the room temperature UTS, in line with the behavior of the as received fiber. The material was tested in tile form within a combustor rig (Figure 26.8). Computational fluid dynamics (CFD) code was used to predict the temperature distribution in the tile (given appropriate thermophysical properties from initial material characterization). The temperature distribution was then applied to an anisotropic finite element (FE) model to determine the stress and strain distribution. A two component model was used to model the non-linear behavior of the composite, assuming a high modulus prior to matrix cracking, and a low modulus after matrix cracking. The model predicted localized matrix cracking around the long edges of the tile and the two air dilution holes.

An Al_2O_3/Al_2O_3 tile was coated with temperature sensitive paint and tested in the combustor for three minutes in order to test the validity of the CFD analysis. In general, the temperature distribution was correct, but the temperature gradients (both through thickness and in-plane) were somewhat underestimated in the calculations. The peak temperature recorded was 1260°C. Figure 26.9 shows the tile after the thermal paint test. It was found to have three cracks (8 to 15mm long, crack lengths shown in round brackets) running into

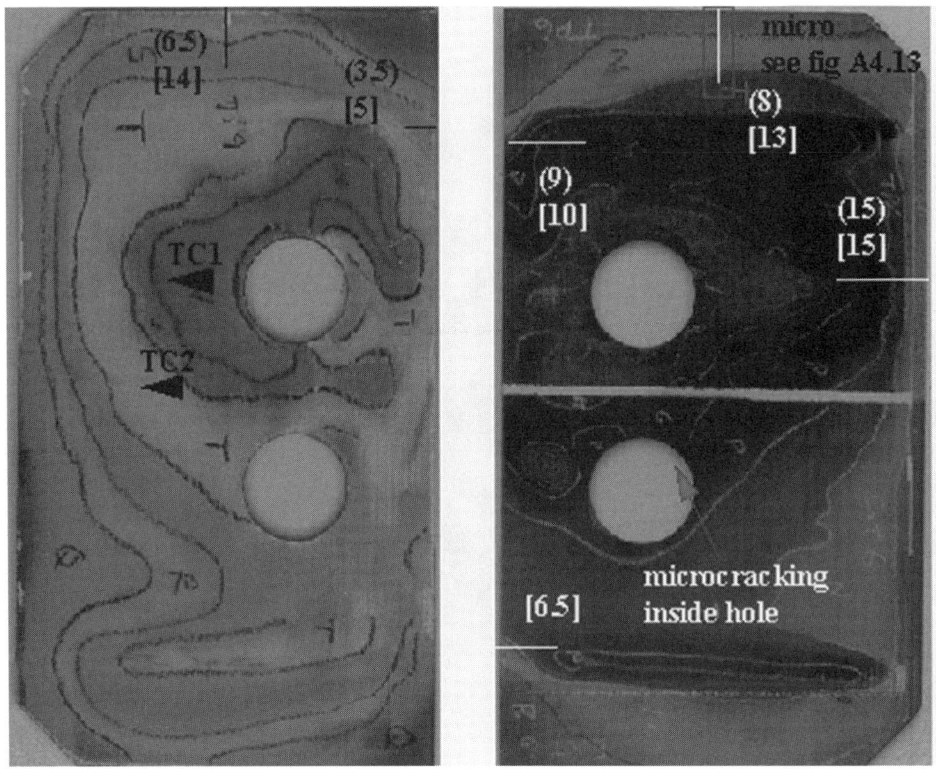

Figure 26.9 Cold and Hot Sides of Tile Post Thermal Paint Test

the tile from the edges. The locations differed slightly from the over-strained regions predicted by CFD/FE calculations. Some microcracking (both delamination and perpendicular cracks) was also observed on the inside of the air dilution holes. The tile was then subjected to cyclic combustor rig testing for 1.5 hours, corresponding to 46 cycles between idling and full load conditions. The cracks observed after the three minutes test had grown between 1 and 5 mm (new crack lengths are shown in Figure 26.9 in square brackets). Some fiber fractures were observed but the major part of the cracks were still bridged by undamaged fiber, only showing matrix damage.

Although the rig testing was considered successful, the material developed can only be regarded as a model, since the cost of the fibers is too high to make it economically viable. The program did however demonstrate that oxide based materials are feasible, and should be considered in the future when suitable fibers become available. The program was supported by the European Commission under the Brite/Euram Framework III program.

ACKNOWLEDGEMENT

All figures in this chapter, except where otherwise indicated, have been provided courtesy of Rolls-Royce plc.

REFERENCES

[1] Beesley, C., "The Application of CMCs in High Integrity Gas Turbine Engines," 1997, Key Engineering Materials, Vols. 127–131, pp. 165–174, Trans. Tech Publications, Switzerland.

[2] Rolls-Royce High Temperature Abradable Seal, US Patent 5 780 146B, 1996.

[3] Holmquist, M., Lundberg, R., Razzell, T., Sudre, O., Molliex, L., and Adlerborn, J., "Development of Ultra High Temperature Ceramic Composites for Gas Turbine Combustors," ASME paper 97-GT-413, presented at the ASME International Gas Turbine and Aeroengine Congress and Exposition, Orlando, Florida, USA, June 2–5, 1997.

[4] Razzell, A.G., Moliex, L., Holmquist, M., and Sudre, O., "Oxide/Oxide Ceramic Matrix Composites in Gas Turbine Combustors," ASME paper 98-GT-30, presented at the ASME International Gas Turbine and Aeroengine Congress and Exhibition, Stockholm, Sweden, June 2–5, 1998.

CHAPTER

27

'CMC-Design' A Brite Euram Program to Develop Design Methodologies for CMC Components for Gas Turbine Engines

Dr. Jun Shi

ABB Alstom Power Technology Centre, Leicester LE8 6LH, United Kingdom

INTRODUCTION

This chapter summarizes a major European R & D project which aims at the development of enabling technologies that would pave the way for a successful introduction of Ceramic Matrix Composites (CMCs) into industrial gas turbine engines. The chapter begins with the main drivers for replacing metallic components in the hot gas path by CMC parts. It then reviews the current technological stumbling blocks, which have hindered the entry of CMCs into the gas turbine market. This follows by a detailed description of a pan European BRITE EURAM program, 'CMC-Design', which intends to address these long-standing issues. Achievements to date and lessons for the future are presented.

Industrial Objectives and Expected Achievements

Gas turbines for both simple and combined cycles had a market in the five years 1991–995 of 25 GW/annum or 8.74 billion euro/year [1]. Gas turbine plants are estimated to have a market of 83.56 GW between 1997–2006 [2]. Within this market, there is tremendous commercial pressure on gas turbine manufacturers to increase the plant efficiency and reduce specific costs whilst continuing to meet the ever more stringent emission limits [3, 4].

In a combined cycle, higher efficiency is obtained from the use of higher turbine inlet temperature in the cycle. However, the current and possibly future metallic materials have reached their limit in terms of strength, oxidation and corrosion resistance at about 1,000°C and the only promising alternative seems to be ceramic matrix composite (CMC) materials. These CMC components will require no cooling air. This air can then be used at higher temperatures thus producing more work from the cycle. Substantial improvements in engine performance such as a 20% increase in power output and a 2.5% rise in simple cycle efficiency could be expected if a CMC is used as the combustor material [5].

The growing environmental concerns have led to stricter and stricter standards for harmful emissions such as NOx and CO. It is well recognized that to strike a balance between these two, an un-cooled combustor is required with primary zone temperature in a narrow band between about 1397°C and 1627°C (1670K and 1900K) [6]. Assuming that no active cooling is needed for a CMC combustor, its wall temperature then will be 200°C to 300°C below the primary zone temperature, which is well beyond the capability of the best alloys available.

The major drivers for advanced gas turbine technology are thus improved efficiency, reduced emissions and lower life cycle costs. This project for the application of CMC materials to gas turbines is in direct response to these drivers. It is believed that successful exploitation of the technologies may lead to a 2.5% increase in efficiency and single digit emissions of NOx and CO.

State-of-the-Art

The ideal approach to a new material development should be concurrent with parallel progress in material processing and characterization, material constitutive law and life prediction, component design and non-destructive evaluation. Unfortunately this has not happened for CMCs. While material processing has progressed over the years and characteristics for certain materials have been established, the modelling and design of CMCs are at such a stage that substantial work is needed [7].

A CMC was first produced by the Atomic Energy Agency (AEA) in the UK in the early 1970s to give an improvement in the toughness and thermal shock resistance of monolithic ceramics and, at the same time retain their high corrosion and oxidation resistance, and high stiffness and strength at elevated temperatures. Over the past two decades, a large number of processing routes have been explored, but only a few of them have emerged with high quality and reproducibility for commercial production [8].

Following the development in CMC processing, a reasonable understanding of the material behavior of CMCs has been obtained for simple fiber architectures and under uni-axial loading. For static loading, the damage sequence of matrix cracking, fiber matrix debonding, fiber fracture and fiber pull-out have been established for most CMCs with a weak interface [9]. These damage modes represent the energy absorbing mechanisms that ultimately provide the toughness of CMCs. Under fatigue loading, it is generally recognized that due to its brittleness, the ceramic matrix does not fatigue in the conventional sense. The fatigue mechanism comes from the changes of interfacial sliding resistance due to frictional wear and reduction of fiber strength due to abrasion, which lead to the decrease of secant modulus and accumulation of permanent strain under stress control [10]. Creep tests have revealed the influence of the relative creep resistance of fiber and matrix and the initial loading rate on creep strength [11].

Despite the knowledge gained from experimental studies, constitutive modelling and life prediction of CMCs are far from adequate for practical design purposes, mainly because of the complicated damage mechanisms caused by material anisotropy. The state of the art of composite design is "little more than empirical rules based on a test database" [12]. This is perhaps acceptable for short life applications such as the Space Shuttle and rocket engines, for which design certification can be achieved solely by component testing. But for components of industrial gas turbine engines, the design of which is usually governed by long-term fatigue and creep, such an approach is apparently too expensive and predictive methods for both deformation and life must be developed. This is particularly true when new CMCs are being developed.

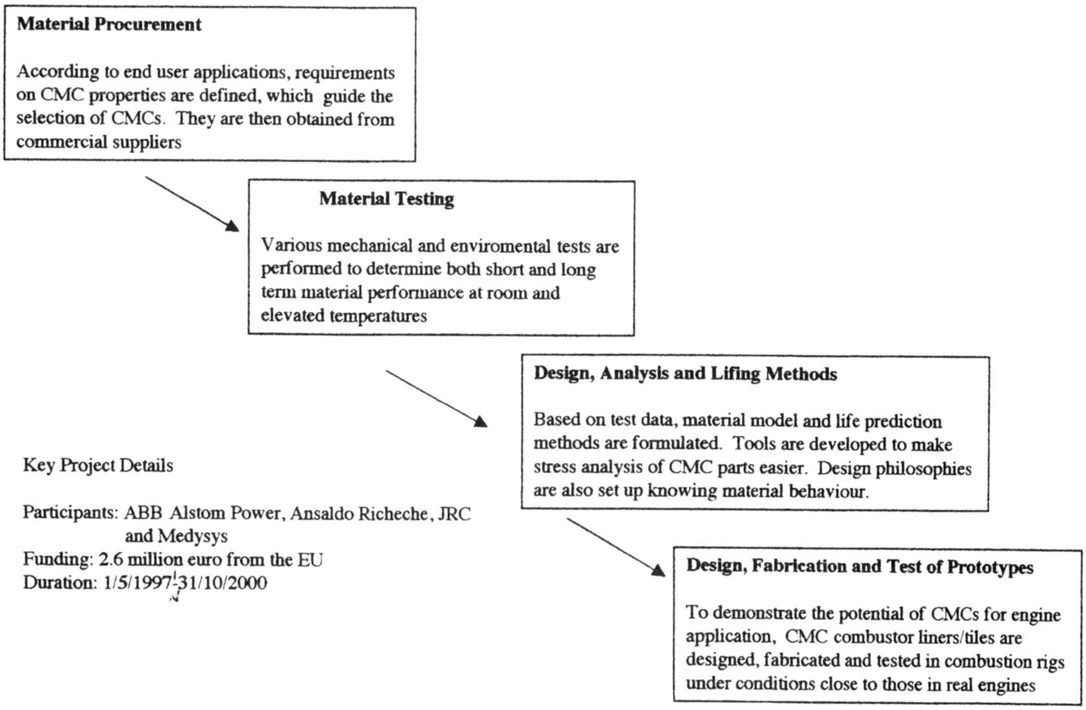

Figure 27.1 Flow Chart of 'CMC-Design', a European R & D Project (Reprinted with permission from Alstom)

Without modelling methods, new product development can be prohibitively expensive due to the amount of testing required.

Very limited design experience of CMC components has been obtained in the past and some of the applications have been mainly space oriented and involve service conditions that are different from those of industrial gas turbine engines, which experience lower temperatures, higher stresses and have much longer lives. Only recently applications for land based gas turbines have emerged that are promising [13, 14].

To address these issues comprehensively, a pan European R & D project was set-up within the Fourth Framework's BRITE EURAM program. Through this, the project is funded nearly 50% by the European Commission. The ABB ALSTOM POWER Technology Centre in the UK leads the project and it has three other partners: (1) ANSALDO Ricerche (Research Centre of the ANSALDO Energy Group) in Italy; (2) the Joint Research Centre of the EU in Holland; (3) MEDYSYS, a consultancy firm in France with expertise in micro mechanics modelling of CMCs. The project plan can be found in Figure 27.1.

The project aims at the development of a new design methodology for CMCs, in particular the capability of applying them to high temperature components in gas turbines. It will address the essential steps towards the industrial application of CMCs (and hence make substantial contributions towards the overall industrial objectives outlined before):

- Establish CMC design data pertinent to the operating conditions of the gas turbine components.
- The development of generic design, analysis and lifing methodologies at *three levels* for CMCs
- The design, fabrication and testing of a prototype CMC combustor can and liner to validate the design, analysis and lifing philosophies developed.

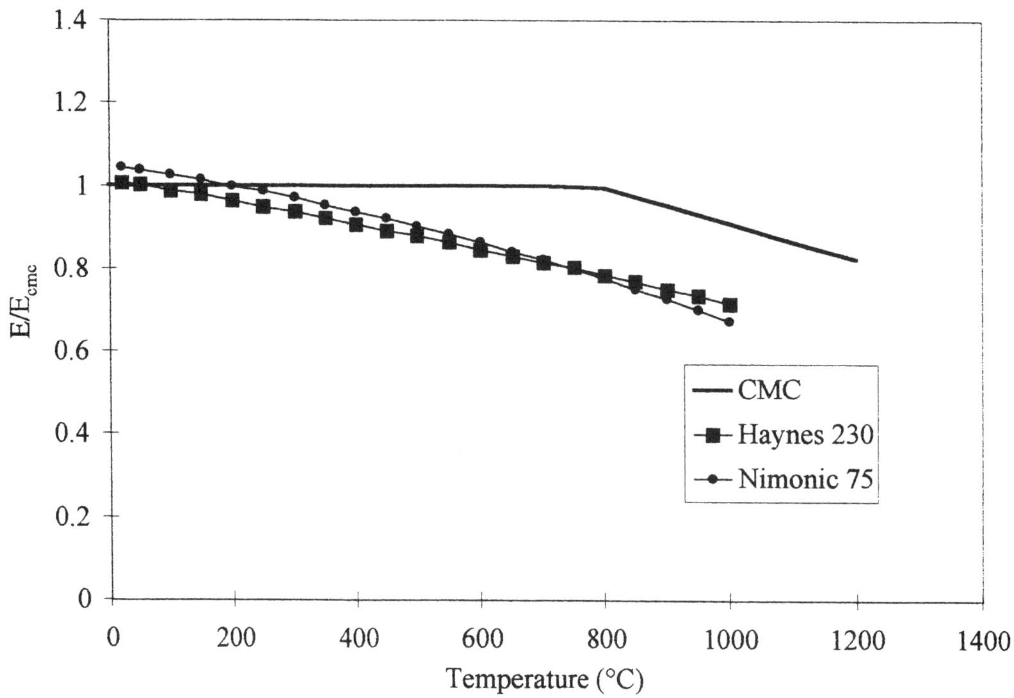

Figure 27.2 Comparison of Relative Young's Modulus (E/E_{cmc}) Between a CMC and Combustor Alloys (Reprinted with permission from Alstom)

PROGRESS TO DATE

Material Testing

Extensive mechanical testing on SiC/SiC composites has been performed, including tensile, fatigue, creep, thermal shock and oxidation testing. Whenever possible, ASTM or CEN test standards were followed closely. The test conditions were specified in such a way that the anticipated engine component loading conditions were fully covered. The test data form the basis for material behavior modelling, component thermal and stress analysis, and life prediction.

The tensile test results [15, 16] are summarised in Figures 27.2 and 27.3, in which data of two combustor alloys are also plotted for comparison. All the properties in these figures are normalized to the RT CMC properties. It is clear that the CMC retains its stiffness and strength up to 1200°C, while the alloys experience serious strength deterioration above 900°C. Therefore, benefits are expected in using CMCs instead of alloys above this temperature.

Creep and mechanical fatigue lives of the CMC can also be compared against those of Haynes 230 (see Figures 27.4 and 27.5). Below 1000°C, the CMC does not creep, while above 1000°C, it still maintains some creep resistance when Haynes 230 loses its creep strength completely.

One important design concern is thermal shock, which was simulated by impinging flames on disk samples intermittently (Figure 27.6). Temperature at the hot surface of the disk was monitored by a pyrometer, while the temperature at the cold surface was monitored by a thermocouple. Strengths before and after a fixed number of cycles were obtained by ball-on-ring tests and they were used as an indicator for the material resistance to thermal shock.

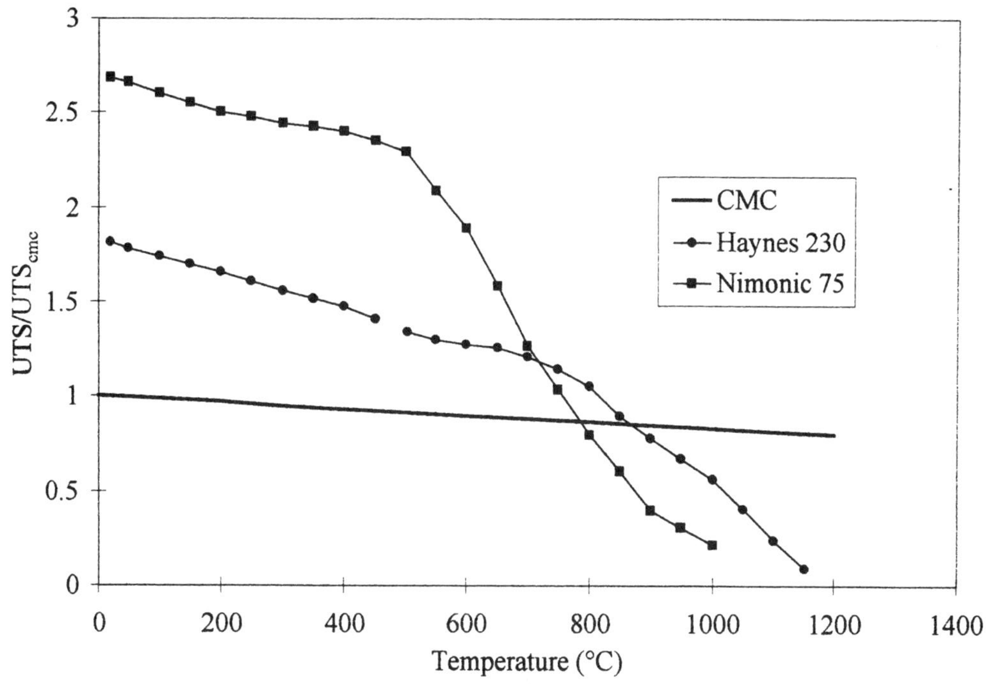

Figure 27.3 Comparison of Relative Ultimate Tensile Strength (UTS/UTS$_{cmc}$) Between a CMC and Combustor Alloys (Reprinted with permission from Alstom)

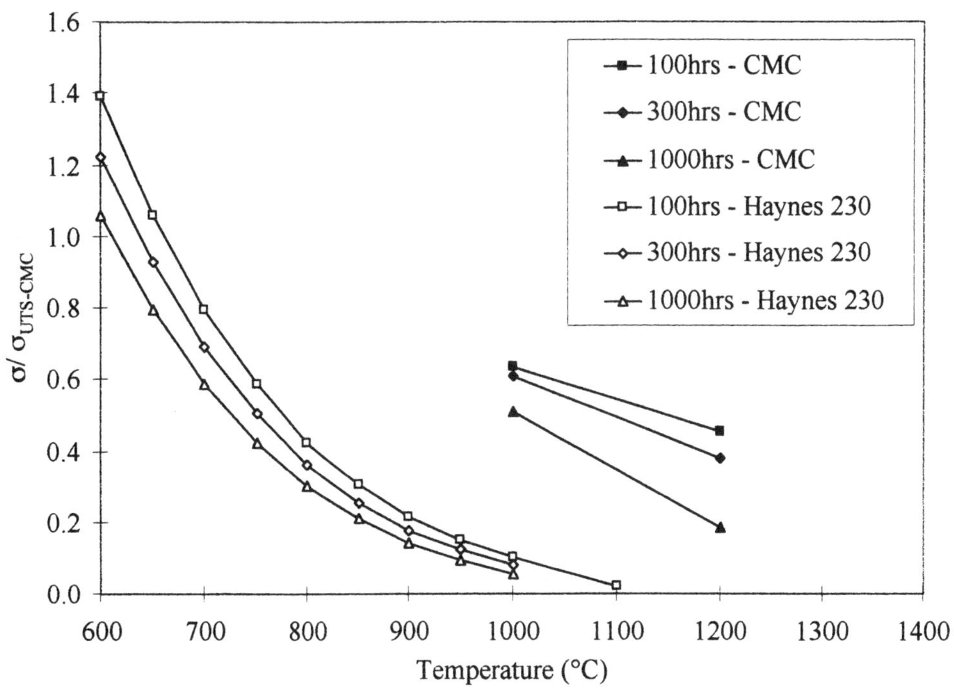

Figure 27.4 Comparison of Relative Creep Rupture Strength (σ/σ_{cmc}) Between a CMC and Haynes 230 (Reprinted with permission from Alstom)

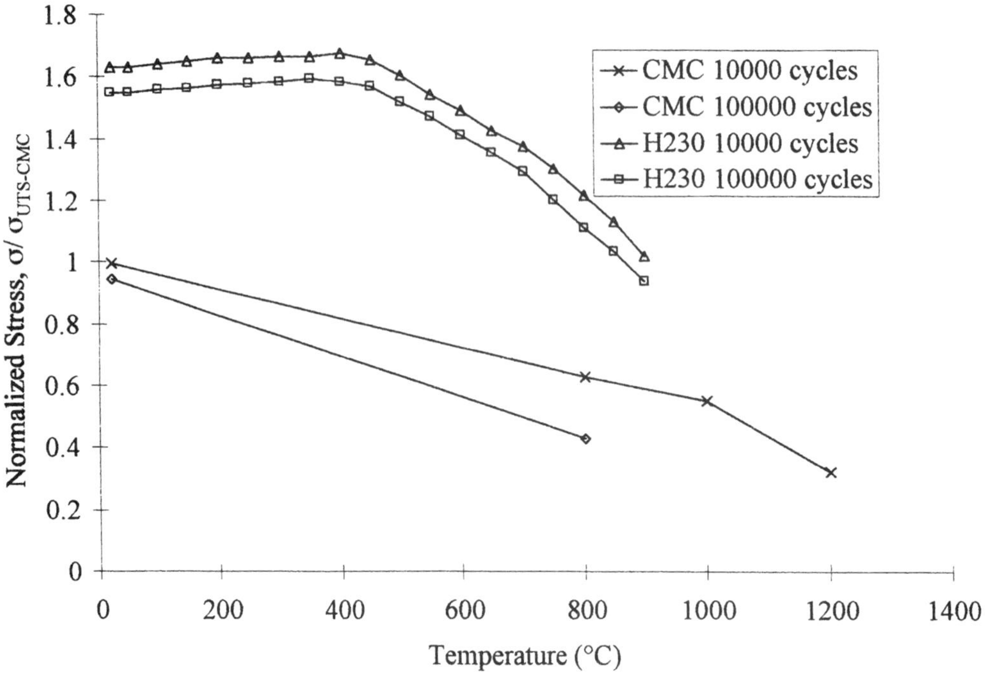

Figure 27.5 Comparison of Cycles to Failure Between a CMC and Haynes 230 (Reprinted with permission from Alstom)

Figure 27.6 Thermal Shock Test of Disk Shapted Specimens (Reprinted with permission from Ansaldo Ricerche s.r.l.)

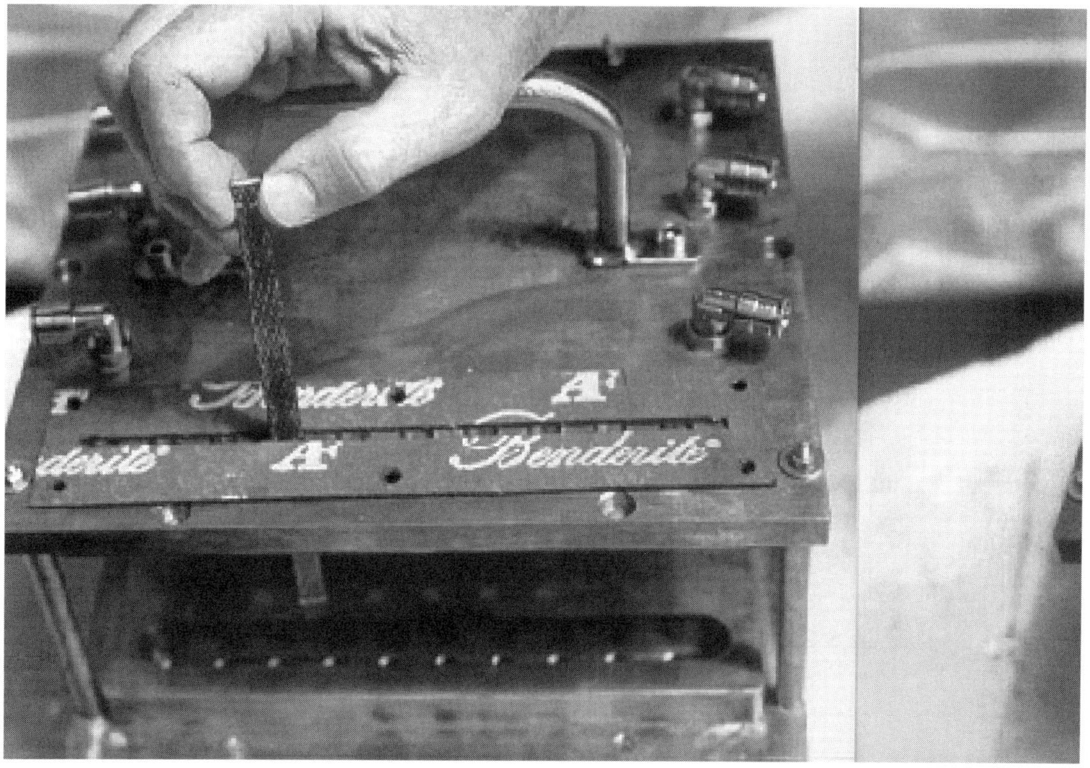

Figure 27.7 Oxidation Test of Flat Specimens (Reprinted with permission from Ansaldo Ricerche s.r.l.)

Oxidation tests (Figure 27.7) were performed by subjecting tensile specimens to combustion gases [17]. They were then tensile tested to find their residual strength after exposure. Reduction in tensile strength is used as a measure of oxidation resistance.

Micro-structural examination by optical and scan electron microscopes was also performed to determine the damage mechanisms which could be useful in building micro-mechanics based models.

Material Modelling

Modelling the non-linear behavior of CMCs is crucial for realising most of their potential. This is so because CMCs often have a low matrix cracking stress and elastic limit. They need to be designed beyond this limit for it to be practical.

Simulating the behavior of CMC material poses two challenges, the first of which stems from the anisotropy in its response caused by its complex fiber architecture. This is particularly so for filament wound components in which fiber directions and layer thickness could be changing continuously, which makes accurate modelling of the component property rather cumbersome. To solve this problem, tools were developed that would automatically generate input decks for a finite element code [18] by reading directly output from a control program of a filament winding machine and thus substantially reduce time and error in pre/post processing associated with a stress and thermal analysis.

The second challenge comes from the multitude of damage mechanisms present inside CMCs and the interaction among these mechanisms. It is believed that for different stages of design, different models of

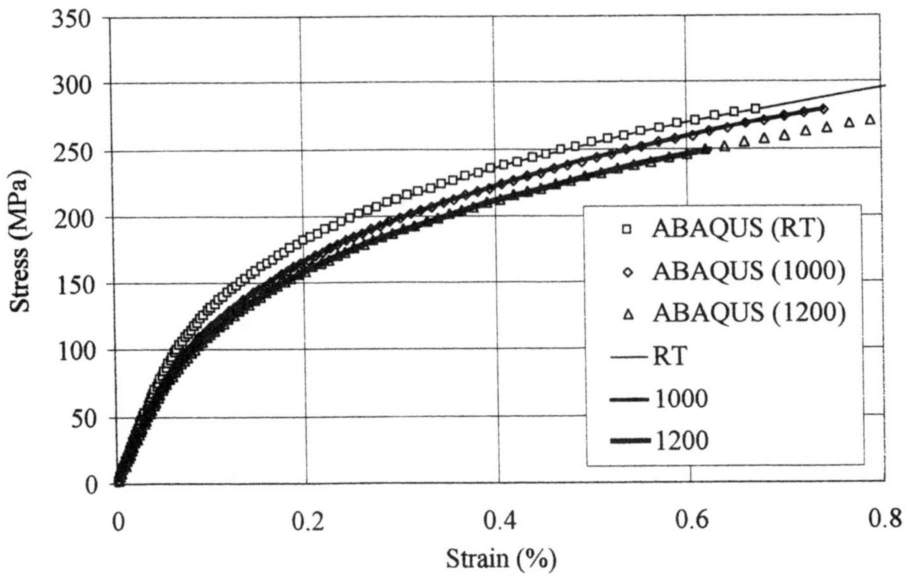

Figure 27.8 ABAQUS Simulation of Mean Tensile Response Based on Ramberge-Osgood Model (Reprinted with permission from Alstom)

varying complexity are required. For most of the engineering design, the material model at the composite level is adequate, while at the detailed design stage, critical areas should be assessed with more elaborate models at the ply or even fiber/interface/matrix levels.

At the composite level, the tensile data were thoroughly analysed and transformed to produce a constant error band. This was achieved by a new master curve concept [19], based on which a constant error band could be obtained and a valid fitting was made. It was found that the well-known Ramberg-Osgood plasticity model fits the test data points quite well. Therefore, it is implemented into the ABAQUS finite element code which can reproduce the mean tensile behavior of the tested CMC precisely (Figure 27.8).

To have a more accurate prediction of CMC behavior, attempts were made to model the ply directly using the theory of continuum damage mechanics [20, 21]. Such a theory could bypass the modelling of cracks explicitly and thus substantially simplify the modelling process. This has been successfully implemented into the ANSYS finite element code, and as can be seen from Figure 27.9, it follows the test data generated within the project closely [22].

The most general model could be built at the fiber, interface and matrix level [20, 21]. Once the in situ fiber, interface and matrix properties are known, they could be utilized to predict the properties of composites made of any fiber, interface and matrix combinations. Two models were produced based on different unit cells (Figure 27.10), both of which reproduce tensile behavior accurately (Figure 27.11) [20, 23]. One of them has been implemented into ABAQUS to assist detailed component stress analysis [20].

To date, the tensile response has been modelled quite successfully. The challenges that remain are how to model the creep and fatigue behavior, and indeed how to predict the life of thermally and mechanically loaded component.

Component Design and Analysis

One of the major challenges in applying CMCs is how to interface a CMC component with its surrounding metallic parts as they have very different coefficients of thermal expansion (CTE). Due to the

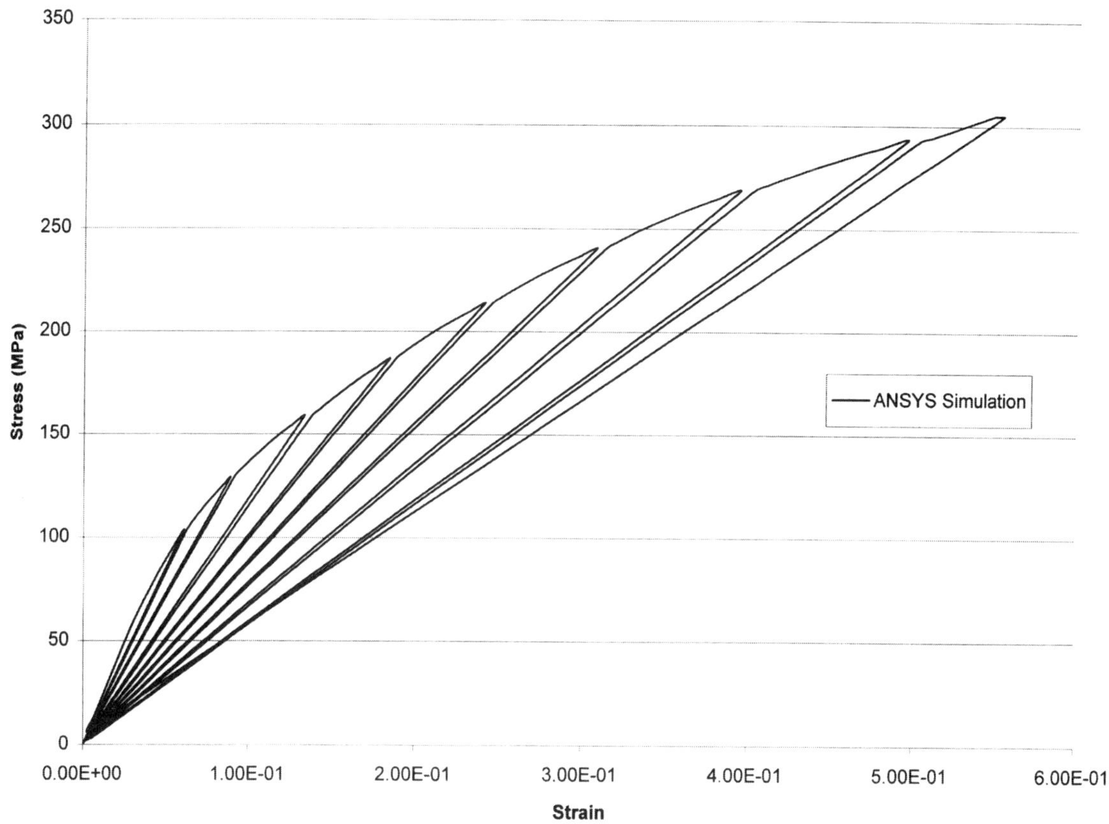

Figure 27.9 ANSYS Simulation of Loading and Unloading of a CMC Specimen (Reprinted with permission from Ansaldo Ricerche s.r.l.)

high operating temperature, such a mismatch in CTE could result in a large differential in thermal deformation. Unless it is assimilated by careful design, structural failure may result caused by unacceptable high thermal stress.

The solutions proposed in the past were either a complete CMC combustor liner [13, 14] or protective tiles [24], both of which have been pursued in the project. The concept based on CMC tiles is more suitable for large engines whose combustor could be of such a size that it exceeds the limit of a production furnace. In addition, the tiles are also cheap and easy to replace when there is localised damage as a CMC part is normally very expensive and no CMC repair technique is known so far. To fix the tiles to the backing structure, mechanical fastening is normally used with refractory metal or ceramic bolts. Such concept could also make it possible to manage stresses in the tile by leaving well-controlled gaps between the tiles.

To avoid excessive thermal stresses, intermediate layers were inserted between the CMC tiles and the metal backing structure (Figure 27.12) [25]. Materials for these intermediate layers need to be chosen carefully in order to minimize the thermal stresses.

The tile concept has the disadvantage of relying on metallic bolts which have temperature limits or ceramic/CMC bolts which have reliability implications. Therefore complete CMC liners are preferable for small gas turbine engines. The final CMC liner design produced in the present project involves a clamping device at the combustor burner end (Figure 27.13) [26]. It is made of a pair of cooled metallic clamping rings. The crucial point is the clamping pressure which should be high enough to give the combustor a firm support, while at the same time it should not be too high to allow some thermal movements between the CMC combustor and the metallic rings.

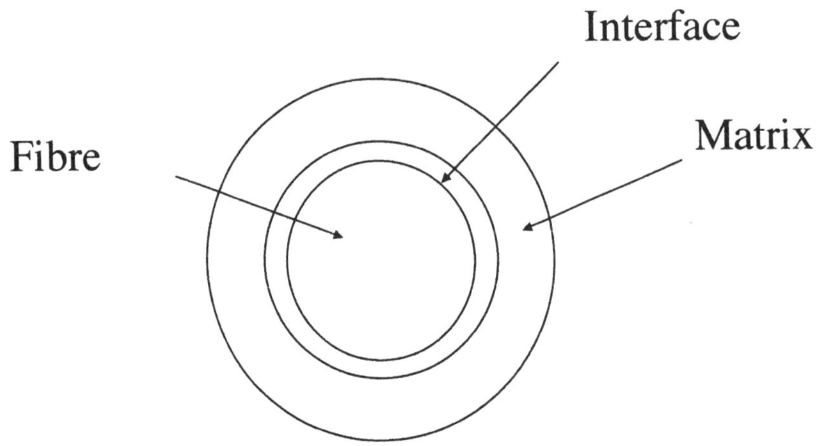

Figure 27.10a **Cylindrical Unit Cell Made of Fiber, Matrix and Interface (Reprinted with permission from Alstom)**

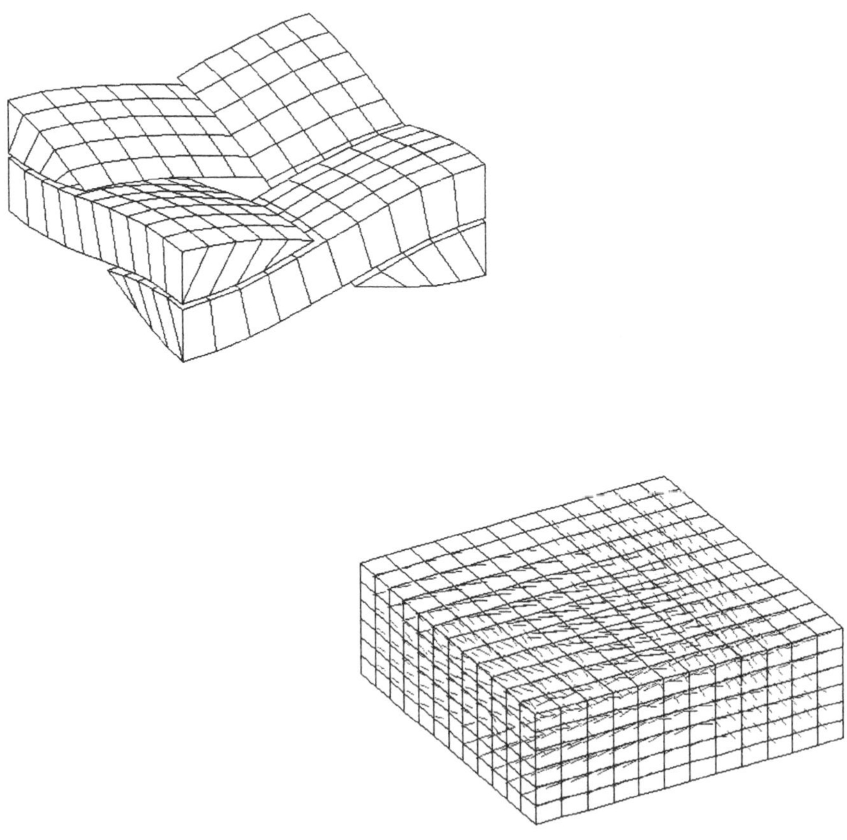

Figure 27.10b **Unit Cells Made of Small Cubes Which May be the Fiber or the Matrix (Reprinted with permission from JRC)**

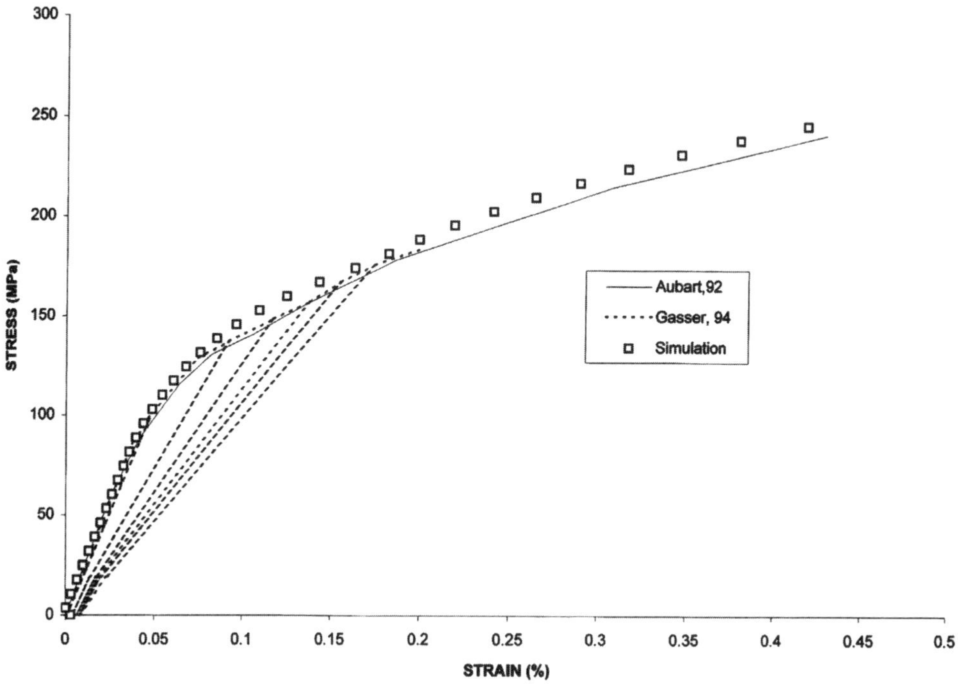

Figure 27.11a Simulation of Tensile Tests Using Micro-model Based on Cylindrical Unit (Reprinted with permission from MEDYSYS)

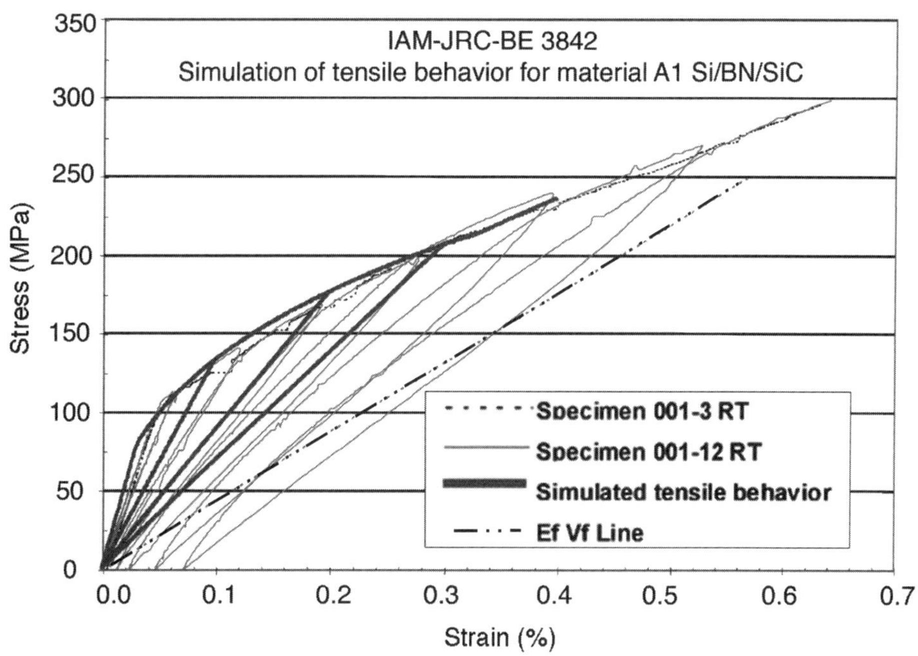

Figure 27.11b Simulation of Loading and Unloading Using Micro-model Based on Cubic Unit-cell (Reprinted with permission from JRC)

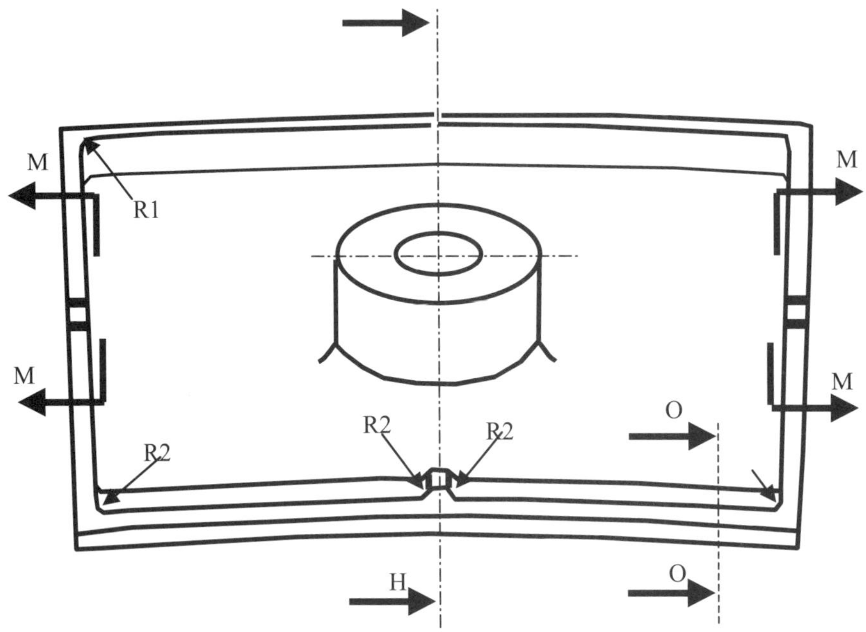

Figure 27.12a CMC Combustor Tile Design (Reprinted with permission from Ansaldo Ricerche s.r.l.)

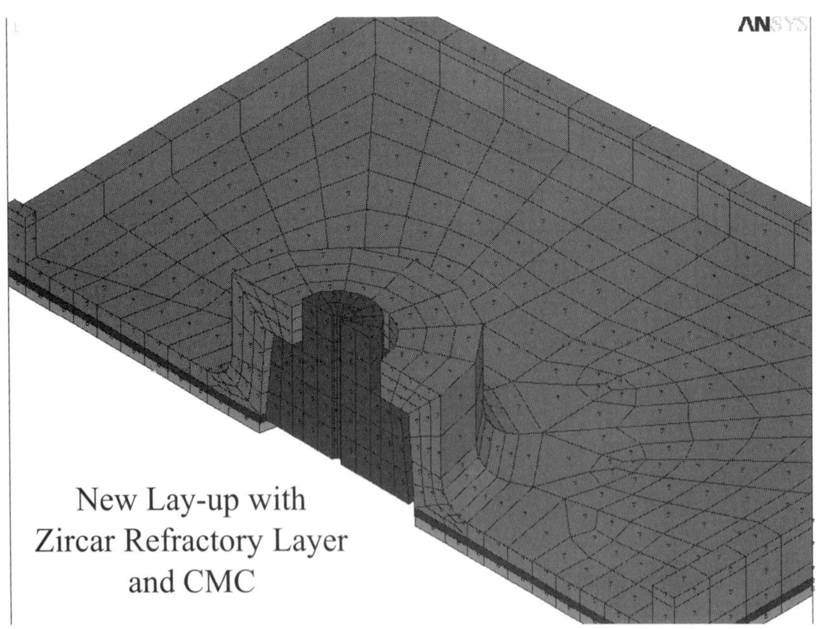

Figure 27.12b Finite Element Model for Thermal and Stress Analysis (Reprinted with permission from Ansaldo Ricerche s.r.l.)

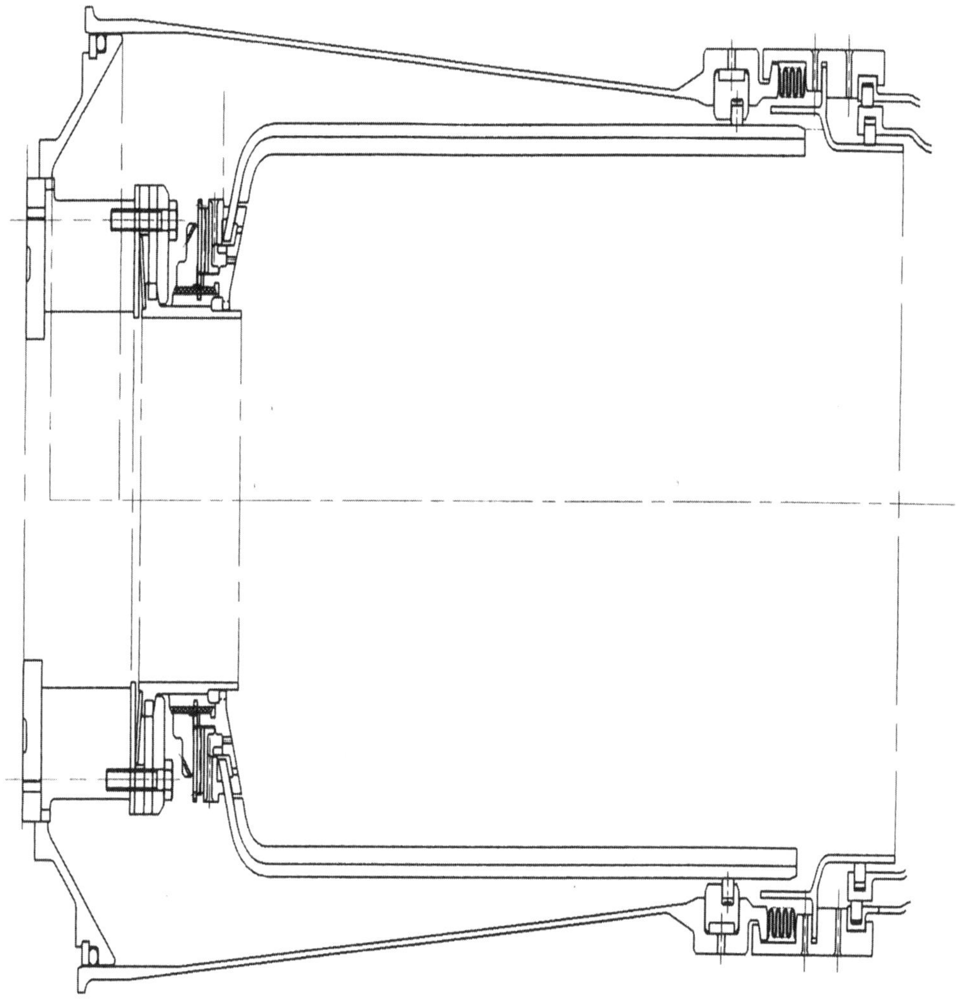

Figure 27.13 Design of a CMC Combustor (Reprinted with permission from Alstom)

In view of the temperature limit of the CMC, a thermal protection system (TPS) of the CMC combustor liner was deemed to be necessary to limit the maximum CMC temperature to below 1200°C, and to protect it from environmental attacks. The TPS is compoosed of short oxide fibers in a highly porous oxide matrix. Because of its high porosity, the TPS was found to be effective in taking most of the temperature gradient through the combustor wall (Figure 27.14). Despite its excellent thermal properties, the current TPS seemed to be weak in mechanical strength.

In addition to thermal analysis, thermal stress, mechanical stress, vibration and buckling analyses were also performed. High cycle fatigue due to combustion acoustic pressure fluctuation is a particular concern. The trend to Dry Low NOx(DLN) combustion often means that the combustion flame operates close to extinction and may present severe pressure fluctuations. It is therefore crucial to design the DLN combustor as stiff as possible so as to move its natural frequencies away from dominant combustion dynamics. Pressure drop across the combustor wall may also cause it to buckle. Due to CMC's high stiffness and low density, the CMC liner was found to be sufficiently strong and not prone to dynamic and buckling failure. The first

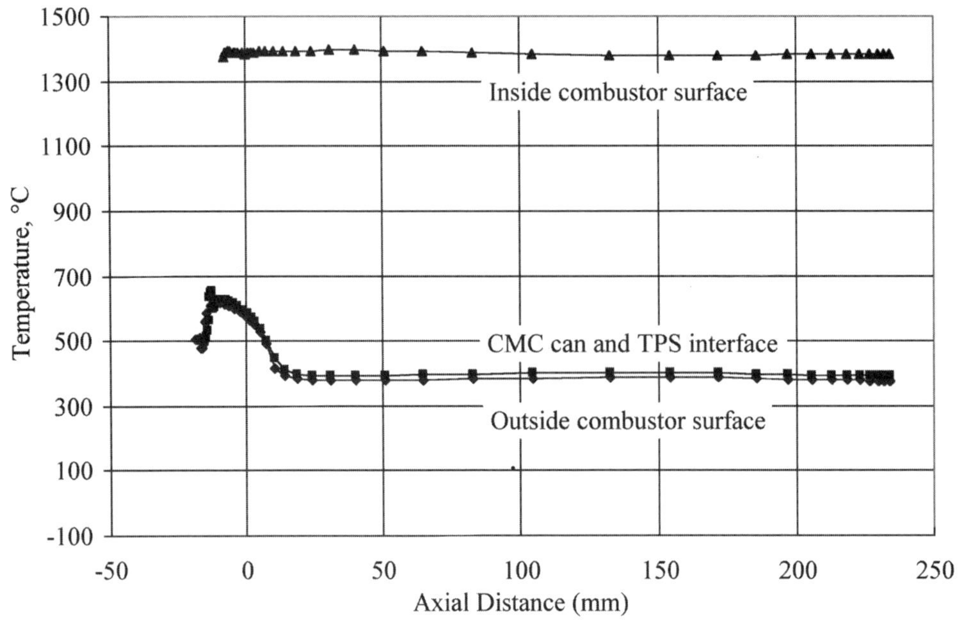

Figure 27.14 Steady-state Temperature in the Combustor Wall (Reprinted with permission from Alstom)

natural frequency of the CMC liner is well above the excitation frequency (see Figure 27.15 for the 1st mode of natural vibration.).

Component Fabrication and Testing

The primary objective in rig testing is to check the interface design that holds the CMC combustor liner in place. These tests were also useful in evaluating the thermal and mechanical integrity of the CMC combustor liner and its TPS. Such tests are invaluable in obtaining information under near engine conditions, which can not be easily duplicated in small laboratory tests. For instance, combined shaking due to engine vibration with a high-speed flame may have significant impact on component integrity that may otherwise be unknown.

CMC combustor prototypes (Figure 27.16) were made by Chemical Vapor Infiltration (CVI) of woven SiC (High Nicalon) Fabrics, a process that seems to have matured over the years as the finished products were of high quality and within reasonably tight tolerance. The interface coating is BN. The delivery time was relatively long but still tolerable if purchase order and production could be better co-ordinated. The supplier, DuPont Lanxide Composites Inc. (now part of General Electric Company), should be congratulated for their excellent job done.

Rings were cut off from the end of the CMC combustors to determine the as-fabricated material properties by subjecting the rings to monotonic and cyclic internal pressure (Figure 27.17). Further pingtests were performed on the combustors to confirm the stiffness properties used in the stress analysis and to serve as a 'signature' of the component before rig testing, which may be used to assess changes in material property after rig tests.

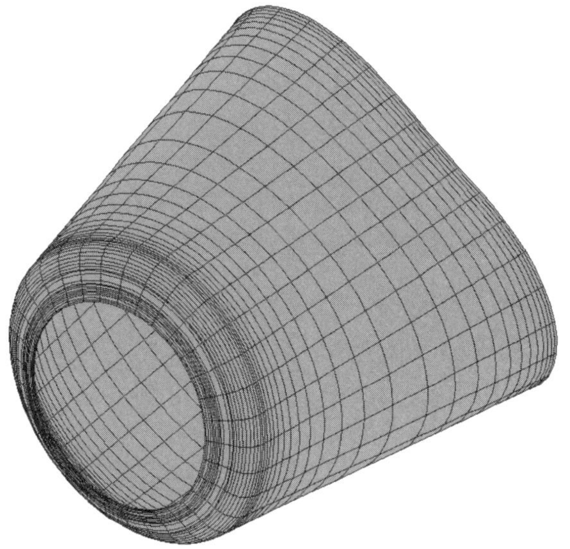

Figure 27.15 First Mode of Natural Vibration is Above the Excitation Frequency (Reprinted with permission from Alstom)

Figure 27.16 A CMC Combustor Can Prototype (Reprinted with permission from Alstom)

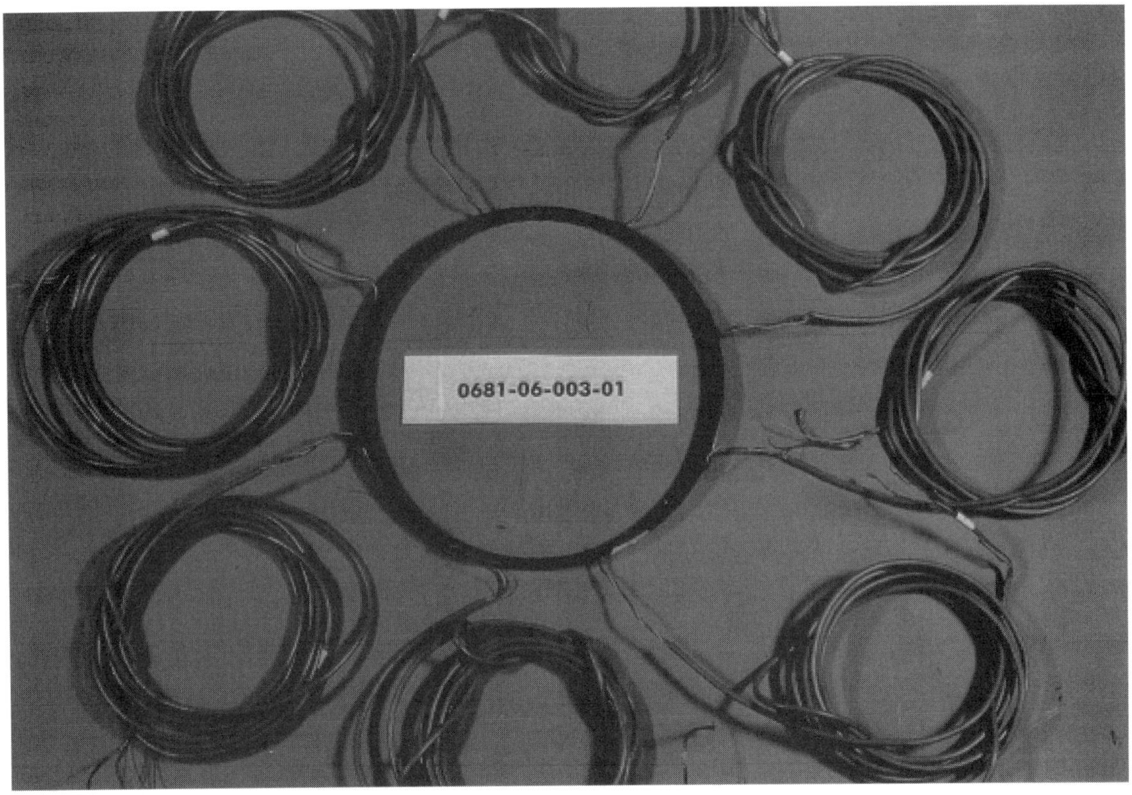

Figure 27.17 Ring Cut from a CMC Combustor is Strain Gauged Before Internal Pressure Test (Reprinted with permission from Alstom)

To minimize the risk, the CMC combustor was tested on an atmospheric pressure rig before testing on the high-pressure rig. During the tests, material temperatures were recorded by thermocouples embedded through the thickness of the combustor wall. The external surface of the CMC combustor was also thermally painted to give extra back up on temperature analysis (Figure 27.18). These temperatures not only provide useful information about material temperatures, but they also may give clues regarding the heat transfer through the wall and the effectiveness of the TPS in reducing the CMC temperature.

It would be nice to have strain gauges on the combustor wall as well since strain information may validate stress calculations and give insight into the deformation of the CMC combustor under real engine operating conditions. However, it was proved extremely difficult to attach strain gauges on the very rough component surface. This can only be achieved by severely dressing the surface, which may damage the substrate. Further suitable high temperature glue could not be found at the time of the rig tests.

The high-pressure rig was also equipped with a gas analyser, which is useful in determining combustion conditions, combustion gas compositions and flame temperature. It gives insight into the level of harmful emissions and water vapor present. The former is linked to one of the main drivers of the project, while the latter is widely recognized as the key factor that controls SiC/SiC component life [27].

After the rig tests, the metallic clamping device and the surrounding combustor can were found to be intact, which gave a tremendous boost in confidence for designing the ceramic-metal interface. In addition, the CMC can seems to have survived the rig tests without any visible damage (Figure 27.18). However, there were problems with the TPS, which was severely damaged. This means that substantial work is needed to

Figure 27.18 Instrumented CMC Combustor After Rig Test (Reprinted with permission from Alstom)

bring the TPS to the status of engine application. The exact cause of TPS failure is not clear at the moment. It could be the result of high thermal stress and low material strength both due to high TPS porosity. Sudden pressure changes inside the combustor could also have played a part in the TPS failure.

CONCLUDING REMARKS

Due to the high risk/cost and the depth/breadth involved in the kind of work reported here, well funded and concerted R & D efforts among gas turbine manufacturers, CMC material/component producers and research establishments are absolutely necessary. It is vital to have available at least a mature material production route with consistent quality and reasonable yields at the beginning of the project. Preferably component fabrication capability at industrial standard is also established. Inferior materials with inadequate component manufacturing would put a project seriously at risk. An early choice of an unsatisfactory CMC (SiC reinforced Alumino Silicate) had put the project in severe jeopardy.

SiC/SiC based CMC is probably the only serious contender for gas turbine combustor applications for the foreseeable future. It has a balanced mechanical strength though oxidation/corrosion in the presence of water vapor still needs to be addressed. Production grade SiC/SiC is now readily available in large quantities and combustor type components have been produced to industrial quality. The CVI process may take a long time but it seems to be acceptable if it is well planned. However, its application is probably limited to the 900°C to 1200°C range, below which current metal components are more competitive and above which the long term stability of the CMC is of major concern unless there is a breakthrough in its oxidation resistance. Before this happens either cooling or some kind of coating is required to protect the CMC from environmental attacks.

The critical part in the CMC combustor design is the interface area where the CMC combustor is connected to metallic structures and where the stress field is three-dimensional and peak stresses are normally found. Consequently, this is the place that is critical for the assessment of structural integrity, which, unfortunately can only be achieved by expensive rig tests.

Instrumentation in rig testing needs to be improved. Ideally, both temperatures and strains should be available to assess the component response under near engine conditions and to validate design calculations.

The CMC prototypes produced are approximately ten times more expensive than their metallic equivalent. Obviously, this is totally unacceptable even taking into account all the potential benefits that the CMCs can offer. Based on possible progress in the current manufacturing technology, it is difficult to envisage a ten-fold reduction in component cost. Therefore high component cost remains a major challenge for the CMC producers until the price comes down to no more than twice that of a metallic component.

ACKNOWLEDGEMENTS

The author likes to thank Drs C. Mao and U. Barberis of Ansaldo Ricerche, Drs M. Steen, P. Bonnel and C. Filio of JRC, Drs C. Rospar and E. Le Dantec of Medysys and Mr. M. Hazell of ABB ALSTOM Power for their valuable contributions to the project. The financial support by the European Commission is much appreciated. He also wants to thank Dr. K. Hatton of the former DuPont Lanxide Composites Inc. for his excellent technical and logistic support during this project, without which it could not have reached its current status.

REFERENCES

[1] Financial Times, "The outlook for the power equipment market," May 16, 1995.

[2] Stein, H., "Electric Power Generation Market Forecast, 1998–2007," *TMI Handbook*, pp 14–21, 1998.

[3] Dodman, K., 1997, "Efficiency will Be the Biggest Issue," International Power Generation, pp. 51–53, Sept. 1997.

[4] European Power News, "Kyoto Calling," p. 6, Dec., 1998.

[5] Cox, N., "Internal Communication," ABB Alstom Power Technology Centre, Leicester, UK, 1999.

[6] Lefebvre, A.H., "Gas Turbine Combustion," 2nd edition, p. 339, Taylor & Francis, Philadelphia, PA, 1998.

[7] Shi, J., "Ceramic Matrix Composites for Gas Turbine Engines," in *Advances in Turbine Materials, Design and Manufacturing*, ed. by A. Strang, et al, Proc. of the 4th Int. Charles Parsons Turbine Conf., 4–6 Nov., 1997, Newcastle Upon Tyne, UK.

[8] Naslain, R.R., "Processing of Ceramic Matrix Composites," Key Engineering Materials, Vols. 164–165, pp. 3–8, Trans Tech Publications, Switzerland, 1999.

[9] Evans, A.G. and Zok, F.W., "The physics and mechanics of the fibre-reinforced brittle matrix composites," J. of Mater. Sci., 29[15], pp. 3857–3896, 1994.

[10] Evans, A.G., Zok, F.W. and McMeeking, R.M., "Fatigue of Ceramic Matrix Composites," Acta Metall. Mater., Vol. 43, No.3, pp. 859–875, 1995.

[11] Holmes, J.W. and Wu, X., "Elevated Temperature Creep Behavior of Continuous Fiber-Reinforced Ceramics," in *High Temperature Mechanical Behavior of Ceramic Composites*, ed. by S.V. Nair and K. Jakus, Butterworth-Heinemann, Newton, MA, 1995. pp.193–259.

[12] Lewin, B., "Why measure?," presented at Composite Components: measurements for design and performance, Inst. of Materials Seminar, March 19, 1996, Airbus, Bristol., U.K.

[13] Price, J.R., Jimenez, O., Faulder, L., Edwards, B., and Parthasarathy, V., "Ceramic Stationary Gas Turbine Development—Fifth Annual Summary," ASME paper 98-GT-181, presented at the

International Gas Turbine and Aeroengine Congress and Exhibition. Stockholm Sweden, June 2–5, 1998.

[14] Nishio, K., Igashira, K., Take, K., Suemitsu, T., "Development of a Combustor Liner Composed of Ceramic Matrix Composites," ASME paper 98-GT-104, presented at the International Gas Turbine and Aeroengine Congress and Exhibition, Stockholm Sweden, June 2–5, 1998.

[15] Bonnel, P., Filiou, C., Fischer, B., and Steen, M., "Room and High Temperature Tensile Tests on SiC/SiC Composites," Brite EuRam Project BE-3842, Progress Report, 20/04/1998.

[16] Filio, C., Bonnel, P. and Steen, M., "Room and High Temperature Tensile Tests on SiC/SiC Composites-Material B," Brite EuRam Project BE-3842, Progress Report, October, 1998.

[17] Mao, C., "Ansaldo Ricerche's Technical Specifications for Oxidation and Thermal Shock Tests," Brite EuRam Project BE-3842, Progress Report, 1999.

[18] Brady, L., Ireson, A., Leggatt, N., and Shi, J., "Two Useful Tools to Generate Composite Finite Element Models," Brite Euram Contract BRPR-CT97-0399, Report, October, 1998.

[19] Shi, J., "Modelling the Tensile Behavior of a SiC/SiC Composites by Master Curves," ASME paper 2000-GT-0066, presented at the 45th ASME International Gas Turbine & Aeroengine Technical Congress, Exposition and Users Symposium, 8–11 May, 2000, Munich Germany.

[20] Rospars, C., Le Dantec E., and Lecuyer F., "A micromechanical model for thermostructural composite," Comp. Sci. Tech., 60, 1095–1102, 2000.

[21] Rospars, C., Le Dantec, E., Lecuyer, F., "CMC damage prediction by micro-macro modelling," *Proceedings of the 12th International Conference on Composites Materials*, 5–9. July 1999, Paris.

[22] Barberis, U., "Damage Macro Model for Ceramic Matrix Composites in Tension. Customized Routines in ANSYS-Rules for their Application," Ansaldo Ricerche s.r.l., T01344UM5000C, 10/03/00.

[23] Bonnel, P., "Method for Modelling the Thermo-elastic Properties of Composites Reinforced with Woven Fabrics," IAM, JRC, report, January, 2000.

[24] Schulenberg, T., "New Development in Land-Based Gas Turbine Technology," *Proceedings Materials for Advanced Power Engineering, Part II, COST*, J. Lecomte- Beckers, F. Schubert, and P.J. Gunig, eds., 1998, p. 849.

[25] Mid-term Report, Contract No. BRPR-CT97-0399, "Development of Design Methodologies for Ceramic Matrix Composites for Industrial Gas Turbines."

[26] Shi, J., "Design, Analysis, Fabrication and Testing of a CMC Combustor Can," ASME paper 2000-GT-0071, presented at the 45th ASME International Gas Turbine & Aeroengine Technical Congress, Exposition and Users Symposium, 8–11 May, 2000, Munich Germany.

[27] Ogbuji, L.U.J.T., "A pervasive mode of oxidative degradation in a SiC-SiC Composite," J. Am. Ceram. Soc., 81[11], 2777–84, 1998.

CHAPTER

28

Siemens KWU – Experience with Refractory Ceramics in Stationary Gas Turbines

Christine Taut, Bernhard Becker
Siemens KWU, Mülheim a.d. Ruhr, Germany

INTRODUCTION AND BACKGROUND

Siemens KWU has been involved in development and application of ceramics for gas turbines since 1960.

This chapter is not intended as a description of the numerous programs for development of ceramics and ceramic gas turbine applications that have been carried out to date. It is merely intended to give an overview of the field experience with ceramic materials subjected to hot gas conditions and about the implications for the practical application of ceramics in gas turbines. Such experience has been gathered by Siemens KWU for two groups of materials: ceramic thermal barrier coatings and refractory materials, the latter of which this contribution focuses on.

Actually most Siemens KWU gas turbines contain between one and two tons of ceramic materials in the hot gas path. Fleet experience includes long-term operation at temperatures of up to 1550°C (1823K): the dreams of generations of ceramists of introducing ceramics in field applications already became reality in Siemens KWU gas turbines years ago. The principal goal was to find technical solutions for the hot gas path that consume as little cooling air as possible. Different solutions were found for parts with low and high aerodynamic loads and with low and high hot gas flow velocities. For items subjected to high aerodynamic loadings such as vanes and blades in the front turbine stages, ceramic thermal barrier coatings on metallic structures are state of the art. Items subjected to low aerodynamic loads such as the combustion chamber wall lining can be realized by two principle concepts: the first is the use of metallic parts, if necessary protected by ceramic thermal barrier coatings, intensely cooled and/or cooled using highly sophisticated cooling technologies. The maximum service temperature of the materials or interface surfaces limits the practicability of this strategy. Currently the maximum long-term service temperature is limited to 1250°C (1523K).

Siemens KWU takes the other approach: the concept of "hot walls" of the combustion chamber. This involves the necessity of using highly insulating materials as the hot wall lining. The main service conditions to which these materials are subjected in the combustion chamber of modern stationary gas turbines are as follows:

- Surface temperature at flame side — 1300 to 1550 °C (1573 to 1823K)
- Steady-state temperature gradient across tile thickness (38mm) — 800 to 1000 K
- Transient thermal gradient (on the flame side) — about 1000 K/20s[1]
- Total pressure — 16 to 17 bar
- Hot gas velocity — 40 to 100m/s
- Inspections performed after — 8000 h[2]
- Typical replacement of tiles after 25,000h — ≤ 50%
- Oxidizing[3] hot gas atmosphere, containing water vapor, Na/Ca/Mg/V/S/Fe and their compounds

The most severe service conditions are due to thermal shock caused by trip at full load. This means that the combustion gas temperature drops from about 1550°C to 450°C (1823 to 723K) immediately. The flame side of the combustion chamber lining is directly exposed to these very rapid temperature changes. Only a few materials can even be considered for such extreme conditions as they must have a high melting point, low coefficient of thermal expansion, low modulus of elasticity, high strength at elevated temperatures and low thermal conductivity. The materials selected should be affected only slightly by water vapor or other corrosive media and should undergo no microstructural changes in the temperature range up to 1550°C (1823K). They should possess high thermal and thermo-mechanical stability as well as erosion resistance and must have favorable crack initiation and propagation behavior. Because of the rapid thermal cycling conditions of gas turbines, thermal shock resistance is a very important aspect of material stability requirements.

It is very clear that these properties are not independent of each other and that the service conditions in a gas turbine sometimes involve opposing requirements. For example, it is necessary to obtain high strength to achieve high resistance to crack initiation, but at the same time, low strength is necessary for ensuring high resistance to crack propagation. Materials with high strength often exhibit a high modulus of elasticity that is disadvantageous in terms of thermal shock resistance. A low thermal conductivity is necessary for wall insulation but a high thermal conductivity would result in lower thermal gradients in the components and consequently lower thermal stresses. For these reasons, an optimum of composition and properties has to be found for each specific application and this requires ranking of the importance of each of the relevant material properties.

HISTORY OF APPLICATION

Siemens KWU began using refractory materials for lining combustion chambers in 1960. More than 8 million hours of operation have been logged in the meantime. Since that time, in-house R&D programs have focused on implementation and further refinement of refractory materials.

Initially, the high-alumina refractories selected were based on a corundum-sillimanite mixture. These materials consisted of alumina with a relatively high silica content. In addition to the silica bound in sillimanite, the microstructure also included some free silica. The material was satisfactory for the first and second-generation machines and is currently in service in most VX4.2 combustion chambers at sites located around the world.

With the introduction of the current generation of machines with higher power output, service temperatures and efficiency, it was necessary to improve the material properties of hot gas path items. Higher performance materials had to be selected, qualified and modified for the planned application. This resulted in

[1] Occurs on trip during operation at full load.

[2] Inspection intervals vary depending on the service conditions (peak load or base load machines).

[3] The term "oxidizing" is meaningless if no reference scale is defined. In this paper the term is used only in a qualitative sense. Oxygen partial pressures equal to or greater than that of air (0.2 atm O_2) are arbitrarily considered to be "oxidizing".

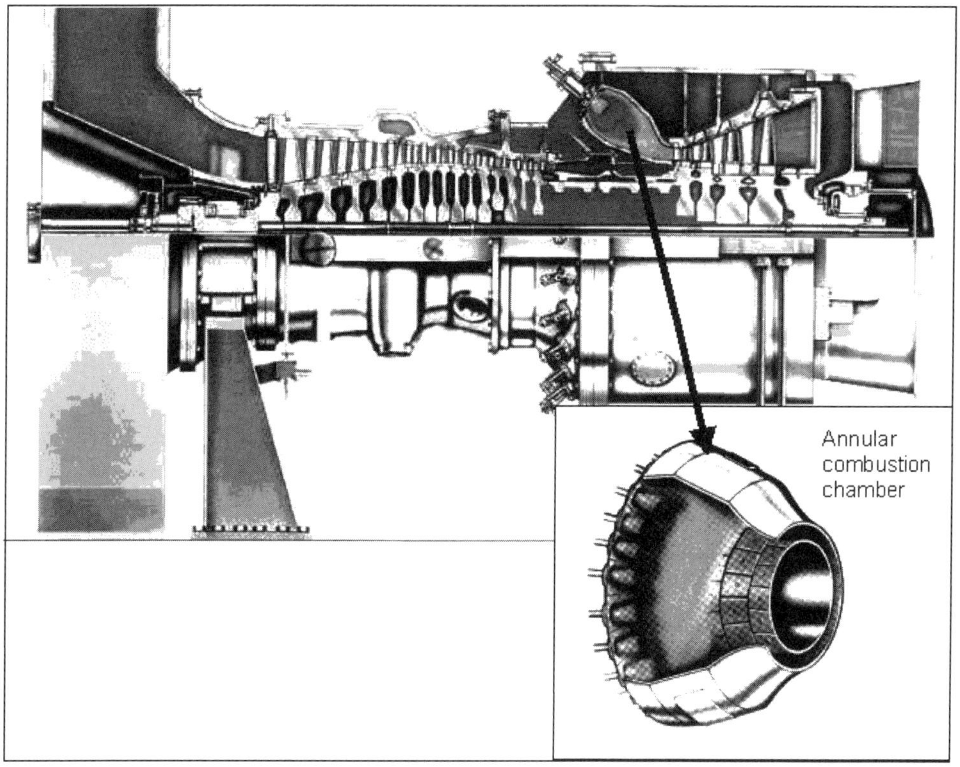

Figure 28.1 Generic Illustration of a Current Siemens KWU Gas Turbine V84.3A(2) with Annular Combustion Chamber, Lined with Ceramic Tiles

the application of higher purity materials with reduced silica contents. For this reason the material selected and presented here is referred to as a "high-alumina" refractory.

A further challenge was encountered with the introduction of the annular type combustor in 1996 with a more complex geometric shape. The first prototype test bed runs with the ceramic-lined annular combustion chamber at the Berlin manufacturing plant demonstrated the feasibility of this application. A wealth of experience with the performance of the high-alumina tiles has been collected in fleet operation. Meanwhile more than 20 machines of this latest type are in service worldwide.

Figure 28.1 is a generic illustration of this type of gas turbine with the annular combustion chamber, equipped with ceramic tiles.

GENERAL CONSIDERATIONS

The refractories discussed here have a very complicated microstructure in contrast to single-phase, homogeneous materials. A sufficiently accurate knowledge and model of this kind of microstructure and its changes during long-term, thermo-mechanical aging is not yet available. This necessitates combining all available knowledge, beginning with details of the production process, the laboratory and prototype test bed results as well as field experience gathered with the installed tiles. In this way it is possible to understand the aging and failure mechanisms of these materials and then derive suitable measures for ensuring reliable operating behavior.

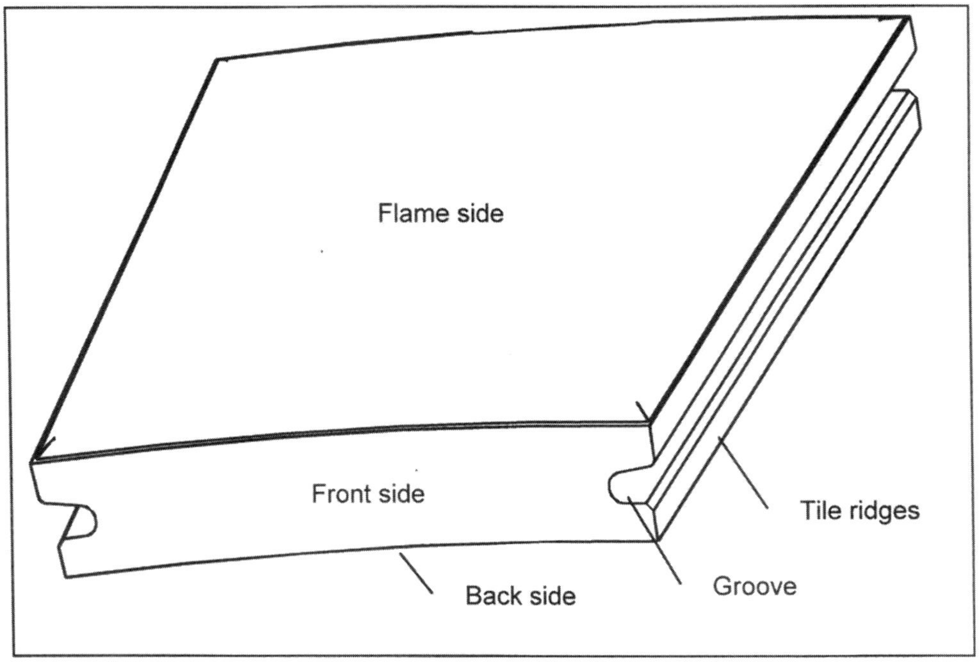

Figure 28.2 Principal Design of the Ceramic Combustion Chamber Tiles

DESIGN APPROACHES

Figure 28.2 shows the principal geometry of the high-alumina tiles. The average size is 200 mm × 200 mm × 38 mm. The tiles have grooves on two opposing sides. Four metallic clips, two on each grooved side, fit into these grooves and hold the tiles in place. It is part of the security concept that each tile is fixed in such a way that if a tile fractures, fragments are not immediately lost. The spring clips fix the tiles in position elastically at the metallic combustion chamber wall. Tiles can be inserted and removed with the upper section of the gas turbine casing in place. A manhole is provided to permit access to the combustion chamber for inspections and service work. Figure 28.3 is a view into a gas turbine combustion chamber showing the location where tiles are installed.

Small gaps around the tiles allow unrestricted thermal expansion of tiles and metallic casing. The gaps are protected against the ingress of hot combustion gas by cooling air at a temperature of about 450°C (723K) fed from the casing side. This results in thermal gradients of up to 500°C (773K) between the edges and the center at the flame side of the tiles and to thermal gradients of up to 727°C (1000K) over the tile thickness.

Free thermal expansion of tiles, clips and the metallic casing wall is the primary design criterion. The thermal insulation of the metallic walls is the secondary criterion and constitutes and optimum between the necessary tile thickness and the induced thermo-mechanical stress level. A further design criterion is the tile holder spring force: the tile holders must hold the tile elastically at the metallic casing wall. They have to be strong enough to hold the tile in place at the wall under all service conditions. The objective is to prevent tile lift-off by sufficiently high tile holder forces but without jeopardizing tile integrity by excessive force. The next criterion is to permit insertion and removal of tiles from the inner ("hot") side of the combustion chamber. And last, but not least, an optimum must be achieved between the overall tile size (i.e. the weight of the tile and the resulting thermo-mechanical stresses) and the total gap area, i.e. the length and width of gaps that have to be protected against the ingress of hot gas. Cooling

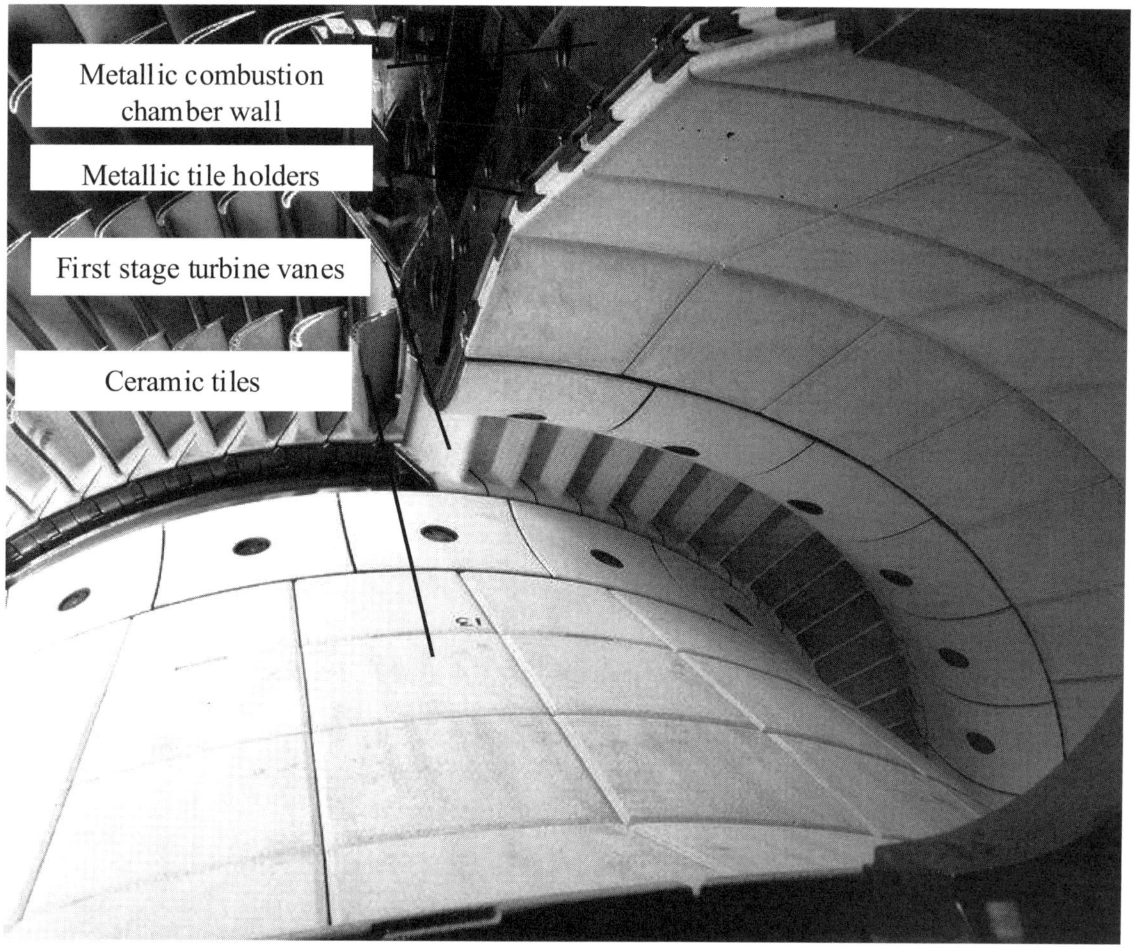

Figure 28.3 View Through a Burner Mount into a V84.3A(2) Annular Combustion Chamber Fitted with Ceramic Tiles

air used in the hot gas path is precious and should be used as sparingly as possible to achieve low temperature, low pollution combustion with high overall efficiency. As a result, the total gap area, and therefore the total amount of seal air required, must be minimized.

FOCUS OF ACTIVITIES/CHALLENGES

Materials Qualification

The procedure used for materials selection and qualification is shown in Figure 28.4. Table 28.1 lists the main properties that must be checked and the information that is obtained on operating behavior. Qualifying a new refractory material, from the initial contact with a potential supplier until the installation of test tiles in a gas turbine, is a process that takes at least two years on average.

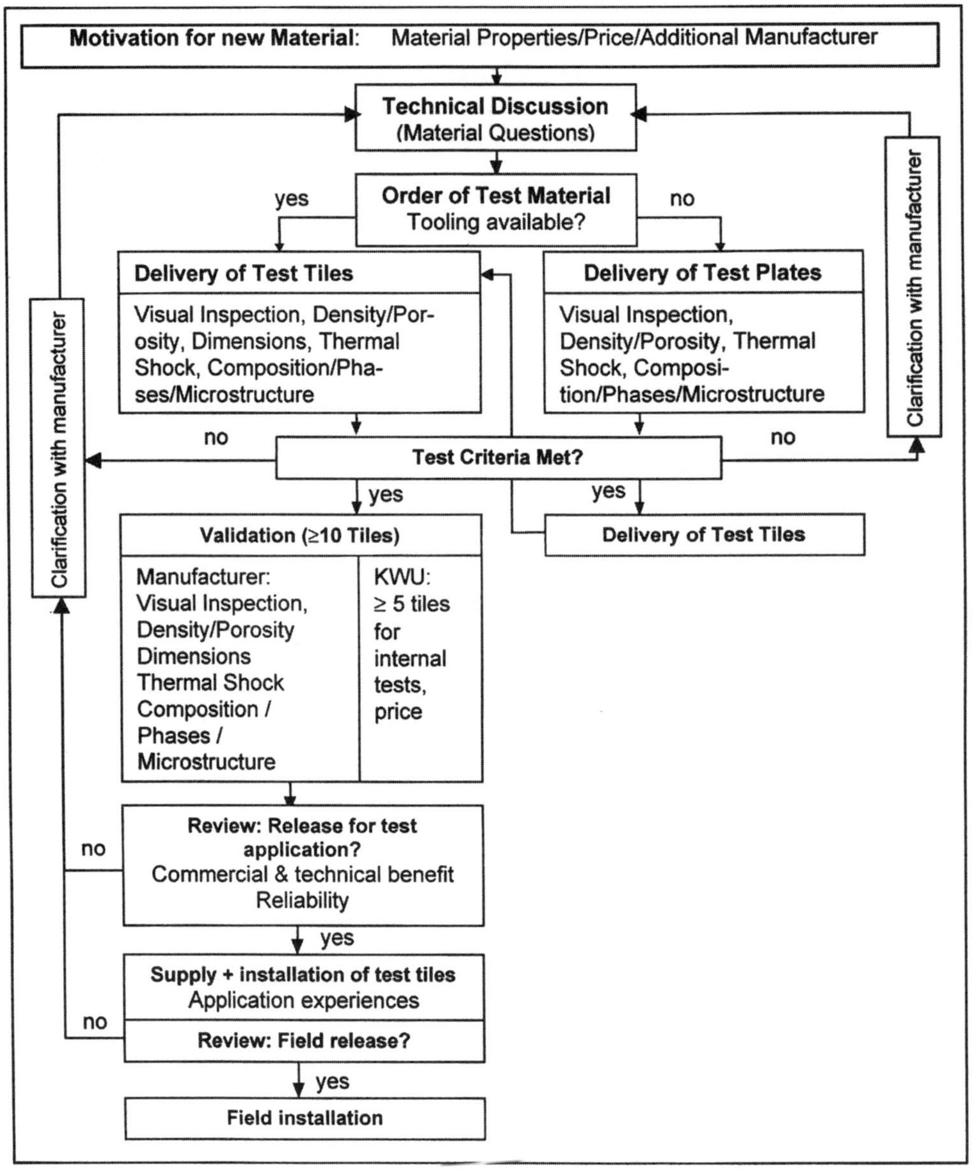

Figure 28.4 Process for Qualification and Validation of New Ceramic Combustion Chamber Lining Materials

During this qualification phase, development effort is required from both the customer Siemens and the potential supplier. This is necessary because of the highly sophisticated demands of the gas turbine application that is by no means a typical refractory application. A central aspect of this process is that the defined material properties must be representative not only for small samples but for entire parts with large volumes and complicated geometries and not only for one part but for the entire production series of many thousands of tiles manufactured per year.

TABLE 28.1 MATERIALS PROPERTIES CHECKED DURING QUALIFICATION AND VALIDATION OF PROSPECTIVE REFRACTORIES FOR COMBUSTION CHAMBER LINING

Material property tested	Method	Corresponding service requirements
Thermal shock behavior	Water quenching from 1000 to 20°C (1273-293K) (at plates 200*200*38mm or whole tiles)	Resistance to a large high number of gas turbine startups, shutdowns and trips
Chemical composition	X-ray fluorescence analysis	Corrosion resistance
Phase composition	X-ray diffraction (Rietfeld analysis)	Microstructure stability
Density/porosity (total amount, distribution)	Water absorption, microscopic cross section analysis	Erosion resistance/insulation properties
Microstructure (grain sizes of each phase, pore size distribution, textures)	Microscopic cross section analysis	Thermo-mechanical resistance
Bending strength, modulus of elasticity at room and elevated temperature	EN and ASTM standard tests	
Thermal expansion parallel and perpendicular to direction of compression	Coefficient of thermal expansion	Narrow scattering range, design features
Hot gas corrosion behavior	Hot gas test	Corrosion resistance
Discoloration, impurity inclusions, pore agglomerations, cracks	Visual check	Low impurity content, high reliability

TABLE 28.2 PRINCIPAL PROPERTIES OF THE HIGH-ALUMINA REFRACTORIES REFERRED TO HERE

Crystalline phases	70% Corundum, 30% Mullite
Density, g/cm_	2.9 … 3.0
Open Porosity, %	18 … 20
Bending Strength, MPa	
20°C (293K)	9 … 12
1200°C (1473K)	7.5
1400°C (1673K)	2.3
Thermal Shock Stability (Water, 1000°C/20°C (1273K/293K))	>30
Modulus of Elasticity$_{static}$, GPa	13 … 16
Thermal Conductivity$_{1200°C(1473K)}$, W/mK	2.5
Heat Capacity$_{1200°C(1473K)}$, J/kgK	1216
Coefficient of Thermal Expansion$_{1200°C(1473K)}$, 10^{-6} K^{-1}	7.2

Materials Characterization

The principal material properties of the high-alumina tiles discussed here are listed in Table 28.2.

Composition. Chemical composition is the main factor influencing the corrosion resistance of the tiles exposed to the atmosphere inside a gas turbine. X-ray fluorescence is used to determine the chemical composition. Here, too, sampling is the main aspect on which the reliability of the analysis results hinges. With maximum grain sizes of the various phases ranging up to several millimeters and microstructural as well as manufacturing process inhomogeneities, the sample volume must be large enough to ensure that representative results are obtained. The type and fraction of the crystalline phases yields information on possible phase

Figure 28.5 Microstructure of the High-alumina Refractory Material, Polished Cross section

changes that must be expected during operation and on the main thermo-mechanical characteristics. These data are usually determined by X-ray diffraction. Phase composition is determined quantitatively by the standard free Rietfeld method. This ensures a high accuracy and reliability of the analysis. The total amount and the distribution of the principal chemical constituents and crystalline phases as well as of trace impurities gives a good indication of the stability of the raw material quality and the robustness of the manufacturing process. The high-alumina tiles currently-used contain no crystalline phases other than the main phases corundum and mullite.

Microstructure. Microstructure evaluation employs scanning electron microscopy of polished cross sections. Figure 28.5 shows typical details of the microstructure of such a high-alumina material. It consists of a fine- and medium-grain corundum matrix containing large grains of mullite. Small amounts of a glassy phase may be included in the grains. Grain sizes vary from the nanometer range up to several millimeters. Both open and closed porosity occur, pore size distribution spans a broad range. A dense network of micro-cracks forms during cooling of the refractory from the sintering temperature. This is the result of either the

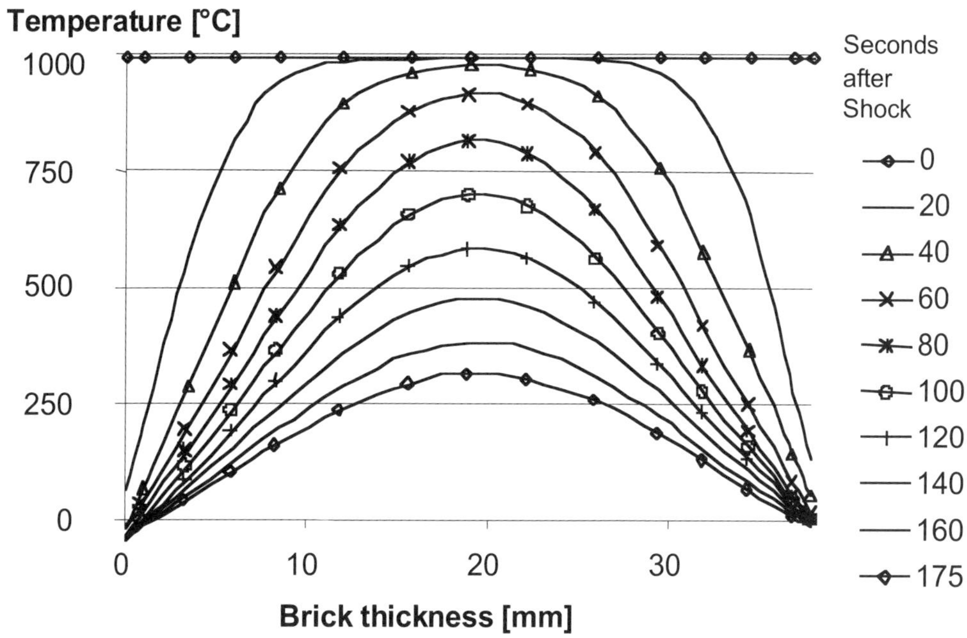

Figure 28.6 Temperature Distribution of High Alumina Brick During Thermal Shock

thermal expansion mismatch of the phases alumina and mullite [1] or the anisotropy of the thermal expansion coefficient exhibited by randomly-oriented grains.

The wide range of inhomogeneities plays a major role in governing the thermo-physical properties. It ensures the excellent resistance to rapid temperature changes and enables a very damage-tolerant tile behavior under most gas turbine operating conditions. This type of microstructure is able to dissipate energy that would first result in cracking and ultimately cause final fast fracture in dense and homogeneous materials.

Mechanical and Thermo-mechanical Properties. A main criterion for the practicability of a material as combustion chamber wall lining is the capacity to resist static and transient thermal stresses. Consequently investigation of the *thermal shock resistance* is a main focal point when evaluating a material. It is also necessary to select a test method that induces somewhat higher stresses than the actual operating conditions as well as a representative sample geometry and volume. Thermal shock resistance is tested with entire tiles under such conditions using the standard multiple-quench test, quenching with water from 1000°C (1273K) to 20°C (293K), by studying the visible crack length and determining the number of cycles to fracture. The quench test is widely used in research and industry [2, 3]. This test characterizes the short-term behavior of the tiles: the resistance against crack initiation at critical temperature differences.

The tests are performed with a temperature difference of nearly 1000°C (1000K), which is comparable to operating conditions. Higher absolute quenching temperatures could improve rather than degrade resistance to thermal shock [4]. The heat transfer coefficients during water quenching are significantly higher than during gas turbine thermal cycling. It follows that these tests induce higher stresses than the thermal shocks sustained during gas turbine starts and trips even though quenching is not performed from the actual operating temperature level.

The Figures 28.6 and 28.7 show typical temperature distributions in a high-alumnia tile from the flame side to the back side during water quenching and on trip of a gas turbine from base load. Whereas during water

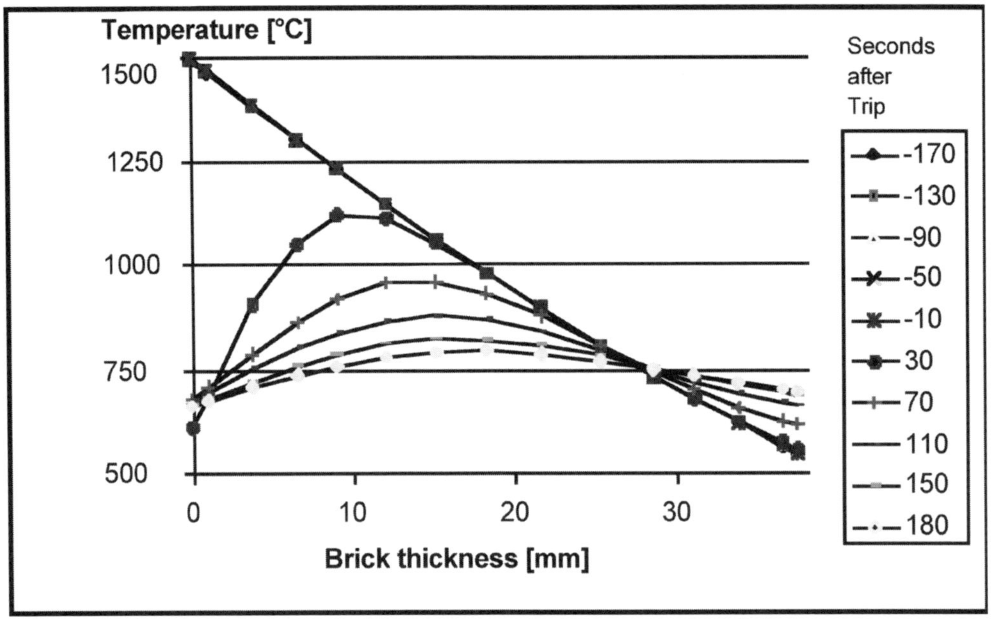

Figure 28.7 Temperature Distribution of High Alumina Brick During Tripping from Base Load at a V94.3A(2) Gas Turbine

quenching thermal shock comes from all sides, the thermal shock on trip comes from one side only. Nevertheless the principal temperature change sequences are similar in both cases. The water quench thermal shock involves steeper temperature gradients—and higher thermal stresses—than the gas turbine trip. In both cases the main temperature decrease occurs during the first 30 seconds after quenching or trip. Stress calculations show that the stresses initiated during such thermal shocks are equal to about 1/0.8 of the stresses sustained during a full load trip of a gas turbine. Test tiles are subjected to at least 30 thermal shock cycles to ensure that the gas turbine can be operated through a full service interval without tile fracture occurring.

The actual high-alumina tiles are subjected to at least 50 such thermal shocks in the testing lab without tile fracture. An example of crack length development as a function of the number of thermal shocks sustained is given is Figure 28.8. Typical "thermal shock cracks" form perpendicular to the tile edges. The crack propagation rate decreases with time and in most cases approaches zero after an initial growth phases during the first 30 to 40 shocks, attaining a maximum visible crack length of about 50 to 60mm. Figure 28.9 shows a typical crack pattern on such a tile after 40 thermal shocks.

A further primary characteristic of potential combustion chamber tile materials is given by their thermo-mechanical behavior at their maximum operating temperatures. These properties include the bending strength, modulus of elasticity, creep behavior, thermal expansion and thermal conductivity.

Typical Weibull-plots of the *bending strength distribution* (4-point bending MOR) sample geometry 10 mm × 13 mm × 90 mm) of the high-alumina material as delivered and after field service for 1100 operating hours are shown in Figure 28.10. The slope of the straight line indicates the scattering of bending strength. The bending strength of the new materials is relatively low (with narrow scattering) because of the high porosity, high micro-cracked microstructure and poor homogeneity of the tiles. The positive effect of this is the high thermal shock resistance of the tiles. It leads to a superior stability when tiles are subjected to the high temperature gradients that occur during gas turbine startup and shutdown. The tiles undergo a decrease in strength and modulus of elasticity and an increase of the scattering of strength during operation at high

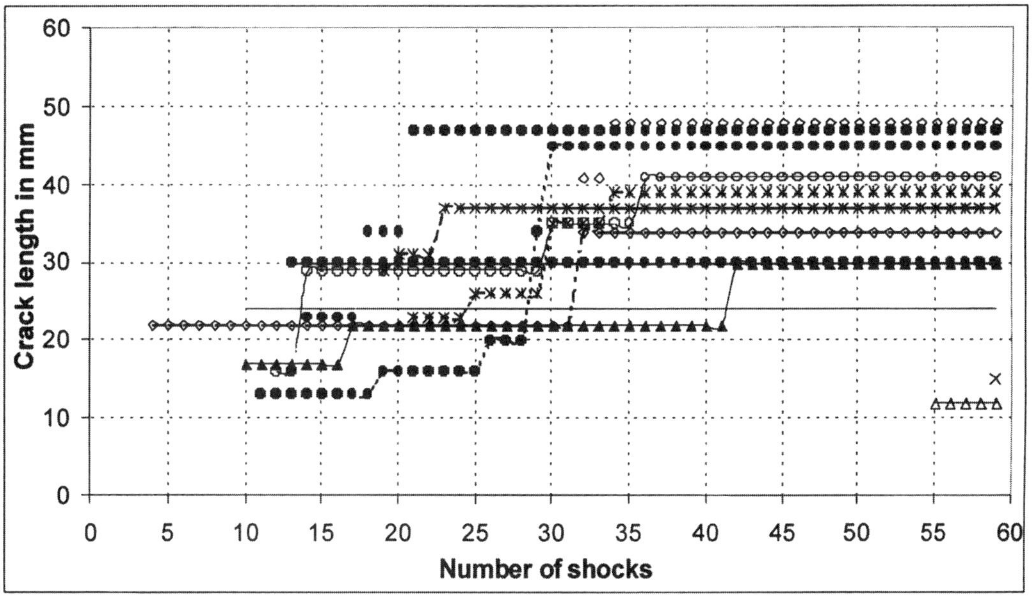

Figure 28.8 Crack Length Development with Number of Thermal Shocks in High Alumina Bricks

Figure 28.9 Crack Pattern at a High Alumina Brick After 40 Thermal Shocks

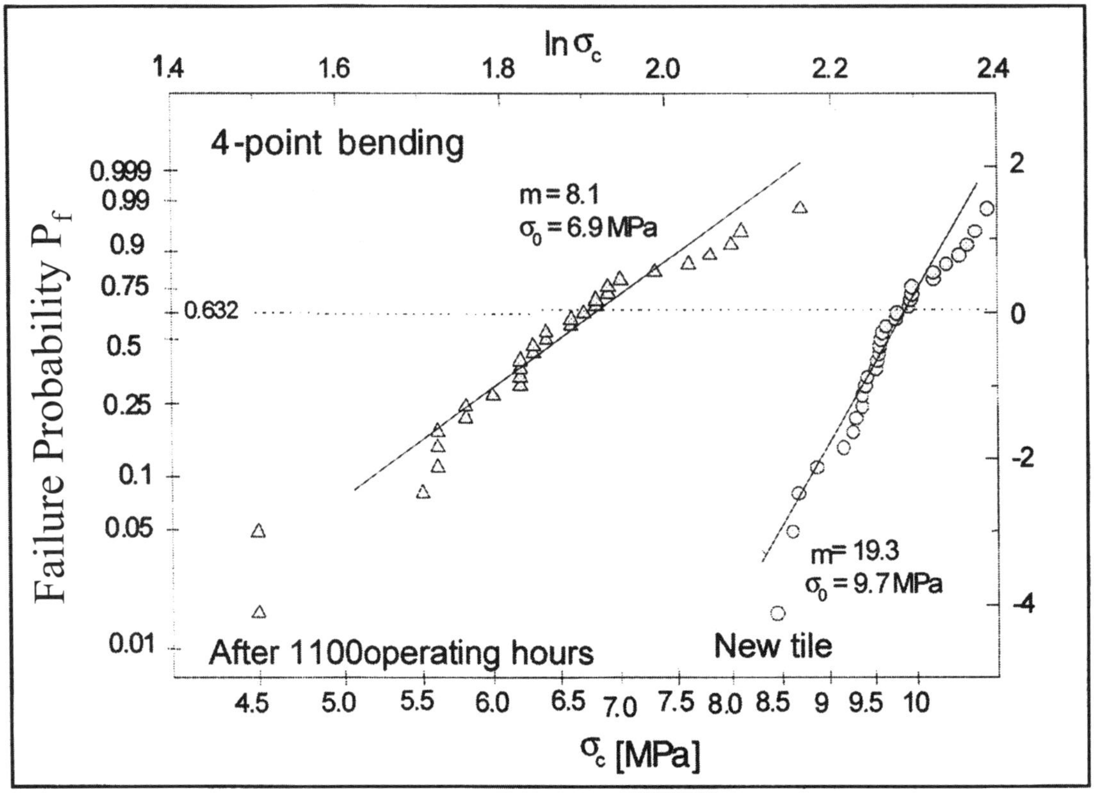

Figure 28.10 Bending Strength/Weibull Plots of a High-alumina Material As-sintered and After 1100 Service Operating Hours

total temperatures, steep temperature gradients and dynamic mechanical loads. Degradation of the mechanical properties involves microstrucutural changes. These microstructural changes occur in the high temperature range, where non-linear-elastic behavior must be allowed for, and are caused by time-dependent effects such as creep or interconnecting of microcracks to form crack networks and ultimately result in the growth of macro-cracks.

The values for the *modulus of elasticity* vary depending on the method of determination: in such highly porous, heterogeneous materials as these high-alumina tiles, the flexural modulus of elasticity can be greater than the modulus of elasticity determined by ultrasonic measurement. This must be considered for calculations of tile behavior under operating conditions.

Creep deformation is a main high temperature consideration for all oxide materials. Creep of the high-alumina material causes stress-induced displacement. This can have the effect that opposing crack flanks no longer fit together when the tile cools, subjecting the crack tip to new stresses. Areas where compressive stresses have been relaxed can be subjected to tensile stresses. Both of these mechanisms can result in slow, sub-critical crack growth. The creep behavior of the high-alumina tile material exhibits only slight temperature dependence up to 800°C (1073K). At higher temperatures, the change from elastic to plastic material deformation already causes creep degradation at slight mechanical stresses.

The temperature dependence of the principal thermal properties of the high-alumina tile material is shown in Figure 28.11 *(thermal expansion)* and 28.12 *(thermal conductivity).* Thermal expansion increases with temperature up to about 1300°C (1573K). At higher temperatures, post-sintering effects cause a

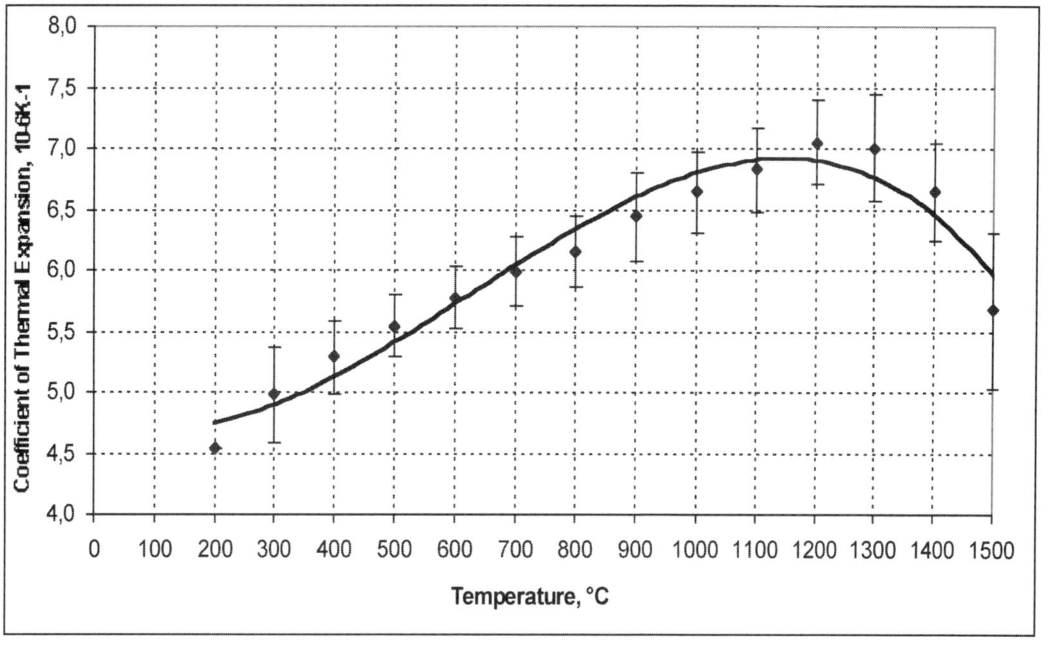

Figure 28.11 Thermal Expansion of High-alumina Tile Material

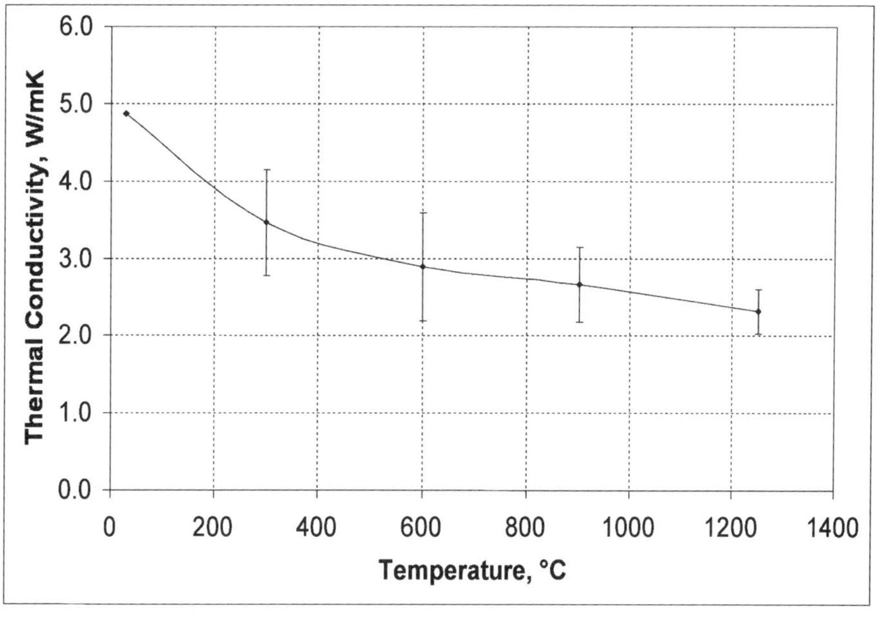

Figure 28.12 Thermal Conductivity of High-alumina Tile Material

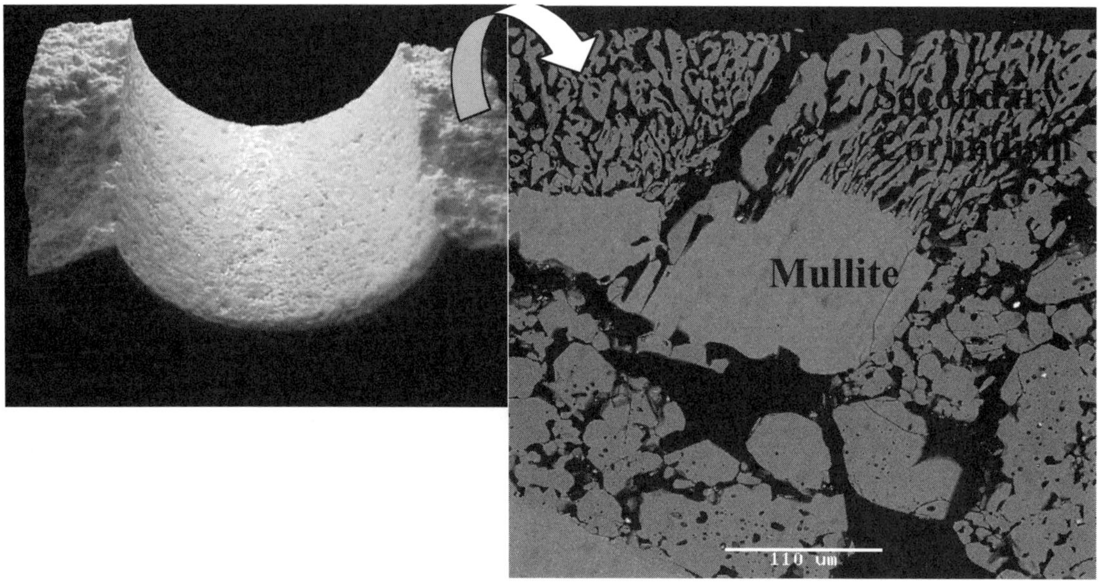

Figure 28.13 Cross Section of a Refractory Containing Alumina and Mullite After Hot Gas Corrosion Test at 1450°C for 100h

decrease in the coefficient of thermal expansion. The thermal conductivity λ of the high-alumina tile was measured using the heated wire technique defined in EN 993-14. Thermal conductivity behavior is similar to λ~1/T, which is reported to be typical for high-alumina refractories at atmospheric pressure [5].

Hot Gas Corrosion. The hot gas corrosion tests are used to simulate corrosion conditions in a gas turbine. It was determined that the thermo-mechanical data typically collected were not sufficiently specific to describe the real behavior of prospective tile materials in gas turbine atmospheres. Consequently, combustion tests were performed in fast flowing combustion gas of up to 1550°C (1823K), at a total pressure of about 5 bar and with water vapor partial pressures of up to 1 bar. These tests very clearly reveal the corrosive effects on test surfaces. Figure 28.13 shows a sample part and the cross section of a typical alumina-corundum refractory material after such a combustion test at a temperature level of 1450°C (1723K). All materials containing mullite show a similar behavior: the mullite decomposes and transforms to a secondary, fine-grained corundum, growing from the former mullite grains. These are no corrosion-affected areas of alumina.

These effects cannot be understood by thermodynamic modeling equilibrium calculations—they are not suitable to describe the high velocity, high pressure flow conditions of the gas turbine combustion atmosphere and to yield a prediction of the reactions under real gas turbine conditions. The corrosion mechanisms of the investigated alumina-mullite mixture are currently not fully understood, however, the effects described are experimentally recognized and confirmed as the principal corrosive effects of this class of high temperature refractories caused by combustion gas.

Field Experience

Damage Tolerance. A key property for determining the suitability of a potential combustion chamber tile material is its damage tolerance. This can be determined by mechanical degradation of the material in

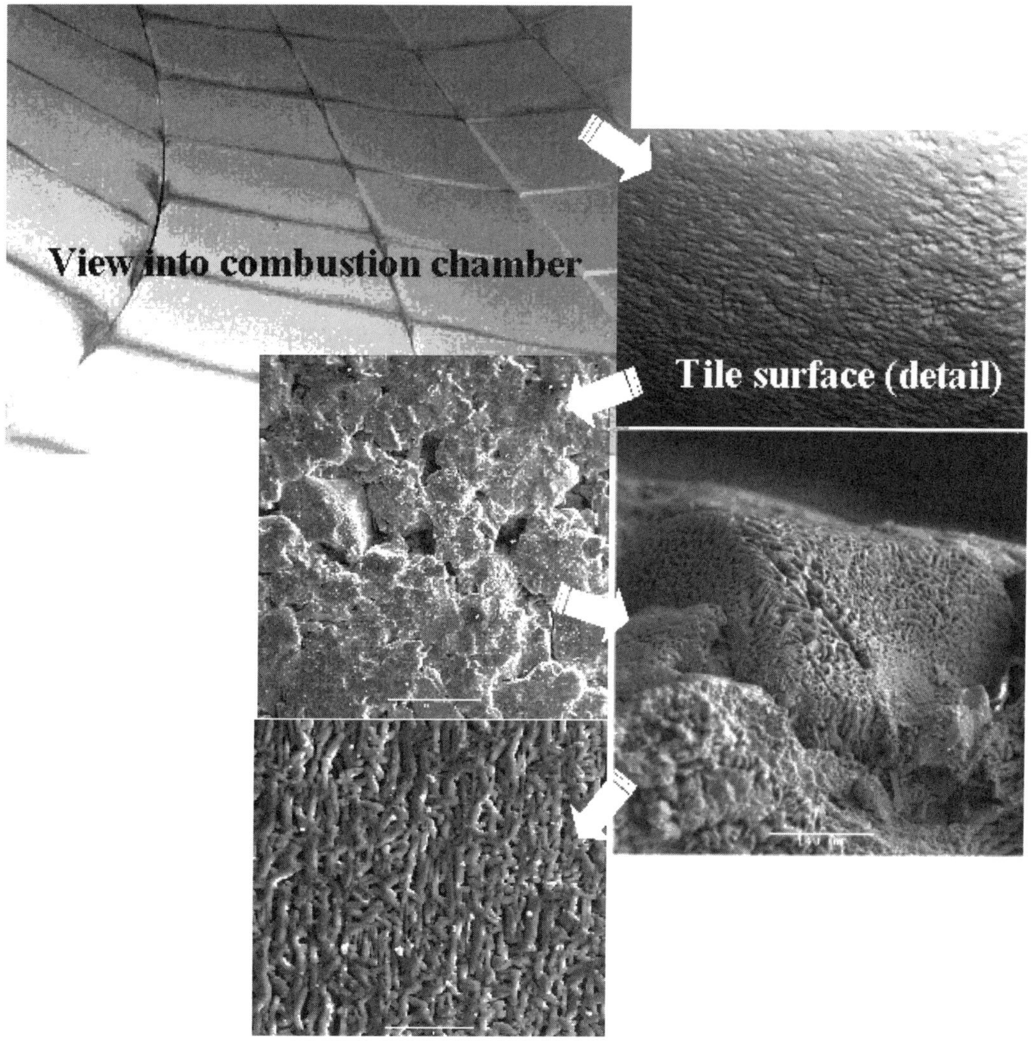

Figure 28.14 Surface of High-alumina Refractory Combustion Chamber tile (Flame Side) After Service

thermal shock tests. The higher the residual strength or residual modulus of elasticity after the initial thermal shocks, the higher the damage tolerance under actual service loads. The damage tolerance of the used high-alumina tiles is caused by two factors: one is a low fracture energy due to the low material strength and the other is the fracture energy consuming effect of reinforcing mechanisms (friction of rough crack surfaces, pull-out of large grains, crack-deflection). These effects cause slow crack growth and slight strength degradation after thermal shock.

The reinforcing mechanisms decline during thermal cycling. Thermally-induced, repeated opening and closing of cracks causes the loss of particles from the crack flanks and consequently a decreased roughness of the opposing crack flank surfaces. As a result, the resistance against crack propagation also decreases. Dislodged particles can collect near the crack tip and if this occurs, these particles can cause a wedge effect that produces tensile stresses at the crack tip when the crack closes and thus contribute to further crack propagation.

Corrosion. The desirable properties of the high-alumina tiles include high resistance to thermal degradation, hot gas corrosion and erosion. In principle high-alumina materials are among the most stable well-known refractories. Water vapor is the only strongly corrosive constituent of the combustion gas. The reaction with water vapor at high temperatures, high pressures and, as in this case, high gas flow velocities is a fundamental problem for all kinds of materials.

Figure 28.14 shows a combustion chamber lined with high-alumina tiles which have been in long-term service. Additional images of the tile surfaces exposed to hot gas show details at various magnification levels. Similar to the findings after the hot gas corrosion tests, the surfaces of the tiles subjected to service loadings have a shiny appearance. The small surface depressions seen here are particularly numerous in the regions which were exposed to the highest temperatures.

Typical microstructural changes involve the decomposition of mullite as described above. Areas affected in this manner have some loose surface material. Small particles can therefore be lost during thermal cycling. Furthermore, the thermal mismatch between the grains of the various phases and/or between the surface regions where corrosion damage has occurred and the unaffected subsurface tile volume causes spalling of small particles during thermal cycling. Both of these mechanisms cause a loss of material which is dependent on time, temperature and the number of thermal cycles and initiates the growth of small surface depressions. This is a very slight effect that leads to a thickness reduction of not more than a few millimeters after several thousand hours of operation.

Dark brown discolorations can be seen near the tile edges (see Figure 28.14). These discolorations are due to cooling air containing iron oxide from upstream regions, for example from the compressor. The brown iron oxide deposits can be found near the "cold" edges of the tiles where the cooling air impinges with the tile surfaces. The temperature of these areas is not higher than about 1000°C (1273K). No significant amounts of iron oxide were detected in the near-surface regions of the tiles. According to [6] high-alumina refractories are able to absorb large amounts of iron oxide under oxidizing conditions without the appearance of a liquid phase even at temperatures of up to 1470°C (1743K).

Cracking. During steady-state operation the ceramic tiles arch slightly on the hot gas side because of the steady-state thermal gradients over the tile thickness. The surface facing the flames expands more than the cool side of the tiles and the uncooled center of the tile expands more than the cooled edges. This results in convex deformation of the tile on the side exposed to the hot gas.

Crack initiation and propagation in ceramic tiles occurs because of thermal or combined thermal and mechanical loadings during gas turbine operation.

Thermo-mechanical Cracking. Thermo-mechanical cracking occurs during steady-state as well as transient operating conditions.

Steady-state conditions: The thermal gradients between the edges and the central surface region on the flame side of the tiles cause thermally-induced tensile stresses at the tile edges. An ideal temperature distribution with uniform temperatures around the tile periphery causes only localized stress peaks near the gaps. The central surface region on the flame side can be considered stress-free. When the stresses near the edges exceed the material's strength, cracks form and grow at the locations with the highest stresses. Crack origins are situated at or near the edges on the flame side (Figure 28.15). These cracks initially grow on the flame side, later the crack surfaces extend around to the flat or grooved sides (see Figure 28.2), perhaps as far as the back side of the tile. When the crack extends beyond the stressed areas on the flame side there is no longer a driving force for further crack growth. Crack growth then stops and the cracks are in a static state.

Transient conditions: This term is used to denote the conditions prevailing during startup, shutdown or gas turbine trip (see Figure. 28.7). The flame side cools down rapidly while the subsurface tile material is still at a high temperature level. This causes tensile stresses on the flame side and compressive stresses in the hot subsurface volume. When the material's strength is exceeded, crack initiation and crack growth occur on the flame side. Crack initiation will occur as early as the very first thermal cycles of gas turbine operation

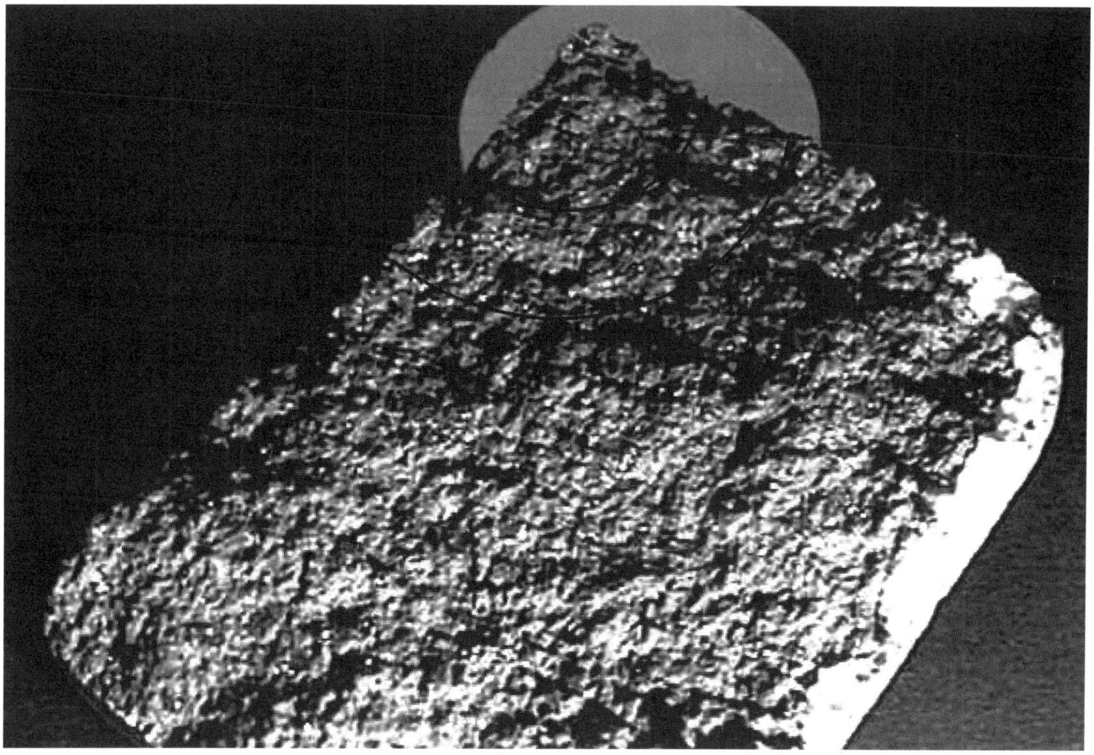

Figure 28.15 Crack Surface with Crack Origin Indicated at the Top

and crack growth occurs at a declining rate. This phenomenon is similar to the thermal shock behavior of the tiles. The origins of these cracks are on the flame side.

These thermo-mechanically-induced cracks (Figure 28.16) grow very slowly and can be evaluated during regular inspections. No extra effort is necessary for such inspections — the crack length can be evaluated with sufficient reliability by inspecting the flame side of the tiles as the greatest crack length is on this side.

Dynamic Mechanical Cracking. Dynamic or mechanically-induced cracking occurs during isolated, off-normal gas turbine operating conditions.

Gas turbine parts are subjected to oscillations. These oscillations are caused by mechanical, thermal or acoustical effects and may increase during isolated, off-normal operating conditions. Such states may lead to severe accelerations at specific frequencies (up to about 300Hz) in the combustion chamber. The system "ceramic tile-metallic clip-metallic casing wall" must be capable of withstanding such pronounced short-term oscillations. When the acceleration forces at the tiles exceed the tile holder forces the tiles lift off from the wall and rebound with a high relative velocity. On impact of the slightly bell-shaped tile against the metallic casing wall, tensile stresses occur at the central region of the edges on the back side. Crack initiation and crack growth occur, starting at the back side of the tiles, when the material's strength is exceeded.

Improvements

Various means are available for enhancing refractory tile performance. These include, but are not limited to:

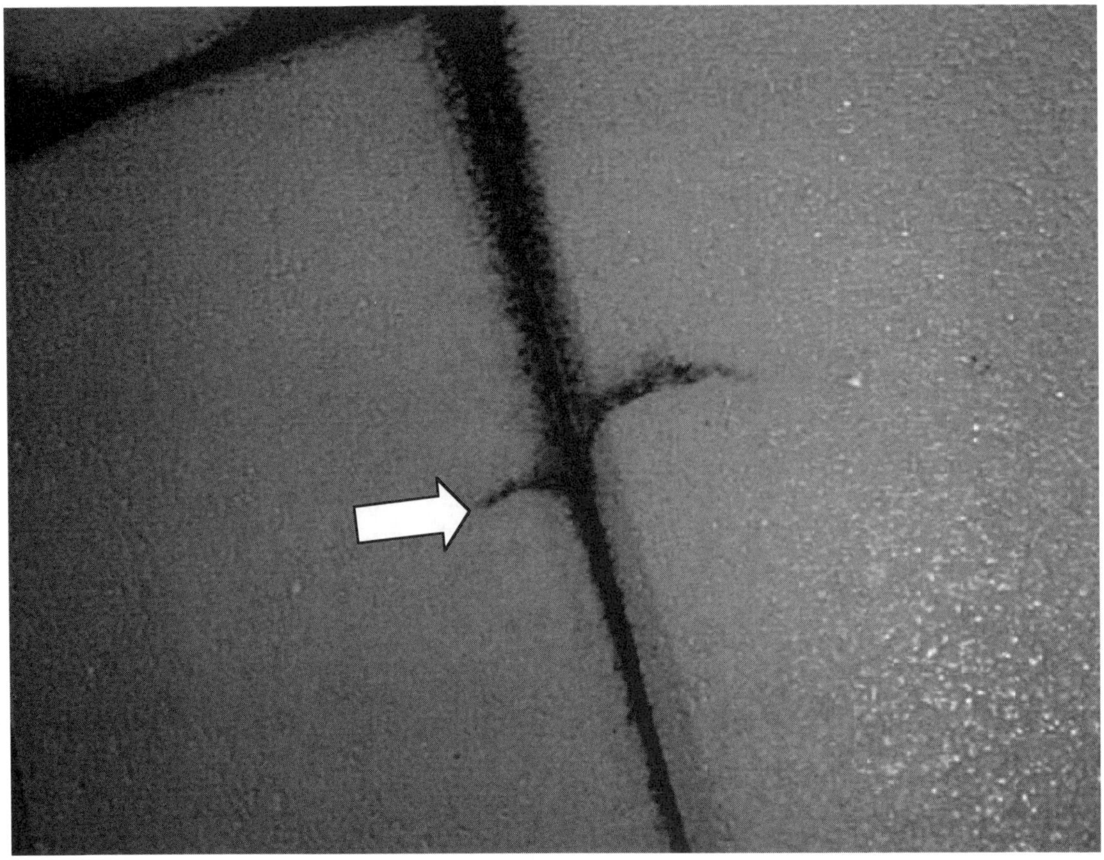

Figure 28.16 "Thermo-mechanical Cracks" Perpendicular to Tile Edges

- Optimization of the microstructure of the tiles for further improvement of damage tolerance,
- Optimization of tile and tile holder design,
- Optimization of composition and flame side surface structure for further improvement of corrosion resistance,
- Development of a quick, reliable, and portable non-destructive method for evaluating cracking in combustion chamber tiles.

These and many other activities contribute to further refinement of tile performance as discussed here to meet current and future demands.

APPLICATION BENEFITS

The use of high-alumina refractory materials in the gas turbine hot gas path involves various benefits:

- The high material service temperature reduces the amount of cooling air required. The only cooling air required in the ceramic lining of the combustion chamber is the air for the gaps between the high-alumina tiles. Cooling air no longer required for reducing material temperatures can be used for "cool" combustion. This technique decreases the flame temperature, thus enabling low-NO_x operation.
- The ceramic "hot wall" design of the combustion chamber enables an improved homogeneity of the temperature distribution in the combustion chamber. This contributes to realization of high-efficiency combustion.

- Service intervals of up to 25,000 operating hours can be realized with high-alumina refractory materials.
- Individual tiles can be replaced without disassembling the entire combustion chamber. Compared to metallic combustion chamber linings this concept has both operational and commercial advantages.
- The specific costs of refractory-lined combustion chamber walls amount to only a small percentage of the costs of metallic-lined combustion chamber walls.

CONCLUSIONS AND OUTLOOK

This paper presents operating experience with refractory ceramics in stationary gas turbine combustion chambers. It was demonstrated that a refractory material is capable of meeting the requirements imposed by applications such as combustion chamber wall linings in modern gas turbines. Discussion also includes the principal material properties that must be investigated for evaluation of ceramic materials under consideration for such applications.

Field experience gathered to date indicates that the capacity of a prospective ceramic material to withstand gas turbine operating conditions cannot be simulated by simple tests alone: the combination of combustion atomsphere, high heat fluxes/high gas flow velocities, high temperature, high pressure and many thermal cycles at various aging states over extremely long terms affects the materials in a manner that cannot be simulated. Realistic prediction of performance under the specified conditions requires a carefully-defined combination of short-and long-term experimental testing using finite element analyses of the anticipated mechanical and thermo-mechanical stresses.

A key property in this context is thermal shock resistance. To demonstrate adequate thermal shock resistance, a material must be subjected to specially adapted thermo-mechanical characterizations. It is absolutely imperative that tests involving conditions approximating gas turbine operation be performed before a prospective material can be considered to have true potential for use in gas turbine applications.

The high-alumina refractory discussed in this paper exhibits the potential for meeting the current demands for mechanical and thermo-mechanical stability, reliability and long-term service stability under most service conditions. It must be pointed out, however, that further potential for improvement in terms of damage tolerance and corrosion resistance has been recognized.

Other material-specific issues that must be still investigated in greater detail include: the mechanisms of thermal and thermo-mechanical fatigue (high-cycle/high-temperature) and of subcritical crack-growth. Based on this knowledge it will be possible to guarantee highly-reliable, long-term service of ceramic combustion chamber linings.

Corrosion resistance should be improved further by physically separating the mullite phase from the combustion gas or by replacing the mullite phase by a more corrosion-resistant phase.

One final key aspect is ensuring a constant raw material composition, thus narrowing this scatter band of the manufacturing processes. This is essential for producing ceramic components with reliable and stable service behavior. Trouble-free operation with ceramic components requires development and implementation of a sophisticated and comprehensive quality control and product evaluation process. This also applies to the production and inspection of ceramic components and their transport, handling and installation at the plant.

ACKNOWLEDGEMENTS

The authors would like to thank their colleagues Mr. Schmahl, Mr. Maghon and Mr. Jeppel (all Siemens KWU Mülheim) for many fruitful discussions and their assistance in all areas of their work and Dr. Kollenberg of Werkstoffzentrum Rheinbach GmbH for his meticulous analytical work and stimulating discussions on refractory materials problems. Valuable suggestions and support received from our colleagues Dr. Rettig, Mr. Oechsner and Dr. Bast (all Siemens ZT) are also gratefully acknowledged. All figures and data tables in this chapter were supplied by Siemens KWU.

REFERENCES

[1] Litovsky, E., Gambaryan-Roisman, T., Shapiro, M., and Shavit, A., "Effect of Grain Thermal Expansion Mismatch on Thermal Conductivity of Porous Ceramics," J. Am. Ceram. Soc. 82[4], 994–1000, 1999.

[2] Peretz, I. and Bradt, R.C., "Andalusite derived mullite matrix refractories," Ceram. Trans. 6, 613–633, 1990.

[3] Tripathi, H.S., Swapan, K.D., and Banerjee, G., "Thermal shock behaviour of high-alumina aggregates derived from sillimanite beach sand with and without Fe_2O_3 doping," Ceram. Internat., 26[1], 1–6, 2000.

[4] Gebauer, J. and Hasselman, D.P.H., "Elastic-Plastic Phenomena in the Strength Behavior of an Aluminosilicate Ceramic Subjected to Thermal Shock," J. Am. Ceram. Soc. 54[9], 468–469, 1971.

[5] Litovsky, E., Ya., Gambaryan-Roisman, T., Shapiro, M., and Shavit, A., "Heat Transfer Mechanisms Governing Thermal Conductivity of Porous Ceramic Materials," in *Trends in Heat, Mass and Momentum Transfer*. pp. 147–167. Research Trends, India, 1997.

[6] Muan, A., "Reactions Between Iron Oxides and Alumina-Silica Refractories," J. Am. Ceram. Soc. 41[8] 275–286, 1958.

CHAPTER

29

DEVELOPMENT OF CERAMIC COMPONENTS FOR UTILITY GAS TURBINES

Axel Kranzmann

Alstom Power Support GmbH, Mannheim, Germany

INTRODUCTION

The R&D group of Asea Brown Boveri (ABB) in Heidelberg has conducted ceramic component development over the past 15 years. Products which originated from this work were an oxygen sensor for combustion engines, the NaS-Battery, an AlN cooler working with water for high current traction electronics, protection tube arrangements for ash vitrification, thermal barrier coatings (up to 1.8 mm thickness) for turbine engines, ceramic coatings on metal-fiber systems for catalysis in the oil & gas business. Nevertheless, the total value of ceramic products is still low in comparison to Ni-base superalloys or steel. Table 29.1 lists the major ceramic activities at ABB since 1965. The major activities show the current research experience.

The main target of ABB's material evaluations was and is the introduction of beneficial components within a design period for a product. The continuing efforts include design tools and development of ceramic systems for gas turbines.

ABB concentrated its efforts in the energy technology on high temperature systems. Regarding the existing TBC technology any new development has to beat the 1300°C surface temperature that is available from TBC technology. The same temperature with just less cooling air can not reach the economical benefit. Thus, 1400°C surface temperature and higher was the goal for the heavy duty gas turbine development.

Within a 500 hour benchmark experiment it was demonstrated that only SiC based materials have potential for uses at 1500°C. Any nitride based ceramic was not able to demonstrate reliability of properties in this benchmark experiment. We followed the rule that a loss of properties at high temperature visible after 500°C disqualified the material. SiC, alumina, and silicon nitride based materials, available in 1993, were tested.

In the design and material R&D work the ABB group worked within boundary conditions which included:

- commercial benefit compared with current superalloy/TBC systems
- co-operation with established manufacturers and use of their material
- probabilistic design tools to reach a minimum failure probability and life time pre-assessment
- accept loss of material by oxidation within a defined limit

TABLE 29.1 THE MAJOR CERAMIC R&D ACTIVITIES AT ABB

Time	Research
1983–1990	Federal Department of education and research, BMB+F, funded project on "Production of homogeneous ceramic"
1988–1992	AMTEC (Alkali Metal Thermal-Electric Converter)
1965–1996	Zirconia oxygen sensor for combustion systems.
1965–1994	Solid oxide fuel cell
1975–1993	Sodium-Sulfur – Battery
1994–1997	AlN cooler for water cooled traction systems
1997	Observations on CMC development
1992–1997	Ceramic combustion chamber heat shield
1995–1998	Ceramic protection tubes for waste combustion ash glasifiers
1994–present	Development of design tools for brittle materials
1991–present	Thick ceramic TBC

- design of a system which offers residual life after impact damage
- use of nondestructive evaluation methods to inspect gas turbine parts
- test of specimens and full scale components
- develop attachment techniques

The main conclusions were:

- Oxidation under gas turbine conditions is crucial for all non-oxide ceramics
- The impact damage failure probability is high because ceramic component pieces destroyed additional components accelerated by the highly turbulent gas in the combustion chamber. A safeguarding system that limits damage after a primary impact needs to be incorporated.
- The loss of mechanical properties justified only designs loosing a minimum of the virgin tensile strength of the material.

PRE-EXISTING EXPERIENCE

The Brown Boveri & Cie. (BBC), and later on the ABB research center Heidelberg were involved in several activities to improve the reliability of ceramic components. Mainly the production of homogeneous ceramic microstructures and the mechanical reliability were investigated. Later brazing technologics and design methods were added.

High Temperature Direct Electricity Converter—AMTEC

The alkali metal thermo-electric converter (AMTEC) is a heat engine which converts heat directly into electricity [1]. In our working group we developed electrodes, metal/ceramic joining, and we designed and operated AMTEC cells containing the ion conducting beta-alumina ceramic tubes, and designed power plant cycles with integrated AMTEC technology. Figure 29.1 shows the brazing geometry between a beta-alumina tube and stainless steel tube. Niobium interconnections were used to overcome the thermal mismatch between the ceramic and the steel. Figure 29.2 gives a view of a beta-alumina test cell in a glass test reactor. This arrangement worked without failure for about 1000 hours at 800°C in direct contact with sodium [2].

The work on the AMTEC module taught the team the importance of the reaction zone between metals and ceramics and gave experience with high temperature brazing. The beta-alumina tubes are of the

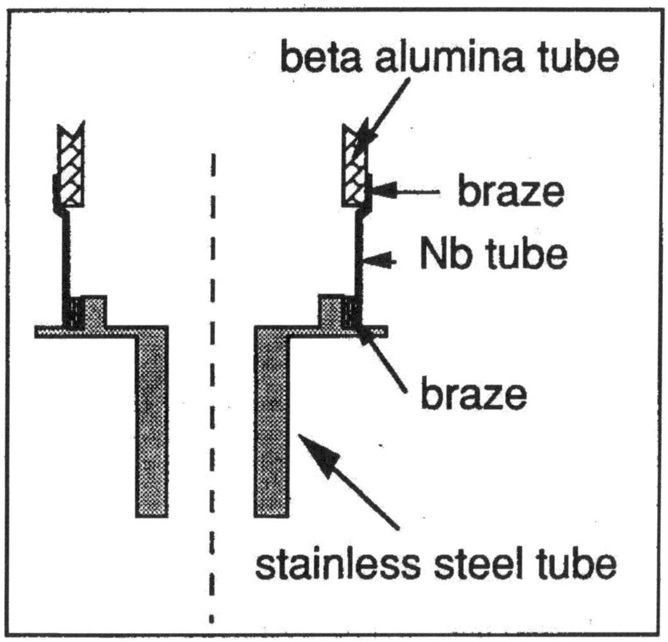

Figure 29.1 Sketch of the Brazing Geometry Between Beta-alumina Tubes and Stainless Steel Tubes (Reprinted with permission from Ref. [2])

same material that was used for the high energy sodium-sulfur battery. It was also concluded that the Ti content of active brazes should not exceed 1 to 2 at. % to avoid further reactions at the interface under working conditions.

Experience From High Energy Battery Development

An early activity of the BBC group was the development work for the sodium-sulfur battery. This battery is still under further development in Japan after ABB sold all knowledge and rights to NGK Insulators, Ltd., Japan. Techniques for the consolidation of ceramic powders and the production of a high number of tubes that were closed on one end were developed. Weibull analysis of the strength distribution and defect analysis tools and experience were developed. One of the battery quality control tools was the exact description of the defect population caused by the production route. It was possible to identify the relation between different production steps and microstructural defects and consequently, the reliability of the ceramic component. Figure 29.3 shows the mass production of beta-alumina tubes for the high temperature battery and the AMTEC project before 1993.

Other Developments—Aluminum Nitride, Silicon Nitride, Oxide Ceramics

AlN itself is not a structural ceramic, when the strength and Weibull modulus are considered, and it is limited to uses below 1000°C, due to its low oxidation resistance. But a lot of work was done on brazing and metallizing the material. The brazing of heater elements was only possible by understanding the fatigue behavior of layered structures prepared from different materials. A heater element of the type shown in

Figure 29.2 View of a Beta-alumina Test Cell in a Glass Test Reactor

Figure 29.4 survived more than 2000 cycles between 350°C and –50°C. From this work it became evident that also in the case of non-oxide materials an active brazing alloy should not contain more than 2 at. % Ti or less. Otherwise the formation of a thick reaction layer (> 10 μm), containing TiAl-phases, would reduce the fatigue life of joints with a high surface area.

Figure 29.4 shows a mini-heater with an edge length of 10 mm and including 3 different materials with different functions. This heater was developed by the author and applied in a Fourier thermal conductivity test equipment. The top layer is a 5% porous alumina with good thermal insulation. The central dark layer is a dense aluminum nitride with high thermal conductivity. This layer includes a number of holes drilled by ultrasonic drilling. The NiCr-alloy heating wire was passed through the holes. The final layer was a copper plate to guarantee a ductile contact between the specimen surface and the heater. The maximum output was 40 Watt. All layers were brazed in one step using Ti containing brazing alloys.

Figure 29.3 Mass Production of Sodium-beta-aluminate Tube Cells Developed at BBC

Figure 29.4 Mini-heater with Aluminum Nitride Elements

TABLE 29.2 TECHNICAL GOALS AND BOUNDARY CONDITIONS

	Combustor	Vane	Blade
Temperature	1400–1500°C	1400°C	1300°C
Stress	< 100 MPa	< 150 MPa	< 200 MPa
Life	> 48,000 hours		
Failure probability during trip conditions	$< 10^{-9}$	$< 10^{-9}$	$< 10^{-9}$
No. of thermal shocks without failure required (gas cooling rate: 900 K/s)	> 500	>500	>500

Silicon Nitride shows excellent oxidation behavior in air up to 1250°C but cannot be used to work for more than 1000 hours at higher temperatures. Also, creep and the rapid deterioration of the mechanical properties are limiting the use of silicon nitride at temperatures in excess of 1250°C.

Oxide ceramics generally show creep, low thermal shock resistance, crack growth and the influence of water on their fatigue behavior [3–5].

TURBINE COMPONENT DEVELOPMENT

Motivated by the work done for smaller gas turbines, e.g. the results on can combustors, (Hitachi and others in Japan), or the small combustors (MTU and Daimler-Benz in Germany), it was decided in the R&D team to develop a bolt/tile concept for a gas turbine combustion chamber lining. Different material concepts such as ceramic foams, filled and reinforced reticular open porous ceramic, sodium-phosphate based zero-expansion materials, and functional gradients were also discussed to design a liner system. A number of qualification tests including thermal shocks in a high temperature gas flow rig were used to select proper materials.

- Alumina foams and reticular ceramics exhibited tremendous thermal shock resistance but the bending strength is below 100 MPa. The gas temperature cooling rate in the thermal shock test was up to 900 K/s. The observed failures after only 60 thermal shock cycles indicated that the requirements were not met (see Table 29.2).
- Filled and reinforced reticular alumina materials became quite heavy. The approach used was very similar to the Russian heat shield system of the Buran space shuttle. This added load to the attachment design. Also, the availability of the material was critical. And in-house production of material was rejected and an external vendor was not identified. Therefore this route was not accepted.
- Zero-expansion materials on the basis of sodium-phosphate alloys were considered to be only useful up to 1200 or 1300°C. Beyond this temperature limit their long term chemical stability was uncertain. Consequently, they were not included in the program.
- Functional porosity gradient materials as a thin layer to substitute for Zirconia-TBC's are still a discussion point in the technical literature. Producing massive combustor parts was considered to be too far away from the target schedule as planned in 1993.
- Commercial silicon nitrides and -carbides passed the thermal shock tests and qualified for a second round of qualification tests.

Boundary Conditions and Material Requirements

The target components under discussion for small gas turbines were vanes, blades and combustion chamber parts.

In case of the utility machines offering 160 to 240 MW power, the ABB annular combustion chamber outer diameter is nearly 4 m, and the sizes of the smallest vanes and blades exceed 20 cm. Because of the size of the engine the surface of the combustor can only be covered by individual tiles. The discussion about different components consequently resulted in potential development goals. A thermal protection system for the combustor, and composite blades engineered from ceramic and metal sub-elements seemed to be of greatest interest.

The final costs of ceramic solutions were estimated using a list of factors concerning gains and losses:

- (–) cost of ceramic part and cost of corresponding metallic part.
- (+/–) lifetime ceramic part versus lifetime metallic + TBC solution.
- (+) savings of cooling air, including savings in the compressor and efficiency increase.
- (+) Maintainability as a goal.
- (+) Service improvements as a goal.

Each factor can be expressed in monetary terms. The sum of all gains and losses was used to calculate the project economics. After completion of the risk analysis it was decided to develop a complete combustor protection system.

The working temperature was defined to be up to 1500°C in flowing combustion gas containing O_2, N_2, H_2O, CO_2, and NOx. Table 29.2 gives a summary of the technical goals and the expected boundary conditions (loads on the components and the atmosphere). The failure probability has to be defined using a system RAMS analysis. The higher stresses for vanes and blades disqualified both for further evaluation.

From Table 29.2 it becomes obvious that the combustion chamber is a good first target in terms of mechanical load. Thermo-mechanical tensile loads higher than 100 MPa were judged to be unrealistic for ceramics. A hard goal was defined for the failure probability. This value should be as low as possible and below 10^{-9}.

Candidate Materials

A number of commercially available materials and R&D activities and the connected vendor companies were screened. It is obvious that in the high temperature combustor regions an oxide such as zirconia, mullite or alumina would be attractive in terms of chemical stability. But as mentioned above in terms of strength and thermal-mechanical fatigue, silicon carbide is superior to these materials. Silicon nitride lacks in oxidation resistance and creep resistance. Finally, alumina-based materials, silicon nitride alloys and pressureless sintered silicon carbide were judged to be the most promising candidates.

The spider diagram in Figure 29.5 shows the comparison between alumina, SiC and Si_3N_4, and the target values at the outer circle. The potential for thermal-mechanical fatigue resistance is described by a small product between the coefficient of thermal expansion (α) and Young's modulus (E), and a high thermal conductivity (κ). To survive thermal stresses the strength (σ) at the component service temperature should be high. The oxidation resistance, described by the parabolic rate constant K_p, was targeted to be smaller than 10^{-8} $g^2m^{-4}s^{-1}$. A small coefficient of thermal expansion is positive for thermal cycling stresses, but increases the difficulty for joining the ceramic to a metallic counterpart or reduces joining to mechanical fixations. The comparison among 4 technical ceramics shows that no material offers a balanced spectrum of properties (Figure 29.5). The nearer the property to the outer circle, the better. Oxide materials show very good stability at high temperature and even the thermal cycling stability can be improved using foam structures to reduce thermal transient stresses but for strength, thermal conductivity and thermal expansion non-oxides are clearly more effective for accommodating transient loads.

Alumina alloys and alumina materials did not survive any thermal shock test near gas turbine conditions. These materials will become useful as soon as the materials engineering community offers reliable high temperature strengthening such as fiber reinforcement.

The chosen silicon nitride and silicon carbide materials were compared in a simple test. From an ordered batch of bending test rods (50×4×5 mm) two half batches were formed. One sub-batch was strength tested at 1500°C. The other half batch was aged for 500 hours at 1500°C in laboratory air, and after this treat-

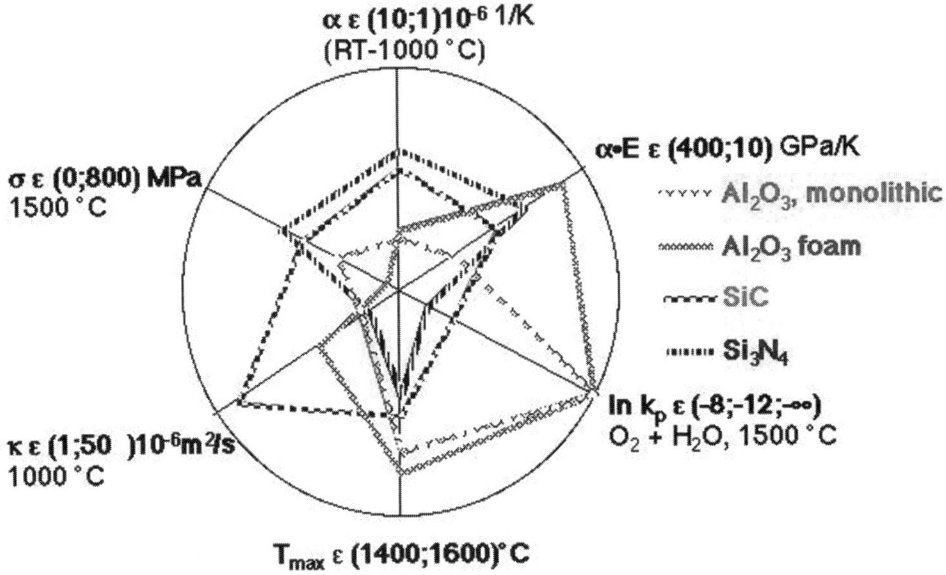

Figure 29.5 Spider Diagram with Properties of Four Different Ceramics

ment strength tested under the same conditions as sub-batch #1. Same condition means: same holder, same engineer, same machine, same crosshead speed, same furnace. This test targeted high temperature fatigue. Figure 29.6 illustrates the test scheme for benchmarking high temperature engineering ceramics.

To get a quick overview of the commercial available materials ABB used a comparison between aged and non-aged specimens from the same batch. 100 Samples were separated using a random selection. The aged and the non-aged batch were tested in 4-point bending (40/20 mm) at 1500 °C in air. Clearly a loss of strength was observed for all nitrides and sintered SiC (SSiC) doped with aluminum carbide.

The result clearly indicated that there was no Si_3N_4 material available on the commercial market that did not show a strength loss of more than 25% between sub-batch #2 and sub-batch #1. It was not exactly clear which mechanisms were responsible. The initial higher strength level of Si_3N_4 materials compared with SiC was completely lost. No indication of the limitation of this loss was observable. This clearly made SiC preferable over Si_3N_4 for the design of combustor heat shield elements. It became obvious that only boron carbide doped SiC shows the expected stability. The result of all tests in 1994 was that Carborundum Hexoloy SA was the most promising material in terms of oxidation resistance and mechanical reliability. Environmental effects on the reliability of silicon nitride and silicon carbide materials were known before and our result confirmed some already published observations [4–6].

COMPONENT DEVELOPMENT

The component development concentrated on a heat shield package that could withstand 1500°C combustion gas temperature. The main goals were:

- a ceramic tile with high reliability
- a functional and durable attachment
- a thermal insulation with a lifetime of more than 6000 hours without a protective tile.

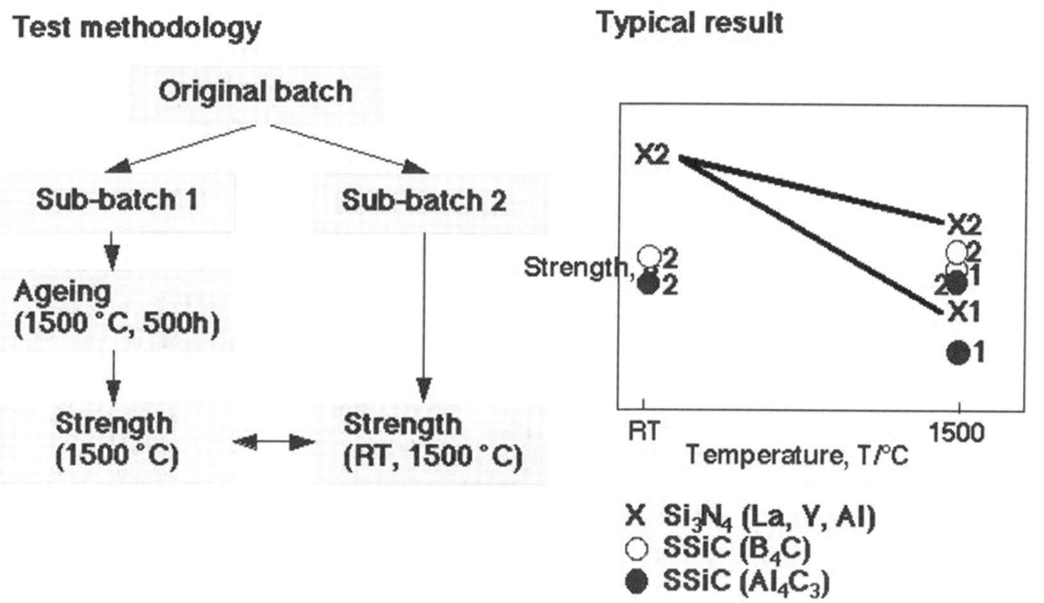

Figure 29.6 Scheme of the High Temperature Benchmark Tests

Ceramic/Metal Integration

In a can combustion chamber, such as that preferred by engines from General Electric Corporation, or in a small gas turbine complete ceramic cans or rings can be considered and clamped into the metallic support. Such a design has been working in an experimental burner rig in our combustion laboratory in Baden-Dättwil without any failure since 1995. In larger gas turbines individual ceramic parts have to be mechanically attached to the metallic outer combustion chamber. A bolt/tile design was developed to attach the ceramic tiles onto a metallic backbone. The bolt was brazed or mechanically attached to a metallic part at the cold end. The positive experience with the brazed AMTEC Ni-β-Aluminate connection gave reason to develop a brazed connection.

The second possibility was the use of SiC/SiC ceramic matrix composite (CMC) and metal screws. These SiC/SiC screws are oxidation resistant up to 1000°C on the inside. The contact area between ceramic and metal never exceeded 600°C. A Ni-superalloy/CMC screw will never join to the tile by oxidation or diffusion. This kind of connection can be opened and fastened several times and is competitive with pure metal screw connections if the maintenance costs are taken into account. IN617 and SSiC were brazed together (see Figure 29.7). Such bolts were tested in a combustion chamber test and in laboratory tests. They survived up to 10^7 mechanical load cycles up to a maximum load of 20 MPa shear stress at 1000°C (see also Figure 29.8) [7].

Figure 29.7 shows the brazed bolt with a direct connection between IN617 and SiC. The connection was tested in a cyclic fatigue test at room temperature and after 1000 hours aging at 1000°C. In the final combustor test the brazed volume was never heated beyond 700°C. No brazed bolt failed in the combustor test or any laboratory thermal shock test. The braze was a Ni-based high temperature braze which was developed at the ABB laboratory in Heidelberg. The braze contained no active element. Such brazes cannot wet a SiC surface. To get a wetting effect the SiC surface was coated with a WC/Co/Mo alloy. This coating adhered very well to SiC, and was excellently wetted by Ni-based high temperature brazes. Figure 29.8 shows the

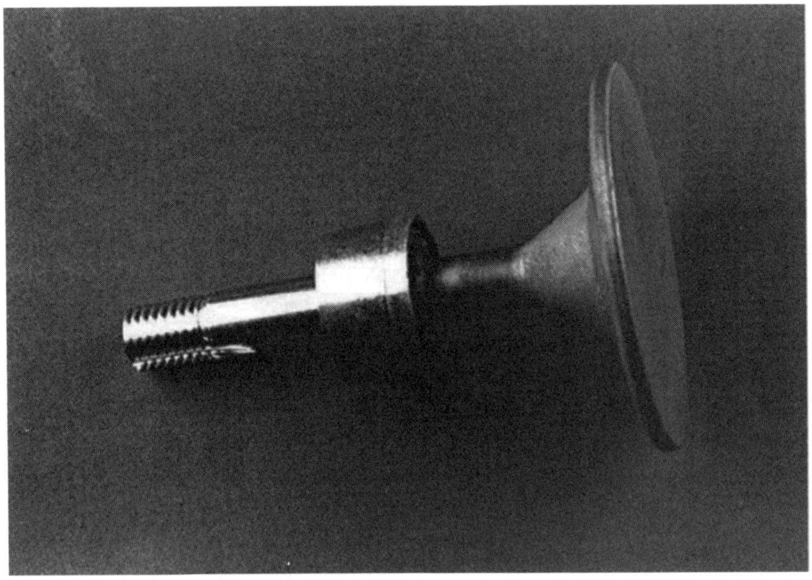

Figure 29.7 Brazed Bolt to Attach the Ceramic Heat Shields to a Metallic Structure

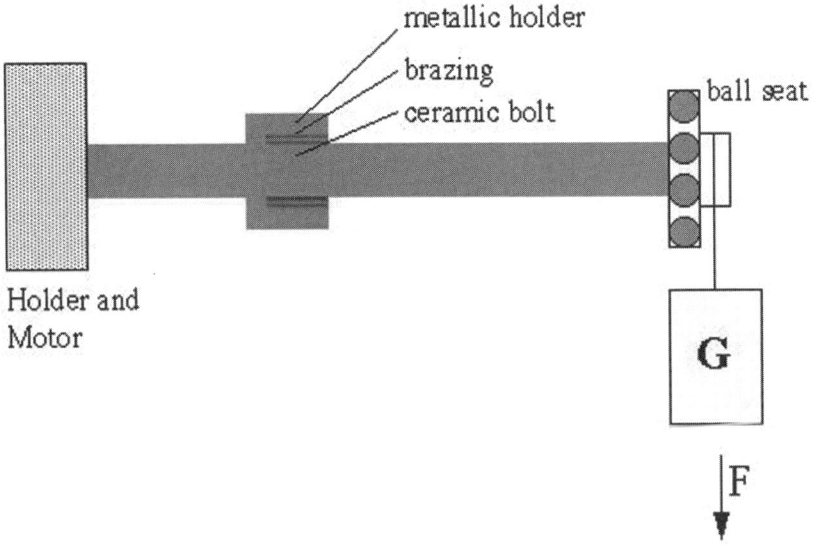

Figure 29.8 Test Set-up for Brazed Bolts

principal test arrangement for the brazed bolts. Bolts of the type shown in Figure 29.7 were tested using a circular shear test. The minumum life of the bolts was defined as 10^7 load cycles.

It became evident that the manufacturing of brazed bolts has to be highly controlled and careful component handling is needed. Otherwise, overloading of the metal-ceramic bolt during installation is very probable. The benefit of this design is that the metallic cold end offers the use of a simple attachment to the

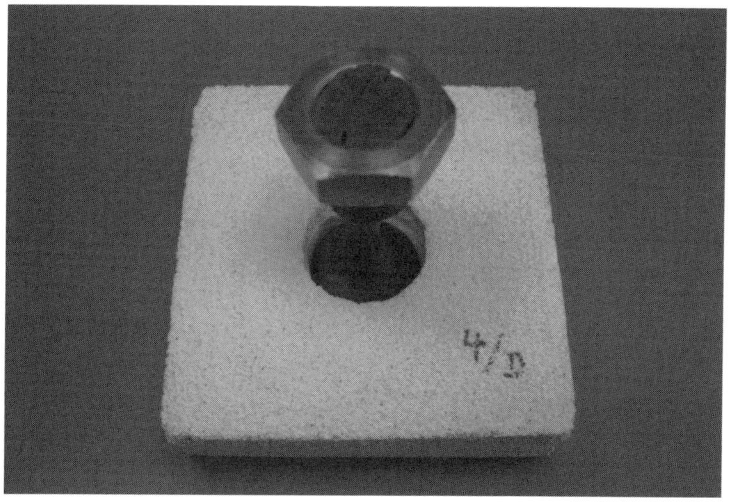

Figure 29.9 Metal/CMC/SiC Attachment Design

combustor housing. Also any known securing system against loss of the connection can be used. Consequently mechanical joining is preferred for the kind of design discussed here.

The metal/CMC/SiC attachment design shown in Figure 29.9 is a purely mechanical attachment. Consequently, it is very robust during handling and well suited for the installation process in an industrial environment. The drawback of this ceramic/metallic design was that it had no obvious possibility for securing the screw against opening. The C/SiC CMC was supplied by MAN Technologie in Munich. Figure 29.9 shows a SSiC bolt and an oxide foam heat shield for laboratory testing. Similar designs were used in a combustion chamber test. In this case not a brazed bolt fixation was tested but a SiC/SiC CMC holder. The holder has an outside thread that can be used to fix the bolt applying a steel nut.

Heat Shield Development

Two different heat shield designs were developed. A SiC tile using a fibrous thermal insulation and a foam concept were designed and compared in laboratory tests. Figure 29.10 and Table 29.3 compare

TABLE 29.3 COMPARISON BETWEEN THE TILE AND THE FOAM DESIGN

	Foam design	Shield design
Relative weight	1.8–2	1
Relative cost	1	2.5
Max. temperature, long term (°C)	1350–1400	1400–1500
Number of thermal shock and component material surface temperature changes of 35–40 K/s	50–60	more than 500
Tensile strength of surface layer (MPa)	20	300
Max. tensile surface stresses of the best design (FE simulation correlated with the observed material cooling rates)	Unknown, because material was not modeled	125
Oxidation damage at 1500°C in combustion atmosphere	no	life restriction to 20,000 hours
Paradigm change of designers in terms of material strength and design rules	very much	Much

Figure 29.10 Illustration of the Parts of the Component in Figure 29.9

both designs. Figure 29.10 shows the parts of the component in Figure 29.9. The CMC half shells were produced by MAN Technologie, Munich. In the laboratory test they survived for more than 5000 hours at a maximum temperature of 650°C and 10^9 mechanical load cycles with a maximum of 30 MPa tensile stress in the bolt shaft.

The ceramic SiC tile was designed applying the principle of minimal probability of failure under trip conditions. The failure probability under trip conditions after more than 4000 hours working time was calculated assuming 25% loss of material strength. STAU II, a modification of STAU, was used to model the failure probability at any time [8]. This calculation included the stress/temperature history and consequently the expected material degradation.

Enabling Design Technologies

The most critical paradigm change is the prediction of failure probabilities with statistical input numbers after aging. The design has to be optimized towards the lowest probability of failure at any point during the duration of service. Therefore, we included oxidation corrosion damage and loss of mechanical strength and creep in our reliability model. Creep damage was not expected in the case of SiC parts and consequently we did not include a creep module into the calculation. Prof. Munz, Prof. Brückner-Foit and Dr. C. Ziegler working at the University of Karlsruhe and the Research Center Karlsruhe were our cooperation partners.

The idea was to get a time-dependent lifetime prediction method on the basis of the Weibull distribution. This time dependent distribution describes the initial Weibull distribution and reflects the change in failure probability with time. All changes in the material and the material properties are due to the aging of the material. Cyclic fatigue effects were discussed separately.

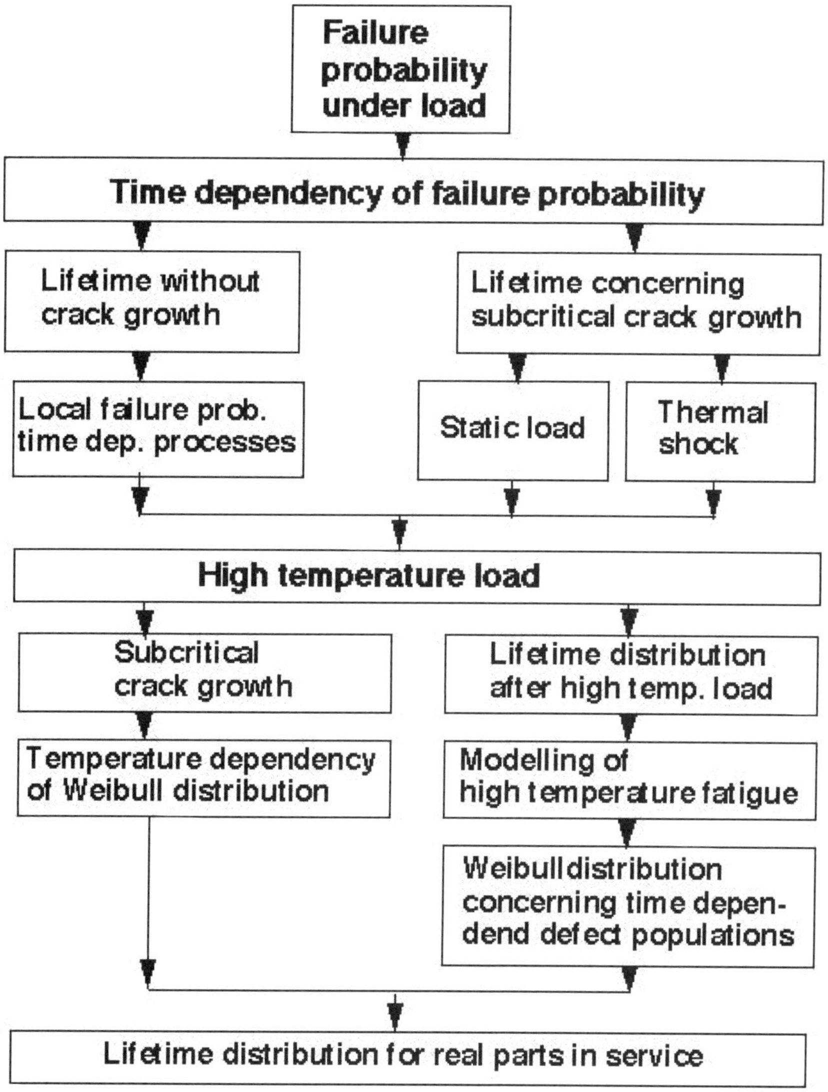

Figure 29.11 Lifetime Assessment Procedure (Adapted with permission from Ref. [8])

Simple analytical assumptions were made:

$$\sigma(t) = \sigma_0 (B - \exp(-A/t)) \qquad m(t) = m_0 (b - \exp(-a/t)) \qquad P_s(t,\sigma,V) = \exp((V(t)/V_0)\bullet(\sigma/\sigma_0(t))^{m(t)})$$

Simply, for each temperature the observed aging of mechanical properties was translated into a time-dependent Weibull distribution. Even the loaded volume will change if a significant part of the component transfers to oxides. We translated the working history into different aging conditions and integrated the complete working time to calculate the state of mechanical properties. Additionally, it was expected, that new defect populations will grow at high temperatures. Such new defect populations may increase the probability of failure.

The final model is summarized in Figure 29.11. The complete load history has to be integrated over time. Alternatively, this history has to be assumed in order to calculate a residual lifetime or the expected lifetime. The basic idea was to define one incident such as a trip and calculate the failure probability under this extreme condition. Under high temperature and the corrosive attack of the combustion gas all changes of the material can be expressed as time-dependent defect populations changing the reliability of the component. The different effects that we took in account are shown in this scheme.

The basic design approach was to minimize the failure probability for any working condition. The trip conditions were assumed and the failure probability during a trip after a history of n hours was calculated. The design with the lowest failure probability under the described conditions was assumed to be the best. Consequently, the end of life is defined by surpassing the upper failure probability limit.

Design Database

The material data base necessary for our design modeling tools for ceramic material application at high temperatures included the mechanical properties at different temperatures, long term oxidation degradation data, creep tests data, and a description of sub-critical crack growth. All these data were used to derive a time- and temperature-dependent Weibull formula describing the failure probability under a defined stress distribution.

The mechanical tests included four point bending tests, tensile tests and component tests. All tests were performed at room temperature and operating temperature. The difference in laboratory atmosphere and wet atmosphere was investigated. The aging tests included atmospheric oxidation and the influence of water and sodium on oxidation. Test data are shown in Table 29.4.

After 4000 hours at 1500°C in air and mechanical testing at 1500°C in air we observed a loss of strength which exceeded 100 MPa. On basis of this observation it became obvious that loads higher than 100 MPa are not acceptable, and any design has to be optimized to be well below 100 MPa maximal tensile stresses. It must be assumed that SiC/SiC CMC systems will also show a degradation of matrix strength.

TURBINE COMBUSTION CHAMBER TESTS

For a test in cooperation with Siemens on an Ansaldo test bed a thermal insulation package and attachment were designed. The brazed bolt discussed previously was used to attach this package to the combustor transition piece. Figure 29.12 shows the insulation package and the attachment feature. The model is for a heat shield system used for the testing within the German turbine program AG Turbo funded by the Federal Secretary of Commerce. All visible layers are a creep resistant alumina to avoid thermal mismatch. The hot outer layer (zirconia/alumina) offers high thermal shock resistance. The other layers afford creep resistance and are highly thermally insulating. The advantage is a higher lifetime of the insulation without protection of the SiC part. The oxide ceramic insulation pieces were mounted in a combustor transition piece of a 20 MW combustor. The total test time in this big rig test was in the range of 50 hours, including 35 hours at full load. A total of 5 trips were performed. No damage was observed after the test. Figure 29.13 shows the principal lining system in the support of the first pressure test. The metallic attachment of the tile can be seen in Figure 29.10. The final heat shields were used in a combustion chamber experiment in co-operation with Siemens. Under the umbrella and with the support of the AG Turbo turbine development program we developed a bolt to fix alumina heat shields onto a metallic backbone support for a combustion chamber experiment [7].

The most sophisticated design included SSiC ceramic tiles fixed on alumina fiber insulation (see Figure 29.14) [8, 9]. These tiles were fixed by metal-ceramic bolts similar to those used for the oxide insulation

TABLE 29.4 MATERIAL DATA PROCESSING PLAN USED TO CHARACTERIZE THE SIC MATERIAL AND COMPONENT DATA

Temp (°C)	Test method	Specimen	Material condition	Load rate (MPa/s) 25	0.25	0.025	0.01	Stress Rupture	Creep	Cyclic fatigue (R= −1)	Combustion Atmosphere
25	Bending	Test bar	As machined	■							
25	Bending	Test bar	as fired	■							
25	Bending	test bar	Aged	■							
25	Double ring	tile	as fired	■							
25	Double ring	tile	Aged	■							
25	Tensile	bolt	as fired	■							
1450	Bending	test bar	As machined	■	■	■	■	■			
1450	Bending	test bar	as fired	■		■	■				
1450	Bending	test bar	Aged	■							
1450	Tensile	rod	As machined					■	■		
1500	Bending	test bar	As machined	■	■	■	■	■			
1500	Bending	test bar	Aged	■		■	■	■			
1500	Bending	test bar	as fired	■							
1500	Tensile	rod	As machined			■		■	■	■	■
1500	Tensile	rod	Aged								■
1500	Double ring	tile	as fired	■						■	
1500	Double ring	tile	Aged	■							
1550	Bending	test bar	As machined	■	■	■	■	■			
1550	Bending	test bar	Aged	■		■	■	■			
1550	Bending	test bar	as fired	■							
1550	Tensile	rod	As machined					■	■		

Figure 29.12 Model of the Heat Shield Used for the German AG Turbo Program

Figure 29.13 Final Heat Shield Design for Combustor Test

Figure 29.14 ABB Heat Shields for Laboratory Combustion Test Equipment (Reprinted with permission from Ref. [8])

packages. This kind of tile can survive 1500°C if only pure oxidation is considered. But the loss of material in a combustor by erosion/volatilization may be as high as 0.1 μm per hour. This amounts to a material loss of 0.8 mm within 8000 hours. The material loss depends on pressure, gas velocity and temperature and the O_2, and H_2O content of the combustion atmosphere.

LIFETIME LIMITATIONS AND FAILURE RISKS

The reliability of ceramic components in gas turbines depends on short term and long term strengths. While short term strength is determined by defect-size distribution at the beginning, long term strength is specified by dynamic degradation effects as defect population shifting (changes of already existing flaws and generation of new flaws), sub-critical crack growth (growth of existing flaws) and creep (material deformation and generation of new flaws). All these long term effects are strongly influenced by oxidation processes.

The model we followed is explained in Figure 29.15. The observed loss of strength during high temperature exposure time was assumed to be result of growing new defect populations. The components that were tested for more than 1000 hours under turbine combustor conditions demonstrated the formation of surface defects. This defect population is covered by the SiO_2 oxidation layer. Testing at low temperatures will show no effect on strength but testing at 1400°C or higher clearly demonstrated the effect in terms of decreased characteristic strength. All cracks started from surface defects under these conditions.

The formation of relevant surface defects was not observed if the high temperature annealing time was less than 1000 hours. An incubation time was observed for all specimens. Also, it seemed from counting pits per area that the number of pits per unit area stays constant over the observation time of 3500 hours after a short formation period in relation to the incubation time (see Figure 29.16). We observed oxidation pits which were as wide as 30 to 50 μm and as deep as 80 μm.

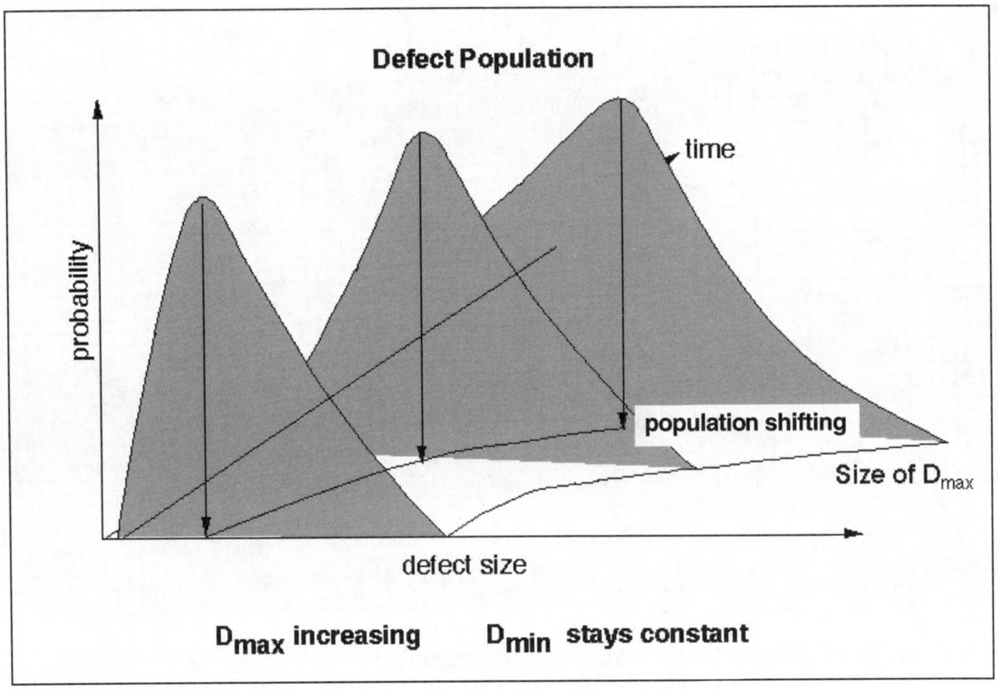

Figure 29.15 Shift of the Defect Population; The New Defect Population Consists Entirely of Surface Defects

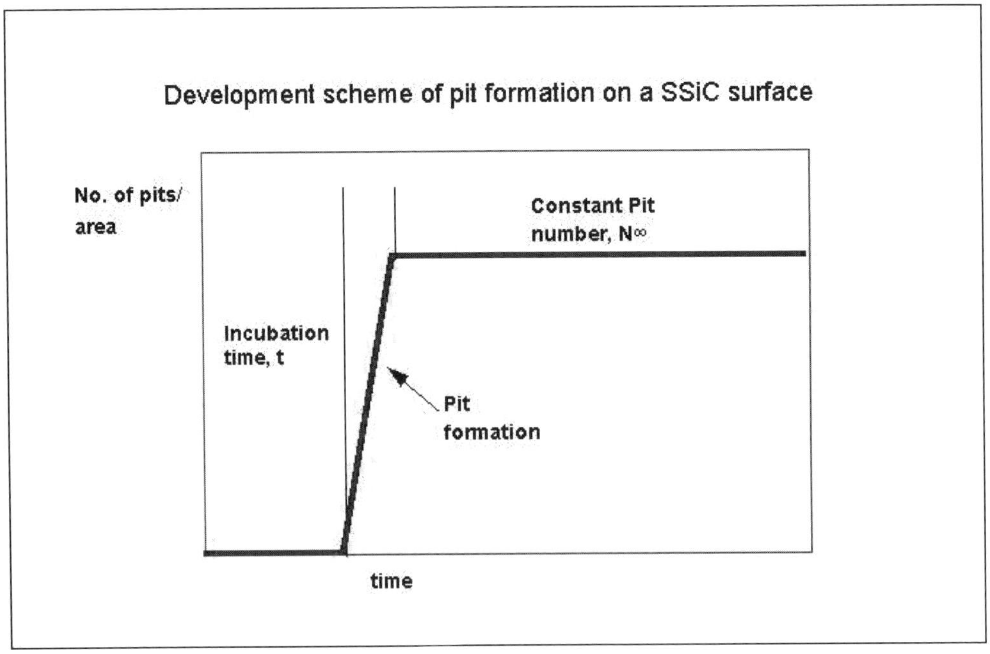

Figure 29.16 Development Scheme of Pit Formation on a SSiC Surface

Once formed pits become deeper but not necessarily larger in diameter. In a numerical description this observation can be expressed as

$$N_{P/A} = 0.5 * N_{\infty} * (1 + \tanh(t_i - t(T))$$

This formula includes the difference between time t at temperature T and the incubation time t_i. The maximum number of pits per unit area $N\infty$ is observable incubation time plus formation time.

The potential mechanical failure mechanisms caused by surface degradation are:

- initiation of crack growth near the surface in volumes under tensile stresses as expected for the combustor tiles
- spontaneous failure if the surface damage is critical

For the second mechanism the pit size distribution must be known. The observed large pits had diameters of about 100 to 500 μm and were up to 100 μm long. This implies that a high diameter coincides with a high radius at the base of the pit. Big pits are rare. Their formation in the Carborundum material (Hexoloy SA) was not observed within 500 hours at 1500°C. The effect of this process is a relative high surface modulus m_s and a low characteristic surface strength σ_s. For that, the pit formation must be included into a Weibull-postprocessor for a surface defects driven failure rate. The calculation of the failure probability of the components was performed within the FEM post-processors CARES or STAU. Both programs showed an increase of more than 10% in failure probability for a given surface near the load. This resulted in a 30% reduction in strength at 1500°C after 3000 hours under combustion conditions.

Oxidation

In case of SiC massive oxidation occurs at temperatures higher than 1400°C. Typically, only the surface is affected but if microstructure defects give a penetration path, internal oxide phase formation can also be observed. Different authors have discussed the influence of oxidation on mechanical properties in the early 1990s or even before [5, 6, 8, 10, 11]. This gave us reasonable facts for the study of oxidation and oxidation effects. Figure 29.17 shows deep penetration of oxide into the microstructure which came in contact with a granulate structure caused during the production. Figure 29.18 shows how the oxidation is penetrating the granulate interface. Such defects reduce the strength of the material at temperatures beyond 1400°C because the SiO_2 phase melts and the result are liquid filled cracks.

Under pressure and the influence of water vapor the oxidation rate is definitely increased. The test conditions in Figure 29.19 show an increase by a factor of 1.8 to 2.1. There is no significant difference in the effect of water vapor on CVD SiC and SSiC (see Figure 29.19). This gave us the evidence for the assumption that second phases or impurities are not predominantly affected by the water vapor but that the SiC itself is attacked by water.

The test atmosphere in Figure 29.19 contained more than 10 vol% water at 1500°C. The flow was 1 1/hour. There was a change of 1% of the atmosphere per hour. Within experimental error no difference in the oxidation behavior was observed between SSiC and the CVD SiC material in the case of the wet atmosphere. An amount of 10% water changes the parabolic constant by as much as a 100°C temperature increase would.

From literature studies it became clear that the oxidation at high pressures and high partial pressure of water is higher than under normal conditions. Because there is still around 16 vol% oxygen in the combustion gas we assumed a reaction like:

$$SiC + 2O_2(g) + 2H_2O(g) \rightarrow Si(OH)_4(g) + CO_2(g)$$

This reaction consumes 4 mol gas and produces 2 mol gas which explains the pressure dependency. This reaction is responsible for fast oxidation at high pressures. Additionally, the hot gas erosion will

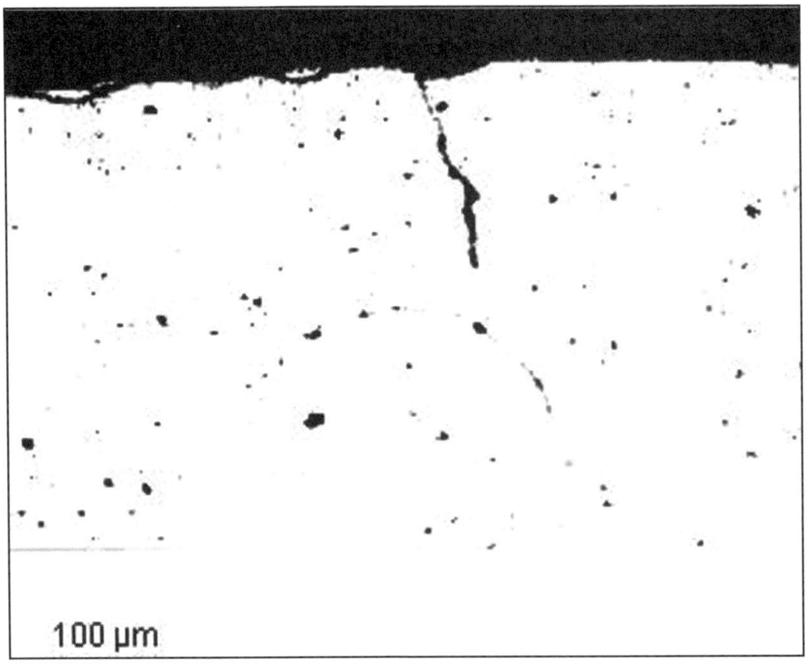

Figure 29.17 Linear Area of High Porosity with Rapid Internal Oxidation

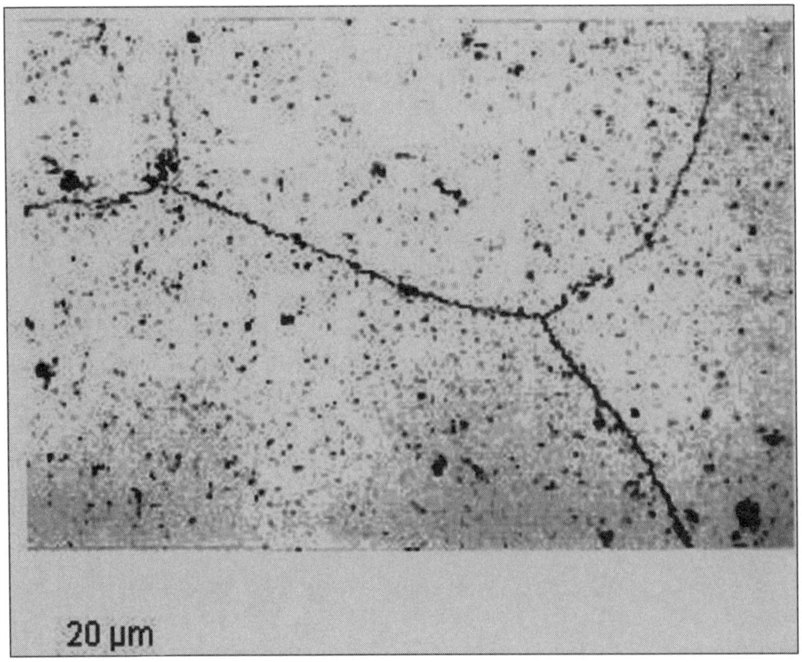

Figure 29.18 Internal Oxidation 200 μm Below the Surface and Deeper

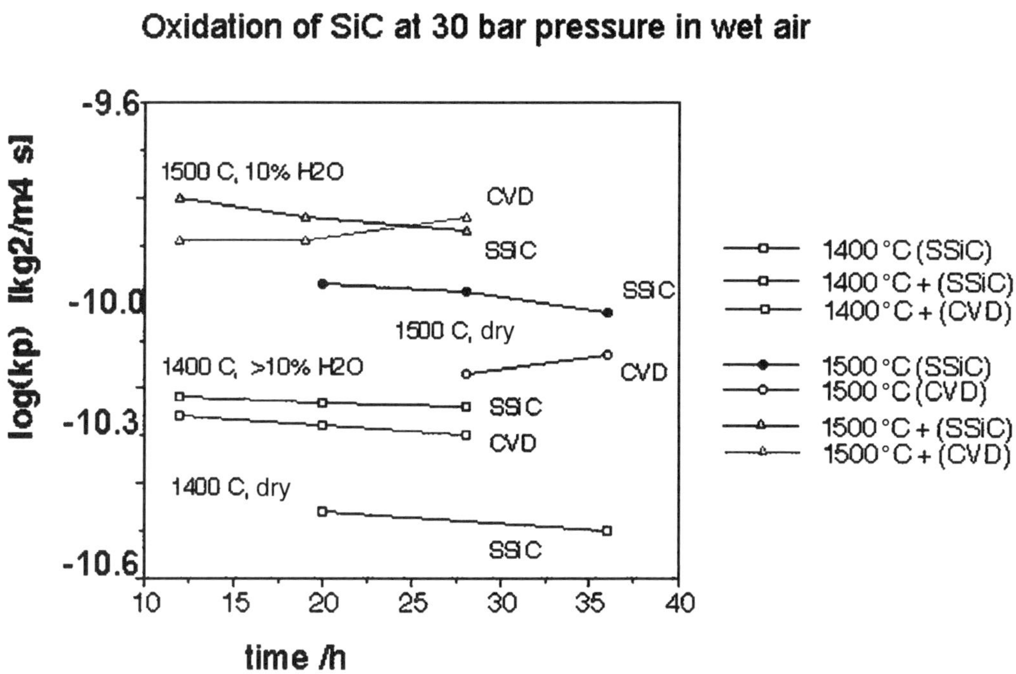

Figure 29.19 Oxidation of SiC at 30 Bars Total Pressure in Wet and Dry Air Showing the Effect of Water Vapor at High Temperatures

destroy an oxidation layer. Based on the experimental experience the oxidation map of Figure 29.20 shows the temperature-pressure region were SiC can be used. This diagram is not accounting for erosion effects.

The oxidation map of Figure 29.20 was prepared as a result of all the in-house experiments and a contract study performed by Prof. Gogozin. In our model the oxygen and the water are part of the reaction at high pressures. The system is able to reduce the pressure and therefore energy by consuming 4 mol gas and producing 2 mol gas in the oxidation reaction. Under the conditions of current gas turbine technology the combustion temperature is already in a critical area for the use of ceramics. It becomes evident that higher material temperatures at higher pressures are not applicable if ceramic components are in the system. Since 1997 models and experimental results were published which supported our idea of water vapor being the constituent primarily responsible for the observed degradation [12, 13].

Reliability of the Protection Tile System Discussed Here

Based on the observations and the STAU model developed and described above we calculated the failure probability of the component during a trip versus time and the cumulative historical damage. Figure 29.21 shows failure probability versus time considering trips, 1500°C steady state operating temperature, starts and stops, and material property degradation. The horizontal line symbolizes a component which is under homogeneous tensile stress but without degradation. All time dependent values were calculated considering a volume under stress equal to the stressed volume and the stress distribution in the real component. The curve symbolized by crosses shows the result including oxidation, loss of mechanical properties and crack growth.

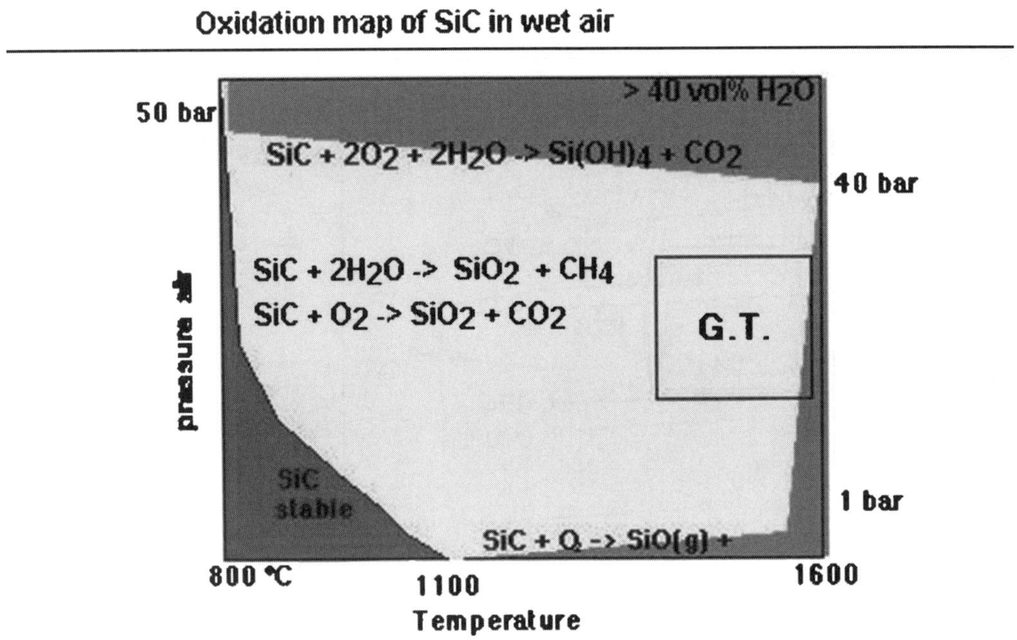

Figure 29.20 Oxidation of SiC in Wet Air

First only the change of stress distribution due to shape changing by oxidation shows an effect but after 30,000 hours the effect of crack growth during trip conditions was observed. For an accepted risk in terms of failure probability the expected life can be calculated in this way. Nevertheless, the example may be optimistic because not all degradation mechanisms may have been considered here. But it is obvious that for a life of 20,000 hours designers have to take a failure probability of 10^{-10} to 10^{-11} per component in this example.

Figure 29.21 represents a model calculation including four steps of loading. Modeled was a SiC heat shield. The failure probabilities during specific conditions of an assumed gas turbine versus time are shown. In the first step the shield is under constant tensile load (rhombohedral symbol). In the second step each rectangle symbolizes a trip condition and the material oxidizes. The triangles are a calculation where we took a loss of strength into account. The final calculation included additional crack growth. The model gives a chance to estimate the increase of risk versus time. The company has to decide which risk can be taken. The loss of reliability directly corresponds to the loss of high temperature strength of SSiC.

The influence of water was expected to accelerate this process because of the faster oxidation. Tensile tests performed under the same environmental conditions showed that aged specimen had crack origins on the surface. Virgin test specimens always showed volume flaws responsible for failure. Long term oxidation tests with high temperature strength tests of the material showed a tremendous loss of mechanical reliability (see Figure 29.22), which is also reflected in the model calculation shown in Figure 29.21.

Not taken in consideration in the model calculation above are impact damages. Impact tests demonstrated that a metallic nut destroys a SiC tile completely if the mechanical impulse exceeds 0.4 kg•m•s^{-1}. In a fast flowing combustion chamber atmosphere much higher velocities and consequently higher impacts of foreign objects can readily be expected.

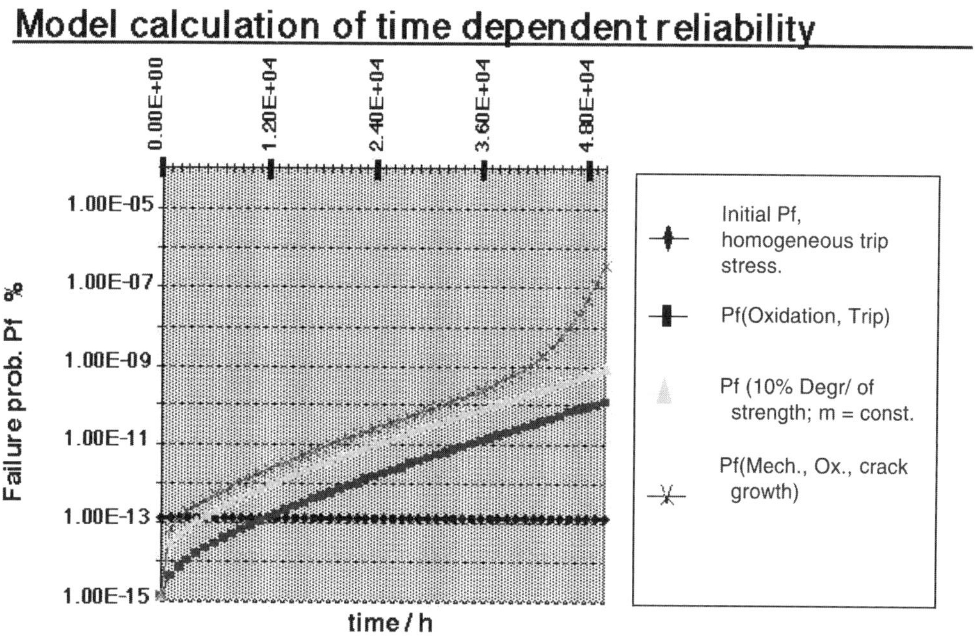

Figure 29.21 Model Calculation of Time-dependent Reliability Obtained from Test Data

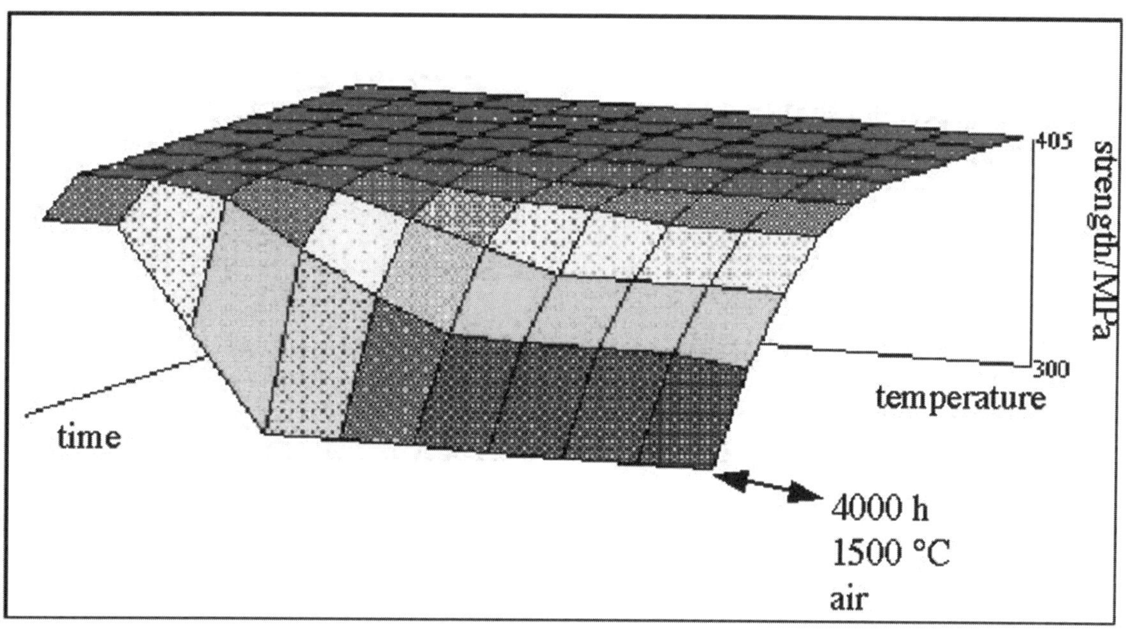

Figure 29.22 Annealed at 1500°C in Air, SSiC had Already Lost 30% of its Bending Strength

SUMMARY

The main result we got was the impact of O_2/H_2O at 1500°C on the properties and consequently the reliability of SiC material. The ABB team got the impression that SiC and may be all non-oxide ceramics are not applicable if the loads include the combined attack of high temperatures, water and tensile stress loads. Caused by oxidation and degradation non-oxide ceramics may be limited to uses below 1350°C in combustion atmospheres. The judgement on oxidation will also include fiber-reinforced materials with currently available non-oxide fibers and matrices, even if the thermal stability of the fiber will be improved some day.

The development of a corrosion and crack growth based lifetime model showed that the discussion of pure mechanical behavior gives an optimistic picture of the material behavior. The potential risk analysis has to include impact failures, corrosion, oxidation and crack growth. The final postprocessor STAU was designed to calculate short and long term effects on the reliability of a component. STAU was used in the design process to optimize the lifetime of the component.

Potential future developments may be protective coatings for CMCs, high temperature oxide fibers and oxide matrices, fiber coatings on the basis of high temperature silicates or other high temperature oxide phases.

Oxide CMC systems will first be included in small commercial engines. Demonstrators may incorporate non-oxide CMCs. Developments being currently discussed where non-oxide CMC material is protected by an environmental barrier coating are expensive and take the additional risk of surface layer failures. Maybe microturbines in the range of some dozen kW up to 150 kW will be the first machines containing CMCs and monolithic ceramic components.

ACKNOWLEDGEMENT

Unless otherwise indicated all figures in this chapter were provided courtesy of Alstom.

REFERENCES

[1] Harbach, F., Knödler, R., Bossmann, H.-P., "Thermoelektrische Festelektrolyt-Generatoren," ("Thermoelectrical solid electrolyte generators,") Physik in unserer Zeit, Vol. 23, pp. 34–41, 1992 (in German).

[2] Knödler, R., Bossmann, H.-P., Kranzmann, A., Harbach, F., "Performance of Single Cells of an Alkali Metal Thermoelectric Converter," Journal of the Electrochemical Society, Vol. 139, No. 11, 3029–3034, Nov. 1992.

[3] Helme, H.F., Lindgren, L.C., Heitman, P.W., and Thrasher, S.R., in *Ceramic Applications in Turbine Engines*, chapter 1, Noyes Publication, ISBN 0-8155-1080-2, 1986.

[4] Komeya, K., "High Temperature Engineering Ceramics," in *Materials Science and Technology*, Vol. 11, eds. R.W. Cahn, P. Haasen, E.J. Kramer, pp. 517–564, chapters 10.1 to 10.5, 1994.

[5] Gogotsi, Y.G., Kaurenko, V.A., *Corrosion of High Performance Ceramics*, chapters 2, 5 and 6, Springer, 1992.

[6] Tressler, R.E., "Environmental Effects on Long Term Reliability of SiC and Si_3N_4 Ceramics," in *Ceramic Transactions*, Vol. 10, Corrosion and Corrosive Degradation of ceramics, R.E. Tressler, M. McNallan eds., Am. Ceram. Soc., pp. 99–124, 1990.

[7] Kranzmann A, Weiler L, "Entwicklung einer Metall/Keramik Lötverbindung für keramische Brennkammern," ("Development of a metal/ceramic brazed joining for ceramic combustors")

Proceedings of the 4th Statusseminar der Arbeitsgemeinschaft Hochtemperaturgasturbine, December 1995, pp. 55–62 (in German).

[8] Ziegler, C., "Bewertung der Zuverlässigkeit keramischer Komponenten bei zeitlich veränderlichen Spannungen und bei Hochtemperaturbelastung," ("Assessment of the Reliability of Ceramic Components Under the Influence of Time-dependent Stresses and High Temperature Loads") PhD Thesis. University of Karlsruhe, 1998 (in German).

[9] Heyer W., Kranzmann A., Pfeiffer A., Reh S., "Keramische Auskleidung," ("Ceramic lining,") European patent No. 97810538.5-2301, 1997 (in German).

[10] Förthmann, R., Naoumidis, A., "Influence of hot gas corrosion on the bending strength of silicon carbide materials," in *High Temperature Corrosion of Technical Ceramics*, R.J. Fordham, ed., 1989, pp. 217–226.

[11] Park, D.S., Danyluk, S., McNallan, M., "Friction and Wear Measurements of Si_3N_4 at Elevated Temperatures in Air, Ar, and Humid Environments," in *Ceramic Transactions*, R.E. Tressler, M. McNallan, eds., Vol. 10, pp. 159–179, 1989.

[12] Robinson R.C., Smialek J.L., "SiC Recession Caused by SiO_2 Scale Volatility under Combustion Conditions: I, Experimental Results and Empirical Model," J. Am. Ceram. Soc., 82[7], 1817–1825, 1999.

[13] Opila E.J., Smialek J.L., Robinson R.C., Fox D.S., Jacobson N.S., "SiC Recession Caused by SiO_2 Scale Volatility under Combustion Conditions: II, Thermodynamics and Gaseous-Diffusion Model," J. Am. Ceram. Soc., 82 [7], 1826–1834, 1999.

CHAPTER

30

EVOLUTION OF PROGRAMS FOR CERAMIC GAS TURBINE ENGINE DESIGN IN THE STATES OF THE FORMER SOVIET UNION

Youry A.Nozhnitsky

Central Institute of Aviation Motors (CIAM), Moscow, Russia

Anatoly V.Soudarev

Research-Technological Institute for Power Engineering (NITIEM), St. Petersburg, Russia

INTRODUCTION

Despite the high level of sophistication achieved so far, gas turbine engine development continues vigorously. The progress in gas turbine manufacture is greatly dependent on the level of development and commercialization of advanced materials superior to the alloys currently applied, in terms of the ultimate operating temperature, and absolute and specific values of strength and toughness.

The application of ceramic materials for the fabrication of reasonably low-stressed small-size rotor and stator elements and, also, hybrid rolling bearings seems to be well within reach in the foreseeable future. Further advancements in ceramic materials processing will be the result of an increase in their fracture toughness, and improvements in ceramic manufacturing technology. These include the homogeneous densification of near-net-shape preforms, increased machining accuracy, enhanced NDE sensitivity, development of methods for diffusive joining of separate ceramic elements into an integral component, as well as ceramic-to-metal joining, etc.. Progress in these areas together with the development of ceramic composite materials engineering is likely to establish the basis for a wide application of ceramics to gas turbine engines. However, the feasibility of applying ceramics must be assessed on a case by case basis through a thorough comparison with the capabilities of alternative technologies.

The ceramic materials could be applied as protective coatings for parts made of metal alloys, ceramics or composite structural materials. Protective coatings have already found widespread application, particularly thermal barrier coatings (TBCs) of partially stabilized zirconia. These TBCs, because of their low thermal conductivity, provide superior oxidation protection to the turbine and combustor hot section components. Additionally, they increase the corrosion resistance of these parts.

The structural ceramic materials have been attracting the attention of the gas turbine engine designers for a long time. For the last three decades, the development of structural ceramics for gas turbine engine applications has been an ongoing activity in many of the developed countries; although the progress with the commercialization of these materials is still not very impressive.

This chapter gives an insight into the experience gained with ceramic gas turbine engines in the former USSR, in Russia and the Ukraine.

CERAMICS APPLICATIONS TO GAS TURBINE ENGINES

The results of an analysis, summarized in Table 30.1, indicate that ceramic structural materials could be applied to a wide range of parts for gas turbines for various applications (aviation, marine and automotive engines, electric power station generators, natural gas blowers, etc).

The use of structural ceramic materials is primarily focused at:

- enhancing cost-effective fuel utilization
- reduction of environmentally undesirable emissions
- reduction of the mass and number of parts
- decreasing the moment of inertia of rotating parts
- increasing life under conditions of high temperature and speed, and aggressive media
- providing a capability to utilize inexpensive fuels, softening of the lubrication requirements, etc.

It is also of vital importance that the development of structural ceramic materials eliminates the need for the utilization of rare alloying elements which are indispensable for the production of high temperature alloys.

MAJOR CERAMIC GAS TURBINE ENGINE PROGRAMS

In the USSR, projects to develop and apply (for different purposes) structural high temperature nonmetallic materials were launched in the early 1970s, primarily with government funding. A number of material science, technological, and research laboratories, and design institutes for aviation, aerospace, defense, automotive, nuclear, electronic, metallurgy, chemical industries, power, refractory materials, as well as academic bodies and universities were involved in these activities. In most cases, though, it was a short-term "attempt" to explore a new attractive direction in research and development.

The major programs in the USSR were focused on the conceptual development of ceramic components for gas turbine engines. Some projects for the production of gas turbine engines for various applications with extensive use of ceramics were also completed.

The principal difference between the Soviet era programs and the programs conducted overseas was that in the USA, Western Europe, and Japan the programs were essentially oriented towards the production of automotive ceramic gas turbine engines while in the USSR these trends (including the ceramic automotive gas turbine engine—GTE) have been quite limited. The only exception seems to be the proposal for the development of a ceramic automotive gas turbine engine in the 150–330 kW output power range, prepared in 1991 by the aircraft engine manufacturing enterprises Central Aviation Motor Manufacture Institute (CIAM) and Rybinsk Motor Design Bureau, jointly with the design bureau of the car plant in Gorky (now Nizhny Novgorod). The proposal was given an award at a the competitive bidding overseen by the State

TABLE 30.1 CERAMIC APPLICATIONS FOR GAS TURBINES

No	Applications	Elements and components of nonmetallic materials	Materials	Objective of application
1	2	3	4	5
(1)	GTEs for automotive application, marine GTEs, stationary GTEs of various applications	Wheels for axial and radial inflow turbines (for engines with power output up to 300 kW)	Monolithic ceramics, ceramic composite materials	Increase fuel efficiency, reduce harmful emissions, reduce mass (time lag), multiple fuel operation capability
		Rotor blades for turbines Nozzle vanes Heat exchangers Combustor liners Hot ducts		
(2)	Engines for unmanned flying vehicles (UFV)	Turbine wheels	Composite materials (carbon-carbon with protection against oxidation, C/SiC, SiC/SiC)	Increase fuel efficiency, reduce mass
		Rotor blades for turbines	Monolithic ceramics, Composite materials	
		Liners		
		Ceramic bearings	Monolithic ceramics carbon-carbon (separator)	Reduce engine mass and volume (less rigid requirements for lubrication and cooling), operate in aggressive media
(3)	Turbopump units for liquid jet engines	Turbine wheels	Composite materials	Increase fuel efficiency, reduce mass
		Rotor blades for turbines	Monolithic ceramics, Composite materials	
		Ceramic bearings	Monolithic ceramics, Carbon-carbon (separator)	Reduce mass, operate in aggressive media
(4)	Aircraft engines	Nozzle elements, liners	Composite ceramic materials	Reduce mass, increase fuel efficiency, life extension
		Low pressure turbine elements (rotor blades, nozzle vanes, casings)	—«—	
		Rotor blades and guide vanes for high pressure compressor	Composite and monolithic ceramic materials	
		Hybrid rolling bearing rollers	Monolithic ceramics	Life extension
(5)	Helicopter engines	Hybrid rolling bearing rollers	Monolithic ceramics	Life extension
(6)	Engines for supersonic passenger aircrafts	Liners, nozzle elements	Composite ceramic materials	Reduce toxic emissions, reduce mass, increase fuel efficiency
(7)	Aerospace engines	Gas-air path elements	Composite ceramic materials	Reduce mass, Increase fuel efficiency
(8)	Engine units.	Rolling contact bearing elements	Monolithic ceramics	Reduce mass (eases requirements for lubrication and cooling), operation in aggressive media, reduce power losses, increase motion smoothness.

Science and Technology Committee, which had been mandated to support the production of environmentally friendly automotive engines, although no funds were allocated to implement the project. Some complex projects for the production of ceramic gas turbines were completed by the State Scientific Center CIAM (named after P.I. Baranov) in Moscow, and the Research-Technological Institute for Power Engineering (NITI EM) in St. Petersburg. The projects were conducted in collaboration with other enterprises which were engaged in the development and implementation of major programs between 1980 and 2000 for the production of ceramic gas-turbine engines (Table 30.2). Regretfully, due either to a lack or termination of funding, these programs were not accomplished during the time frame planned. At present, only one program, N 2, with NITI EM as the major contractor is still in progress, although the period of performance was stretched out. The results of the research and development programs researches are surveyed in this volume (chapters 31 and 32) and also in the 2nd volume (chapters 12 and 21). The participation of the Rybinsk Motors Design Bureau (currently the joint stock company "Rybinsky Motory", Rybinsk, Yaroslavl region—see chapter 33 in this volume) and the Research Engine Institute (Moscow) also deserve mention in conjunction with the development and study of ceramic gas turbine components.

For the development of structural ceramic materials, the State Scientific Center of Russia, namely the Research-Manufacture Merger "Tekhnologhia" (Obninsk, Kaluzhskaya region) and the Kiev Institute of the National Academy of Sciences in the Ukraine (the Institute for Material Science problems and the Institute for Superhard Materials) should be mentioned first. The Institute for Strength Problems of the National Academy of Sciences in the Ukraine (Kiev) contributed essentially to the development of the facilities and methods for the investigation of the mechanical properties of the ceramic materials and the methods of life prediction for the parts made of these materials. The expertise of "Tekhnologhia" of Obninsk and that of the Kiev Institutes is discussed in chapters 12 and 20 of Volume II of this text.

THE LESSONS LEARNED AND CONCLUSIONS

The main lessons learned during the multi-year projects can be summarized as follows:

(1) Likely applications of the ceramic materials are shown in Table 30.1. However, to assess the requirement for the application of the structural ceramic materials, a multi-factor analysis should be conducted on a case-by-case basis. One should take into account the impressive progress recently achieved with the development of alternative technologies such as new heat resistant alloys, including single crystal alloys, thermal barrier and anticorrosion coatings, high efficiency cooling schemes, the development of low-NO_x combustors, magnetic and gas bearings, etc.
When selecting a design and technology solution key factors must be identified, the multiple restrictions associated with each must be considered, and a risk analysis performed. Such an objective analysis is rather complicated. In addition, the analysis must be repeated and refocused at each new level of development of the competitive technologies which can result in a change of priorities.

(2) Impressive progress is observed with the development of the structural ceramic materials. However, insufficient plasticity of the current ceramic materials prevents their application for engine parts where their application would likely cause failure. In many cases, the competitiveness of the current ceramic materials is limited by their ultimate operating temperature, which for the majority of the ceramic materials is still too low for the manufacture of uncooled nozzle vanes for advanced engines that must operate at the gas temperature at a turbine inlet temperature (TIT) ≥ 1327°C (1600K).

(3) Design of ceramic components must be performed while taking into account the deformation and degradation behavior of these materials. While segmenting a ceramic component into smaller sized elements reduces the likelihood of critical flaws, due to the scale factor effect, and generally lowers the thermal stresses it does not increase the strength reliability. Reducing the absolute dimensions of the parts essentially improves their microstructure by decreasing the relative number of defects, etc.

TABLE 30.2 MAJOR CERAMIC GAS TURBINE PROGRAMS IN THE USSR, RUSSIA AND THE UKRAINE (1980–2000)

No	Program	Year start	Year finish	Objective	Approved by	Notes
(1)	R&D program for the production and commercialization of components of heat resistant ceramic and metal-ceramic materials for automotive gas-turbine engine and internal combustion engine applications	1983	1990	Manufacture of 1st production ceramic engines and their tests		
(2)	R&D program for the production at MGO "Energomash"*) and Mintyazhmash**) of the USSR of gas turbine and gas-pumping units and diesel engines with heat resistant components of structural ceramic and composite materials for the 1988–1995 period	1988	1995	Manufacture of gas-turbine gas-pumping units (GPU) with efficiency not below 42% Conducting of complex tests on laboratory test rigs and at pipeline compressor stations	The Council of Ministers of the USSR as of Sept. 9, 1988 N 1816p.	*) Interbranch State Merger "Power Engineering" **) Ministry of heavy engineering industry
(3)	Program of concerted activities of MGO "Energomash" and defense Industry Enterprises for the production of high efficiency, environmentally friendly gas turbine engines with parts and components of structural ceramic materials intended for natural gas and oil production and transportation (Neftegazceramica-95)	1990	1995	Manufacture and supply of 1st production ceramic GTEs of 2.5, 6 and 10 MW power to sites of gas industry		Coordinated by the ministries of Gas, Oil, Engineering, and power engineering industries

However, at the same time, new challenges arise such as joining the elements into larger components. This requires the development of techniques for diffusive bonding of ceramics, or ceramics-to-ceramics and ceramics-to-metal soldering, etc.

Using a material with a high thermal conductivity and a low coefficient of linear thermal expansion will minimize thermal stresses. The management of thermo-physical properties of ceramics (in particular, increasing their thermal conductivity) has not been given sufficient attention yet. The reduction of the elastic modulus is often useful for reducing thermal stresses. It is especially important to avoid significant stress concentrators, including contact and thermal stresses in ceramic structures, and if required, to locate these stress concentrators within a low stress zone.

In design one must consider that ceramics are stronger in compression than in tension. This inherent property of ceramic materials, is still only seldom used by designers who, as a rule, tend to rely on conventional engineering approaches typically utilized for the development of metal structures. Fundamentally different design strategies should be used for components that operate under predominantly compressive stress conditions. In this case, it is often much more useful to utilize hybrid structures consisting of ceramic and metallic elements, making sure that the predominantly compressive stresses are located in the ceramic elements. Sometimes, assemblies in which the ceramic parts operate in compression have a higher strength compared to structures in which the ceramic parts operate in tension, but they may be inferior in other aspects (complexity, cost, etc).

For the development of hybrid components with ceramic and metallic parts, one must take special care to lower the contact stresses and to accommodate the difference in the coefficients of linear thermal expansion. This latter factor must be taken into account in the analysis for potential leaks in the components.

Attention should be paid to the development of special dampers and other means to lower the dynamic and vibration loads in the ceramic parts.

(4) The ceramic part strength reliability is intimately related to the quality of manufacturing.

The following factors are especially important:

- a high material density and lack of internal defects (pores, inclusions, etc.);
- high precision machining, exclusion of surface defects;
- development of reliable nondestructive evaluation (NDE) methods

The high density (e.g. hot pressed) materials typically have a high rigidity which facilitates achieving a high machining accuracy, although at the same time, they have a high hardness which necessitates the use of expensive diamond tools. Hence, zero-shrinkage ceramic materials which can be machined as "green," not yet ceramized preforms, appear to be especially attractive. After the final geometry is obtained, the preform is converted by high temperature synthesis into the ceramic part. This process significantly reduces the cost of the ceramic parts, and eliminates the emergence of surface stress concentrators after machining.

To manufacture complex shaped components (e.g. turbine blades), it is useful to utilize materials that can be densified after machining. The application of zero-shrinkage ceramic materials may help in this respect. Furthermore, in many cases complex shaped parts should be segmented, and manufactured from several smaller parts with simpler configurations that are subsequently joined using state-of-the art soldering-welding technologies (or in general, diffusion bonding).

The NDE methods that have been developed for ceramic components are either insufficiently effective or too expensive, and are not suitable for 100% quality control of serial production parts.

(5) The most complex challenge is the verification of ceramic component life. It is necessary to evaluate life limits both for slow crack growth and creep. Statistical methods similar to those used for metallic alloys cannot be applied because of the lack of reliable data for the distribution of defects. At the same time the characteristics of the crack growth rate in ceramics usually have substantial scatter, which makes it difficult to precisely define a regime of validation tests. The necessity of conducting validation tests under significant overloads restricts the life and strength of the ceramic components. Data that accounts for the influence of creep on crack growth in ceramics are not available.

The ceramic composite materials have higher fracture toughness than monolithic ceramics; although, their high temperature strength is still insufficient for the manufacture of highly stressed parts. Models for deformation and failure of ceramic composites have not yet been properly developed either. Such factors as anisotropy of properties, structural heterogeneity, fiber and matrix creep, scatter of mechanical properties, and flaws in the material must be taken into consideration in these models.

(6) The ceramic materials should be primarily employed for the fabrication of relatively low-loaded small-size rotor and stator parts, such as liner segments, turbine shrouds, nozzle flaps, etc (composite ceramic materials) and, also for hybrid rolling bearings for high speed rotors (hot-pressed silicon nitride). Further progress with ceramic material development will likely result in the application of these materials for the manufacture of some other parts including relatively high stressed turbine blades for small engines. As it is more difficult to cool small parts than large parts, the use of uncooled ceramic components may be advantageous in this case.

(7) The challenge of resistance of the ceramic components to impact from mechanical particles, foreign objects, etc. which may enter the gas stream still remains to be solved. Design solutions that can eliminate the "material loss" which causes catastrophic failure of the engine hot section still have to be developed. Advanced approaches could involve the use of composite ceramic materials, the arrangement of traps in the stator elements to catch the debris of the rotor parts, the development of the structural ceramic materials that convert to dust on impact, etc.

(8) For widespread implementation of ceramic materials in gas turbine gas engines further progress is essential, both in material and process development, structural fabrication technologies, and in the improvement of methods for designing and substantiating ceramic part strength reliability and life.

CHAPTER

31

Ceramic Gas Turbine Component Development Experience at the Central Institute of Aviation Motors

Youry A. Nozhnitsky, Youlia A. Fedina,
Anatoly D. Rekin and Nickolai I. Petrov
Central Institute of Aviation Motors (CIAM), Moscow, Russia

INTRODUCTION

At this time monolithic ceramic materials (silicon nitride, silicon carbide, and sialons), carbon-carbon composite materials and ceramic composites can be referred to as structural heat-resistant non-metallic materials.

The Central Institute of Aviation Motors (CIAM) participated in several programs for the development of gas turbine engines (aircraft engines and gas turbine mechanical drives for various purposes) made of heat-resistant non-metallic materials [1]. These programs were mainly government funded. The projects were initiated at the beginning of the 1970s. Unfortunately, because of financial constraints the funding of these projects was abruptly reduced after 1990.

CIAM's responsibility includes the following activities:

- Coordination of activities related to the application of non-metallic heat-resistant materials in aircraft engines
- Development of materials requirements and investigations into their structural strength
- Development of design methodology and test methods for engine components made of these materials
- Assessment of the effectiveness of non-metallic materials applications for gas turbine engines
- Calculations and experimental investigations, including tests at laboratory facilities and in experimental engine gas generators
- Development and optimization (jointly with other organizations) of novel design strategies.

CIAM developed test techniques in support of these activities. Unique equipment to investigate the structural strength of non-metallic materials over a wide temperature range (see also Volume 2, Chapter 21) was designed or modified for engine component tests (both at laboratory facilities and in experimental engine gas generators). The software was also adequately upgraded.

To carry out all necessary calculations the software used for the calculation of the stress-strained state (both universal software such as ANSYS, etc., and software developed at CIAM such as DVUKON, etc.) were supplemented with programs for the calculation of reliability, durability and proof test regimes based on the weak link model and fracture mechanics methodology [2, 3]. Several institutes participated in the projects. Materials developed by the State Scientific Center "Technologiya" (Obninsk), the All-Union (now Russian) Central Institute of Aviation Motors (Moscow), the Institute of Materials Problems, and the Institute of Superhard Materials of the National Academy of Sciences of the Ukraine (Kiev), and the Research Institute of Graphite-based Materials (Moscow), and others, were investigated.

Experts of the Research Institute of Engines Technology (Moscow) participated in the development of fabrication techniques for ceramic engine parts. The Research Institute of Engines (Moscow), "Rybinsk Motors," and a number of other design bureaus contributed to the development and studies of engine parts. The Institute of Bearing Industry (Moscow) participated in development of bearings with ceramic elements.

Initially, the investigations focused on two areas, monolithic ceramics and composite carbon-carbon (C/C) materials. C/C materials have a number of unique mechanical properties. The absolute and specific strength and stiffness properties do not decrease up to around 2000°C. The low linear coefficient of thermal expansion allows small stable gaps between rotor and stator, essential for optimal efficiency, and also reduces thermal stresses in C/C parts. However it complicates the development of assemblies consisting of metallic and C/C parts. When designing C/C parts it is also necessary to take into consideration anisotropy of material properties. The time consuming process of the manufacture of C/C parts and their high cost also needs to be mentioned. The problem with C/C parts is the requirement for high-temperature oxidation protection. Without going into a detailed discussion of the results of the C/C materials studies, it should be noted that a number of turbine, combustor liner and nozzle elements made of C/C materials were developed and tested, and significant progress was made in the development of coatings for oxidation protection. However, it was difficult to ensure reliable operation of coated C/C parts over the expected long service lives, and their high cost was an impediment for C/C material applications in stationary gas turbine equipment. To ensure reliable service life, requiring protection against oxidation, ceramic matrix composites are more attractive

Extensive experience has been gained from investigations of gas turbine engine components made of monolithic ceramic materials. The maximum operating temperatures of the developed ceramic materials is 1400°C for Si_3N_4-based materials and 1600°C for SiC-based materials. However, taking into consideration creep and other modes of deterioration of material properties during gas turbine operation, the required life between overhauls (no less than 10,000 hours), the component temperature must be kept 200°C to 300°C below the maximum operating temperatures (the actual allowable temperature depends on the stresses in a part). The mechanical properties of ceramic materials are essentially dependent on their density, but high density materials typically have high hardness. So they need costly machining which in turn determines the high cost of parts made of these materials. Inherent brittleness is the main obstacle for the use of monolithic ceramics: it limits the allowable stresses in the materials and, therefore, limits the potential range of applications.

Due to their insufficient plasticity, ceramic materials should be applied, first and foremost, for the manufacture of parts whose failure have no hazardous effects. For component design one should consider that these materials function better in compression than in tension. Stress concentrators, including those of a thermal and contact origin, should be eliminated, or located in regions of low nominal stresses. The scale factor effect on strength reliability is reflected in the limitations on ceramic part dimensions.

The durability of ceramic parts can be limited by crack growth, first of all from material flaws, but also from creep, high temperature oxidization, and degradation of properties [4, 5]. Cracking due to initial defects is of special importance, because the low plasticity of the ceramic material in conjunction with the statistical defect distribution will result in reduced durability. In some cases hot isostatic pressing was used to reduce internal porosity. Component nondestructive evaluation was conducted by fluorescent dye penetrant inspection, and X-ray and ultrasonic methods. However, presently no reliable data on the defect distribution in a ceramic part are available that will permit the use of fracture mechanical modeling, recently developed for metallic alloy part life evaluation [6]. So a preliminary short-term proof test for each part is the only way of ensuring the required strength reliability and service life. A proof test under overload conditions will indicate minimum durability. At the same time the component must be carefully inspected for flaw growth or other undesirable changes in the structure after the proof test.

Analysis, design and calculations have shown that ceramic materials can be considered for the manufacture of a wide range of parts for gas turbine engines for different applications (see Table 30.1 in chapter 30). Besides, despite the differences in part dimensions, operational conditions and service lives, most of the problems with ceramic part development are of a common nature. For detailed analysis of these problems and possible ways of solving them, the efforts at CIAM concentrated on the development of ceramic parts for a small gas turbine engine, ceramic elements for a combustion chamber, and rolling element bearings.

In the first phase of the project, engine designs specially intended to accommodate ceramic component application were studied, but subsequently the design of individual ceramic parts and components were evaluated [1–3, 7–13]. Testing of these designs was conducted at laboratory facilities and in specially modified engine gas generators. The basic results of this work are given below.

CERAMIC TURBINE PARTS

The strength of integrally manufactured ceramic nozzle vanes is essentially inferior to that of a segmented design consisting of separately manufactured nozzle vanes and rings. However, the design with separately manufactured vanes has high stresses in the vane edges due to temperature induced strains. It was demonstrated that these strains could be decreased by using airfoils with bowed edges, decreasing chords, making edges thinner, and reducing the thickness of the vane platforms [12–14]. Nevertheless, after the analysis the ceramic nozzle vane design was excluded from consideration [10, 11, 13]. First, replacement of cooled nozzle vanes of the 1st stage of the turbine with uncooled vanes did not appear to be very effective. Differences in the coefficients of thermal expansion between the ceramic and superalloy components of the assembly cause leakages, which need to be eliminated. Secondly, the maximum operating temperature of the existing structural ceramic materials is not sufficient for the 1-st stage uncooled nozzle vanes of an advanced high-temperature turbine, taking into account "hot spots" resulting from the combustor pattern factor and radial profile.

The mechanical properties of state-of-the-art ceramic materials seem to be insufficient for ensuring strength reliability of the highly loaded turbine disk or blisk. The task of manufacturing a turbine wheel with a metal disk and inserted non-metal blades appears to be less complicated. The application of uncooled ceramic blades increases the efficiency by 2% and reduces the rotor mass by about 30%. To increase the strength reliability of a ceramic blade, it would be appropriate to use the design approach in which the ceramic part of a blade operates in compression, but the known rotor designs with such blades are, as a rule, less effective. For a conventional rotor with uncooled ceramic blades that operate in tension due to centrifugal forces, the major problem is to ensure reliability of a lock and metal disk connection.

Various locks of the "hinge" and "dovetail" types were examined. The calculations and experimental work indicated the need for a "dovetail" type lock of larger height (in comparison with metallic prototypes) and a compliant layer between the surfaces of the ceramic blade and the metal disk to redistribute the contact stresses. Variations of the modulus of elasticity and coefficient of friction of the compliant layer material (made of a nickel gauze, or solid nickel or platinum), within a certain range, is of little importance for the redistribution of contact stresses. But if the coefficient of friction is significantly elevated, the compliant layer ceases to be useful. The availability of the compliant layer also results in a substantial decrease in vibration stresses, although it still does require the insertion of special dampers beneath the blade platforms. It is also possible to reduce vibrations using a ring attached to the face surfaces of the blades roots. However, as mentioned, if the coefficient of friction rises significantly, the compliant layer ceases to be useful.

Figure 31.1 Ceramic Turbine Blades Being Prepared for Thermocycling Tests

The calculations and experimental investigations have been performed on blades made of different materials and manufactured by the State Scientific Center "Technologhia" (Obninsk), the Institute of Materials Problems and Institute of Superhard Materials of the National Academy of Science of the Ukraine (Kiev) and other organizations. In particular, the reaction-bonded nitride based materials (OTM-904 and others), densified reaction-bonded silicon nitride based (OTM-913 and others) materials, hot pressed silicon nitride based materials (OTM-914, Tisinit and others). The properties of some materials are described in Chapter 12 of Volume 2.

The reaction-bonded silicon nitride blades were manufactured without machining. The blades made of densified reaction-bonded or sintered material required minimum machining. The hot pressed material blades were made of billets of a bar type and it required a great amount of expensive diamond machining. It was demonstrated that only hot pressed or densified reaction-bonded silicon nitride could be used for the manufacture of highly loaded turbine blades. Applying semi-finished blade densification may cut down the expenses associated with diamond machining. State-of-the-art silicon carbide-based materials should not be used to manufacture the turbine blades as they are inferior to silicon nitride-based materials in strength and slow crack growth resistance. At the same time, the physical properties of silicon carbide result in higher thermal stresses compared with those of similar components made of silicon nitride-based materials.

For the experimental investigation, a number of test rigs including a thermal cycling rig, spin rigs, and experimental engine gas generators were updated. Blades being prepared for thermal cycling testing can be seen in Figure 31.1. The test facility allows a maximum gas temperature of 1450°C, a maximum pressure 6.0 Pa, and a maximum temperature rate of 350°C/s. One of the spin rigs (see Figure 31.2) allows the testing of a running wheel with blades under atmospheric conditions (not in a vacuum chamber) using friction between the blades rotor and air to heat the blades. Spin rigs have a maximum rotational velocity of 70,000 rpm. When testing a metal disk with several inserted ceramic blades the temperature was kept below 800°C because of the risk of failure of the uncooled metal blades or disk rim. It is also possible to excite blade vibrations by special turbolators or air nozzles (Figure 31.3). A holographic technique was used to investigate frequency spectra and modes of natural vibrations, while high-cycle fatigue was explored with vibrorigs in which the vibrational modes were excited by a modulated compressed air jet. The interferograms of some oscillation

Figure 31.2 Spin Test Rig

modes of a turbine ceramic blade are shown in Figure 31.4. An air vibro-bench for fatigue testing at frequencies up to 20 kHz is shown in Figure 31.5. Vibrations are excited by a modulated compressor air jet.

Because of the high cost of blade machining and the criticality of the strength of the blade root for blade performance, tests of blades fabricated from hot pressed material subjected to centrifugal forces were conducted in several stages. Initially, i.e. prior to machining of the blade airfoil, the blade attachment strength was verified in the cold spin rig. Subsequently, a completely finished blade was tested in the spin rig (at speeds in excess of the design value) and in the gas generator (see Figure 31.6).

The sintered and reaction-bonded material blades were finally manufactured and tested at the spin rig and in the engine gas generator. The total amount of the tested blades was approximately 300. A significant number of the blades were tested at "Rybinsk Motors"

Tests with stepwise increase in temperature, centrifugal loads, number of cycles, and operation at the most severe conditions have substantiated the potential performance of blades fabricated of hot pressed and densified reaction-bonded silicon nitride. However, to demonstrate the durability of high-stressed blades made of a low-toughness ceramic, it is necessary to conduct proof tests at overload conditions. Invariably this leads to a large number of blades being rejected in the proof test, and a high rejection rate is unacceptable. The different lots of the blades were tested in the spin rig at various coefficients of overloading. To reduce the number of failed blades during these tests the coefficient of stress overloading did not exceed 1.2, and a significant number of the blades were tested in the spin rig without overloading, i.e. at the maximum design operating stresses. In view of these considerations, it appears that state-of-the-art monolithic ceramic materials can be considered only for production of turbine blades with relatively low stresses (not higher than 250 MPa).

A significant portion of the tests in the gas generator were conducted at maximum stresses that were essentially lower than the maximum design stresses for the engine in operation. This was done to minimize

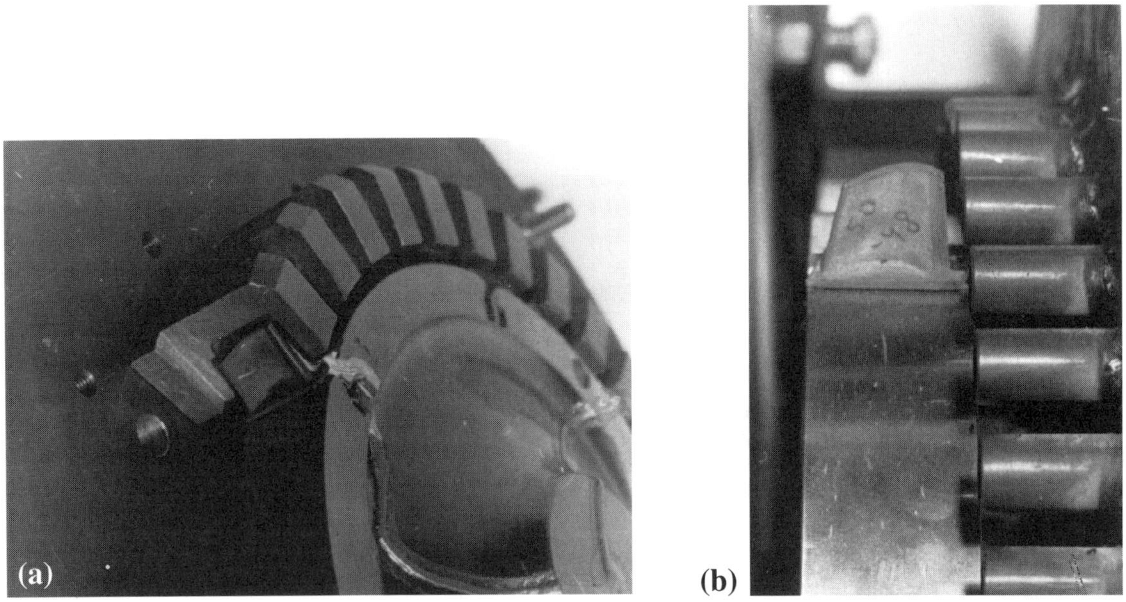

Figure 31.3 Turbolators (a) and Air Nozzles (b) Exciting Blade Oscillations in the Spin Test Rig

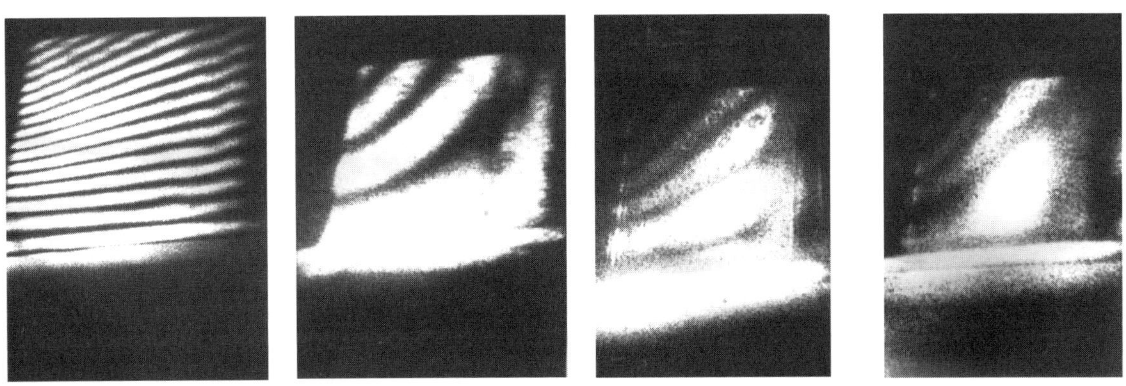

17359 16901 11454 20952

Figure 31.4 Some Vibrational Modes of a Ceramic Turbine Blade

the possibility of failure of the blades not tested at the required overloading stress in the spin rig. The maximum material temperature at the tests in the gas generator was ~ 1200°C. The maximum duration was about 35 hours. The timing of a ceramic blade failure was recorded with an optical pyrometer that measured the temperature of the rotating blades. During these tests there were cases of blade failures as a result of a slow crack growth due to initial defects. These failures likely occurred because these blades had not been tested at the required overload (because of the expected high level of blade rejection). It seems that it is also nec-

Figure 31.5 An Air Vibro-bench

Figure 31.6 Mounting of Specimen with Machined Ceramic Blade

essary to improve nondestructive control methods to evaluate the probability of part failure at the known defect distribution instead of conducting proof tests at overload conditions.

There are various ways for increasing the toughness of ceramics. While a modest improvement in fracture toughness can be achieved by SiC whisker reinforcement, more substantial toughness increases are achievable by using fiber-reinforced (primarily, SiC) structural ceramics. Unfortunately, the strength of the state-of-the-art SiC fibers tends to decrease substantially at a temperature of 1200°C. So it is necessary to continue development of ceramic materials that combine a higher fracture toughness than current monolithics with better strength than SiC-reinforced ceramic matrix composites. A material with $K_{IC} \geq 15$ MPa.m$^{1/2}$, flexural strength of about 1000 MPa, and excellent environmental resistance would be required for highly stressed turbine blades. This task seems to be rather complicated. As far as fracture toughness is concerned the best silicon nitrides have K_{IC} of about 8–9 MPa.m$^{1/2}$ (e.g. AS-800), whisker-reinforced silicon nitrides and silicon carbides probably have K_{IC} of about 6 MPa.m$^{1/2}$, and the best SiC/SiC CMC's have K_{IC} of about 25 MPa.m$^{1/2}$ (see Volume 2).

The influence of creep on the blade microstructure was not evident after the tests. But a thorough analysis of the effects of creep and oxidation on blade durability is also essential for the design of ceramic turbine blades for engines with long service life [5, 15]. Component temperature is a key factor for creep, and creep will need to be considered if temperatures exceed 1200°C. Oxidation is very much dependent on temperature, total pressure, partial pressure of water vapor formed in the combustion process, and gas velocity. So all relevant factors need to be considered. Unfortunately, the amount of data on ceramic material creep is extremely limited now.

Reliable operation of the dovetail blade with intermediate compliant layer (thin metal foil, sprayed coating, etc.) at the root over the service life of the engine [16] and optimization of the structural damping are also necessary. The vast body of experience with the optimization of structural damping for metallic blades, accumulated at CIAM and other companies, is available. In some cases the tip speed of the ceramic turbine blade had to be limited because of edge failure as a result of the impact of hard particles [17]. This would require additional study although our preliminary investigations have shown excellent resistance of superhard ceramic materials (in particular, hot pressed silicon nitride) to abrasive erosion. In particular, during an erosion test with blowing quartz sand with particle dimensions up to 300 μm at particle velocities up to 300 m/s the erosion resistance of hot pressed silicon nitride materials is essentially higher than that of nickel alloys.

CERAMIC COMBUSTOR PARTS

The application of ceramics instead of metal alloys for combustor liners can provide reliable liner operation with low cooling air flow. As a result, combustor mass and HC,CO and NO_x emissions can be decreased substantially. Air saved can be redirected to lean out the flame which lowers NOx [10, 11, 18].

In some projects ceramic integral liners are made from a solid billet. The development of liners of a segmented design constitutes an advanced trend for combustors with improved durability. [10, 11]. The main merits of the segmented combustor compared to a single-piece combustor are the reduction of thermal stresses, and the possibility of using heat-resistant materials, in particular ceramics, for the segments. The studies at CIAM have focused on segmented ceramic liners since they were believed to be more reliable than integral ceramic liners. Reducing the thickness of the segments lowers the thermal stresses, and also results in less damage to the turbine in case of a segment failure. Since 1991 some combustors with liners consisting of square ceramic segments, 1-1.2 mm thick, were developed at CIAM [6]. These segments were secured to the metal wall by a "dovetail type" attachment placed in the center of each segment (Figure 31.7).

To identify the coefficient of convective heat transfer between the segments and the cooling air and, also, between the segments and the heated air stream, tests were carried out with a flat panel with segments exposed to the open air stream. The temperature distribution over the external surfaces of the segments at

Figure 31.7 Hot Pressed Silicon Nitride Tiles

sudden cooling air flow changes was determined using the thermovision system AGEMA-780, and the coefficients of convective heat transfer were calculated according to the heat rate.

The segments were tested in a high-temperature combustor at steady-state and transient heating conditions (Figure 31.8). The maximum pressure at the combustor inlet ranged from $P_C{}^*$ = 300–1500 kPa, the maximum air temperature at the combustor inlet was $T_C{}^*$ = 327–527°C (600–800 K), and the excess air coefficient in the combustor varied between α_{Cmax} = 4.5–6.5 to 1.2 which corresponds to near stoichiometry.

In other cyclic tests the cooling air flow in the model combustor was varied stepwise from 15% to 3.5%. The corresponding air flow in a conventional annular combustor varied from 37% to 9%. The duration of the cyclic tests was 40-60 s. The segment temperature was recorded using thermocouples. The experimental results were close to those predicted by three-dimensional calculations. The maximum segment temperature was 1220–1300°C. The best results in the preliminary tests were obtained using segments made of hot-pressed silicon nitride, developed by the State Scientific Center "Technologhia" (Obninsk). The tiles of different reaction-bonded, sintered and hot pressed materials developed by the State Scientific Center "Technologhia" (Obninsk), the Institute of Materials Problems and Institute of Superhard Materials of the National Academy of Science of the Ukraine (Kiev) and other organizations have been investigated. The properties of some investigated materials are shown in Volume II, chapter 12.

Fractographic studies, performed according to MIL HANDBOOK 790, indicated that the failure of silicon carbide segments (of various manufacturers) was the result of fabrication defects. After optimization of the cooling system of the segments which led to a reduction in thermal stresses, and an improvement in the quality of the fabricated parts, the silicon carbide segments managed to survive the tests. Total times of testing of silicon nitride segments and silicon carbide segments at mentioned conditions were about 6 and 4 hours, respectively. After testing, the color of the silicon carbide segment "hot" surface became grayer, but the color of the silicon nitride segment surface remained almost unchanged.

3–D calculations for targeted service operation indicated that the thermal stresses in the design were low (≤ 25 MPa), except for those in the attachment, where the temperature (1350°C) and compressive thermal stresses (75 MPa) are at their maximum. The analysis indicates that special cooling of the attachment (for example additional impingement cooling around the dovetail) can decrease the thermal stress and assure

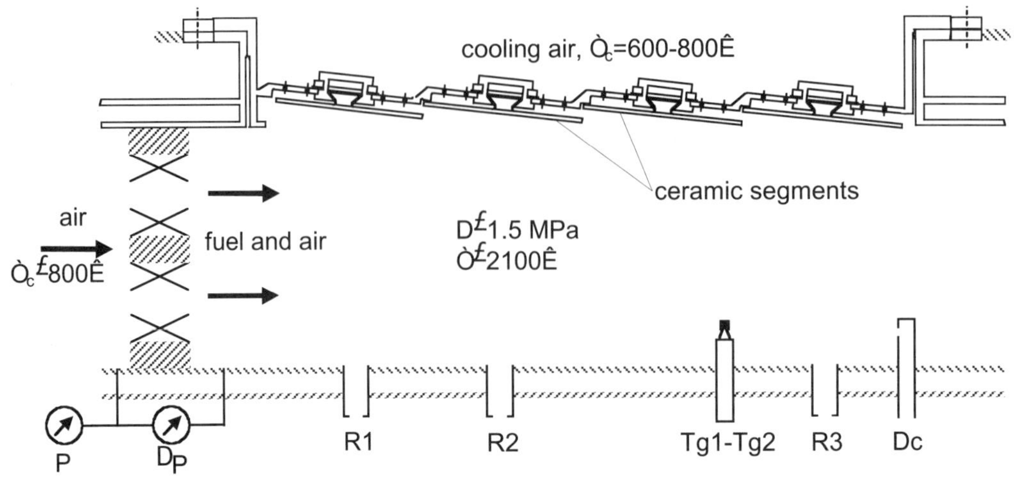

Figure 31.8 Experimental Model of Combustion Chamber with Ceramic Segments R_1, R_2, R_3—Radiometers, Tg_1, Tg_2—Platinum/Rhodium Thermocouples

acceptable life for the combustor with the ceramic segments. Also, the analysis demonstrates that ceramic matrix composite materials, in particular SiC/SiC, have properties that are adequate for segment fabrication. The results obtained can also be used for the development of other stator parts, in particular for ceramic segmented turbine shrouds and nozzle elements.

CERAMIC MATERIALS FOR ROLLING ELEMENT BEARINGS

The interest in the development and application of bearings with ceramic rolling elements and steel rings (hybrid bearings) derives from the realization that component life targets can be achieved at higher speeds by decreasing the centrifugal loads imposed by the rolling elements made of a relatively low density ceramic material [10, 11, 19].

The calculations and experimental studies for hybrid bearings with hot-pressed silicon nitride balls, supplied by the Research-Manufacture Enterprise "Technologhia" in Obninsk, were performed at CIAM. The properties are given in Volume II (chapter 12). The tests performed at $d_m \cdot n = 3{,}05 \cdot 10^6$ mm rpm (d_m is the diameter for rolling body centers in mm; n is the speed in rpm), at an axial load of 200–250N and radial load of 35–50N, validated the feasibility of the hybrid bearings for high speed rotors. The durability of the hybrid bearings is currently under investigation. The high cost of machining of the ceramic balls and the necessity of a reliable nondestructive quality control method remain challenges which have not been resolved yet.

Designers have great interest in ceramic bearings (with ceramic rings and rolling bodies) because the requirements for lubrication and cooling are much less severe [10, 11, 19]. These bearings are 2.5 times lighter than metallic bearings and are capable of operating under aggressive and cryogenic environmental conditions. Ceramic bearings with balls and rings made of OTM-914 (hot pressed silicon nitride—State Scientific Center "Technologiya," Obninsk, Russia) were investigated at CIAM.

Preliminary testing of the ceramic bearings revealed a number of other challenges. To compensate for the thermal expansion difference between the ceramic and steel bearing parts, the bearing assembly must have special elastic elements. To solve this problem compensating elastic elements were installed between the ceramic and steel parts of the bearing assembly. The ceramic bearing with a compensator was success-

Figure 31.9 Bearing with Ceramic Rings, Balls and Carbon/Carbon Cage

fully tested at $d_m \cdot n=2{,}99x10^6$mm rpm under and axial load of 1700N and a radial load of 350N. Duration of the operation of the ceramic bearing at maximum revolutions without special lubrication supply was 1 hour.

The tests also demonstrated that the ceramic bearings do not survive vibration and dynamic loads well. In particular, the transition through a critical speed, usually associated with higher vibrational loads, can result in destruction of the ceramic bearing parts. Therefore, shifting the critical frequency to a lower speed is desirable for decreasing the vibrational loads and for reducing the number of load cycles during the transition through the critical frequency. Dampers with a certain elasticity must be installed in the rotor supports to provide a shift of the rotor critical frequencies from the operational speed range.

The choice of the cage material and design is another challenge for the development of the ceramic bearings. The tests of bearings with a ceramic cage have demonstrated its ineffectiveness under high speed conditions. Metal and fluoroplastic cages do not ensure reliable operation of the bearing at high temperatures. Metal (Figure 31.9a) and fluoroplastic cages do not ensure reliable operation of the bearing at high temperatures. The use of a carbon/carbon cage appears promising (Figure 31.9b). Lack of funding to-date has not allowed the completion of this work. So, a number of intricate problems must be solved before ceramic bearings can be successfully implemented.

SUMMARY

It was shown that in the near term uncooled small turbine blades, segments of combustor liners, and some other stator components, and bearing balls will be the most desirable ceramic materials applications for gas turbine engines. These components may be able to meet the expected performance targets. However, in order to ensure adequate and reliable operating life for high-stressed structures (foremost for turbine blades) improved fracture toughness will be required. Development of composite ceramic materials is a promising way to solve this problem.

ACKNOWLEDGEMENTS

The important contributions of Prof. I.A. Birger, Prof. K.M. Popov, Prof. T.D. Karimbaev, Dr. I.Y. Kelina, Mr. E.G.Kouznetsov and Mr. A.N. Menshikov for the work reported here are acknowledged. The authors are grateful to a large number of CIAM experts for their participation in various investigations, especially to Mr. V.G. Seleznev for providing the investigation of ceramic blade vibration modes using a holographic technique. All figures in this chapter have been provided courtesy of CIAM.

NOMENCLATURE

C/C	carbon/carbon
d_m	distance between centers of balls in rolling bearing
°C	degree (centigrade)
K	Kelvin
K_{1c}	fracture toughness
kPa	kiloPascal
mm	millimeter
MPa	megaPascal
N	Newton
n	revolution
Pa	Pascal
P_c*	total pressure before combustor
rpm	revolutions per minute
s	second
SiC	silicon carbide
SiC_f	silicon carbide (fiber)
Si_3N_4	silicon nitride
T_c*	total temperature before combustor
α_{cc}	coefficient of air excess in combustor.

REFERENCES

[1] Nozhnitsky, Y.A.,Ogorodnikov, D.A., Birger, I.A., Popov, K.M. "Perspectives of the Application of Ceramics for Gas Turbine Engines," *The Abstracts of the XIIth All Union Scientific Conference "Design and Technologies of Non-metal Parts Manufacture,"* Obninsk, 1990, part 1, p. 4 (in Russian).

[2] Boutourlinova, N.A., Karimbaev, K.D., Kouznetsov, E.G., Navrotsky, V.V., Nozhnitsky, Y.A., Fedina Y.A., and Shechtman, Y.V., "The Evaluation of the Stress-Strain State, Strength and Reliability of Ceramic Parts," The XXIIIth All-Union Scientific Meeting on Engine Strength, 1990, Moscow, p. 48 (in Russian).

[3] Navrotsky, V., Nozhnitsky, Y., Shechtman, Y., Boutourlinova, N., Fedina Y., and Chyiaston, E., "Designing Gas Turbine Ceramic Elements," *Proceedings of the 4th International Symposium on Ceramic Materials and Components for Engines*, Göteborg (Sweden), 10–12 June, 1991. pp. 1035–1041.

[4] Evans, A.G. and Wiederhorn, S.M., "Proof Testing of Ceramic Materials—an Analytical Basis for Failure Prediction," Int. Journal of Fracture, Vol. 10, pp. 379–392, 1974.

[5] Nemeth, N.N. and Gyekenyesi, J.P., "Probabilistic Design of Ceramic Components with the NASA/CARES Computer Program," in *Engineered Materials Handbook*, Vol. 4. Ceramics and glasses, ASM International, Metals Park, OH, USA, pp. 700–708, 1991.

[6] Domas, P. A., "Probabilistic Life Prediction Methods—Meeting the Challenges for Practical Application," *Proceedings of the Sixth International Fatigue Congress*, Vol. 3, May 6–10, 1996, Berlin, Germany, Pergamon Press.

[7] Nozhnitsky, Y.A., Kouznetsov, E.G., Zudin, N.N. "The Experience of Ceramic Blade Development," *Abstracts of the XIIth All-Union Scientific Conference "Design and Technologies of Non-metal Parts Manufacture,"* Obninsk, 1990, Part 1, p. 4 (in Russian).

[8] Nozhnitsky, Yu.A., Markov, A.S., Sakovich, V.N. Smirnov, L.E., and Fedina, Yu.A., "Investigation of the Strength of Ceramic Rotor Components," in *Ceramics Today—Tomorrow's Ceramics, Proceedings of the 7th International Meeting on Modern Ceramics Technologies (7th CIMTEC—World Ceramics Congress)*, Montecatini-Terme, Italy, June 24-30, 1990. P. Vincenzini, ed., Elsevier Science Publishers B.V. Amsterdam, The Netherlands, pp. 2717–2724.

[9] Nozhnitsky, Y.A., Markov, A.S., and Menshikov, A.N., "The Experimental Investigation of non-metal heat resistant rotor element strength," The Union scientific meeting of engine strength problems, Academy of Science of USSR, 1990, pp. 122–123. (in Russian).

[10] Nozhnitsky, Y.A., Fedina, Y.A., Rekin, A.D. and Petrov, N.I., "Development and Investigation of Ceramic Parts for Gas-Turbine Engines," ASME paper 97-GT-157, presented at the International Gas Turbine and Aeroengine Congress and Exposition, Orlando, FL, USA, June 2–5, 1997.

[11] Nozhnitsky, Y.A., Fedina Y.A., Kouznetsov, E.G., Navrotsky, V.V., Rekin, A.D. and Petrov, N.I., "The Evaluation of Gas Turbine Engine Ceramic Element Strength and Durability," The XXVIth International Scientific Meeting on Engine Dynamics and Strength, June, 1996, Samara, (in Russian).

[12] Boutourlinova, N.A., Fedina, Y.A., Navrotsky, V.V., and Kouznetsov, E.G., "Assesment of Strain-Stress Condition, Strength and Reliability in Details Made of Ceramic Materials," *Proceedings of the European Ceramic Society Second Conference*, Augsburg, Federal Republic of Germany," Euroceramics II," G. Ziegler and H. Hauser, eds., 11–14 September, 1991. Vol. II, pp. 1131–1136.

[13] Fedina Y.A., "Gas Turbine Ceramic Part Strength Reliability," *The International Scientific Conference "The Reliability of Machines, Mechanisms and Equipment,"* February 15–17, 2000, Slavskoe, the Ukraine, p. 119, (in Russian).

[14] van Roode, M., Price, J.R., Gates, S., Waslo, S. and Hoffman, P., "Ceramic Stationary Gas Turbine Development—Opportunities and Challenges" *Proceeding of the 7th International Gas Turbine Congress*, Kobe, Japan, November 14–19, 1999. Vol. I, pp. 65–72, 1999.

[15] DiMascio, P.S., Orenstein, R.M. and Rajiyah, H., "Reliability of a Conceptual Ceramic Gas Turbine Component Subjected to Static and Transient Thermomechanical Loading," ASME paper 97-GT-284, presented at the International Gas Turbine and Aeroengine Congress and Exposition, Orlando, FL, USA, June 2–5, 1997.

[16] Kinney, T.W. and Easley, M.L., "Ceramic Gas Turbine Technology Development," ASME paper 97-GT-465, presented at the International Gas Turbine and Aeroengine Congress and Exposition, Orlando, FL, USA, June 2–5, 1997.

[17] Takehara, I., Tatsumi, T., and Ichikawa, Y., "Development Summary of CGT302 Ceramic Gas Turbine," *Proceedings of the 7th International Gas Turbine Congress*, Kobe, Japan, November 14–19, 1999. Vol.I, pp. 57–64, 1999.

[18] Sikorski, R.L., "Thermal Mechanical Fatigue and its Effect on Combustion Design," AGARD CP-569, paper presented at the Conference on Thermal Mechanical Fatigue of Aircraft Engine Materials, Banff, Canada, October 2–4, 1995.

[19] Zaretsky E.V. "Ceramic Bearings for Use in Gas Turbine Engines," ASME paper 88-GT-138, presented at the Gas Turbine and Aeroengine Congress, Amsterdam, The Netherlands, June 6–9, 1988. *Transactions of the ASME, J. for Gas Turbines and Power*, 111 [1], pp. 146–157, Jan. 1989.

CHAPTER

32

RESEARCH-ENGINEERING CENTER "CERAMIC HEAT ENGINES (NIZ KTD) AT THE RESEARCH-TECHNOLOGICAL INSTITUTE FOR POWER ENGINEERING (NITI EM)"

Anatoly V. Soudarev and Vitaly Yu. Tikhoplav

St.Petersburg, Russia

INTRODUTION

Because of their excellent thermal stability ceramics are considered as structural materials for gas turbines, and it is believed that ceramic gas turbine components may raise this equipment to a higher technological level [1, 2]. To attain this goal, a variety of challenges are being addressed by NIZ KTD. The most pressing of them are the following:

- development and demonstration of fabricated structural shrinkage-free ceramics, high-temperature ceramic components and a ceramic gas turbine engine (CGTE)
- improvement of structural ceramics to meet requirements of reliability and fabricability so that they can be applied as hot section components for CGTEs
- development, manufacture and testing of components and assemblies of ceramic high temperature GTEs taking into account the specific properties and behavior of ceramic materials

To meet the above challenges, NIZ KTD, has been conducting a program to develop and demonstrate a low-NO_x high efficiency hybrid metal-ceramic 2.5 MW GTE for power generation applications. Due to funding problems following the political changes associated with the transition from the USSR

TABLE 32.1 PROGRAM SCHEDULE FOR THE NIZ KTD 2,5 MW METAL-CERAMIC GTE (CGTE-2.5) (REPRINTED FROM REF. [3])

Description	1991	1992	1993	1994	1995	1996	1997	1998	1999	2000	2001	2002
Development of techniques of manufacture of ceramic parts	■	■	■	■								
Elaboration of gas-generator contractor design	■	■	■									
Refinement of techniques of optimum screening and evaluation of ceramic components strength	■	■	■	■								
Tests of subscale combustor		■	■	■								
Research over module structures of shell-tube ceramic recuperators		■	■	■								
Tests of rotor blade models			■	■								
Tests of full-scale module of recuperator					■	■						
Complex tests of full-scale rotor blade					■	■						
Complex tests of full-scale guide vanes					■	■	■					
Tests of full-scale combustor						■	■					
Tests of full-scale stator elements of high temperature paths						■	■					
Manufacture of demonstration unit assemblies							■	■	■	▒		
Tests of demonstration unit									■	▒		
Manufacture of gas-generator components										▒	▒	
Test of complete gas-generator											▒	▒

to Russia in the late 1980s and early 1990s, the program had to be delayed. The current program is funded primarily by the Russian Joint Stock Company "GAZPROM" and there is some foreign funding, mostly negotiated through the French National Aerospace Research Institute (ONERA—Office National d'Etudes et Recherches en Aerospatiales). The current schedule for the four phases of the program is shown in Table 32.1 [3].

By the time the program was underway (Table 32.1), foreign companies in Japan, the USA, in Germany, etc, had already gained (over 20 years and more) experience with the production of structural ceramic materials (SCMs) and with ceramic piston and gas-turbine engine development. By 1990, around a hundred of comparatively promising silicon nitride- and carbide-based and also metal oxide-based SCMs had been developed by these foreign companies. These SCMs have a unique combination of properties which enables manufacture of high-temperature GTE parts. Notwithstanding the impressive progress in fabrication of ceramic components, long term ceramic engine tests have only recently been conducted by companies such as Honeywell Engines and Systems, Solar Turbines Incorporated, Kawasaki Heavy Industries, Ltd., and Rolls Royce Indianapolis [4–7].

TABLE 32.2 PHYSICAL AND MECHANICAL PROPERTIES OF THE ALUMINO-BORON-NITRIDE CERAMIC "SUGRAV"

NAME AND DESIGNATION OF SIZE DIMENSION	Units	Numerical value
Density	g/cm^3	2.6–2.8
Maximum operating temperature	°C	1600
Coefficient of linear thermal expansion	K^{-1}	$9.0x10^{-6}$
Thermal conductivity	W/(m.K)	14.0–16.0
Specific heat capacity	kJ/(kg.K)	1.1
Ultimate bending strength (20°C-1350°C-1500°C)	MPa	120-60-60
Ultimate compressive strength	MPa	400–450
Modulus of longitudinal elasticity (20°C-1200°C-1400°C)	GPa	130-90-100
Weibull's modulus, 20°C	–	14
Fracture toughness, 20°C	$MPa.m^{1/2}$	4.1

CERAMIC COMPONENT MATERIAL

Our analysis indicated that the main causes of failures with the development of reliable ceramic GTEs result from two principal factors:

- First, complete copying of conventional metal designs for ceramic GTE parts. In the realm of ceramic gas turbine development representatives of various disciplines are collaborating with different backgrounds, and engineering assumptions. The design engineers on many programs were very knowledgeable in designing metal components but had very little experience with the design of ceramic components and the limitations and benefits of ceramics. Many programs had a characteristic learning curve where the constraints of working with ceramics had to be discovered in rig and (costly) engine testing.
- Secondly, insufficient consideration for fabrication requirements which is actually true for all SCMs known by 1990. This is attributable in the first place to the essential shrinkage (up to 10-18%) of ceramic parts during processing, and, consequently, the need for expensive finishing operations such as machining of ceramic parts using diamond tools, and also the lack of techniques for welding and diffusive joining of parts of structural ceramic components.

As a result, a basically new structural alumino-boron-nitride ceramic material "SUGRAV" was developed and a technique for diffusive joining of parts of this material was verified [8–10]. The main properties of the ceramic "SUGRAV" are shown in Table 32.2 [9]. The same reference provides a detailed description of the main features of this material, techniques of manufacture of parts and components of the high-temperature GTE path made of this material, the current studies of improved modifications of the alumino-boron-nitride composition-based SCMs. The important points to be made are that the material is shrink-free, requires little finishing, and has a use temperature in the 1500°C to 1600°C range. The material also has low flexure (bending) strength, but that may be OK since the design philosophy is based on placing the components in compression, and since the compressive strength is much higher.

The development of the structural ceramic capable of operating reliably without cooling at temperatures up to 1600°C provided the solution to two basic challenges for the gas turbine engine manufacturer. The first one is "how to raise the thermal efficiency of the GTE," and the second one is "how to develop a GTE that is environmentally friendly."

Addressing the second problem first, it is worth noting that for a conventionally cooled metal combustor the liner requires up to 30% of the total air flow from the compressor. Consequently, this cooling air is no longer available to lean out and cool the flame in the primary zone which is a critical requirement for a low-NO_x lean premixed combustor. The replacement of the cooled metal liner with an uncooled structural ceramic material conserves cooling air which can now be redirected to the combustor primary zone where it leans out and cools the flame resulting in a substantial reduction of thermal NO_x emissions. Note, apart from lowering NO_x a hot ceramic wall also helps lowering CO. The reason is that the increased wall temperature (typically from 850°C–900°C to 1000°C–1200°C) shifts the formation of CO to lower values. The CO contribution comes mainly from the areas near the combustor wall where air enters the combustor through the cooling holes in the liner creating localized low temperature conditions.

As for the higher cost-effectiveness of the ceramic GTE, the reasons for this phenomena are well known. Replacement of the metal cooled high-temperature gas turbine with a virtually uncooled one with its flow path incorporating hybrid ceramic-metal components enables the TIT to be up to 1500°C to 1600°C with an attractive reduction of fuel flow. Consequently, CO_2 emissions are also reduced which lowers the burden on the environment.

CERAMIC COMPONENTS TECHNOLOGY

All the ceramic elements (shells of guide vanes and rotor blades, heat protection platforms, liner, recuperator matrix, etc) were manufactured of upgraded alumino-boron-nitride "SUGRAV" ceramics, not requiring the use of a diamond tool for manufacture. The manufacture of the ceramic product is a two-stage process (Figure 32.1). First, a component is fabricated from a "green" mold using conventional tools in accordance with a drawing, then it is sintered at a temperature of 1000 to 1100°C in air and thermally treated at 1400°C. The dimensions of the components do not actually change during this operation. The yield of high quality products is up to 90%. Also, the technique of diffusion welding of the ceramic elements was developed with the strength of the welded joint not being inferior to that of the parent material.

To confirm the feasiblity of using "SUGRAV" ceramics, thermal and thermo-cyclic tests of ceramic model parts and full-scale components were performed. The heat-resistance tests were carried out on 50 × 80 × 10 mm plates at temperatures of 1250°C, 1350°C, 1400°C, and 1450°C for 7.5, 15 and 19 hours, and at 1600°C for 3 hours in an oxidizing atmosphere (in air). No changes in the structure, composition and configuration of the specimens were observed. The thermo-cyclic tests were carried out on the same plates and, also, on full-scale ceramic components (rotor blade shell elements). Heating from 20°C to the maximum temperature was implemented by insertion of the cold specimens into a preheated furnace. Cooling was conducted by blowing cold air on the specimens. No degradation of the specimens was observed after 20 thermal cycles.

More detailed descriptions of the manufacturing process for alumino-boron-nitride-based products, also, of the diffusive joining process for ceramic elements into a monolithic part have been published [9, 10].

THE COMPLEX CYCLE AND THE GENERAL LAYOUT OF THE CGTE-2.5 ENGINE

The gas-turbine unit developed under contract from the Joint Stock Company "GAZPROM" is meant for application by the gas industry for power generation or mechanical drive. The main design parameters of the engine, based on the cycle shown in Figure 32.2, are presented in Table 32.3 (The microturbo data are for a 1 MW demonstration unit fabricated for another customer). Here, the engine efficiency will not be below 45% which is around 1.5 times as high as that for existing engines of the same output power (Figure 32.3). The abbreviations used in the description and figures are as follows: LPC, LPT—low pressure compressor and low pressure turbine; HPC, HPT—high pressure compressor and high pressure turbine; CC—combustion chamber; IC—air intercooler; ER—end regenerator; PT—power turbine; LPA—low pressure assembly; HPA—high pressure assembly; DU—driven unit.

At the same time, the temperature of the working media at the inlet of the power turbine of such engine will be over 1060°C which will necessitate the application of cooled blades to this turbine. The latter complicates the turbine's design and makes the engine more costly. Given the above, and in view of the level

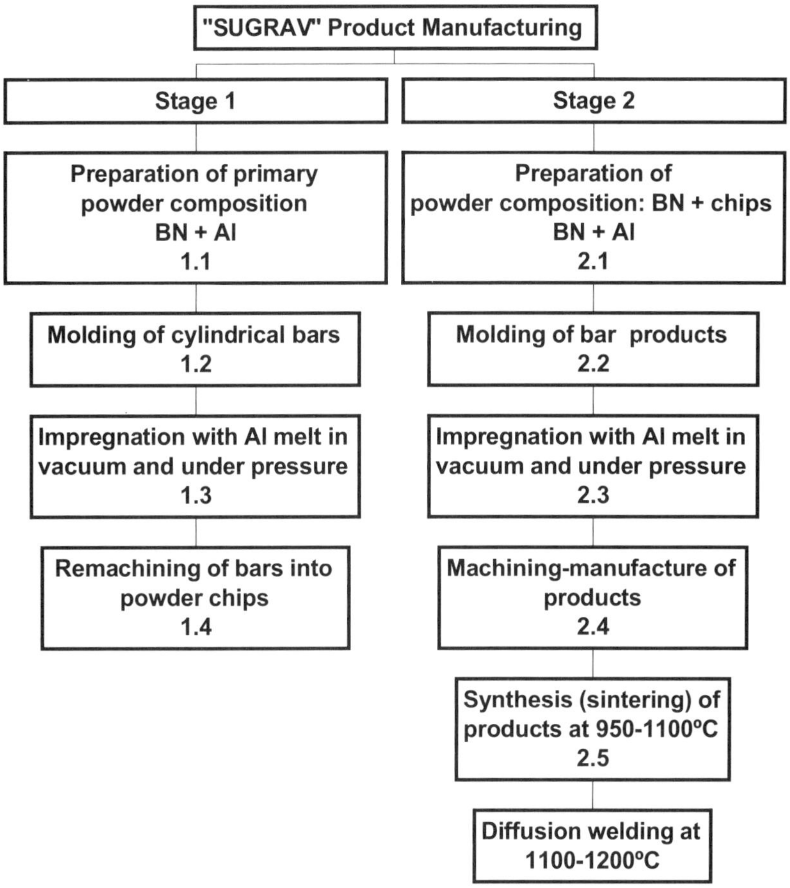

Figure 32.1 Flowchart Showing Manufacturing of "SUGRAV" Products (Adapted from Ref. [10])

of efficiency specified by the customer, namely 42%, it was decided to lower the temperature for the pilot GTE at the inlet of the power turbine by installing an inter-gas heater (Figure 32.2) with a regeneration ratio of 0.41.

The preliminary layout of the CGTE-2.5 with intermediate and end recuperators-airheaters is shown in Figure 32.4. The engine gas generator includes two assemblies (low and high pressure ones), and the air intercooler is mounted in between them. The high and low pressure gas generator assemblies have a similar design.

Demonstration Unit for the CGTE-2.5

The demonstration unit manufactured as scheduled for the 3rd phase of the program (Table 32.1) is indispensable for carrying out joint control tests for the metal-ceramic engine components which are mostly mechanically and thermally loaded. These are the turbine assembly, combustor, and intermediate airheater. The demonstration unit is actually a high-temperature part of the GTE gasgenerator and it excludes the high assembly and end airheater. Design and thermal parameters of the demonstration unit are presented in Table 32.3, the heat cycle and parameters in representative points are shown in Figure 32.5. All main elements of

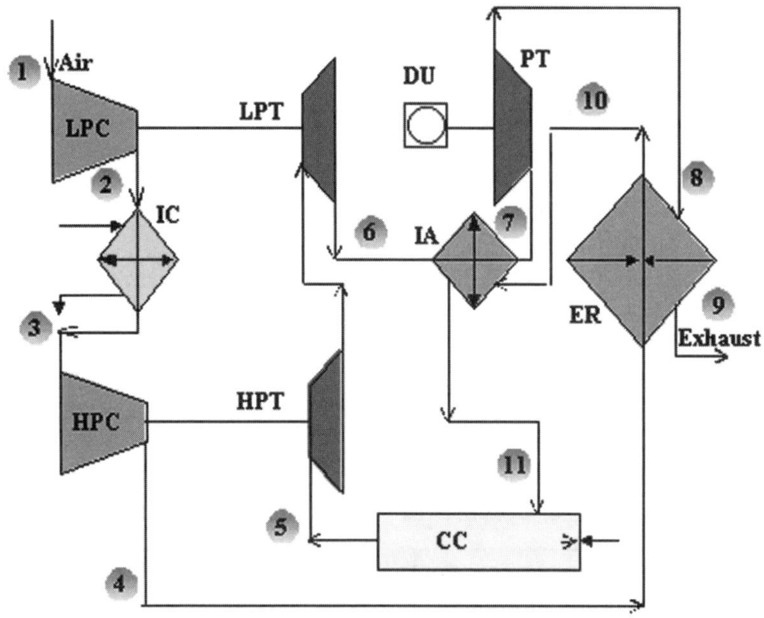

N	T(°C)	P (MPa)
1	25	0.1013
2	150	0.329
3	40	0.317
4	208	1.17
5	1350	1.08
6	1062	0.419
7	850	0.407
8	569	0.106
9	271	0.103
10	519	1.14
11	745	1.113

Figure 32.2 Engine Cycle (Adapted from Ref. [3])

the demonstration unit have already been manufactured and tested. At present, manufacture of the full-scale components of the demonstration unit is underway; with some of them undergoing (or having undergone) the rig tests.

The **turbounit** (Figure 32.6) assembly (see also Figure 32.4) includes the single-stage centrifugal compressor and axial single-stage metal-ceramic turbine with gas-static bearings. The main metal components of the **turbounit,** namely the centrifugal compressor wheel (1), shaft (2), turbine disk (3), and compressor diffuser (4) were manufactured in mid-2000. The demonstration unit turbine includes the following ceramic and metal-ceramic components: nozzle vanes, rotor blades, scroll, inlet and outlet pipe branches.

The **nozzle vanes** (Figure 32.7, see also Figure 13.7 in Ref. [9]) are all ceramic and include the external (1) and internal (2) rings made up of separate elements. Solid vanes (3) are mounted loosely between the elements, the rings are secured by bosses, with the external ring being inside the groove of the metal sleeve 5 of the bearing housing. This design ensures the required mechanical strength and free thermal expansion of all the elements. This design, developed on the basis of tests of the initial metal-ceramic design [11], is more ideal in terms of aerodynamics; it is thermally elastic and does not need cooling air consumption.

TABLE 32.3 MAIN DESIGN PARAMETERS OF ENGINE CGTE-2.5 AND DEMONSTRATION UNIT

	CGTE-2.5		
NAME	CGTE-2.5	Demonstration unit	GGTE Microturbo
Gas temperature at high pressure turbine inlet (TIT, °C)	1350	1350	1350
Gas flow Gg, kg/s	7.0	1.03	2.48
Pressure ratio in:			
LPC	3.25	—	—
HPC	3.69	3.69	3.73
Total pressure ratio	12	3.69	3.73
Speed of:			
low pressure assembly shaft ηLP, rpm;	22000	—	—
high pressure assembly shaft ηHP, rpm	23000	23000	50000
Regeneration ratio, %	86	41	—
Maximum peripheral velocity (m/s)	377	377	450
Cooling air flow, %	3	1.5	2
Running time, h	1000	1000	5
Number of thermal cycles	1000	1000	2
Cooling air temperature, °C	208	170	194
Maximum blades ceramic shell	–40	–40	–200
stress, nozzle metal core	180	180	850
MPa vanes ceramic shell	50	50	50

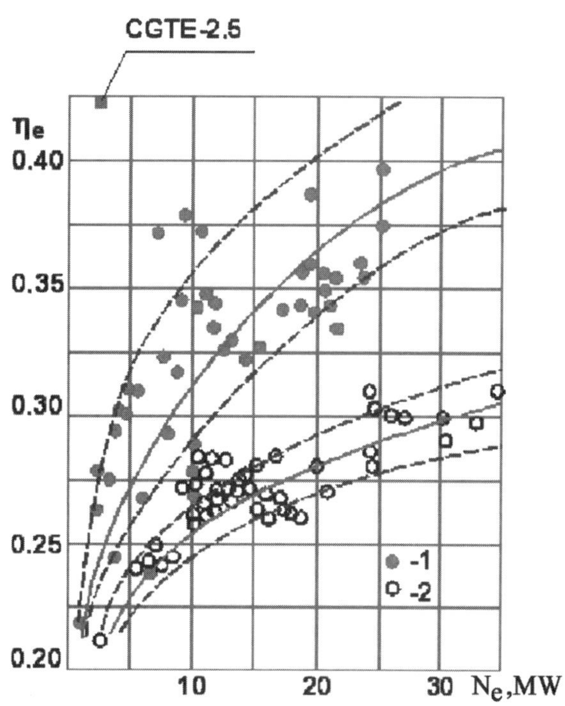

Figure 32.3 GTU Efficiency vs Its Power for: Regenerative and Aircraft GTUs (1); simple cycle GTU (2) (CGTE-2.5) (Reprinted from Ref. [3])

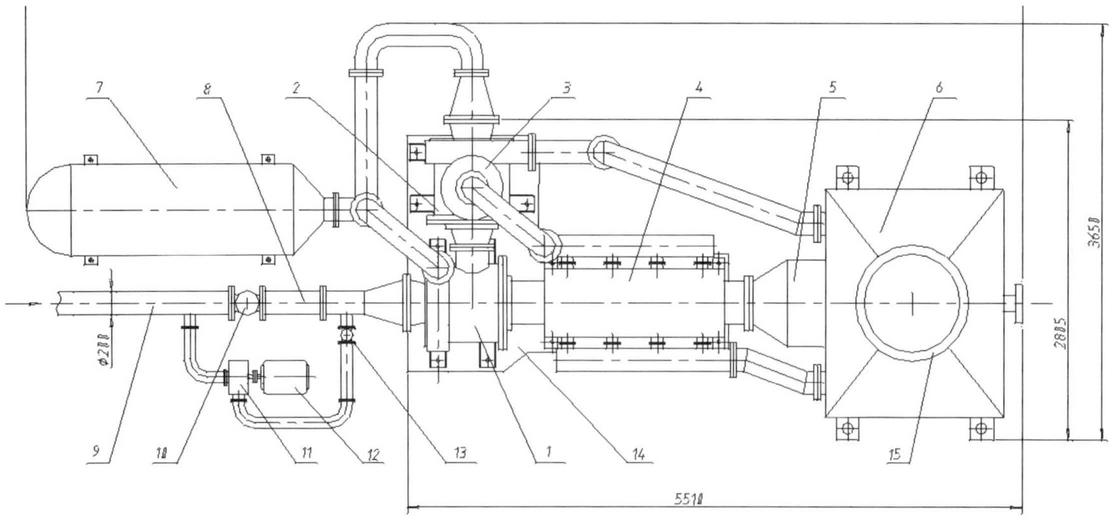

(1) LP turbocompressor
(2) HP turbocompressor
(3) Combustor
(4) Intermediate recuperator
(5) Power turbine
(6) End recuperator
(7) Intermediate cooler
(8) Damper
(9) Inlet duty gate valve
(10) Start-up blower
(11) Blower engine
(12) Start-up gate valve
(13) Frame
(14) Exhaust

Figure 32.4 Lay-out of 2.5 MW Plant CGTE-2.5 with Intermediate Cooler and Intermediate and End Recuperators (Reprinted from Ref. [3])

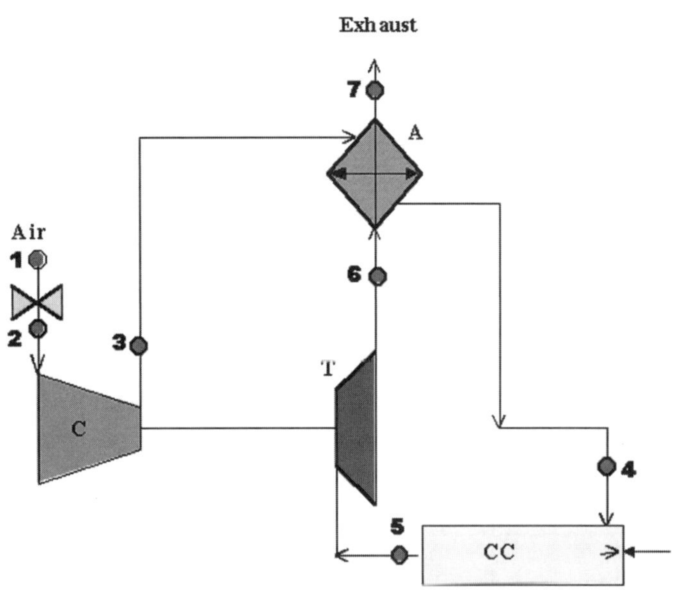

N	T (°C)	P (MPa)
1	15	0.1013
2	15	0.0505
3	170	0.1785
4	656	0.1731
5	1350	0.1679
6	1210	0.1061
7	783	0.1029

Figure 32.5 Heat Cycle and Parameters in Representative Points of Demonstration Unit; C—compressor; T—turbine; CC—combustion chamber and A—airheater (Adapted from Ref. [3])

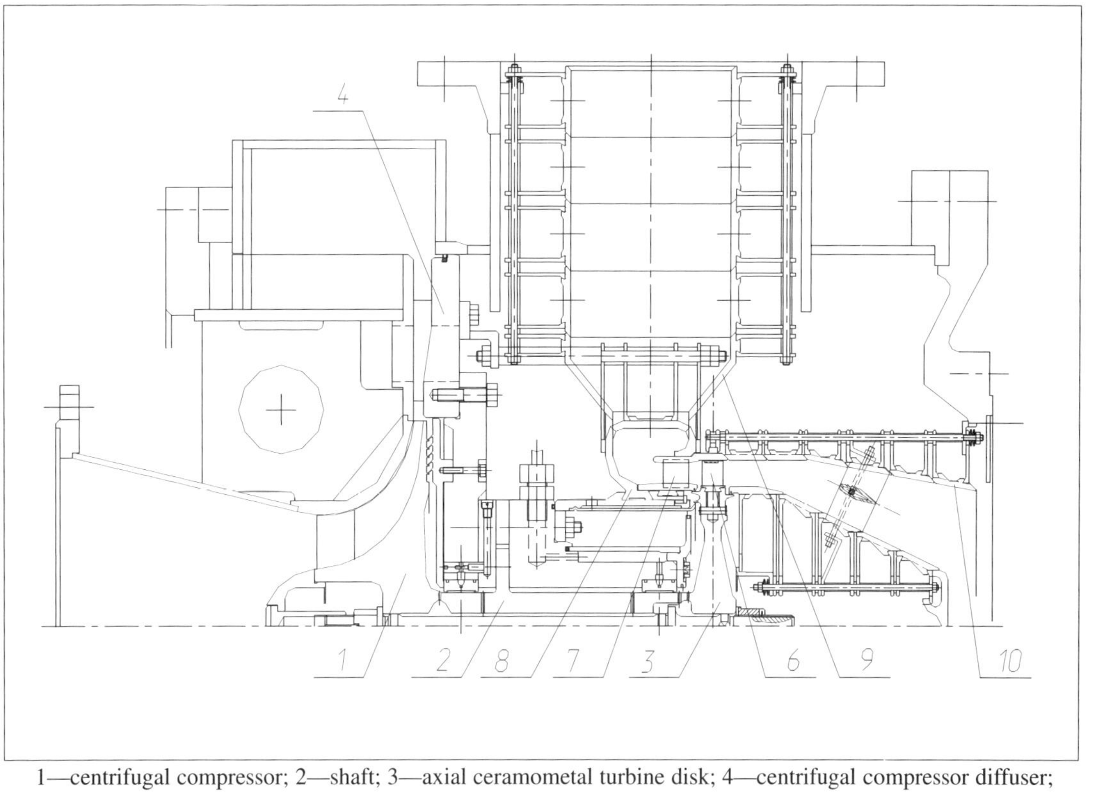

1—centrifugal compressor; 2—shaft; 3—axial ceramometal turbine disk; 4—centrifugal compressor diffuser; 5—casing; 6—rotor blade; 7—nozzle vanes; 8—scroll; 9—inlet ceramic branch pipe; 10—outlet ceramic branch pipe.

Figure 32.6 Longitudinal Section of Demonstration Unit Turbine CGTE-2.5 (Reprinted from Ref. [3])

The **rotor blade** design is shown in Figure 32.8. The rotor blades are knock-down and consist of the metal carrier core (6), ceramic shell (5), loosely mounted on that core, and the metal screen-deflector (4), installed between the ceramic shell and the metal core. The cross section configuration of the metal core fits into the internal surface of the rotor blade. The carrier core is made integral with the blade tail where the thread is provided for securing in the disk (1). When the disk rotates the ceramic shell is engaged by the centrifugal forces acting freely upon the core which rests with its upper end upon the metal core platform. Thus, a compressive stress is generated in the ceramic shell during rotation. Cooling air flows only between the core and the screen-deflector (4) with the semi-spherical recesses ensuring a point contact with the outer ceramic shell. The lock nut (2) fixes the blade angle and serves as a channel for the cooling air flow. The disk is protected against contact with the hot gas by the ceramic platform inserted over the disk (3). When the ceramic uncooled shell is in contact with the gas, 1.5% air flow is sufficient to ensure an admissible temperature state of the rotor blade elements.

As of the middle of 2000 all the ceramic rotor parts including the rotor blade shells, and lining elements had been manufactured (Figure 32.9), and the assembly of the turbine wheel was underway. Some of the parts had already undergone the static mechanical tests and, also, cold spin testing. (see Figure 13.4 in Ref. [9]).

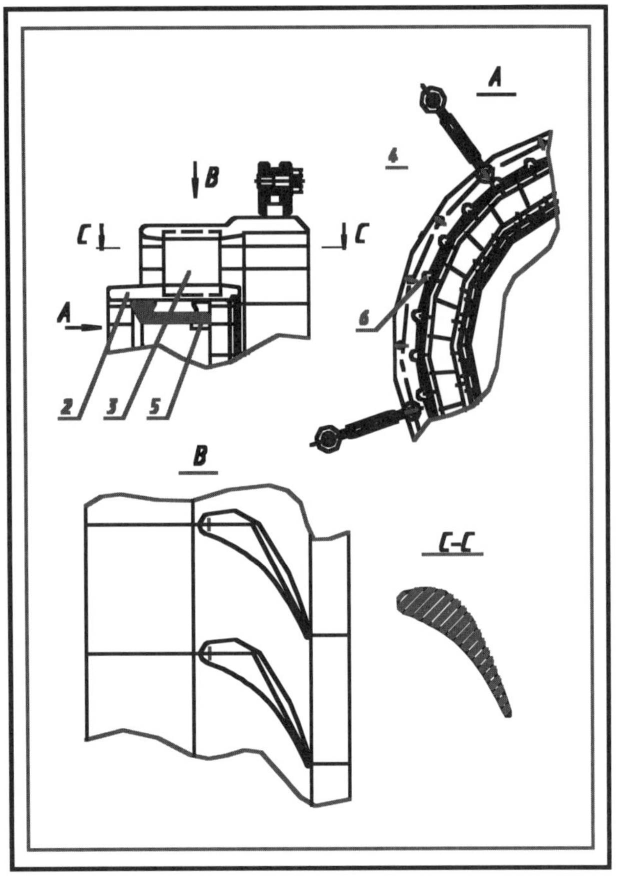

1—Outer Ring; 2—Inner Ring; 3—Vane; 4—Membrane; 5—Sleeve of Bearing Housing; 6—Cap.

Figure 32.7 Ceramic Nozzle Vanes (Reprinted from Ref. [3])

The hot section of the turbounit (Figure 32.6) for the demonstration plant includes, besides the guide vanes and rotor blades, the inlet branch pipe (Figure 13.6 in Ref. [9]), scroll and discharge annular diffuser (Figure 13.8 in Ref. [9]). The above components were fabricated using a common design concept, i.e. the ceramic shell which has direct contact with the hot gas consists of separate elements secured in the pre-compressed state on the bearing thermoelastic metallic membranes, thereby ensuring both the air-tightness and the free thermal expansion. At the same time, another design of the scroll made as a single monolithic component (Figure 32.10) comprising separate sections joined by a diffusive welding technique was being prepared for testing in 2000.

Gas-static thrust and journal **bearings** are used in the demonstration unit turbine in order to decrease fire and explosion hazards of the engine. Furthermore, this design also has simplified the gas generator lubrication system since there is no need for an oil tank, filters, oil coolers, oil heaters, pumps, etc. The tests of the air thrust and journal supports were carried out on special test beds with disk-simulators for the centrifugal compressor and axial turbine [7] with a step-by-step increase in the specific load on the journal surfaces.

The design of demonstration unit **combustor** (Figure 32.11, Table 32.4) is essentially the same as that of the full-scale combustor of the CGTE-2.5. The analysis of the initial test results on model components

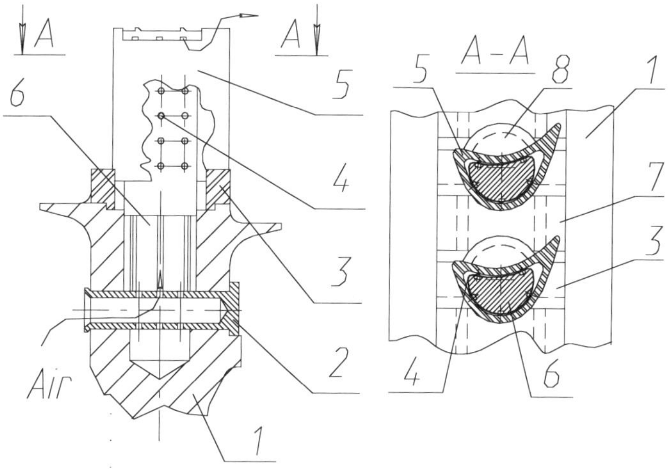

1—Turbine disk; 2—Lock; 3, 7, 8—Thermally Protective Ceramic Elements; 4—Deflector; 5—Ceramic Shell of Rotor Blade; 6—Carrier Core.

Figure 32.8 Metal-Ceramic Turbine Blade Design (Reprinted from Refs. [3, 11])

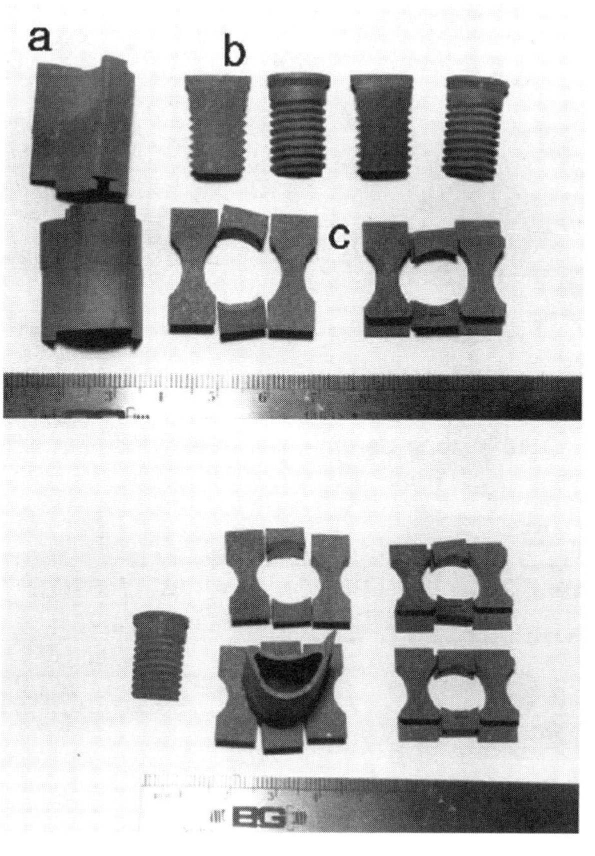

Figure 32.9 Ceramic Elements of Rotor Blades: Shells (a), Cores (b), Thermally Protective Inserts (c) (Reprinted from Ref. [3])

Figure 32.10 Ceramic Scroll for Demonstration Unit (Reprinted from Ref. [3])

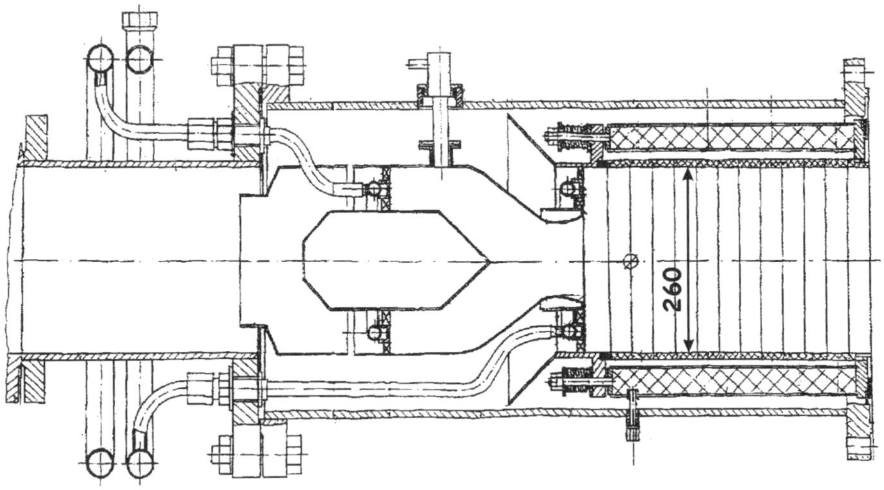

Figure 32.11 Double-zone Ceramic Combustion Chamber for CGTE-2.5 (Reprinted from Ref. [3])

TABLE 32.4 MAIN DESIGN CHARACTERISTICS OF COMBUSTION CHAMBER

	Values	
NAME OF PARAMETER	CGTE-2.5	Microturbo CGTE
Air flow at inlet, G_A, kg/s	6.50	2.4
Fuel flow, G_F, kg/s	0.117	0.382
Fuel	Natural gas	aviation kerosene TRO-JP8
Gas temperature at outlet, T_G, °C	1350	1350
Gas pressure at outlet, P_G, MPa	1.08	0.365
Air temperature at inlet, T_A, °C	745	194
Air pressure at inlet, P_A, MPa	1.113	0.405
Excess air coefficient α_o	3.5	2.0
NO_x emissions, mg/Nm^3 (15% O_2)	<25	—
CO emissions, mg/Nm^3 (15% O_2)	<50	—

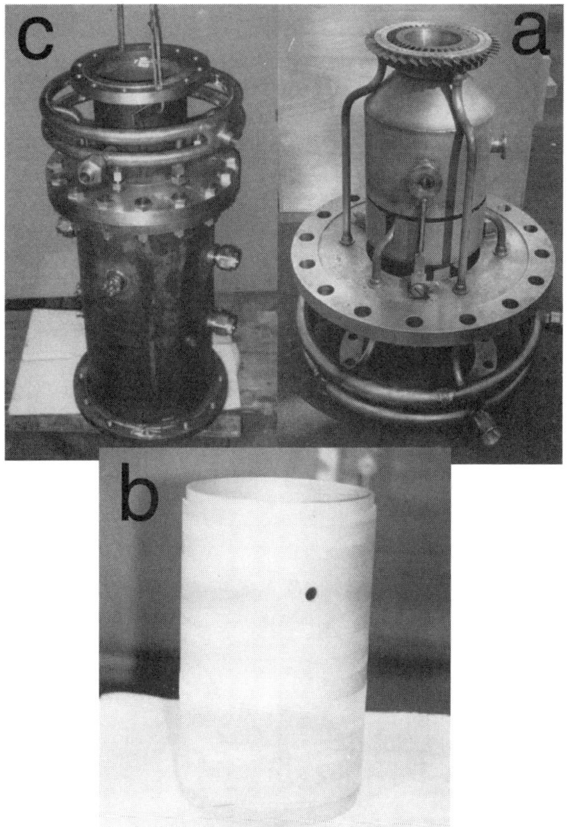

Figure 32.12 Double-Zone Burner (a), Ceramic Liner of the 2nd Zone (b) and Low-NOx ceramic Combustor of Demonstration Unit as an Assembly (c) (Reprinted from Ref. [3])

with liner diameters of 140 mm [12] and 250 mm [13], respectively, allowed the introduction of some corrections into the design of the attachment and co-fixation of the double-zone burner (Figure 32.12a) in the 2nd zone (Figure 32.12b) of the metal casing. In the investigated full-scale combustor, the 2nd zone liner is made as a series of connected uncooled ceramic rings of 260 mm diameter pressed by 4 springs.

To decrease the undesirable NO_x emissions from the combustor, a combination of a cascade air supply with microflame fuel burner was applied [12–14]. Two series of parallel connection domes of similar design (Figure 32.11) are envisioned in the combustor, each of these domes forming a fuel combustion zone. Each dome includes a flame-holder which, at the same time, operates as a gas manifold, and vanes placed on either side of the latter. The vanes swirl the air stream in a mutually opposite direction under a 45° angle. The fuel gas is delivered into the recirculation zone downstream of each flame-holder through a system of holes.

The 1st zone dome is meant to produce a fuel/air mixture which is directed towards the middle part of the combustor through a cylindrical nozzle. The similar 2nd zone device is utilized to produce a stream of high-temperature particles, igniting the mixture supplied from the 1st zone nozzle. There is no mixer in the combustor, and all air is delivered only through the domes.

The recuperated airheater (A, Figure 32.5) for the demonstration unit is designed in such a way that the data is applicable to both the intermediate and end regenerators of the CGTE-2.5 during its testing (Table 32.5). So, the airheater incorporates two sections of parallel arrangement, namely the metallic section consisting of 4 modules and the single-module ceramic one. The metallic modules are of the tubular cassette

TABLE 32.5 MAIN DESIGN CHARACTERISTICS OF INTERMEDIATE AND END REGENERATORS

	Values		
	End regenerator	Intermediate regenerator	
PARAMETER	CGTE-2.5	CGTE-2.5	Demonstration unit
Gas flow, G_G, kg/s	6.90	6.70	1.008
Gas temperature at inlet, T_G, °C	569	1062	1200
Gas pressure at inlet, P_G, MPa	0.106	0.419	0.1061
Air flow, G_A, kg/s	6.50	6,70	1.015
Air pressure at inlet, P_A, MPa	1.17	1.14	0.1785
Air temperature at inlet, T_A, °C	208	519	170
Air temperature at outlet, T, °C	519	745	656
Regeneration ratio μ	0.86	0.41	0.47

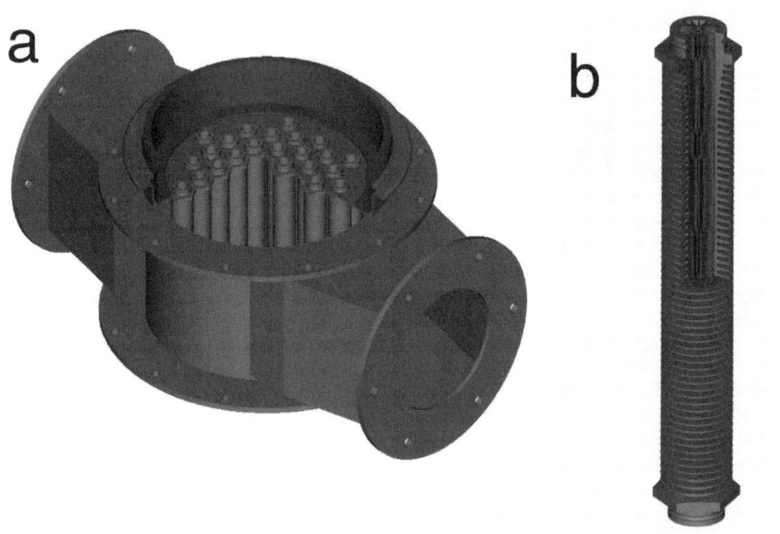

Figure 32.13 Design of Module (a) of Air Heater of Demonstration Unit with Matrix of Two-sided Finning (b) (Reprinted from Ref. [3])

type, highly compact (the coefficient of compactness is up to 1000 m^2/m^3) and lightweight. Their design and also the complex test data are provided in detail in [15].

The ceramic section module is a full-scale cell of the intermediate CGTE-2.5 recuperator since it's structure is mostly responsible for ensuring reliability of the large-size ceramic products [16]. It is known that the sensitivity of ceramics to stress concentrators tends to increase because of structural defects of the material. Our experience indicated that the seriousness of ceramic defects depends on part size and that the thermo-mechanical characteristics improve as the part size decreases.

The size of the ceramic recuperator can be reduced by a design based on modules that are connected in parallel (Figure 32.13), and by enhancing the heat exchange path. An increase in the convective heat exchange can be achieved by incorporating longitudinal or cross finning on the outside tubes (Figure 32.14a, b). For inside tubes, it is achieved by a longitudinal split finning, welded from the inside to the

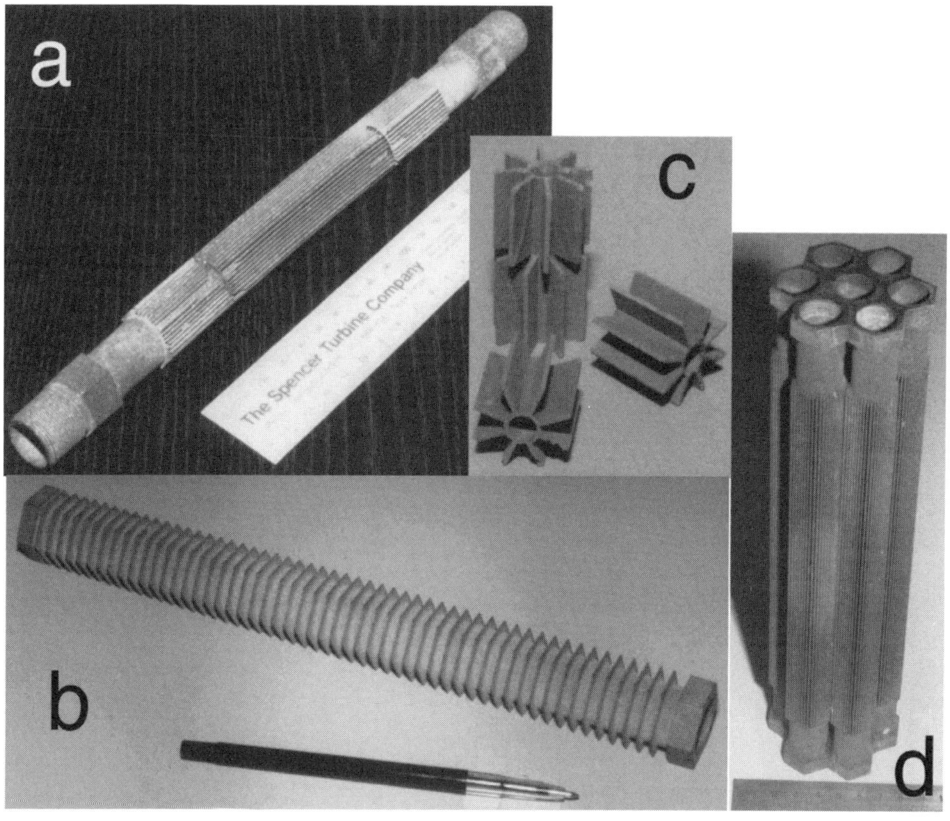

Figure 32.14 Tube Fin Elements of Ceramics (a, b, c) and 7-tube Model (d) of Heat Exchanger Module Matrix (Reprinted from Ref. [3])

tube and using the radiation effect in the central part of the inserts (Fig.32.14c, d), heated to the hot gas temperature [17–19].

NITI EM/ONERA COLLABORATIVE CGTE STUDY

An impressive scientific and manufacture CGTE project was undertaken by NIZ KTD in collaboration with the French aerospace agency ONERA (Office National d'Etudes et Recherches en Aerospatiales). The objective of this joint effort was to develop both the materials and technologies, as well as the advanced design approaches for hot section components for automotive ceramic GTEs. Since 1995, research and development has been conducted to develop a ceramic turbine stage (Table 32.6) for the high-load low-life Microturbo production GTE [20, 21] aimed at enhancing its thrust while at the same time, keeping the overall dimensions unchanged. In the axial metal-ceramic turbine stage developed, all the basic elements in contact with the hot gas were made of uncooled ceramic materials, while the load-bearing elements are made of cooled metal materials (Figures 32.15a, b).

Figure 32.16 shows the **nozzle** of this turbine stage. The load-bearing shroud rings of the nozzle set (2, 7) have two openings each to accommodate the carrier cores (8). The semi-spherical recesses to mount the nuts (5) of the carrier cores are made in the outer ring. The peripheral (4) and root (1) ceramic inserts are secured to the load-bearing shroud rings. The ceramic shell comprising five similar elements (3) threaded on

TABLE 32.6 TYPES OF CERAMIC COMPONENT TESTS FOR THE CGTE AND TEST FINDINGS

CGTE	NAME	Tests	Objective	Test conditions	Test findings	References
ONERA (Microturbo)	Guide vanes. Rotor blades.	Aerodynamic	Identification of power loss	M=0.775, Re=38000, M=0.76;	Figure 32.19 Re=350000	[11, 22]
		Thermal	Temperature of elements. Cooling air flow.	TIT=800–1400°C	Figure 32.20	
		Thermal cycling	Verification of heat shock resistance at startups and shutdowns of engine.	78–190°C/s, 10 cycles	No defects	
		Life	Operation at constant and rated load.	5 hours	—	
	Combustor	Thermal	Ignition and stabilization of flame, temperature of elements, combustion completeness, hydraulic resistance.	Fuel—aviation kerosene; air pressure—0.11 MPA; rating—6h	Burn out completeness 99.9%, resistance 9.17%, reliable startup and flame stabilization.	[24, 25]
		Thermal cycling	Operation at transients, startup and shutdown conditions.	Not less than 2 cycles	4 cycles, no defects	
	Rotor blades	Overspeed	Operation at stresses 1.5–2 times over the rated values.	70–90 MPa	Figure 32.21	[11]
CGTE-2.5	Combustor	Thermal	Combustion completeness, startup, flame stabilization, hydraulic resistance, temperature of elements.	Fuel – natural gas, air pressure—0.11 MPa, gas temperature down-stream of combustor—up to 1500°C.	Reliable startup and flame stabilization, combustion completeness at rating 99.9%. Hydraulic resistance 3.3%. Temperature of metal elements not above 900°C.	[12, 18]
		Thermal cycling	Transients, startup-shutdown cycles.	200 cycles	200 cycles accomplished, no defects	
		Life	Operation at rating	Not less than 400 hours	The tests were carried out for 400 hours, no defects.	
		Environmental	Identification of NO_x and CO emissions within the 75 to 100% operating range.	Air pressure – 0.11 MPa	NO_x (15% O_2) at rating (conversion with high pressure accounted for)—not above 20 ppm CO (15% O_2) at 75% load – not less than 100 mg/Nm^3.	
CGTE-2.5	Inlet branch pipe, scroll, exit diffuser	Life	Operation reliability: – at transients	Gas temperature up to 1450°C air pressure – 0.11 MPa	Test in progress (2000)	
		Thermal cycling	– at rated load	Up to 200 startup-shutdown cycles		
	Recuperator for the demonstration unit	Thermal	Regeneration ratio, pressure loss, reliability	Rated values of air and gas temperatures and velocity, media pressure – around 0.11 MPa		
		Life	Operation reliability at rated load.			
		Thermal cycling	Operation at transients	Not less than 200 "startup-shutdown" cycles.		
Automotive CGTE	Recuperator-airheater	Thermal, thermal cycling	Regeneration ratio, pressure losses	Hydrodynamic similarity for conventional design engine operating conditions. Air and gas pressure – 0.11 MPa.	Desirable regeneration ratio provided at specified pressure losses by gas and air and specified constraints for mass and overall dimensions.	[19, 20, 26, 27, 28]

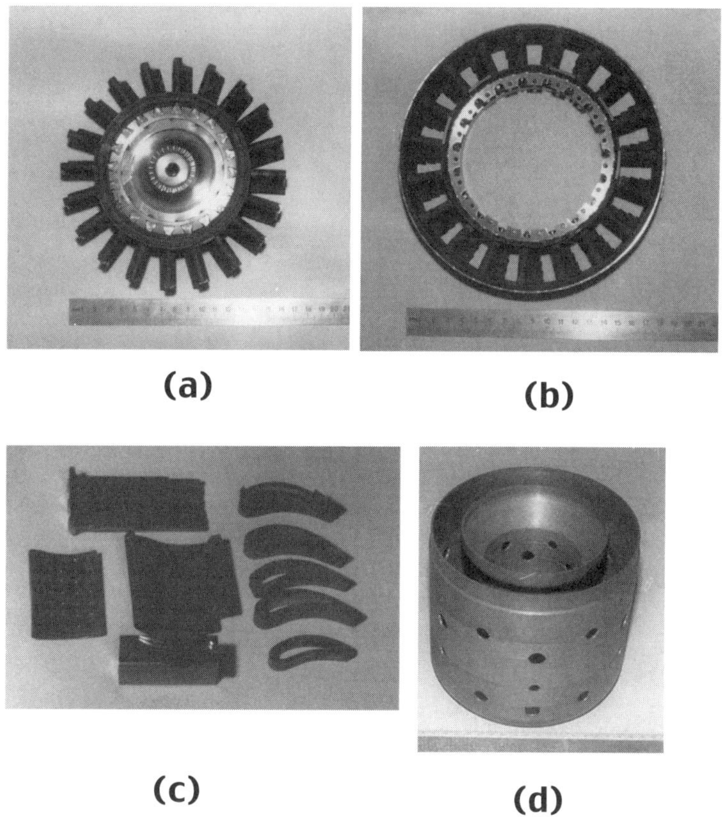

Figure 32.15 (a) Turbine Wheel, (b) Nozzle Vanes, (c), Rotor Blade Elements, and (d) Ceramic Liner for Annular Combustor (Fuel-aviation Kerosene) of Metal-Ceramic Axial Turbine CGTE for ONERA (Reprinted with Permission from NIZ KTD, NITI EM)

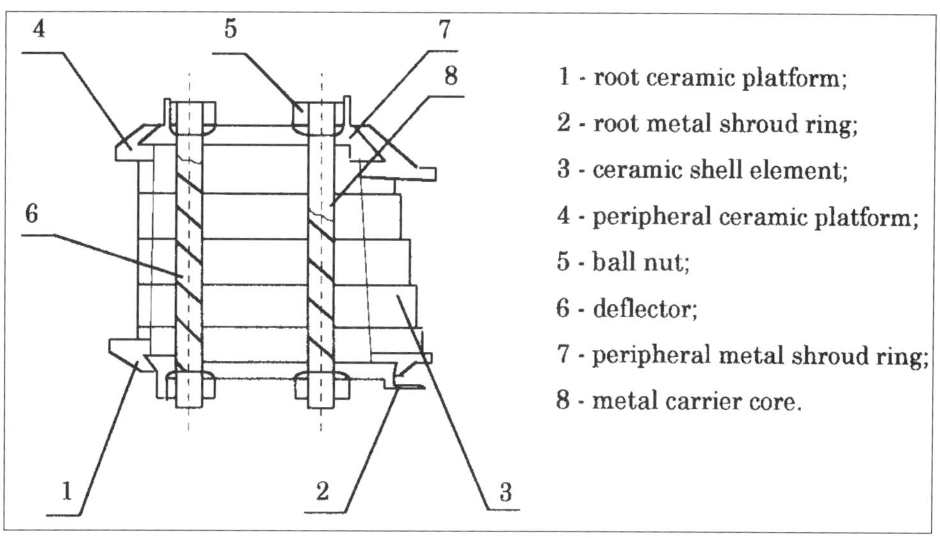

Figure 32.16 Nozzle of Microturbo CGTE (Reprinted with permission from NIZ KTD, NITI EM)

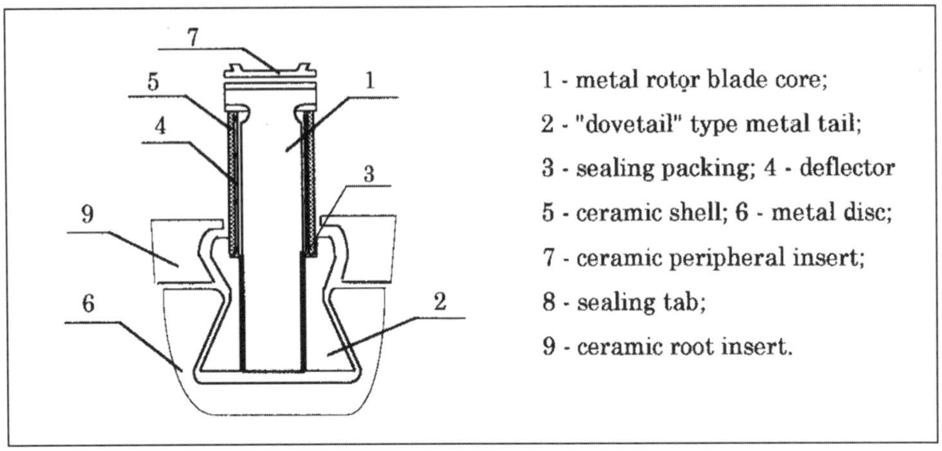

Figure 32.17 Microturbo CGTE Rotor Blade Assembly (Reprinted with permission from NIZ KTD, NITI EM)

two carrier cores has a two-step twist on a radius over the entire height of the airfoil. Securing of the cores to the external and internal shroud rings and tightening of the shells are implemented using ball nuts. Spaces to supply air for cooling the carrier cores are provided for. Air flows between the cores (cross section as a circle with two flats) and around the deflectors. The number of ceramic peripheral and root platforms protecting the shroud rings against a contact with the hot gases is equal to the number of nozzle vanes. The ceramic platforms of the nozzle vane and the shroud rings are butt-joined with the trapezoidal grooves and bulges. To preclude mutual sliding and turning, and, also, to ensure the required accuracy at assembly, two upper and three bottom elements are rigidly joined using diffusion welding. Thus, the shell comprises two independent sections which ensure essentially lower thermal stresses [22].

The **rotor blade** design is shown in Figure 32.17. The carrier core (1) of the profiled section is soldered with a high temperature braze into a "dovetail" root (2) at an angle to balance the bending stresses due to the gas forces and a mismatch of the centres of gravity of the ceramic shell and the root section of the core. Prior to brazing, the screen (4) with the semi-spherical craters providing a point contact with the external ceramic shell (5), is mounted. The ceramic shell consists of five similar airfoil elements joined by diffusion welding and is supported by the carrier core platform (1), i.e. it operates in "compression" during rotation. Cooling air flows only between the core and the screen-deflector (4). In this event, with the uncooled ceramic shell in contact with gas, it is enough to have 1.5% of the air flow to ensure an acceptable temperature state of the metal elements of the rotor blade. Separate metallic (core, tail, deflector) and ceramic (shell elements) parts of the rotor blade are shown in Figure 32.15c.

An annular combustion chamber [23, 24] for the same Microturbo engine with a ceramic dome and liners was designed, manufactured and tested using kerosene as fuel. The initial fairly complex and contradictory specifications were the basis for designing the ceramic combustor. The key requirements for the combustor design were:

- Outer and fitting sizes constrained by the size of the existing metal chamber
- Limitation of thickness of the ceramic liner wall because the combustor's effective volume could not be reduced
- Availability of the air supply holes in the walls of the ceramic liner which are critical areas since they cause considerable thermal stresses in the ceramic
- Providing the combustor alignment in the engine casing and reliable movable joints for the liner and the metal parts of the fuel nozzles, the igniter and the circular seals of the turbine nozzle guide vanes
- Providing thermal protection for the engine casing against the heat radiation of the liner

The ceramic annular combustor developed and manufactured with consideration of the above demands is shown in Figure 32.15d. The combustor has an outer metal shroud with 6 sliding (movable) metal sleeves in the end face semi-torroidal area to accomodate the fuel nozzles. The holes are drilled on the internal and external cylindrical areas of the shroud to supply air to the combustor's working zone. To diminish the thermal stresses, the liner is segmented into a number of separate elements. At the butt joints between the creamic rings, the gaskets of a sheet material made of crystalline silicon carbide fibers are mounted. A high-temperature insulation material is placed between the ceramic and metal parts. The combustor design ensures bilateral heat movements of the ceramic liner and the metal shroud which is achieved by appropriate heat gaps between the ceramics and the metal and, also, by the thermoelastic design of the end face semi-torroidal area of the metal shroud. Furthermore, spring elements imposing a constant compressive force on the ceramic sectional liner are placed in the end face area which eliminates air overflows between the liner rings.

The collaborative activities of NITI EM and ONERA for the development of high-temperature recuperators for automotive CGTE's are described in detail in chapter 23 of this volume [25]. It contains both the description and specific features of design of the hybrid recuperators which consist of two sections, namely the high-temperature (>600°C) ceramic and low-temperature metallic sections, and results of the thermotechnical and thermocyclic tests of the models of these heat exchangers. Therefore, we provide here only the photos (Figure 32.18) of the manufactured and tested full-scale recuperators [18, 19, 26, 27].

Screening and Proof Tests of Ceramic Parts

All the ceramic parts underwent 100% multi-stage screening and proof tests. The screening test is specific for each part and is based on a certain probability of failure for each for a specified service life. For instance, four-stage screening of the ceramic shell of the rotor blade involves:

- Process control during the manufacture of a set of the ceramic shell elements with the geometry and manufacture quality control maintained up to the final sintering
- Static screening of the least loaded sintered elements on a press with loads ensuring a specified reliability of the remaining parts
- Thermal cycling screening intended to detect significant internal flaws in the part structure after static screening
- Conducting proof tests of a set of welded ceramic shells on a cold spin rig at overspeed.

When carrying out the static screening of the ceramic elements of the shell, no break-downs of the least loaded lower elements were observed, while around 35% of the upper (the highest loaded) shells failed. The subsequent overspeed tests confirmed the effectiveness of the screening performed. On the whole, the screening and proof tests of both the rotor and stator components confirmed necessity and effectivenes of these tests. Table 32.6 shows the results of some of the rig tests of the ceramic components and parts of the CGTE performed during the R&D projects [3, 11, 12, 21, 23, 24].

To date, there are clear cut complex challenges in ceramic gas turbomachinery awaiting solutions. These include:

- Nondestructive control and screening technologies for ceramic parts that have shown problems during development, and the use of effective technologies for utilization of rejected ceramic products in order to reduce the CGTE cost
- "Impact problems" due to carry-over by gas of debris of a destructed ceramic part in the flow passage with the resultant destruction of the GTE by this debris
- Challenges associated with the development and commercialization of high-temperature test rigs and the development of a measuring technology for physical parameters of the working media and ceramic elements of GTE's operating with a the high-temperature stream of gas.

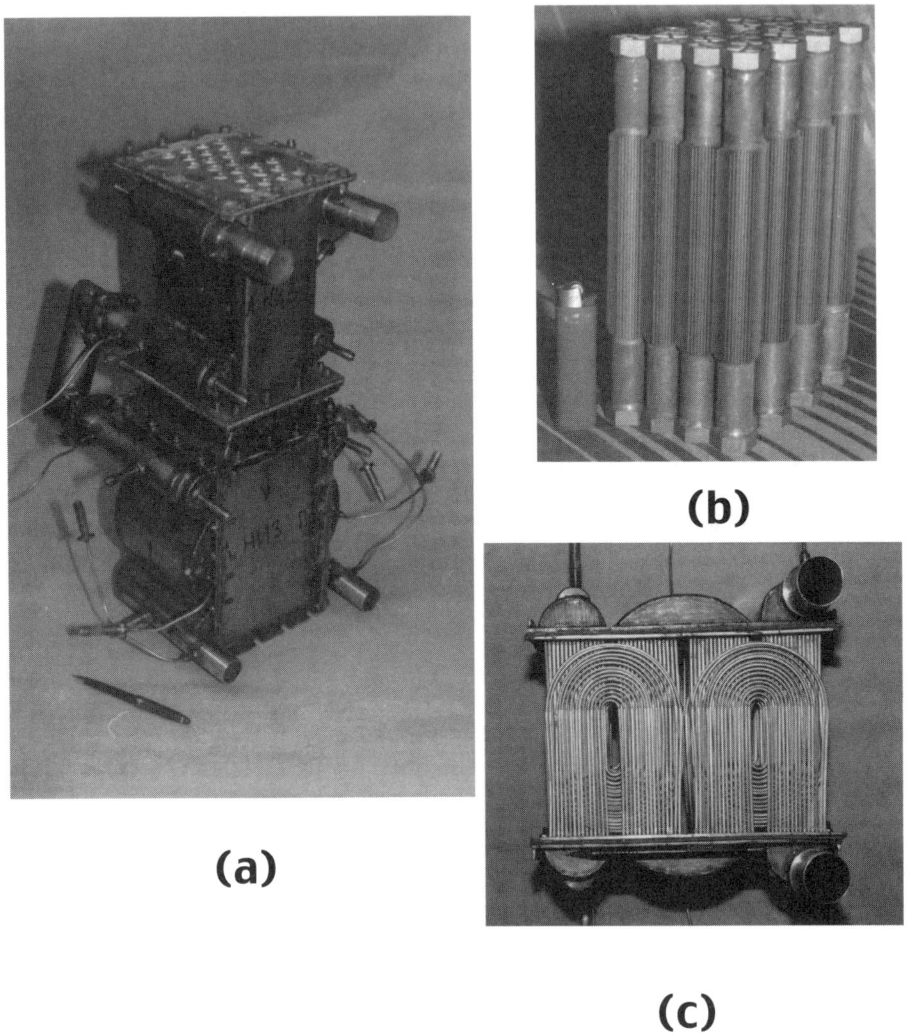

Figure 32.18 Module of Metal-ceramic Air Heater (a) of Ceramic (b) and Metal (c) Sections. Heat Exchanger is Intended for Automotive Ceramic GTE, Made for ONERA (Reprinted with permission from NIZ KTD, NITI EM)

MAIN LESSONS LEARNED

(1) A shrinkage-free ceramic material is the easiest to manufacture, the least expensive and the most competitive, and, therefore, the most promising.

(2) Advancements in the development of the ceramic "SUGRAV" and the components fabricated from this material are attributable to the development, demonstration, and implementation of basic concepts or ceramic gas turbine design. The most important concepts of CGTE design are the following:
   - The concept of modular design
   - The concept of the assembly of an integral design using a set of ceramic parts

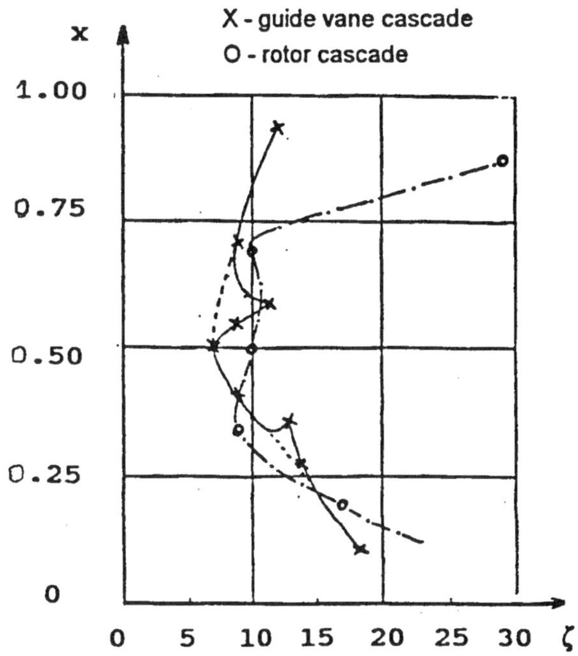

Figure 32.19 Energy Loss Coefficient Variation vs. Blade Height (Reprinted with permission from NIZ KTD, NITI EM)

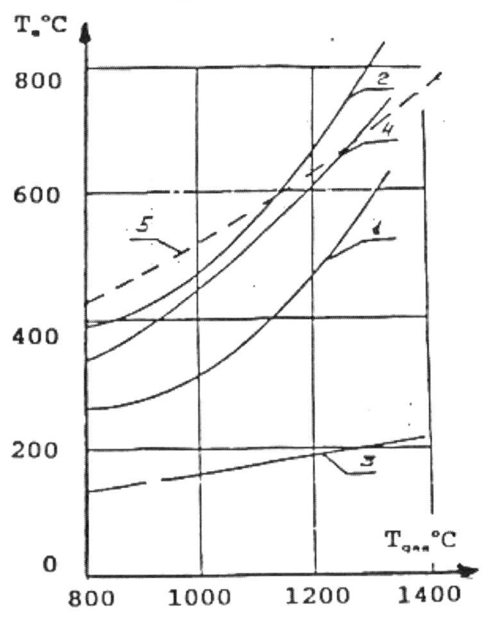

1—Temperature Under Platform in Rotor Blade Core; 3—Temperature of Cooling Air;
2, 4—Temperature in Platform of Rotor Blade (Sealed and Unsealed Versions);
5—maximum Temperature in the Center of Carrier Spokes of Guide Vane.

Figure 32.20 Thermal State of Elements of Rotor Blades and Guide Vanes (Reprinted with permission from NIZ KTD, NITI EM)

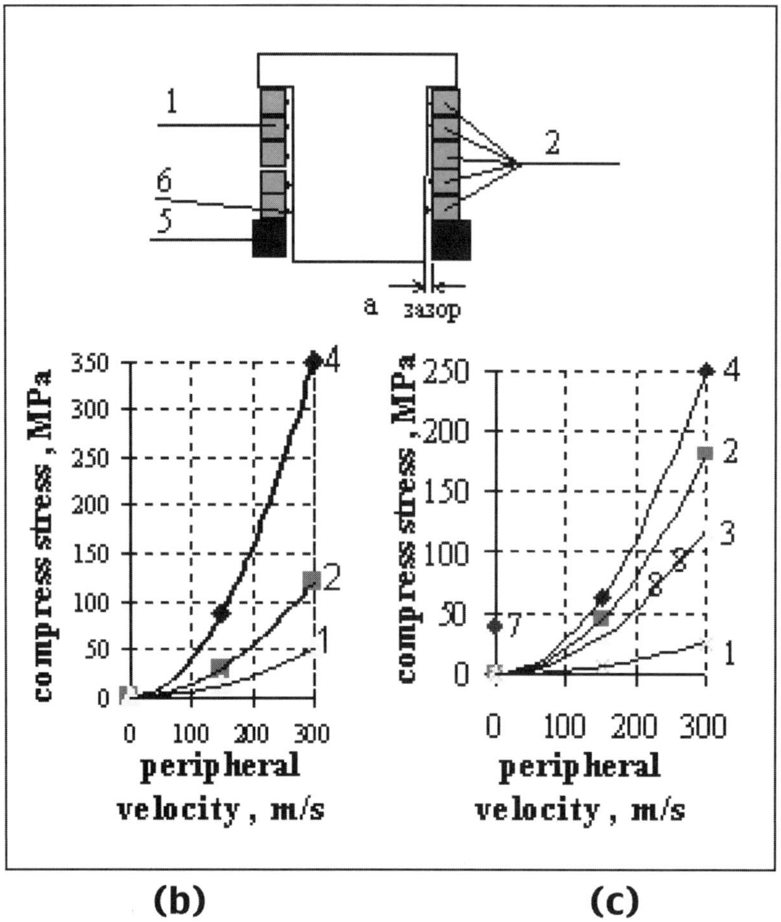

1—Shell Element; 2—Sectional Shell; 3—Shell Element with Additionally Loaded Washer; 4—Sectional Shell with Additionally Loaded Washer; 5—Additionally Loading Washer; 6—Deflector; 7–Operating Stresses in Engine; 8—Shell Destruction at Gap of 0.2–0.3 mm.

Figure 32.21 Overspeed Tests of Ceramic Model Shell (a-Test Scheme; b-Cylindrical Model Shells; c-Full-scale Airfoil Shells) (Reprinted with permission from NIZ KTD, NITI EM)

Modular design ensures:

- Standardization of components
- Reduction of fuel consumption
- Improved environmental-friendliness of the GTE operating at part load (by switching off individual power modules of a gas turbine plant), common techniques for repair, etc.
- An integrated component assembly using a set of ceramic elements for the fabrication of hybrid ceramic-metal and ceramic-ceramic components for high-temperature GTEs taking into account scale factors and implementing quasi-adiabaticity of heat cycles in the GTE by a substantial reduction in cooling, thereby eliminating external heat exchange.

CONCLUSIONS

(1) A number of heat-resistant shrinkage-free alumino-boron-nitride structural ceramic components that were easily fabricated, were developed and successfully tested. These ceramic components were designed to operate predominantly in compression. Green parts are easily machined, and "welded," and do not need finish machining.
(2) The alumino-boron-nitride ceramic material was incorporated in a number of components of a high-temperature GTE, and survived complex tests, including the thermodynamic and strength tests at speeds in excess of the design speed in a test rig.
(3) High-temperature alumino-boron-nitride ceramic elements for a demonstration plant of engine CGTE-2.5 were fabricated. A prototype gas turbine designed for short life with hybrid ceramic-metal nozzles and rotor blades, and an annular ceramic combustor was fabricated and tested at a combustor gas temperature of 1350°C. The gas turbine can potentially be operated at a combustor gas turbine temperature of 1500°C to 1600°C.
(4) The fabrication and testing of a prototype package for a 2.5 MW GTE operating under high pressure for a natural gas blower are currently in progress (year 2000). All the ceramic parts of the GTE, fabricated from the alumino-boron-nitride ceramic material will operate essentially in compression.
(5) The basic development concepts of the quasi-adiabatic high-temperature GTE and the structural alumino-boron-nitride ceramic material are the platform for the advanced high-temperature gas turbine manufacture in Russia.

ACKNOWLEDGEMENTS

This chapter would not have been possible without the important contributions of our colleagues A.A. Souryaninov, V.Ya. Podgorets, A.S. Molchanov, Yu. I. Zakharov, and E.D. Vinogradov.

REFERENCES

[1] Kirillov, I.I., Soudarev, A.V., Reznikov, A.G., "Ceramic in High-temperature GTUs (Survey)," Prom. Teplotekhnika, Volume 12, [4], pp. 67–89, 1990 (in Russian).

[2] Soudarev, A.V., "Prospects of Development of Environmentally-friendly Ceramic Gas-turbine Engines for Heavy Power Applications," Energhetyka i transport, Volume 38, M., RAN, 49–59, 1992 (in Russian).

[3] Soudarev, A.V., Souryaninov, A.A., Grishaev, V.V., Podgorets, V.Ya., Tikhoplav, V.Yu. and Soudarev, B.V., "Ceramic Gas-Turbine 2.5 MW Engine for Gas Industry Application in Russia," ASME paper 2000-GT-0074, presented at the 45th International Gas Turbine and Aeroengine Congress and Exhibition, Munich, Germany, May 8–11, 2000.

[4] Schenk, B., "Ceramic Turbine Engine Demonstration Project—A Summary Report," ASME paper 2000-GT-0052, presented at the 45th International Gas Turbine and Aeroengine Congress and Exhibition, Munich, Germany, May, 8–11, 2000.

[5] Price, J.R., Jimenez, O., Parthasarathy, V., Miriyala, N. and Leroux, D., "Ceramic Stationary Gas Turbine Development Program—Seventh Annual Summary," ASME paper 2000-GT-0075, presented at the 45th International Gas Turbine and Aeroengine Congress and Exhibition, Munich, Germany, May 8–11, 2000.

[6] Takehara, I., Tatsumi, T. and Ichikawa, Y., "Summary of CGT302 Ceramic Gas Turbine Research and Development Program," ASME paper 2000-GT-0644, presented at the 45th International Gas Turbine and Aeroengine Congress and Exhibition, Munich, Germany, May 8–11, 2000.

[7] Wenglarz, R.A. and Kouns, K., "Ceramic Vanes for a Model 501-K Industrial Turbine Demonstration," ASME paper 2000-GT-0073, presented at the 45th International Gas Turbine and Aeroengine Congress and Exhibition, Munich, Germany, May 8–11, 2000.

[8] Soudarev, A.V., Grishaev, V.V., "Material application aspects of development of stationary ceramic gas turbine engine," Turbines and Compressors, St.Petersburg, NIKTIT, Volume 8–9 [1, 2], pp. 82–92, 1999 (in Russian).

[9] Soudarev, A.V., Grishaev, V.V., "Structural Alumino-Boron-Nitride Ceramic Material For Gas-Turbine Applications," chapter 13 in *Progress in Ceramic Gas Turbine Development, Volume II—Ceramic Gas Turbine Component Development and Characterization*, Mark van Roode, Mattison K. Ferber, David W. Richerson, eds., ASME Press, New York, USA, 2003.

[10] Grishaev, V.V., Soudarev A.V., and Okatova, G.P., "Research into Mechanism of Formation of Alumino-Boron-Nitride Ceramics. A Structural Material for GTE Parts," ASME paper 97-GT-323, presented at the 47th International Gas Turbine and Aeroengine Congress and Exposition Orlando, Florida, USA, June, 2–5, 1997.

[11] Soudarev, A.V., Soudarev, B.V., Molchanov, A.S., Grishaev, A.V., Podgorets, V.Ya. and Souryaninov, A.A., "Program of Development of Metal-Ceramic Gas-Turbine 2.5 MW Unit and the Initial Results of its Implementation," ASME paper 97-GT-356, presented at the 42nd International Gas Turbine and Aeroengine Congress and Exposition, Orlando, Florida, USA, June 2–5, 1997.

[12] Sudarev, A.V., Zakharov, Yu.I., Vinogradov, E.D., Butovsky, L.S., and Granovskaya, E.A., "Process studies as applied to ceramic gas turbine engine low-emission double-zone micro-diffusion combustion chamber," ASME paper 94-GT-445, presented at the 39th International Gas Turbine and Aeroengine Congress and Exposition, the Hague, Netherlands, June 13–16, 1994.

[13] Sudarev, A.V., Zakharov, Yu.I., Ljubchik, G.N., Butovsky, L.S. and Granovskaya, E.A., "Development and Bench Test of High-Temperature Combustion Chamber with Structural Ceramic Components," ASME paper 91-GT-315, presented at the 36th International Gas Turbine and Aeroengine Congress and Exposition, Orlando, Florida, USA, June 3–6, 1991.

[14] Sudarev, A.V., "Low-Toxic Method of Fuel Firing in Gas Turbine Micro-Flame Combustors," ASME paper 90-GT-101, presented at the 35th International Gas Turbine and Aeroengine Congress and Exposition, Brussels, Belgium, June 11–14, 1990

[15] Soudarev, A.V., Soudarev, B.V., Soudarev, V.B., Kondratiev, A.A., and Avran, P. "Application of Cassette-tube Airheaters as a Way to Improve the Mass-size Characteristics of Regenerative Gas-turbine Units," ASME paper 99-GT-331, presented at the 44th International Gas Turbine and Aeroengine Congress and Exposition, Indianapolis, Indiana, USA, June 7–10, 1999.

[16] Tikhoplav, V.Yu., Soudarev, A.V., and Orlov, A.P., "Option for Structure of High-Temperature Gas-turbine Plant with Ceramic Elements," Prom. Teplotechnika, Volume 12 [4], 77–89, 1990 (in Russian).

[17] Soudarev, A.V., and Sudarev, B.V., "High-temperature compact airheater for GTE. Turbines and compressors," St.Petersburg: NIKTIT, Volume 5 [1], 21–27, 1998 (in Russian).

[18] Soudarev, A.V. and Soudarev, B.V., "Heat Exchange and Hydraulic Resistance Inside the Inner Path of the Ceramic Airheater for GTU," 6th UK National Conference on Heat Transfer Book Abstracts, pp. 503–508, 15–16 September, 1999, Edinburg, UK.

[19] Soudarev, A.V., Soudarev, B.V. and Grishaev, V.V., "Development and Study of Compact High-temperature Ceramic Tubular Airheaters," *Fourth ISHME-ASME Heat and Mass Transfer Conference and Fifteenth National Heat and Mass Transfer Conference Proceedings*, pp. 1033–1038, January 12–14, 2000, Puna, India.

[20] Soudarev, A.V., Soudarev, B.V., Grishaev, V.V., Molchanov, A.S., Podgorets, V.Ya., and Souryaninov, A.A., "History of Development of Metal-Ceramic Axial Turbine Stage for Highly Effective Gas Turbine Engine," VDI Berichte 1280, Entwicklungslinien der Energie- und Kraftwerkstechnick, Tagung Siegen, Germany, 10–11 September, 1996, pp. 527–541.

[21] Avran, P., Soudarev, A.V., Grishaev, V.V., and Podgorets, V.Ya., "Complex test of thermal-stressed elements in high temperature gas turbine," *Aerothermodynamics of internal flows III, Proceedings of the 3 ISAIF*, Beijing, China, 1–6 September, 1996, pp. 281–288.

[22] Nakakado, K., Machida, T., Miyata, H., and Mori, N., "Development and reliability evaluation of ceramic stator vane for industrial gas turbines," *4th International Symposium on Ceramic Materials and Components for Engines, Proceedings*, Goteborg, Sweden, 10–12 June, 1991, pp. 985–1000.

[23] Soudarev, A., Vinogradov, E., Zakharov, Yu., Avran, P., and Grishaev, V., "Test Bed Development of Design of Ceramic Combustor for Gas-Turbine Engine," *Proceedings of the International Gas Turbine Congress 1999 Kobe*, 14–19 November, 1999, Kobe, Japan, Vol.II, TS-57, pp. 679–683, November, 1999.

[24] Soudarev, A.V., and Grishaev, V.V., "Studies over high-temperature long-term behaviour of alumino-boron-nitride ceramic "SUGRAV," in oxidizing media," *Proceedings conference on materials in ceramic environment*, Vol.II, Lisbon, Portugal, 22–24 July, 1998, pp. 43–49.

[25] Avran, P., and Boudigues, S. "Ceramic Development at ONERA," chapter 23 in *Progress in Ceramic Gas Turbine Development, Volume I, Ceramic Gas Turbine Design and Test Experience,* Mark van Roode, Mattison K. Ferber, David W. Richerson, eds., ASME Press, New York, USA, 2002.

[26] Soudarev, A.V., Soudarev, B.V., and Kondratiev, V.V., "Convective Heat Transfer and Hydraulic Resistance at Air Flow Through Ceramic Tubes with Heat Exchange Intensifiers," *Proceedings Fourth International Symposium on Experimental and Computational Aerothermodynamics of Internal Flows*, Volume II, Dresden, Germany, August 31—September 2, 1999, pp. 208–216.

[27] Sudarev, A., Sudarev, B., Avran, P., and Sudarev, V., "Unsteady Operating Conditions as Applied to Ceramic Recuperative Air Heater," *Proceedings of the International Gas Turbine Congress 1995 Yokohama*, 22–27 October, 1995, Yokohama, Japan, Vol. II, pp. 271–274.

CHAPTER

33

Ceramic Component Development at Rybinsk Motors Design Bureau

Mark van Roode

Solar Turbines Incorporated, San Diego, California, USA

INTRODUCTION

This chapter is a brief summary of past ceramic gas turbine activities at Rybinsk Motors Design Bureau, a Russian national design bureau in the former USSR, responsible for the development of gas turbines that power military and commercial aircraft. The company is the successor to a national design bureau which was originally based in the Moscow area, but which was relocated to Rybinsk in 1939. Rybinsk Motors Design Bureau (RMDB) has been involved with the development of turbojet and turboshaft helicopter engines, turboprop engines, aircraft diesel engines up to 200 hp, and 1-90 MW ground mobile power stations. Table 33.1 summarizes some of the engines developed by Rybinsk Motors Design Bureau [1]. One of its products was the engine for the Tupolev-144, the Russian equivalent of the French-British Concorde supersonic aircraft.

The author visited RMDB in 1996 as part of an information exchange visit sponsored jointly by the US Department of Energy and Solar Turbines Incorporated of San Diego, California, USA. The findings of this visit have been reported [2]. This chapter summarizes a few key aspects of RMDB's ceramic gas turbine activities based on the information obtained during the visit and discussions with RMDB staff.

A major ceramic gas turbine project at RMDB was funded by the Ministry of Aviation Industry and included the development of the design methodology and manufacturing techniques for hot section components of aircraft and aeroderivative ceramic gas turbines. The ceramic manufacturer Technologhia in Obninsk was one of the suppliers of monolithic ceramic components. Tekhnologhia's ceramic gas turbine related activities have been described in chapter 12 of Volume II of this text. The ceramic gas turbine project lost momentum in the 1990s after the breakdown of the USSR because of lack of funding.

RMDB CERAMIC COMPONENT DEVELOPMENT

RMDB's engineers were interested in Si_3N_4– and SiC-based materials because of their low specific weight, an important factor for aircraft engines. However it was realized that their thermal stability and

TABLE 33.1 AIRCRAFT ENGINES DEVELOPED BY RYBINSK MOTORS DESIGN BUREAU [1]

Engine	Type	Performance specifications	Applications
VD-4K	Aviation piston engine	4300 HP SFC 185 g/HP.hour Weight; 1900 kg	TU-85 aircraft
VD-7B	Single-shaft turbojet engine	9500 kgf Take-off Thrust	3M aircraft
RD-7M2	Single-shaft turbojet engine with afterburner	16500 kgf Take-off Thrust	TU-22R and TU-22K aircraft
RD-36-41	Single-shaft turbojet Engine with Afterburner	16000 kgf Take-off Thrust	T-4 aircraft
RD-36-51A	Supersonic single-shaft turbojet engine	20000 kgf Take-off Thrust	TU-144 supersonic aircraft
RD-38	Single-shaft turbojet engine	3250 kgf Take-off Thrust	YAK-38 aircraft (vertical take-off)
RD-38K	Single-shaft turbojet engine		Samolet A-40 water-based aircraft
TVD-1500	Turboprop engine	SFC 179 g/HP.hour 1300 HP Take-off Power	AN-38 and S-80 commercial aircraft
RD-600	Turboshaft engine	SFC 209 g/HP.hour	KA-62 helicopter and amphibian yamal aircraft
DN-200	Turbosupercharged diesel	SFC 170 g/HP.hour 200 HP max.	Aircraft, water Transport, power plants

durability in the gas turbine environment did not allow their application for long term service at gas temperatures exceeding 1300°C to 1500°C. Practically, 1300°C may well be the limit because of degradation in the gas turbine environment which is accelerated by the water vapor in the combustion atmosphere (see Volume II, Chapter 26). Metal components in aircraft and aeroderivative engines operate with turbine inlet temperatures (TIT) in the 1300°C to 1500°C range because these parts are fabricated from advanced alloys and are kept at an acceptable temperature with sophisticated cooling techniques.

From a commercial perspective the above temperature range is still attractive for small engines. Another area of interest were the peripheral velocities of the turbine wheel blade rims which exceed 525-530 m/s. A program was therefore initiated to develop components for the combustor and turbine of a small-size engine of the TVD-1500 power class (see Table 33.1). The principal target of study was the gas generator of this twin-shaft small-size engine. For this class of engines the design parameters are close to critical if the blade rims are not cooled. The gas generator is representative of the engine and it was tested in a rig which was capable of fully simulating the engine operating conditions. The combustor gas collector components, turbine nozzle vanes of various designs, ceramic turbine rotor blades, and the thermal state of the support systems were investigated under various lubrication and cooling conditions.

The best results were obtained with nozzle vanes fabricated from a composite which consisted of carbon/carbon and silicon carbide (C/C + SiC). In a relatively simple design, integral nozzle assemblies of this material survived both the static strength tests and tests in the hot gas stream without structural degradation, while monolithic ceramic components of a similar design revealed cracks at the same loads. The cracks were observed both on the structural elements that were load bearing and on those that did not carry any critical mechanical and thermal loads. In particular, damage at the interfaces between the ceramic flanges and metal elements was observed even at rather moderate nominal bearing stresses along the interfaces. Combustor gas collector casing rings made of monolithic Si_3N_4-based ceramic materials showed similar results. The analysis of the mechanical and thermal stress levels did not indicate any significant stresses within the damaged zones. Satisfactory test results were achieved only with the C/C + SiC material.

In the design of the gas generator turbine rotor an analysis of the structural ceramic silicon nitride- and carbide-based materials was performed. The analysis indicated that with a Weibull modulus m = 7–10 it was impracticable to produce an all-ceramic rotor for operation under the specified conditions using slip casting as the fabrication route.

A hybrid design was therefore pursued based on inserted dovetail blades in a metal disk. For the dovetail blade root a 60° angle was selected in order to achieve the optimum density of the blade rim of the turbine

Figure 33.1 Silicon Nitride Blades and Nozzle Shroud Rings

rotor and an acceptable stress level of the metal disk locking mechanism at the design temperature range of 500°C to 550°C. The normal bearing stresses along the interface were on the order of 334 MPa. An evaluation of the turbine rotor blades produced by slip casting indicated that this fabrication process was unable to meet the dimensional accuracy required for insertion into the metal turbine disk at the design peripheral velocities. Warping of the airfoil and the blade attachment would result in greater bending and contact loads in the blade root area. To meet the specified tolerances machining of the blades was required. A technique for diamond tool profiling to perform profile grinding of the blade root surfaces and manufacturing of special-purpose formwheels to shape the blade was developed. With this technique the turbine rotor blade could be precision machined to a dimensional accuracy which was equivalent to that of ceramic preforms produced by any molding and densification technology (including hot pressing and hot isostatic pressing) which produced maximum strength of the material. The facilities and the grinding technology established provided the tool durability adequate for the serial production of commercial quantities of blades.

Figure 33.1 shows several fully machined silicon nitride blades produced by hot pressing. The blades were proof tested individually or in pairs. A fully bladed disk was cold spin tested at 130% of design speed. The blades that survived the proof test were hot spin tested at a temperature that simulated the temperature of the disk shroud at the location of the blade attachment. A fully bladed rotor is shown in Figure 33.2.

The test data led the RMDB researchers to draw the following general conclusions:

- The ceramic materials were adequate for the blade design. During the testing the blades rarely failed along the airfoil notwithstanding the relatively low K_{IC} of the materials under study. This led RMDB to believe that the ceramic composites which had superior fracture toughness (compared to monolithics)

Figure 33.2 Axial Rotor with Silicon Nitride Ceramic Blades and Radial Metal Compressor Rotor

would provide reasonably acceptable blade strength under simulated thermal and impact conditions characteristic for the operation of an aeroderivative gas turbine.

- The majority of failures occurred along the blade dovetail. Analysis indicated that the damage occurred at the blade attachment – disk slot interface. The analysis also showed that the stress level and the location of the zone of maximum stresses depended on whether the surface contact was through sliding or sticking. During sliding, the stresses in the ceramic material were substantially lower than during sticking. The type of contact depended on how the contact zone was designed and manufactured. It is highly unlikely that sliding occurs if the quality of machining of the interfaces is poor. The studies performed indicated that in the case of an erroneous selection of the materials within the contact zone extremely detrimental processes associated with degradation of the ceramic material strength are likely to occur, even if sliding along the contract surfaces takes place.

An assembly of nozzle segments held together between inner and outer shroud rings is shown in Figure 33.3. The shroud rings can also be seen with the blades in Figure 33.1. The nozzles were evaluated in static rigs and in an engine at temperatures up to 1500°C.The studies of the functionality of the ceramic bearings under the critical conditions of low lubrication oil consumption indicated that they operated normally. The operation of the series 206 bearings was thermally stabilized with only air cooling (and without oil cooling); while for metal bearings of the same series and similar D^*_n parameters oil cooling in addition to air cooling was required.

ACKNOWLEDGEMENT

The author is indebted to the researchers at the Rybinsk Motor Design Bureau for the information summarized in this chapter. The photographs were taken by the author during his visit to RMDB in 1996.

Figure 33.3 Silicon Nitride Nozzle Assembly

REFERENCES

[1] Dynkii A.P., "Aircraft starts with the engine," Zametki o Rybinskom Konstructorskom Buro Motorostroenya, Pubinskoye Podvorie, 1995 (in Russian).

[2] van Roode, M., and Glezer, B., "Status of Ceramic Gas Turbines in Russia," *Proceedings of the Advanced Turbine Systems Annual Review Meeting*, Abbie Layne and Patricia Hoffman, eds., Washington, D.C., November 7,8, 1996, pp. 671–681.

INDEX

A

ABAQUS 602
Abradable seal 115
Acoustic tiles 248
Advanced Material Gas Generator Program 307
Advanced Turbine Systems (ATS) 126
AGATA 448, 501, 523, 532
AGT-5 123
AGT 100 117
AGT101 86
AIN 637
Air permeability 490
AiResearch 78
AiResearch Casting Company (ACC) 82, 90
Airfoil 585
Al_2O_3/Al_2O_3 591
AlliedSignal Engines 94
Allison Engine Company 111
Allison Gas Turbine Division,
  General Motors Corporation 111
Alstom Power 446, 597
Alumina fiber insulation 648
Alumina foams 640
Annawerk 455, 470
Annular 159
Annular combustor 617, 701
Annular combustion 640
Annular combustion chamber 617
Ansaldo 597
ANSYS 602, 603, 670
APU 78, 94
AS800 169
AS800 silicon nitride 100, 128, 250
Asahi Glass, C600 389
Asahi Glass Co., Ltd. 286
Asea Brown Boveri (ABB) 447, 635
ASEA Cerama 448, 499
Atmospheric rig test 470
Attachment 103
ATTAP 92, 123, 124
Automotive 100kW CGT program 280, 281, 285
Automotive ceramic gas turbine (CGT) 283
Automotive Gas Turbine Engine 117, 123
Axial 123, 284, 319, 365, 499, 574, 663
AY6 486

B

Ball bearings 241
Ball turbolators 527
Beta-alumina 636
BF Goodrich Aerospace (BFG) 165
Biaxial stress 84
Blade 640
Blisk 274, 339
BMFT program 446
BNFL 462
Boostec Industries 449
BN 608
Borescope 184
Boron nitride 113, 171
Brayton Cycle 2
Brite/Euram 592, 595, 597
Burner rig testing 263
Burst testing 490

C

C600 394
C600B 394
Car 2000 program 446, 479, 481, 483, 485
Carbon/carbon 446
Carborundum 114
Carborundum Hexoloy SA 642

CARES 146, 344, 369, 653
CARES/LIFE 146, 166, 169
Cascade tests 389, 390, 391, 398, 416
Casing 585
Castable 195
Castable zirconia 264
Catalytic combustor 207, 449, 502, 532
Catalytic igniter supports 242
CATE 113
C/C 583, 565, 670
C/C combustor 270
C/C composites 242
C/C rotor 272
Central Research Institute of Electric Power Industry (CRIEPI) 278
Ceradyne Inc. 158
Ceralloy 147A 158
Ceramic bearings 712
Ceramic blisk 104, 107
Ceramic bolt 644
Ceramic coatings 394
Ceramic combustor 56, 556, 676
Ceramic foam 645
Ceramic gas turbine (CGT) 277
Ceramic heat exchanger 449, 487
Ceramic honeycomb 504, 532
Ceramic Matrix Composite (CMC) 226, 242, 253, 286
Ceramic nozzles 78, 165, 322, 464
Ceramic nuts and bolts 411
Ceramic rolling element bearings 678
Ceramic rotor design 66
Ceramic shell 541
Ceramic sleeve 429
Ceramic stator 59
Ceramic stator vanes 196
Ceramic tiles 617, 642, 643, 648
Ceramic turbine blade 101, 311, 313
Ceramic turbine nozzles 94, 101, 137
Ceramic turbine wheel 136
Ceramic turbocharger 285
Ceramic vane 198
Cèramiques et Composites 449, 522
Cerasep™ 448, 585
CERITS-L 557
CFCC 165, 182, 214, 228
CFD code 591
CGT301 280, 305
CGT302 280, 331
CGT303 280, 361
Characteristic strength 31
Chemical Vapor Infiltration (CVI) 608
CIAM 664
Clearance 323, 375, 446, 669
CMC 446, 448, 551, 595
CO 456, 535, 560, 596
Cold isostatic pressing 543
Cold spin testing 172
Coil springs 339, 340
Combustor 7, 271, 288, 293, 294, 316, 341, 640, 648, 692, 368, 385, 386, 390, 405, 456
Combustor flame 364
Combustor liner 158, 164, 188, 229, 245, 254, 293, 4545
Combustor test rig 46
Combustor tile 447, 589, 641
Commercialization 96
COMPGLAS® 242
Compliant insulation 164
Compliant layer 106, 171, 175, 202, 312, 314, 374, 391, 476, 561, 671
Compliant materials 323
Compressor 7, 288, 585
Compressor airfoil 586
Compressor case 586
Compressor inner shrouds 243
Computed Tomography 165
Contact stresses 84, 89, 97, 176, 204
Continuum damage mechanics 602
Convective cooling 575
Cooled ceramic nozzle 397, 398
Cooled ceramic vanes 249
Cooled nozzle 448, 565
Cooling 3
Cordierite 207, 523, 524
Corrosion 394, 630
Corundum 622
Corundum-silimanite 616
Cost 188, 227, 243
Crack growth 656
Creep 431, 455, 648
Creep resistance 507, 648
CRIEPI 281, 404
Cruise missiles 261
CSGT 159
C/SiC 565, 645

CSN 101 504, 505, 507, 510
CTEDP 96
Curvic coupling 65
CVD 564
CVD SiC 653
CVI process 254
Cyclic fatigue 643, 646

D

Daimler-Benz 446, 453
Degradation 182
Degussa 455, 470
Design codes
DFVLR 454
Diffusive joining 685
Dimox™ 228, 585
DIMOX process 215
Directed metal oxidation 585
Disks 454
Divergent flaps 242
Divergent seals 242, 248
Dornier 449
Dovetail 103, 113, 169, 202, 671, 676
Duo-density 24
Duo-density ceramic rotor 472
DuPont Lanxide Composites (DLC) 165
DVUKON 670

E

EC-152 310
EC-152 Si_3N_4 293
EC-155 Si_3N_4 293
EC-157 Si_3N_4 293
Efficiency 3, 161, 180, 196, 217, 226, 301, 309, 324, 326, 327, 332, 352, 353, 361, 403, 406, 426, 557, 596, 686, 687
Emissions 180, 461, 505, 535, 590
End walls 238
Endurance test 354
Engine field demonstration 130
Engine testing 68, 84, 89, 105, 120, 158, 173, 243, 269, 299, 323, 346, 375, 461, 484
Enhanced SiC/SiC 182
Environmental barrier coatings (EBCs) 130, 182, 233, 256, 422
Erosion resistance 676
ESK Feldmühle 455
Eureka R&D program 448
European Gas Turbines (EGT) 446
Exhaust diffuser 364
Exhaust guide vanes 446, 589

F

Failure 84, 93, 117, 175
Failure analysis 274
Failure modes 139
Failure probability 505, 557, 569, 574, 640, 641, 655
Fatigue 637, 641
FE 467, 469, 475, 591
Fiber-Reinforced Composite hybrid component 336
Fibrous monolithic ceramic 264
Field testing 94, 105, 132, 180
Fine Ceramics 278, 306
Finite Element Modeling (FEM) 291, 299, 311, 315, 385, 505
Flame tube 448
Flameholders 446, 583
Flat wall model 572
Flexure strengths 462, 544
Fluorescent Penetrant Inspection (FPI) 122, 437
Foreign Object Damage (FOD) 9, 10, 95, 107, 122, 135, 141, 229, 301, 344, 351, 366, 392, 394, 484
Ford Motor Company 17
Fracture probability 432
Fracture toughness 507
Fraunhofer IzfP Saarbrücken 448
Fuel efficiency 454

G

Garrett Turbine Engine Company 77, 78
Gas-pressure-sintered 100
Gasifier rig 484
Gasifier rotor 125
GAZPROM 684
GE Power Systems 225
German Aerospace Institute 448
German Aerospace Research Establishment 454
GFICES 395
Glass ceramic composites 585

Glass-ceramics 242
Glass encapsulation 507
Glass encapsulation method 499
Glow plugs 279
GN-10 silicon nitride 92
Green machining 543
GT110 501
GTE 447, 486

H

Hamilton Sundstrand Power Systems 133
HASSMAP 588
HC 535
Heat exchanger 156, 288, 318, 320, 454, 499, 510, 513
Heat shield 438, 645, 648
Heloxy SA silicon carbide 166, 411, 418, 447
Hi-Nicalon 182
High-alumina refractory 617
High-alumina tiles 447, 618
High cycle fatigue 431
High-Performance for Severe Environments Program 307
HIP encapsulation technology 448
Hitachi Ltd. 278, 281, 382, 390, 404
Honeywell Ceramic Components 100
Honeywell Engines, Systems & Services 77
Hot gas rig 461, 466, 470
Hot gas spin rig 482
Hot isostatic pressed (HIPped) 100
Hot isostatic pressing (HIP) 119, 351, 507
Hot pressed silicon nitride 67, 446, 672
Hot rig tests 456
Hot spin rig testing 52, 269, 366
Hot wall 164, 195, 615
HPSN 158, 446, 472, 475, 483
HS-130 Si_3N_4 197
HXL 293
Hybrid ball bearings 511
Hybrid metal-ceramic nozzle 447, 539
Hybrid metal-ceramic rotor 474
Hybrid metal-ceramic turbine blades 550
Hybrid nozzle 157
Hybrid rotors 81
Hybrid SiC blade 447
Hybrid turbine 307, 327
Hybrid type rotor 390, 429

I

Image enhancement 84
Impact resistant 93
Impingement cooling 543, 575
Injection molding 59, 118, 464, 472, 487, 505
Integral rotor 479
Interface 68, 313
Interface coating 608
Interface friction 103
Intermediate duct 372, 373
Ishikawajima Harima Heavy Industries (IHI) 278, 280, 281, 301
Isothermal forging 240

J

Japan Automobile Research Institute (JARI) 280, 281, 286
Japan Industrial Standards (JIS) test specimens 293
Joining 236, 455
Joint Research Centre 449, 597
J-phase 178

K

Kawasaki Heavy Industries (KHI) 278, 280, 281, 331
Keramische Bauteile für Fahrzeug-Gasturbinen 454
Kyocera 90, 100, 331
Kyocera Corporation 361, 433, 447, 487

L

Lanxide™ 585
LAS 293
Leakage 522
Lean premix 161
Lenticular pad 476, 478
Lessons learned 69, 88, 94, 187, 197, 204, 208, 214, 218, 220, 231, 274, 300, 377, 529, 664, 702
Life assessment 143
Life prediction 256
Limited life engines 261
Lithium aluminosilicate 243

Low cycle fatigue 431
Low emission combustors 456
Lubricants 68, 106, 113

M

Malden Mills 186
MAN Technologie 645
MAS 293
Matrix cracking 215, 228
Max Planck Institut 448
Mechanical Engineering Laboratory 366
MEDYSYS 597
Melt Infiltrated-Ceramic Composite (MI-CMC) 227
Metal pads 390, 391, 392
Microfocus x-ray radiography 84
MICROTURBO 514, 535
Ministry of International Trade and Industry (MITI) 277, 280, 285, 331
Mitsubishi Heavy Industries, Ltd. (MHI) 278, 281, 382, 388
Mitsubishi Motor Corp. 279, 281, 283, 285
Mixed flow turbine 148
Mixed flow turbine rotor 92
MK1 499
Modulus Of Rupture (MOR) 32
Monkman-Grant 144
Monorotor 144
Moonlight Project 331, 361
MTU 446, 539, 453
Mullite 622
Multiple-quench test 623

N

NASA CARES/Life Code 291
NASALife 256
NC132 202, 240, 484
NC-132 HPSN 81
NC-350 RBSN 81
NC 430 158
NDE 23, 84, 165
NDE standards 84
Near net-shape processing 487
NEDO 8,000 kW Program 356
New Energy and Industrial Technology Development Organization (NEDO) 280, 361
New Sunshine Project 331, 361
NGK Insulators, Ltd. 166, 307, 390
NGK Locke 90
NGK Spark Plug, Ltd. 286, 307, 361, 389
Nicalon 215, 242
Nissan Motor Corp. 279, 281, 283, 284, 286
NITI EM 449, 527, 535, 664
NIZ KTD 683
Nondestructive testing 455
Norton Advanced Ceramics (NAC) 100
Norton Company 81
Norton NC 132 158
Nose cone 26, 58, 322, 325, 454, 462, 466
NOx 161, 180, 301, 316, 317, 327, 332, 354, 355, 388, 412, 446, 456, 502, 535, 560, 590, 596
Nozzle 385, 388, 390, 697
Nozzle petals 583
Nozzle ring 461, 464
Nozzle vane 8, 688
NT154 silicon nitride
NT164 silicon nitride 100
Nuovo Pignone 449

O

ONERA 449, 505, 513, 684, 697
ORNL 182
OTM-904 672
OTM-913
OTM-914 672, 678
Oxidation 59, 130, 181, 197, 215, 274, 394, 455, 640, 641, 645, 648, 653
Oxidation/corrosion 98, 464
Oxidation pits 651
Oxidation resistance 301
Oxide-based CMC 214
Oxide foam 645
Oxide/oxide 446, 589
Oxide-oxide CFCCs 218
Oxide/oxide CMC 215

P

Partially Stabilized Zirconia (PSZ) 541
Pattern factor 409, 671
Petroleum Energy Center (PEC) 280, 285
Pinned root 169
PIP process 254

Porosity gradient materials 640
Porous ceramic 563
Powder technology 455
Pressureless sintering 543
Probabilistic 29
Probability of Failure (POF) 30, 473
Probability of Survival (POS) 481, 482, 548
Proof test 84, 99, 167, 172, 268, 295, 301, 462, 481, 701, 712
Proof testing 129, 311, 395
PWT 110 483, 484, 486

R

Radial 86, 117, 134, 284, 319, 362, 574, 663
Radial profile 671
Radial turbine rotor 88
Radial turbine wheel 502
Rapid prototyping 479
RBSC 22
RBSN 22, 57, 59, 84, 240, 446, 464, 469, 470, 472, 475, 488, 490
Reaction-bonded silicon nitride 672
Recuperator 156, 341, 449, 502, 513
Recuperator effectiveness 522
Refel 458, 460, 461, 464
Refel silicon carbide 56
Regenerator 86, 117, 126, 280, 293, 361, 364, 492
Reliability 67, 431, 487
Retainer ring 314
Reticular ceramics 640
Rig testing 54, 84, 263
Rim seal 172
Road tests 455
Rocker arm pads 279
Rolls-Royce 111, 446, 583
Rosenthal 455, 470, 490
Rotary regenerator 288, 488
Rotating blades 454
Rotor 385, 390, 391
Rotor attachment 27
Rotor blades 39, 202, 429, 435, 691, 700
Rotor failure 68
Rotor stress 35
Rybinsk Motors Design Bureau 664, 709

S

SC-21 394
Scroll 125, 158, 284, 292, 293, 340, 364, 369, 448, 574, 692
Seal damper 314
Seal rings 78
Sector rig 251
Segmentation 557, 577
Segmented combustor 676
SEP 448
Shell-and-tube type recuperator 514
Shroud 292, 340, 364, 446, 466, 541
Shroud seal 588
Si_3N_4 126, 519
Sialons 449, 520
SiC 519
SiC-1 386, 392
SiC/Al_2O_3 446, 585, 587, 588
SiC/SiC 583, 585, 589
SiC/SiC CFCC 165, 565
SiC/SiC screws 643
SiC tile 645
Siemens KWU 447, 615
Siemens Westinghouse Power Corporation (SWPC) 193
SIGRI 455, 462
Silicon carbide 116
Silicon nitride 205, 250
Siliconized SiC 158
SIN-E 433
Sintered α-silicon carbide 543
Sintered silicon nitride 90
SiSiC 448, 462, 464, 465, 469, 470, 557
Slip casting 118, 464, 465, 487
Slow Crack Growth (SGC) 67, 291
SN84 319, 390
SN84HD 309
SN88 309, 394, 389, 391, 398
SN88M Si_3N_4 293
SN88 Si_3N_4 166, 293
SN90 Si_3N_4 293
SN91 Si_3N_4 293
SN220 90, 293
SN220M 119, 331
SN250M 90
SN252 124, 148, 293, 343, 347, 389, 391, 408, 411, 433

SN253 293
SN281 169, 178, 293, 343, 351, 366, 394, 398
SN282 100, 178, 250, 343, 351
Solar Turbines Incorporated 155
SOUGRAV 449
Spallation 184, 186
Specific thrust 262
Spin rig 535
Spin test 45, 295, 347, 391, 481, 507, 551, 673
Spin testing 41, 268, 437, 473, 476, 478, 481
SPSLIFE 142, 166
SRBSN 478
SSiC 449, 462, 464, 540, 544, 551, 559, 642, 648, 652, 653
SSiC bolt 645
SSN 464, 486
ST-1 310
Static fatigue 34
Stationary ceramic components 294
Stationary components 293
Stationary vanes 454
Stator blades 427
Stator vane 39, 405
STAU 646, 653
Stressed oxidation 265
Structural ceramics program 446, 486
Strut 322
Sub-critical crack growth 648
SUGRAV 527, 685
Sumitomo Precision Products 331
Surface defects 653
Support springs 364
Swirl chamber 279
Syalon 446

T

TARC 535
TBC 227
Technical University in Berlin 448
Technical University in Clausthal-Zellerfeld 448
Technical University of Karlsruhe 448
Technologhia 672
Tekhnologhia 664
Tensile stress 571, 648
Tensile surface stresses 645
Texaco 184
Thermal Barrier Coatings (TBCs) 155, 547
Thermal conductivity 233, 490, 523, 559, 572, 616, 626, 627
Thermal cycling test 368
Thermal efficiency 331, 349, 352
Thermal expansion 7, 626, 627
Thermal imaging 165
Thermal insulation 642, 645, 648
Thermal-mechanical fatigue 641
Thermal protection system (TPS) 446, 607, 641
Thermal shock resistance 464, 490, 523, 559, 616
Thermal shock rig 267, 299
Thermal shock test 44, 347, 481, 570, 629
Thermal shocks 490, 640, 645
Thermo-mechanical cracking 630
Thermo-mechanical stress 618
Thermo-mechanical tensile loads 641
300kW CGT program 280, 281
Three-layer airfoil wall 566
Thrust 261
Tiled combustor 195
Time-dependent lifetime prediction 646
Tip clearance 340
Tip rubs 231
Tip seal 215
Tip seal ring 8
Tip shroud 62, 115
TJ-50 thrust engine 144
Tokyo Electric Power Co., Inc. (TEPCO) 278, 281, 382
Toshiba Corporation 278, 281, 381, 382, 386, 425
Toyota Motor Corp. 278, 281, 283, 286
Trailing edge ejection 575
Transition piece 386, 407, 461, 648
Turbine 7, 288
Turbine blades 364
Turbine Inlet Temperature (TIT) 3, 6, 26, 84, 86, 92, 134, 158, 196, 262, 294, 300, 301, 323, 325, 327, 352, 354, 664, 710, 383, 397, 453, 456, 479, 480, 547, 566
Turbine nozzle 136, 150, 272, 314, 336, 364
Turbine rotor 149, 272, 293, 294, 311
Turbine Rotor Inlet Temperature (TRIT) 156
Turbine shrouds 229
Turbine stationary components 314
Turbine stator 293

Turbine support ring 246
Turbine wheel 505
Turbocharger rotor 279
Turbofan 261
Turbogenerator 501

U

United Technologies Pratt & Whitney 237
United Technologies Research Center (UTRC) 237
United Turbine 448

V

Vane 640
Vane cascade test 544
Valve 447
Vibration 84, 106
Volkswagen AG 446, 453
Volvo 448, 499

W

Waspaloy 172
Wave rings 339
Weibull 30, 309, 481, 646, 653
Weibull modulus 31, 432, 462, 544
Weibull-plots 624
Weibull statistical analysis 142
Westinghouse Electric Corporation 193
Williams International 261
Wind tunnel test 433

Y

Yanmar Diesel Engine Co., Ltd. (YDE) 280, 281, 361